Fracture Mechanics of Ceramics

Volume 12

Fatigue, Composites, and High-Temperature Behavior

Fracture Mechanics of Ceramics

Volume 12

Fatigue, Composites, and
High-Temperature Behavior

Edited by

R. C. Bradt
University of Alabama
Tuscaloosa, Alabama

D. P. H. Hasselman
Virginia Polytechnic Institute and State University
Blacksburg, Virginia

D. Munz
University of Karlsruhe
Karlsruhe, Germany

M. Sakai
Toyohashi University of Technology
Toyohashi, Japan

and

V. Ya. Shevchenko
High Tech Ceramics Scientific Research Centre
Moscow, Russia

SPRINGER SCIENCE+BUSINESS MEDIA, LLC

ISBN 978-0-306-45379-3   ISBN 978-1-4615-5853-8 (eBook)
DOI 10.1007/978-1-4615-5853-8
Library of Congress Catalog Card Number 83-641076

Second part of the proceedings of the Sixth International Symposium on the Fracture Mechanics of Ceramics,
held July 18 – 20, 1995, in Karlsruhe, Germany

© 1996 Springer Science+Business Media New York
Originally published by Plenum Press, New York in 1996

PREFACE

The volumes 11 and 12 of Fracture Mechanics of Ceramics constitute the proceedings of the 6th International Symposium on Fracture Mechanics of Ceramics held at the Research Center Karlsruhe, Germany, July 18, 19, 20, 1995. As in previous conferences the state of the art of the failure behaviour of monolithic engineering ceramics and of reinforced ceramics was discussed. The 90 papers by over 200 authors and co-authors address the recent developments in the understanding and the modelling of the fracture processes in brittle materials. The main topics are R-curve behaviour, toughness determination, surface effects, composite materials, high temperature behaviour, ceramic-metal joints, fatigue.

The program chairmen gratefully acknowledge the financial support of the Deutsche Forschungsgemeinschaft (German Science Foundation) which made possible the participation of scientists from East Europe. We also thank Mrs. Eva Schröder from the Research Center Karlsruhe, the local organization committee (Dirk Hertel, Claus Petersen, Dr. Franz Porz, Rainer Weiß) and the conference secretary Mrs. Lucia Borchardt for their conscientious and efficient organization of all details of the conference, Mrs. Isabella Daubenthaler and Mrs Natascha Rothweiler for preparing the booklet of abstracts and Dr. Theo Fett for his help in editing the proceedings.

R.C. Bradt   D.P. Hasselman   D. Munz
Tuscaloosa, USA   Blacksburg, USA   Karlsruhe, Germany

M. Sakai   Y.V. Shevchenko
Toyohashi, Japan   Moscow, Russia

July 1995

CONTENTS

FRACTURE MECHANICS OF FATIGUE OF STRUCTURAL CERAMICS[†]

I-Wei Chen, Shih-Yu Liu[*] , David S. Jacobs[§] and Mehernosh Engineer

Materials Science & Engineering
University of Michigan
Ann Arbor, MI 48109-2136

INTRODUCTION

Stable crack growth under cyclic loading has been observed in many structural ceramics which exhibit a rising toughness (R-curve) behavior.[1-7] Although the growth is sometimes enhanced by the environmental factors, it can also proceed in vacuum where the environmental assistance is absent.[6] This phenomenon has been extensively documented in recent years so that it is now possible to give a fairly general description of its mechanisms and kinetics. The purpose of this article is to incorporate the main body of the phenomenology into a micromechanical framework, within which the current understanding of the R-curve behavior and fatigue is fully reconciled.

For ceramics which have no capacity for crack tip blunting, stable crack growth under a monotonically increasing load must imply the existence of an R-curve.[8] Applying the same instability consideration in fracture mechanics to cyclic loading conditions of identical peak stress intensity factor, K_{max}, and demanding incremental crack advance in every cycle when the stress intensity factor reaches K_{max}, we can conclude that stable fatigue crack growth likewise implies the existence of an underlying R-curve. On the other hand, stable crack growth under the latter conditions also implies a degradation of fracture resistance during unloading and reloading, since cyclic loading in ceramics cannot cause blunting and resharpening of the crack as it does in metals. Thus, a mechanistic understanding of the R-curve behavior and shielding degradation in ceramics holds the key to the problem.

The most general cause for the R-curve behavior in ceramics is crack wake bridging due to frictional pull out of grains.[9] The magnitude of crack tip shielding from this contribution depends on the population of participating grains, the friction, the sliding distance, and geometrical factors such as the aspect ratio of the grain.[10] Among the above factors, friction is obviously amenable to degradation by wear during cyclic loading. Direct physical evidence of wear under a cyclic load is vividly displayed in Fig. 1,[11] which shows accumulation of wear debris at the sliding interfaces under cyclic but not monotonic loading conditions. This mechanism is adopted in the present analysis which will further explore the interplay between cyclic loading, interfacial friction, and crack tip shielding.

[†] This research was supported by the Air Force Office of Scientific Research, Grant No. AFOSR-F49620-95-1-0119.
[*] Now at Westinghouse Electric Corp., Pittsburgh, PA.
[§] Now at Saint Gobain Norton, Northboro, MA.

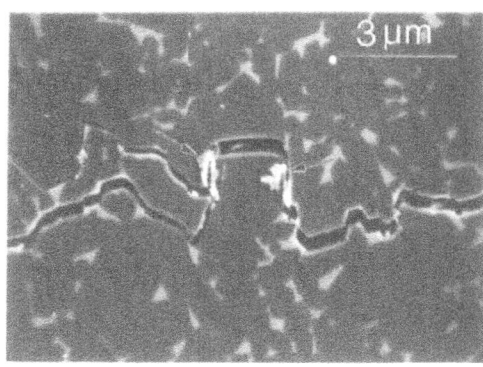

Figure 1. Crack profile in Si_3N_4 after cyclic fatigue; wear debris at the sliding interfaces is indicated by arrows.

An important distinction should be made at the outset between low temperature and high temperature concerning the kinetics of friction, wear and crack advancing processes.[12] Both frictional sliding and crack tip advancement are essentially athermal at low temperature but thermally activated at elevated temperature. In many structural ceramics, a critical temperature can be identified that separates the two temperature regimes. This is the softening point of the grain boundary phase which may lie several hundred degrees below the "intrinsic" softening of "clean" ceramic grain boundaries that rarely exist except in high purity materials. The analysis below implicitly assumes the dominance of some grain boundary phases on the decohesion and sliding of grains both in the wake and at the tip of an advancing crack.

MECHANICS OF TOUGHENING AND FATIGUE

We first present a simple dimensional analysis that outlines the interplay between toughness, friction, temperature and fatigue. Consider the frictional work contributions of grains which, on one end, are separated, intergranularly, from the matrix. The steady state toughness increment can be written as[10,12]

$$\Delta G = V_f \tau L \left(\frac{L}{R}\right) \tag{1}$$

where V_f is the volume fraction of grains participating in pull out, τ is the intergranular friction, L is the length of the pulled out grains, and (L/R) is their aspect ratio which appears as a stress concentration factor determined by the shear-lag model. Of the above parameters, V_f, τ and L can all be affected by either cyclic loading or temperature. Two broad cases thus apply.[12]

(a) Low temperature

At low temperature, V_f is determined, apriori, by the population of grains that are debonded through crack deflection but accidentally left in the "socket" constrained by asperities and residual stresses. A typical value for V_f is quite low, only around 10%, which is not surprising in view of the "accidental" circumstances required for its existence.[10] Although V_f is sensitive to the material properties, such as toughness and residual stresses of the grain boundary, it is not affected by cyclic loading. It will probably decrease, however, with temperature as the residual stress due to thermal expansion mismatch decreases. In a similar way, L is determined, apriori, by crack deflection. It is probably reasonable to regard L to be about the same as the average half length of grains ℓ, and no effect of temperature or cyclic loading on L is expected. Fracture toughness thus increases with τ, as schematically illustrated in Fig. 2a.

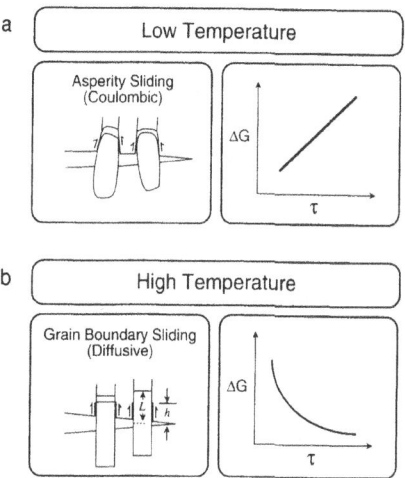

Figure 2. Grain bridging and pull out processes (left) and their associated toughness-friction relations (right). (a) low temperature; (b) high temperature.

(b) High temperature

As grain boundary phases soften thermally with increasing temperature, decohesion of grain boundaries due to deflection of the main crack is increasingly unlikely. Grains bridging the main crack begin to slide over a distance h, which scales as $R(\sigma_f/\tau)$ because the stress concentration due to shear lag $\tau(h/R)$ must not exceed σ_f, the fracture strength of the grain. If h reaches the triple junction, then stress concentration at the triple junction may cause decohesion of the transverse grain boundary, allowing grains to be pulled out eventually making a large contribution to the work of fracture. V_f in this case is thus determined by the probability of a triple junction lying within a distance h from the crack plane, i.e., $V_f = h/\ell$. Meanwhile, obviously, L = h. Substituting V_f and L into Eq (1), we obtain

$$\Delta G = (h/\ell)\, \tau\, h(h/R) = \sigma_f R\, (R/\ell)\, (\sigma_f/\tau)^2 \tag{2}$$

Therefore, fracture toughness decreases with τ^2 as schematically illustrated in Fig. 2b.

We now consider the dependence of τ on temperature and cyclic load. First, for asperity sliding, we may assume τ to follow Coulomb's law and be proportional to the normal compressive stress, σ_{rr}, which is in turn assumed to be caused by a misfit normal strain ε_{ms}. Thus, $\tau = c\mu E\varepsilon_{ms}$, where c is a dimensionless constant, μ is the coefficient of friction, and E is Young's modulus. The effect of the cyclic load, which produces cyclic displacement, u, is to wear off the material and hence reduce the misfit strain. Thus,

$$\frac{d\varepsilon_{ms}}{du} = -\frac{k\sigma_{rr}}{R} = -\frac{k\tau}{\mu R} \tag{3}$$

In the above, we have assumed a linear wear law and a wear constant k that states that the rate of material removal per sliding displacement is proportional to the normal pressure (which can in turn be related to τ, as before, using Coulomb's law). Combining Eq (2) and the expression for τ, we obtain

$$\frac{d\tau}{du} = -\frac{kcE}{R}\tau \tag{4a}$$

or

$$\tau = \tau_o \exp\left(-\frac{kcE}{R}u\right) \tag{4b}$$

That is, τ decreases exponentially with sliding distance.[11] In the above, τ_o is the initial friction which is dependent on μ and ε_{ms}. Considering thermal softening that affects μ and thermal expansion mismatch that affects ε_{ms}, we expect τ_o to decrease with temperature and τ to further decrease with the number of cycles of cyclic loading.

Wear considerations of the above kind probably also apply at high temperature. Moreover, regarding grain boundary sliding as a diffusion-controlled process, we also expect its resistance to increase with grain boundary roughness, which may be smoothed out by repeated back-and-forth sliding. Therefore, the basic trend of τ in response to temperature and cyclic load is similar in both low temperature and high temperature regimes.

We can now appreciate the effect of cyclic loading on toughness increments and crack growth resistance. At low temperature, cyclic loading decreases τ and hence ΔG. The

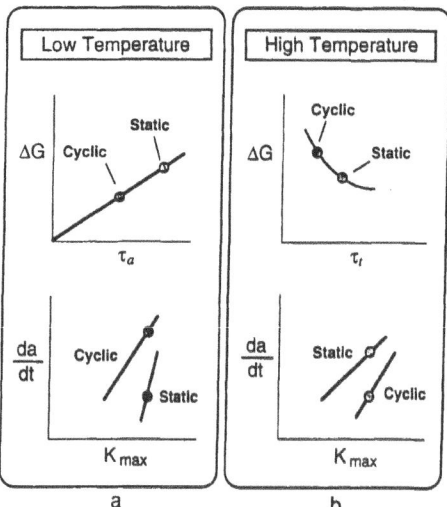

Figure 3. Effect of degradation of interfacial shear stress, due to cyclic loading, on fracture energy and crack growth rate. (a) low temperature; (b) high temperature.

degradation of crack growth resistance causes crack growth rate to increase, under cyclic loads, as schematically illustrated in Fig. 3a. At high temperature, cyclic loading again decreases τ but increases ΔG, therefore, the increase of crack tip crack resistance causes crack growth rate to decrease, as schematically illustrated in Fig. 3b. Following the negative temperature dependence of τ and applying Eq (1) and (2) at low temperature and high temperature, respectively, we also envision that the toughness-temperature relation undergoes a reversal, with the peak toughness obtained near the softening temperature.

The above simple analysis essentially captures the major trends exhibited by many glass-containing ceramics under cyclic loading at different temperatures.[13-18] In the following, a more detailed examination of some related aspects concerning fatigue crack propagation is made to elucidate their mechanics and kinematics.

SHIELDING DEGRADATION AND ACCUMULATION IN FATIGUE

To understand and to model the evolution of crack wake shielding in cyclic loadings, it is convenient to first consider a single grain, modeled as a short fiber embedded in a matrix, with the far end (in the matrix) at a distance L from the surface, free of normal traction. The following steps are considered. In step I, the fiber is initially pulled out by a tensile stress, the stress-distance relation follows

$$\sigma_b = \frac{2\tau}{R}\left(\frac{L}{2} - u\right) \tag{5}$$

In step II, the tensile stress is reversed, and the fiber undergoes reverse elastic sliding,

$$u = u_{max}\left(1 - \frac{1}{2}\left(1 - \frac{\sigma_b}{\sigma_{max}}\right)^2\right) \tag{6}$$

where

$$-1 \le \frac{\sigma_b}{\sigma_{max}} \le 1 \tag{7}$$

and the subscript "max" refers to the stress-distance state at the end of step I. In step III, the fiber is pushed in under a compressive stress, and the stress-distance relation is just the opposite of I.

$$\sigma_b = -\frac{2\tau}{R}\left(\frac{L}{2} - u\right) \tag{8}$$

In step IV, the compressive stress is reversed, and the fiber elastically reloads all the way back to a configuration that is the same as the starting one, prior to step I.

$$u = \frac{u_{max}}{2}\left(1 + 2r^* - \frac{2r^*\sigma_b}{\sigma_{max}} + \left(\frac{\sigma_b}{\sigma_{max}}\right)^2\right) \tag{9}$$

where

$$r^* = \frac{\sigma_b}{\sigma_{max}} \tag{10}$$

and subscript "max" refers to the stress-distance state at the end of step III. The above sequence is schematically shown in Fig. 4.

The above relations were obtained following a similar analysis as that given by Marshall et al. for the instrumented fiber push-in experiment.[19] Mathematically, the sequence repeats itself precisely without backlash under repeated loading/unloading cycles, provided the friction remains unchanged. In the crack wake, of course, the friction decreases with wear and thus the sequence will evolve with cyclic loading. The important point is that, a hysteresis loop always exists which contributes to the work of fracture under cyclic loading conditions.

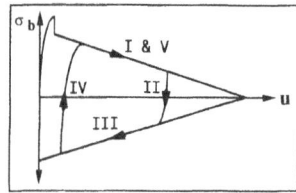

Figure 4. Schematic representation of a stress-distance plot for a single grain, modelled as a short fiber embedded in a matrix undergoing an unloading-reloading process.

Modeling crack wake with bridging fibers that undergo loading/unloading cycles of the above type can proceed using standard methods of either integral equation or superposition of discrete influence functions. At any given point, the stress-displacement distribution in the wake must be determined in a self-consistent manner to satisfy the pertinent stress-displacement condition shown above. As an example, we show in Fig. 5 the crack profile at full loading $(R = 1)$ and at subsequent unloading $(R < 1)$. For comparison, a crack

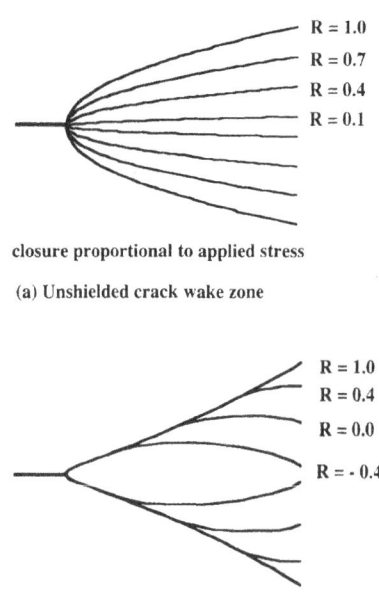

closure proportional to applied stress

(a) Unshielded crack wake zone

crack closure more at the mouth due to fiber push in

(b) Shielded crack wake zone

Figure 5. Crack profile during an unloading process; (a) unshielded; crack closure is proportionate to applied stress, (b) shielded crack wake zone; crack closure is more at the mouth.

which is free of bridging in the wake is also shown. Clearly, the crack free of bridging unloads proportionally and maintains a self-similar (parabolic-shaped) profile at all times.[8-9] In contrast, the shielded crack experiences a much larger closure displacement at the mouth, but maintains a nearly fully open profile in between the tip and the mouth. This feature can be understood by inspection of the stress-displacement curve of individual fibers at various locations in the wake. The fiber close to the crack tip, which has very little displacement in the first place, is found to simply unload elastically following Eq (6) without any push in. The fiber in the center of the wake zone, which has been pulled out considerably, is found to elastically unload fully and then to push in a bit. It maintains a relatively high compressive load because most of the distance L is still in the matrix experiencing a friction. The fiber close to the crack mouth, which has been almost completely pulled out, is found to elastically unload fully and then to push in the most. Yet, the stress is quite low on this fiber because only a very short segment still remains in the matrix to exert the friction. It is then clear that the fiber in the mouth is "plastically soft" and hence, able to close more during unloading. It is this larger reverse sliding that causes wear of the fiber and the consequent reduction of friction, per Eq (4), that results in the degradation of shielding. When the problem is modeled using discrete fibers distributed at regular spacing in the wake, the process of shielding degradation can be clearly seen in Fig. 6. It shows the results of a calculation that incorporates cyclic sliding, friction degradation, crack tip advance and inclusion of new fibers. With repeated load/unloading, crack-wake shielding degraded almost linearly. When the shielding is sufficiently reduced, so that the stress intensity factor at the crack tip, K_{tip}, increases to reach K_o, the crack growth resistance of the material, the crack advances by a distance of fiber spacing. The incorporation of a new fiber then raises the shielding to the previous level, and the process repeats itself. Following this picture, steady state crack propagation can be envisioned as an intermittent process that consists of short bursts of crack advances, each burst preceded by an incubation period during which crack tip shielding is being degraded.

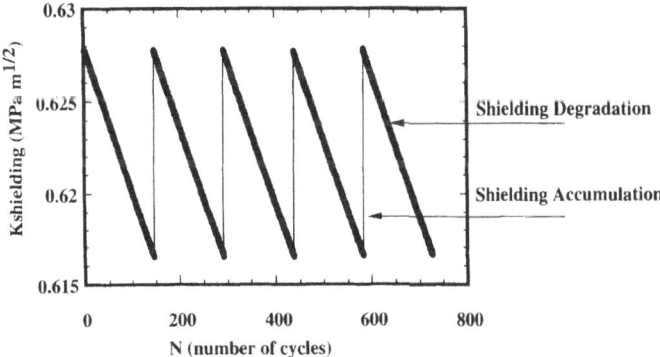

Figure 6. Crack shielding degradation and accumulation at steady state crack growth under cyclic loading.

We have examined the effect of two mechanical parameters, K_{max}, the maximum stress intensity factor applied, and ΔK, the difference between K_{max} and K_{min}, the minimum K in the cycle, on crack growth rate. The results of some calculations are shown in Fig. 7. Obviously, crack growth rate (da/dt), increases with both K_{max} and ΔK, but their effects are much different. The dependence of (da/dt) on K_{max} is much stronger than that on ΔK. In the case of Fig. 7, the dependence may be written as (da/dt) $K_{max}^{11.4} \Delta K^{3.7}$. This is in accord with the experimentally observed fatigue kinetics at lower temperature which all follow

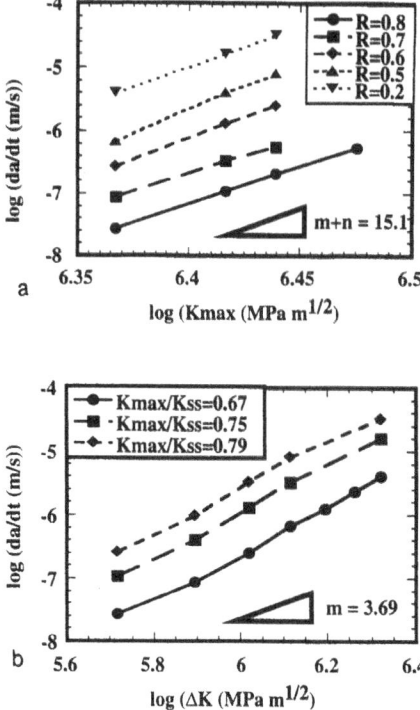

Figure 7. Dependence of fatigue crack growth rate $(\frac{da}{dt})$ on (a) K_{max} and (b) ΔK.

$$\frac{da}{dt} = A K_{max}{}^{n} \Delta K^{m} \qquad (11)$$

with an n which is exceedingly large, typically between 10 and 50, but m lying between 2 and 4, and A is a constant.[4,11,20-21]

The dependence of $(\frac{da}{dt})$ on K_{max} is a unique feature of fatigue in ceramics. In metals, the crack growth rate is essentially independent of K_{max} and dependent on ΔK only, following the so-called Paris law $(\frac{da}{dt}) = A \Delta K^{m}$ where $m \sim 4$.[22] For such kinetics, fatigue testing can be conducted either by setting K_{max} constant and varying K_{min} (i.e., only ΔK varying), or by varying K_{max} and K_{min} at the same time but keeping their ratio R $= K_{min}/K_{max}$ constant (i.e., both K_{max} and ΔK, equal to (1-R) K_{max}, are varying). Tests of the latter kind would yield for ceramics a seemingly strong dependence on ΔK with an exponent of n + m if the effect of varying K_{max} is not accounted for explicitly. This interpretation, which was adopted in the earlier literature on ceramic fatigue, is incorrect.

A simple physical explanation of the unique feature of separate ΔK and K_{max} dependence has been proposed that draws upon the characteristics of the R-curve characteristics of ceramics.[11] We demand, for an R-curve material in the athermal regime, that crack tip satisfy an equilibrium condition $K_{tip} = K_{o}$. If degradation and accumulation of shielding occur, then at equilibrium these processes must balance so that the net change in shielding, dK_{sh}, is zero

$$0 = dK_{sh} = \frac{\partial K_{sh}}{\partial a} da + \frac{\partial K_{sh}}{\partial t} dt \qquad (12)$$

In the above, the first term on the right hand side refers to the shielding accumulation and the second term refers to shielding degradation. By rearrangement, we obtain crack growth rate

$$\frac{da}{dt} = \frac{-\frac{\partial K_{sh}}{\partial t}}{\frac{\partial K_{sh}}{\partial a}} \qquad (13)$$

Thus, under steady state conditions, this equilibrium naturally leads to a steady state growth rate. One can associate the shielding accumulation, $(\partial K_{sh}/\partial a)$, to the slope of the R-curve. Since R-curve rises initially and saturates eventually, the slope decreases rapidly with K and may be represented, parametrically as

$$\frac{\partial K_{sh}}{\partial a} \propto K_{max}^{-n} \qquad (14)$$

In the above, K_{max} has been identified with K on the R curve and the value for n is probably very large in view of the shape of the R-curve. This leads, via Eq (13), to the dependence of (da/dt) on K_{max}^{n}, given in Eq (11). On the other hand, $(\partial K_{sh}/\partial t)$ can be written, through chain rule, into

$$\frac{\partial K_{sh}}{\partial t} = \frac{\partial K_{sh}}{\partial \tau} \frac{\partial \tau}{\partial u} \frac{du}{dt} = \frac{\partial K_{sh}}{\partial \tau} \frac{\partial \tau}{\partial u} \, f \, \Delta u \qquad (15)$$

where f is the frequency and Δu is the amplitude of cyclic displacement per cycle. Of the above terms on the right hand side, only the last term has a direct dependence on ΔK. Further theoretical consideration and calculations indicate that the ΔK dependence in Δu is relatively weak, with $\Delta u \propto \Delta K^2$ to ΔK^4 This, via Eq (13), explains the ΔK dependence of (da/dt) as expressed in Eq (11).

We can now conclude that the unique kinetics of fatigue in ceramics is directly related to the R-curve behavior of these materials, which gives large increments of shielding at low K but increasingly smaller increments of shielding as the saturation K is reached. Since shielding increment is the source of fatigue resistance, the characteristics of R-curve gives rise to an exceedingly strong dependence on K_{max} in (da/dt). A dependence of (da/dt) on ΔK, which is similar to the Paris law, also arises but is from an entirely different origin. It is related not to crack tip blunting/sharpening due to dislocation plasticity but to crack mouth wear and weakening due to intergrain sliding. Together, this simple analysis has provided direct and useful insight into the origin of fatigue in ceramics.

R-CURVE AND SHORT CRACK BEHAVIOR

Transient behavior marked by accelerating or decelerating growth rates upon changes of K_{max}, K_{min}, or R ratio is a common observation in fatigue testing of ceramics. The origin of it can be easily appreciated now that the role of shielding degradation and R-curve in determining fatigue growth rate is known. Specifically, changing K_{max} necessitates climbing up the R-curve which must be accompanied by some crack advancement, whereas changing K_{min} or R ratio alters the amplitude of cyclic displacement and hence the wear rate. Both disturb the mechanical equilibrium between shielding accumulation and degradation and thus trigger a transient evolution toward the new steady state. It seems reasonable to expect the distance of crack advance during the transient to be (inversely) determined by the wear rate of the interface. The details of the transient kinetics, however, are complicated and must be resolved by either experimental investigation or numerical simulation.

When a fresh crack is cyclically loaded, the transient is especially pronounced because of the need to establish a steady state wake zone of bridging grains. As the wake zone develops, the shielding builds up and the crack growth rate decelerates. If a constant peak load is maintained, however, eventually the longer crack length causes K_{max} to increase gradually. The latter effect eventually overcomes the shielding accumulation, which approaches a steady state, so that accelerating crack growth is observed at a later stage. This sequence of events gives rise to a sigmoidal shape to the a-t curve, or equivalently, a V-shape to the (da/dt) - K_{max} curve. The latter relation is reminiscent of the short crack behavior which often arises because of variations in residual stresses or microstructures over the dimension of the crack length.[4,22-23]

A particularly interesting set of observations of short-crack behavior has been reported for high temperature crack growth under static and cyclic loading conditions.[24] For a silicon nitride which maintains a strength of 800 MPa up to 1400°C, the short crack behavior was found more prominent at higher temperature, at lower initial K_{max} (K_i) and at lower R-ratio (or larger cyclic loading component). These results can be numerically simulated using a grain bridge model which allows friction to degrade in cyclic loading but not in static loading. For the crack growth rate, it can be assumed that at high temperature

$$\frac{da}{dt} = AK_{max}^{(tip)n} \qquad (16)$$

where $K_{max}^{tip} = K_{max} - K_{sh}$ and both A and n are functions of temperature. (A increases but n decreases with temperature in keeping with the phenomenology of crack growth at high temperatures.) As a crack propagates under a cyclic load that maintains the same peak value, K_{max} increases with crack length while K_{sh} increases with wake bridging. This in turn causes (da/dt) to vary with time. Using the parameters listed in Table I, we have calculated the crack growth rate as a function of K_{max} for test conditions shown in Fig. 8.

Table 1. Parameters used in calculation of crack growth rate for test conditions shown in Fig. 8.

Temperature	Static Fatigue			Cyclic Fatigue		
	τ	h	D	τ	h	D
(°C)	(Mpa)	(μm)	(μm)	(Mpa)	(μm)	(μm)
1300	100	0.60	2.0	75	0.80	1.50
1350	80	0.75	1.6	62.5	0.96	1.25
1400	75	0.80	1.5	60	1.00	1.20

The observed trend and the magnitude of the growth rates are in good agreement with the experimental data. Also shown in Fig. 8 is an envelope in the R-K_i-T space within which the short crack behavior is observed. Under a constant peak load, the crack extension curve as a function of time (a vs. t) has a sigmoidal shape within the envelope and a concave-upward shape outside the envelope.

A simple rationalization of the above trend can be obtained once again by appealing to the fracture mechanics of the R-curve behavior.[24] In general, for wake-shielding mechanisms, the length of the wake bridging zone is comparable to the distance of crack advancement required for the crack growth resistance to approach saturation (Δa_{ss}). At saturation, the wake bridging reaches a steady state in which one grain at the mouth becomes fully pulled out (thus carrying friction no more) when the crack advances by one increment bringing in a new bridging grain at the crack tip. At high temperature, the length of the wake bridging zone can be approximately estimated by setting the crack opening displacement (COD) at the mouth to equal the sliding length of the grains, h. Approximating COD using linear elastic fracture mechanics

$$COD = C\frac{K\sqrt{x}}{E} \qquad (17)$$

where x is the distance from the crack tip, and C is a dimensionless constant, and substituting h from Section II(b) for COD above, we obtain

$$\Delta a_{ss} = x = \left(\frac{E\sigma_f R}{C}\right)^2 \left(\frac{1}{K\tau}\right)^2 \qquad (18)$$

where Δa_{ss} is equated to x for a crack that has propagated by Δa_{ss}. According to this relation, we can now predict that Δa_{ss} increases with decreasing initial K_{max} and with decreasing τ (either due to higher temperature or fatigue wear). A longer transient and hence more pronounced short crack behavior are thus expected at lower K_i, higher temperature, and lower R, in agreement with the results shown in Fig. 8.

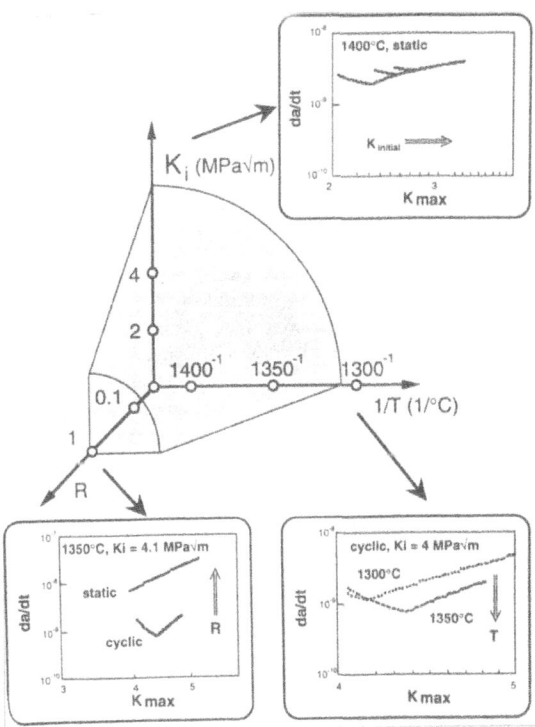

Figure 8. Fatigue crack growth rate as a function of K_{max} for various test conditions lying inside and outside of the envelope delinetated in the center.

CONCLUSIONS

The toughness of ceramics is generally derived from the crack wake zone in which grains are pulled out against an intergrain friction. This process has essentially different characteristics at low temperature and at high temperature. At low temperature, the pull out of a small fraction of bridging grains, which were completely debonded due to prior deflection of the main crack, accounts for the toughening. Friction degradation due to cyclic interfacial wear reduces the frictional work and hence the crack growth resistance under cyclic loading. At high temperature, sufficient softening of grain boundary phases allows the sliding zones to extend to the triple points, allowing a majority of grains that lie normal to the crack plane to separate at the ends provided their grain length is shorter than their sliding length. Friction degradation due to cyclic interfacial wear or a temperature rise increases the sliding length and hence triggers more pull out. Opposite fatigue effects on crack growth rate at low and high temperature are thus expected.

Unlike the case in metals in which fatigue causes crack tip blunting, distortion, and possibly resharpening, fatigue in grain bridging ceramics does not disturb the crack tip as much as it disturbs the crack mouth. A large cyclic displacement is experienced at the crack mouth due to the relatively low traction therein, which is a result of the very short segment of grain length that remains constrained in the matrix. The cyclic interfacial wear of these grains and the attendant degradation of crack wake shielding is proportional to ΔK^m. At low temperature, this shielding degradation due to fatigue is countered by shielding accumulation of new, undamaged bridging grains entering at the advancing crack tip. At the steady state, an equilibrium is reached between shielding degradation and accumulation and the crack tip stress intensity factor is always maintained at the value equal to the intrinsic toughness (K_o) of the material. At high temperature, subcritical crack growth is thermally activated and follows a power law dependence on K_{tip}. At the steady state, mechanical equilibrium is not maintained but the crack growth is kinetically constrained by the thermal activation.

Transient crack growth behavior emerges naturally from the rising wake-shielding and the ΔK sensitivity of wake-wear as crack advances and cyclic loading continues. When tested under a constant load, a growing crack can also exhibit short crack behavior, with initial deacceleration and later acceleration, giving a V-shaped ($^{da}/_{dt}$) -K relation. Such behavior can be rationalized by the evolution of shielding accumulation and degradation, as well as the development of the crack wake zone. It is especially pronounced at elevated temperature because of the decreased sensitivity to K_{max} in subcritical crack growth and increased shielding near the softening temperature.

It is further concluded that at low temperature, fatigue crack growth resistance under the same loading conditions should improve with increasing wear resistance, increasing steepness of R-curve, and a higher saturation toughness. At high temperature, some fatigue retardation of crack growth rate eventually emerges in ceramics containing a substantial amount of glassy phase. The magnitude of this retardation is time dependent and a unique ($^{da}/_{dt}$) -K relation does not exist under most conditions. To the extent that crack opening displacement, hence wake shielding and the R-curve, are all dependent on the specimen geometry and loading configurements, certain caution is warranted in applying fatigue test data to design applications.

REFERENCES

1. R.H. Dauskardt, W. Yu, and R.O. Ritchie, Fatigue crack propagation in magnesia-partially-stabilized zirconia ceramics, *J. Am. Ceram. Soc.*, 70:10:C248-252 (1987).

2. L. Ewatt and S. Suresh, Crack propagation in ceramics under cyclic loads, *J. Mater. Sci.*, 22:1173-92 (1987).

3. S.Y. Liu and I-Wei Chen, Fatigue of yttria crack-stabilized zirconia, I. fatigue damage, fracture origin and lifetime prediction, *J. Am. Ceram. Soc.*, 74:6:1197-205 (1991).

4. S.Y. Liu and I-Wei Chen, Fatigue of yttria crack-stabilized zirconia, II. crack propagation, fatigue striations, and short-crack behavior, *J. Am. Ceram. Soc.*, 74:6:1206-16 (1991).

5. S. Lathabai, J. Rodel, and B. Lawn, Cyclic fatigue from frictional degradation at bridging grains in alumina, *J. Am. Ceram. Soc.*, 74:6:1340 (1991).

6. D.S. Jacobs and I-W. Chen, Mechanical and environmental factors in the cyclic and static fatigue of silicon nitride, *J. Am. Ceram. Soc.*, 77:5:1153-61 (1994).

7. See References 1-33 quoted in 3.

8. See, for example, J.W. Hutchinson, Nonlinear fracture mechanics, Technical University of Denmark (1979).

9. See, for example, B. Lawn, Fracture of brittle solids, Cambridge University Press, New York (1993).

10. G. Vekinis, M.F. Ashby, and W.R. Beaumont, R-curve behavior of Al_2O_3 ceramics, *Acta Metall. Mater.*, 38:6:1151-62 (1990).

11. D.S. Jacobs and I-W. Chen, Cyclic fatigue in ceramics: a balance between crack shielding accumulation and degradation, *J. Am. Ceram. Soc.*, 78:3:513-20 (1995).

12. I-W. Chen, S-Y. Liu and D.S. Jacobs, Effects of temperature, rate and cyclic loading on the strength and toughness of monolithic ceramics, *Acta Metall. Mater.*, 43:4:1439-46 (1995).

13. T. Fett, G. Himsolt and D. Munz, Cyclic fatigue of hot pressed Si_3N_4 at high temperatures, *Adv. Ceram. Mater.*, 1:2:179-84 (1986).

14. M. Masuda, T. Soma, M. Matsui, and I. Oda, Fatigue of ceramics (Part 3), cyclic fatigue behavior of sintered Si_3N_4 at high temperatures, *J. Ceram. Soc. Jpn., Int. Ed.*, 97:601-07 (1989).

15. L.X. Han and S. Suresh, High temperature failure of an alumina-silicon carbide composite under cyclic loads: mechanisms of fatigue crack-tip damage, *J. Am. Ceram. Soc.*, 72:7:1233-38 (1989).

16. L. Ewart and S. Suresh, Elevation-temperature crack growth in polycrystalline alumina under static and cyclic loads, *J. Mater. Sci.*, 27:5181-91 (1992).

17. C-K.J. Lin and D.F. Socie, Static and cyclic fatigue of alumina at high temperatures, *J. Am. Ceram. Soc.*, 74:7:1511-18 (1991).

18. S-Y. Liu, I-W. Chen, and T.Y. Tien, Fatigue crack growth of silicon nitride at 1400°c: a novel fatigue-induced crack tip bridging phenomenon, *J. Am. Ceram. Soc.*, 77:1:137-42 (1994).

19. D.B. Marshall and W.C. Oliver, Measurement of interfacial mechanical properties in fiber-reinforced ceramic composites, *J. Am. Ceram. Soc.*, 70:8:542-48 (1987).

20. R.H. Dauskardt, M.R. James, J.R. Porter, and R.O. Ritchie, Cyclic fatigue crack growth in a silicon carbide whisker-reinforced alumina ceramic composite: long- and small-crack behavior, *J. Am. Ceram. Soc.*, 75:4:759-71 (1992).

21. R.H. Dauskardt, B.J. Dalgleish, D.Yao, and R.O. Ritchie, Cyclic fatigue crack growth in a silicon carbide whisker ceramic composite: role of load ratio, *J. Mater. Sci.*, 28:12:3258-66 (1993).

22. See, for example, S. Suresh, Fatigue of materials, Cambridge University Press, New York (1991).

23. M.V. Swain and V. Zelizko, Comparison of static and cyclic fatigue on Mg-PSZ alloys, in Advances in Ceramics, Science and Technology of Zirconia III, Vol. 24B. S. Somiya, N. Yamamoto, and H. Hanagida, eds., American Ceramic Society, Westerville, OH (1988).

24. S.Y. Liu and I-W. Chen, "High Temperature Crack Growth in Silicon Nitride under Static and Cyclic Loading: Short-Crack Behavior and Brittle/Ductile Transition," in press, *Acta Metall. Mater.* (1995).

FATIGUE BEHAVIOR OF SINTERED Al₂O₃ UNDER ROTARY BENDING AND STATIC FATIGUE

H. N. Ko

Nakanihon Automotive College
Sakahogi-cho, Kamo-gun, Gifu-ken, Japan 505

INTRODUCTION

The fatigue behavior of ceramics is an important factor in the application of ceramics, particularly for structural components, since the fatigue strength is generally lower than the static strength. The fatigue behavior has already been examined in the various types of loading, and it is known for some ceramics that the fatigue strength depends on the loading condition such as the applied stress state [1]. However, the basic data on the effect of cyclic loading are of limited availability for plain specimens [2,3]. More information on factors affecting fatigue strength seems to be necessary to clarify the fracture mechanism of ceramics and their reliability as structural materials. The effect of grain size on the fatigue strength of sintered Al₂O₃ has not been studied much in comparison with that on the static strength, and elementary data are not sufficient [4,5]. The fatigue strength, both static and cyclic, seems to be dependent on grain size, but elementary data on this problem are of limited availability.

Therefore, both rotary bending and static fatigue tests were performed at room temperature on sintered Al₂O₃ plain specimens with different grain sizes to study the effects of loading condition and grain size on the fatigue behavior. The specimens with different grain sizes were prepared controlling the sintering temperature. The two fatigue tests were performed for a fairly long period to examine the existence of a knee. The rotary bending test was performed on the specimens with three grain sizes within the range of 10^1 to 10^9 stress-cycles using an Ono's rotary bending fatigue testing machine operating at 3420 cycles per minute. The static fatigue test was performed on the specimens with two grain sizes for less than 10^7 seconds using the non-rotating fatigue machine. The test results were compared with the static bending strength measured by the non-rotating fatigue machine, and the fractured surfaces were examined.

Based on the results, the difference in the two fatigue types, the effect of grain size on the fatigue behavior, a micro crack growth behavior and fatigue fracture features were discussed.

EXPERIMENTAL PROCEDURE

The material used was sintered Al₂O₃ obtained from Kyocera, Japan, with an alumina content of 99 %. The material was fabricated with SiO₂ and MgO, etc., as

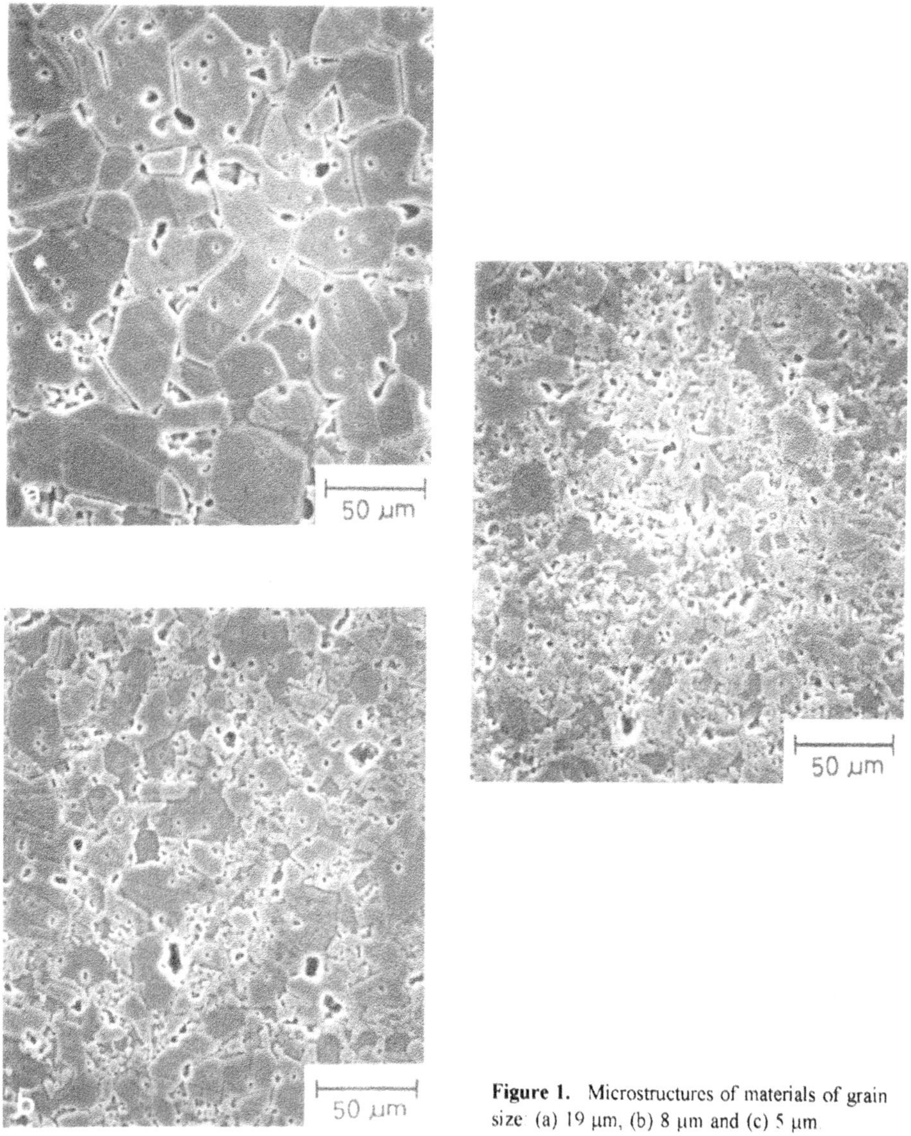

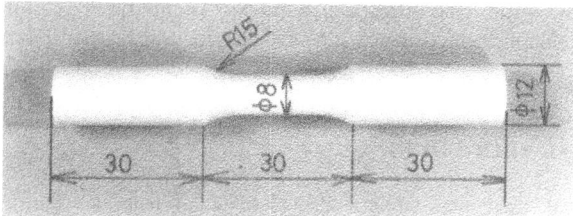

Figure 1. Microstructures of materials of grain size: (a) 19 μm, (b) 8 μm and (c) 5 μm.

Figure 2. Specimen geometry (dimensions in mm).

additives [6]. Three kinds of specimens with different grain sizes were prepared changing the sintering temperature slightly by 25 °C. Each microstructure of material is shown in Fig.1. Average grain sizes, determined by the linear intercept method, were about 5, 8, and 19 μm, respectively. Material characters such as fracture toughness and density were almost the same. The specimen geometry is shown in Fig.2. The specimen had a cylindrical shape. The diameter of the middle part was 8 mm, and each end of the specimen had a larger diameter of 12 mm for chucking. The straight length of the middle part was 15.4 mm, and the two parts of different diameters were connected by a smooth curved surface without generating stress concentration. The specimen was ground perpendicularly to its axis to make the finished surface smooth.

The machine used was an Ono's rotary bending fatigue testing machine operating at 3420 cycles per minute. The loading type of the machine was four-point bending; the stress state of the specimen under rotary bending was reversed bending. The static test and static fatigue test were performed using the non-rotating machine. The fractured surfaces were examined by scanning electron microscopy (SEM) and scanning laser microscope (SLM).

RESULTS AND DISCUSSION

Rotary Bending Fatigue Strength

The rotary bending test was performed on the specimens with three different grain sizes. The test was carried out generally within the range 10^4 to 10^5 stress-cycles. Some specimens were tested at cyclic numbers more than 10^5 to examine the existence of a knee; 10^5 stress-cycles is equivalent to about 20 days under the present test condition. Test points for samples with different grain sizes are plotted together on a semilogarithmic graph shown in Fig.3. The arrowed points indicate points for which testing was stopped. It is known that the fatigue strength of finer-grained material is higher than that of larger-grained material. It can be considered that the grain size is the primary factor for increased fatigue strength, although some characters of materials such as grain boundaries may be different [7]. The life of each material increases remarkably as the stress amplitude decreases, and each material does not have a distinct knee at cyclic numbers less than 10^7; some specimens are fractured at near 2×10^7 stress-cycles. However, from the figure, the knee seems to exist at cyclic numbers more than 10^5. The assumed fatigue limits are about 70, 100 and 120 MPa, respectively. A similar S-N curve, indicating the existence of a knee, was obtained in the previous study [4] for

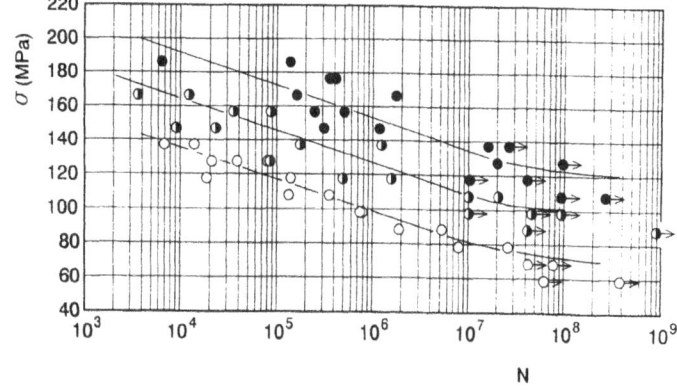

Figure 3. S-N curves for materials of grain size: (○) 19 μm, (◑) 8 μm and (●) 5 μm

sintered Al₂O₃ with grain sizes of 10 and 28 μm. These rotary bending results suggest that fatigue testing should be carried out at least for more than several tens of days to confirm the basic fatigue strength of sintered Al₂O₃, as well as sintered Si₃N₄ [8-10].

Test points are plotted on a logarithmic graph shown in Fig.4. The straight line in the figure was obtained by the least squares method from the data except the arrowed points. Each S-N curve can be approximately represented by a straight line up to about 10^8 stress-cycles and can be expressed by

$$\sigma^n N = \text{constant} \tag{1}$$

where σ is the stress amplitude and N is number of cycles to failure. The exponent n was obtained from the slope of the straight line. The exponents were 13.8 for 19 μm grain size, 19.8 for 8 μm grain size and 22.1 for 5 μm grain size, respectively. As already pointed out in the previous studies [4,5], the value of n for finer-grained material is higher than the value for larger-grained material.

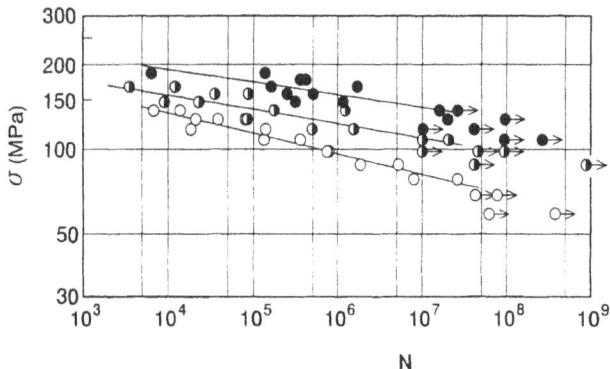

Figure 4. S-N curves for materials of grain size: (○) 19 μm, (◑) 8 μm and (●) 5 μm.

Static Bending Strength

The static bending test was performed using an Ono's rotary bending testing machine. After the specimen was attached to the machine, the applied stress was increased gradually without rotating the machine until it was fractured. Each mean static bending strength obtained from two specimens and the assumed fatigue limit are shown in Fig.5; the results [4], obtained on sintered Al₂O₃ with grain sizes of 10 and 28 μm, are also shown in the figure. The static strength of finer-grained material is higher than that of larger-grained material. The strength increment with grain size was in agreement with reported data [11,12].

Comparing the assumed fatigue limit with static bending strength, the ratio of the fatigue limit to the static strength decreased as the grain size increased, as shown later in Table 1. The effect of grain size seemed to be stronger on the fatigue strength than on the static strength. A similar result was obtained on sintered Al₂O₃ used in the previous study [4]; the ratios were 0.32 for material of 10 μm grain size and 0.15 for 28 μm grain size. These results show that the ratio is high as the material is strong. A similar tendency is seen in the results of sintered Si₃N₄ [8-10]. The ratio is important from the practical view point, so it seems necessary to obtain more information on ceramics.

As it appeared possible to correlate static bending strength with fatigue strength, their test results were plotted together on a logarithmic graph shown in Fig.6. Each straight line in the figure was obtained by the least squares method from the data except the arrowed points. Each S-N curve, represented by a straight line up to about 10^5 stress-cycles, can be expressed by Eq.(1). Each straight line in the figure is similar to that shown in Fig.4. It seemed that datum point of the straight line could be taken as the mean static strength. It is considered that the S-N curve, including the static strength, is useful for estimating the fatigue strength of ceramics under rotary bending, since the fatigue strength can be estimated by the static strength and the value of n.

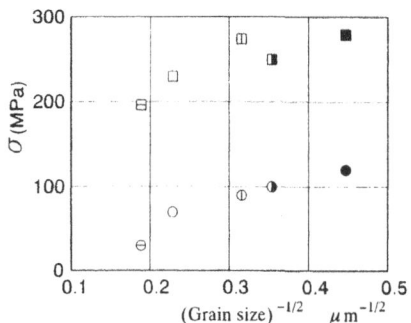

Figure 5. Static bending strength and assumed fatigue limit plotted against grain size for present and previous materials.

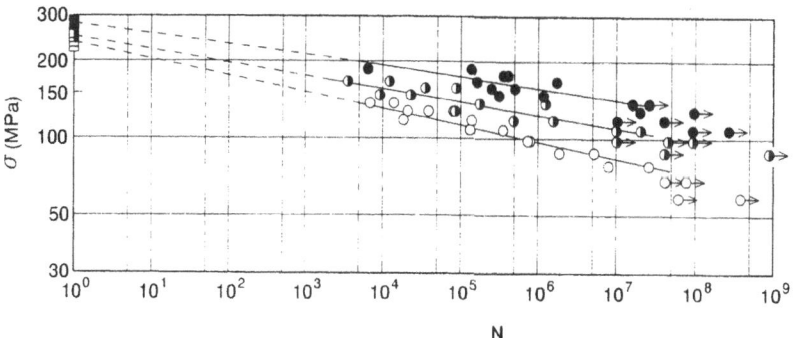

Figure 6. S-N curves, including static bending strength, for materials of grain size: (○) 19 μm, (◑) 8 μm and (●) 5 μm.

Static Fatigue Strength

The static fatigue test was performed on the specimens with two different grain sizes using a non-rotating fatigue machine. The time holding a constant load was measured with a hand made apparatus capable of displaying time digitally. The test was carried out generally for less than 10^7 seconds, and test points were plotted on a logarithmic graph shown in Fig.7. The solid and dashed lines were obtained by the least squares method from the data except the arrowed points. The S-t curve for 19 μm grain size can be approximated by a straight line up to about 10^7 seconds, and can be expressed by

$$\sigma^n t = \text{constant} \qquad (2)$$

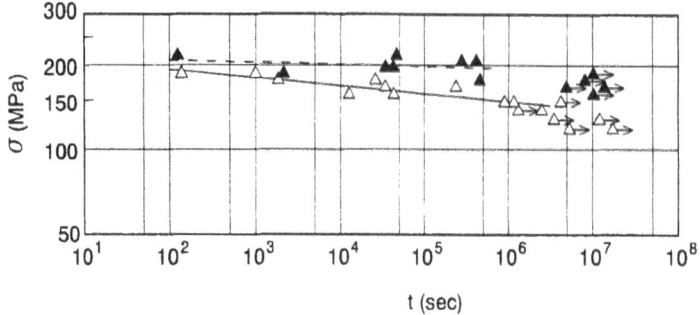

Figure 7. S-t curves for materials of grain size: ($\triangle$) 19 μm and ($\blacktriangle$) 5 μm.

where σ is the applied stress and t is the time to failure. The exponent n was 34.6, and was much larger than the value under rotary bending. The exponent n for the dashed line could not be determined exactly due to large scattering, but from the figure the value for 5 μm grain size seemed to be larger than the value for 19 μm grain size.

The static fatigue limit, defined as the stress level below which prolonged service life expected, seems to exist at the time more than 10^6 seconds. The assumed static fatigue limits are about 130 and 170 MPa, respectively. The strength increment with grain size is much smaller under static fatigue than under rotary bending.

Effect of Grain Size on Fatigue Behavior

The test results obtained are summarized in Table 1. The static strength is much higher than the fatigue limit or static fatigue limit. The effective volume V_E of the specimen for rotary bending is different from that for static fatigue although the specimen geometry is the same. Taking the size effect [13] into account, the fatigue limit σ_{RB} is related to the static fatigue limit σ_{SF} by the following equation

$$\frac{\sigma_{RB}}{\sigma_{SF}} = \left(\frac{V_{E.SF}}{V_{E.RB}} \right)^{1/m} \tag{3}$$

where m is the Weibull modulus. The effective volumes of circular specimens under rotary bending and static fatigue are obtained as follows:

$$V_{E.RB} = \frac{2V}{m+2} \tag{4}$$

$$V_{E.SF} = (2V/\pi) \int_0^{\pi/2} \sin^m\theta\cos^2\theta d\theta \tag{5}$$

where V is the volume of the middle part of the specimen in which the fracture occurred. Taking m=10 and applying Eqs.(4) and (5), the ratio of σ_{RB} to σ_{SF} in Eq.(3) was 0.81. As this value is larger than the experimental value 0.54 or 0.71, it can be said that the fatigue limit under rotary bending is lower than that under static fatigue, even taking the size effect into account. The initiation stress for the crack growth from the initial flaw is believed to be lower under rotary bending than under static fatigue. It should be emphasized that the ratio σ_{RB}/σ_{SF} is different due to grain size. The ratio is 0.54 for 19 μm grain size and 0.71 for 5 μm grain size, since the fatigue strength changes more remarkably with grain size under rotary bending than under static fatigue. This indicates that the effect of grain size is stronger on the rotary bending strength than on the static fatigue strength.

Table 1. Assumed fatigue limit and static strength.

Test conditions				Ratios			n	
Grain size d μm	Rotary Bending σ_{RB} MPa	Static Fatigue σ_{SF} MPa	Static Test σ_S MPa	$\dfrac{\sigma_{RB}}{\sigma_S}$	$\dfrac{\sigma_{SF}}{\sigma_S}$	$\dfrac{\sigma_{RB}}{\sigma_{SF}}$	RB	SF
5	120	170	280	0.43	0.61	0.71	22.1	
8	100		250	0.40			19.8	
19	70	130	230	0.30	0.57	0.54	13.8	34.6

Table 2. The values of n for three sintered Al_2O_3 specimens under cyclic and static fatigue at room temperature.

n	Experimental method		Alumina content (%)	Material	Reference
14	P	Cyclic fatigue (reversed)	99	Present material	
35	P	Static fatigue	99	Present material	
29	P	Cyclic fatigue (reversed)	99.6	Material A	2
40	P	Static fatigue	99.6	Material A	2
21	P	Cyclic fatigue (reversed)	99.5	Material B	22
37	P	Static fatigue	99.5	Material B	22
27	C	Cyclic fatigue (unidirectional)	99.5	Material C	26
44	C	Static fatigue	99.5	Material C	26
21	C	Cyclic fatigue (reversed)	99.6	Material A	2
38	C	Static fatigue	99.6	Material A	2
18	CT	cyclic fatigue (unidirectional)	99.9	Material D	19
60	CT	static fatigue	99.9	Material D	19
14	CT	cyclic fatigue (reversed)	99	Material E	27
33	CT	static fatigue	99	Material E	27

P   Plain specimen

C   Specimen pre-cracked by Vickers or Knoop indentor.

CT Compact tension specimen

It became clear that from Eqs.(1) and (2) the value of n obtained under rotary bending whose stress state was reversed bending was small compared with the value under static fatigue. The similar tendency is seen in the results [3] obtained on sintered Si_3N_4 plain specimens under the present test conditions. In Table 2 the values of n for sintered Al_2O_3 are listed for plain specimens, pre-cracked specimens and compact tension specimens, respectively. The difference in n for plain specimens is similar to those for pre-cracked specimens and compact tension specimens. It can be said that the value of n is smaller under cyclic fatigue than under static fatigue; the difference between the fatigue behavior in reversed bending and in repeated unidirectional bending will be the subject of a future study. As seen from Table 1, the value of n for finer-grained material is higher than the value for larger-grained material under rotary bending. This can be seen in the data (Fig.7) under static fatigue. A similar tendency is seen in the results obtained under dynamic fatigue [14,15]. It is considered that in the same applied stress state the value of n for sintered Al_2O_3 is different due to grain size.

The differences in the fatigue behavior, corresponding to loading condition and grain size, seem to be caused by the bridging effect [16] and the residual stress [15], but the quantitative analysis is needed in a future study.

Assumed Fatigue Crack Growth Behavior Corresponding to Loading Condition and Grain Size

The value of n indicates the fatigue resistance of materials, and it also seems to indicate the exponent in the following relation between the subcritical crack growth rate V and the stress intensity factor K_I

$$V = AK_I^n \tag{6}$$

where A is a material constant. The value of n, obtained from Eq.(1) or (2), was smaller under rotary bending than under static fatigue, and the value was higher for finer-grained material than for larger-grained material. The difference in n corresponding to loading condition or grain size, obtained on the present plain specimens, is believed to stem from the subcritical crack growth behavior from the initial flaw [17]. Although the mechanism is not clear, the applied stress state or the grain size seems to have direct effect upon the crack growth behavior, viz. crack initiation stress and n value.

Based on the results of sintered Si_3N_4 plain specimens [3], the failure of the present specimen seemed to occur when the size of the subcritical crack growth from the initial flaw reached a critical value dependent on the applied stress. The subcritical crack growth region looks semicircular and the size seems to be less than several hundreds μm; the fracture toughness of the material was 3.9 $MN/m^{3/2}$ which was measured by the Vickers indentation method [18]. Since it is difficult to obtain directly the relationship between K_I and V for such a micro crack, some simplified relationship must be assumed to explain the result.

The assumed schematic diagram of the micro crack growth behavior under rotary bending and static fatigue is shown in Fig.8, based on the assumption in which the crack growth behavior is treated in the simple way [17]. The abscissa of the coordinate axis can be estimated by

$$\frac{\sigma_{RB}}{\sigma_{SF}} = \frac{K_{Ith,RB}}{K_{Ith,SF}} \tag{7}$$

where K_{Ith} is the threshold stress intensity factor for the crack growth initiation, and σ_{RB} and σ_{SF} are the fatigue limits under rotary bending and static fatigue, respectively. The straight line in Fig.8 is simply based on the analogy that a micro crack growth is expressed by Eq.(6). The separation between the two lines in the figure becomes smaller for finer grain size since the ratio σ_{RB}/σ_{SF} increases as the grain size decreases. The material with much finer grain size may not take effect of cyclic loading.

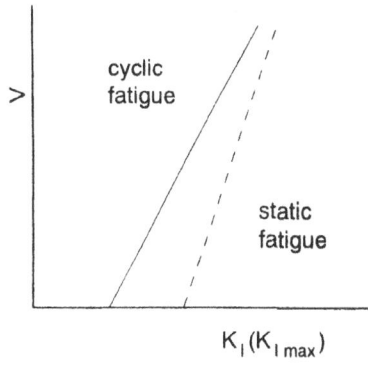

Figure 8. Schematic diagram of assumed fatigue crack growth behavior corresponding to loading condition.

The tendency of the difference corresponding to loading condition, shown in Fig.8, is similar to that [19] measured on compact tension sintered Al_2O_3 specimens under repeated unidirectional bending and static fatigue. The data [19] show that the crack growth rate under repeated unidirectional bending is higher than that under static fatigue at the same maximum stress intensity factor, and as shown in Table 2 the value of n obtained from Eq.(6) is smaller for the former than for the latter. The bridging effect seems to decrease in cyclic loading for a macro crack, and a similar effect may exist for a micro crack.

According to the assumption in which the crack growth behavior is treated in a simple way, the general tendency of the difference in the crack growth behavior under cyclic and static fatigue can be understood well. However, considering the effect of grain size on a crack growth behavior under the same loading condition, the micro crack growth behavior seems to differ with the macro crack behavior measured on such compact tension specimens. The crack growth data [19] of compact tension sintered Al_2O_3 specimens, under repeated unidirectional stress, show that the crack growth rate is lower for larger grain size (19 μm) than for finer grain size (1 μm) at the same maximum stress intensity factor. The larger-grained material is expected to have a greater resistance to the macro crack growth in alumina. On the other hand, the data [20] measured on a small crack whose size is about 340 μm show that the crack growth rate is higher for larger grain size than for finer grain size. The crack growth behavior of ceramics, as well as metals, seems to be different due to the crack size.

Fig.9 shows the assumed schematic diagram of the cyclic fatigue crack growth behavior corresponding to grain size. The abscissa of the coordinate axis can be estimated by

$$\frac{\sigma_{RB}}{\sigma_s} = \frac{K_{Ith,RB}}{K_{Ic}} \qquad (8)$$

where K_{Ith} is the threshold stress intensity factor for the crack growth initiation and σ_s is the static strength. The straight line in the figure is simply based on the analogy that a micro crack growth is expressed by Eq.(6). The assumed micro crack growth behavior should be similar to the above small crack growth behavior since the fracture toughness K_{Ic} is almost the same for the present materials. The difference in the static fatigue crack growth behavior corresponding to grain size can be treated in the similar way. Since the ratio σ_{SF}/σ_s increases less remarkably than the ratio σ_{RB}/σ_s as the grain size decreases, the separation between the two lines in the figure becomes smaller under static fatigue.

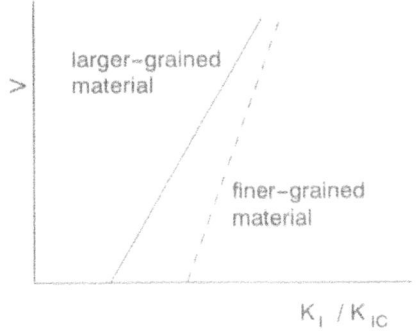

Figure 9. Schematic diagram of assumed fatigue crack growth behavior corresponding to grain size

Based on the basic fatigue data and the assumption in which the micro crack growth behavior is expressed by Eq.(6), the effect of grain size on the crack growth behavior can be understood well. The accumulation of fatigue data is needed hereafter for further explanation.

Fracture Features of Fractured Surface

The typical appearance of the fractured specimen is shown in Fig.10a and b. The fractured surface is perpendicular to the specimen axis. In many cases the last failure parts of specimens were fractured with branching (Fig.10b), whereas some specimens were fractured straight (Fig.10a), i.e. without branching. Sintered Si_3N_4, the fatigue strength of which was much higher than that of the present material, had a fractured shape similar to an inverted T or Y [8,10]. The fractured shape seems to be different because of the strength of the material or the applied stress. The present specimens were fractured with clearer branching as the fracture stress was high.

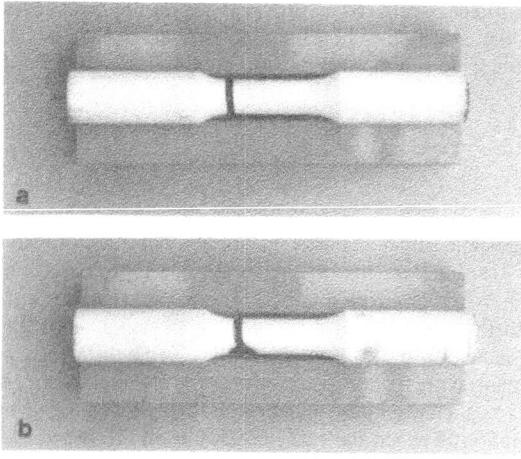

Figure 10. Fractured specimens after fatigue test: (a) without and (b) with branching.

Macroscopic observations showed that each overall fractured surface after the two fatigue tests was uniform and smooth in comparison with that after static test. An example of fractured whole surface is shown in Fig.11. The fracture initiation region is the top in the figure. The fracture seemed to propagate radially from a certain part near the specimen surface; this matter is confirmed as described later. Mirrors were not revealed on the fractured surfaces. It is pointed out by Kirchner and Gruver [21] that large grain material does not have mirrors. To observe mirrors it may be necessary to prepare much finer-grained material, viz. 1-2 μm, considering the results of sintered Si₃N₄ [8,10] and the reported observation [22].

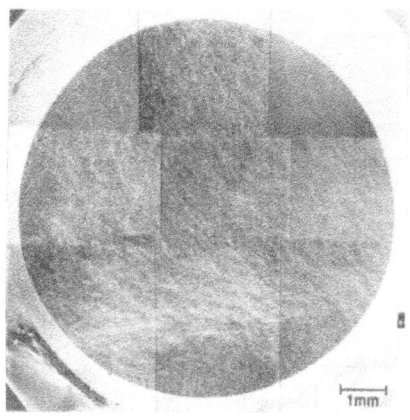

Figure 11. Fractured whole surface after static fatigue test (d=19 μm, σ=167 MPa, t=3.39 × 10⁴ sec).

Fig.12a-e show typical micrographs near the fracture initiation region of the fractured surface after the two fatigue tests. The fracture initiation region had the fracture origin such as a pore although there were some cases where the fracture origin could not be detected clearly. The fracture occurred mainly as a transgranular process and the fracture features were not different in the two fatigue tests. The fracture features were similar to those near the middle part of the fractured whole surface. Moreover, microscopic observations on the fractured surfaces after two fatigue tests were similar to those after static test and the fracture features after the three tests could not be distinguished. Within the limits of the experiment, the observations on finer-grained material were similar to those on larger-grained material, and no remarkable difference could be found; many pores with various shapes and sizes were observed, and the pore size for finer-grained material seemed to be a little smaller than that for larger-grained material. These observations show that no characteristic signs corresponding to the differences in the crack growth behavior were present.

The material in which the fracture occurred mainly as a transgranular process had the effect of cyclic loading. On the other hand, it has been reported elsewhere [23] that cyclic fatigue occurs only in the materials in which the intergranular-fracture takes place. This view is based on pre-cracked specimens made by Vickers indentor and may not be essential to existence of the effect of cyclic loading on plain specimens.

The information on fracture origins is necessary to clarify the fracture mechanism of ceramics. The fracture origin seemed to be a pore for the present specimens, but it was not easy to confirm this matter. Since these materials did not have clear mirrors, it was not easy to detect the fracture origin or crack initiation region clearly. Rice [24] points out that the combination of larger or a greater number of pores combined with one of a few larger grains is typically the fracture origin, based on his minute observation on Al₂O₃ samples used in the present study.

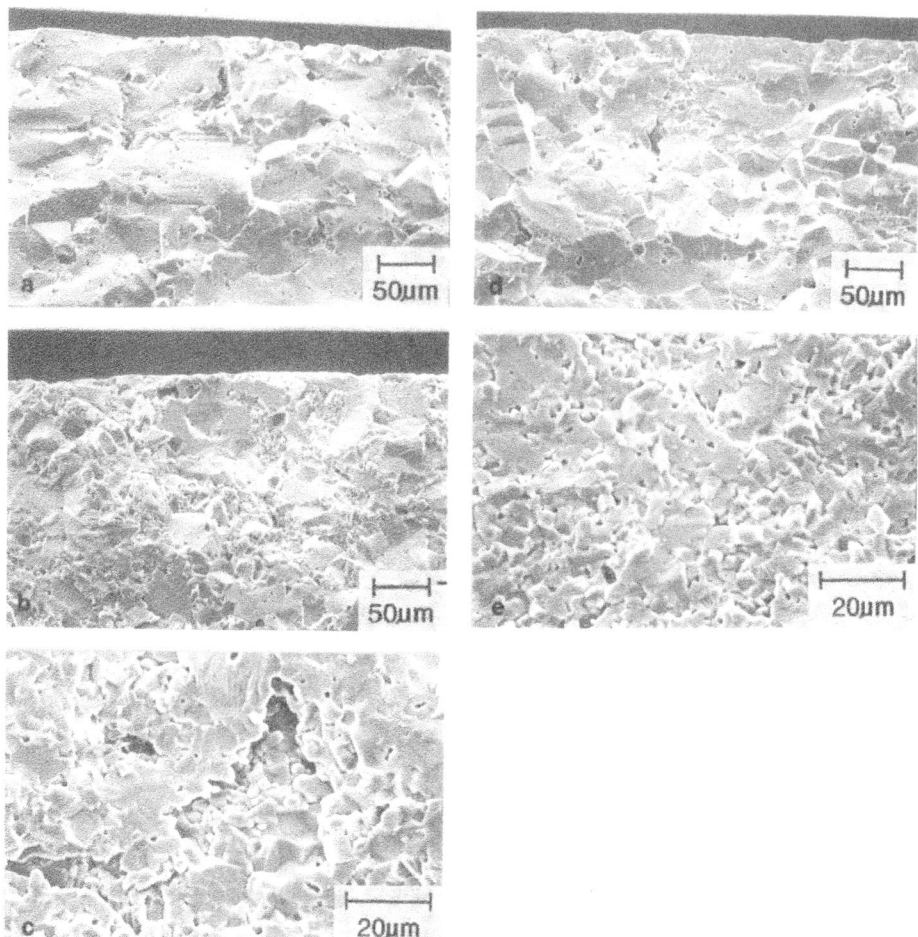

Figure 12. Typical micrographs near the fracture initiation region of the fractured surface after:
(a) rotary bending test (d=19 μm, σ=78 MPa, N=2.62 X 10^7)
(b) rotary bending test (d=8 μm, σ=108 MPa, N=2.06 X 10^7)
(c) rotary bending test (d=5 μm, σ=127 MPa, N=2.01 X 10^7)
(d) static fatigue test (d=19 μm, σ=167 MPa, t=2.37 X 10^5 sec)
(e) static fatigue test (d= 5 μm, σ=206 MPa, t=2.77 X 10^5 sec).

In the process of the microscopic observations, it was found that a pore had a string-like decoration [25] which seemed to indicate the fracture direction. Many decorated pores were found although some pores had no decorations. An example of a decorated pore is shown in Fig.13. The string-like decoration, named the tail, seemed to be the crack misfit due to the interaction of crack growth with pore. The directions of decorated pores were measured by SLM with the measuring instruments. The distribution of the directions, measured on the fractured whole surface, showed that the fracture initiation region or fracture origin could be detected inverting the tail direction. In Fig.14 the directions of decorated pores near the fracture origin are shown. From the figure it can be seen that the fracture propagates radially from the fracture origin; some decorated pores, which have conflicting directions, may be affected by the grain microstructure. It is generally difficult to detect the fracture origin for materials having no clear mirrors even by the minute observation. Thus, analyzing the tail direction is an useful method for detecting the fracture initiation region or fracture origin for materials having decorated pores. It is considered that the decorated pore is an useful sign, as well as mirrors, for clarifying the fracture features of ceramics.

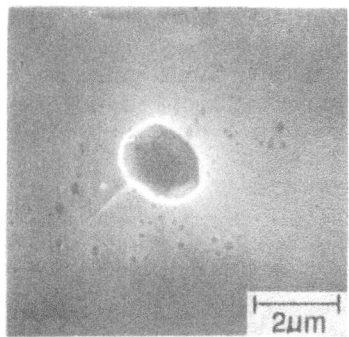

Figure 13. Decorated pore.

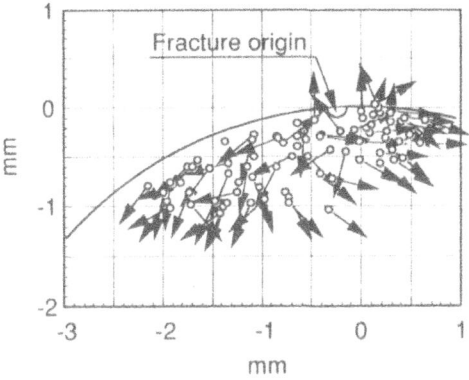

Figure 14. Tail directions near fracture origin

CONCLUSIONS

Rotary bending and static fatigue tests were carried out at room temperature on sintered Al_2O_3 plain specimens with different grain sizes. The results obtained are summarizes as follows:

1. The fatigue strength of finer-grained material was higher than that of larger-grained material. The effect of grain size seemed to be stronger on the cyclic fatigue strength than on the static fatigue strength.

2. The life of each material increased as the stress amplitude or the applied stress decreased, and the knee seemed to exist at cyclic numbers more than 10^5 or at the time more than 10^6 seconds, respectively. The assumed fatigue limit under rotary bending was lower than that under static fatigue. The ratio of the assumed fatigue limit to the static bending strength decreased as the grain size increased.

3. The value of n under rotary bending was much smaller than the value under static fatigue, in the expressions of $\sigma^n N$=constant and $\sigma^n t$=constant, and the exponent n for finer-grained material was higher than that for larger-grained material. The value of n was believed to be different due to not only the applied stress state but also the grain size.

4. The fatigue microcrack growth behavior seemed to be different corresponding to the grain size even under the same loading condition. The difference in the fatigue behavior due to the loading condition seemed to become smaller for the material with finer grain size.

5. Mirrors were not revealed clearly on the fractured surfaces. The fracture occurred mainly as a transgranular process and microscopic observations on finer-grained material were similar to those on larger-grained material. No remarkable difference could be found between fatigue fractured morphology and static fractured one.

6. The pore had a string-like decoration which seemed to indicate the fracture direction. Observing the decorated pores it could be seen that the fracture propagated radially from the fracture initiation region.

ACKNOWLEDGMENT

Sincere thanks should be presented to Dr. O. Kamigaito of Toyota Central Research and Development Laboratories for valuable advice and suggestions. Thanks should be also presented to Kyocera Kagoshima Factory for their support.

REFERENCES

1. R.O.Ritchie and R.H.Dauskardt, J. Cer. Soc. Japan, 99:1047 (1991).
2. T.Fett, G.Martin, D.Munz and G.Thun, J. Mat. Sci., 26:3320 (1991).
3. H.N.Ko, "Fracture Mechanics of Ceramics 9," Plenum Press, New York, 517 (1992).
4. H.N.Ko, J. Mat. Sci. Lett., 8:1438 (1989).
5. H.N.Ko, J. Mat. Sci. Lett., 11:1711 (1992).
6. Kyocera, private communication.
7. R.W.Davidge, "Fracture Mechanics of Ceramics 2," Plenum Press, New York, 447 (1974).
8. H.N.Ko, J. Mat. Sci. Lett., 6:175 (1987).
9. H.N.Ko, Proceedings of the MRS International Meeting on Advanced Materials 5, Material Research Society, Pittsburgh, 43 (1989).
10. H.N.Ko, J. Mat. Sci. Lett., 10:545 (1991).
11. R.R.Matheson, Ceram. Age, 79:54 (1963).
12. O.Johari and N.N.Parikh, "Fracture Mechanics of Ceramics 1," Plenum Press, New York, 399 (1974).
13. D.G.S.Davies, Pro. Brit. Cer. Soc., 22:429 (1973).
14. E.M.Rocker and B.J.Pletka, "Fracture Mechanics of Ceramics 4," Plenum Press, New York, 725 (1978).
15. A.J.Gesing and R.C.Bradt, "Fracture Mechanics of Ceramics 5," Plenum Press, New York, 569 (1983).
16. R.W.Steinbrech, "Fracture Mechanics of Ceramics 9," Plenum Press, New York, 187 (1992).

17. H.N.Ko, J. Mat. Sci. Lett, 12:1866 (1993).

18. K.Niihara, R.Morena and D.P.H.Hasselman, J. Mat. Sci. Lett., 1:13 (1982).

19. A.Ueno, H.Kishimoto, H.Kawamoto and S.Ogawara, Paper presented at the 20th Symposium on Fatigue, J. Soc. Mat. Sci. Japan, 153 (1990).

20. A.Ueno, H.Kishimoto, S.Ogawara, T.Kondou and A.Yamamoto, J. Soc. Mat. Sci. Japan, 43:183 (1994).

21. H.P.Kirchner and R.M.Gruver, "Fracture Mechanics of Ceramics 1," Plenum Press, New York, 309 (1974).

22. F.Guiu, M.J.Reece and D.A.J.Vaughan, "Fatigue of Advanced Materials," Materials and Component Engineering Publication Ltd., 193 (1992).

23. S.Horibe and R.Hirahata, Acta Metall. Mater., 39:1309 (1991).

24. R.W.Rice, private communication.

25. H.N.Ko and A.Ueno, J. Mat. Sci. Lett.; in press.

26. T.Kawakubo and A.Goto, J. Soc. Mat. Sci. Japan, 37:939 (1988).

27. M.J.Reece, F.Guiu and M.F.R.Summur, J. Amer. Cer. Soc., 72:348 (1989).

FATIGUE OF NOTCHED ALUMINA SPECIMENS

D. Hertel, T. Fett, and D. Munz

Universität Karlsruhe, Institut für Zuverlässigkeit und Schadenskunde im Maschinenbau, and
Forschungszentrum Karlsruhe, Institut für Materialforschung II
Postfach 3640, 76021 Karlsruhe, Germany

ABSTRACT

Subcritical crack growth data are necessary for lifetime predictions of ceramic components. Although, their transferability is limited, subcritical crack growth data are commonly obtained from simple specimen geometries and transferred to complex ceramic structures.

In this context, the influence of notches on the cyclic crack growth behavior was investigated for two different types of alumina. v-K_I curves were determined by a statistical method based on strength and cyclic tension-compression lifetime measurements of notched bending bars. The reliabilities of the notched specimens were evaluated using the multiaxial Weibull theory and compared with the experimental results.

The cyclic v-K_I curves were found to be dependent on the notch geometry. It will be shown that pure cyclic compression of notches can cause a degradation in strength, depending on the notch geometry. The same damage process is considered to be active during the cyclic tension-compression tests and is thus related to the geometry dependency of the v-K_I curves. Furthermore, it will be shown that the contribution of R-curve behavior to an increase in inert and fatigue strengths is also dependent on the notch geometry.

INTRODUCTION

Reduced cyclic lifetimes of ceramic materials have long been attributed to the same crack growth mechanism as observed under static loading[1]. In the last decade, however, several studies unequivocally proved the existence of a mechanical cyclic fatigue effect in various ceramic materials[2-10]. A summary of these studies can be found, e.g., in ref. 11. The authors also give a review of fatigue mechanisms which have been proposed so far.

The major mechanism acting in alumina is believed to be a reduction of crack-tip shielding due to the degradation of bridging grains in the wake of the crack. In situ observations showed an accumulation of „wear" debris, indicating a loss of traction at the junction of bridging grains[12,13]. Other mechanisms may be the formation of a microcrack-zone in front of the crack-tip or a wedging effect due to debris particles or surface asperities. Ewart and Suresh[2] reported on mode-I crack propagation in front of notches subjected to fully compressive far-field cyclic loads. This effect is reported to be due to the formation of a microcracking zone in front of the notch. Frictional sliding and crack closure of these microcracks led to residual tensile stresses within this zone upon unloading from the compressive stress[14].

The latter observations indicate that cyclic lifetimes of notched components cannot necessarily be predicted directly from experimental results obtained from smooth specimens, since different damage mechanisms may be involved. The aim of this study is to elucidate the influence of the notch geometry on cyclic life.

EXPERIMENTAL AND ANALYSIS PROCEDURES

Materials and Mechanical Properties

Two different qualities of alumina were investigated. Table 1 shows characteristic parameters of the materials. Material A is a fine-grained alumina with a higher flexural strength and a higher fracture toughness than material B.

Table 1. Characteristic parameters of the materials (according to manufacturers)

Material	Mean Grain Size [μm]	Flexural Strength (4-point) [MPa]	Weibull Modulus	Fracture Toughness [MPa$\sqrt{m}$]
A	≈2.5-3	510	10	4.3
B	≈20	300	10	3.5

The R-curve behavior was determined by controlled fracture of notched bending bars. The 3-point-bending test arrangement used is described in ref. 15. Figure 1 shows the crack resistance over crack extension for both materials. The R-curve behavior is more pronounced for the coarse-grained alumina (B) than for the fine-grained alumina (A).

Table 2. Specimen geometry

Material	Specimen Dimensions W×T×L [mm]	Notch Length l [mm]	Notch Radius r [mm]	Stress Concentration Factor K_t
A	4.0×3.0×45	1	0.05	5.9
A	4.0×3.0×45	1	0.8	1.79
B	4.5×3.5×45	1	0.05	6.26
B	4.5×3.5×45	1	0.8	1.89

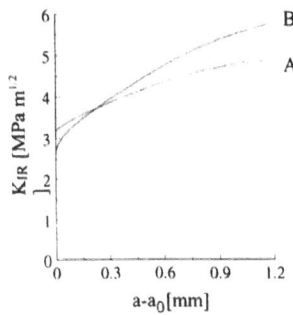

Figure 1. R-curves of the fine - (A) and the coarse -grained (B) alumina, initial notch length: a_0=2.2mm.

Sample Preparation

For each material, single edge notched specimens were prepared from the same batch of bending bars. The geometry of the notches was varied by changing the notch radius (Figure 2). The specimen and notch dimensions with the corresponding stress concentration factors are listed in Table 2. The stress concentration factors were calculated by finite-element analysis. Comparable surface qualities in the notch root were obtained by grinding the notches with the same feed and coarseness of the grinding wheels. All samples were annealed at 1150°C for 5 hours.

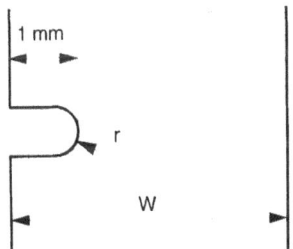

Figure 2. Notch geometry.

Fatigue Testing

Two different types of bending fatigue tests were performed (Figure 3). In the first type of test, the samples were subjected to fully reversed step shaped loading with stress amplitudes of $\pm\sigma_{max}$. For the second type of test, the notches were subjected to fully compressive loading at a minimum stress of $-\sigma_{max}$ and a maximum stress of $-0.1\sigma_{max}$. All tests were carried out at a frequency of 0.5Hz. A fatigue limit was defined at 500 hours. Samples exceeding a lifetime of 500 hours were defined as "survivors".

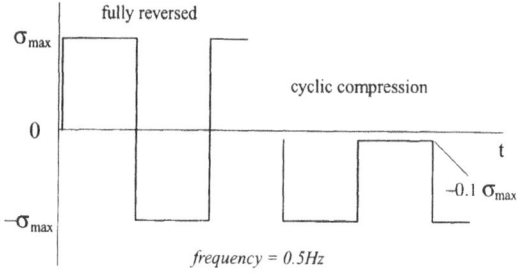

Figure 3. Loading history of the fatigue tests performed.

All tests were carried out with the cantilever arrangement[17] illustrated in Figure 4. The cyclic load was generated by the magnet systems of two loudspeakers and was transferred to the sample by a cantilever. The samples were fixed at their ends in brass tubes using an epoxy resin. The brass tubes were clamped into the fixing brackets. The bending moment was measured by strain gauges applied on the upper fixing bracket. At the moment of failure, a silver strip on the specimen was interrupted and a time counter was stopped.

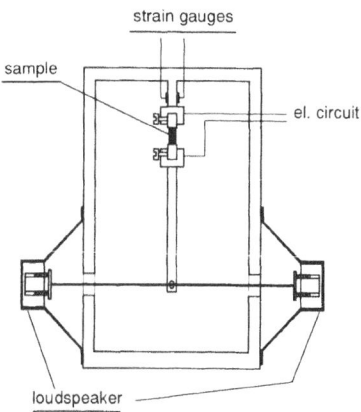

Figure 4. Testing device[17].

Inert Strength Testing

For comparison, the inert strengths were determined with the same testing device as described above. The load was monotonically increased by increasing the loudspeakers' voltage input. Stress rates of about 3000MPa/s were obtained.

NUMERICAL METHOD TO EVALUATE RELIABILITY OF NOTCHED BARS

The strength of a ceramic component is controlled by the most dangerous flaw. The scatter of strength can be attributed to the scatter of the flaw size. Since the most dangerous flaw may be located in a region of comparatively low stress, any stressed volume in the component contributes to the risk of rupture. Hence, for strength predictions, the stress distribution of the entire component has to be considered.

A powerful tool to evaluate the reliability of a component subjected to an arbitrary stress state is given by the multiaxial Weibull theory[18-21]. The basic idea of this theory is to consider the flaws as planar cracks. The cracks are assumed to be of random size, of random orientation and of random location with respect to the principal stress axes. The most dangerous flaw with the most unfavorable combination of size, location and orientation will cause failure. The reliability of a component is determined via an integral over the stress field. This integral may be regarded as the effective volume of the component:

$$V_{eff} = \frac{1}{4\pi} \int_V \int_\Omega \left(\frac{\sigma_{eq}}{\sigma^*} \right)^m d\Omega dV \tag{1}$$

where V denotes the volume of the component, Ω the surface of a unit sphere, $d\Omega = \sin\alpha \, d\alpha \, d\beta$ a surface element of a unit sphere with the polar angles α and β, m is the Weibull modulus and σ^* is a reference stress characterizing the loading of the component. σ_{eq} is an equivalent stress taking into account that the cracks are subjected to mixed mode loading, depending on their orientation relative to the principal stress axes. The equivalent stress has to be derived from a suitably defined multiaxiality criteria[22].

The effective volume is independent of the applied load level and is hence a convenient tool to characterize the effect of the stress distribution on the failure probability, particularly if two components with different geometries are compared. At a given level of reliability, the allowable stress levels $\sigma^{(1)*}$ and $\sigma^{(2)*}$ for two different geometries are related by:

$$\frac{\sigma^{(1)*}}{\sigma^{(2)*}} = \left(\frac{V_{eff}^{(2)}}{V_{eff}^{(1)}} \right)^{\frac{1}{m}}. \tag{2}$$

Equation (2) was applied to predict the failure stress of one notch geometry from the experimentally determined failure stress of the other geometry. Therefore, the stress fields in the notched bars were determined using the finite-element method, and the effective volumes were calculated by numerically integrating over the stress fields using the postprocessor STAU[23]. The mode I failure criterion was chosen as a multiaxiality criteria, leading to $\sigma_{eq} = \sigma_n$, where σ_n denotes the stress normal to the crack plane.

RESULTS AND DISCUSSION

Inert Strengths and Strength Predictions

Figure 5 shows the results of the bending strength tests in a typical Weibull-plot. Please note that the abscissa displays the failure stress at the notch root. The fitted lines were calculated using the maximum-likelihood-method[24]. The bold lines show the predictions of the strength for the sharp notches based on the results of the blunt notches. The experimental data clearly show that for both materials, the sharp notches tolerate higher notch stresses than the blunter ones. This is in agreement with the notion that a component with a smaller effective volume tolerates higher stresses (see equation (2)). The stress gradient in front of the sharp notch is steeper, compared to that of the blunter notch. Therefore, the highly stressed volume and, thus, the effective volume are smaller.

The strength predictions were calculated according to the procedure described above. For material A, very good agreement between experiment and prediction was obtained.

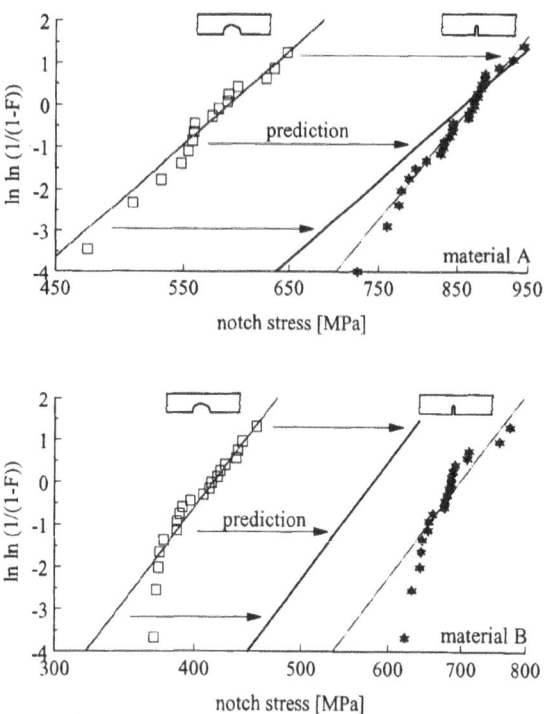

Figure 5. Flexural strength for both notch geometries and strength predictions for the sharp notches based on results of the blunt notches. Please note: figures indicate the stress at notch root.

However, for material B, the measured strength of the sharp notches significantly exceeded the prediction. It will be qualitatively shown below that the reason for this discrepancy is the pronounced R-curve-behavior of material B (Figure 1) which was not considered in calculating the predictions. Figure 6 illustrates the effect of the R-curve on the strength of the two notch geometries. The bold lines display the stress intensity factors of an edge crack emanating from the notch root. The stress intensity factors were calculated by boundary-element analysis. The dashed lines qualitatively show the R-curves of materials A and B, respectively, as well as the constant crack resistance curve (denoted as K_{I0}) of a material without R-curve behavior. Considering the latter, catastrophic failure occurs when the stress intensity factor of the initial crack a_i reaches the fracture toughness (K_{I0}). Hence, failure occurs at a stress σ_1 for the blunt notch and σ_2 for the sharper notch, respectively. However, considering the R-curve behavior, the crack may propagate in a stable manner until the tangent criterion of instability

$$\left(\frac{\partial K_{Iappl}}{\partial a} \right)_{\sigma=\sigma_c} = \frac{dK_{Ibr}}{d(\Delta a)}$$

is fulfilled. It becomes obvious that the pronounced R-curve-behavior of material B leads to a significant increase in strength of the sharp notch geometry. The failure stress increases from σ_2 to σ_3. For the blunt notch geometry, the R-curve does not significantly influence the strength because the tangent criterion is satisfied at about the initial crack size a_i. Analogous considerations for the moderate R-curve of material A reveal that for the sharp notch geometry very little, and for the blunt notch no R-curve controlled increase in strength can be expected. Therefore, only the strength of the specimens of material B with the sharp notches is significantly increased by the R-curve behavior. This is in agreement with the observed discrepancy between prediction and experiment. The sharp notches reach higher strength values than predicted because the failure is controlled by the R-curve behavior. Finally, it should be mentioned that another source of inaccuracy in predicting the strength of the sharp notches may be the steep stress gradient in front of the sharp notches. The steep stress gradient may lead to the violation of one basic assumption of the Weibull-theory, where only a moderate change in stress is allowed over the size of a crack.

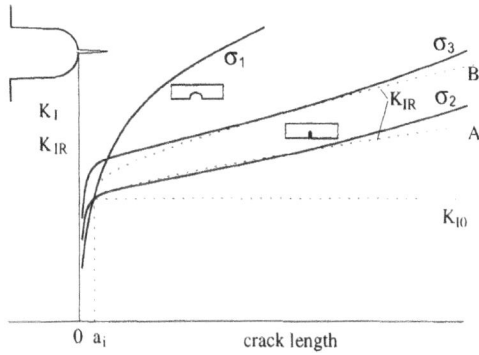

Figure 6.   R-curve-effect on strength for the two notch geometries.

Cyclic Lifetimes

Figure 7 shows the results of the cyclic lifetime experiments in a double logarithmic stress/lifetime plot. Again, the data are displayed with respect to the stress at the notch root. On the right hand side of the diagram, specimens exceeding a lifetime of 500 hours are listed as survivors. Similar to the inert strength, the sharp notches obtain higher fatigue strengths for both materials. The scatter in lifetime at a fixed stress level generally decreases with increasing lifetime. This can be appreciated when analyzing the failure probabilities. The lines connecting the failure probabilities of 10%, 50%, and 90% converge with increasing lifetime. Fett and Munz[27] related the reduced scatter of cyclic lifetimes, compared to the scatter in static tests, to the vanishing shielding stress intensity factor under cyclic loading. For the cyclic tests, the authors considered the limit case of a totally vanished shielding stress intensity factor. Since the bridging stresses between the crack surfaces, and thus, the shielding stress intensity factor, decrease with increasing number of cycles, it is consistent that the scatter gradually decreases with increasing lifetime.

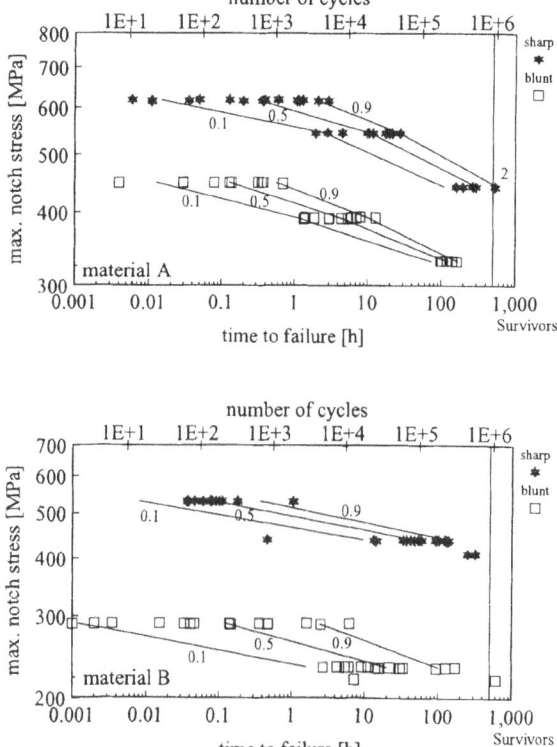

Figure 7. Cyclic lifetimes; straight lines indicate failure probabilities (10, 50 and 90%) at fixed stress levels.

The influence of notch geometry on cyclic lifetimes can be discussed using a normalized stress-lifetime representation. If the same flaw population is responsible for strength and fatigue failure, normalization of the applied stresses by the inert strength of the corresponding notch geometry will lead to a nearly unique stress-lifetime curve. It will be independent of the specimen geometry. Figure 8 shows the normalized representation of the lifetime data, where the mean strengths of the corresponding geometry was obtained from the data given in Figure 5. The slopes of the fitted lines yield the crack growth exponents n_c for a power law description of cyclic crack growth. The lines were fitted through the mean values of lifetime of each stress level.

Material A shows the expected unique normalized stress-lifetime-curve for both notch geometries. The obtained crack growth exponents are in good agreement with each other. However, for material B, we do not obtain a unique curve. The crack growth exponent of the sharp notch is significantly larger. Extrapolation of the fitted lines indicates that for the sharp notches a 10^6-cycle "fatigue limit" approaches approx. 60% of their inert strength, whereas for the blunter notches it approaches approx. 46% of the corresponding inert strength.

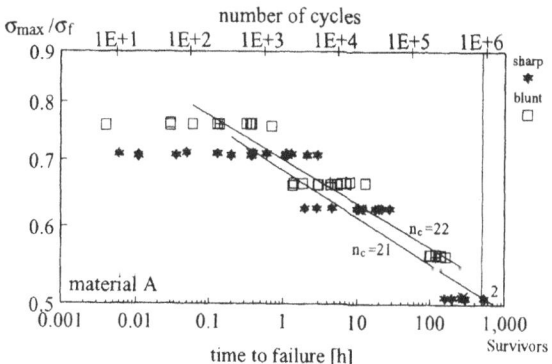

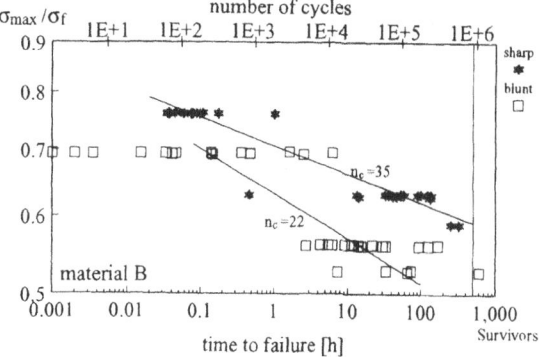

Figure 8.   Cyclic lifetime at normalized maximum stress, frequency = 0.5 Hz.

Evaluation of v-K Curves

A more sophisticated procedure of evaluating the $v(K_I)$-relationship from the stress/life data is given by the modified lifetime method[25]. This procedure is based on the scatter of natural cracks. The resulting scatter of the inert strength is related to the scatter of the lifetime at a fixed stress level. This method has the advantage that no assumption has to be made on the type of the $v(K_I)$-relationship (e.g., power law).

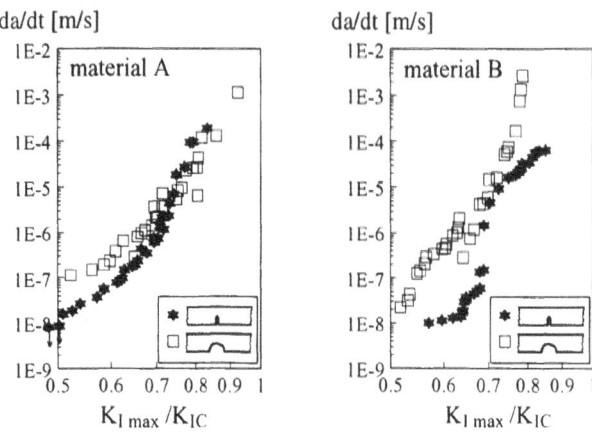

Figure 9. v-K curves.

Figure 9 shows the v-K_I-relationships for the two materials obtained by the modified lifetime method. For material A, a deviation of the v-K_I curves of the two notch geometries can be observed for a region of relatively small loading ($K_{Imax}/K_{IC} < 0.7$). The crack growth rates for the blunt notches are by up to one order of magnitude higher than those of the sharp notches. For material B, the deviation of the two curves is more pronounced. The slope of the v-K_I curve of the sharp notch geometry is significantly steeper compared to that of the blunt notch. This agrees with the diverging crack growth exponents n_c found in Figure 8. (The jagged shape of the curve of the sharp notches is caused by the pronounced R-curve behavior of material B. Further details on this effect can be found in ref.[26].) It can be concluded from these plots that the crack growth rates are influenced by the notch geometry. The influence is stronger the lower the loading and it is more pronounced for material B.

Role of Cyclic Compression on Fatigue Behavior

One indication for understanding the role of the notch geometry in cyclic fatigue is the formation of a damage zone in front of notches subjected to cyclic compression (see Introduction). In order to examine the amount of damage introduced by the compressive part of the (fully reversed) cyclic lifetime tests, the notches were subjected to pure cyclic compression (Figure 3) prior to strength testing. In the testing procedure basically the same cyclic experiments were repeated, except that the tensile part of the cycles was set zero and that the tests were stopped after the corresponding lifetime. Figure 10 compares the residual strengths, after cyclic compression for a given time and stress, to the corresponding inert strengths. The stress levels in the cyclic compression experiments are quantified in terms of the ratio of the absolute value of the minimum stress σ_{min} to the inert strength of the corresponding notch geometry σ_f:

compressive loading:
$$\frac{|\sigma_{min}|}{\sigma_f} 100\%$$

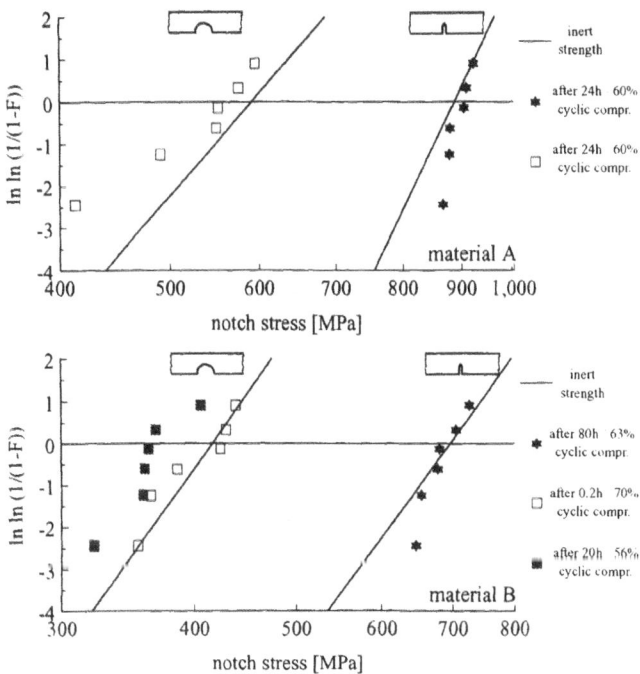

Figure 10. Degradation of strength by cyclic compression; 100% in the figure means |minimum stress| = strength.

Figure 10 (top) shows the results of material A. Both notch geometries were subjected to cyclic compression at a "stress level of 60%" for a period of 24 hours. For the sharp notch geometry, no degradation in strength could be observed. However, the residual strength of the blunter notch geometry appears to be somewhat lower, compared to the inert strength. For material B (Figure 10, bottom), the blunt notches were examined at two different stress levels: Cyclic compression at 70% for 0.2 hours did not show any degradation in strength. However, a significant degradation was observed after cyclic compression at 56%

for 20 hours. Finally, for the sharp notches no degradation was observed after cyclic compression at 63% for 80 hours.

The results indicate: 1) For both materials, the blunt notches are more susceptible to strength degradation by cyclic compression than the sharp notches: Cyclic compression at the same normalized stress level and for the same period of time revealed a degradation in strength for the blunt notches only (material A). For material B, cyclic compression of the sharp notches at a "stress level of 63%" for 80 hours caused no degradation, whereas cyclic compression of the blunt notches at a lower stress level and for a shorter period of time (56%, 20 hours) led to a significant decrease in strength. 2) The degradation in strength is strongly controlled by the loading period. This becomes obvious if one compares the residual strength of the blunt notches (material B) cycled at 70% for 0.2 hours with those cycled at 56% for 20 hours. The latter samples showed a stronger degradation in strength, even though they were subjected to smaller compressive stresses. The longer loading period obviously overcompensated the lower compressive stresses.

It is believed that the dependency of the v-K_I curves on the notch geometry is caused by an additional damage process introduced by the compressive cycles of the tension-compression tests. The increased crack growth rates observed for the blunt notches can be related to the accumulation of damage caused by cyclic compression. The amount of damage introduced was shown to be higher the longer the loading period lasted thus leading to increased crack growth rates at lower stress levels. A comparison of the results of the two materials makes obvious that material B is the material more susceptible to damage under cyclic compression, leading to a more pronounced reduction in residual strength and in cyclic lifetimes.

CONCLUSIONS

The main topic of this report is the influence of notches on the cyclic subcritical crack growth behavior of alumina under cyclic loading. Strength and cyclic fatigue tests were carried out on single edge notched bending bars with two different notch geometries. Two different types of alumina were examined. The influence of notches on strength was discussed and compared to the influence on the crack growth behavior. The main results are:

•Sharper notches with steeper stress gradients generally tolerate higher notch stresses, not only in strength but also in fatigue testing. This is because the failure is controlled by the spatial stress distribution and not by the maximum stress in the component.

•Strength predictions on the basis of the multiaxial Weibull theory were found to be in good agreement with the experimental results as long as failure was caused by unstable crack extension of the initial crack. For the coarse-grained alumina (material B), it was shown that the R-curve behavior can cause distinct stable crack extension prior to catastrophic failure, leading to a significant increase in strength. This increase in strength depends on the notch geometry and could not be taken into account in the strength predictions.

•Cyclic tension-compression lifetime tests revealed a dependency of the v-K_I curves on the notch geometry. In comparison with the sharp notches, the blunt notches showed higher crack growth rates. This tendency was stronger the lower the stress level, i.e., the higher the number of cycles was. Furthermore, this effect was found to be more distinct for the coarser-grained alumina (material B).

•The dependency of the v-K_I curves on the notch geometry was related to an additional damage process introduced by cyclic compression of the notches during the cyclic tension-compression tests. In analogy to the enhanced crack growth rates, the amount of damage introduced by cyclic compression of notches was shown to be controlled by the same parameters, such as the loading history (stress level and loading period), the shape of the notch, and the microstructure of the material.

REFERENCES

1. Evans, A.G., Fuller, E.R.: Crack Propagation in Ceramic Materials under Cyclic Loading Conditions. Metall. Trans., 5 [1] 27-33. (1974).

2. Ewart, L., Suresh, S.,: Crack Propagation in Ceramics under Cyclic Loads. J. Mater. Sci. 22 [4] 1173-1192 (1987).

3. Reece, M.J., Guiu, F., Sammur, M.F.R.: Cyclic Fatigue Crack Propagation in Alumina under Direct Tension - Compression Loading. J. Am. Ceram. Soc., 72 [2] 348-352, (1989).

4. Fett, T., Martin, G., Munz, D., Thun, G.: Determination of da/dN-ΔK Curves for Small Cracks in Alumina in Alternating Bending Tests, J. Mat. Sci. 26, 3320-3328, (1991).

5. Dauskardt, R.H., Yu, W., Ritchie, R.O.: Fatigue Crack Propagation in Transformation-Toughened Zirconia Ceramic. J. Am. Ceram. Soc., 70 [10] C248-252, (1987).

6. Sylva, L.A., Suresh, S.: Crack Growth in Transforming Ceramics under Cyclic Tensile Loads. J. Mat. Sci. 24, 1729-1738, (1989).

7. Masuda, M. Soma, T., Matsui, M., Oda, I.: Cyclic Fatigue of Sintered Silicon Nitride. Ceram. Eng. Sci. Proc. 9 [9-10] 1371-1382 (1988).

8. Masuda, M. Soma, T., Matsui, M.: Cyclic Fatigue Behavior of Silicon Nitride Ceramics. J. Eur. Ceram. Soc. 9 [4] 253-258 (1990).

9. Kishimoto, H., Ueno, A., Kawamoto,H.: Crack Propagation Behavior of Silicon Nitride under Cyclic Loads, in Fatigue of Advanced Materials, in press, ed. R.O. Ritchie, R.H. Dauskardt, B.N. Cox.

10. Jacobs, D.S., Chen, I-W.: Cyclic Fatigue in Ceramics: A Balance between Crack Shielding Accumulation and Degradation, J. Am. Ceram. Soc., 78 [3] 513-20 (1995).

11. Ritchie, R.O., Dauskardt, R.H.: Cyclic Fatigue of Ceramics: A Fracture Mechanics Approach to Subcritical Crack Growth and Life Prediction., J. Ceram. Soc. Japan. 99 [10] 1047-1062 (1991).

12. Lathabai, S., Rödel, J., Lawn, B.R.: Cyclic Fatigue from Frictional Degradation at Bridging Grains in Alumina, J. Am. Ceram. Soc., 74 [6] 1340-48, (1991).

13. Frei, H., Grathwohl, G.: The Fracture Resistance of High Performance Ceramics by in situ Experiments in the SEM, Beitr. Elektronenmikroskop. Direktabb. Oberfl. 22, 71-78 (1989).

14. Brockenbrough, J.R., Suresh, S.: Constitutive behavior of a microcracking brittle solid in cyclic compression, J. Mechanics and Physics of Solids 35, 721-42.

15. Fett, T., Thun, G.:Vorrichtung zur Durchführung von kontrollierten Biegebruchversuchen, German device patent G9207089.2.

16. Fett, T.: Contributions to the R-curve Behavior of Ceramic Materials, KfK-Bericht 5291, 138-143 (1994).

17. Fett, T., Thun, G.: German Device Patent G 9107645.5.

18. Batdorf, S.B., Crose, J.G.: A Statistical Theory for the Fracture of Brittle Structures Subjected to Nonuniform Polyaxial Stresses, J. Appl. Mech., 41, 459-61 (1974).

19. Batdorf, S.B., Heinisch, H.L.:Weakest Link Theory Reformulated for Arbitrary Fracture Criterion, J. Am. Ceram. Soc., 61 [7-8] 355-58 (1978).

20. Evans, A.G.: A General Approach for the Statistical Analysis of Multiaxial Fracture, J. Am. Ceram. Soc., 61 [7-8] 302-308 (1978).

21.   Matsuo, Y.: A Probabilistic Analysis of the Fracture Loci under Bi-axial Stress State, Bull.JSME, 24, 290-94 (1981).

22.   Thiemeier, T., Brückner-Foit, A., Kölker, H.: Influence of the Fracture Criteria on the Failure Prediction of Ceramics Loaded in Biaxial Flexure, J. Am. Ceram. Soc. 74 [1], 48-52 (1991).

23.   Heger, A., Brückner-Foit, A., Munz, D.: STAU-ein Programm zur Berechnung der Ausfallwahrscheinlichkeit mehrachsig beanspruchter keramischer Komponenten als Post-Prozessor für Finite-Elemente-Programme, Internal Report, Institute for Reliability and Failure Analysis, University of Karlsruhe, Germany, (1991).

24.   Kreyszig, W.: Statistische Methoden und ihre Anwendungen, Vandenhoeck & Rupprecht Göttingen, 7. Auflage (1979).

25.   Fett, T., Munz, D.:Determination of v-K_I-curves by a modified evaluation of lifetime measurements in static bending tests, Commun. of the Amer. Ceram. Soc. 68, C213-C215 (1985).

26.   Fett, T., Munz, D.: Subcritical Crack Growth of Macro- and Microcracks in Ceramics, Fracture Mechanics of Ceramics [9], 219-233 (1991).

27.   Fett, T., Munz, D., Differences between Static and Cyclic Fatigue Effects in Alumina, J. Mater. Sci. Let. 12, 352-354 (1993).

STATIC AND CYCLIC FATIGUE OF ZIRCONIA MATERIALS
MEASURED BY DOUBLE TORSION

J. Chevalier, C. Olagnon, and G. Fantozzi

INSA-GEMPPM, URA 341
F69621 Villeurbanne Cedex France

ABSTRACT

A study of the static and cyclic fatigue behaviour of 3Y-TZP has been conducted with the Double Torsion (DT) method. In order to obtain precise results, we have first focused on the DT specimen itself. An evolution of the stress intensity factor with crack length has been observed, and taken into account in a modified K_I expression. A strong crack accelerating effect has been observed in cyclic fatigue compared to static fatigue. Power laws were shown to describe well data for both static and cyclic measurements with an exponant lower in cyclic loading. Thresholds under which no propagation occured were measured. They were lower under cyclic solicitation, and depended on loading conditions.

I. INTRODUCTION

The use of Yttria-Stabilized Zirconia Polycrystal (Y-TZP) ceramics for hip prosthesis applications is growing because of their unique advantages at room temperature[1]. Y-TZP presents a high toughness associated with an high strength, as consequences of transformation toughening. Moreover, its fine grain microstructure leads to very good wear properties. For hip prosthesis applications, the evaluation of the lifetime is of major interest and must integrate crack propagation laws under service conditions. Hip protheses are subjected to both static and cyclic solicitations. Therefore, subcritical crack growth laws under static and cyclic loadings are required.

Subcritical crack growth (SCG) under static loads has been extensively studied in the past decades. Theoretical and experimental investigations have shown that SCG was due to stress corrosion at the crack tip[2]. Crack extension rate (V) is controlled by the reaction between the ceramic and the corrosive molecules and by the applied stress intensity factor K_I. It has been observed for a wide range of ceramic materials that three propagation stages were present at ambient temperature[3]. The first stage is mainly governed by the reaction rate at the crack tip, while the second is ruled by the diffusion of the corrosive molecules to the crack tip. The third stage stage takes place near the toughness K_{IC} and leads to fast fracture. Figure (1) shows a typical V-K_I diagram in a log-log scale, where the three stages are present. In the first propagation stage, i.e for crack rates less than 10^{-4} m/s^2, V can be related to K_I by the expression:

$$V = A * K_I{}^n \qquad (1)$$

where A and n are material properties.
K_{I0} is a threshold value under which no propagation occurs. It has sometimes been observed in ceramic materials[4] and is of prime importance in practice because it defines unlimited lifetime for ceramic components.

In the past decades, the traditionnal view of ceramics as brittle materials prohibited the admittance of specific cyclic mechanisms (as observed in metals) operative under cyclic loading conditions. The refuted existence of mechanical fatigue was based on the absence of plasticity. It was shown for glass and porcelain that SCG under cyclic loading could be explained by the same environnemental and corrosive processes responsible for SCG under steady loads[5]. However, recently, following the studies of Dauskardt et al. for Mg-PSZ[6], several investigations have shown reduced lifetimes for both transforming[7-12] and nontransforming ceramics[13-16] under cyclic loading conditions.

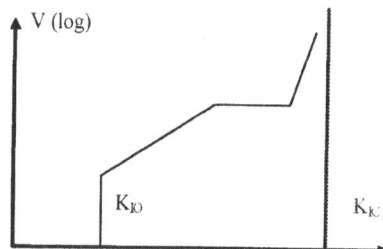

Figure 1 : Typical V-K_I diagram in ceramics (log-log scale).

In nontransforming ceramics such as Al_2O_3 and Si_3N_4, crack-bridging occurs in static loading conditions and imparts for a rising toughness (R-curve)[17,18]. The R-curve results from the accumulation of shielding by crack bridging ligaments with crack extension. A decrease of crack-tip shielding has been observed in cyclic fatigue from in situ observations of the crack-bridging ligaments degradation, due to repeated abrasion[19]. Gilbert et al.[20] and more recently Jacobs et al.[21] have modelled cyclic crack propagation in this way, and a good agreement with experimental results was obtained. In transformation-toughened zirconia ceramics, crack-tip plasticity occurs due to the martensitic transformation from tetragonal to monoclinic symmetry at the crack front[22,23]. A compressive transformed zone surrounds the crack and results in crack-tip shielding, thus in a rising toughness. It is to note that crack bridging can superimpose to transformation toughening. Recently, Tsai et al.[24] have investigated the origin of cyclic fatigue in a Ce-TZP/Al_2O_3 ceramic. From their results, it appears that cyclic fatigue effects are due to the transformation zone shielding degradation. Transformation zones formed during cyclic loading showed lower compressive residual stresses at the crack tip than the zones formed in static loading. Observations of compressive transformed zones revealed a different morphology in cyclic loading as compared to sustained loading conditions. As for nontransforming ceramics, cyclic fatigue effects are therefore linked to the degradation of crack-tip shielding. In our mind, cyclic fatigue can therefore be expressed in terms of stress corrosion or/and crack-tip shielding degradation.

SCG under sustained loading in 3Y-TZP has been studied by Knechtel et al.[25] and by Chevalier et al.[26] and was shown to be due to stress corrosion. Knechtel et al.[25] suggested the presence of a threshold stress intensity factor K_{I0}, but the few results presented are to be confirmed. Independently, cyclic fatigue of the same material has been investigated by Liu et al.[7]. The authors proposed a propagation law in cyclic fatigue for short cracks, strongly dependent on the maximum stress intensity factor of the cycles. Cyclic fatigue effects were attributed to microcraking around the crack front, but no microstructural observations addressed this issue. No threshold K_{I0} under cyclic loads was noted, because of the relatively high crack rates investigated ($V > 10^{-9}$ m/s). The purpose of the present investigation was to determine precise SCG laws in both static and cyclic fatigue, and to compare them in terms of shielding degradation. Threshold intensity factors were investigated from the characterisation of low crack rates ($< 10^{-10}$ m/s). Macroscopic flaws were characterized with the Double Torsion (DT) method, and results were compared with those reported by Liu et al.[7] for microscopic cracks (Vickers-induced cracks).

II. EXPERIMENTAL PROCEDURE

II.1. Material

The material studied in this work was a 3%mol Yttria-stabilized Zirconia (3Y-TZP) sintered at 1450°C (PROZIR, Ceramiques Techniques Desmarquest). A density of 6.06 g/cm³ was measured by Archimed method and the average grain size was 0.58 μm. 3Y-TZP exhibits a steeply rising R-Curve over few microns (~ 10 μm), and a subsequent plateau toughness[7]. Therefore, DT study using specimens containing large cracks in the present investigation was not perturbed by a rising toughness.

II.2. Specimen geometry

The geometry of the DT specimen is shown in figure (2). It consists of a plate of dimensions 20*40*2 mm³.

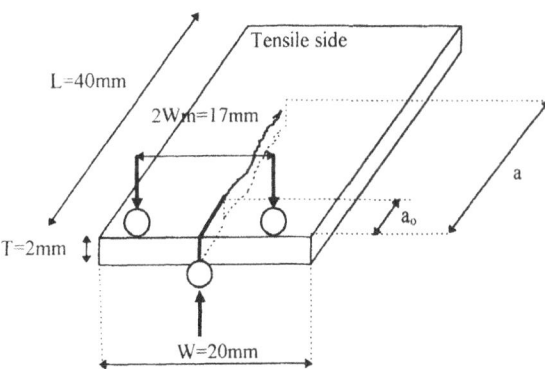

Figure 2 : Double Torsion specimen and loading configuration.

The tensile surface was polished down to 1 μm to observe the cracks with a precision of ± 2μm with an optical microscope. The notch of dimensions : length a_o=10 mm and root ρ=0.1mm was machined with a diamond saw. Subsequent precracking induced a sharp 'natural' crack of length of ~13mm in each test specimen. A good alignment in the jig, associated with a good control of the specimen parallelism, allowed to obtain a straight propagation through the specimen.

Most of the studies conducted by Double Torsion published in the literature were carried out on specimens with a groove to guide the crack. However, it has been shown that such a groove has a strong influence on the results[3], leading to significant errors on crack velocities. Therefore, no groove was machined on the tensile side in order to avoid any residual stress intensity factor.

II.3. Testing methods

Tests were conducted using an universal electromechanic testing machine (Model 8562, Instron) in ambient air, water, Ringer solution (characteristic of the human body) or Silicon oil. Cyclic fatigue frequency was limited to 1 Hz by the testing machine.

For the determination of the evolution of K_I with crack length and for the study of SCG under static loading, experimental work consisted of constant loading and relaxation tests. In constant loading tests, specimens were subjected to static loads (P) for varying durations. Crack speed in each case was calculated as a ratio of crack increment to the loading time. The load (P) was related to the stress intensity factor. One test therefore allowed us to obtain a "crack growth rate-applied stress intensity factor" couple. SCG laws were recorded in the graphical form of well-known (V-K_I) relation. In the relaxation tests the pre-cracked specimens were subjected to fast loading, followed by subsequent stopping of the crosshead at a given load. The crack propagation induced a load relaxation. The load versus time curve allowed the determination of the V-K_I curve[28]. Cyclic fatigue tests consisted in submitting the DT specimens to sinusoïdal loads at constant P_{max} and ΔP. Data were presented in terms of V-K_{max} or V-ΔK (where K_{max} and ΔK are respectively the maximum and the range of the applied stress intensity factor during a cycle).

III. STUDY OF THE DOUBLE TORSION SPECIMEN

III.1. Introduction

The Double Torsion method was chosen because of its simple geometry and the possibility of conducting extensive stable crack propagation in a wide range of velocities under constant loading or in the stress-relaxation mode[27-29]. Moreover, studies based on a compliance analysis show that the stress intensity factor is independent of the crack length[29]. Therefore, the DT test theoretically offers a great advantage in comparison with other methods. However, the effective constancy of K_I over the specimen length has often been questionned[30]. From a finite element stress analysis, Trantina[31] observed a variation of K_I with crack length strongly dependent on the crack profile. Chevalier et al.[32] have shown the important role of contact stresses generated on the uncracked compressive side of the specimen. Therefore, in this study, we have first studied the evolution of K_I with crack length, in order to further obtain accurate crack propagation laws. A corrective formula, including the variation of crack length is proposed to obtain SCG parameters during static and cyclic fatigue.

The well-known analytical expression of the stress intensity factor for the DT specimen is based on a compliance analysis[29].

A number of assumptions is required for the K_I determination :

The first is that compliance C, given as the ratio of load-point displacement to the load, varies linearly with crack length. The crack front is assumed to be straigth through the thickness of the specimen, and invariant with crack length. Finally, failure is assumed to occur in mode I. Under these restrictive assumptions, K_I can be calculated with the following expression :

$$K_I = H * P \quad (2)$$

$$\text{with} \quad H = \frac{W_m}{T^2} * \left[\frac{3*(1+\nu)}{\psi(T/W)*W} \right]^{1/2} \quad (3)$$

where W_m, T, W are geometrical parameters defined in figure (2) and ν is the Poisson ratio. $\psi(T/W)$ is a geometrical function[28].

We must note that the analytical stress intensity factor is therefore independent of the crack length. However the numerous assumptions for establishing this formula have to be checked. The first work has consisted in verifying the linear dependance of the compliance (C) with crack length (a) in the range of crack lengths investigated (10-30 mm). In this range of crack lengths, C(a) was defined as a straight line. therefore, possible changes in K_I with crack length could therefore not be attributed to a change in compliance formulation.

III.2. Constant loading tests

Constant loading tests were performed in air, in order to investigate the relationship between crack length and crack velocity during subcritical crack extension in the linear region of compliance. For this purpose, an unique specimen was loaded alternatively by three loads (100 N, 105 N and 110 N) and crack velocity V was recorded as a function of crack length for each load. The results are presented in a log (V) versus log (a/a_o) plot in figure (3). It appears that crack velocity increases with crack length for the three applied stresses, traducing an increase in K_I. Indeed, slow crack velocity under static loading in the investigated range (10^{-9} - 10^{-5} m/s) is directly related to K_I in the first stage of SCG (cf. equation 1). Therefore, expression of K_I with equation (2) seems to be unappropriate.

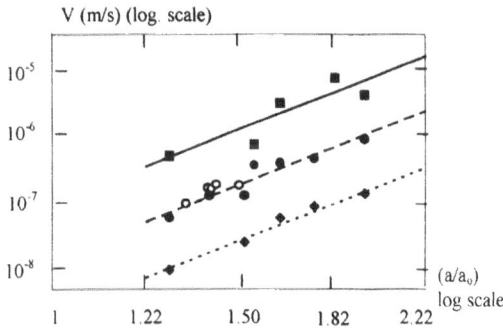

Figure 3 : Crack kinetic versus ratio of crack length to notch length for three applied loads.
■ : P=110 N   ● and ○ : P=105 N (● : a_o=10mm, ○ : a_o=21mm)   ◆ : P=100 N.

On a log-log plot, the experimental results define three parallel straight lines.
Therefore, in the range of crack velocities investigated and for a given applied load, the velocity V appears to be linked to the crack length by :

$$V = F(P)*(a / a_o)^m \qquad (4)$$

Where a_o is the notch length, F(p) depends on the applied load and m appears to be independent of P.
We therefore defined a corrected velocity Vcorr only function of the applied load P, independent of the crack length. Vcorr is related to V by :

$$Vcorr = V*(a_o / a)^m \qquad (5)$$

III.3 : Multi relaxation tests

In order to investigate the relationship between P and Vcorr, multi relaxation tests were performed in ambient air on a single specimen for crack velocities in the range of 10^{-7} - 10^{-5} m/s (again in the first stage of SCG).
The results are presented in the form of log (V) or log (Vcorr) as a function of log (P) in figure (4). The effect of crack length is obvious from the shift of log (V) vs log (P) curves, although all of them show similar slopes. The increase of initial crack length in the relaxation test results in a shift to lower values of P, due to the influence of crack extension on K_I. Interestingly, the plot of log (Vcorr) vs log (P) do not present this shift, as the effect of crack length is hereby taken into account. Note that the slope of log (Vcorr) vs log (P) is different from those of the log (V) vs log (P) curves.
Vcorr can therefore be expressed as a function of P by :

$$Vcorr = L*(P)^k \qquad (6)$$

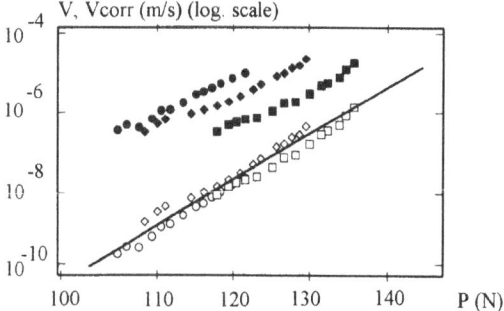

Figure 4 : V (full symbols)and Vcorr (empty symbols) versus P for three successive relaxations on a sample.
■ and □ : Final crack length af=15.9mm, ◆ and ◊ : Final crack length af=20.0mm,
● and O : Final crack length af=27.2mm.

III.4. A modified expression of K_I for the Double Torsion test

Combination of equations (5) and (6) gives :

$$V = L * P^k * \left(\frac{a}{a_o} \right)^m \qquad (7)$$

Under the assumptions that (i) both the relaxation tests and the constant loading tests were performed in the first stage of the subcritical crack propagation law (crack velocity typically on the 10^{-9}-10^{-4} m/s range), that is to say for velocities given by the power law relationship :

$$V = A * K_I^n \qquad (1)$$

and (ii) the stress intensity factor is related to P and a by an empirical relationship of the form :

$$K_I = H * P * \left(\frac{a}{a_o} \right)^x \qquad (8)$$

Replacing the value of K_I from equation (8) in (1) and comparing (8) with (7), we have :

$$n=k \quad \text{and} \quad x=m/n$$

Thus equation (8) can be rewritten in the form :

$$K_I = H * P * \left(\frac{a}{a_o} \right)^{m/k} \qquad (8.b)$$

Hence K_I can be conveniently calculated from constant loading and multirelaxation tests. The values of m and k can be easily obtained using figure (3) and (4), resulting in k=32 and m=6. Therefore, the stress intensity factor is correctly given by :

$$K_I = H * P * \left(\frac{a}{a_o} \right)^{6/32} \qquad (8.c)$$

This expression of K_I will be preferred to the expression (2) for the static and the cyclic fatigue analysis in the following sections.
In a more complete study about the Double Torsion method, Chevalier et al.[32] have studied crack geometries in the DT specimen. A curve crack front (already mentionned by a number of authors) was observed, and the compressive side of the specimen remained often uncracked.
Therefore, the authors showed that the increase in K_I was mainly due to the curved crack front, resulting in an uncracked region with increasing residual stresses on the compressive side of the specimen. The classical expression of K_I with equation (2) is therefore not accurate for precise SCG parameters measurements, because these stresses are not taken into account in the standart compliance analysis.

IV. CRACK PROPAGATION RATES UNDER STATIC LOADING

IV.1.Static fatigue in air

Tests performed in the preceeding section were analysed in terms of (V-K_I) for slow crack growth behaviour of the material in air. Additional stress relaxation and constant loading tests were performed at higher loads in order to determine the complete V-K_I curve of the material (stages II and III of the SCG law), and the presence of a threshold K_{I0} was investigated with the constant load technique at low stress levels.The results of both relaxation and constant loading tests are presented in figure (5). The overall shape of the curve is that of a standart 3-stage graph. The first stage parameters A and n were calculated, and we obtained :
A=1,8.10^{-26} and n=31,5.

A well defined threshold K_{I0} is present, under which crack propagation is presumed dormant. The presence of this threshold value at 3,5 MPa√m, already suggested by Knechtel et al.[25], is of great importance for hip prosthesis applications.

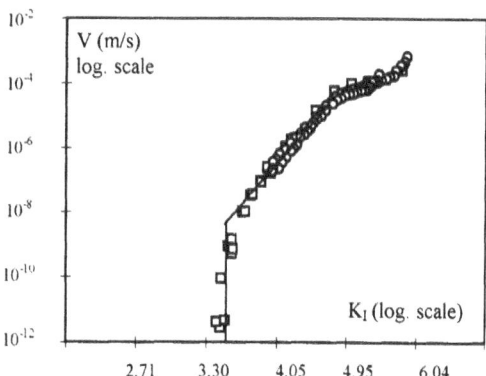

Figure 5 : Complete V-K_I curve of 3Y-TZP.
□ : Relaxation method, with correction of K_I, O : Constant load technique, with correction.

IV.2. Influence of the media on the crack propagation behaviour under static loading

Relaxation tests were performed in water, in a Silicon oil and in Ringer solution to understand the effect of environnement on SCG.

The results are presented in figure (6). In all the liquids, an unique stage is present, because no saturation in corrosive species is reached. The first propagation stage in air is identical to the propagation law in water or Ringer's solution (same slope). The V-K_I law in oil is very different from the others, crack growth appearing only very near K_{IC}. The difference between the Silicon oil and the other media is the absence of H_2O molecules. Therefore, SCG is probably due to the stress assisted corrosion of the ceramic by H_2O molecules. It is to note that the different salts present in the Ringer's solution do not promote SCG. This result is of importance since SCG measurements can be conducted in air or water to simulate behaviour in Ringer's solution.

V. CYCLIC FATIGUE BEHAVIOUR

Liu et al.[7] and Jacobs et al.[21] have found that a modified Paris Law of the form :

$$da / dt = A*\left(K_{max}\right)^{m}*(\Delta K)^{m'} \qquad (9)$$

described well cyclic data for microscopic flaws in a 3Y-TZP, with m=19 and m'=2, and in a Silicon Nitride with m~22 and m'~3.
We therefore hereby analyse our results in terms of V-K_{max} and V-ΔK dependence.

Some authors[33,34] have sometimes questionned the validity of results obtained for macrocracks because of the effect of rising crack growth resistance with extension. In 3Y-TZP, however, such effects must be neglectable due to the fact that the plateau toughness is reached at the early stage of propagation. It will be interesting to compare our results on macroscopic flaws with those of Liu et al.[7] on microcracks.

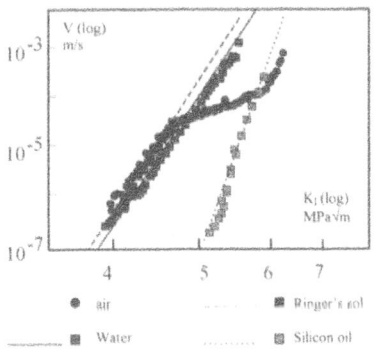

Figure 6 : Influence of the media on the stress corrosion of 3Y-TZP.

V.1. ΔK dependency of da/dt

For a fixed frequency (1 Hz) and for different R ratio (defined as the ratio of K_{max} to K_{min} in the cycle), we have recorded the crack speed versus the stress intensity range ΔK.

The results are reported on figure (7), in a log (da/dN) versus log (ΔK) diagram. It appears that V is only weakly dependent of ΔK, because of the strong importance of the R ratio.

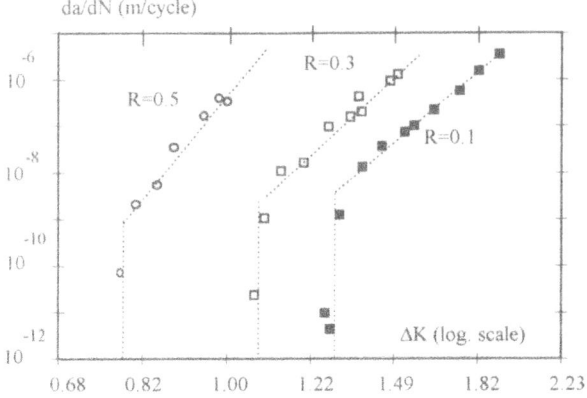

Figure 7 : V-ΔK curves for three R ratios, during cyclic fatigue at 1 Hz.

V.2. Kmax dependency of da/dt

The same results have been analysed in terms of V-K_{max} for the different R ratio, and reported in figure (8). This time, the results are more confined, although a sligth effect of R is observed. We also note threshold values K_{IOmax} for each R ratio. In order to take into account the weak dependence of the V-K_{max} curves with R, we have represented the results in a V/(1-R^2)-K_{max} graph. The results reported in figure (9) define a single line of slope m=18 for crack speed superior to 10^{-9} m/s. Under 10^{-9} m/s we note a strong decrease of the crack rates, and thresholds depending of R are measured.

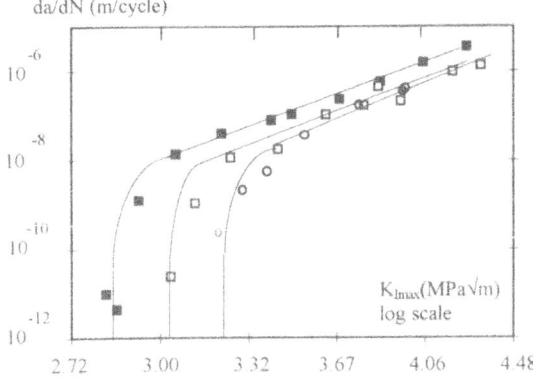

Figure 8 : V-K_{max} curves for the three R ratios considered, during cyclic fatigue at 1 Hz.

We therefore propose a growth law in cyclic fatigue for macroscopic cracks :

$$da \,/\, dt = A * \left(K_{max} \right)^{m} * \left(\Delta K \right)^{2} \qquad (9.b)$$

with m=18, and for $K_{max} > K_{I0max}$.
K_{I0max} is dependent of R, and we have :
K_{I0max}=3.2 MPa√m for R=0.5, K_{I0max}=3.0 MPa√m for R=0.3, K_{I0max}=2.8 MPa√m for R=0.1

VI. DISCUSSION

A comparison of the static fatigue and cyclic fatigue results (for R ratio of 0,1) is reported in figure (10) in terms of crack rate versus K_I or K_{Imax}.

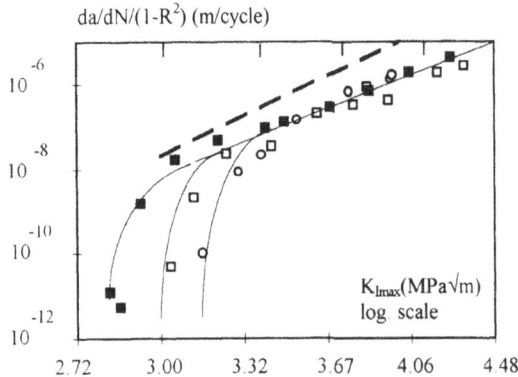

Figure 9 : (da/dN)/(1-R²) versus K_{max} curve, comparison with the results of Liu et al on microcracks.

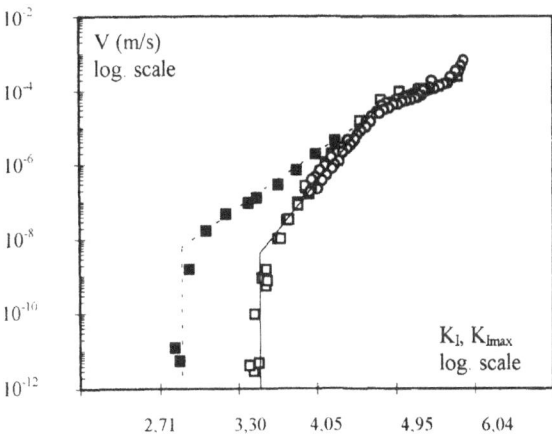

Figure 10 : Comparison of the SCG behaviour of 3Y-TZP in static fatigue (□,○ : V-K_I graph) and in cyclic fatigue (■ : V-K_{max} for R=0.1).

It appears that :
- Crack propagation rates are higher during cyclic fatigue.
- Stress exponent m =18 is much lower than exponent n=32.
- Threshold value K_{IOmax} (2,8 MPa√m) in cyclic fatigue is lower than K_{IO} in static fatigue (3,5 MPa√m).

The comparison of crack rates under static and cyclic loading in figure (10) is indirect. Indeed, it is more appropriate to compare experimental crack velocities under cyclic fatigue with the integration of the SCG law over the cycles.

If stress corrosion was the unique phenomenon responsible for crack extension, crack rate per cycle would be obtained by :

$$da \, / \, dN = \int_{0}^{T} A * K_{I}^{n} = A * K_{max}^{n} * \int \left[\left(\frac{1+R}{2} \right) + \left(\frac{1-R}{2} \right) \sin(\omega t) \right]^{n} = \Omega * K_{max}^{n} \qquad (10)$$

It is to note that the sole presence of stress corrosion during cycling loading would lead to a crack exponent m equal to n. The comparison of the experimentals results of cyclic fatigue for R=0.1 and the simulation by the integration of SCG law over the same conditions is presented on figure (11). The cyclic crack accelerating effect is even pronounced in figure (11) than in (10). Stress corrosion is therefore far from being the sole crack extension factor during cyclic loading.

Shielding degradation during cyclic fatigue can explain this cyclic crack accelerating effect :

Liu et al. have expressed cyclic fatigue behaviour of a Si_3N_4 in air in terms of a mechanical damage of crack tip shielding. It was assumed that a crack would propagate in cyclic fatigue when the stress intensity at the crack tip exceeded the intrinsic toughness of the material K_O. Because of crack tip shielding due to crack bridging, the applied K_I needed to cause crack advance, noted K_{Iappl} is given by :

$$K_{Iappl} = K_O + K_{sh} \qquad (11)$$

In cyclic fatigue, shielding degradation leads to a decrease of shielding stress intensity factor K_{sh} in comparison with the static loading case. The authors showed on the Si_3N_4 that after a

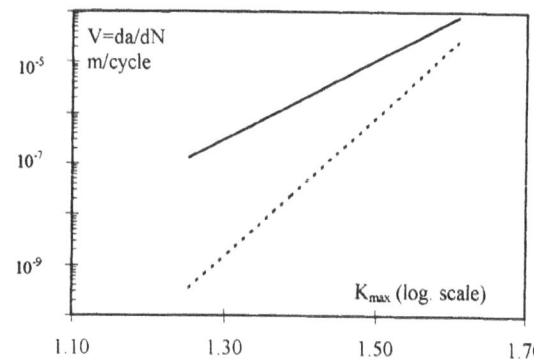

figure 11 : V-K_{max} curve for R=0,1 in the 10^{-9}-10^{-5} m/s crack velocity range.
————— : from experimental data.   · · · · · · : from integration of stress corrosion over the cycles.

first decrease of K_{sh}, a steady state was reached, so that the crack grew at a constant rate for a constant cycling loading configuration. At this equilibrium, accumulation and degradation of shielding balanced. Cyclic fatigue was therefore expresssed in terms of shielding degradation $\Delta K_{sh}=(K_{sh,static}-K_{sh,cyclic})$ between quasi static and cyclic fatigue tests.

Therefore, for a Si_3N_4, where we remember that stress corrosion is unsignificant[16], SCG in cyclic fatigue was accurately explained by a mechanical equilibrium between accumulation and degradation of crack tip shielding.

For 3Y-TZP, (and other ceramic oxides) sensitive to stress corrosion, one can expect that such an equilibrium exists, but that crack propagation occurs when the stress intensity factor at the crack tip reaches the value needed to cause crack propagation by stress corrosion, i.e below K_{IC}.

Crack velocity could therefore be expressed by :

$$da / dt = A * K_I^n = A * (K_{Iappl} + \Delta K_{sh})^n \qquad (12)$$

where ΔK_{sh} represents the crack tip shieding degradation at equilibrium. Therefore, the crack increment per cycle da/dN can be calculated by :

$$da / dN = \int_0^T A * K_I^n * dt = A * \int \left[K_{max}\left[\left(\frac{1+R}{2}\right)+\left(\frac{1-R}{2}\right)\sin(\omega t)\right] + \Delta K_{sh} \right]^n * dt \qquad (13)$$

With the experimental value of da/dN measured in cyclic fatigue (section V), and the SCG parameters A and n measured with constant loading tests, it is possible to determine ΔK_{sh} for a given R value and for different K_{max} values. This has been conducted by numerical calculations of expression (13) for the three R ratios. For a given K_{max} value, expression (13) has been fitted to the experimental da/dN by varying ΔK_{sh}. The determination of ΔK_{sh} for various K_{max} allows the determination of three ΔK_{sh} versus K_{max} curves, reported in figure (12).

It appears that shielding degradation is decreased with increasing K_{max} (increasing crack velocities) and R (decreasing stress intensity range for a given K_{max}). The most interesting

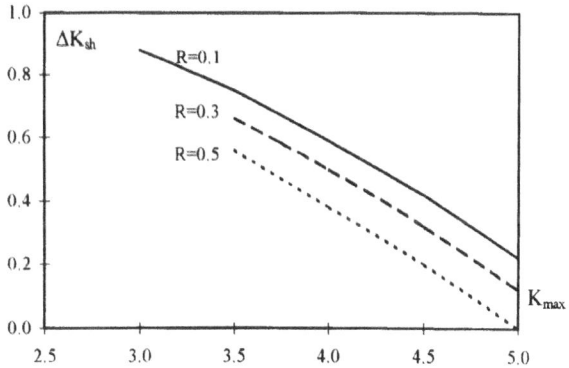

Figure 12 : ΔK_{sh} versus K_{max} curves for the three R ratios.
———— : R=0,1   — — — : R=0,3   · · · · · : R=0,5.

values of ΔK_{sh} are the one calculated for K_{max}=2,8 MPa$\sqrt{m}$ and R=0.1 (ΔK_{sh}~0.85), K_{max}=3 MPa$\sqrt{m}$ and R=0.3 (ΔK_{sh}~0.65), K_{max}=3,1 MPa$\sqrt{m}$ and R=0.5 (ΔK_{sh}~0.55). Indeed, they correspond quite well to the difference between K_{IO} and K_{IOmax} (noted ΔK_{IO}) for the three R ratios (R=0.1, ΔK_{IO}=0.7 - R=0.3, ΔK_{IO}=0.5 - R=0.5, ΔK_{IO}=0.4). The three thresholds in cyclic fatigue are therefore accurately related to the threshold stress intensity factor in static fatigue K_{IO} and to the crack tip shielding degradation by cyclic loading.

In conclusion, cyclic fatigue of 3Y-TZP seems to be governed by stress corrosion and crack tip shieding degradation, the two processes acting in series. Of course, this has to be confirmed by further studies. We currently work on cyclic fatigue in vacuum and in inert atmospheres, where stress corrosion is inoperant, to confirm this assumption.

VII. CONCLUSION

A study of the 3Y-TZP crack propagation behaviour has been conducted with the Double Torsion method.

1. In order to obtain precise crack propagation laws in both static and cyclic fatigue, the evolution of K_I with crack length has been investigated. The results show that the stress intensity factor is significantly dependent of the crack length, which is not taken into account in the usual compliance analysis. We therefore propose a modified expression that takes into account the evolution of K_I with crack extension.

2. Static fatigue behaviour is due to stress assisted corrosion at the crack tip by the H_2O molecules. Three stages are observed in a V-K_I curve, and the first one is well described by a power law :

$$V = A * K_I^{\,n}$$

with n=32. A well defined threshold intensity factor K_{IO}=3,5MPa$\sqrt{m}$ has been measured.

3. Under cyclic solicitations, a modified Paris law proposed by Liu and Chen[7]of the form :

$$da / dt = A * \left(K_{max}\right)^{m} * \left(\Delta K\right)^{2}$$

with m=18 correctly fits the experimental results. Our results on macroscopic flaws are in good agreement with those of Liu and Chen for indentation cracks. Threshold values K_{IOmax} depending on the (K_{Imin}/K_{Imax}=R) ratio were also measured.

4. Crack extension rates are much higher and threshold intensity factors lower during cyclic fatigue than during static loading. This has been interpreted in terms of shielding degradation ΔK_{sh} at the crack tip. ΔK_{sh} has been calculated by simple numerical simulations for different cyclic loading configurations, and reaches a value on the order of 1 MPa$\sqrt{m}$ for R=0,1. Thresholds in cyclic fatigue correspond to K_{IO} in stress assisted corrosion. Their lower values are explained by shielding degradation. These arguments are however to be confirmed by further studies, particularly with tests in inert atmosphere.

REFERENCES

1. Cales B. "Zirconia ceramic for improved hip prosthesis -a review-" Proceeding of the 6th Biomaterial Symposium, september 1994 - to be published.
2. Wiederhorn S. M. "Subcritical crack growth in ceramics", in : Fracture Mechanics of ceramics, vol.2, Microstructure, Material and applications, 1974, New York, edited by Bradt R.C., Hasselman D.P, Lange FF, 1974, N.Y.

3. Sudreau F. "Methodes d'analyse et de simulation de la fatigue thermique de céramiques thermomécaniques", Ph D. thesis, INSA Lyon, 1992, France, pp 273.

4. Fett T, Germerdonk K, Grossmüller A, Keller K, Munz D "Subcritical crack growth and threshold in borosilicate glass", *Journal of Materials science*, vol.26, 1991, pp 253-257.

5. Evans E. G. & Fuller R. "Crack propagation in ceramic materials under cyclic conditions", *Metallurgical transactions*, 1974, volume 5, pp 27-33.

6. Dauskardt R. H. & Ritchie R. O. "Fatigue crack propagation in transformation toughned zirconia ceramic".
   J. Am. Ceram. Soc. ,1987, volume 70, n°10, pp 248-252.

7. Liu S. Y. & Chen I.W "Fatigue of yttria-stabilized zirconia : II, Crack propagation, fatigue striations, and short-crack behavior". *J. Am. Ceram. Soc.* ,1991, volume 74, n°6, pp 1206-1216.

8. Liu T., Matt R., Grathwohl G. "Static and cyclic fatigue of 2Y-TZP ceramics with natural flaws", *J. Europ. Ceram. Soc.*, 1993, volume 11, pp 133-141.

9. Grathwohl G. & Liu T. "Crack resistance and fatigue of transforming ceramics : I. Materials in the ZrO2-Y2O3-Al2O3 System", *J. Am. Ceram. Soc.* ,1991, volume 74, n°2, pp   318-325.

10. Grathwohl G. & Liu T. "Crack resistance and fatigue of transforming ceramics : II, CeO2-Stabilized Tetragonal ZrO2", *J. Am. Ceram. Soc.* ,1991, volume 74, n°12, pp 3028-3034.

11. Dauskardt H, Marshall D, Ritchie R "Cyclic fatigue-crack propagation in Mg-PSZ ceramics", *J. Am. Ceram. Soc.* ,1990, volume 73, n°4, pp 893-903.

12. Tzai J., Yu C., Shetty D. "Fatigue crack propagation in Ceria-Partially-Stabilized Zirconia (Ce-TZP)-Alumina composites". *J. Am. Ceram. Soc.* ,1990, volume 73, n°10, pp 2992-3001.

13. Guiu F., Reece M. J., Vaughan D. A. J. "Cyclic fatigue of ceramics", *Journal of Materials Science*, 1991, volume 26, pp 3275-3286.

14. Fett T., Martin G., Munz G., Thun G. "Determination of da/dN-ΔKI curves for small cracks in alumina in alternating bending tests", *Journal of Materials Science*, 1991, volume 26, pp 3320-3328.

15. Dauskardt R, James M, Porter J, Ritchie R "Cyclic fatigue crack growth in a SiC-Wiskers-Reinforced Alumina composite : long- and small-crack behavior". *J. Am. Ceram. Soc.* ,1992, volume 75, n°4, pp 759-771.

16. Horibe S "Fatigue of Silicon Nitride Ceramics under cycling loading", *J. Eur. Ceram. Soc.*, 1990, n°6, pp 89-95.

17. Vekinis G, Ashby M, Beaumont P "R-Curve behaviour of Al_2O_3 ceramics", *Acta metall. mater.*, 1990, vol 38, n°6, pp 1151-1162.

18. Rödel J. Kelly J. Lawn B "In situ measurements of bridged crack interfaces in the scanning electron microscope", *J. Am. Ceram. Soc.* ,1990, volume 73, n°11, pp 3313-18.

19. Lathabai S., Rodel S.& Lawn B. R. "Cyclic fatigue from frictional degradation at bridging grains in alumina". *J. Am. Ceram. Soc.* ,1991, volume 74, n°6, pp 1340-1348.

20. Gilbert C. Petrany R. Ritchie R, Dauskardt R, Steinbrech R "Cyclic fatigue in monolithic alumina : mechanisms for crack advance promoted by frictional wear of grain bridges", *Journal of Materials Science*, 1995, volume 30, pp 643-654.

21. Jacobs D. Chen I "Cyclic fatigue in ceramics : a balance between crack shielding accumulation and degradation", *J. Am. Ceram. Soc.* ,1995, volume 78, n°3, pp 513-520.

21. Porter D. Heuer A "Mechanisms of toughening Partially Stabilized Zirconia", *J. Am. Ceram. Soc.* ,1977, volume 60, n°3-4, pp 183-184.

23. Marshall D. Shaw M, Dauskardt R, Ritchie R, Readey M, Heuer A "Crack-tip transformation zones in toughened Zirconia", *J. Am. Ceram. Soc.* ,1990, volume 73, n°9, pp 2659-2666.

24. Tsai J, Shetty D "Cyclic fatigue of Ce-TZP/Al2O3 composites : Role of the degradation of transformation zone shielding". *J. Am. Ceram. Soc.* ,1995, volume 78, n°3, pp 599-608.

25. Knechtel M., Garcia D., Rödel J., Claussen N. "Subcritical crack growth in Y-TZP and Al2O3 -Toughned Y-TZP", *J. Am. Ceram. Soc.* ,1993, volume 76, n°6, pp 2681-2684.

26. Chevalier J, Olagnon C, Fantozzi G, Cales B "Crack propagation behaviour of Y-TZP ceramics". *J. Am. Ceram. Soc.* - accepted for publication.

27. Fuller E "An evaluation of Double Torsion testing - Analysis", *in :* Fracture Mechanics Applied to Brittle Materials, proceedings of the 11th symposium on fracture mechanics, part II. ASTM STP 678,pp 3-19. editor S.W. Freiman.

28. Plekta B, Fuller E, Koepke B "An evaluation of Double Torsion testing - Experimental", *in :* Fracture Mechanics Applied to Brittle Materials, proceedings of the 11th symposium on fracture mechanics, part II. ASTM STP 678, pp 19-38 editor S.W. Freiman.

29. Williams D, Evans A. "A simple method for studying slow crack growth", *Journal of testing and evaluation*, Vol. 1, N°4, July 1973, pp. 264-270.

30. Shetty, D.K., Virkar A.V., Harward M.B, *Journal of The American Ceramic Society- Discussions and Notes*, Vol. 62, 5-6 1979, pp 307-309.

31. Trantina G.G. "Stress analysis of the Double-Torsion specimen", *Journal of The American Ceramic Society*, Vol.60, 7-8 1977, pp 338-341.
32. Chevalier J, Saadaoui M, Olagnon G, Fantozzi G "Experimental investigations on the Double Torsion specimen with a 3Y-TZP", *Ceramics International* - Submitted to publication.
33. Fett T, Munz D, Keller K "determination of subcritical crack growth on glass in water from lifetime measurements on Knoop-cracked specimens", *Journal of Materials Science*, 1988, volume 23 , pp 798-803.
34. Fett T, Munz D "Subcritical crack growth of macro- and microcracks in ceramics", *in* : Fracture Mechanics of Ceramics, volume 9, Plenum Press, New York, 1992.

THE FATIGUE BEHAVIOUR OF Mg-PSZ AND ZTA CERAMICS

M.M. Nagl, L. Llanes, R. Fernández and M. Anglada

Departamento de Ciencia de los Materiales e Ingeniería Metalúrgica
Universidad Politécnica de Cataluña, E.T.S.I.I.B., Avda. Diagonal 647
08028 Barcelona, Spain

ABSTRACT

The static and cyclic fatigue behaviour of partially stabilized zirconia (Mg-PSZ) with two different microstructures and zirconia toughened alumina (ZTA) containing 5, 15 or 30% zirconia has been studied. In both materials, the tetragonal zirconia was contained within a matrix material of different structure. In the former it precipitated during cooling from the coarse-grained cubic zirconia matrix, whereas in the latter it formed a dispersed phase in a fine-grained alumina matrix.

The microstructural characteristics and the fatigue mechanisms were studied using X-ray diffraction, optical, scanning and transmission electron microscopy techniques. Fatigue tests were carried out using 1) smooth specimens containing only natural flaws, 2) samples having small indentation induced cracks, only in the case of ZTA, and 3) single edge notched bend specimens with large cracks produced by compression-compression fatigue.

All the materials showed a clear cyclic fatigue effect, i.e. the cyclic crack growth rates were higher and the cyclic fatigue lives shorter than the corresponding static fatigue crack growth rates and fatigue lives. Such effect was discussed in terms of degradation of the active toughening mechanisms. The operation of one or another of these mechanisms was closely related to microstructural parameters. For the two Mg-PSZs, differences in the mechanical and fatigue properties could be associated with microstructural changes, which in turn also changed the main toughening mechanism from crack tip shielding by transformation to crack deflection. Similarly for ZTA, the reduction of grain size as a result of the zirconia addition changed the toughening mechanism from grain bridging to transformation toughening. However, in this case such microstructural differences hardly affected the mechanical and fatigue resistance of these materials.

INTRODUCTION

High technology ceramics are potential candidates for structural applications where critical design requirements include high stiffness to weight ratio, chemical stability or high temperature stability. Therefore, it is not surprising to find that research focused on understanding the relationship between mechanical properties and processing parameters has been one of the most highly publicized area of ceramics in the last fifteen years.

In terms of mechanical properties, strength and toughness are those which have received the greatest attention. The exceptional strength of ceramics is usually undermined by their intrinsic low toughness and considerable interest has been focused on the development of tough ceramics. One of the most successful approaches has proven to be "transformation toughening" of zirconia. Transformation toughening, first identified by Garvie et al.[1], is achieved by the shielding effect on a propagating crack, that results from the stress-induced transformation of metastable tetragonal particles to the stable monoclinic crystal form. An extensive investigation, mainly microstructurally-driven, has led to optimal processing routes and as a result to dramatic increases in the relative toughness of these ceramics. However, within a more general framework, such toughness improvement may not be considered as completely beneficial under fluctuating loads.

In a recent survey on the cyclic properties of engineering materials Fleck et al.[2] have pointed out that material developments to improve fracture toughness generally make the material more sensitive to fatigue. Several experimental investigations on the fatigue behaviour of advanced ceramics (e.g. Refs. 3,4) have confirmed this trend. Therefore, in structural applications involving both toughened ceramics and cyclic loading conditions a specific understanding of the relationship between microstructure and fatigue mechanisms is another critical requirement for safe design. Although there is a relatively large number of studies on the fatigue behaviour of advanced ceramics, most of them have been mainly concerned in demonstrating and explaining the mechanical side of cyclic fatigue in these materials[*]. On the other hand, only a few works have been focused on the description and interpretation of the influence of microstructural parameters on the fatigue behaviour of toughened ceramics. Thus, it is the aim of this work to investigate and correlate microstructural changes with the relative fatigue behaviour of two transformation toughened ceramic systems.

The study was initiated with a detailed microstructural characterization of the two systems to be investigated: one a MgO-partially stabilized zirconia-based system (Mg-PSZ) with two different microstructures and the other, a system formed by a high modulus alumina matrix that contained three different volume fractions of Y_2O_3-stabilized zirconia particles (ZTA). Then, the fatigue behaviour of the materials with natural or processing flaws as well as with small and long cracks was investigated and discussed. Here, in order to meet the above objective, attention was focused on describing relative cyclic to static fatigue effects as well as fatigue crack propagation mechanisms, under cyclic and constant applied loading conditions, and correlate them with relevant microstructural features.

EXPERIMENTAL PROCEDURE

Materials

The Mg-PSZ used for this study was a commercial material containing about 3 wt% MgO[**]. The material was isostatically pressed into plates and then sintered at 1800 °C. It was supplied in the form of rectangular bars (dimensions 50x5x8 mm).

Two microstructures were investigated. One microstructure corresponded to the as-received condition which may be described, following a detailed transmission electron microscopy (TEM) study, in terms of a large volume fraction (> 30 vol%) of tetragonal precipitates homogeneously distributed in a cubic matrix. The mean grain size of the matrix was 65 μm. A few monoclinic particles in regions adjacent to grain boundaries, assumed to form by heterogeneous nucleation and transformation during cooling, were also observed. The size of the tetragonal precipitates ranged from 100 to 200 nm (Figure 1a). This latter value is close to the critical one required for spontaneous transformation to monoclinic structure at room temperature[5]. Hence, this material is referred to as peak-aged (PA) in this investigation.

The second microstructure was obtained by ageing the above material at 1320 °C during 8 hours. This material is referred to as eutectoid-aged (EA) in the present study, because the ageing temperature was slightly below the eutectoid temperature of 1400 °C[6]. The microstructural changes resulting from such ageing included: 1) coarsening and transformation of the tetragonal precipitates into the stable monoclinic structure (Figure 1b), and 2) eutectoid decomposition of the cubic matrix into monoclinic zirconia and MgO mainly at grain boundaries. The eutectoid product had a cellular-like character (Figure 1c), similar to that observed by Farmer et al.[7]. The eutectoid decomposition promoted pronounced microcracking at the grain boundaries, as indicated in Figure 1d. Such microcracking was due to the significant thermal expansion mismatch between the cubic matrix and the eutectoid product and played a

---

[*] In the ceramics literature the term "cyclic fatigue" is commonly used to distinguish the true mechanical fatigue of that associated with environmentally assisted crack growth or "static fatigue".

[**] Type FZM, Friatec AG, Mannheim, Germany

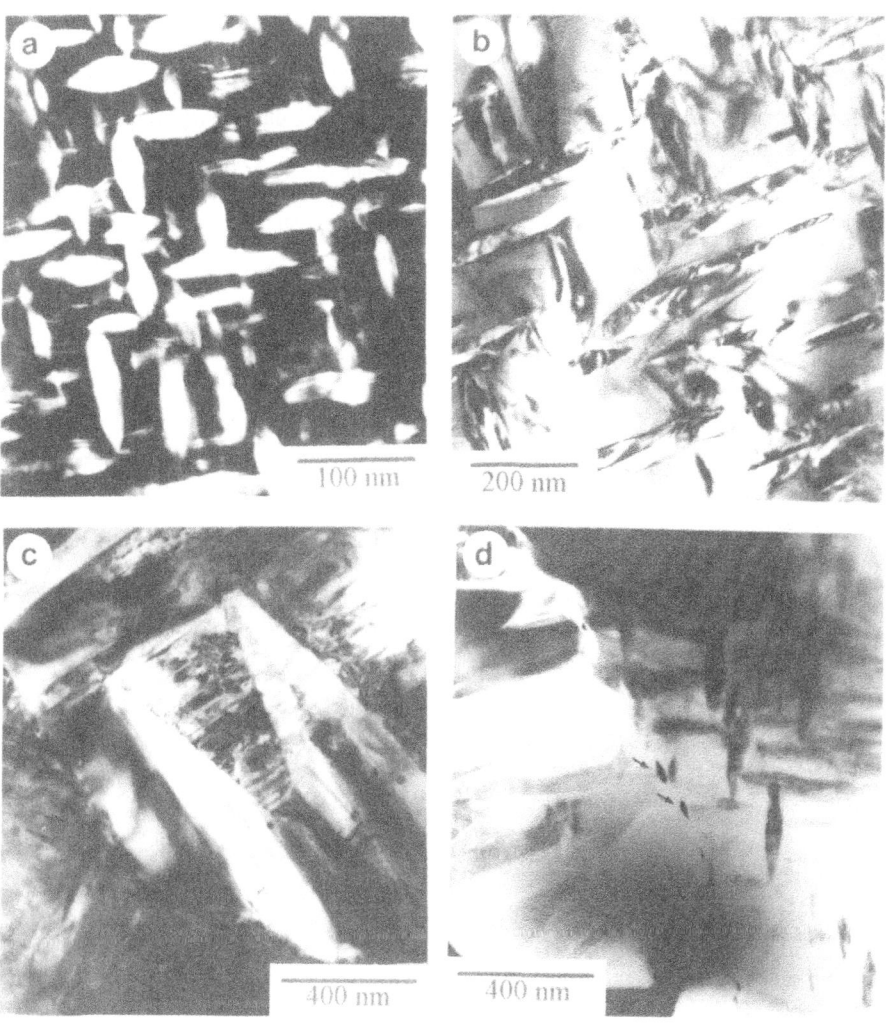

Figure 1 TEM micrographs showing microstructural characteristics of the Mg-PSZs studied: **a)** PA-PSZ, tetragonal precipitates within the cubic matrix (dark field image); **b)** EA-PSZ, large transformed monoclinic precipitates; **c)** EA-PSZ, cellular aspect of eutectoid decomposition product; and **d)** EA-PSZ, microcracks at grain boundaries as a consequence of the eutectoid-like ageing.

very important role in explaining the mechanical behaviour of the EA-PSZ material, as will be seen later.

On the other hand, three ZTAs with 5, 15 and 30 vol% of stabilized zirconia (from now on termed A5Z3Y, A15Z3Y and A30Z3Y respectively, according to following nomenclature $A(ZrO_2$ in vol.%$)Z(Y_2O_3$ in mole%$)Y)$ and 99.7 % pure alumina were studied[*]. The latter formed also the base material of the ZTAs. Like the PSZs, these materials were isostatically pressed into plates at room temperature using a pressure of 200 MPa, subsequently sintered for 2 hours at 1600 °C and finally cut in to bar shaped specimens (dimensions 45x5x8 mm). Figure 2 shows a scanning electron microscopy (SEM) image of the microstructure of the

[*] Produced by the Instituto Tecnológico de Materiales, Asturias, Spain

A15Z3Y material after thermal etching for 30 min at 1350 °C. The distribution of the zirconia particles was fairly uniform within the alumina matrix. Some clusters of zirconia were found, and obviously the number and size of the clusters were larger when zirconia content was higher. The grain size diminished with increasing zirconia addition (Table 1) since the zirconia impedes the grain growth of the alumina matrix[8]. Additional TEM studies showed that some of the largest zirconia grains transformed to monoclinic symmetry on cooling. This was mainly observed within zirconia clusters and the critical grain size for this spontaneous transformation on cooling was about 0.6 - 0.8 μm, which compared well with other studies on ZTA[9,10]. In all cases the monoclinic content at the surface of the polished specimens was still below 1%, as confirmed by X-ray diffraction (XRD). Finally, in a few instances microcracks were found around zirconia particles.

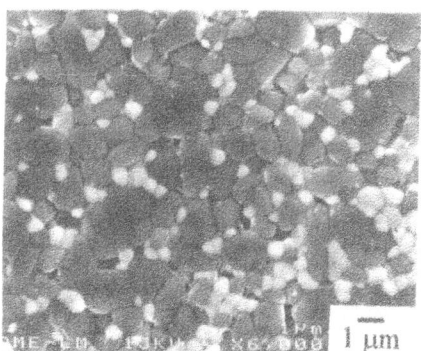

Figure 2. Secondary SEM mircograph showing the microstructure of A15Z3Y.

Table 1. Average and maximum grain sizes for alumina and ZTA.

		Al_2O_3	A5[3Y	A15Z3Y	A30Z3Y
Al_2O_3	d_{avg} [μm]	3	1.6	1.3	1
	d_{max} [μm]	5-7	3-4	2-2.5	1.5-2
ZrO_2	d_{avg} [μm]	-	0.4	0.5	0.43
	d_{max} [μm]	-	0.5-0.7	0.8-1	0.8-1

Measurement of Fracture Strength, S-N and S-t Curves

The specimens were first polished on the surface which was later subjected to the maximum stress in bending. Polishing was conducted using diamond pastes of 30 and 6 μm and colloidal silica as a final step. The edges of the polished specimens were slightly chamfered for the test with natural flaws. A 294 N Vickers indentation was placed in the centre of the polished surface, for the tests with small indentation cracks. The indents were aligned in such a way that one pair of the indentation cracks was perpendicular to the specimen central axis with a maximum error of ±3°. Such tests were only performed with ZTA and alumina. They were not possible in Mg-PSZ because indents did not produce a well defined crack system in the PA-PSZ, even for small loads. Similar observations have also been made by Marshall and Swain[11] in another commercial Mg-PSZ. Furthermore, in the case of A15Z3Y some of the specimens were annealed for 1h at 1100 °C to eliminate the residual stress field produced by an indentation. This was done in order to study the effect of such stress field on the fatigue behaviour.

A fully articulating four-point bend test jig (inner span 20 mm, outer span 40 mm) was used for all tests. Flexural strength and static fatigue tests were conducted in an electro-mechanical testing machine (Instron, model 8562). Cyclic fatigue tests were carried out in a resonant testing machine (Rumul, model Mikrotron) using a load ratio of 0.2. The working frequency varied between 225 and 240 Hz depending on the specimen geometry and the Young's modulus.

The cyclic fatigue limit was evaluated as the maximum applied cyclic stress at which the specimens did not fail in 10^7 cycles. Likewise, the static fatigue limit corresponded to the stress which the specimens could sustain for 15 hours without failure, i.e. the time to perform a cyclic fatigue test of 10^7 cycles.

Fracture Toughness and Crack Growth Rate Measurements

Fracture toughness and crack growth rates were measured using single edge-notched bend (SENB) specimens with a crack length-to-specimen width ratio (a/W) of ~0.35. In this case, the side surfaces of the specimens were polished to a 6 µm surface finish to facilitate crack growth measurements. The samples were prepared by cutting a 2.5 mm deep notch in the centre of the specimens using a 150 µm thick diamond disk. Then, they were subjected to cyclic compression in a servo-hydraulic testing machine (Instron, model 1341), in order to induce a pre-crack at the notch tip. These tests were conducted under load control with a sinusoidal waveform at a testing frequency of 20 Hz and a load ratio of 10. Pre-cracks were observed at minimum compressive stress of 200 to 400 MPa. Typical final pre-crack lengths, i.e. length of the crack in front of the notch tip after 10^5 - 10^6 cycles, varied between 50 and 150 µm. In ZTA and alumina the cracks were worn out and in subsequent fatigue tests they started to grow at rates $>10^{-5}$ m/cycle, i.e. they did not act as real fine pre-cracks. Therefore, in these materials the cracks were extended another 100 - 200 µm beyond the pre-crack by cyclic bending. Afterwards the specimens were annealed for 1 hour at 1100 °C. Such treatment was performed in order to invert any transformation which might have occurred at and around the crack tip in these materials during crack extension as well as to remove possible residual stresses produced by the compressive loading. On the contrary, in Mg-PSZ no such crack extension was necessary since generally the compressive pre-cracks started to grow at relatively low rates in the subsequent fatigue test, i.e. they acted as sharp pre-cracks. Likewise, these materials were tested without any further annealing treatment.

The actual fracture toughness, static and cyclic fatigue tests were conducted in a servo-hydraulic testing machine (Instron, model 8511) in four point bending. In all cases crack growth was measured *in situ* with an accuracy of ±5 µm using a long-distance optical microscope (Questar, model QM100). The fracture toughness was determined using the maximum crack extension obtained under stable crack growth at constant load together with the following stress intensity factor expression, from Tada *et al.*[12], for SENB geometry:

$$K_{IC} = \sigma\sqrt{\pi a}\left[1.12 - 1.38\left(\frac{a}{W}\right) + 7.33\left(\frac{a}{W}\right)^2 - 13.08\left(\frac{a}{W}\right)^3 + 14.0\left(\frac{a}{W}\right)^4\right] \quad (1)$$

where σ is the applied stress, a is the crack length and W is the specimen height.

The cyclic fatigue tests were conducted at a frequency of 2 Hz under load control using a sine waveform and a load ratio of 0.2. For comparison environmentally assisted or static fatigue crack growth was measured with the specimens subjected to constant applied load. All tests were carried at room temperature and humidity of about 55%.

RESULTS AND DISCUSSION

Strength and Fracture Toughness

The mean flexural strength and the fracture toughness of the EA-PSZ were 251 MPa and 5.30 MPa m$^{1/2}$, respectively. Both were significantly lower than the corresponding values measured for the PA material, i.e. 474 MPa and 8.64 MPa m$^{1/2}$, respectively. Still, these values for the PA-PSZ seemed to be rather low compared to those reported for other commercial Mg-PSZs (e.g. Ref. 13). It is believed that a small increase of the mean tetragonal precipitate size (to about 200 µm) would improve the mechanical response of the material studied. On the other hand, it is interesting to note that, besides the large difference in the fracture parameters measured, both materials showed a relatively pronounced R-curve. Similar findings have also been made by Swain and Hannink[14]. As it is already well known, the R-curve behaviour observed in the PA-PSZ is a result of its intrinsic transformation toughening capability. On the other hand, the increase of crack growth resistance with crack extension in the EA-PSZ was found to be rather related to the existence of "grain-size-scale" crack deflection mechanisms. These mechanisms were a direct consequence of the prominence of cracks to nucleate at and propagate along frail grain boundaries. Such grain boundary weakness was explained by the microcracking which resulted from the localized deposition of eutectoid decomposition products on them[5]. Hence, the differences in the mechanical properties between PA- and EA-PSZ can be associated with the clearly different crack-microstructure interactions controlling the mechanical response of each of them.

For the ZTAs the fracture toughness increased only moderately with increasing zirconia content, as can be seen from Figure 3. This was probably due to 1) the limited amount of tetragonal-to-monoclinic transformation during fracture (~5% for all ZTAs) as measured by XRD and 2) the smaller bridging effect in the ZTAs compared to alumina as a consequence of the smaller grain size. On the other hand, the fracture strength increased considerably compared to pure alumina (Figure 3), due to the decrease in processing flaw size with zirconia content. Taking into account the observed change in fracture toughness, the increase in fracture strength of A30Z3Y compared to alumina corresponds to a change in flaw size by a factor of about 2, which is in good agreement with the measured variation in void size.

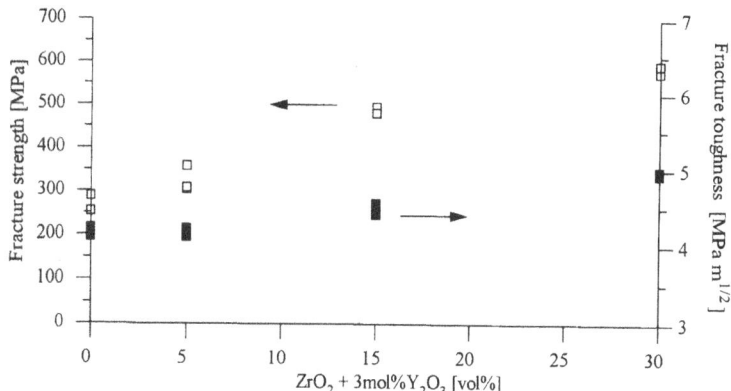

Figure 3. Fracture toughness and fracture strength as a function of zirconia content for ZTA and alumina.

Fatigue Life

All materials showed fatigue, i.e. they failed at stresses well below the fracture strength, both under constant and cyclic loading conditions. Furthermore, evidence of a cyclic fatigue limit was observed in all the materials studied.

In the case of Mg-PSZ, similarly to the mechanical response observed under monotonic loading, the fatigue resistance of the PA-PSZ is much higher than that of the EA one. However, the differences between cyclic and static fatigue, i.e. the real cyclic fatigue effects were dependent upon microstructure. This is observed in the S-t curves shown in Figure 4, where it is obvious that real cyclic fatigue effects are smaller in the EA-PSZ than in the PA one.

In ZTAs the fatigue limits for specimens with natural flaws increased with increasing zirconia content, but to a lesser extend than the fracture strength. This indicates that ZTAs suffer more fatigue than pure alumina.

The cyclic fatigue lives of indented specimens were at least one or two orders of magnitude smaller than the static fatigue lives under the same applied maximum stress, as shown in Figure 5 for the A30Z3Y material, confirming that all materials suffered real cyclic fatigue. In absolute values, the static fatigue limit was very similar for all materials including alumina. The cyclic fatigue limits increased only slightly (~10 MPa) for the materials with higher zirconia content, even though the fracture strength of the indented specimens was 30 MPa higher for A30Z3Y compared to pure alumina. This shows that most of the strength increase caused by zirconia addition was compensated by the higher susceptibility of ZTAs to static fatigue.

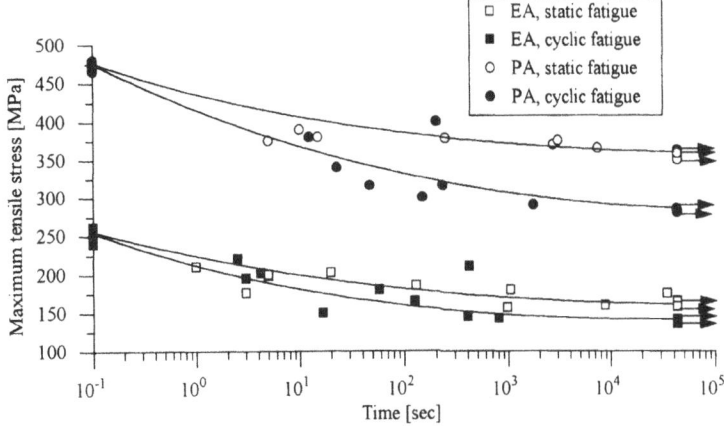

Figure 4. S-t curves for PA- and EA-PSZ. The data points at 10^{-1} seconds correspond to the fracture strength. Specimens which did not fail in 5×10^4 seconds are plotted with an arrow indicating that their fatigue lives would be even higher than such value.

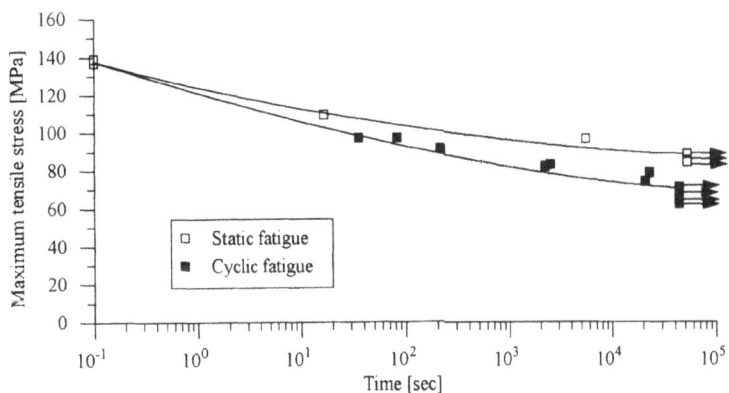

Figure 5. Comparison of the static and cyclic fatigue lives for A30Z3Y using Vickers indented specimens. The data points at 10^{-1} seconds correspond to the fracture strength of indented specimens. Specimens which did not fail in 5×10^4 seconds are plotted with an arrow indicating that their fatigue lives would be even higher than such value.

Fatigue Ratios

In order to allow a more objective comparison of the results obtained under cyclic loading conditions with different materials, it is common to normalize basic fatigue parameters by the mechanical properties. One of such correlations, usually made for S-N and S-t-curves (i.e. for un-notched specimens) is the fatigue ratio, which is defined as the ratio between the fatigue limit and the flexural strength, σ_E/σ_F. Hence, the fatigue ratios were calculated for the tests with natural flaws and indentation cracks. In the latter, they corresponded to the fatigue limit divided by the fracture strength of similarly indented specimens.

The cyclic and static fatigue behaviour of Mg-PSZ displayed different trends depending upon microstructure. Two aspects may be pointed out. First, the cyclic fatigue ratios for the materials studied are very similar (0.59 and 0.56). Second, the static fatigue ratio of EA-PSZ is smaller (0.66) than that of PA-PSZ (0.76), i.e. EA-PSZ is more prone to static fatigue than the PA one. Hence, it is suggested that the differences in the cyclic to static fatigue behaviour or the real cyclic fatigue effect, are due to the differences in the individual static fatigue behaviour of each material. Such finding may be rationalized in terms of the larger environmental sensitivity of the eutectoid products which formed in the EA-PSZ.

In ZTA and alumina, the cyclic fatigue ratios showed a slight trend towards smaller values with increasing zirconia content (Figure 6). On the other hand, the static fatigue ratios for the ZTAs were considerably lower compared to pure alumina. In other words the real cyclic fatigue effect was smaller in the ZTAs compared to alumina. This led to the conclusion that adding zirconia to alumina affects primarily the static fatigue mechanisms, and that ZTAs suffer less real cyclic fatigue than alumina. The higher static fatigue effect in ZTA can be explained by the intrinsic susceptibility of Y-stabilized zirconia to static fatigue. Similarly Becher[9] found that adding zirconia to alumina decreases the static fatigue threshold.

As expected, the indented and annealed A15Z3Y specimens needed much higher loads for failure than those without any additional treatment after indentation. However, the measured cyclic fatigue ratio was very close to that obtained with as-received and indented specimens, i.e. specimens with residual stresses induced by the indentation (Figure 6). This suggests that the residual stress field underneath the indent did not affect the fatigue mechanisms. More importantly, the cyclic fatigue ratios obtained with specimens containing natural flaws and indentation cracks were also similar, i.e. they were not affected by the size of the initial void. Hence, it may be concluded that the slow growth of the intrinsic processing flaws determines fatigue and that specimens with small indentation cracks simulate quite well the actual fatigue behaviour of specimens containing random processing flaws. On that basis, the fatigue ratios measured with indentation cracks might be used to predict the fatigue behaviour of samples or actual parts produced under similar conditions and which only contain processing flaws, at least in the case of alumina and ZTA. In other words, the fatigue ratio could be measured using only a few indented specimens instead of a much larger batch which is necessary to determine the fatigue limit when using specimens with processing flaws. The fatigue ratio obtained in this way might then be used for a design based on fatigue thresholds, below which cracks are believed to be dormant.

Crack Growth Rates

The measured crack growth rates using SENB specimens are shown in Figures 7, 8 and 9 for the Mg-PSZs, alumina and A30Z3Y, respectively. In all cases the load was increased in

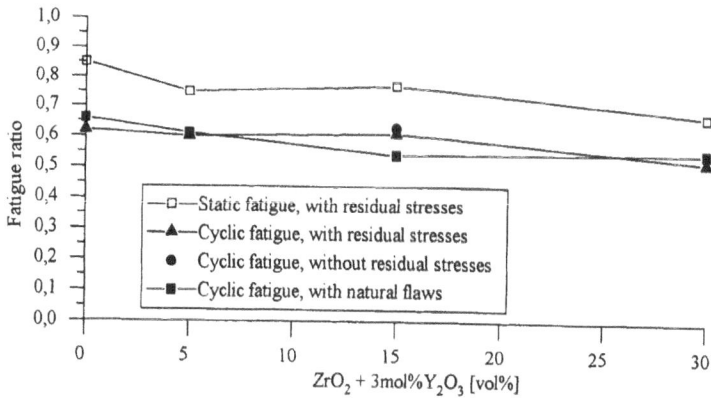

Figure 6. Summary of the fatigue ratios measured using specimens with natural flaws and indentation cracks as a function of zirconia content.

0.1 MPa m$^{1/2}$ steps until crack growth was detected. In most cases, the cracks arrested or slowed down during initial crack extensions associated with relatively small stress intensity factors[*], i.e. values within the threshold regime. Thus, all the materials showed short crack behaviour. Such behaviour was observed for different crack extensions depending upon the material. These crack extensions were up to 30 and 50 μm for ZTA and alumina, respectively, but of up to 70 μm for PA-PSZ and 250 μm for the EA one. In general, this behaviour was associated with either 1) the gradual formation of a transformation or microcracking zone at the crack tip (in ZTA and PA-PSZ), or 2) the slow development of a bridging zone within the crack wake zone (in alumina), or 3) deflection mechanisms (in EA-PSZ). The feasibility of observing one or more of these mechanisms is discussed in detail later. In this context, it should be pointed out that the small crack behaviour observed in the PSZs was not as pronounced as previously reported for another "peak-aged"-like Mg-PSZ[15].

The crack propagation curves were qualitatively very similar for all materials. Besides the referred short crack regime three distinct stages, as usually found for metals, could be identified. A first stage, i.e. the threshold regime, where crack growth rates depended strongly on the applied ΔK. Then, a second or Paris regime in which crack growth rates showed a power-law dependence on the applied ΔK. And finally, a third stage whose onset was marked by crack growth instability and therefore characterized by a rapid ascending crack growth rate with increasing crack extension.

In Mg-PSZ, large differences in the crack growth behaviour under static and cyclic loading conditions are found for both materials. These differences were mainly observed for crack growth rates below 10^{-5} m/s. At crack growth rates about 10^{-5} m/s the curves corresponding to static and cyclic loading conditions tended to join each other. As observed for other mechanical parameters, the corresponding crack growth thresholds for the PA-PSZ material were significantly higher, about twice, than those for the EA one. Within the second regime the slopes or Paris exponents associated with static crack growth were much larger, as expected (e.g. Ref. 3), than those corresponding to cyclic crack growth. However, it should be indicated that the slopes corresponding to crack growth under cyclic loading were similar for the PA and EA-PSZs.

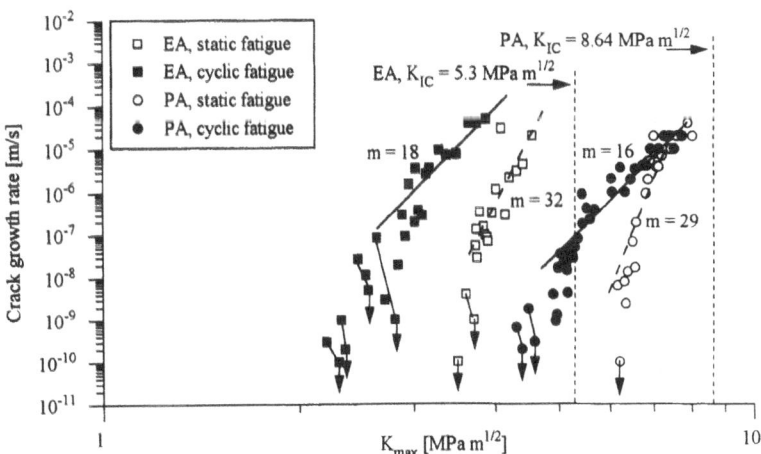

Figure 7. Comparison of measured static and cyclic crack growth rates as a function of maximum applied stress intensity factor for PA and EA-PSZ. The corresponding Paris coefficients are given by m.

---

[*] In this work, the cyclic fatigue data have been presented as a function of K$_{max}$ for comparison purposes with static fatigue data. In fact, the dependence (slope) of da/dt upon K$_{max}$ or ΔK is the same, because all cyclic fatigue tests were conducted under constant load ratio conditions. Thus, when appropriate, cyclic fatigue effects were also discussed in terms of ΔK.

For ZTAs, the static and cyclic fatigue growth rates showed considerable differences at crack growth rates below 10^{-8} m/s. With increasing crack velocity the differences become smaller and above 10^{-6} m/s cracks grew at similar rates in static and cyclic fatigue tests in all materials. Hence, it seems that cyclic loads have only an effect at very low crack growth velocities. The static crack growth thresholds were very similar and of the order of 3.1 to 3.3 MPa m$^{1/2}$ for all materials. In comparison, the cyclic fatigue thresholds were lower (between 2.7 and 2.9 MPa m$^{1/2}$) but also similar for all materials, as in the case of the fatigue limits. Likewise, the Paris exponents were very similar for all materials (cf. Figures 8 and 9). This confirms that the toughness increase as a result of the zirconia addition seems to be compensated by the higher fatigue effect in these materials. In other words, the progressive weakening of the shielding effect due to cyclic or static loading leads to very similar fatigue thresholds, which might be related to the intrinsic fracture toughness of the alumina matrix.

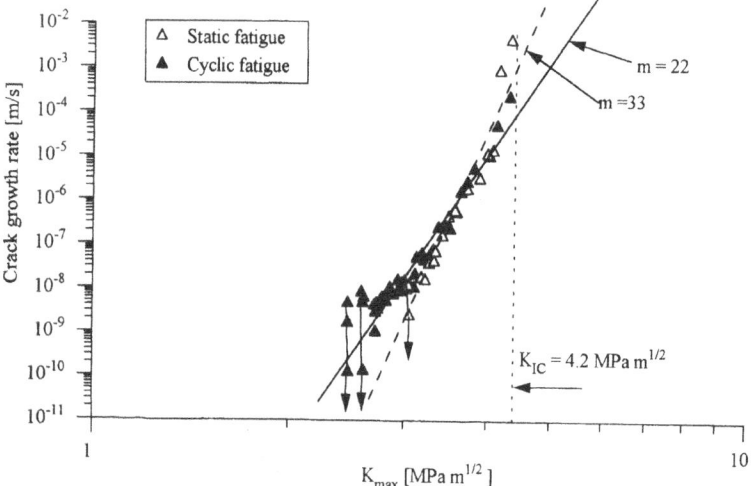

Figure 8. Measured static and cyclic crack growth rates as a function of maximum applied stress intensity factor for alumina. The corresponding Paris coefficients are given by m.

Fatigue Threshold-Fracture Toughness Coefficients

As for the S-N and S-t-curves, the crack growth results may be used to calculate the ratio between the threshold ΔK (or corresponding K_{max} for a given load ratio) and the fracture toughness, $\Delta K_{th}/K_{IC}$, which is a measure of the relative sensitivity of pre-cracked materials to fatigue[2]. Such ratio is referred to as fatigue coefficient in this study. It facilitates the comparison of the results obtained with the different materials. Together with the fatigue ratio it enables a comparison of the fatigue data obtained with natural flaws, small and long cracks.

In this context, the determination of the fatigue threshold is an important aspect. As can be seen in Figures 7 to 9, all materials showed small crack behaviour, i.e. crack started to grow at relative low stress intensity factor values, which could be defined as short crack or microstructural thresholds[16]. In many instances, these cracks were arrested either due to the crack tip shielding produced by transformation, microcracking and bridging or due to microstructural barriers such as large grains which forced the cracks to experience deflection mechanisms. In these cases, the load had to be increased in steps in order to produce further crack growth. As a result of this short crack behaviour a second or mechanical threshold can be defined at stress intensity factor values where crack growth starts to be continous. The

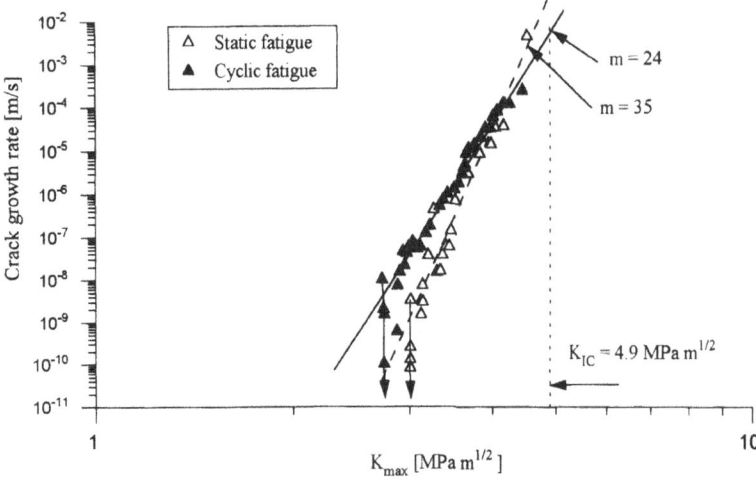

Figure 9. Measured static and cyclic crack growth rates as a function of maximum applied stress intensity factor for A30Z3Y. The corresponding Paris coefficients are given by m.

difference between mechanical and microstructural threshold represents the distance which a crack has to grow in order to overcome the shielding effect or the major microstructural barrier[16]. In this investigation, the mechanical threshold was used to calculate the fatigue coefficients since only continuous growing flaws lead to failure in fatigue life testing.

Using this mechanical threshold, the corresponding fatigue coefficient and fatigue ratio were almost identical for alumina, ZTA and Mg-PSZ, which shows that the fatigue coefficients obtained with long cracks might be used to estimate the fatigue ratios of specimens with natural flaws as in the case of indentation cracks (see above). Hence small and long crack data might be used for a design based on fatigue thresholds.

Crack Growth and Fracture Mechanisms

Fracture paths were studied either *in situ* using a long-distance optical microscope or by examining fracture surfaces as well as unbroken specimens, which contained fatigue cracks, through conventional optical microscopy and SEM.

Significant differences in crack growth and fracture mechanisms were observed in the two Mg-PSZs. On the other hand, for a given material, the fracture surfaces corresponding to specimens tested under monotonic, constant and cyclic loading conditions were very similar in agreement with previous reports[3,15,17].

PA-PSZ was mainly characterized by transgranular-like fracture, independently of the applied ΔK. Thus, and as expected from the TEM features found in this material, crack shielding by transformation toughening was induced through the interaction between transgranularly distributed precipitates and the propagating crack. Besides this mechanism, crack branching and low-angle crack deflection were also often observed (Figure 10). These mechanisms are well known to increase the fracture toughness under monotonic loading[18]. Cyclic fatigue of "PA-like" Mg-PSZ materials has usually been attributed to damage accumulation, in regions ahead of the crack tip, associated with a wide range of mechanisms: purely intrinsic (i.e. mechanically induced phenomena such as microcracking or mode II and III cracking, for example)[3], transformation plasticity[19], bridging degradation[20], etc. Extensive TEM and SEM examination of crack paths in the PA material studied here allowed to speculate that wedging effects and implicit mode II and III cracking due to the intrinsic roughness of the transgranular crack path as well as bridging degradation (bridges left after crack branching) played an important role in cyclic fatigue. Nevertheless, other mechanisms like reverse transformation or microcracking were either not studied or very difficult to evaluate, and therefore they must also be considered as possible mechanisms in accounting for the cyclic fatigue effect measured.

The fracture surface of EA-PSZ presented an intergranular-like aspect. Here, the intergranular mode was clearly delineated by the eutectoid decomposition products formed at grain boundaries. Crack growth mechanisms were mainly correlated to propagation along "weakened" grain boundaries. Swain and Hannink[14] made similar observations in an overaged Mg-PSZ. Thus, "grain-size-scale" crack deflection was prominent within the crack path, as shown in Figure 11. Such mechanism promoted a large retardation effect and the existence of pronounced short crack behaviour at stress intensity factor values close to that of the threshold (see above). At higher stress intensity factors the propagating crack showed a less intergranular-like mode and other retardation mechanisms such as crack branching were also found. As discussed before for the PA-PSZ, and following Dauskardt et al.'s ideas[3], the cyclic crack growth mechanisms in the EA material are suggested to be mainly associated with mechanically induced phenomena. However, two aspects must be pointed out. First, in that work on cyclic fatigue of "EA-like" material a transgranular fracture mode was found and crack deflection mechanisms were relatively scarce[3]. Because of the "overaged" nature of these materials, the reason for such observation is uncertain. Second, the observed fracture mechanisms in the EA-PSZ indicated that there was already an intrinsic damage within the material, before testing. This resulted in a diminished intrinsic resistance of the material that, in addition to the higher environmental sensitivity of the eutectoid product, would imply in EA-PSZ a less real cyclic fatigue than in the PA one. The results found here confirm such a trend.

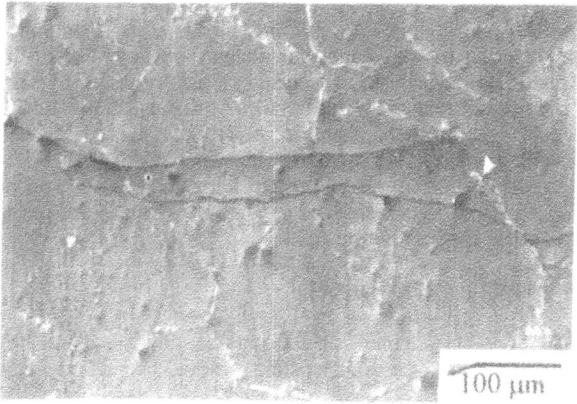

Figure 10. Optical micrograph indicating crack branching in PA-PSZ.

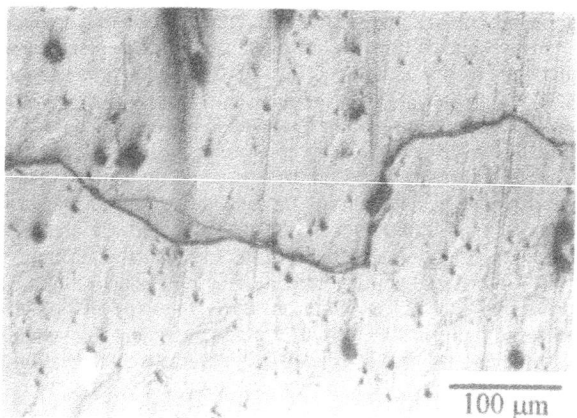

Figure 11. Optical micrograph showing pronounced crack deflection in EA-PSZ.

Two mayor observations were made in alumina and ZTA and these were independent of the type of test, either cyclic or static fatigue conditions. First, grain bridges or interlocking grains were found in the crack paths of unbroken samples (Figures 12 and 13). The size and the number of the bridges decreased with increasing zirconia content according to the reduction in grain size observed. Some of the bridges appeared already quite degraded, especially those further away from the crack tip which formed at the beginning of the tests. Likewise, small scale crack deflections around grains was found, but obviously this crack deflection played only a secondary role due to the small grain size of these materials, i.e. it can be hardly compared to the large scale deflections observed in EA-PSZ where the grain size is 15 - 30 times larger.

Second, differences were found in the fracture surfaces of alumina when stable crack growth occurred. In tests with indentation cracks, the fracture surfaces appeared more intergranular next to the indentation crack where stable growth took place. Much further away, in the region of unstable crack growth, the fracture surfaces showed a mixture of trans- and intergranular failure, which was also typical for fracture strength tests in this material. A very

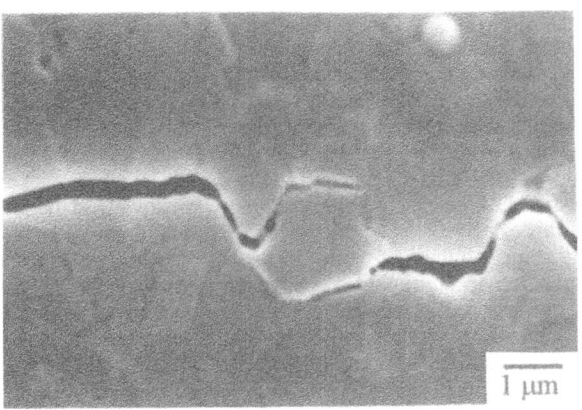

Figure 12. Secondary SEM micrograph showing a grain bridge in the trajectory of an indentation crack in pure alumina after cyclic fatigue testing.

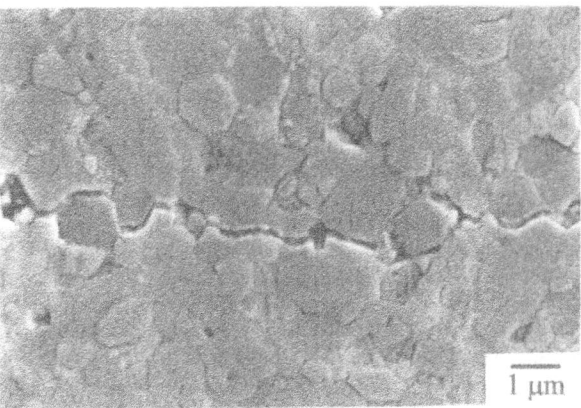

Figure 13. Secondary SEM micrograph showing small scale crack deflection and grain bridges in the path of an indentation crack in A30Z3Y after cylic fatigue testing.

similar behaviour was found in the tests with SENB specimens, the cracks propagated predominantly in an intergranular mode next to the pre-crack, i.e. at the beginning of the crack growth experiment when stable crack growth occurred. With increasing crack extension, i.e. at higher crack velocities, more and more transgranular failure was found. In the region of unstable crack growth or fast failure, the fracture surfaces showed once more a mixture of transgranular and intergranular fracture mode. In the A5Z3Y material, these differences could still be observed. However with increasing zirconia content it was increasingly difficult to observe any differences. This can be explained in conjunction with the first observation on crack paths. The more intergranular failure mode during slow crack growth was a result of the formation and subsequent degradation of the grain bridges by the environment in a static fatigue test and cyclic loads in a cyclic fatigue test. Hence, the cracks grew mainly along the grain boundaries and the fracture surfaces appeared more intergranular-like during stable growth in alumina and also to some extent in A5Z3Y. With increasing zirconia content fewer and much smaller bridges were found along the crack path due to the smaller grain size. In these finer microstructures no obvious differences could be found and the crack surfaces were similar during stable and unstable crack growth in A15Z3Y and A30Z3Y.

FURTHER DISCUSSION

The mechanical response observed under constant applied loading indicates that all materials experience static fatigue. Such effects may be explained by the intrinsic susceptibility of the different materials to environmental attack or stress corrosion cracking. The eutectoid decomposition product which forms along the grain boundary of the EA-PSZ appears to be particularly sensitive to environmental attack or static fatigue, and as a consequence: 1) this material suffers more static fatigue compared to the PA-PSZ; and 2) cracks propagate mainly along grain boundaries. Likewise, ZTAs are much more sensitive to static fatigue compared to pure alumina and this can be explained by the higher susceptibility of ytria stabilized zirconia (Y-TZP) to show static fatigue[21]. In both cases, microstructural changes as result of either eutectoid decomposition or addition of second phase particles gave rise to additional phases. This changed the intrinsic static fatigue resistance of the materials and therefore, it may be pointed out as an important feature in promoting static fatigue effects on ceramics.

In all materials, cyclic loading results in: 1) failure at stress values lower than the observed modulus of rupture; 2) fatigue lives shorter than those measured under constant load for a given maximum applied stress; and 3) crack growth thresholds lower than those observed under static fatigue conditions. Hence, a real cyclic fatigue effect is also evident. Such cyclic damage may be explained by means of the active toughening mechanisms and their degradation during cyclic loading which make toughened ceramics more susceptible to fatigue.

In PA-PSZ, stress induced transformation of the zirconia precipitates in addition to crack branching were the main toughening mechanisms and their cyclic degradation is suggested to be at least partly responsible for fatigue in these materials. On the other hand, crack deflection played the most important role in determining the toughening and cyclic fatigue behaviour of the EA material. It is very interesting to note that in both cases a similar slope in the da/dt-K_{max} curve was found. That would mean that completely distinct toughening mechanisms could have a very similar impact on the fatigue properties of the materials, depending upon their effective magnitude.

In pure alumina, the main toughening mechanism was grain bridging, as recently shown by Li and Guiu[22,23]. The grain bridges were degraded by cyclic loads in fatigue tests. With only 5% of zirconia added as in the case of A5Z3Y, the crack growth mechanisms hardly changed and grain bridges could still be found. Obviously these bridges were much smaller in size compared to pure alumina since the grain size decreased. As the amount of zirconia was increased further, fewer and even smaller bridges were found. According to Steinbrech et al.[24], the associated bridging stress is expected to decrease proportionally with the grain size for cracks with similar crack opening angle. In other words, bridging had less effect in ZTA and another mechanism such as transformation must have been active in order to maintain the same level of toughness as in A5Z3Y or increase it as in the materials with 15 and 30% of zirconia. As the main toughening mechanism changes, the cyclic fatigue characteristics are altered too, i.e. the main fatigue mechanism is expected to be cyclic degradation of the amount of transformation toughening when the zirconia percentage is increased. In addition, other mechanisms such as crack surface friction, crack closure and microcracking might have been active. However, in contrast to the Mg-PSZ materials, these mechanisms were believed to be of second importance due to rather negligible crack roughness and deflection as a result of the small grain sizes in the ZTAs.

CONCLUSIONS AND OUTLOOK

The above summary of the toughening and fatigue characteristics of the different materials studied illustrates very well that microstructural modifications may have a profound impact on both mechanical and fatigue behaviour of advanced ceramics. It is misleading to think that the different toughening mechanisms are additive and that the more distinct mechanisms are active at the same time, the better the material. In fact, toughening mechanisms may be even exclusive as observed in ZTA. In this case, the principal toughening mechanism was changed from bridging in alumina to transformation toughening in A30Z3Y. Nevertheless their properties and fatigue crack growth curves were very similar, because one mechanism excluded the other. An effective way of improving the fracture toughness of zirconia toughened alumina seems to be the introduction of zirconia agglomerates instead of dispersed particles. As shown by Wang and Stevens[25], in this case crack deflection, crack bridging, microcracking and bridging might occur at the same time, which also appears to improve the fatigue resistance of these materials[4].

Similarly, it is quite surprising to observe that EA-PSZ, i.e. a material with weak grain boundaries, shows qualitatively similar cyclic fatigue baheviour to PA-PSZ. Thus, microstructural tailoring such that both crack deflection and intergranular transformation toughening mechanisms are promoted would be an excellent choice for improving the mechanical behaviour of PSZ materials. However, and as pointed out by Suresh[26] in the case of metals, the prominence of crack deflection or intergranular crack growth mechanisms usually results from microstructural changes that imply a deleterious effect on the intrinsic resistance to fatigue. This is clearly observed in the mechanical response of the EA material studied here, and seems to indicate that the above ideas might also be valid for ceramics.

ACKNOWLEDGMENTS

This research has been conducted as part of the projects number MAT93-0328 and MAT94-0431, both financed by the CICYT (Comisión Interministerial de Ciencia y Tecnología). M. M. Nagl would like to thank the European Union for the financial support provided under the Human Capital and Mobility program (Contract No. ERBCH I CT930701). Likewise, R. Fernández would like to express his gratitude for the scholarship received from the Instituto de Cooperación Iberoamericana (ICI). Special thanks also to C. Domínguez at the Instituto Tecnológico de Materiales in Asturias (Spain) for producing the ZTA specimens as part of the above mentioned project and to M. Marsal for assistance on the SEM studies.

REFERENCES

1. R.C. Garvie, R.H. Hannink and R.T. Pascoe, Ceramic steel, *Nature.* 258:703 (1975).
2. N.A. Fleck, K.J. Kang and M.F. Ashby. The cyclic properties of engineering materials, *Acta metall. mater.* 42:365 (1994).
3. R.H. Dauskardt, D.B. Marshall, and R.O. Ritchie, Cyclic fatigue-crack propagation in magnesia-partially-stabilized zirconia ceramics, *J. Am. Ceram. Soc.* 73:893 (1990).
4. K. Duan, Y-W. Mai and B. Cotterell, Crack growth in a sintered Al_2O_3/ZrO_2 composite subjected to monotonic and cyclic loading, *in:* "Fracture of brittle disordered materials: Concrete, rock and ceramics," G. Baker and B.L. Karihaloo, eds., E&FN Spon, London (1995).
5. D.L. Porter and A.H. Heuer, Microstructural development in MgO-partially stabilized zirconia (Mg-PSZ), *J. Am. Ceram. Soc.* 62:298 (1979).
6. C.F. Grain, Phase relations in the ZrO_2-MgO system, *J. Am. Ceram. Soc.* 50:288 (1967).
7. S.C. Farmer, A.H. Heuer and R.H.J. Hannink, Eutectoid decomposition of MgO-partially-stabilized ZrO_2, *J. Am. Ceram. Soc.* 70:431 (1987).
8. D.J. Green, R.H.J. Hannink and M.V. Swain, "Transformation toughening of ceramics," CRC Press Inc., Boca Raton (1989).

9. P.F. Becher, Slow crack growth behaviour in transformation toughened Al_2O_3-$ZrO_2(Y_2O_3)$ ceramics, *J. Am. Ceram. Soc.* 66:485 (1983).
10. B.L. Karihaloo, Contribution of t-m phase transformation to the toughening of ZTA, *J. Am. Ceram. Soc.* 74:359 (1991).
11. D.B. Marshall and M.V. Swain, Crack resistance curves in magnesia-partially-stabilized zirconia, *J. Am. Ceram. Soc.* 71:399 (1988).
12. H. Tada, P.C. Paris and G.R. Irwin, "The stress analysis of cracks handbook," Del Research Corporation, Hellertown (1973).
13. R.H.J. Hannink and M.V. Swain, Magnesia-partially stabilized zirconia: The influence of heat treatment on thermomechanical properties, *J. Aust. Ceram. Soc.* 18:53 (1982).
14. M.V. Swain and R.H.J. Hannink, R-curve behaviour in zirconia ceramics, *in:* "Advances in ceramics, Volume 12: Science and Technology of Zirconia II," N. Claussen, M. Rühle and A.H. Heuer, eds., The American Ceramic Society Inc., Columbus, Ohio (1984).
15. L.A. Sylva and S. Suresh, Crack growth in transforming ceramics under cyclic tensile loads, *J. Mater. Sci.* 24:1729 (1989).
16. K.J. Miller, Materials science perspective of metal fatigue resistance, *Mat. Sci. & Technol.* 9:453 (1993).
17. R.H. Dauskardt, W. Yu and R.O. Ritchie, Fatigue crack propagation in transformation-toughened zirconia ceramic, *J. Am. Ceram. Soc.* 70:C248 (1987).
18. H.P. Kirchner, R.M. Gruver, M.V. Swain and R.C. Garvie, Crack branching in transformation-toughened zirconia, *J. Am. Ceram. Soc.* 64:529 (1981).
19. I.-W. Chen and S.-Y. Liu, Constitutive relations for mechanical fatigue in zirconia ceramics, *in:* "Proc. of the Eng. Found. Int. Conf. on Fatigue of Advanced Materials," R.O. Ritchie and R.H. Dauskardt, eds., Mat. & Component Eng. Publ. Ltd., UK (1991).
20. M.H. Jørgensen and B.L. Karihaloo, Cyclic fatigue of Mg-PSZ - Theoretical and experimental study, *in:* "Fatigue 93," J.-P. Bailon and J.I. Dickson, eds., EMAS, UK (1993).
21. H. Yin, M. Gao and R.P. Wei, Phase transformation and sustained load crack growth in ZrO_2 + 3mol% Y_2O_3: Experiments and kinetic modeling, *Acta metall. mater.* 43:371 (1995).
22. M. Li and F. Guiu, Subcritical fatigue crack growth in alumina - I. Effects of grain size, specimen size and loading mode, *Acta metall. mater.* 43:1859 (1995).
23. M. Li and F. Guiu, Subcritical fatigue crack growth in alumina - II. Crack bridging and cyclic fatigue mechanisms, *Acta metall. mater.* 43:1871 (1995).
24. R.W. Steinbrech, A. Reichl and W. Schaarwächter, R-Curve behaviour of long cracks in alumina, *J. Am. Ceram. Soc.* 73:2009 (1990).
25. J. Wang and R. Stevens, Toughening mechanisms in duplex alumina-zirconia ceramics, *J. Mat. Sci.* 24:3421 (1989).
26. S. Suresh, "Fatigue of materials," Cambridge University Press, Cambridge (1992).

THE INFLUENCE OF PROCESSING TECHNIQUES ON THE FATIGUE PROPERTIES OF YTTRIA STABILIZED ZIRCONIA WITH DIFFERENT GRAIN SIZES

Ralf Matt[1] and Georg Grathwohl[2]

[1] Institut für Keramik im Maschinenbau, University of Karlsruhe, Germany
[2] Keramische Werkstoffe und Bauteile, University of Bremen, Germany

Introduction

In recent years strong research efforts were made to reduce the inherent brittleness of structural ceramics. One promising material is transformation toughened ZrO_2, where the stress induced tetragonal to monoclinic (t-m) transformation can be used to increase strength and fracture toughness. With the use of different additions (stabilizers) such as Y_2O_3, CeO_2, MgO etc. the t-m transformation behaviour can be varied, leading to transformation zone sizes in the range of a few μm (e.g. Y_2O_3) up to mm (e.g. CeO_2). While Y_2O_3 stabilized ZrO_2 ceramics (Y-TZP) reveal rather limited toughness values due to small transformation zones, pronounced R-curve behaviour can be obtained for ZrO_2 materials stabilized with MgO and CeO_2.

Several studies have shown that transformation toughened zirconia ceramics are susceptible to cyclic fatigue [1-8]. The dominant cyclic fatigue mechanisms of ZrO_2 are still not completely understood. It is supposed that connections between the R-curve behaviour and the cyclic fatigue effects exists. For Si_3N_4 or Al_2O_3 which can also exhibit significant R-curve effects such interrelations between the toughening mechanisms and the cyclic fatigue effects are known [9,10]. The cyclic fatigue mechanisms discussed can be classified in two categories (extrinsic and intrinsic mechanisms)[11]. While extrinsic mechanisms are leading to higher crack-tip stress intensities due to diminished shielding processes, intrinsic mechanisms are causing microstructural changes (damaging processes) ahead of the crack-tip. Intrinsic mechanisms are favoured so far since the direct experimental evidence of a cyclic effect on the transformation behaviour (e.g. variation of the size and shape of the zone or the monoclinic volume fraction within the zone) was not given. However, recent studies [12] indicate that extrinsic mechanisms also have to be considered to understand the cyclic fatigue phenomena.

The mechanical behaviour of sintered Y-TZP can be improved by a post-densification process (hot isostatic pressing (hip)). The influence of this hip-process on the fatigue properties of Y-TZP is unknown. Therefore the static and cyclic fatigue properties of two materials with different transformation behaviour due to different grain sizes are studied. Moreover, the influence of a post-hip process on the fatigue properties was investigated and compared to the as sintered condition using sample series with large numbers of specimens with natural defect populations. It is then the aim of this work to describe the cyclic fatigue properties of Y-TZP as being influenced by the state of sintering and the microstructural parameters.

Experimental Procedures

Materials

A commercial ZrO_2-powder containing 2 mol% Y_2O_3 (Tosoh, Corp., Tokyo, Japan) was used for these experiments. After a compaction with a uniaxial die at 6 MPa and isostatic pressing at 400 MPa the green compacts were sintered at two different temperatures (1400 and 1500°C) to obtain microstructures with different grain sizes. Some sintered plates were additionally hot isostatic pressed (hip) at 1300°C with 150 MPa in argon atmosphere to reduce the defect sizes (to enhance the initial strengths). Bend specimens with the dimensions 3,5x4,5x45 mm^3 were cut from the plates and ground. The residual stresses caused by machining were released by a heat treatment at 1100°C for 2h in air. One series of specimens was tested in the as ground condition without this additional heat reatment i.e. with machining induced residual stresses still being present. The phase composition of the materials was determined by X-ray diffraction technique. The densities were determined by the water displacement technique. Large numbers of samples were used (up to 24) for the determination of the initial bending strength. A high loading rate of 1000 N/s was chosen to avoid subcritical crack growth effects on the initial strengths. The inner and outer span width of the loading fixture was 20 and 40 mm, respectively. The fracture origins of the specimens were investigated by SEM systematically. The fracture toughness K_{Ic} was measured by the bridge method [13].

Fatigue experiments

The static experiments were performed as four-point bending tests at room temperature with a maximal duration of loading of 400 h for the sintered and 200 h for the hip-materials. The cyclic fatigue tests (under load control) were carried out with two stress ratios (R=0,2 and -1) using an electromagnetic resonance machine. The frequency of about 70 Hz was kept constant. As for the strength tests a four-point bending fixture was used for the fatigue experiments. The alternating bending tests (R=-1) required a special fixture, where the sample was fixed in two brass cylinders with a mixture of adhesive (for details see ref.[14]). The outer fibre stress was controlled by strain gauges attached on both sides (tensile and compression surface) of the specimens. The maximum number of cycles was fixed at N_L=2· 10^6 cycles for the sintered and 10^7 for the hip-variants. For each material series with large numbers of samples were tested in static and cyclic experiments at a single stress level. The number of specimens used for the fatigue tests depended on the scattering of the initial strengths. With a higher scattering of the strengths a larger number of specimens was used for each fatigue test to assure an acceptable statistical accuracy.

On the base of the critical flaw size distribution in a specimen series a variation of the initial stress intensities (K_{Ii}) or the range of the initial stress intensities ΔK_{Ii} is provided by applying a constant stress σ_s and stress range $\Delta\sigma$. With the distribution of the initial strength (σ_c) and the lifetime (t_f) or numbers of cycles to failure (N_f) it is then possible to calculate [15, 16] the crack velocity as a function of K_{Ii}/K_{Ic} or $\Delta K_{Ii}/K_{Ic}$ using equation (1) and (2):

$$v\left(\frac{K_{Ii}}{K_{Ic}}\right) = -\frac{2 \cdot K_{Ic}^2}{t_f \cdot \sigma_c^2 \cdot Y^2}\frac{d[\log(\sigma_s/\sigma_c)]}{d[\log(t_f \cdot \sigma_s^2)]} \qquad (1)$$

$$v\left(\frac{\Delta K_{Ii}}{K_{Ic}}\right) = -\frac{2 \cdot K_{Ic}^2}{N_f \cdot \sigma_c^2 \cdot Y^2} \cdot \frac{d[\log(\Delta\sigma/\sigma_c)]}{d[\log(N_f \cdot \Delta\sigma^2)]} \qquad (2)$$

The geometrical factor Y is kept constant in this calculation.

Results and analysis

Material characterization

The sintered material 2Y-TZP-I shows due to the lower sintering temperature and sintering time a significant lower grain size compared to the material 2Y-TZP-II (table 1); however, the densities of both sintered materials are similiar. The hip-process after sintering did not change the grain size but reduced the porosity leading to almost dense materials. The K_{Ic}-values are rather moderate. The tendency of a higher fracture toughness of TZP-II can be explained by the larger grains and therefore the higher potential of the t-m transformation.

The initial bending strengths of all materials are shown in figure 1.

Table 1. Sintering conditions, grain size, density and fracture toughness of the 2Y TZP-ceramics. Some specimens were redensified by post-hot isostatic pressing (h)

Materials	Sintering conditions	Mean Grain size [μm]	Density [% theor.density]	K_{Ic} [MPa$\sqrt{m}$]
2Y-TZP-I			99,7	
	1400°C,	0,2		5.1
2Y-TZP-I(h)	0,5 h		>99,9	
2Y-TZP-II			99,4	
	1500°C,	0,6		5.3
2Y-TZP-II(h)	2,5h		>99,9	

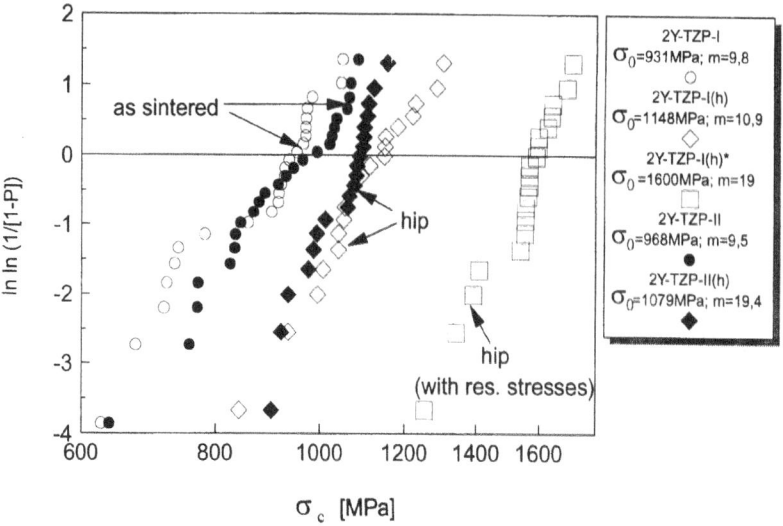

Figure 1   Weibull distribution of initial strength of the tested materials

The sintered materials reveal similiar strength distributions with rather low values of the Weibull modulus m. For the hip materials enhanced initial strengths are observed, what is especially pronounced for the material 2Y-TZP-I(h)* with residual stresses. The hip-process and the following thermal treatment to remove the residual stresses lead to similiar strength disributions for 2Y-TZP-I(h) and 2Y-TZP-II(h) at least up to a failure probability of about 60 percent (lnln (1/[1-P]~0). For higher failure probabilities the fine grained 2Y-TZP-I(h) shows higher strengths and a lower Weibull modulus m compared to 2Y-TZP-II(h). The effect of the residual stresses is very strong leading to an additional strength increase by more than 400 MPa and a higher Weibull modulus. Analysis of the residual stresses of the ground surface revealed compressive stresses of about 350 MPa (parallel) and 450 MPa (perpendicular to the grinding direction), resp. The contribution of these near-surface stresses to the strength seems to be surprisingly high since the zone thickness influenced by machining is in the range of a few μm only. With SEM the fracture surfaces of the specimens were investigated to identify the type of failure origins and also the location of the critical defects. This analysis shows that for the sintered and hiped samples processing defects were relevant and not flaws caused by machining. The strength limiting defect populations observed for the sintered materials were pores and agglomerates mostly located near the surface. In some cases (for higher failure probabilities) also volume flaws with a distance of more than 50 μm from the surface were found. However, the post-hip materials revealed a changed flaw distribution. For these materials not only a reduction of the defect sizes (as it is obvious due to the enhanced strengths) but also a change of the relevant defects occurred. In this case the most frequent defects were agglomerates. Furthermore the defects were mostly located at the surface where they are influenced by the residual compressive stresses.

Table 2 Results of the static and cyclic fatigue tests of the sintered and post-hip materials
with and without residual stresses

Materials	Loading condition and maximum stress (MPa)	number of specimens		
		initial fracture	fatigue failure	survivor
as sintered:				
2Y-TZP-I	static (690)	1	7	15
	R=0,2 (600)	1	8	15
	R=-1 (600)	2	21	0
2Y-TZP-II	static (710)	1	8	15
	R=0,2 (620)	0	9	15
	R=-1 (550)	0	18	6
post-hip and thermal treatment:				
2Y-TZP-I(h)	static (700)	0	9	9
	R=0,2 (650)	0	16	2
2Y-TZP-II(h)	static (800)	0	7	8
	R=0,2 (800)	2	13	0
post-hip without thermal treatment:				
2Y-TZP-I(h)*	static (1200)	0	5	10
	static (1320)	3	11	1
	static (1400)	4	8	0
	R=0,2 (1200)	3	9	0

Fatigue experiments

The stress levels and some results of the fatigue tests are presented in table 2.
In table 2 three types of specimens are distinguished depending on their performance during
the test: a) Specimens which fail before the final stress level of the fatigue experiment is
reached (initial fractures), b) specimens which fail during the fatigue tests (fatigue fractures) and c) specimens which did not fail before the end of the test (survivors). It can be
clearly seen from table 2 that all materials reveal strong fatigue effects. Some first state-
ments can then be drawn from table 2:

- At comparable stress levels (static stress and maximum stress in the cyclic test) the
  number of specimens showing fatigue failure is higher in cyclic tests compared to
  static tests.
- Compared to the higher initial strength of the post-hip materials the level of stresses
  to be actualized in the fatigue experiments is rather similar to the sintered materials.
- A higher sensitivity to fatigue effects is then observed for the fine grained post-hip
  material 2YTZP-I(h).

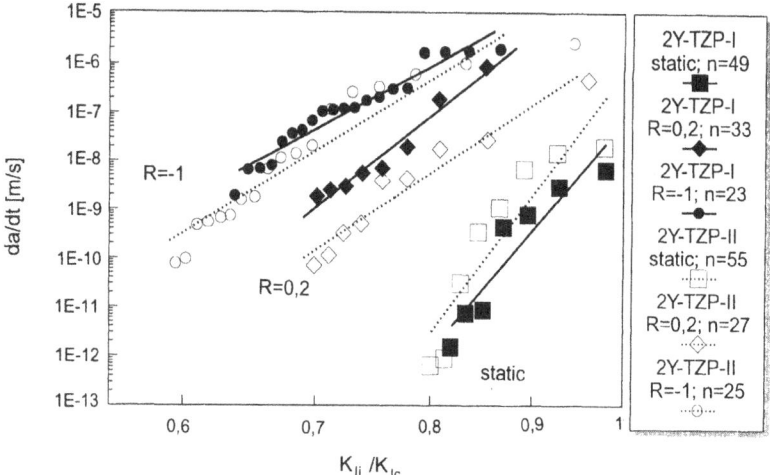

Figure 2   v-K_I-curves of the sintered materials under static and cyclic loading

The v-K_I-curves of the sintered materials 2Y-TZP-I and 2Y-TZP-II calculated from the results following eq. (1) are represented in figure 2.

The much higher crack velocities under cyclic loading (especially under alternating loading (R=-1)) demonstrate the pronounced cyclic fatigue effect for the tested materials, whereby the crack velocity curves for both materials can be described by the power law (da/dt=A $(K_{Ii}/K_{Ic})^n$). Furthermore the lowest n-values (slopes of the v-(K_{Ii}/K_{Ic})-curves) were found for R=-1 (obtained by a least square fit).The larger effective surface of the specimens in alternating bending tests compared to the specimens tested by static and pulsating loading was taken into account. The corresponding strength for a specimen with the larger effective surface $\sigma_{c,2}$, which is relevant for the alternating bending tests can be calculated with the relation $P(\sigma_{c,2})=1-[1-P(\sigma_c)]^2$ (see ref.[16]). Despite the relatively high scattering and a rather small number of static fatigue fractures to calculate the v-K_I curve higher n-values for the static loading condition can be stated. Figure 2 shows also a slight tendency of higher crack growth rates (static) for the coarser grained material 2Y-TZP-II. However, this is not the case in the cyclic tests.

The measured residual strength distributions of the static and cyclic survivor specimens are shown in figure 3a and 3b together with the Weibull distribution of the initial strength. Beside these survivor specimens (with corresponding higher failure probabilities) also fatigue fractures and in some cases initial fractures (with lower failure probabilities) have to be considered; the distributions appear then as truncated distributions.

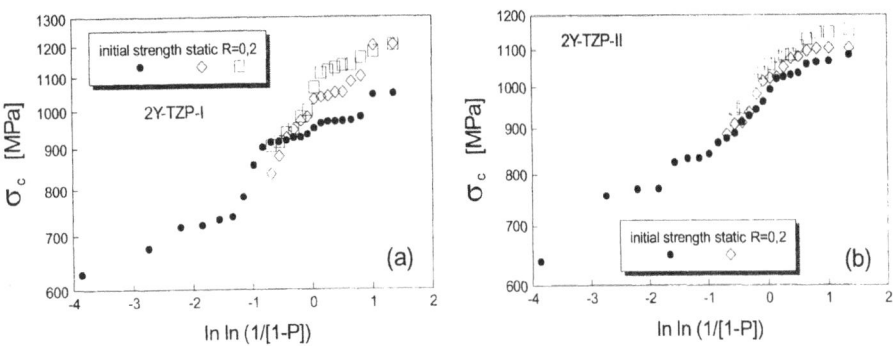

Figure 3   Weibull distributions of the initial strength and the residual strength of survivor
specimens (truncated) after static and cyclic loading of the sintered materials

The comparison of the residual strength and initial strength values (with corresponding
failure probabilities) indicates the existence of a strengthening effect, esp. for 2Y-TZP-I. It
is interesting to see that the strengthening effect seems to be more pronounced under cyclic
loading. Note that the stress levels of the cyclic loading were lower and also the loading
time of the fatigue experiments. The residual strength distribution of 2Y-TZP-II (static)
shows an almost identical distribution as the initial strength distribution. It can then be
stated that during the loading time not only fatigue effects but also strengthening effects are
active.

The influence of the post hip-process on the fatigue properties is demonstrated in figure 4
and 5. In both cases the specimens were subjected to a thermal treatment to remove the
residual stresses.

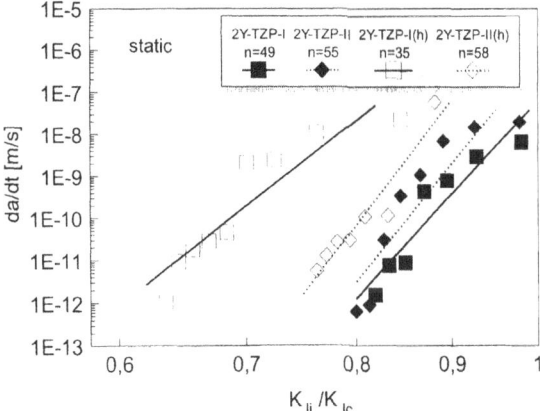

Figure 4   v-K_I-curve (static) for the sintered and post-hip materials with different grain size

Figure 4 shows that the static fatigue behaviour of 2Y-TZP-I(h) was changed drastically after the post-hip-process. This fine grained material reveals the lowest static fatigue resistance (highest crack growth rates) and the lowest n-value (n=35). In contrast only small differences between the crack velocities of the sintered and the post-hip 2Y-TZP with n-values of 55 and 58, resp. were determined.

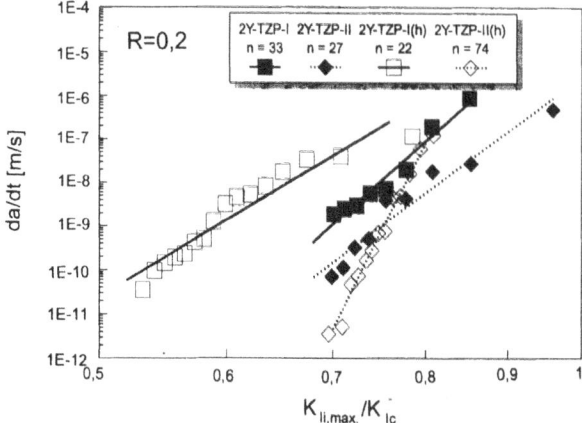

Figure 5   v-K_I-curve (R=0,2) for the sintered and post-hip materials with different grain size

A similiar situation is observed in the case of pulsating stresses (figure5). Again the fine grained 2Y-TZP-I(h) shows the highest crack growth rates and the lowest n-values (n=22). A very high n-value (74) was found for 2Y-TZP-II(h). The crack velocities are similiar for the sintered and post-hip TZP-II. While for the sintered materials lower crack velocities of the coarser grained material can be stated, these differences are extremely high for the post-hip materials.

The measurement of the residual strength distributions of the survivor specimens of 2Y-TZP-I(h) are shown in figure 6(a) for 2Y-TZP-I(h) and figure 6(b) (2Y-TZP-II(h)).

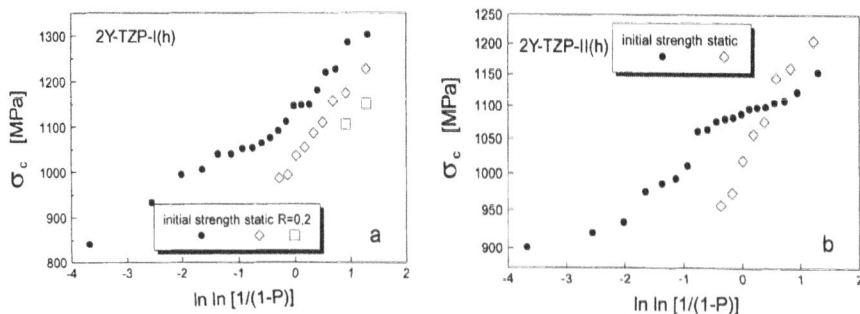

Figure 6   Weibull distributions of the initial strength and the residual strength of the survivor specimens (truncated) after static and cyclic loading of the post-hip-materials

The residual strength distribution of 2Y-TZP-I(h) confirms the fatigue effect of this material (figure 7a). The survivor specimens reveal significant lower strengths compared to the initial strength values at corresponding failure probabilities. This is in opposite to the results of the sintered materials (figure 4a). This effect is much less pronounced in both states of 2Y-TP-II, i.e. in the sintered and in the post-hip state (figure 3b and 6b).

It has been demonstrated that the fine grained post-hip material 2Y-TZP-I(h) reveals the lowest static and cyclic fatigue resistance. However, for the specimens which were not subjected to the thermal treatment (2Y-TZP-I(h)*) and therefore revealed residual stresses at the surface, the applicable stresses of the fatigue tests were very high (see table 2). The crack velocity curves for 2Y-TZP-I(h)* confirm the high fatigue resistance of the material in this state (figure 7a and 7b).

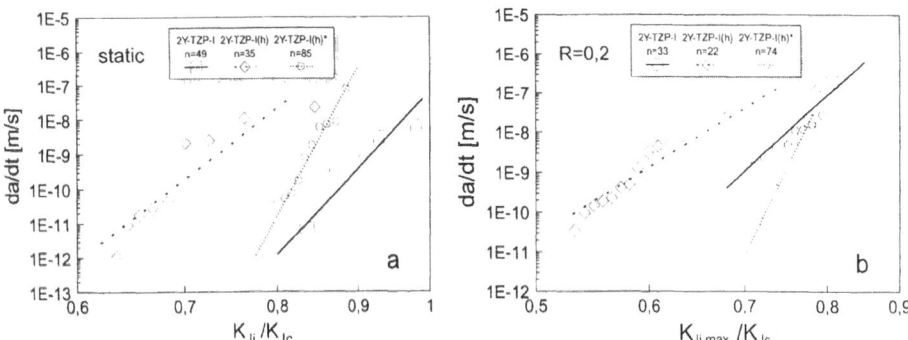

Figure 7   v-K_I -curves for static (a) and cyclic loading (b) of 2Y-TZP-I in three states: sintered, sintered and post-hip, sintered and post-hip with residual stresses.

In order to investigate the influence of the thermal treatment and its effect on the residual stresses and the fatigue behaviour a series of 2Y-TZP-I(h) and -II(h) specimens was polished to remove the residual stresses. 8 specimens were polished and additionally subjected to the same thermal treatment as the ground specimens. The initial strength distribution of these specimens are presented in figure 8.

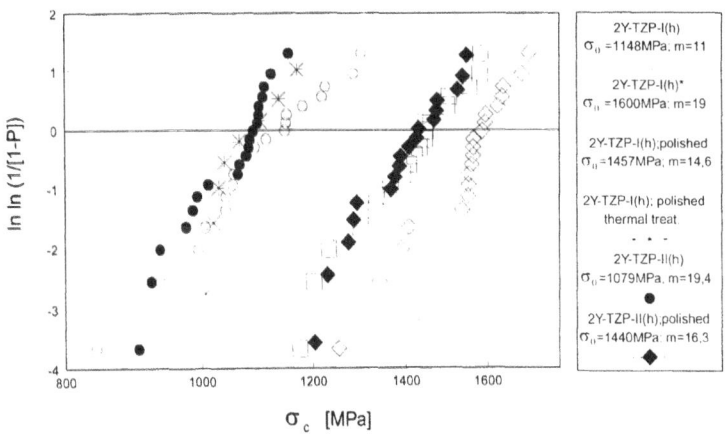

Figure 8   Initial strength distribution of polished and thermal treated specimens

Figure 8 makes clear that the initial strength distributions of the polished and thermal treated specimen series of 2Y-TZP-I(h) and 2Y-TZP-II(h) exhibit significant differences. Assuming that both, the thermal treatment and the polishing, lead to a removal of the residual stresses at the surface the initial strength distributions should be comparable. However, a large difference between both states was found.

The high σ_0-level in the as-ground state (1600 MPa) is reduced by thermal treatment to σ_0 =1148 MPa while polishing after grinding leads to a moderate reduction in strength (σ_0 =1457 MPa). If, however, the polishing procedure is followed by a thermal treatment the same final strength is reached as in the case of grinding and annealing.This indicates that the thermal treatment causes not only a removal of the the residual stresses but also micro-structural changes which reduce the initial strengths. Moreover, the static and cyclic fatigue resistance of the material (2Y-TZP-I(h) was dramatically changed by annealing as shown before.

The polished 2Y-TZP-I(h) and 2Y-TZP-II(h) exhibit almost an identical strength distribution with σ_0-values of about 1450 MPa.

Due to the observations presented above polished specimens of both materials were also tested in cyclic fatigue experiments (R=0,2) with a maximum applied stress of 1050 MPa. In both cases 17 specimens were tested. All 17 specimens of 2Y-TZP-I(h) tested failed during the loading time (10^7 cycles), while 3 specimens of 2Y-TZP-II(h) did not fail during this high loading. The higher fatigue resistance of the coarser grained 2Y-TZP-II(h) under cyclic loading is also demonstrated by the calculated v-K_I curve (figure 9). In this figure the crack propagation rates are compared to the crack growth of the thermal treated materials. It can be seen that the crack velocities of the fine grained polished material (2Y-TZP-I(h)) are significantly reduced. Moreover, a higher n-value (n=41) was found after polishing. For the coarser grained 2Y-TZP-II(h) only slight differences in the crack growth rates of the polished and ground samples (without residual stresses) were found. Very high n-values (74 and 76, resp.) were measured in both states. This result shows that the cyclic fatigue beha-viour of 2Y-TZP-II(h) is not affected by the thermal treatment as it is the case for 2Y-TZP-I(h). The comparison of the crack growth behaviour of both polished materials demonstrates again the lower fatigue resistance of the fine grained 2Y-TZP-I(h).

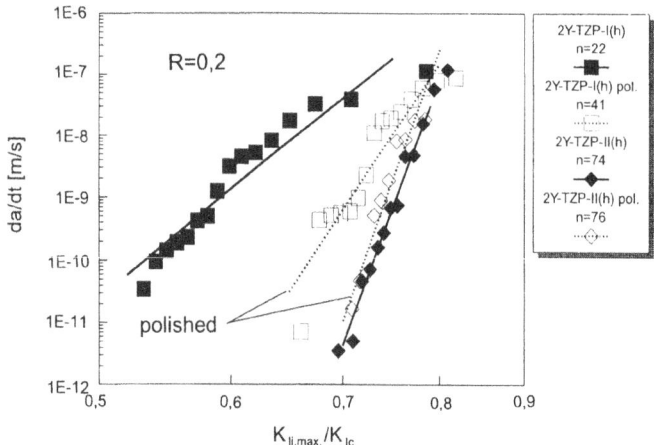

Figure 9   v-K_I-curves of ground (and thermal treated) and polished (without thermal treatment) materials

Discussion

The sintered and post-hip 2Y-TZP materials reveal both, static and cyclic fatigue effects with a more pronounced cyclic fatigue effect under alternating loading, which lead to decreased lifetimes (in the order of several magnitudes). The v-K_I-curves determined by the modified lifetime method using natural flaws can be described by a power law. Note, that this method does not pre-assume a power law dependance of the crack velocity. The cyclic fatigue effect is also obvious due to the enhanced crack growth rates compared to the static loading condition. Moreover, the n-values (slopes of the curves) are decreased under cyclic loading, with lowest values under alternating loading (23-25). The crack growth rates for highest stress intensities ($K_{Ii}/K_{Ic} \rightarrow 1$) under cyclic loading (especially R=-1) are higher ($\sim 10^{-7}$ to 10^{-6} m/s) compared to the velocities under static loading ($\sim 10^{-8}$ m/s). Following the v-K_I-curve evaluation by the modified lifetime method the shortest lifetime of a fatigue test serie is correlated with the lowest initial strength (taken from the initial strength distribution with the same failure probability), leading to corresponding high crack growth rates for high initial stress intensities (K_{Ii}/K_{Ic}-ratios). Since the loading rates to reach the final stress levels of the static and cyclic fatigue tests were not identical the influence of this effect on the measured lifetime should be considered. For extreme short lifetimes (a few seconds) a stronger influence has to be expected compared to longer lifetimes, where this effect is neglectable. In the case of cyclic fatigue tests the loading rate is lower compared to the static case and especially under alternating loading (R=-1), some specimens failed with very short lifetimes. Thus, the measured lifetimes of these specimens seem rather underestimated leading to overestimated crack velocities.

Only minor grain size effects on the fatigue behaviour can be found for the sintered materials 2Y-TZP-I and -II, with a tendency of slightly higher crack growth rates (lower fatigue resistance) for the fine grained material under cyclic loading (figure 2).

The longterm loading condition leads, beside the strength degradation (crack propagation) and failure, to strengthening effects of the survivor specimens (figure 3). It is believed, that this effect is connected with the stress induced t-m transformation in the process zone around defects. It is interesting to see that under cyclic loading strengthening seems to be more pronounced compared to the static loading condition. However, the loading time as well as the applied stress levels of the static and cyclic tests are different. Despite this fact, these results indicate that also under the cyclic loading condition with its strong fatigue effects significant strengthening can be observed. These phenomena are studied in detail in current investigations.

The post-hip process leads to a rather complex mechanical behaviour (initial strength and fatigue properties) depending on a thermal treatment. The initial strengths of the post-hip materials are enhanced compared to the sintered states. Based on the results of the initial strength distributions (figure 8) the contribution of the post-hip process to enhanced strengths seems to be different for the thermal treated and the polished materials, because the polished samples revealed about 300 MPa higher strengths compared to the heat treated materials. In both cases the residual stresses at the surface were removed. Furthermore, SEM investigations indicated that in both cases processing flaws were limiting the strength. Therefore, an additional effect is caused by the thermal treatment, which leads to further decreased initial strengths. This observation will be discussed later.

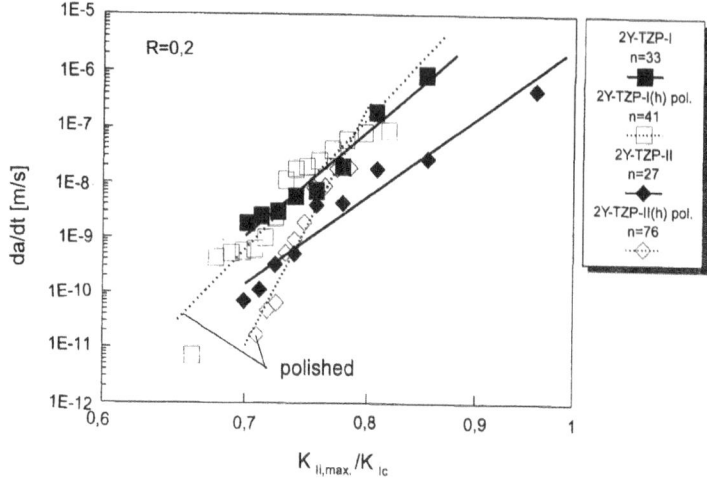

Figure 10 v-K_I-curves (R=0,2) of the sintered and post-hip materials

The effect of the post-hip process on the cyclic fatigue behaviour is shown figure 10. In this v-K_I-curve the cyclic crack growth behaviour of the sintered and post-hip (polished) materials are compared. In both cases the influence of surface residual stresses due to the machining were removed.

It can be seen that the fine grained material (TZP-I) reveal similiar crack velocities in both states. Due to the higher initial strengths (σ_0= 1457 MPa) the applied cyclic stress level was much higher with a maximum applied stress of 1050 MPa compared to the stress level of the sintered material (max. stress of 600 MPa). Thus the post-densification allows to increase the applied loading of a fatigue test (due to reduced critical defect sizes) to get equivalent initial stress intensities. The observation of similiar v-K_I-curves of the sintered and post-hip materials indicate then an almost unchanged cyclic crack growth behaviour. Only slight differences of the crack velocities between the sintered and post-densified coarser grained material TZP-II can be found, with a high n-value of 74 for the post-hip material. For both processing techniques, the sintering and the post-densification, slightly reduced crack velocities can be observed for the coarse grained materials. The higher fatigue resistance of the coarse grained and polished 2Y-TZP-II(h) is also expressed by the fact that 3 specimens withstand the high cyclic loading (survivors) and all specimens of the fine grained material failed at the same stress level of 1050 MPa, despite the almost identical initial strength distribution of the polished materials (figure 8).

Concerning the effects occuring due to the thermal treatment two main effects should be discussed. Firstly, a strength degradation of the hip and thermal treated materials compared to the polished sample series of 2Y-TZP-I(h) and 2Y-TZP-II(h) (figure 8). Secondly, a dramatically reduced static and cyclic fatigue resistance for the fine grained material 2Y-TZP-I(h) was found, when the specimens were subjected to the thermal treatment. In contrast, the polished samples (without this heat treatment but also with removed residual stresses) revealed a much better cyclic fatigue behaviour (figure 9). The causes of these phenomena are not completely understood. In both cases a connection with the thermal treatment is supposed. This assumption is confirmed by the fact that the only polished and the polished and additionally thermal treated samples revealed the same strength differences of about 300 MPa as it was found for the ground materials (figure 8). Intensive SEM investigations revealed no difference of the fracture origins of these materials and also no significant differences of the microstructure around the defects could be detected. Masaki [17] reported that Y-PSZ materials which were prepared by hot isostatic pressing under argon and nitrogen atmosphere degraded greatly in strength after ageing in air at temperatures of 800 - 1200°C for 50 -1500 h. Masaki explained this effect with the formation of cavitations which were produced by the oxidation of carbon due to CO-portions in the gaseous atmosphere. Although the thermal treatment lasts only 2 h at 1100°C, a similiar effect for the materials 2Y-TZP-I(h) and 2Y-TZP-II(h) can not be ruled out, since a graphite heater was used for the post-densification process. However, this phenomenon requires further investigations.

The second effect of a reduced fatigue resistance of the fine grained 2Y-TZP-I(h) seems also to be connected with the thermal treatment. If this material is only polished to remove the residual stresses (caused by machining) a high cyclic fatigue resistance can be achieved (figure 9), in contrast to the low fatigue resistance obtained when the specimens were subjected to a thermal treatment. Despite the observation that both, the fine and coarse grained materials experienced a strength degradation with similiar initial strengths after the thermal treatment the fatigue behaviour of these materials are drastically different. For 2Y-TZP-II(h) the phenomenon of a higher fatigue susceptibility could not be found. Since the defect populations for 2Y-TZP-I(h) and 2Y-TZP-II(h) are comparable a microstructural influence seems to exist, which may involve the different grain size. It is believed that the coarse grained microstructure of 2Y-TZP-II(h) with its higher transformability is responsible for the improved fatigue properties of 2Y-TZP-II(h). The fatigue behaviour of this material and the sintered state is similiar.

The grain size dependance of the t-m transformation is reflected by results of phase analysis, where a clearly higher monoclinic phase content after grinding (15 %) and at the fracture surfaces (~55%) for the coarse grained 2Y-TZP-II was detected, compared to the fine grained material, where only 10 % (as ground) and about 40 % (fracture surfaces) could be measured.

The method used to calculate the v-K_I curves requires a material without R-curve behaviour and a single defect population for the tested sample series. Moreover, it is assumed that the lifetime is determined by crack propagation. The assumption of a single defect population is best fulfiled for the post-hip materials where agglomerates at the surface were detected as fracture origins. However, the transformation toughened 2Y-TZP materials used in this study have a limited potential for R-curve behaviour. The shape of the ("short-crack") R-curve for the critical processing defects which control the lifetime is unknown. The fact that only very thin transformation zones (several μm) are generated at the fracture surface, which are not visible with an optical microscope, strongly limits the potential of a rising crack resistance. This is also confirmed by the low K_{Ic}-values measured by the bridge method. Additional tests were made to investigate the "validity" of the parameters A and n of the crack velocity curve calculated under these assumptions. Static fatigue tests were performed to compare the predicted (using the parameter A and n of the crack velocity curve of 1320 MPa and calculated with the conventional lifetime equation) and the measured lifetimes (at stress levels of 1200 and 1400 MPa). The result is shown in figure 11 where the calculated (predicted) and measured lifetimes are in reasonable accordance.

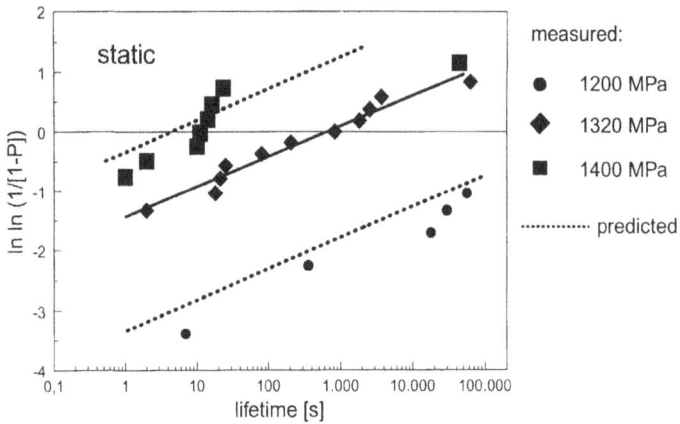

Figure 11 Measured and predicted lifetimes for static loading conditions (2Y-TZP-I(h)*)

This comparison shows at least for this material, that despite the assumptions made to allow a calculation of the v-K_I-curve, esp. the assumption of absent R-curve behaviour, a lifetime prediction with an acceptable accuracy is possible.

The results presented here, demonstrate that the static and cyclic fatigue behaviour is influenced by the microstructure (grain size), processing and the preparation of the specimens. A higher static and cyclic susceptibility was found for the fine grained materials. This means that a higher potential for the t-m transformation (and therefore enhanced R-curve behaviour) lead to improved static and cyclic fatigue behaviour.

Conclusions

A post-densification (hot isostatic pressing) leads to enhanced initial strengths (up to 1700 MPa (4-point-bending) and a lower scattering (m-values up to 19) compared to the sintered state.
The post-hip-process caused no significant changes of the fatigue properties. However, if the post-densified materials are subjected to a thermal treatment (1100°C, 2h) the initial strengths are drastically reduced (of about 300 MPa) and the fatigue behaviour of the fine grained material variant is dramatically changed revealing a much higher static and cyclic fatigue susceptibility for this material. It is assumed that the reduced potential of the stress induced t-m transformation (due to the smaller grain size) in combination with the effects caused by the thermal treatment are responsible for this observation.
Only a slight influence of the microstructure (grain size) on the fatigue properties is found for the sintered and post-hip materials, with a higher fatigue resistance for the coarse grained materials.

The strongest cyclic fatigue effects and lowest n-values (23-25) were found under alternating loading (R=-1).

The parameters of a v-K_I-curve determined by a fatigue test at a single stress level are used to conduct a lifetime prediction for other stress levels. The comparison of the predicted (calculated) and measured lifetimes are in a reasonable accordance, demonstrating the "validity" of the parameters despite some assumptions made to allow the calculation of the v-K_I-curve.

References

[1] R. H. Dauskardt, W. Yu and R. O. Ritchie: "Fatigue Crack Propagation in Transformation-Toughened Zirconia Ceramic"; J. Am. Ceram. Soc., 70 [10] C-248-C-252 (1987)

[2] G. Grathwohl and T. Liu: "Crack resistance and fatigue of transforming ceramics: II, CeO_2-stabilized tetragonal ZrO_2"; J. Am. Ceram. Soc., 74 [12] 3028-3034 (1991)

[3] S.-Y. Liu and I.-W. Chen: "Fatigue of yttria-stabilized zirconia: I, Fatigue damage, fracture origins and ifetime prediction"; J. Am. Ceram. Soc., 74 [6] 1197-1205 (1991)

[4] S.-Y. Liu and I.-W. Chen: "Fatigue of yttria-stabilized zirconia: II, Crack propagation, fracture striations and short-crack behaviour"; J. Am. Ceram. Soc., 74 [6] 1206-1216 (1991)

[5] J.-F. Tsai, C.-S. Yu and D.K. Shetty: "Fatigue Crack Propagation in Ceria-Partially-Stabilized Zirconia (Ce-TZP)-Alumina Composites"; J. Am. Ceram. Soc., 73 [10] 2992-3001 (1990)

[6] T. Liu, Y.-W. Mai and G. Grathwohl: "Cyclic Fatigue Crack Propagation Behaviour of 9Ce-TZP with Different Grain Size"; J. Am. Ceram. Soc., 76 [10] 2601-606 (1993)

[7] T. Liu, Y.-W. Mai and M.V. Swain: "Cyclic Fatigue Behaviour of Eutectoid Aged Mg-PSZ Ceramics ith Processing Flaws"; J. Europ. Ceram. Soc. 12 (1993) 221-226

[8] S.-Y. Liu and I.-W. Chen: "Plasticity-Induced Fatigue Damage in Ceria-Stabilized Tetragonal Zirconia Polycrystals"; J. Am. Ceram. Soc., 77 [8] 2025-35 (1994)

[9] H. Frei and G. Grathwohl: "New test methods for engineering ceramics - in-situ microscopy investigation"; cfi/Ber. DKG 68 (1991) No.1/2 (1991)

[10] F. Guiu, M. Li and M. J. Reece: "Role of Crack-Bridging Ligaments in the Cyclic Fatigue Behavior of Alumina"; J. Am. Ceram. Soc., 75 [11] 2976-2984 (1992)

[11] R. H. Dauskardt, D.B. Marshall and R.O. Ritchie: "Cyclic Fatigue-Crack Propagation in Magnesia-Partially-Stabilized Zirconia Ceramic"; J. Am. Ceram. Soc., 73 [4] 893-903 (1990)

[12] J.F. Tsai, J. Belnap and D.K. Shetty: "Crack Shielding in Ce-TZP/Al$_2$O$_3$ Composites: Comparison of Fatigue and Sustained Load Crack Growth Specimens"; J. Am. Ceram. Soc., 77 [1] 105-117 (1994)

[13] R. Warren and B. Johannesson: "Creation of stable cracks in hard-metals using bridge" indentation"; Powder Metall. [27] 25-29 (1984)

[14] R. Matt, T. Liu and G. Grathwohl: "Microstructural and environmental influences on the fatigue properties of 2Y-TZP ceramics with processing flaws", Materials& Design 14 [3] 159-168 (1993)

[15] T. Fett and D. Munz: "Determination of v-K$_I$-curves by a modified evaluation of lifetime measurements in static bending tests"; J. Am. Ceram. Soc., 68 (1985) C213-C215

[16] T. Fett, G. Martin, D. Munz and G. Thun: "Determination of da/dN-ΔK$_I$ curves for small cracks in alumina in alternating bending tests"; J. Mat. Sci. [26] (1991) 3320-3328

[17] T. Masaki: "Mechanical Properties of Y$_2$O$_3$-Stabilized Tetragonal ZrO$_2$ Polycrystals After Ageing at High Temperature"; J. Am. Ceram. Soc., 69 [7] 519-522 (1986)

FATIGUE CRACK INITIATION AND PROPAGATION IN CERAMICS

Akira Ueno and Hidehiro Kishimoto

Toyota Technological Institute
2-12-1, Hisakata, Tempaku, Nagoya, 468 Japan

ABSTRACT

Fatigue crack initiation and propagation in plain specimens of a sintered Si_3N_4 have been investigated. The crack initiation was observed by the *in situ observing system* equipped with a scanning laser microscope. Micro-crack propagation behaviors of the plain specimens are compared with those of long cracks and semi-elliptical surface cracks. Main results obtained are as follows:

(1) Cracks in the plain specimens can propagate with low stress intensity factors, which are far lower than the threshold stress intensity of the long cracks.

(2) In some cases, a crack stops propagation or propagates very slowly after the crack initiation. In these cases, the crack propagation rate is not determined uniquely by the stress intensity factor, but varies significantly.

(3) At the final stage of the fatigue life, the relationship between K_{Imax} and da/dN of micro cracks approaches that of small semi elliptical surface cracks. This means that the power low is not applicable, for a large part of the fatigue life.

INTRODUCTION

Ceramic materials have been expected as new promising materials because of their many excellent characteristics including light weight and high heat-resistance. Many reports on cyclic fatigue and creep behavior have been published to study the mechanism of cyclic fatigue, because strength of some ceramics are weakened by cyclic load. As crack propagation has been thought to be the most important in cyclic fatigue, pre-cracked specimens such as a compact tension specimen[1-6] and a specimen with artificial small crack[7-12] have been used. It is not enough to study the cyclic fatigue using pre-cracked specimens only, because ceramic components of commercial use are supplied without a crack.

We investigated the crack initiation and propagation behaviors of micro-cracks originated from defects in plain specimens. Then we compare their propagation behaviors with those of long cracks or artificial small cracks.

EXPERIMENTAL PROCEDURE

The material used in this study was a sintered silicon nitride doped with Al_2O_3 and Y_2O_3. Average grain size of the material was about 0.48μm and average aspect ratio of the elongated grains was 6.8[15]. Mechanical properties are shown in Table 1. Fatigue specimens were machined into the shape shown in Fig.1, after which the specimen surface was buff-finished using Al_2O_3 powder. In order to compare the propagation behavior of micro-cracks with that of long-cracks, fatigue specimens were cut out from the used CT specimens[3,4] as shown in Fig.2.

Fatigue tests were carried out by a newly designed ball-screw type compact 4-point bending fatigue testing machine, as shown in Fig.3. The testing machine was placed under an objective lens of a scanning laser microscope (SLM, Lasertec Corp., Japan, Type 1LM21H), and the "*in situ observing system*" was constructed as shown in Fig.4. A cyclic fatigue tests were performed under a triangle wave form of f=0.1Hz, R=0.1. The crack length was measured using a SLM of high (up to $\times$6000) magnification. Stress intensity factor K_I was calculated by the Raju-Newman's equation[14], assuming that the crack shape was semi-circular.

Table 1 . Mechanical properties of a sintered Si_3N_4.

Bending strength σ_{3b} (MPa)[*1]	Young's modulus E (GPa)	Fracture toughness K_c (MPa$\sqrt{m}$)[*2]	Hardness HV[*3]
740	310	5.70	1270

*1 3-point bending
*2 IF method using Niihara's equation
*3 Vicker's hardness

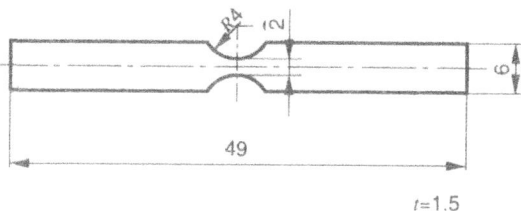

t=1.5

Stress concentration factor α=1.037

Figure 1. Configuration and dimensions of fatigue specimen.

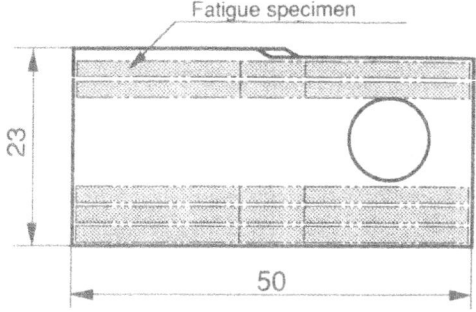

Figure 2. Extracted position from CT specimen.

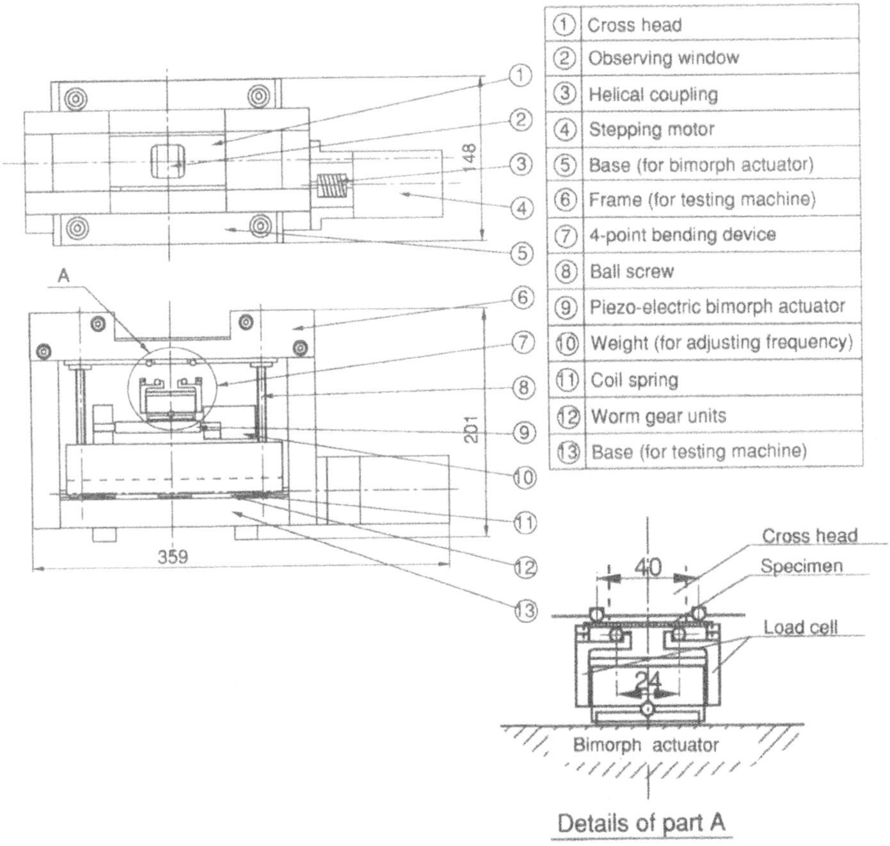

① Cross head
② Observing window
③ Helical coupling
④ Stepping motor
⑤ Base (for bimorph actuator)
⑥ Frame (for testing machine)
⑦ 4-point bending device
⑧ Ball screw
⑨ Piezo-electric bimorph actuator
⑩ Weight (for adjusting frequency)
⑪ Coil spring
⑫ Worm gear units
⑬ Base (for testing machine)
A
148
201
359
40
24
Cross head
Specimen
Load cell
Bimorph actuator
Details of part A

Figure 3. 4-point bending type fatigue testing machine for *in situ* observation.

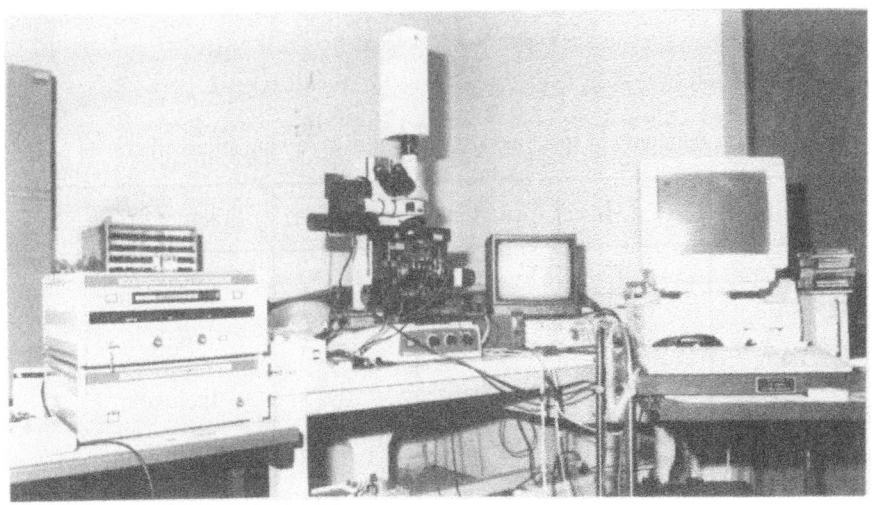

Figure 4. Whole view of the "*in situ observing system*".

EXPERIMENTAL RESULTS

S-N Properties

Figure 5 shows the relationship between σ_{max} and N_f, where σ_{max} is the maximum stress, N_f is the number of stress cycles to failure. In the figure, symbols with arrows indicate truncated specimens. Data points ($\square$) on $N_f = 1$ cycle indicate results obtained from the monotonic 4-point bending tests. Average bending strength was 773 MPa. Although the fatigue life decreases as the applied stress increases, the scatter in the fatigue life is very large. Defects such as pores or inclusions were found at the center of *mirror* region in SEM observation to the fracture surface.

Crack Growth Behavior of Micro-cracks

Test results are listed in Table 2, where N_i is the crack initiation life. In the table, $N_i = 0$ denotes that a fatigue crack initiated during the 1st loading. Except for specimen No. 3, fatigue cracks initiated early in the fatigue life, because of the high maximum stress. The crack initiation and crack propagation of each specimen are summarized as follows:

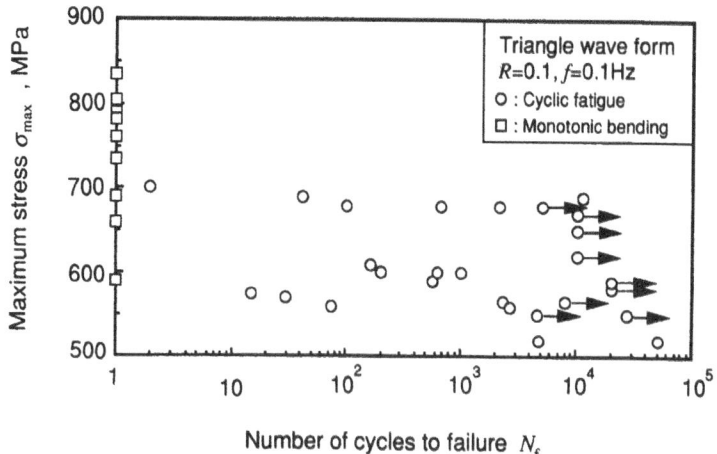

Number of cycles to failure N_f

Figure 5. S-N properties.

Table 2. Summary of specimens succeeded in observing micro-cracks.

Specimen	σ_{max} (MPa)	N_i (cycles)	N_f (cycles)	Fracture origin
1	660	0	1	Inclusion
2	560	0	74	Pore
3	520	50501	50976	Inclusion
4	520	140	4700	Inclusion
5	680	5	678	Pore

Specimen No. 1: A fatigue crack of about 3μm in length initiated from an inclusion on the specimen surface at the middle of the 1st stress cycle (Fig. 6(a)). SEM micrograph of the inclusion is shown in Fig. 6(b). Diameter and depth of the inclusion are about 66μm and 86μm, respectively.

Specimen No. 2: A fatigue crack initiated from a large pore on the specimen surface at the 1st stress cycle. Although the crack did not grow until 70 cycles (Fig. 7), the crack suddenly started to propagate. After rapid crack growth, the specimen failed at 74 cycles.

Specimen No. 3: As any crack did not initiate until 50500 cycles, the fatigue test was truncated. Successive fatigue test was done. At the first cycle, which was 50501th cycles from the beginning, a fatigue crack initiated from an inclusion on the specimen surface (Fig. 8). Although this crack did not grow until 50976 cycles, the specimen suddenly failed at 50976 cycles. The micro-crack, shown in Fig. 8, was determined not to be a main crack, because the above mentioned inclusion could not be found around the fracture starting point in SEM observation. This is an example that plural cracks initiate in a specimen.

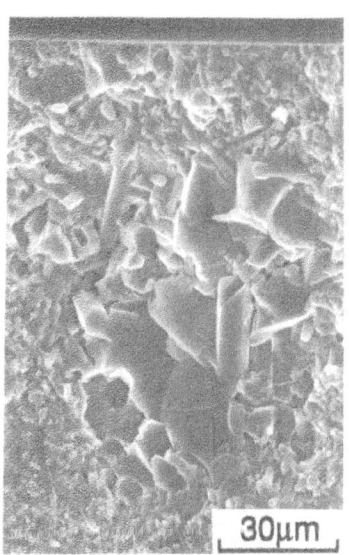

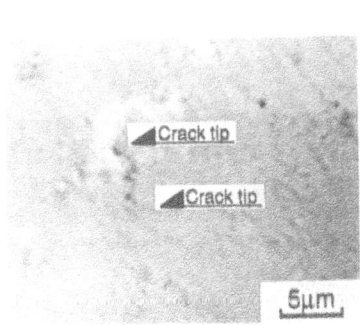

(a) SLM micrograph of a surface crack.   (b) SEM micrograph of fracture surface.

Figure 6. Micrographs of specimen No. 1.

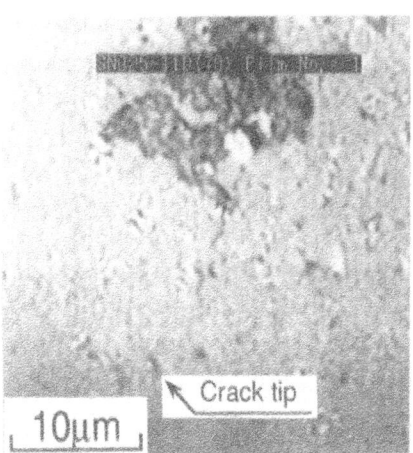

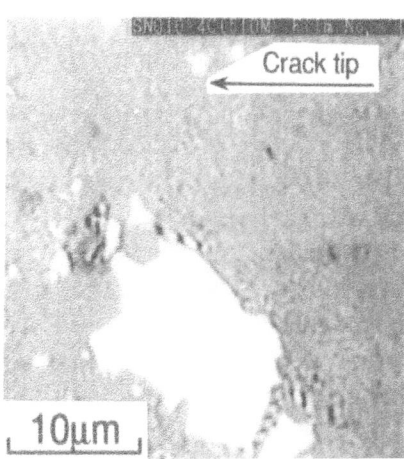

Figure 7. Micrographs of specimen No. 2.   **Figure 8.** Micrographs of specimen No. 3.

Specimen No. 4 : A crack initiated in a large inclusion of 34 μm in diameter and 108 μm in depth at the middle of the 1st stress cycle, and it started propagation into the matrix at 140th cycle. The specimen failed at 4700 cycles after subcritical crack growth. An example of SLM micrograph of the crack and SEM micrograph of the inclusion on the fracture surface are shown in Figs. 9(a) and 9(b), respectively.

Specimen No. 5 : A crack initiated from a large pore of 22 μm in diameter at 5th cycle. The specimen failed at 678th cycle after some amount of crack growth. An example of SLM micrograph of the crack and SEM micrograph of the pore on fracture surface are shown in Figs. 10(a) and 10(b), respectively.

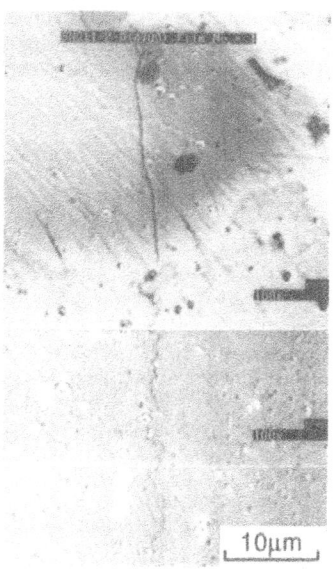
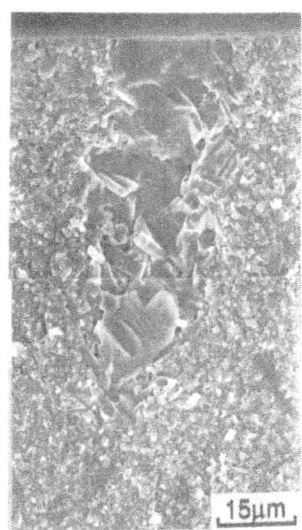

(a) SLM micrograph of a surface crack.   (b) SEM micrograph of fracture surface.

Figure 9. Micrographs of specimen No. 4.

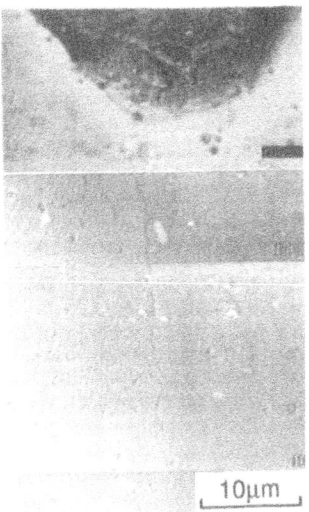
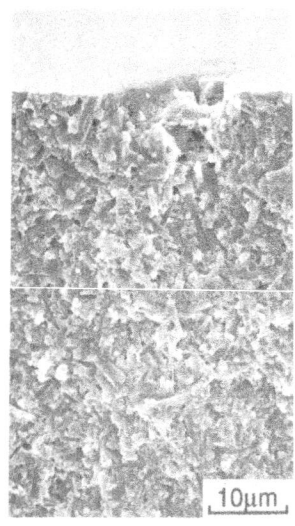

(a) SLM micrograph of a surface crack.   (b) SEM micrograph of fracture surface.

Figure 10. Micrographs of specimen No. 5.

Figure 11 shows the relationship between crack length $2a$ and normalized stress cycles N/N_f in specimens No. 4 and No. 5. The crack length does not increase continuously as stress cycles rise. The growth behavior of micro-cracks can be divided into three successive regions;

Region 1: The period when a crack grows rapidly right after the initiation.
Region 2: The period when a crack grows slowly or stops growing.
Region 3: The period when it grows rapidly again, just before failure ($N/N_f > 0.9$).

K_{Imax}- da/dN Properties

Figure 12 shows the relationship between maximum stress intensity factor K_{Imax} and crack propagation rate da/dN in specimens No.4 and No. 5. A solid line and a hatched area in Fig.12 indicate a behavior of a long crack in CT specimen[3] and that of a small artificial semi-elliptical surface crack[11], respectively. The crack repeats propagation and non-propagation, especially in the *Region 2* as shown in Fig. 11. The da/dN are denoted in two ways; one is an average growth rate with symbols of $\bigcirc$ and $\bullet$, the other is a scatter band taking account of the crack propagation and non-propagation. It is obvious that the propagation behaviors of micro-cracks are different from those of small or long cracks. The characteristics of the micro-crack behavior in a sintered Si_3N_4 are summarized as follows:

1. The micro-cracks grow with lower stress intensity factor (approximately 2.7 MPa√m̄) than the estimated threshold stress intensity factor K_{th} (approximately 3.2 MPa√m̄) of a long crack and propagate with a relatively high crack propagation rate.
2. When a crack grows up, da/dN of the micro-crack approaches gradually to da/dN of a small artificial semi-elliptical surface crack, rather than that of a long through crack in CT specimen.

Similar behaviors have been observed in a gas-pressed Si_3N_4 and a polycrystalline Al_2O_3.

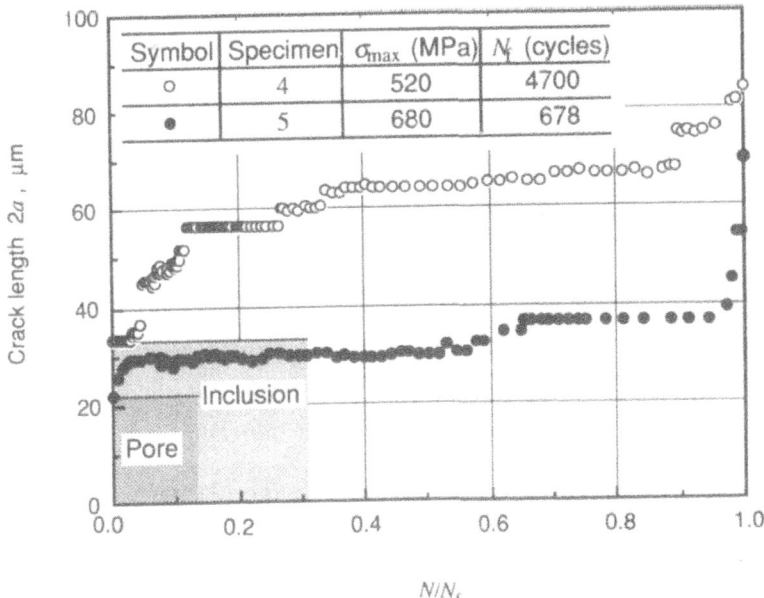

Figure 11. Relationship between crack length $2a$ and the normalized stress cycles N/N_f.

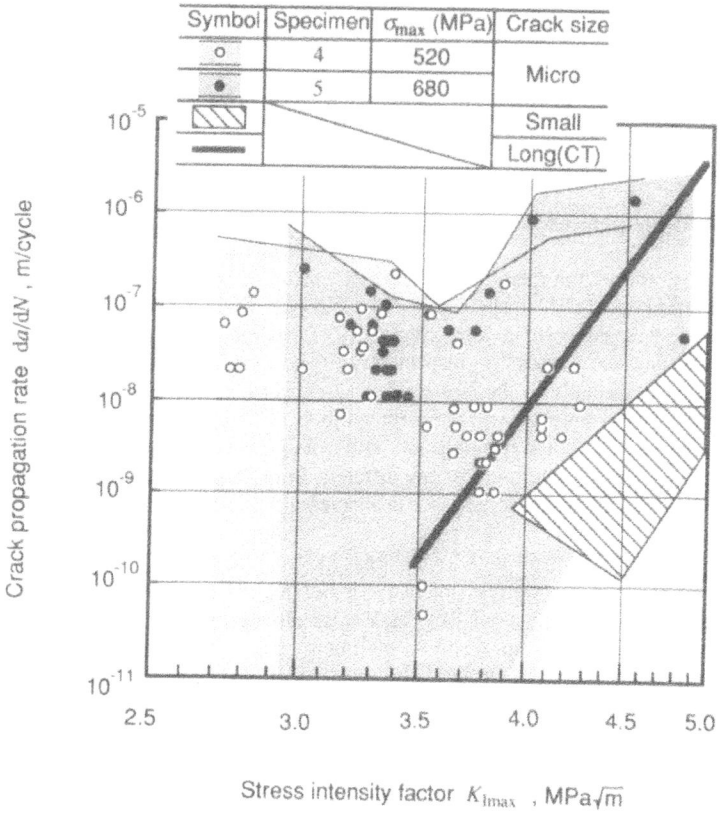

Symbol	Specimen	σ_{max} (MPa)	Crack size
○	4	520	Micro
●	5	680	
			Small
			Long(CT)

Figure 12. Relationship between maximum stress intensity factor K_{Imax} and crack propagation rate da/dN.

DISCUSSION

Above mentioned behaviors are close to those of short fatigue cracks of *metallic materials* [15]. Although the mechanics of the short fatigue cracks have been studied extensively, it is obvious that the proposed mechanics of the short fatigue cracks in *metallic materials* is not applicable to the micro-cracks of *ceramic materials*, because ceramic materials do not possess the ductility at room temperature. Therefore, it is necessary to clear up the causes of those behaviors in terms of characteristics of ceramics, for instance, *grain bridging* [16-17], *residual stress caused by difference in coefficient of thermal expansion* [19], and so on [20]. Probable causes of the rapid growth just after the initiation and an applicability of the *Linear Elastic Fracture Mechanics* (LEFM) to micro-cracks of ceramics are discussed in the following sections.

The causes of the rapid growth just after crack initiation

Figures 13(a) and 13(b) show examples of the crack propagation rate da/dN representing the effect of crack path. The crack propagates along a zigzag path from the microstructural point of view, as shown in Fig. 13(b). However, no evidence which is concerned with the propagation and non-propagation behavior around the fracture origin, is observed. On the other hand, it seems that cracks always slow down its growth rate at the site approximately $10 \sim 20\mu m$ away from the edge of defects.

The manufacturing defects, such as pores and inclusions, originated from admixture of contaminations. T. Tanaka et al.[21] detected impurity elements at the fracture origins. They concluded that the size of defects together with segregated impurity elements are successfully determined by the outer bounds of segregation region.

The areal inspections around the fracture origin were conducted through EPMA, in order to investigate the diffused area of the impurity elements. Figures 14(a)~14(d) show SEM micrograph and X-ray images of the inclusion on the fracture surface of specimen No. 4. X-ray images of Si, N and Fe are shown in Figs. 14(b), (c) and (d), respectively. The Si-rich inclusion has a considerable Fe-segregation at its edge. The segregation area of these impurity elements almost coincides with the profile of the inclusion. No segregations are observed around the pore on the fracture surface of specimen No. 5. It is concluded that the rapid crack growth just after the crack initiation from defects is not caused by the segregation of impurity elements.

On other hands, it have been reported that the stress field of a crack in the vicinity of a defect are strongly affected by the stress field associated with the defect. The residual stress is caused by coherency between the inclusion and the matrix[19], by difference in coefficient of thermal expansion[20] and so on. The effects of stress concentration around a defect should be considered.

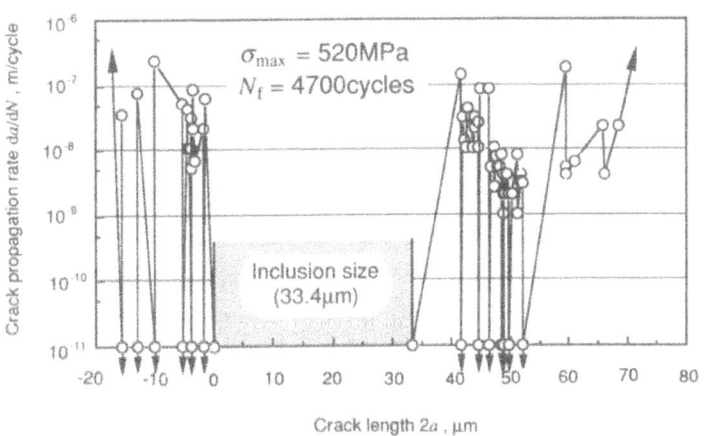

(a) Change in da/dN around inclusion.

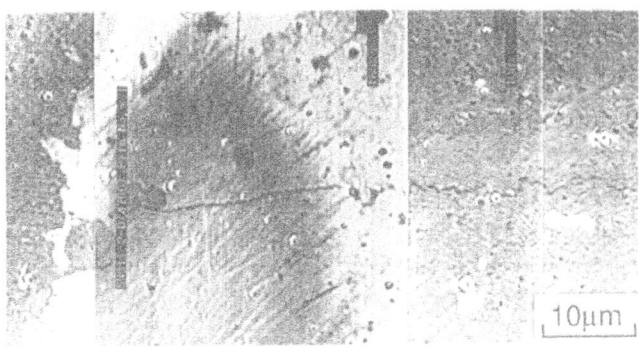

(b) SLM micrograph of crack path.

Figure 13. Crack propagation rate da/dN representing effect of crack path.

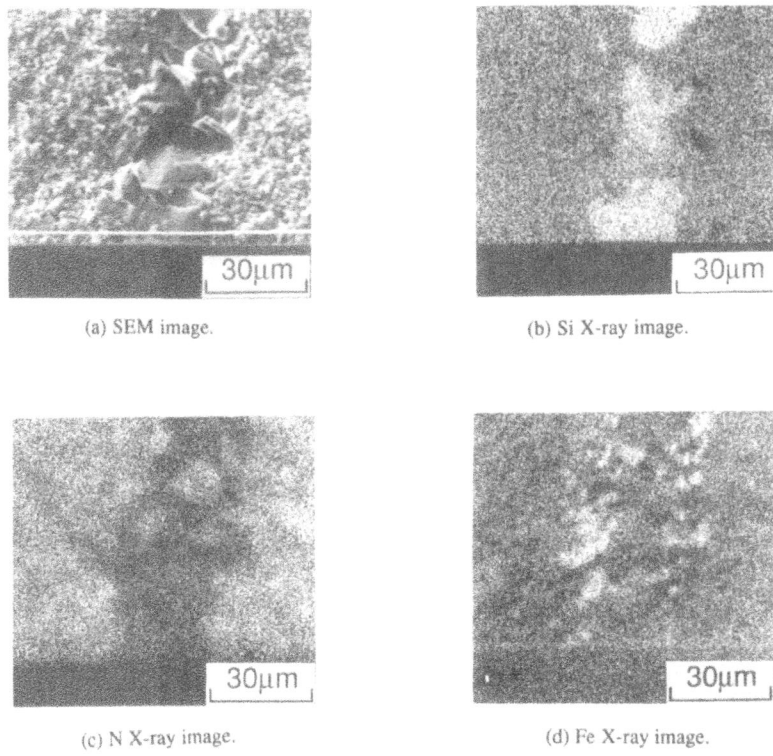

(a) SEM image.

(b) Si X-ray image.

(c) N X-ray image.

(d) Fe X-ray image.

Figure 14. Examples of X-ray analysis near inclusion.

Applicability of the LEFM

The stress intensity factor K of the *LEFM* is suitable to discuss the fracture of cracked specimen. For instance, a power low of the stress intensity, $da/dN = CK^n$, can be applicable to the long crack growth behavior in *ceramic materials*, as well as the crack growth behavior in metal fatigue. It is not yet made clear whether *LEFM* can be successfully applicable to as small crack as natural crack of several dozens of micrometer. As the sizes of measured micro-cracks are very short, less than 100 µm, applicability of the *LEFM* must be examined.

The stress σ_y at the distance r from a through crack in an infinite plate is expressed by eq. (1)[22].

$$\sigma_y = \frac{\sigma_0 (a + r)}{\sqrt{r(2a + r)}} = \frac{K_I}{\sqrt{2\pi r}} \frac{1 + \dfrac{r}{a}}{\sqrt{1 + \dfrac{r}{2a}}} \qquad \cdots \quad (1)$$

where, σ_0 is an applied nominal stress, a is a half-length of the crack and $K_I (= \sigma_0 \sqrt{\pi a})$ is a stress intensity factor of mode I type loading. The one-term approximation $\sigma_y{'}$ that neglects below the 2nd term in Taylor expansion of eq. (1) for $r \ll a$, is expressed as eq. (2).

$$\sigma_y{'} = \frac{K_I}{\sqrt{2\pi r}} \qquad \cdots \quad (2)$$

It is well known that eq. (2) causes significant error for $r \geq 0.1\,a$. Figure 14 shows the relationship between distance r from the crack tip and non-dimensional stress $\sigma_y/\sigma_y{}'$. σ_y and $\sigma_y{}'$ can be calculated from eqs (1) and (2). The stresses σ_y and $\sigma_y{}'$ in the vicinity of a crack tip are almost equal and the difference between σ_y and $\sigma_y{}'$ increase as the distance r rises. Therefore, the disagreement between the growth behavior of micro and small cracks is not explained by the stress at the crack tip.

It is known that the strength of a micro-cracked specimen is well explained by using the σ_y at the point where is proper distance ahead of the crack tip. For instance, S. Usami et al.[23] have proposed the fracture criterion by using a *grain-fracture model*. They derived the specific size r_0 in terms of the average grain size d [*1]. By using r_0, the apparent reduction of the plain specimen strength is successfully described. On the other hand, J. Kitazumi et al.[24] have proposed another fracture criterion using a *hypothetical crack length model*. They derived a hypothetical crack length l_0 [*2] and modified the relation between the strength and the flow size by adding l_0 to the original equivalent crack length. In order to examine the deviation of $\sigma_y{}'$ from σ_y in relation to distance from the crack tip, two specific distance, r_0 and l_0, are applied to the present results. Figure 15 shows the relationship between the crack length $2a$ and normalized stress $\sigma_y/\sigma_y{}'$. Although similar tendency are obtained by both criterion, the $\sigma_y/\sigma_y{}'$ increases as the crack size decreases. It is concluded that *LEFM* is not applicable to a micro-crack growth behavior, and the inapplicability of *LEFM* is one of causes for the disagreement between the growth behavior of micro-cracks and small ones.

Fatigue life estimation methods and proof testing methods of the plain specimen have been proposed on the assumption that the fracture of the plain specimen is dominated by the crack propagation process expressed by the equation $da/dN = C \cdot K_{Imax}^n$ [25]. However, the crack propagation rate of cracks initiated from the plain specimens can not be well expressed by the equation, as seen in Fig. 12. K_{Imax} has some difficulty in expressing the stress of a small crack. More study is necessary for the fatigue life estimation and proof testing of the plain specimen to be treated rationally.

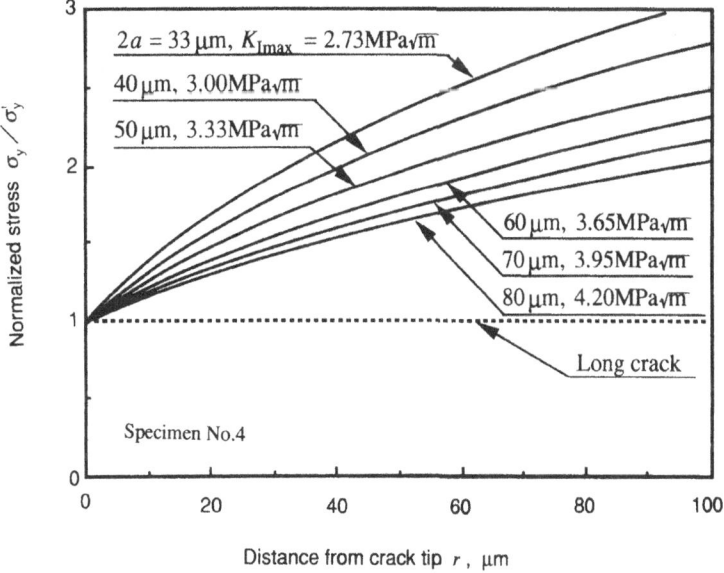

Figure 14. Relationship between distance r from crack tip and normalized stress $\sigma_y/\sigma_y{}'$.

---

*1 $r_0 = 2d$ for a semicircular surface defect.

*2 $l_0 = 10\mu m$ for a silicon nitride.

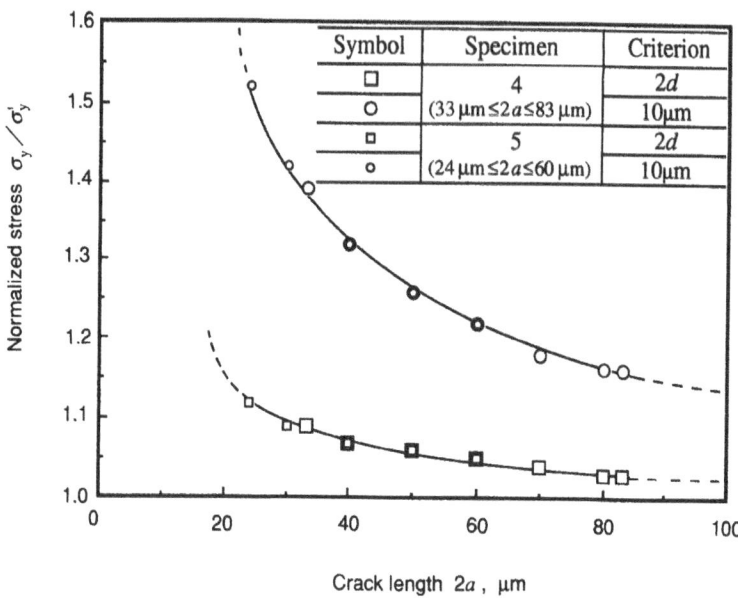

Figure 15. Relationship between crack length $2a$ and normalized stress σ_y/σ_y'.

CONCLUSIONS

Initiation and growth behavior of a micro-crack emanated from manufacturing defects were investigated by using plain specimens. The major results obtained are summarized as follows:

1. Crack length of a micro-crack initiated from manufacturing defects did not increase continuously as the stress cycles rose. This growth behavior of the micro-crack can be divided into three successive regions:

Region 1: The period when a crack grows rapidly right after the initiation.
Region 2: The period when a crack grows slowly or stops growing.
Region 3: The period when it grows rapidly again, just before failure ($N/N_f > 0.9$).

2. The propagation behaviors of micro-cracks are different from those of small artificial cracks or long cracks. The characteristics of the micro-crack behavior in a sintered Si_3N_4 are summarized as follows:

   (1) Micro-cracks grow with lower stress intensity factor (approximately $2.7\ MPa\sqrt{m}$) than the estimated threshold stress intensity K_{th} (approximately $3.2\ MPa\sqrt{m}$) of a long crack and propagate with relatively high crack propagation rate, da/dN.

   (2) When micro-cracks have grown up, their da/dN approaches gradually to da/dN of a small artificial semi-elliptical surface crack, rather than that of a long through crack in CT specimen.

3. It is noted that *LEFM* is not applicable to a micro-crack growth behavior, and the inapplicability of *LEFM* is one of causes for the disagreement between the growth behavior of micro-cracks and small-cracks.

ACKNOWLEDGMENTS

The authors are grateful to Dr. M. Okumiya of Toyota Technological Institute for assistance with the EPMA analytical works, and Prof. Y. Murakami of Kyusyu Univ. for his instructive advice concerning to the applicability of the Fracture Mechanics.

REFERENCES

1. H. Kishimoto, A. Ueno and H. Kawamoto, Crack Propagation Characteristics of Sintered Si_3N_4 under Static and Cyclic Loads, *J. of the Society of Materials Science, Japan*, **3 6**:1122 (1987).
2. H. Kishimoto, A. Ueno, H. Kawamoto and Y. Fujii, The Influence of Wave Form and Compressive Loads on the Crack Propagation Behavior of a Sintered Si_3N_4 under Cyclic Loads, *J. of the Society of Materials Science, Japan*, **3 8**:1212 (1989).
3. H. Kishimoto, A. Ueno and H. Kawamoto, Crack Propagation Behavior of Sintered Silicon Nitride under Cyclic Load - Influence of Differece in Materials -, *JSME International Journal (Series I)*, **3 4**:361 (1991).
4. A. Ueno, H. Kishimoto, H. Kawamoto and M. Asakura, Crack Propagation Behavior of Sintered Silicon Nitride under Cyclic Load of High Frequency and High Stress Ratio, *J. of the Society of Materials Science, Japan*, **3 9**:1570 (1990).
5. R. H. Dauskardt, D. B. Marshall and R. O. Ritchie, Cyclic Fatigue-Crack Propagation in Ceramics: Behavior in Overaged and Partially-Stabilized MgO-Zirconia, in: *"Proceedings of the MRS International Meeting on Advanced Materials, Vol.5"*, M. Doyama, S. Somiya and R. P. H. Chang, ed., Materials Research Society, Pittsburgh (1989).
6. H. Kishimoto, A. Ueno, S. Okawara and H. Kawamoto, Crack Propagation Behavior of Polycrystalline Alumina under Static and Cyclic Load, *J. Am. Ceram. Soc.*, **7 7**:1324 (1994).
7. T. Tanaka, N. Okabe and Y. Ishimaru, Fatigue Crack Growth and Crack Closure of Silicon Nitride under Wedge Effect by Fine Fragments, *J. of the Society of Materials Science, Japan*, **3 8**:137 (1989).
8. T. Hoshide, T. Ohara, T. Yamada, Fatigue Crack Growth from Indentation Flow in Ceramics, *Int. J. Fract.*, **3 7**:47 (1988).
9. M. Yoda, Subcritical Growth of As-Indented Pyramid and Knoop Cracks in Soda-Lime Glass, in: *"Euro-Ceramics, Vol.3"*, G. de With, A. Terpstra and R. Metselaar, ed., Elsevier Applied Science, (1989).
10. A. A. Steffen, R. H. Dauskardt and R. O. Ritchie, Cyclic Fatigue-Crack Propagation in Ceramics: Long and Small Crack Behavior, in: *"Proc. of the 4th International Conference on Fatigue and Fatigue Thresholds, Vol.II"*, H. Kitagawa and T. Tanaka, ed., Materials and Component Engineering Publications Ltd., Birmingham (1990).
11. A. Ueno, H. Kishimoto and H. Kawamoto, Effects of Crack Size on Crack Propagation Behavior and Experimental Verification of Cyclic Fatigue Mechanism of Sintered Silicon Nitride, *Fracture Mechanics of Ceramics*, **9**:423 (1992).
12. A. Ueno, H. Kishimoto, S. Okawara, T. Kondo and A. Yamamoto, Crack Propagation Behavior of Small Crack of Polycrystalline Alumina and Effects of Cyclic Load and Grain Size on Bridging, *J. of the Society of Materials Science, Japan*, **4 3**:183 (1994).
13. H. Kishimoto and A. Ueno, Fractographic Analysis of Fracture Surface of Sintered Silicon Nitride Fractured under Static or Cyclic Load, *J. of the Society of Materials Science, Japan*, **4 0**:695 (1991).
14. I. S. Raju and J. C. Newman, Jr., Stress-Intensity Factor for a Wide Range of Semi-Elliptical Surface Cracks in Finite-Thickness Plates, *Eng. Frac. Mech.*, **1 1**:817 (1979).
15. K. Tanaka, Mechanics of Small Fatigue Cracks, in: *"Proc. of the 4th International Conference on Fatigue and Fatigue Thresholds, Vol.I"*, J. -P. Bailon and J. I. Dickson, ed., Engineering Materials Advisory Services Ltd., West Midlands, 355 (1993).

16. R. O. Ritchie, Mechanisms of Fatigue Crack Propagation in Metals, Ceramics and Composites: Role of Crack Tip Shielding, *Mater. Sci. and Eng.*, **A103**:15 (1990).

17. G. Grathwöhl, Ermudung von Keramik unter Schwingbeanspruchung, *Mat.-wiss. u. Werkstofftech.*, **19**:113 (1988).

18. S. Lathabai, J. Roedel and B. R. Lawn, Cyclic fatigue from Degradation at Bridging Grains in Alumina, *J. Am. Ceram. Soc.*, **74**:1340 (1991).

19. J. C. Swearengen, E. K. Beauchamps and R. J. Eagan, Fracture Toughness of Reinforced Glasses, *Fracture Mechanics of Ceramics*, **4**:973 (1978).

20. J. S. Nadeau and R. C. Bennet, Some Effects of Dispersed Phases on the Fracture Behavior of Glass, *Fracture Mechanics of Ceramics*, **4**:961 (1978).

21. T. Tanaka, N. Okabe, A. Sakaida and H. Nakayama, Relationship of Tensile Fracture Strength and Evaluation of Inherent Defects through EPMA Observasion for Sintered Silicon Nitride, *J. of the Society of Materials Science, Japan*, **37**:1197 (1988).

22. H. M. Westergaard, Bearing Pressures and Cracks, *Trans. ASME J. Appl. Mech.*, **61**:A-49 (1939).

23. S. Usami, H. Kimoto, I. Takahashi and S. Shida, Strength of Ceramic Materials Containig Small Flaws, *Eng. Frac. Mech.*, **23**:745 (1986).

24. J. Kitazumi, Y. Taniguchi, T. Hoshide and T. Yamada, Characteristics of Strength and Their Relations to Flaw Size Distribution in Several Ceramic Materials (Part 1 : Static Strength), *J. of the Society of Materials Science, Japan*, **38**:1254 (1989).

25. R. W. Davidge, *"Mechanical Behaviour of Ceramics"*, Cambridge University Press, Cambridge (1979).

ON FATIGUE AND FRACTURE BEHAVIOR
OF Si-ALLOYED PYROLYTIC CARBON

Ling Ma,[1] George Sines,[1] and C. Barclay Gilpin[2]

[1]Department of Materials Science & Engineering
University of California, Los Angeles
CA 90095-1595
[2]Department of Mechanical Engineering
California State University, Long Beach
CA 90840

INTRODUCTION

A chemical vapor deposited, silicon alloyed, isotropic pyrolytic carbon (PyC) has been widely regarded as the primary candidate for the structural material for components of mechanical heart valve prostheses. The biocompatibility of PyC, in particular its thromboresistance, together with its resistance to degradation in the biological environment, has made it the material of choice[1]. The design of the valve and the choice of material must provide for survival for lifetimes of the order of 10^9 cycles.

Fatigue Behavior of PyC

Although PyC has been used successfully in mechanical heart valve prostheses for decades, very little data has been published on its strength and its resistance to cyclic stressing. Some early studies have demonstrated that this form of carbon possesses a fatigue strength that is within the scatter band of the fracture strength [2-5].

Though three studies have shown that the cyclic fatigue strength of PyC is within the scatter band of the static strength, only one of them has data approaching the cyclic life of interest. In Figure 1, the work of Kepner et al. [4-5] shows seven specimens that endured 10^8 cycles without failure at a strain of 11,530 $\mu\varepsilon$ which is near to the middle of the scatter band of the static strength. If PyC were a typical metal, and since the nature of the fatigue curves for metals are well known, one would not hesitate to extrapolate the data of Figure 1 to longer lifetimes; a strain of 10,100 $\mu\varepsilon$ would be reasonable for 6 x 10^8 cycles.

Several very important questions remain to be answered: Will a crack in PyC, extrinsic flaws for example, grow under cyclic stress? Is there a true threshold crack size for no growth under cyclic stress? Will cracks initiate in PyC specimens under the service cyclic stress at the lifetimes of interest ($\sim 10^9$)?

Fracture Mechanics of Ceramics, Vol. 12
Edited by R.C. Bradt *et al.*, Plenum Press, New York, 1996

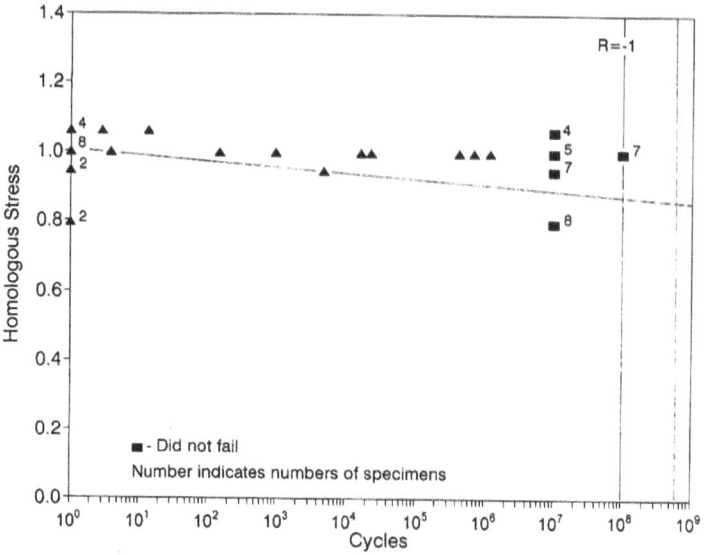

Figure 1. Rotating beam fatigue test of PyC.

Fatigue of PyC with Statistical Probability of Survival

Static and dynamic finite element analysis, using boundary conditions determined from experiments on models and prototypes of bileaflet valves, show that a conservative upper bound for the elastic strain is 250 $\mu\varepsilon$ ($\mu\varepsilon = 10^{-6}$). Although this strain is a small fraction of that one would predict to be safe from extrapolation to longer lives of the existing fatigue data (10,100 $\mu\varepsilon$ from Kepner's study for example), sufficient data must be taken to ensure an extremely high probability of survival because of the critical nature of the application of PyC in heart valves and because of the paucity of information of the fatigue behavior of PyC. In fact, the US Federal Drug Administration (FDA) has set the requirement that a fatigue strength having a probability of 90% survival with 95% confidence be achieved on uncracked specimens at 6 x 10^8 cycles with a safety factor of two on the strain level. New methods have had to be developed to reach the extreme cyclic lifetime and obtain the extensive replication of tests needed for statistical confidence.

EXPERIMENTAL - FATIGUE TEST

Material

The PyC used in this study was supplied by Carbon Implants, Inc. (now known as Medtronic Carbon Implants, Inc.) of Austin, Texas. The material is a pyrolytic carbon made by chemical vapor deposition in a fluidized bed on graphite substrate to give an isotropic coating. The carbon was alloyed with 5 to 8% silicon. The Young's modulus

Figure 2. Specimen tested in Ringer's solution.

of PyC is 31.2 GPa with standard deviation of 1.9 [6]. The Poisson's ratio ranges from 0.22 to 0.30. The sandwich specimens used in fatigue tests had a 0.25 mm coating on each side of a 0.30 mm graphite substrate. The specimens measured 6 x 30 x 0.8 mm.

Testing Conditions and Environment

The strain levels of the fatigue tests used in this work were set at 500 $\mu\varepsilon$, 1,000 $\mu\varepsilon$ and 2,000 $\mu\varepsilon$; thus making the testing strain level higher than the estimated peak service strain by a factor of two, four and eight. The ratio of the maximum strain and the minimum strain (R) was set at 0.05 and at -1. The loading level is given in terms of elastic strain instead of stress, because elastic strain was used to set the testing machines and monitor the tests. Since PyC does not flow plastically, strain and stress control should be essentially equivalent.

A means has been developed to provide a special aqueous environment for some of the specimens in the fatigue test. A transparent plastic capsule was used to cover the specimen (see Figure 2). The solution was introduced into the capsule that was then sealed. Ringer's saline solution was used to simulate the environmental conditions of the valve. All tests were conducted at room temperatures.

Fatigue Testing Machines and Loading Beams

The conventional Krouse plate bending fatigue machine was modified to accommodate as many as twenty specimens at a time on a triangular loading plate. The loading plate is shaped so that it experiences the same stress throughout when it is bent as a cantilever beam (see Figure 3). The ratio of the minimum stress to the maximum stress (R) is adjusted by vertically positioning the vise. The testing speed is about 1,725 rpm.

To avoid fracturing the aluminum loading beam from fatigue damage during the long life test, the strain level applied to the loading plate had to be kept low; however the testing strain level was really limited by the strength of the adhesive system. Difficulties were encountered when specimen and adhesive system were subjected to a strain above

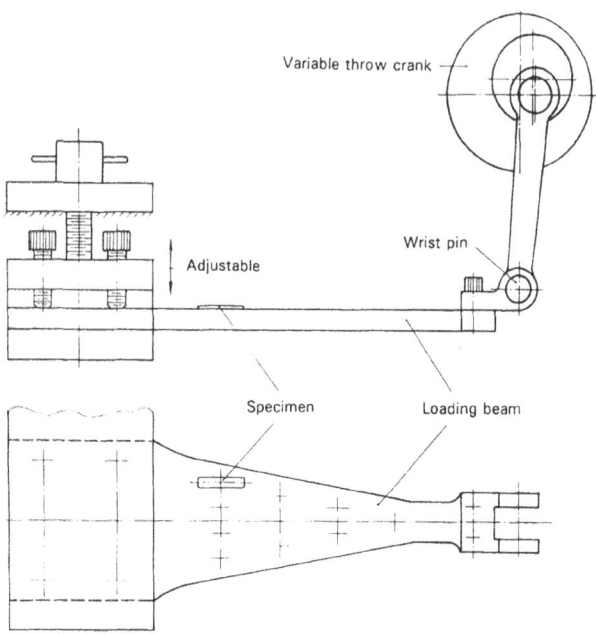

Figure 3. Modified Krouse plate bending machine with triangular loading beam.

1,000 $\mu\varepsilon$. To test the PyC at higher strain levels, a special specimen shape was developed. A strain amplification of two was attained in the test section of the specimen illustrated in Figure 4.

To obtain increased testing speed, Krouse cantilever rotating beam fatigue machines, which run up to 10,000 rpm, were modified to test up to four specimens at a time. The 8 mm square-sectioned rotating loading beam was made so that a specimen

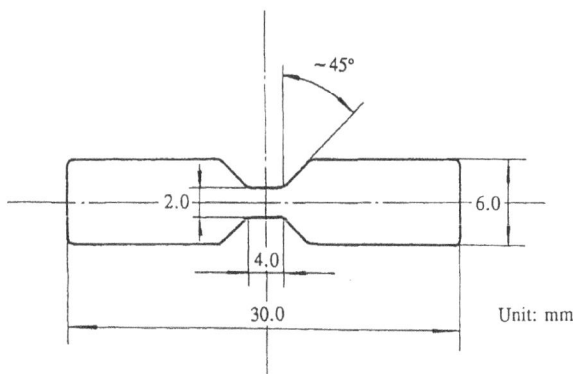

Figure 4. Specimen to amplify strain by a factor of two.

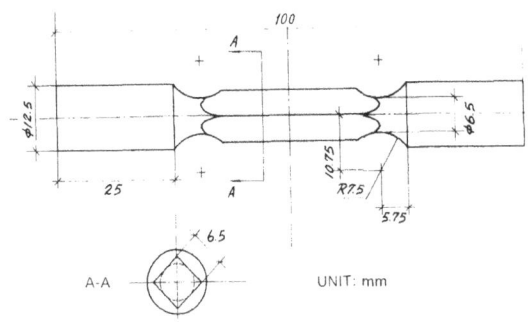

Figure 5. Square-sectioned rotating loading beam.

could be affixed to each flat surface (see Figure 5). The specimens were also installed by adhesives in the same way as was used for the plate machine to give uniaxial loading. A titanium alloy, 6V-4Al in the fully heat treated condition, was chosen for the beams after several fatigue failures of aluminum ones occurred. The limited number of specimens on a machine was compensated by using eight testing machines.

Adhesive System

The M-bond AE10/15 and GA-61 electric resistance strain gage adhesives manufactured by Measurement Group, Inc. at Raleigh, North Carolina were used in all the tests. The selection was made after numerous attempts to find a suitable adhesive system for this study.

The specimens were fastened by adhesive to the plate in a way that the center portion (6 mm x 6 mm) of the specimen was free from the plate to ensure that the strain was predominantly uniaxial. This is achieved by inserting a piece of cellophane tape in between the specimen and the plate. The cellophane tape does not bond with the adhesive and therefore the area covered by the tape is free from bonding (see Figure 6). In order to level the unevenness due to the insertion of the tape, two pieces of paper with the same thickness as the tape were laid on both sides of the tape. Despite these measures, a tensile strain of about 150 $\mu\varepsilon$ was measured on the top surface of one of the specimens; thus the maximum strain value may be somewhat higher than the nominal reported, thereby creating a more severe testing condition.

After several incidents of adhesive failure occurred, a strain gage was installed to every specimen at the end of each test to check whether the bonding was still intact and whether the loading was still at the desired level.

Uncracked Sandwich Specimen under Cyclic Stress

Thirty seven uncracked sandwich specimens were tested under uniaxial stress at several strain levels and R values. The tests were conducted with Krouse plate bending machines and with Krouse cantilever rotating beam machines. In a spot test, one specimen was tested in Ringer's solution.

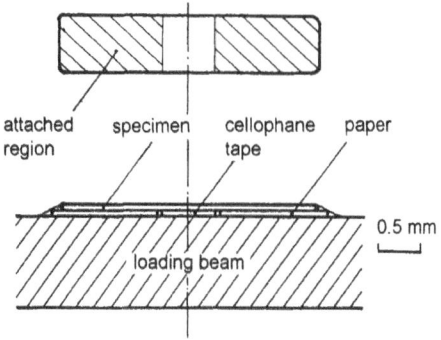

Figure 6. Specimen installation.

Pre-cracked Sandwich Specimens under Cyclic Stress

Forty sandwich specimens, identical to the ones described above, had cracks induced by a Vickers diamond indenter with loads from 89 N to 133 N (20 lb to 30 lb). The longest crack was 2.44 and the shortest crack was 0.60 mm. The mean value of the crack length was 1.24 mm with the population standard deviation of 0.46 mm. The crack length was measured with a B&L metallograph with a resolution of 10^{-5} m.

In most cases when inducing cracks by the Vickers diamond indenter, the corner cracks emanated from the impression longitudinally and transversely to the axes of the specimen and were well suited for this study because they all had a very sharp tip. Under the impression secondary cracks also form; a pyramidal crack and within the pyramid conical cracks. Several pre-cracked specimens were sectioned and observed under an optical microscope. It was found that the crack penetrated the top PyC coating and was arrested at the PyC-graphite interface (see Figure 7).

The pre-cracked specimens were tested at strain levels of 500 and 1,000 $\mu\varepsilon$ with R=0.05 and -1. In a spot test, one pre-cracked specimen with a 1.58 mm crack was tested in Ringer's solution. The installation and testing methods of the specimens were the same as those of their uncracked counterparts.

RESULTS - FATIGUE TEST

Un-cracked Sandwich Specimens under Cyclic Stress

Thirty seven uncracked sandwich specimens were tested at strain levels of 500, 1,000, 1,500 or 2,000 $\mu\varepsilon$ with R of 0.05 or -1. They were tested up to and beyond 6 x 10^8 cycles. No crack initiation was observed in any specimen. All tests were terminated with no failure to the specimens. The testing strain levels, R values and cycles at which tests were terminated are tabulated in Table 1.

The accelerated fatigue test of PyC was conducted at 1,725 and 10,000 rpm. We realize that for metals there may be an effect of frequency when there is corrosion, elevated temperatures, or when the frequency is above 30,000 rpm. However this type of

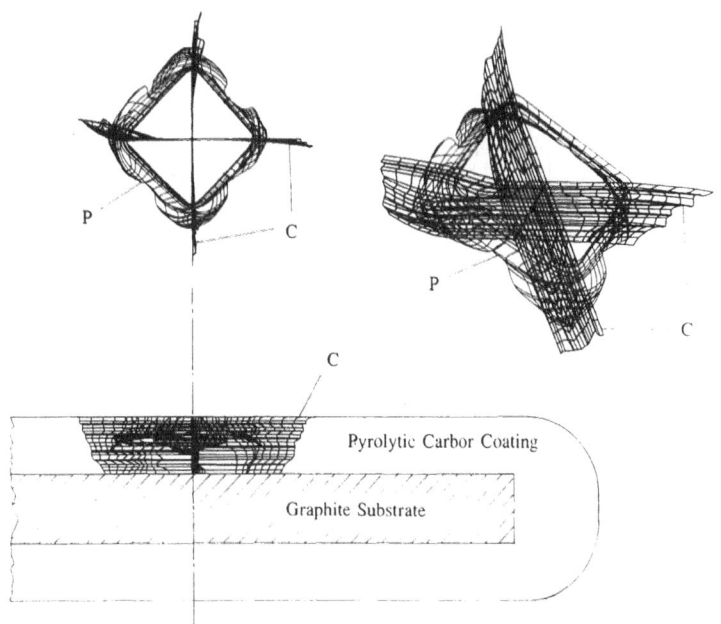

Pyrolytic Carbor Coating

Graphite Substrate

Figure 7. Cracks in PyC made by the Vickers diamond indenter. C represents corner cracks; P represents the pyramid crack. Some secondary cracks are omitted.

Table 1. Fatigue test of un-cracked sandwich specimens

Number of specimens	Strain ($\mu \varepsilon$)	R	Test terminated at cycles
1[A)	500	-1	1.34×10^9
1[A)	500	-1	7.86×10^8
2[B)	500	-1	2.40×10^9
1[B)	500	-1	6.36×10^8
2[A)	1,000	0.05	$8.10 \times 10^{8,\,C)}$
1[A)	1,000	0.05	6.00×10^8
8[B)	1,000	-1	6.00×10^8
8[B)	1,000	-1	6.10×10^8
4[B)	1,000	-1	6.13×10^8
4[B)	1,000	-1	6.25×10^8
2[B)	1,000	-1	2.20×10^9
1[A)	2,000	0.05	6.00×10^8
1[B)	1,500	-1	6.31×10^8
1[B)	2,000	-1	6.31×10^8

A) Krouse plate bending machines.
B) Krouse cantilever rotating beam machines.
C) One specimen tested in Ringer's solution.

113

PyC is very resistant to corrosion; the testing temperature is low; and the testing speed is below where frequency might be a problem. No difference in behavior of PyC was observed when the specimens were tested at speeds of 1,725 rpm and 10,000 rpm. These facts lead us to believe that the accelerated cyclic tests are valid.

Thirty seven uncracked specimens survived 500 $\mu\varepsilon$ (with R=-1) or more severe conditions for 6×10^8 cycles and beyond, with no crack initiation or failure, giving 92.2% probability of survival with 95% confidence [7]. This exceeds the requirement of the FDA.

Pre-cracked Sandwich Specimens under Cyclic Stress

The results of the fatigue test of pre-cracked specimens are listed in Tables 2 and 3. The K_1 was calculated by the following equation [8]

$$K_1 = \frac{E v_o (\pi a)^{1/2}}{h} \tag{1}$$

where E is the Young's modulus, v_o is the normal displacement, a is the half crack length and h is the effective half length of the specimen. The tabulated K_1 values have been modified for the sample width [8, 9]. The cracks were modelled as a through, central crack under constant displacement. The tip to tip length of the corner cracks was regarded as the crack length 2a in the fracture mechanics calculations. The results in Table 2 and 3 were collected by conducting the accelerated test at speeds of 1,725 rpm and 10,000 rpm. We found no difference in results from the testing speeds of 1,725 rpm and 10,000 rpm.

Table 2. Fatigue test of pre-cracked specimens (A)

No. of specimens	strain $(\mu\varepsilon)^{C)}$	R	Crack 2a (mm)	K_1 max (MPa$\sqrt{m}$)	ΔK_1 (MPa$\sqrt{m}$)	No failure[D]
28[A), B]	500	-1	0.60-1.84	0.50-0.90	1.00-1.80	$\geq 6\times10^{8,\ E)}$
1[A]	1,000	0.05	1.63	1.69	1.61	6×10^8
1[A]	1,000	0.05	1.58	1.66	1.58	$8.1\times10^{8,\ F)}$

[A] Krouse plate bending machines.
[B] Krouse cantilever rotating beam machines.
[C] Strain given is the maximum of the cycle.
[D] Tests terminated with no failure.
[E] One specimen exceeded 2.4×10^9 cycles
[F] Tested in Ringer's solution.

Table 3. Fatigue test of pre-cracked specimens (B)

No. of specimens	strain $(\mu\varepsilon)^{C)}$	R	Crack 2a (mm)	K_1 max (MPa$\sqrt{m}$)	ΔK_1 (MPa$\sqrt{m}$)	Failure cycles
1[A]	1,000	0.05	2.44	2.14	2.03	<100
1[A]	500	-1	2.29	1.03	2.06	5×10^6
1[A]	500	-1	2.04	0.95	1.90	4×10^6
7[B]	1,000	-1	0.70-1.88	1.08-1.83	2.16-3.66	1

[A] Krouse plate bending machine.
[B] Krouse cantilever rotating beam machine.
[C] Strain given is the maximum of the cycle.

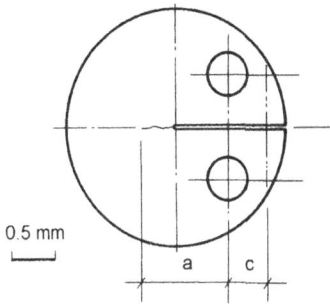

0.5 mm

a   c

Figure 8. Modified disk-shaped compact specimen.

Thirty pre-cracked specimens survived 500 $\mu\varepsilon$ (with R=-1) or more severe conditions for 6 x 10^8 cycles and beyond, with no crack propagation or failure, giving 90.5 % probability of survival with 95% confidence if all the specimens had the least crack length of 0.60 mm [7]. Here the FDA requirement is met with pre-cracked specimens.

EXPERIMENTAL - CRACK GROWTH RATE

The specimens for crack growth rate were in the form of a modified ASTM E399 disk-shaped compact specimen DC(T). The specimens were slightly different in that there was no flat portion, as measured by the "c" dimension (see Figure 8). Five monolithic PyC and four PyC graphite sandwich specimens were tested. The thickness of the monolithic PyC specimens was 0.69 mm. The sandwich specimen has a 0.27 mm graphite substrate with a 0.30 mm PyC coating on each side. The machined crack was 4.8 mm long, 0.2 mm wide with a tip radius of 0.1 mm. The machined crack was sharpened on a MTS machine under the displacement control mode. All tests were conducted with a MTS model 810 closed loop hydraulic test machine. Crack length was measured on the specimen surface with a microscope having a resolution of 0.02 mm.

The crack lengths were reported every 0.20 mm or 0.25 mm along with the number of cycles at the crack length. Data collection began at a crack length of 6 mm with a load sufficient to give a crack growth rate in the low 10^{-11} m/cycle range. Data was collected at this same load until crack growth was in the low 10^{-7} m/cycle range. The load was lowered to give a crack growth rate in the low 10^{-11} m/cycle range. This same load was used until again the crack growth was in the low 10^{-7} m/cycles range. In this manner, two curves were obtained from most samples.

RESULTS - CRACK GROWTH

Plots of da/dN versus ΔK are shown in Figure 9. It is to be noted that the data in Figure 9 shows very little scattering when compared with other reported results from the similar tests [10]. This can be attributed to the different data acquisition procedures. It is also noted that the specimens tested in air and in Ringer's solution exhibited virtually the same crack growth rate.

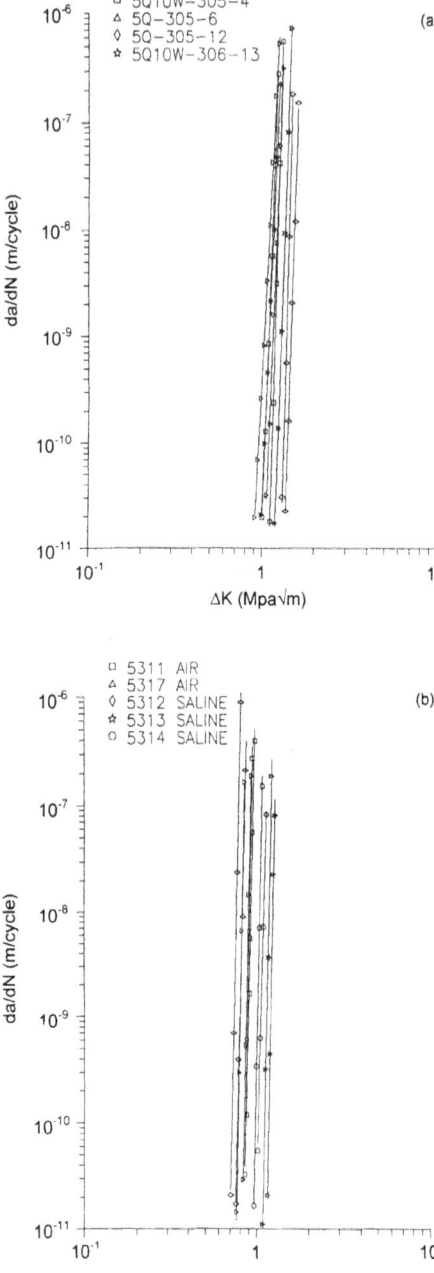

Figure 9. Figure crack growth of (a) sandwich specimens and (b) monolithic PyC specimens.

EXPERIMENTAL - K_{1C} MEASUREMENT

Measurement of K_{1C} of Monolithic PyC by Tensile Test

Three pieces of monolithic PyC strips were prepared by coating graphite substrates and then splitting the midplane of graphite. The graphite was subsequently removed leaving monolithic PyC samples 0.47 mm thick. Doublers were attached to the ends for tensile testing (see Figure 10). The monolithic PyC strips had cracks induced by the Vickers diamond indenter with loads of 578 N (130 lb). The specimens measured 12 x 24 x 0.47 mm. The test was performed on an Instron at a crosshead speed of 0.25 mm/min.

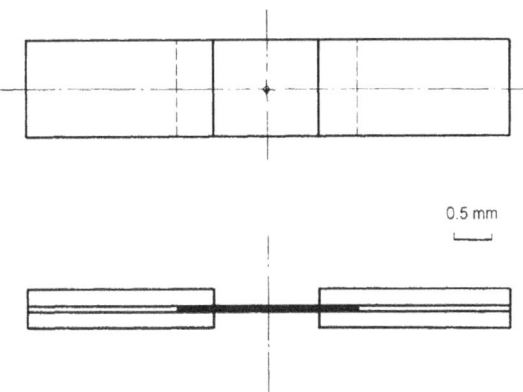

0.5 mm

Figure 10. Pre-cracked monolithic PyC specimen for K_{1C} measurement.

RESULTS - K_{1C} MEASUREMENT

K_{1C} Measurement by Tensile Test

The crack was modeled as a central through crack in the calculation, even though the crack did not penetrate to the back side, because we did not know shape of the crack. The K_{1C} of the monolithic PyC specimens is listed in Table 4. The following equation was used to calculate K_{1C}

$$K_{1C} = \sigma[\pi a \, \sec(\pi \frac{a}{w})]^{1/2} \tag{2}$$

where σ is the applied stress, a is the half crack length and w is the specimen width.

Table 4. K_{IC} of monolithic PyC.

Specimen number	490-4A	490-15A	490-17A
Width (mm)	12.27	11.64	11.77
Thickness (mm)	0.46	0.47	0.47
Fracture stress on gross area (MPa)	19.81	22.33	20.42
Crack length measured before test, 2a (mm)	2.20	2.34	2.26
Measured K_{IC} (MPa$\sqrt{m}$)	1.19	1.39	1.25
Average K_{IC} (MPa$\sqrt{m}$)		**1.28**	

K_{IC} Measurement by Compact Disk Specimen Test

The mean value of K_{IC} of the five monolithic PyC specimens of Fig. 9 (b) was 1.17 MPa$\sqrt{m}$ with standard deviation of 0.17. The mean value of K_{IC} of the four sandwich specimens of Fig. 9 (a) was 1.84 MPa$\sqrt{m}$ with the standard deviation of 0.22. The procedures of ASTM E399 were followed in the calculation of the critical stress intensity factors.

DISCUSSION

Crack Growth Rate and da/dN Curve Extrapolation

This work on the cyclic stress crack growth rate of PyC, which used improved data collection techniques, shows very little scatter in contrast to the large scatter in the earlier work of Ritchie et al. [10]. For the cyclic lifetimes of interest, which are of the order of 10^9 cycles, the order of magnitude for a threshold crack growth rate would be of the order of 10^{-15} m/cycle. Even though the crack growth curves have very little scatter, it is not prudent for this critical application to extrapolate four orders of magnitude, from the slowest measured rate of 10^{-11} m/cycle, to a threshold value.

Detailed study of the crack growth rate is of less importance to this application than establishing whether a true threshold value of the stress intensity factor range exists for no crack growth. The initial crack size for the specimens tested in crack growth rate measurement were approximately 6 mm. The fatigue tests conducted above using crack lengths less than 2.5 mm at stresses twice the nominal service stress are closer to the reality of the application. The testing methods developed in this work permitted sufficient replication to obtain statistical information without extrapolation of crack growth rates or of lifetimes.

Fracture Mechanics Calculations

When calculating the crack growth rate under cyclic stress, it is a common practice to use only the positive portion of the cycle for reversed strain conditions. The justification of doing so is that the crack closes during the negative portion of the cycle; however we feel strongly that this may not be the case for the pyrolytic carbon. During the fatigue test, it was observed that carbon powder was pushed out to the specimen surface from the crack. This shows that the crack does not close during the negative portion of the cycle, therefore, the full range of strain should be used. Tables 2 and 3 give the full range of the strain in ΔK_1.

It is demonstrated in Tables 2 and 3 that all the specimens that survived with no crack growth had ΔK_1 values less than 1.80 MPa$\surd$m, while all specimens that failed had ΔK_1 values greater than 1.90 MPa$\surd$m.

The calculated critical stress intensity factor K_{IC} for the sandwich specimens by disk-shaped compact specimen was 1.84 MPa$\surd$m with the standard deviation of 0.22. Ritchie et al. reported K_{IC} of 1.90 MPa$\surd$m on a similar material [10].

It was observed during the compact disk test that long transverse cracks (6 to 8 mm) in the PyC of sandwich disks had grown under ΔK_1 of 0.89 to 1.29 MPa$\surd$m. It was not clear whether the crack in the PyC layer was leading that in the graphite or vice versa. However direct comparison between the stress intensity factors for the graphite coated carbon sandwich materials obtained by the compact tensile specimen and the Vickers indented strip may not be feasible because of the overly simplified models used in their calculation. It is to be noted that calculations in this work were based on the assumption that the specimens were isotropic and homogeneous; despite the fact that the modulus of elasticity of the two materials is quite different. Better models to analyze cracks in layered composite materials are needed.

Threshold Crack Size in PyC

In spite of the difficulties encountered in analyzing the cracks in layered composite materials, we would like to bring attention to our experimental observations. As it is demonstrated in Tables 2 and 3, cracks longer than 2 mm grew and those less than 1.80 mm did not grow at a strain level of 500 $\mu\varepsilon$ with R=-1. This indicates that there is a threshold size for cyclic fatigue crack non-propagation. We did not detect any crack growth for all the pre-cracked specimens in Table 2 up to 6×10^8 cycles and beyond at the strain levels that are two and four times higher than the estimated peak service strain. For the cracks that survived 6×10^8 cycles without growth, the growth rate had to be less than 10^{-14} m/cycle because the accuracy of the crack measurement was 10 microns. This is to *assume* that in the worst case the crack propagated up to our limit of detection of 10 microns by the end of 6×10^8 cycles.

The existence of the threshold size of fatigue crack non-propagation of pyrolytic carbon is of importance in terms of service reliability within the designed lifetime of the heart valve and the quality control during the manufacturing process. When the flaw size is less than the threshold size, the flaw would not propagate during the service lifetime provided that the valve is operated under the designed stress. The threshold size of fatigue crack non-propagation makes the proof test of the components and the whole assembly for the valve very effective to screen out the flaws. A comparison of the elastic strain to failure under static loading of 11,530 $\mu\varepsilon$, to the maximum nominal strain in the application of 250 $\mu\varepsilon$, shows that a proof test can be performed at 5,000 $\mu\varepsilon$ to screen out any intrinsic flaws greater than 20 microns.

Extrapolation of the data of Kepner et al. shows that at 6×10^8 cycles a cyclic strain of 10,100 $\mu\varepsilon$ would be acceptable for common engineering applications. It is not surprised that a cyclic strain of 250 $\mu\varepsilon$ (or 1/40 of Kepner's) would give a probability of survival greater than needed for this critical medical application.

ACKNOWLEDGMENT

This work was supported by Carbon Implants, Inc. of Austin, Texas.

REFERENCES

1. J. C. Bokros, Carbon biomedical devices, *Carbon*, 15:355 (1977).
2. F. J. Schoen, On the fatigue behavior of pyrolytic carbon, *Carbon*, 11:413 (1973).
3. H. S. Shim, The behavior of isotropic pyrolytic carbon under cyclic loading, *Biomaterials, Medical Devices and Artificial Organs*, 2(1):55 (1974).
4. J. C. Bokros, A. D. Haubold, R. J. Akins, L. A. Campbell, C. D. Griffin and E. Lane, The durability of mechanical heart valve replacements: past experience and current trends, *in*: "Replacement Cardiac Valves," E. Bodnar and R. W. M. Frater, editors, Pergamon Press, New York (1991).
5. J. Kepner, A. D. Haubold, L. A. Beavan, Cyclic fatigue testing of pyrolytic carbon, presented at the 41th Pacific Coast Regional Meeting of the American Chemical Society, San Francisco, October (1990).
6. L. Ma and G. Sines, Threshold size for cyclic fatigue crack propagation in a pyrolytic carbon, *Materials Letters*, 17:49 (1993).
7. P. F. Packman, S. J. Klima, R. L. Davies, J. Malpani, J. Moyzis, W. Walker, B. G. W. Yee and D. P. Johnson, Reliability of flaw detection by nondestructive inspection, "Metals Handbook: Nondestructive Inspection and Quality Control," 11:414, American Society for Metals, Metals Park, OH, (1976).
8. D. P. Rooke and D. J. Cartwright, "Compendium of Stress Intensity Factors," Her Majesty's Stationary Office, London (1976).
9. M. Isida, Effect of width and length on stress intensity factors of internally cracked plates under various boundary conditions, *International Journal of Fracture Mechanics*, 7(3):301 (1971).
10. R. O. Ritchie, R. H. Dauskardt and Weikang Yu, Cyclic fatigue-crack propagation, stress-corrosion, and fracture-toughness behavior in pyrolytic carbon-coated graphite for prosthetic heart valve applications, *Journal of Biomedical materials Research*, 24:189 (1990).

AN ANALYSIS OF CYCLIC FATIGUE EFFECTS
IN CERAMIC MATRIX COMPOSITES

Dietmar Koch and Georg Grathwohl

University of Bremen
28359 Bremen
Germany

INTRODUCTION

Within the last years many efforts have been made to improve the properties of ceramic materials with the aim to cover new demands in the design of structural elements. Beside of low density and high temperature resistance ceramics ought to be damage tolerant to be used not only for simple applications but also for critical use in air and space projects as well as in the gas turbine industry. One of the most promising attempts to realize high fracture energy and non catastrophic failure is the development of ceramic matrix composites which do not fail in a brittle manner compared to monolithic ceramics.

Components made of ceramic matrix composites are loaded not only monotonically but also – especially in the above defined working fields – in a cyclic manner. Fatigue of ceramic matrix composites induced by cyclic loading, however, is not investigated very intensively, yet. Therefore, the present paper examines cyclic fatigue behavior of ceramic matrix composites which depends on loading history, load amplitude and applied frequency. Tensile tests should be preferred to bending tests because the composites are built up with laminates and therefore inhomogeneous in cross section. A new developed testing equipment has been used for fatigue testing in tension with homogeneous uniaxial loading of the composites. For monitoring fatigue tests a data acquisition system has been designed which makes it possible to interprete fatigue behavior by quantifying single load cycles. As these stress-strain loops are resolved with a high amount of values, parameters as hysteresis area (loss energy) or the tangent of the hysteresis curves can be used to characterize the development of microstructural damage processes. The sensivity of the materials to the loading parameters is evaluated and the causes of the fatigue behavior are investigated. This especially includes the degradation of internal sliding interfaces as well as the failure of fiber and matrix during cycling which leads to the so-called S-curve behavior of the composites, i.e. an increase of the tangent modulus at the stress reversal points. The results are analysed on the basis of a single fiber model which describes the decrease of fiber matrix frictional forces; the S-curved shape of the stress-strain-cycles of the composites can then be explained.

FATIGUE OF CERAMIC MATRIX COMPOSITES

Fatigue of materials is characterized by a loss of strength caused by microstructural processes due to cyclic stresses. In case of metals fatigue effects are observed even if they are loaded below the proportional limit. Fatigue can be accompanied by strengthening or weakening depending on the process of microplastic deformation and crack initiation and propagation.

Monolithic ceramics without any particular reinforcement react linear-elastic if they are loaded under tension up to failure which is induced by critical crack growth. Fatigue is supposed to play a rather unimportant role in these highly brittle materials. However, ceramics with R-curve behavior which is defined as an increase of crack resistance with increasing crack length reveal clear fatigue effects. This can be explained as follows: induced by the number of load cycles and the applied stress amplitude the crack resistance of the material is lowered e.g. by a reduction of bridging stresses due to internal friction and wear along the crack surfaces. Thus, fatigue, i.e. a decrease in tolerable maximum stress with an increasing number of load cycles, is observed in several R-curve ceramics [1, 2, 3, 4, 5].

In contrast to monolithics the mechanical behavior of ceramic matrix composites (CMC) under tensile conditions seems to be similar in some aspects to the behavior of metals. Initially the stress-strain response is linear followed by a decrease in the tangent modulus when the load exceeds the so-called proportional limit. The decrease in the modulus is induced by matrix cracking, fiber matrix debonding processes and with further loading also by fiber failure; it is, however, in contrast to metals, not influenced by plastic deformation.

If composites are reinforced in a two-dimensional way two distinct changes in the slope of the stress-strain curve can be observed. In case of glass ceramic composites (SiC-fibers in CAS-matrix) the first proportional limit occurs below the limit of the unidirectional composite [6]. This is explained by the onset of matrix cracking in 90° plies and delamination cracks. Several investigations are available which correlate the matrix crack initiation with the first non-linear stress-strain behavior. By measuring the change of the electrical resistivity of a sputtered gold film on the surface of the specimen the initiation stress for matrix cracking is measured and can be well compared to the proportional limit [7, 8]. Other investigations of the matrix cracking stress using surface replicas and acoustic emission showed that the first matrix cracks already occured below the proportional limit [9, 10]. Thus, before the first non-linearity of the stress-strain curve is detected by the load cell and the strain gauges the composite may already be damaged by cracks.

Consequently, it can be followed that cyclic loading around the proportional limit also leads to a change of internal composite properties (e.g. load transfer between fiber and matrix or fiber matrix interface bonding). A degradation of internal frictional forces or interface parameters would then lead to further damage which results in fatigue of the composites. Thus, matrix cracking is not only a measure to characterize the microstructural damage processes under quasistatic loading but also under cyclic loading.

Karandikar and Chou [11, 12] correlated matrix crack density with measured modulus reduction in quasistatic and cyclic tests. By sinusoidal loading, the matrix crack density increases and equals the crack density measured in static and quasistatic tests. Several other authors also showed that the determination of crack density in glass ceramic composites is reliable to quantify the damage state of fatigued specimens [6, 13].

However, as the ceramic matrix composites investigated in this study are produced by the polymer process and chemical vapor deposition they contain already matrix microcracks in the as-fabricated condition. Thus, it is necessary to use other methods than counting the number of matrix microcracks for quantifying the fatigue behavior.

One suitable way to determine the degradation of ceramic matrix composites caused by cyclic loading is the measurement of the temperature increase in the specimen which is induced by internal friction of crack surfaces [14]. In glass ceramics (CAS with SiC-fibers) and C/SiC-composites it is found that at the beginning of the cyclic test the temperature of the specimen strongly increases. After a certain amount of cycles the heating of the specimen levels out and with further cyclic loading the temperature of the specimen even slightly decreases [15]. By varying the testing frequency it is observed that with increasing testing frequency the heating of the specimen strongly increases [16, 17]. Furthermore, the increase in testing frequency leads to failure of the specimens at a lower number of cycles compared to lower frequencies. In case of graphite/epoxy composites the produced heat is also a function of the testing frequency and leads to earlier crack propagation and a lower number of cycles to failure [18]. It is reasonable to interprete the frequency dependence of fatigue life by the degradation of the interfacial shear stress τ of the fiber matrix interface and the dependence of the composite microstructure on the temperature [19].

Another possibility to determine the fatigue of CMCs is the evaluation of the stress-strain loops as a function of the number of load cycles [11, 14, 19]. However, the hysteresis area and the actual stiffness of the specimen can only be measured at high testing frequencies if a fast high resolution data acquisition system is available. This evaluation method is often restricted to low testing frequencies (e.g. 10Hz) or low resolution of single stress-strain loops has to be accepted. Therefore, this paper is focussed on both the measurement of the heating of the specimen and the quantitative evaluation of hysteresis loops at high testing frequencies using an advanced data acquisition system.

TEST PROCEDURE

To realize fatigue tests of ceramic matrix composites under tensile conditions it is necessary to use a gripping system which allows an alignment of the specimen axis and the load axis of the testing machine and, furthermore, a reduction of bending strains. It should also be capable to transmit tension and compression forces. As commercial high precision gripping solutions for ceramic matrix composites were not available, a tensile testing equipment has been developed.

This system as shown in figure 1 is not self-aligning but can be adjusted and fixed with minimal bending moments being present in the specimen using strain gauge measurement. It is therefore suitable for tensile and compression tests. All the specimens which are rectangular in cross section are equipped with strain gauges on all four sides to measure the bending strains in the gauge length. The specimen was clamped between wedge grips where it was fixed both with glue and with clamping forces. The glue which was fatigue proved transmit the forces from the machine to the specimen and reduces stress peaks at the end of the grips. It was then ensured that the specimens failed within the gauge length and not at the grip ends. The alignment was adjusted using spherical bearings in the upper and lower alignment fixture. The spherical bearings and the possible lateral displacement of the fixture allowed the optimal alignment of the specimen with the force axis as well as they reduced the bending (ratio of bending strain to tensile strain) below 2% by controlling the strain gauges mounted on the specimen surface. The grips were fixed in a servohydraulic testing machine which allows testing frequencies up to 1000Hz. The data acquisition system which is explained in detail below enables an evaluation of the cycles after having finished the test (fig. 1).

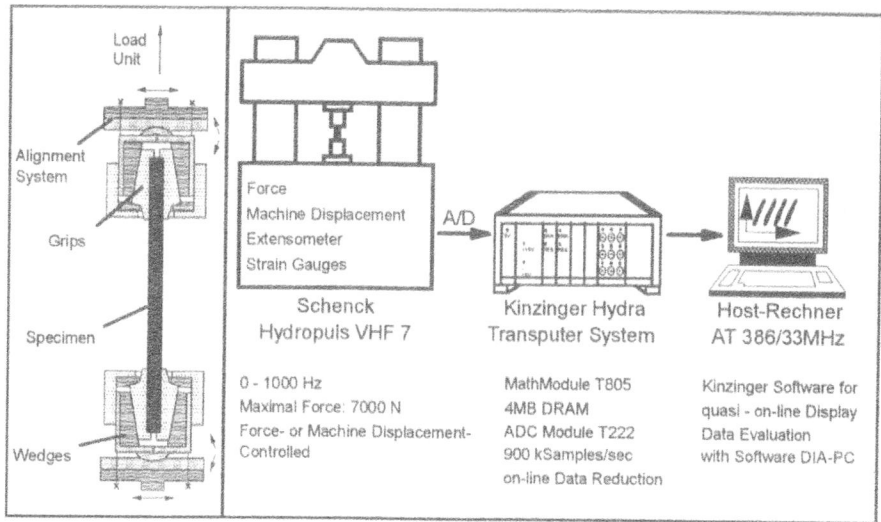

Figure 1. Gripping technique with alignment fixture (left) and experimental test arrangement (right)

As it has been pointed out in the previous section, the determination of fatigue effects of CMCs with as-processed matrix cracks should be investigated by measurement of the change of the stress-strain behavior during cyclic loading. Therefore, a data acquisition system has been developed for high frequency testing. The system is transputer based and allows a 12bit high resolution data sampling with up to 33000 data points per channel and second. If every cycle is stored a large amount of data is produced during fatigue tests. Therefore, event triggering has been established which reduces the data without loosing important information about the fatigue process. The function of the event triggering can be explained as follows (fig. 2): all the data of four channels are sampled but they are stored only if a trigger condition is fulfilled. When testing in displacement control the force signal is the trigger channel. A changing stress-strain behavior of the material leads to a change of the force maximum. If the maximum differs from the first maximum by a definite value a given number of cycles before and after the event are stored in the computer. The triggered maximum is now defined as the new trigger level and the next cycles are only stored if the force maximum now leaves the shifted trigger band. This kind of data storing has two main advantages. Firstly all the cycles where damage emerges are stored; e.g. the specimen failure is detected as a large drop of the applied maximum load and the cycles before are stored and can be evaluated after the test. Secondly, the acquisited cycles are resolved with a high amount of data points, the realizable resolution are 66 points per cycle at a testing frequency of 500Hz. This data acquisition system typically leads to a test recording as it is shown in figure 2 where the data blocks are not stored in equal but in variable time intervals.

All the tests were conducted in the displacement control mode. After two quasi-static loading ramps the cyclic test started and was finished either by specimen failure or by reaching 1,000,000 cycles. The specimens which had not failed were finally loaded up to failure strength. For the interpretation of the fatigue properties the maximum applied stress was varied at a testing frequency of 100Hz , in tension-tension with a R-ratio of 0.1 (i.e. the quotient of minimum applied to maximum applied stress). To

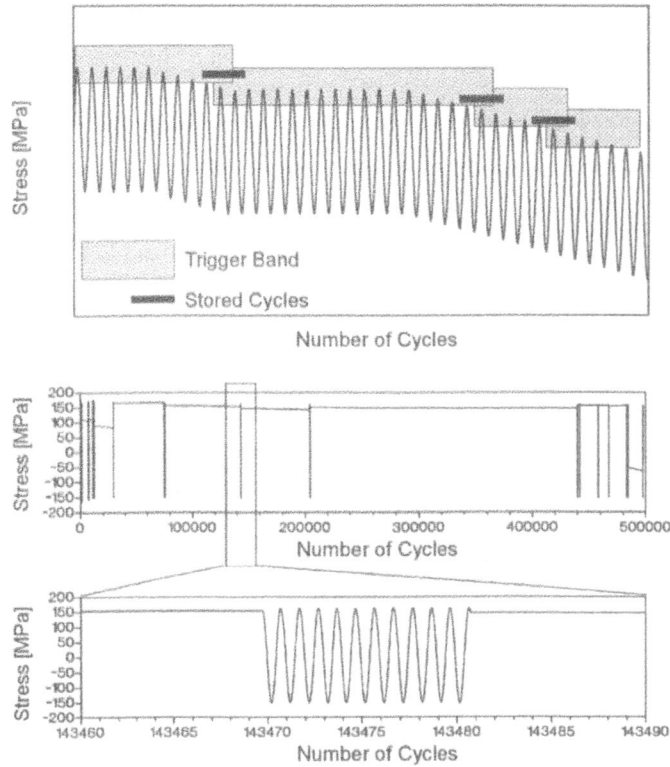

Figure 2. Event triggering for reduced but high resolution data storing (above) and a typical test record (below)

investigate the specimen's temperature increase the testing frequency was changed with R=0.1 and constant maximum load in a second test series.

To evaluate the tests the cyclic stress-strain curves were quantified by defining a medium curve which was calculated by $\sigma(\varepsilon) = \frac{1}{2} \cdot (\sigma_{ascending}(\varepsilon) + \sigma_{descending}(\varepsilon))$. This curve defines the actual stiffness of the specimen dependent on the number of cycles. The slope of the medium curve (MTM = medium tangent modulus) of the medium curve is used to interprete the change in stiffness at the stress reversal points. Since the existence of a hysteresis gives an indication of energy dissipation processes the area of the stress-strain loop, i.e. the hysteresis area, was also determined and defined as loss energy.

TENSILE TESTS

The specimens investigated are carbon fiber (*T800*) reinforced SiC and SiC fiber (*Tyranno*) reinforced SiC. The 0°/90° cross-ply-reinforced C/SiC was produced by the polymer pyrolysis process, whereas the bidirectional woven SiC/SiC was manufactured via chemical vapor infiltration (CVI). Their tensile stress-strain behavior differs significantly (fig. 3). C/SiC remains almost linear-elastic up to the failure strength of 255MPa. In the case of SiC/SiC, however, an extensive quasiplastic range is observed after reaching the proportional limit at about 80MPa. Above the proportional limit the tangent modulus decreases continuously up to failure at about 280MPa.

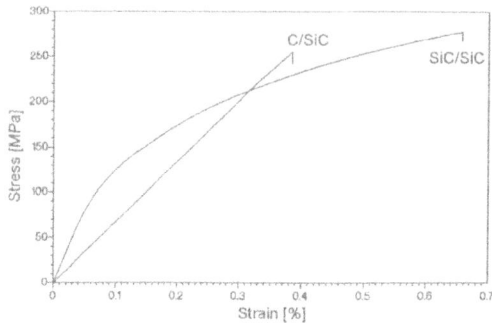

Figure 3. Tensile stress-strain graphs of C/SiC (polymer pyrolysis, 0°/90° cross-ply-reinforced) and SiC/SiC (CVI, bidirectional woven)

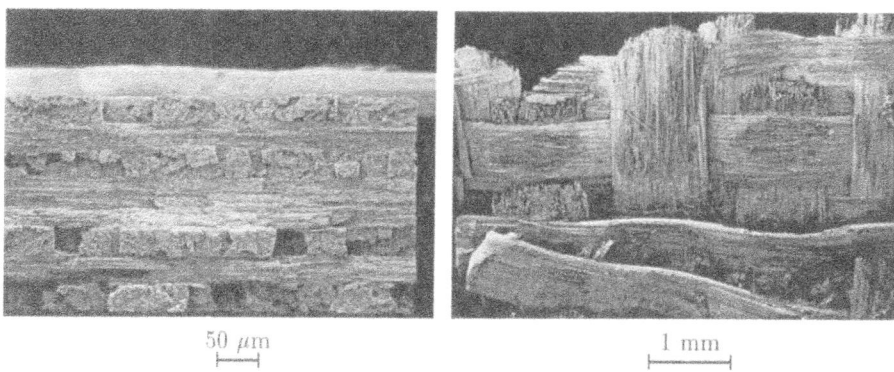

$50~\mu\mathrm{m}$ 1 mm

Figure 4. Tensile specimens of C/SiC (left, fracture surface) and SiC/SiC (right, side view) after tensile tests

The tensile specimens reveal the difference between the two materials (fig. 4). C/SiC failed more or less in one single plane perpendicular to the stress direction with some fiber bundle pull-out taking place along as-processed matrix cracks, as shown in figure 4, left. However, the fracture surface of SiC/SiC is characterized by a scattering of the fracture process with very different plateaus of failure of the woven bundles.

EVALUATION OF FATIGUE EFFECTS

The tensile stress-strain behavior of the CMCs already revealed significant differences between C/SiC and SiC/SiC. It may be expected that C/SiC is not affected by significant fatigue effects since its tensile stress-strain behavior is almost linear elastic up to failure strength. However, when C/SiC is loaded cyclically at stresses beyond a specific stress level hysteresis curves could be measured. Figure 5 shows typical hysteresis curves of a cyclic test with a maximum stress of 225MPa at the beginning of the test. The hysteresis curves are shifted in the diagram to allow an overview of the development of the hysteresis parameters as hysteresis area and the actual stiffness of the specimen. In figure 6 the loss energy which corresponds with the hysteresis area is plotted as a function of the number of cycles at different maximum stress levels.

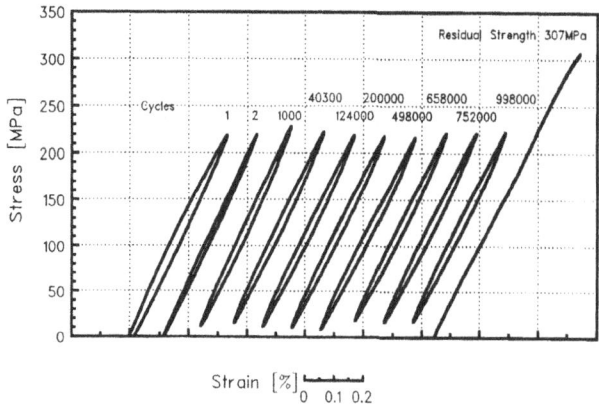

Figure 5. Cyclic tensile stress-strain curves C/SiC at 100Hz, σ_{max}=225MPa and R=0.1

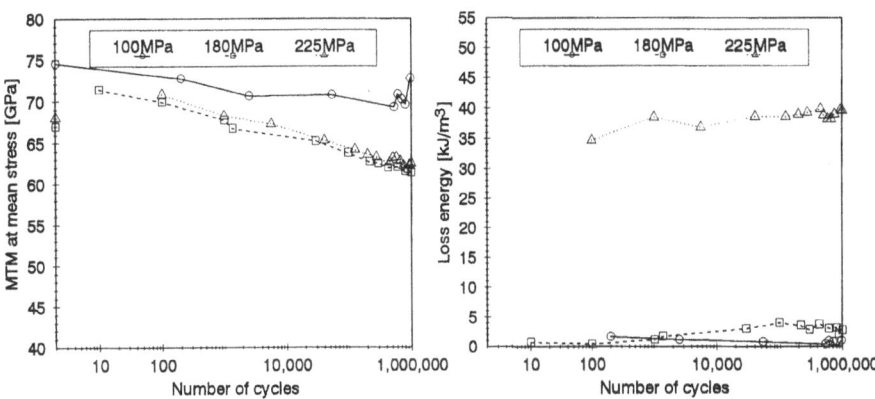

Figure 6. Stiffness change (MTM_m) and loss energy of the C/SiC-specimens dependent on maximum applied load and number of load cycles (f=100Hz, R=0.1)

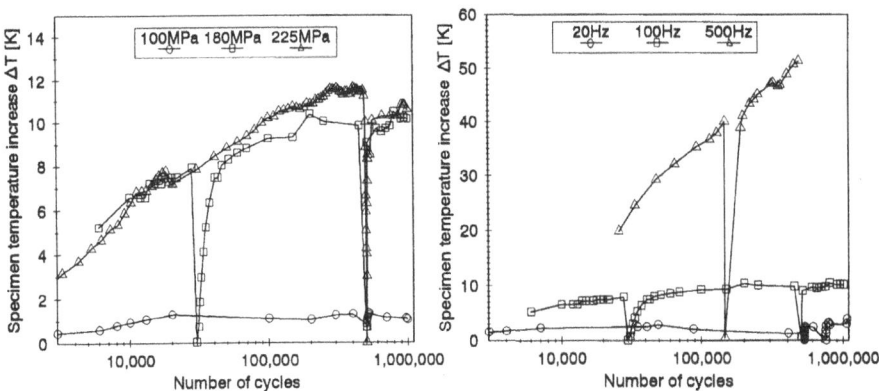

Figure 7. Temperature increase of the C/SiC-specimens dependent on load cycles and maximum applied load (f=100Hz, R=0.1), left, and on testing frequency (maximum applied stress 225MPa, R=0.1),right

It should be emphasized that at stress amplitudes up to 180MPa only small or no hysteresis curves exist whereas at high stress amplitudes (225MPa) compared to the failure strength significant hysteresis loops are observed. The MTM_m, i.e. the tangent on the medium curve at mean stress is also recorded which represents the development of the mean stiffness of the specimen during cycling. Firstly, in figure 6, it can be observed that with an increase of the maximum applied load the level of the MTM_m at the beginning of the test drops. Secondly, even at low stress amplitudes where no hysteresis is measured the MTM_m decreases slightly with the number of cycles. Thus, even if no energy dissipation can be measured the specimen is susceptible to fatigue effects. Finally, it has to be pointed out that the residual strength of C/SiC is higher than the original tensile strength even when the specimen was cyclically loaded at high amplitudes (fig. 5).

By measuring the temperature increase of the specimens it was observed that there is no direct correlation between the loss energy and the heating of the specimen (fig. 7). Although the specimen loaded with a maximum stress of 180MPa shows the same temperature increase as the specimen loaded with 225MPa the stress-strain loops of both specimens differ strongly from each other. On the other hand, the change of the MTM_m can be compared in a qualitative way with the temperature change of these specimens. This means that the loss energy (as being determined from the loop area) is not an adequate measure to characterize the state of fatigue in case of C/SiC.

CVI-SiC/SiC behaves very different to C/SiC as it was already expected after the tensile tests. A typical test result is shown in figure 8 where the different behavior to C/SiC seems to be limited mainly to the first cycle. There, an extensive quasiplastic non-linear stress-strain response is observed. The following cycles up to 1,000,000 loops are at the first look rather similar to C/SiC. But in detail, there are significant differences. Evaluating the stress variation tests at a testing frequency of 100Hz (fig. 9) it is obvious that up to high maximum stresses (230MPa) the increase of loss energy and the reduction of MTM_m level out with increasing number of cycles, suggesting microstructural stability against fatigue at lower maximum stress amplitudes. Above this level (230MPa) the specimens are damaged continously which can be quantified by a steady decrease of the MTM_m-values. In the same way, the hysteresis area changes which leads to a continuous increase in energy dissipation and finally, just before failure, to a strong rise until fracture.

Regarding the temperature increase in the SiC/SiC-specimens during cycling the temperature raises with the number of cycles, reaches a maximum and decreases slightly approaching the end of the cyclic tests (fig. 10). Depending on the applied stress amplitude the heating of the specimens is shifted to higher temperatures at high stress levels. Maximum values of about 100K are observed which are significantly higher compared to the C/SiC-specimens where at high loads only a heating of the specimens by about 12K was measured.

In the case of SiC/SiC the loss energy and the temperature increase could be well correlated in the range where the specimen temperature remained nearly constant. Considering radiation, convection and conduction [19, 20] the heat loss of the specimen is calculated and compared to the loss enery calculated from the hysteresis area. The heat loss per cycle corresponds in tendency with the loss energy but is lower by a factor of 3 to 4 (e.g. at a stress amplitude of 200MPa loss energy is 20kJ/m^3 compared to heat loss calculated to 5kJ/m^3).

With C/SiC, the influence of the testing frequency on the fatigue behavior was evaluated. The temperature increase at an applied amplitude of 225MPa in tension-tension (R=0.1) was measured as being dependent on the testing frequencies (fig. 7).

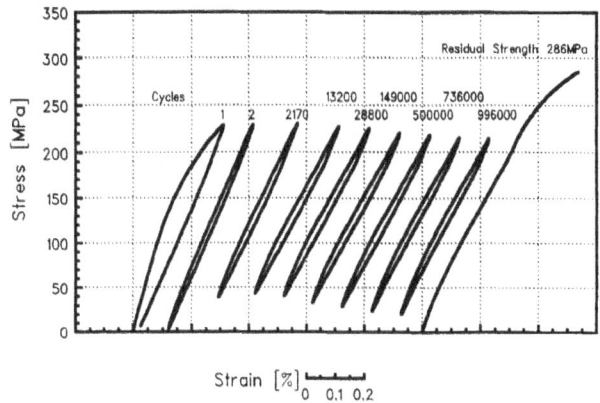

Figure 8. Cyclic tensile stress-strain curves SiC/SiC at 100Hz,σ_{max}=230MPa and R=0.1

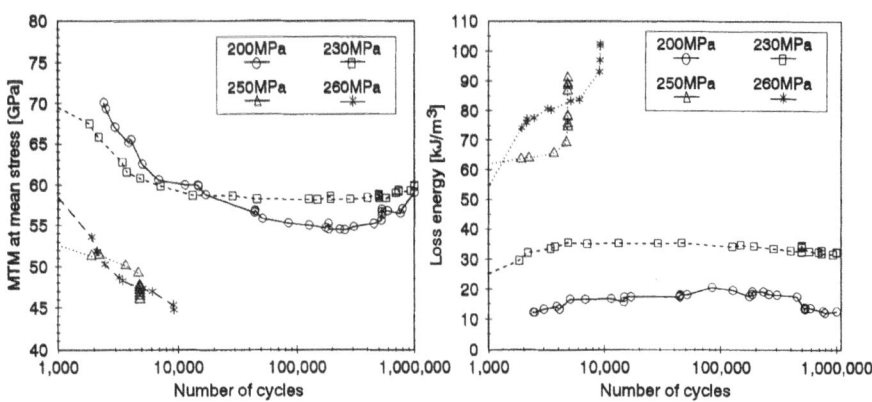

Figure 9. Stiffness change (MTM_m) and loss energy of the SiC/SiC-specimens dependent on maximum applied load and number of load cycles (f=100Hz, R=0.1)

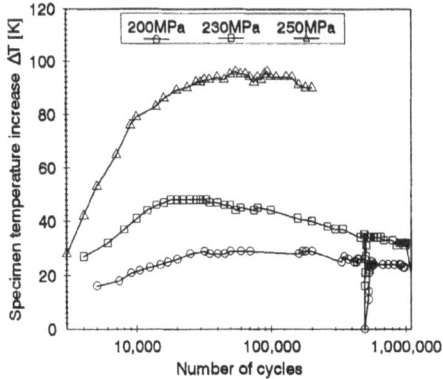

Figure 10. Temperature increase of the SiC/SiC-specimens dependent on maximum applied load and load cycles (f=100Hz, R=0.1)

Whereas up to 100Hz a plateau of the heating of the specimen was measured, the temperature increase did not level out at a testing frequency of 500Hz and the specimen failed before reaching 1,000,000 cycles.

The higher the testing frequency the higher the specimen heating and also the damage accumulation leading to premature fatigue failure. In fig. 7 and 10 it can also be seen that the specimens cooled down to room temperature when the tests were interrupted and regained very quickly the same temperature level where the tests have been interrupted when they were restarted. Thus, the specimen temperature level can be used as an indicator for the state of the fatigue process.

S-CURVE MODEL

In the previous section the hysteresis development and the stiffness of CMCs have been evaluated. Apart from these fatigue effects there is another remarkable change in the stress-strain curve which arises during cyclic fatigue testing and is called S-curve behavior. The S-curve stands for different slopes of the stress-strain curve during loading and unloading. This can be quantified by using the above defined MTM which represents the tangent on the medium curve. In figure 11 three characteristic slopes of the medium curve are drawn, the MTM at the upper (MTM_u) and lower (MTM_l) stress reversal points and at the mean applied stress (MTM_m). The specimens were loaded at $f=100$Hz with a stress amplitude of 225MPa ... 230MPa in tension-tension. The value MTM_m which was already used to characterize the change of stiffness of the specimen during cycling is obviously always lower than the values of the MTM at the lower and upper stress reversal points. In the case of C/SiC the tendency of the MTM's to decrease during cycling is identical for the three characteristic values. Thus, C/SiC looses stiffness but the shape of the loops is not changed particularly.

At the applied stress amplitude of 230MPa the MTM_m of SiC/SiC levels out with the number of cycles (see also fig. 9). The tangent MTM_u at the upper stress reversal point also decreases and levels out near the end of the fatigue test but is always higher than the MTM_m. The tangent at the lower stress reversal point, however, differs significantly from the others as it increases with the number of cycles.

To explain the differences of the slopes of the loops depending on the applied

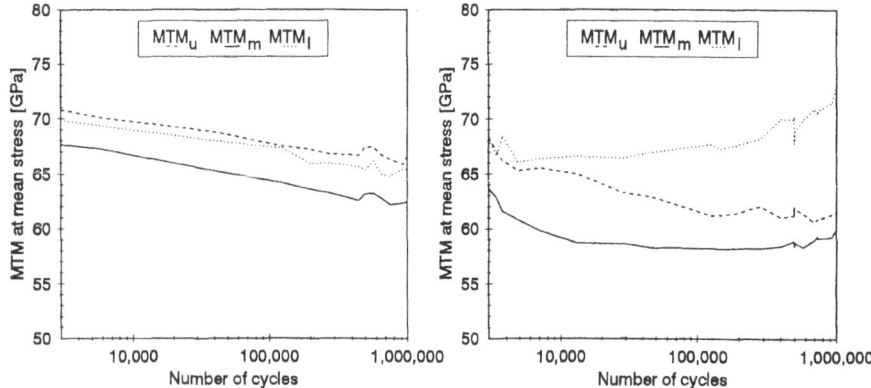

Figure 11. Changes of the stiffness of a C/SiC-specimen (left) and a SiC/SiC-specimen (right) at the upper (MTM_u) and lower (MTM_l) stress reversal points as well as at the mean applied load (MTM_m) $(f=100$Hz, $R=0.1$, applied stress amplitude 225MPa ... 230MPa)

stress the so-called S-curve model has been established (fig. 12). This model is based on the work of Marshall et al. [21, 22] who used a single fiber model to describe the displacement between fiber and matrix using two different fiber matrix interface rules (constant shear stress and Coulomb friction). When a fiber is pulled with an applied stress σ_a the fiber matrix interface is debonded up to a debond length $l_{Debonding}$ dependent on the interfacial fracture energy and the shear stress between fiber and matrix. By unloading and reloading the displacement δ between fiber and matrix can be calculated by integration of the stress profile along the fiber axis (shaded area in fig. 12) and leads to a hysteresis loop with constant slope of the medium stress-strain curve.

The basic idea leading to the S-curve model is the fact that during loading and unloading the fiber is moved relatively to the matrix. The relative sliding displacement between fiber and matrix is not constant along the debonded interface; it reaches minimum values at the debonding crack tip and increases within the debonding zone. The variation of the relative displacement depends on the actual state (e.g. roughness) of the interface. By repeated back and forward sliding it can be assumed that the interface would be damaged by friction and wear leading to a drop of the interfacial shear stress τ between fiber and matrix.

The S-curve model describes the drop of τ by defining two areas with different values of τ. The shear stress τ_1 gives the value for the undamaged interfacial area whereas τ_2 characterizes the smoothed fiber matrix interface. The equations for reloading and unloading of the Marshall model are then varied and can be written as [20]:

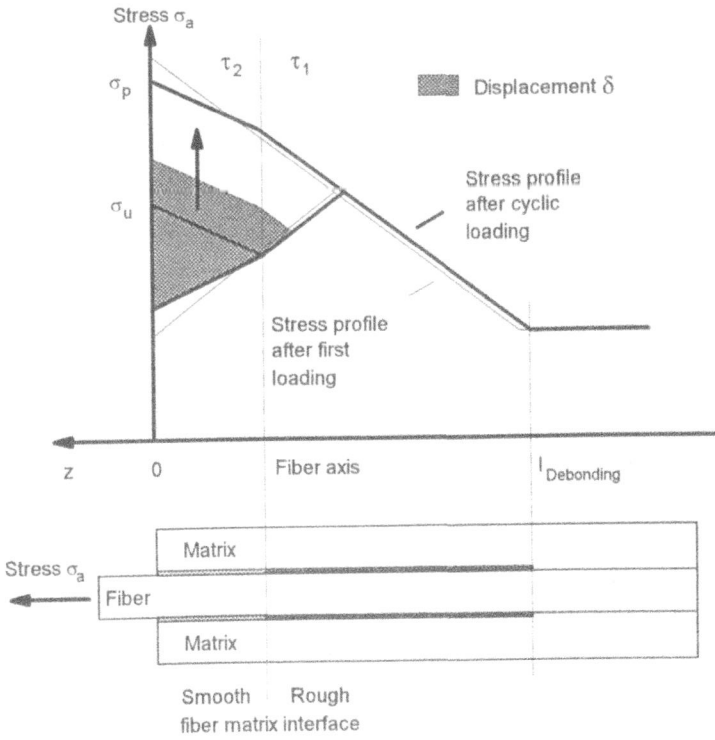

Figure 12. S-curve single fiber composite model considering the smoothing of the fiber matrix interface induced by cyclic relative movement between fiber and matrix

reloading :

$$
\delta = \begin{cases} \left[\frac{(b_2+b_3)(1-a_1V_F)^2 r_F}{8E_M}\right] \frac{\sigma_a^2}{\tau_2} & : \ \sigma_a < \sigma_u \\[12pt] \left[\frac{(b_2+b_3)(1-a_1V_F)^2 r_F}{8E_M}\right] \left(\frac{\sigma_a^2}{\tau_2} + (\sigma_a^2-\sigma_u^2)\left(\frac{1}{\tau_1}-\frac{1}{\tau_2}\right)\right) & : \ \sigma_a > \sigma_u \end{cases} \tag{1}
$$

unloading :

$$
\delta = \begin{cases} \delta_p - \left[\frac{(b_2+b_3)(1-a_1V_F)^2 r_F}{8E_M}\right] \frac{(\sigma_p-\sigma_a)^2}{\tau_2} & : \ \sigma_a > \sigma_p-\sigma_u \\[12pt] \delta_p - \left[\frac{(b_2+b_3)(1-a_1V_F)^2 r_F}{8E_M}\right] \left(\frac{(\sigma_p-\sigma_a)^2}{\tau_2} + (\sigma_p-\sigma_u-\sigma_a)^2\left(\frac{1}{\tau_1}-\frac{1}{\tau_2}\right)\right) & : \ \sigma_a < \sigma_p-\sigma_u \end{cases} \tag{2}
$$

These equations are used to calculate the hysteresis loops. The shape of the unload-reload cycles changes by increasing the difference between τ_1 and τ_2 to a S-shaped curve leading to an increase in the tangent at the stress reversal points compared to the applied mean load and also to a drop of the absolute stiffness.

The behavior of a unidirectional composite can be predicted if the composite is divided in single fiber matrix composite cells with a characteristic stress strain behavior and compliance. There are three possible types of cells: first is an intact composite, second a composite with a matrix crack and a fiber crack, and third a S-curve cell with changing interfacial shear properties, i.e. an intact fiber bridging a matrix crack. If these cells are linked in parallel and serial way to a unidirectional multi-fiber composite the stress-strain loops of this composite can be calculated as a function of the arrangement of the cells and as a function of the reduction of the shear stress. In figure 13 the hysteresis of a unidirectional composite is simulated (66% cells with S-curve properties and 33% intact cells) with a reduction of the interfacial shear stress by a factor of 10 compared to the virgin value of $\tau_1=10$MPa. This value seems to be realistic for SiC/SiC-composites whose fiber and matrix properties are used to calculate the stress-strain curves. If τ remains constant along the debonded area – as it is the case in the Marshall model – the MTM's remain constant as well. If the ratio of the interfacial shear stresses τ_1/τ_2 increases the hysteresis is increasingly S-shaped. It can be concluded

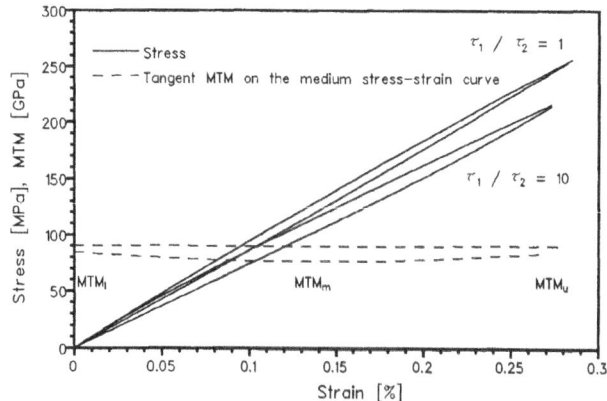

Figure 13. Simulation of the hysteresis curve of a unidirectional reinforced fiber matrix composite consisting of single fiber matrix composite cells with intact fiber matrix bonding (33%) and with reduced values of interfacial shear stress (66%)

that the S-curve model is suitable to explain the observed stress-strain behavior of real composites.

The measured increase of the stiffness of SiC/SiC at the lower stress reversal point, however, cannot be explained by this theory. It is supposed that matrix debris blocks the matrix cracks and therefore leads to a higher stiffness when the cracks are closed before reaching stress minimum (crack closure). As the Young's modulus of the matrix of SiC/SiC is much higher than that of C/SiC this effect can only be observed at SiC/SiC.

CONCLUSIONS

The interpretation of cyclic fatigue of CMCs with as-processed matrix cracks requires a testing technique with a quantitative evaluation of single load cycles. With a powerful data acquisition and evaluation system the change of stiffness and energy dissipation can be quantified. The composites C/SiC and SiC/SiC which differ in the reinforcement and production process exhibit very different behavior under cyclic loading conditions. Whereas the compliance of C/SiC is continuously reduced with the number of stress loops the fatigue process of SiC/SiC is stabilized resulting in a steady stiffness after a certain amount of cycles. The specimen heating is a further suitable measure for the state of the fatigue of the specimen and with higher frequencies the temperature increases dramatically which results in premature failure.

An effective explanation of the observed phenomena of an increase in the tangent on the hysteresis curves induced by cyclic loading is given by the S-curve model. The model consists of single fiber composites which do have a characteristic stress-strain behavior and compliance assuming changes in the interfacial shear stress due to interfacial friction and wear. These cells are combined to a multi-fiber composite which can describe the S-curved load cycle.

ACKNOWLEDGMENTS

The support of Deutsche Forschungsgemeinschaft within the framework of the "Graduiertenkolleg Technische Keramik" is greatfully acknowledged. Furthermore the authors would like to thank Dornier GmbH and MAN Technologies for providing the composite materials.

REFERENCES

[1] R.O. Ritchie & R.H. Dauskardt. Cyclic Fatigue of Ceramics: A Fracture Mechanics Approach to Subcritical Crack Growth and Life Prediction. *J. Ceram. Soc. of Japan*, 99, 1991.

[2] D.A. Krohn & D.P.H. Hasselman. Static and Cyclic Fatigue Behavior of a Polycristalline Alumina. *J. Am. Ceram. Soc.*, 55(4), 208–211, 1972.

[3] T. Fett & D. Munz. Lifetime Prediction for Ceramic Materials under Constant and Cyclic Loading. In *ASTM STP 1201, Life Prediction, Methodologies and Data for Ceramic Materials, Cocoa Beach, Fl.*, page 161–174. Philadelphia: American Society for Testing and Materials, 1993.

[4] G. Grathwohl. Ermüdung von Keramik unter Schwingbeanspruchung. *Mat.-wiss. und Werkstofftechnik*, 19, 113–124, 1988.

[5] G. Grathwohl. Crack Resistance and Fatigue Limits of Structural Ceramics. *Powder Met. Int.*, 24(2), 98–105, 1992.

[6] D.S. Beyerle, S.M. Spearing & A.G. Evans. Damage Mechanisms and Mechanical Properties of a Laminated 0/90 Ceramic/Matrix Composite. *J. Am. Ceram. Soc.*, 75(12), 3321–30, 1992.

[7] M.W. Barsoum, P. Kangutkar & A.S.D. Wang. Matrix Crack Initiation in Ceramic Matrix Composites. Part I: Experiments and Test Results. *Compos. Sci. Technol.*, 44, 257–269, 1992.

[8] A.S.D. Wang, X.G. Huang & M.W. Barsoum. Matrix Crack Initiation in Ceramic Matrix Composites. Part II: Models and Simulation Results. *Compos. Sci. Technol.*, 44, 271–282, 1992.

[9] L.P. Zawada, L.M. Butkus & G.A. Hartman. Tensile and Fatigue Behavior of Silicon Carbide Fiber-Reinforced Aluminosilicate Glass. *J. Am. Ceram. Soc.*, 74(11), 2851–2858, 1991.

[10] C. Cho, J.W. Holmes & J.R. Barber. Distribution of Matrix Cracks in a Uniaxial Ceramic Composite. *J. Am. Ceram. Soc.*, 75(2), 316–324, 1992.

[11] P. Karandikar & T.-W. Chou. Microcracking and Elastic Moduli Reductions in Unidirectional Nicalon-CAS Composite under Cyclic Fatigue Loading. *Ceram. Eng. Sci. Proc.*, 13, 881–888, 1992.

[12] P. Karandikar & T.-W. Chou. Characterization and Modelling of Microcracking and Elastic Moduli Changes in Nicalon/CAS Composites. *Compos. Sci. Technol.*, 46, 253–263, 1993.

[13] A.W. Pryce & P.A. Smith. Matrix Cracking in Unidirectional Ceramic Matrix Composites under Quasi-Static and Cyclic Loading. *Acta metall.*, 41(4), 1269–1281, 1993.

[14] J.W. Holmes. Fatigue of Fiber Reinforced Ceramics. In *Ceramics and Ceramic Matrix Composites*, page 193–238. New York: ASME, 1992.

[15] J.W. Holmes, X. Wu & B.F. Sørensen. Frequency Dependence of Fatigue Life and Internal Heating of a Fiber-Reinforced Ceramic Matrix Composite. *J. Am. Ceram. Soc.*, 77(12), 3284–3286, 1994.

[16] J.W. Holmes, C. Cho & J.R. Barber. Estimation of Interfacial Shear in Ceramic Composites from Frictional Heating Measurements. *J. Am. Ceram. Soc.*, 74(11), 2802–2808, 1991.

[17] S.F. Shuler, J.W. Holmes, X. Wu & D.H. Roach. Influence of Loading Frequency on the Room Temperature Fatigue of a Carbon-Fiber SiC-Matrix Composite. *J. Am. Ceram. Soc.*, 76(9), 2327–36, 1993.

[18] A. Rotem. Load Frequency Effect on the Fatigue Strength of Isotropic Laminates. *Compos. Sci. Technol.*, 46, 129–138, 1993.

[19] J.W. Holmes & C. Cho. Experimental Observations of Frictional Heating in Fiber-Reinforced Ceramics. *J. Am. Ceram. Soc.*, 75(4), 929–938, 1992.

[20] D. Koch. *Fortschrittberichte VDI, Reihe 18, Nr.172, Analyse des Ermüdungsverhaltens faserverstärkter keramischer Werkstoffe*. Düsseldorf: VDI-Verlag, 1995.

[21] D.B. Marshall. Analysis of Fiber Debonding and Sliding Experiments in Brittle Matrix Composites. *Acta metall. mater.*, 40(3), 427–441, 1992.

[22] D.B. Marshall, M.C. Shaw & W.L. Morris. Measurement of Interfacial Debonding and Sliding Resistance in Fiber Reinforced Intermetallics. *Acta metall. mater.*, 40(3), 443–454, 1992.

FATIGUE LIFE PREDICTIONS OF PZT USING CONTINUUM DAMAGE MECHANICS

AND FINITE ELEMENT METHODS

Tze-jer Chuang, Zhengdong Wang, Michael Hill and Grady White

Ceramics Division
National Institute of Standards and Technology
Gaithersburg, Maryland 20899

ABSTRACT

Owing to their high dielectric and piezoelectric constants, piezoelectric ceramics have been used for sensor and actuator applications. The reliability of these materials under service conditions is of major concern. The fatigue life of a poled PZT specimen under mechanical cyclic loading conditions in four-point bending is predicted by the finite element method, using a continuum damage mechanics approach. The damage parameter is defined as a scalar in proportion to the microcrack density in a representative volume comparable to a typical finite element size. The material is modelled as transversely isotropic elastic to reflect the 6mm poled crystalline texture, with elastic and piezoelectric constant tensors all functions of the damage parameter. The damage-dependent constitutive laws and energy-based damage evolution law were incorporated into the finite element program to calculate the history of stress, strain, electric potential and damage fields for the four-point loaded beam specimen in an iterative manner as number of loading cycle increases. The results indicated that the macroscopic cracks initiate at loadpoint and inner span which link together to cause failure. Comparisons between the predictions of the total cycle numbers to failure and the experimental stress amplitude-life cycle data at a frequency of 20 Hz showed good agreement. Implications of the results with regards to damage mode, mechanical failure criteria, and its potential applications are discussed.

INTRODUCTION

In recent years, piezoelectric ceramics such as lead zirconate titanate (PZT) are finding increasing use for manufacturing transducers, micropositioners, smart sensors and quiet actuators owing to their strong electro-mechanical coupling effects and their prompt responses. In these applications, severe mechanical/electrical stressing often occurs in service, mostly in the form of low-frequency cyclic loading. As a consequence, mechanical and electrical properties deterioration and premature failure owing to fatigue damage are introduced which result in unreliable service as well as added maintenance cost from excessive replacements[cf. 1-2]. Fatigue behavior of PZT is of major concern to both engineering designers and end users who like to be assured of reproducible outputs from those components within the expected designed life. The reliability issue of PZT materials under fatigue conditions is the focal point of the present paper.

Historically as revealed by a literature review, the fracture problem of PZT under sustained loads has been treated mainly in the context of fracture mechanics in linearly (isotropic or

Fracture Mechanics of Ceramics, Vol. 12
Edited by R.C. Bradt *et al.*, Plenum Press, New York, 1996

anisotropic) elastic materials. Parton [3] in 1976 published an elementary fracture mechanics theory of piezoelectrics where a slit-like crack with vanishing electric boundary conditions subjected to a remote uniaxial tensile stress state was considered. Deeg [4] extended the analysis to cover other major defects including cracks, dislocations and inclusions embedded in piezoelectric solids and presented solutions of the associated forces and fields produced. Special analytical tools [5-6] such as complex potential technique, eigenfunction expansion method etc. have been applied to solve the electromechanical field equations for the transverse isotropic piezoelectric media containing elliptical cavities. McMeeking [7] worked out the electrostrictive stress fields for conducting cracks. Suo, et. al. [8] presented a general theory of crack mechanics. Crack extension forces or J-integral expressions were derived by a number of researchers [9-10] under a variety of boundary conditions or loading modes. Experimentally, indentation cracks were used to determine the fracture toughness of piezoelectric materials [cf. 12]. The problem of fatigue crack growth in ferroelectric ceramics was considered both theoretically and experimentally [13-15] for pre-existing cracks.

However, for as-received PZT containing a fixed amount of initial porosity under cyclic loading conditions, the mode of fatigue failure normally is not dictated by a dominant crack, rather by a sequence of damage generation and accumulation leading to final electric breakdown (arcing) or mechanical fracture[3-4]. Recent studies by electron microscopy [16-17] on post fatigue failed PZT specimens indicated that the major forms of fatigue damage are 90 deg. domain switching and microcracks at the triple-point junctions. Those observations imply that, in real applications, the above-mentioned approaches are likely not to be quite useful in describing the global fatigue behavior for a PZT component, *albeit* they are relevant to predicting the field parameters in the localized area surrounding the microcracks. Thus, we believe that a new approach employing discipline of continuum damage mechanics may have to be adopted from the perspective of fatigue life prediction, since a majority of lifetime is consumed by damage accumulation on a global scale, not by subcritical fatigue crack growth. Furthermore, because of the complexity involved in the geometry and interactions between mechanical and electrical field parameters induced by fatigue, numerical approaches such as finite element methods may have to be implemented.

The present paper focuses the attention on the issue of reliability of PZT materials as influenced by fatigue effects. The major goal is to ascertain service life of a component (transducer, actuator, resonator, etc.) limited by fatigue failure, so that unexpected service interruptions can be avoided. Towards this goal, an algorithm is hereby outlined which incorporates the damage mechanics concept into the finite element formulations. The methodology, as will become clear later, is capable of estimating lifetime of a PZT component under a variety of mechanical/electrical cyclic loading conditions. The program of the current paper is as follows: we start in the next section to describe the fundamental constitutive laws for a class of poled PZT piezoelectric solids. The forms essentially follow the classical expressions for piezoelectric media, except the elastic and electrical constants in the equations are no longer fixed. Rather, they are functions of damage, which in turn are history-dependent (i.e., depending on number of cycles in service). In this way, the property deteriorations can be monitored from the begining to the end of service life. Next, we will briefly summarize the necessary governing equations for finite element computations. Those are formulated on the basis of each individual element. An algorithm will be constructed wherein an iterative numerical scheme is implemented to solve incrementally the electroelastic field parameters including damage within the body as number of loading cycles increases until damage reaches a critical value. In this way, the total cycle number to failure can be estimated. Example of a four-point bend bar subjected to a variety of mechanical load amplitudes at a frequency of 20 Hz is given to demonstrate the application of this methodology. Experimental procedure paralleling this case study is described. Fatigue tests on a grade of PZT-8 manufactured by Morgan Matroc Inc. [25] at room temperature were performed. Fatigue life data are then compared with the theoretical predictions. Results and their implications are discussed along with a recommendation list for future research.

DAMAGE DEPENDENT CONSTITUTIVE LAW

Since PZT belongs to a class of piezoelectric ceramic materials, it is appropriate to employ the constitutive laws in the theory of piezoelectricity wherein mechanical and electrical effects are coupled. Moreover, since we are primarily interested in estimating fatigue life of a PZT part in service where the range of the cyclic loading frequency tends to be in the low end, we may justifiably neglect dynamic and thermal effects, and treat the system as quasi-static and isothermal at room temperature. Under these basic assumptions, there are four important field parameters that can be identified to describe the coupling effects, namely, two mechanical variables: stress (σ_{ij}) and strain (ε_{ij}) tensors; and two electrical variables, electric field (E_i) and electrical displacement (D_i) vectors. It is arbitrary to treat one set as independent variables and another set as dependent. However, in keeping with the spirit of conventional finite element theory, we will treat strain tensor and (negative) electrical potential field vector as independent parameters. Then the constitutive laws as given in the theory of piezoelectricity take the following forms for the two dependent variables σ_{ij} and D_i:

$$[\sigma_{ij}] = [C_{ijkl}] [\epsilon_{kl}] + [e_{ijk}] \{-E_k\} \tag{1}$$

where C_{ijkl} is the elastic stiffness tensor and e_{ijk} is the piezoelectric coefficient tensor, and the symbols [•] and {•} designate tensorial and vectorial quantities, respectively. As dinstinct from the conventional elasticity theory, due to high piezoelectric constants of PZT, a term representing additional stress generated due to the presence of electric field must be added to the expression for the total stress. In the case of electric displacement,

$$\{D_i\} = [e_{kli}]^T [\epsilon_{kl}] - [d_{ij}] \{-E_j\} \tag{2}$$

where $[d_{ij}]$ is dielectric constant tensor and superscript T on [•] stands for transpose of the matrix representation of the tensor. Again, elastic strain will introduce electrical flux in addition to electric field, because of high coupling coefficients. In the SI system, the units of [C], [e] and [d] are Nm^{-2}, C/m^2 and $C/(V•m)$. respectively. Those materials property data are measurable and initial values were obtained by tests. For a general isotropic elastic solid, [C] can be simplified to a matrix containing only two independent constants. However, for a poled PZT, the elastic properties become highly anisotropic, at least for the poled direction as related to other directions. Although the individual PZT grains possess a tetragonal perovskite-type crystal structure, and thus are highly anisotropic at the microscopic level, the properties will be statistically averaged to yield isotropy, except in the polarization direction. Thus, the elastic properties of PZT can be characterized as transversely isotropic elastic solid. In this case, adopting a rectangular Cartesian coordinate system, 1-2-3 with Axis-3 being the poling axis, then the matrix [C] takes the following form:

$$[C] = \begin{pmatrix} C_{11} & C_{12} & C_{13} & 0 & 0 & 0 \\ . & C_{11} & C_{13} & 0 & 0 & 0 \\ . & . & C_{33} & 0 & 0 & 0 \\ . & . & . & C_{44} & 0 & 0 \\ . & . & . & . & C_{44} & 0 \\ . & . & . & . & . & \frac{1}{2}(C_{11}-C_{12}) \end{pmatrix} \tag{3}$$

137

Since [C] is a symmetric matrix, we replace the lower side elements with dotted symbols, signifying they are identical to the upper side. As can be seen, there are five independent elastic constants to be measured in a transversly isotropic solid, and all must be real. On the other hand, the piezoelectric coefficient tensor is not symmetric. Furthermore, of the 5 nonvanishing elements, only three are independent and their values may be negative. This feature is described explicitily in the following equation (4):

$$[e]^T = \begin{pmatrix} 0 & 0 & 0 & 0 & e_{15} & 0 \\ 0 & 0 & 0 & e_{15} & 0 & 0 \\ e_{31} & e_{31} & e_{33} & 0 & 0 & 0 \end{pmatrix} \tag{4}$$

where $e_{31} = e_{32}$ and $e_{15} = e_{24}$. Thus, only e_{31}, e_{33} and e_{15} need be measured. Finally, since electrical field and displacement are both aligned with crystallographic axes, the dielectric constant tensor must be diagonal with zero off-diagonal terms. If we also assume the electrical properties of PZT are transversly isotropic, then [d] takes the following simple form:

$$[d] = \begin{pmatrix} d_{11} & 0 & 0 \\ 0 & d_{11} & 0 \\ 0 & 0 & d_{33} \end{pmatrix} \tag{5}$$

where $d_{11} = d_{22}$, and only two dielectric constants are needed to characterize the electric properties of PZT.

So far, we pretty much follow the principle of the classical continuum mechanics in formulating the constitutive laws for PZT. Now we begin to deviate the approach by injecting the concept of damage mechanics. Here we define a damage matrix [M] containing damage elements which are functions of a damage parameter ω (a scalar quantity). The damage parameter ω could be a function of microcrack density and/or domain orientation. At this point we leave open the explicit expression for ω. We demand, in principle, that all coefficients in the constitutive laws, Eqs. (1-2) are no longer fixed; rather, they become functions of ω. In this way, the history of property degradation can be followed. Accordingly, [C] and [e] in Eqs. (1-2) are now modified to have a form in matrix notation:

$$[c(\omega)] = [M]^{-1} [c] \tag{6}$$

and

$$[e(\omega)] = [M]^{-1} [e] \tag{7}$$

where superscript (-1) means inverse operation, and the damage matrix takes the following form:

$$[M] = \begin{pmatrix} m_{11} & m_{12} & m_{13} & 0 & 0 & 0 \\ \cdot & m_{11} & m_{13} & 0 & 0 & 0 \\ \cdot & \cdot & m_{11} & 0 & 0 & 0 \\ \cdot & \cdot & \cdot & m_{44} & 0 & 0 \\ \cdot & \cdot & \cdot & \cdot & m_{44} & 0 \\ \cdot & \cdot & \cdot & \cdot & \cdot & m_{44} \end{pmatrix} \qquad (8)$$

The form of [M] must obey the following requirements: (1) it must be symmetric, otherwise the resulting [c] will become unsymmetric which is not permissible since this will violate the existence of strain energy; and (2) when $\omega=0$, [M] must reduce to the indentity matrix [I]. The latter requirement assumes the as-received material contains no damage. It is noteworthy that because of the modification on [C] by [M], the initially isotropic solid may become anisotropic after damage is introduced into the solid. Explicit expressions for m_{ij} should be derived by appropriate micromechanics modeling. If we assume damage is dominated by a distribution of microcracks, then the classical Taylor's model [20] depicts that $m_{11}= m_{22}= m_{33}= 1 + \pi (\omega + 0.373 \omega^2)$; $m_{12}= m_{13}= m_{23}= 0.041 \pi\omega^2$; and $m_{44}= m_{55}= m_{66}= 1+ \pi (2\omega + 0.664 \omega^2)$. Those expressions are derived from the stress field solution induced by the interactions of randomly oriented and distributed microcracks. Cyclic loading may change the dielectric constants of the material by stress induced de-aging, in which case the constant will increase slightly [18]. On the other hand, microcracks have been reported to decrease the constant [19]. Accordingly, for dielectric constant matrix [d], we assume, for the time being, no degradations will occur during the entire fatigue life.

FINITE ELEMENT FORMULATIONS

Finite element methods (FEM) are a branch of numerical techniques widely employed by modern engineers since the 1960s to search approximate solutions for field parameters within a medium existing in the real world. The power of this technique is that it is capable of dealing with a realistic structure with complex geometry loaded by complicated boundary conditions. The idea of this method is to divide the structure fictitiously into a finite number of regions called elements using an appropriate mesh design. The boundary lines of the element are connected by nodal points. Instead of solving the field parameter (e.g., stress, strain, temperature and electric potential etc.) at an arbitrary point, the FEM seeks approximate solutions at the discrete nodal points. If it becomes necessary to find solutions somewhere else, interpolation from the neighboring nodal solutions can alway be utilized using shape functions. Since the FEM is basically a continuum approach at a macroscopic global scale, the size scale of a typical element must be, as a rule of thumb, at least two orders of magnitude larger than the characteristic size scale of heterogeneity (e.g., grain size). In the case of PZT materials with a typical grain size of 1 μm considered in the present paper, the appropriate finite element sizes should be in the range of 100 μm to 1 mm. The general finite element theory is documented in a number of textbooks, and commercially available general purpose codes such as ANSYS[1] are well developed.

---

[1]ANSYS is a trade mark of a general purpose finite element computer code developed by Swanson Analyis Systems, Inc..in Houston, PA. Its use should not be construed as endorsement by the National Institute of Standards and Technology.

Here we only summarize the essential formula in piezoelectricity used by the current investigation. Let us consider an arbitrary element and let the unknown displacement and electric potential vectors at its nodal points be represented as $\{u\}$ and $\{V\}$. Using shape functions N to interpolate, the displacement and the scalar potential at any point within the element can be obtained by $\{u(x)\} = [N^u]^T \{u\}$ and $V(x) = \{N^e\}^T \{V\}$, respectively. Here the superscripts u and e designate shape functions for displacement and potential, respectively. The elastic strains and electrical field at the nodal points are related to nodal displacement and potential vectors by $\{\varepsilon\}=[B^u]\{u\}$ and $\{E\}= -[B^e]\{V\}$, respectively, where $[B]=\text{grad} \cdot [N]$. The variational principle leads to the following equilibrium equations for the unknowns $\{u\}$ and $\{V\}$:

$$
\begin{pmatrix} [K] & [K^z] \\ [K^z]^T & [K^d] \end{pmatrix} \begin{Bmatrix} \{u\} \\ \{V\} \end{Bmatrix} = \begin{Bmatrix} \{F\} \\ \{L\} \end{Bmatrix} \tag{9}
$$

where $\{F\}$ is the nodal mechanical force vector, $\{L\}$ is the nodal charge flux vector; and the overall stiffness matrix can be decomposed into 3 submatrices: $[K]=\int_v [B^u]^T[C][B^u]\, dV$, $[K^z]=\int_v [B^u]^T[e][B^e]\, dV$ and $[K^d] = \int_v [B^e]^T[d][B^e]\, dV$, where $[C]$, $[e]$ and $[d]$ are the elastic, piezoelectric and dielectric constant tensors given by the constitutive laws. Since those constants are dependent on the current damage state, a damage evolutionary law is required. This is described in the following subsection.

Damage Evolution Law

Ideally, a damage evolutionary law to give damage increment per cycle of loading should be based on physics-based micromechanics modeling. For example, in the case of creep damage, Chuang, et. al. [21] have proposed a creep microcrack growth model that can be used as input to a macro-damage mechanics model to estimate creep life. The resulting lifetime expression consists of measurable physical constants, in addition to the external loading parameters. No empirical parameters or curve fitting methods are involved. In the case of fatigue damage, however, micromechanics theories are scarce, although some attempts in this area have been made[22]. Lacking a credible theory, investigators then rely on some type of energy-based empirical law. The rationale is that as energy is pumped into each element per cycle, fatigue damage must be monotonically increased in proportion to the energy release. The most popular functional expression for $d\omega/dN$ adopted is in the power-law form[23]:

$$
\frac{d\omega}{dN} = \alpha\, G^n \tag{10}
$$

where N is number of loading cycles, α is the proportionality constant to be determined by empirical method, the exponent constant n =6.1 is found to be applicable to a variety of materials[24], and G is the energy release due to damage: $G = (\partial U/\partial\omega)$, U is the total energy per element. It is interesting to observe that succesful cases have been reported based on this damage evolution law for a wide variety of materials and applications[23-24]. Hence, we propose using this law as well, extending the total energy to cover other energies: $U = U^E+U^d+U^z$, where U^E is the elastic strain energy defined by $U^E = \frac{1}{2}\{u\}^T[K]\{u\}$, U^d is the dielectric energy defined by $U^d = \frac{1}{2}\{V\}^T[K^d]\{V\}$ and U^z is the coupling energy given by $U^z = \frac{1}{4}[\{u\}^T[K^z]\{V\}+\{V\}^T[K^z]\{u\}]$.

FATIGUE LIFETIME PREDICTIONS

Having presented the constitutive laws involving damage with PZT ceramics and the finite element formulations, we are now in a position to construct a methodology for calculating reliably the service life of a PZT structure limited by fatigue. Figure 1 is a flow chart showing the finite element algorithm by which the fatigue life prediction of a PZT component can be made.

For a PZT structure with a given set of boundary and initial conditions on its environment and mechanical/electrical loading as schematically drawn in Fig. 1, the first step is to discretize the structure into finite elements by a proper mesh design. A preprocessor can accomplish this task. Once a finite element model is constructed, constitutive laws with the initial data on the coefficients as described previously are input to the program, and field parameters such as stress, strain, electrical potential and damage are solved. At this point in time, a question must be posed, namely, has the damage reached the critical level as defined by the material property? For example, does $\omega=1$? If not, the number of loading cycle is allowed to increase, and the damage evolution law as given by Eq. (10) is called in to compute the increase in damage. New damage-dependent constitutive laws

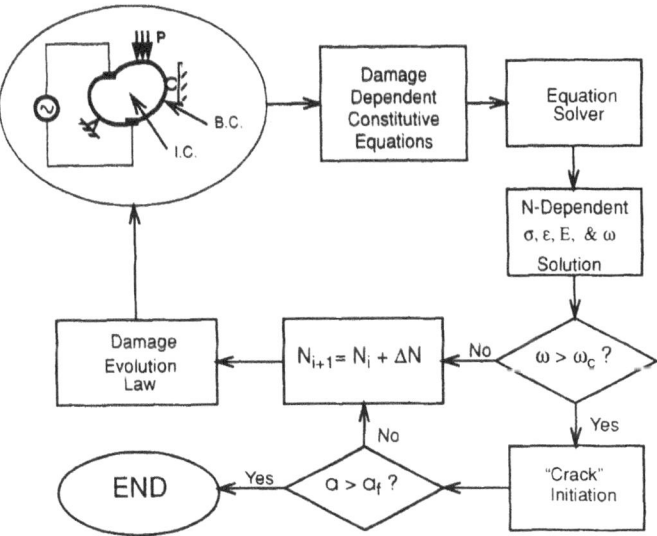

Figure 1. Finite element algorithm for fatigue life prediction of PZT

must be re-calculated using the new coefficients. New solutions on the field parameters must be solved, this process repeats until the damage level reaches the critical value. Then, a "crack" is initiated at this stage, and although physically dominated by crack growth, the procedure is allowed to continue until the linkage of cracks occurs that defines the failure and the computation stops. In this way the total number of loading cycles can be recorded to come up with the fatigue life. Based on this numerical iterative procedure, a computer code was developed in cooperation with the commercial code ANSYS. An example of predicting fatigue life of a 4-point bend PZT bar subjected to a variety of stress amplitudes at 20 Hz is given in the following section to demonstrate its application.

Case Study on a Four-Point Bend Beam

For the sake of demonstration, we select the simplest case of a PZT-8 four-point bend beam with dimensions L= 16 mm, B= 3 mm and H= 4 mm, subjected to only mechanical loading. The applied stress is in a sine wave form of the type $\sigma_a = \sigma_o + \Delta\sigma \sin(2\pi ft)$ generated by loading in the vertical direction y, where $\Delta\sigma$ is one-half of the stress amplitude, σ_o is fixed at a level (depending on maximum $\Delta\sigma$ applied) to insure that tension side remains tension and compression side remains always at compression, the frequency f is fixed at 20 Hz and t is time. A two-dimensional finite element model is constructed within the general purpose code ANSYS for one half of the beam due to the symmetry conditions. The model consists of 200 rectangular elements in the x-y plane (x-y-z corresponding to 1-2-3) where z-axis is parallel to the poling direction (see Figure 2). The input to the program includes stress amplitude $\Delta\sigma$ and the initial PZT-8 material constants in the constitutive equstions Eqs. (1-2), namely, the data as supplied by the vendor [25] for the versions of Eqs. (3-5) now read as :

$$[C] = \begin{pmatrix} 146.9 & 81.1 & 81.0 & 0 & 0 & 0 \\ . & 146.9 & 81.0 & 0 & 0 & 0 \\ . & . & 131.7 & 0 & 0 & 0 \\ . & . & . & 31.3 & 0 & 0 \\ . & . & . & . & 31.3 & 0 \\ . & . & . & . & . & 32.9 \end{pmatrix} [GPa]$$

$$[d] = 10^{-9} \begin{pmatrix} 11.42 & 0 & 0 \\ 0 & 11.42 & 0 \\ 0 & 0 & 8.85 \end{pmatrix} [C/(V \cdot m)]$$

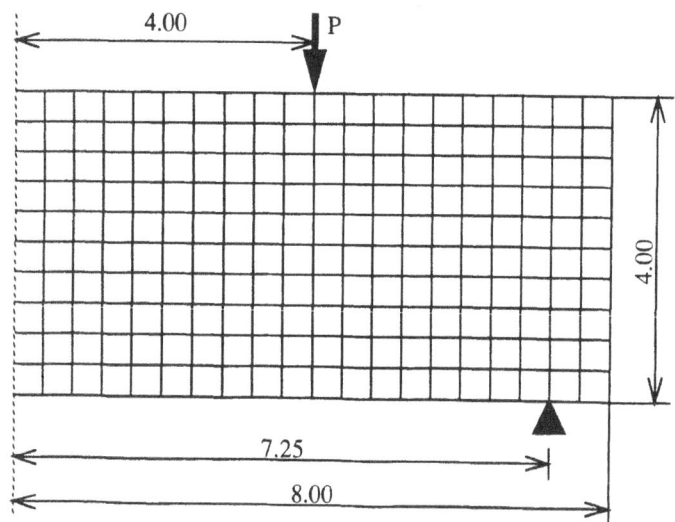

Figure 2.   Two dimensional finite element model for one-half of the beam (unit in mm).

$$[e]^T = \begin{pmatrix} 0 & 0 & 0 & 0 & 10.34 & 0 \\ 0 & 0 & 0 & 10.34 & 0 & 0 \\ -3.88 & -3.88 & 13.91 & 0 & 0 & 0 \end{pmatrix} \; [C/m^2]$$

An iterative scheme based on the algorithm described in the previous section was applied to solve the unknown stress, strain, potential and damage distributions within the half-beam as the number of loading cycles increases. The solutions of a specimen for the stress amplitude of 20 MPa (maximum stress 80 MPa) are reported as follows for two selected N: N=1,000 cycles representing the earlier stage of fatigue life, and N= 40,869 cycles representing a later stage of the service life.

Stress   Figure 3 shows the contour plots for the stress (σ_{xx}) distributions at two stages. When N=1,000 cycles, the maximum stress in tension occurs at the edge of the tension side with a value of 100 MPa, whereas the minimum stress (i.e., max. stress in compression) has a value of 170 MPa at the loading point. Notice that at the lower loading point, the stress is 50 MPa in compression, which results from the superposition of localized contact compressive stress and far-field tensile stress. As the number of cycles increases to N=40,869, the location of maximum stress shifts towards the mid-span somewhat, although maintaining at about the same magnitude. On the other hand, the location of minimum stress remains at the upper loading point, although the level reduces somewhat to about 162 MPa in compression. The stress at the lower loading point stays at the 50 MPa level.

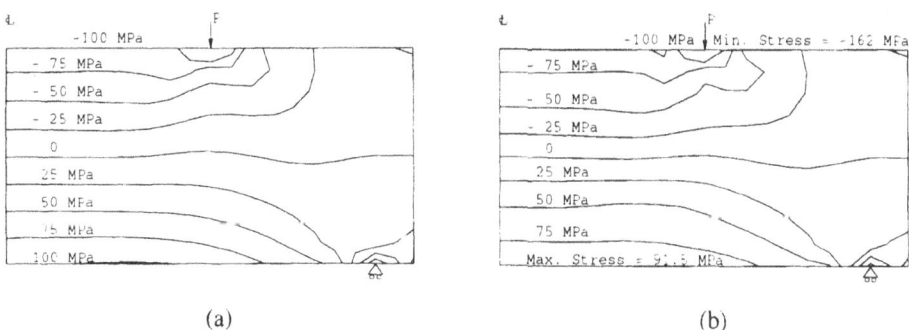

(a)   (b)

Figure 3.   Contour plots of stress fields for the half- beam. (a) at N=1000; (b) N=40869.

Strain   The solutions of strain fields are shown in Figure 4 in contour plots within the half-beam specimen. Understandably, the patterns of the strain contour plots are similar to that of stress solutions. At N=1,000, the maximum tensile strain occurs at about one-third of the half-span in the lower edge at approximately 0.1%. As shown in Fig. 4(b), when the number of cycles reaches N=40,869, the strain seems to saturate at about 0.2% in that region, together with its surrounding area. It is likely that damage accumulation increases the compliance of the material leading to increased anelastic tensile strains. On the other hand, the effect of damage in compressive strain is not significant. For example, the maximum compressive strain at the upper loading point is about 0.2% when N=1,000. It increase a little to 0.25% when N reaches to 40,869 cycles (Fig. 4b). Again, due to compressive contact, the lower contact point creates a lower tensile zone surrounded by a half-circle compressive ring area.

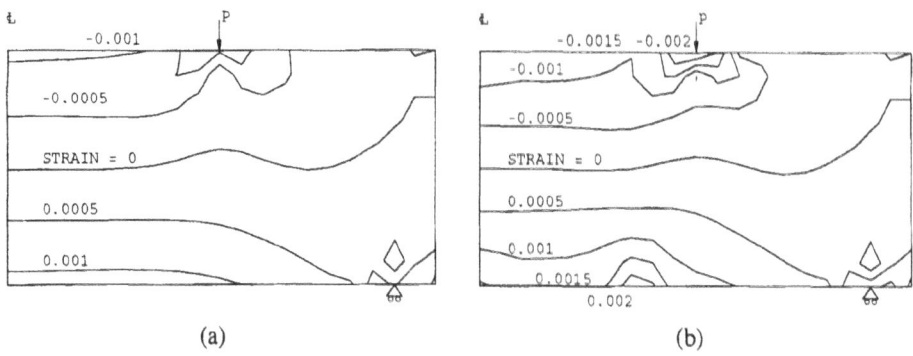

Figure 4. Contour plots of strain fields for the half-beam. (a) at N=1000; (b) N=40869.

Electric Potential   The solutions for the electric potential V (E=B•V) are presented in Figure 5 in terms of contour plots for the half-beam. Fig. 5a shows V-distribution at N=1,000 cycles. A maximum potential of 336 volts occurs at the upper loading point. On the other hand, a negative minimum potential of -207 volts present at the lower loading point. When N increases to 40,869 cycles (Fig. 5b), the maximum voltage at the upper loading point has increased to 499 volts, whereas the potential at the lower loading point lowers to -291 volts. Furthermore, it is observed that a severe potential gradient approaching the dielectric breakdown strength has developed at the upper loading point, indicating a possible spot for electrical failure (arcing).

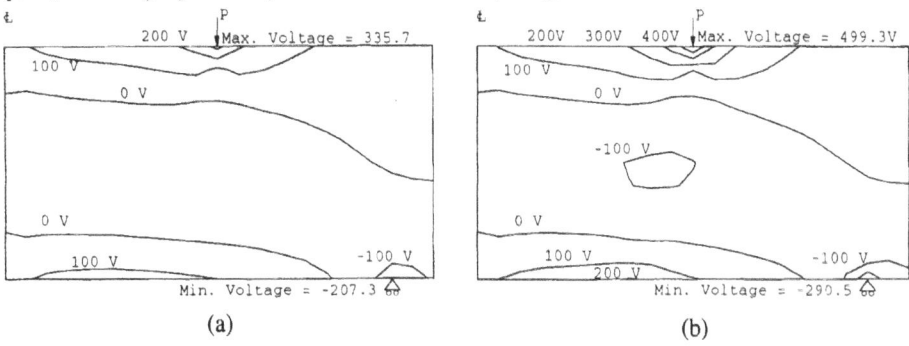

Figure 5. Contour plots of potential fields (V) for the half-beam. (a) at N=1000; (b) N=40869.

Damage State   The distributions of the damage parameter ω at a typical early and late stages in the fatigue life are given in Figure 6. As shown in Fig. 6a when N=1,000 cycles, the levels of damage everywhere are low (almost undamaged), except an enhanced value of 0.02 is observed at the upper and lower edges of the middle span. When the loading cycle increases to 40,869, we observe that high level (0.08) of damage develops at the loading points and at the one-third of the lower edge. Clearly those are the potential sites for crack nucleation. To have an in-depth understanding of the damage evolution, we plot in Figure 7 the damage values ω for the four critical elements located at the potential sites (see the inset of Fig. 7) as functions of N. It is seen that for the majority of the fatigue life, the value of ω for all elements increases only mildly. It is towards the end of the fatigue life, ω suddenly takes off and reaches the critical value (set at unity) almost at the same time. This kind of behavior allows us to estimate the total number of cycles to failure with ease. For example, it is estimated that this particular specimen will fail by linkage of two macrocracks from the bottom to the top at approximately 41,000 cycles of mechanical loads. In this way, fatigue life vs. stress amplitude can be predicted. A companion experimental program for the four-point bend fatigue tests on the PZT-8 was developed earlier and is described in the following section for comparison with the theoretical predictions.

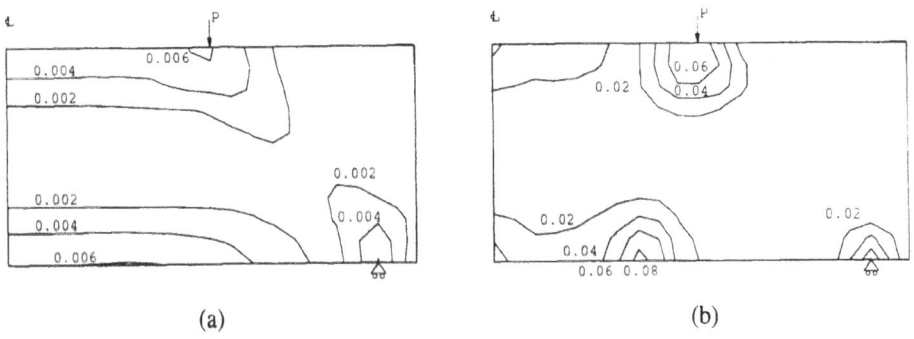

Figure 6. Contour plots of damage fields for the half-beam. (a) at N=1000; (b) N=40869.

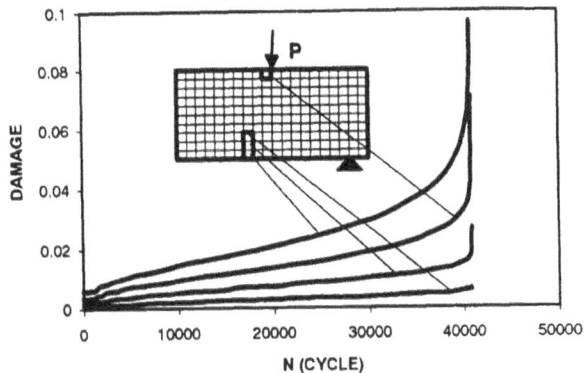

Figure 7. Damage evolution for the 4 critical elements as a function of N.

Experimental Procedure

Material   The material used in this experimental program was PZT-8 with a 300°C Curie Temperature[25]. Detailed description of the microstructure and physical properties were given elsewhere [17]. Here we only briefly summarize the relevant information. Essentially the microstructure of the as-received PZT-8 consisted of equiaxed grains of 1-2 μm in diameter, with second phases segregated at grain boundaries and triple point junctions. The material was received in the form of poled 6.5 mm x 12.4 mm x 70 mm with eletrodes attached. These billets were sectioned into 24 samples, each had dimensions 3x4x17 mm, by a diamond saw and the poling direction was maintained at the 3 mm long direction. The electrode surfaces were polished off and the edges of the specimens chamfered with 600 grit SiC paper.

Mechanical Loading   Poled samples were mounted in a semi-articulating fixture and tested, using an Instron 8502 testing machine[26], in a cyclic four-point bending configuration with the poling direction normal to the load train axis. To determine the stress limits for the cyclic loading conditions, failure strengths of 15 samples were measured and a failure stress of 101 MPa with a standard deviation of 7 MPa was determined. For the mechanical loading, a sinusoidal stress of $\Delta\sigma$ was superimposed upon a constant offset stress σ_0 of 61.5 MPa such that $\Delta\sigma \leq \sigma_0$. In this way, the tensile surface is assured to remain tensile always, and compressive surface remains under compression inside the beam throughout the fatigue test. The time-dependent stress at the tensile surface was: $\sigma_a = \sigma_o + \Delta\sigma \sin(2\pi ft)$, where $f = 20$ Hz., a condition exactly identical to the cases analyzed above. A test sample with a thermocouple attached was used to monitor the change, if any, in the sample temperature during mechanical cycling. Due to low cycle fatigue testing, no significant temperature rise was observed. Overall, a total of 23 fatigue life data were collected as a function of maximum tensile stress applied (or stress amplitude applied) and comparison with the finite element predictions is presented in the following section.

Comparison between Theory and Experiment

To compare the measured data on the fatigue life for PZT-8 at 20 Hz with that predicted by the continuum damage mechanics and finite element calculations, we plot stress amplitude versus number of cycles to failure, the so called S-N curves in Figure 8. Open circles are the 23 data points measured, whereas 5 solid square points are theoretical predictions made by the finite element methods, using the damage evolutionary law of Equation (8) with $\alpha=0.125$ and $n=6.1$. The three open triangular data represents unfailed specimens interrupted prematurely. It is seen that the life data scatter a lot. This is not surprising and quite a normal phenomenon, since any life data as reported in the literature, regardless of "in-use" conditions, are bound to scatter due to a perturbation of as-received conditions and the stochastic nature of materials microstructure and properties and "in-service" loading conditions. Nevertheless, even with a great uncertainty involved, we are still able to obtain fair agreement between the theory and experiment.

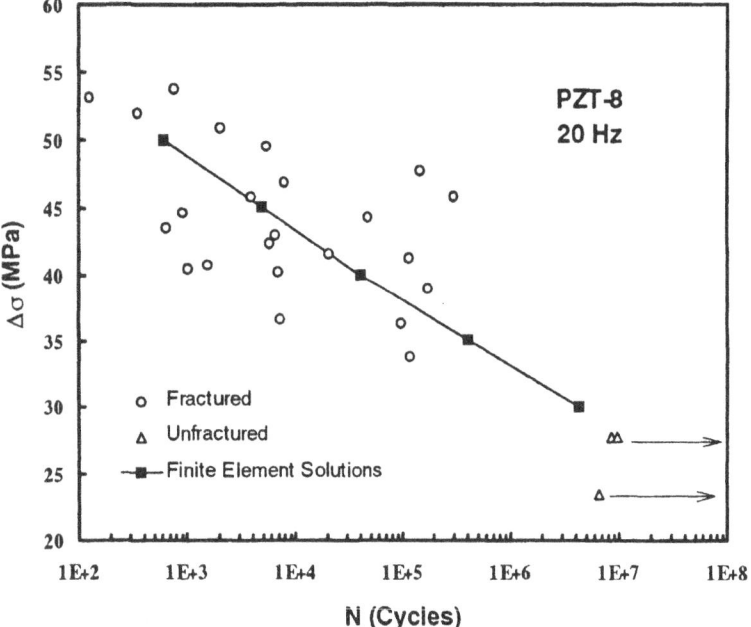

Figure 8.   Comparison of fatigue life on PZT-8 between theory and experiment.

SUMMARY

We have presented an innovative numerical algorithm using damage mechanics concept incorporated into the finite element techniques to predict fatigue life of a general piezoelectic experimentally obtained S-N curve using a sinusoidal wave type loading at a frequency of 20 Hz. It was shown that the theoretical predictions agree fairly well with the fatigue life data.

In the analysis, we assume a damage parameter ω exists as an internal variable in the context of irreversible thermodynamics. We leave it open the precise expression, although micrcracking and domain switching have been suggested as the dominating attributes [14-15]. The coefficients of the constitutive equations were then modified by ω, resulting in damage-dependent functionals, and are no longer constants. In this way, the property degradations or deteriorations can be predicted or monitored at each stage of service life.

Another important ingredient in the current theory is the adoption of a power-law function of energy release due to damage for the fatigue damage evolution law. Although the use of energy as a driving force for fatigue damage is somehow justifiable based on physics argument, the determination of the two constants involved in the evolution law remains empirical, and must rely on curve-fitting methods. It would be desirable to formulate the law from the micromechanics stand point so that the law consists of no curve-fitting parameters. Investigations along this lines have been attempted[22].

Recently, extension of the analysis to cover the whole beam was performed[27]. The results indicated that, in reality, a crack-like damaged element initiates first at either side of the beam (say, on the right hand side) at a later stage of fatigue life. Once initiated, the critical stresses on the other side (say, left-hand side) relax somewhat. But as N continues to increase, the stresses over there eventually had reached to a level high enough resulting in the initiation of the second "crack". Although the damage events are different in sequence and are essentially asymmetrical in character as opposed to the symmetric conditions assumed in the present analysis, the predicted fatigue life, however, was not altered significantly.

Finally for PZT subjected to cyclic electrical loading particularly at the resonance frequency, earlier work seemed to indicate energy conversion into heat, leading to temperature enhancement[16-17]. For pure mechanical loading, however, little temperature increase has been observed [14-16]. So it is alright to neglect thermal effects in the present analysis. Nevertheless, for general loading conditions involving a combined mechanical/electrical load, additional thermal analysis must be invoked to analyze the coupled effects induced by the temperature field.

ACKNOWLEDGMENT

It is a pleasure to thank Office of Naval Research, Solid Mechanics Program on Adaptive Quiet Structures with Active Materials, Dr. Roshdy S. Barsoum, for partial financial support under Interagency Contract No. N00014-94-F-0016. Helpful discussions with Drs. C. K. Chiang and Edwin R. Fuller, Jr. are hereby acknowledged.

REFERENCES

1. Q. Jiang, E.C. Subbarao and L.E. Cross, "Grain Size dependence of electric fatigue Behavior of Hot Pressed PLZT Ferroelectric Ceramics," **Acta Metall. Mater.**, 42 [11] 3687-94 (1994).

2. Q. Y. Jiang and L. E. Cross, "Effects of Porosity on Electric fatigue Behavior in PLZT and PZT Ferroelectric Ceramics," **J. Am. Ceram. Soc.** 28 [12] 4536-43 (1993).

3. V. Z. Parton, "Fracture Mechanics of Piezoelectric Materials," **Acta Astronautics,** 3 [2] 671-683 (1976).

4. W. F. J. Deeg, "The Analysis of Dislocation, Crack, and Inclusion Problems in Piezoelectric Solids," Ph. D. Thesis, Stanford University, Stanford, CA (1980).

5. H. Sosa, "Plane Problems in Piezoelectric Media with Defects," Int. J. Solids Structures, 28 [4] 491-505 (1991).

6. H. Sosa, "On the Fracture Mechanics of Piezoelectric Solids," *ibid*, 29 [10] 2613-22 (1992).

7. R. M. McMeeking, "Electrostrictive Stresses near Crack-like Flaws," ZAMP, 40 [9] 615-627 (1989).

8. Z. Suo, C.-M. Kuo, D. M. Barnett and J. R. Willis, "Fracture Mechanics for Piezoelectric Ceramics," J. Mech. Phys. Solids, 40 [4] 739-765 (1992).

9. Y. E. Pak, "Crack Extension Force in a Piezoelectric Material," J. Appl. Mech. 57 [9] 647-653 (1990).

10. R. M. McMeeking, "A J-Integral for the Analysis of Electrically Induced Mechanical Stress at Cracks in Elastic Dielectrics," Int. J. Engng. Sci, 28 [7] 605-613 (1990).

11. T.-Y. Zhang and J. E. Hack, "Mode III Cracks in Piezoelectric Materials," J. Appl. Phys. 71 [12] 5865-5870 (1992).

12. S. W. Freiman, L. Chuck, J. J. Mecholsky, D. L. Shellman and L. J. Storz, "Fracture Mechanisms in Lead Zirconate Titanate Ceramics," in *Fracture Mechanics of Ceramics*, Vol. 8, pp. 175-185, Eds. R. C. Bradt, A. G. Evans, D. P. H. Hasselman and F. F. Lange, Plenum Press, New York, (1986).

13. C. S. Lynch, L. Chen, W. Yang, Z. Suo and R. M. McMeeking, "Crack Growth in Ferroelectric Ceramics Driven by Cyclic Polarization Switching," to be published.

14. H. Cao, M. Y. He and A. G. Evans, "Electric Field-Induced Fatigue Crack Growth in Ferroelectric Ceramics," to be published.

15. S. R. Winzer, N. Shankar and A. P. Ritter, "Designing Cofired Multilayer Electrostrictive Actuators for Reliability," J. Am. Ceram. Soc. 72 [12] 2246-57 (1989).

16. G. S. White, A. S. Raynes, M. D. Vaudin and S. W. Freiman, "Fracture Behavior of Cyclically Loaded PZT," J. Am. Ceram. Soc. 77 [10] 2603-2608 (1994).

17. M. D. Hill, G. S. White, C.-S. Hwang and I. K. Lloyd, "Cyclic Damage in PZT," J. Am. Ceram. Soc. in press.

18. H. H. A. Kruger, "Stress Sensitivity of Piezoelectric Ceramics: Part 2 Heat Treatment," J. Acous. Soc. Amer. 43 [3] 576-582 (1981).

19. M. Kahn, Private Communication (1995).

20. J. W. Ju and T.-M. Chen, "Effective Elastic Moduli of Two-Dimensional Brittle Solids with Interacting Microcracks, Part II: Evolutionary Damage Models," J. Appl. Mech. 61 [6] 358-366 (1994).

21. T.-J. Chuang, J.-L. Chu and S. Lee, "Diffusive Crack Growth at a Bimaterial Interface," *ibid*, in press.

22. Y. Zhang and Q. Jiang, "Microcracking in Ferroelectric Ceramics, Part 2: Intergranular and Transgranular Cracking" ONR Report No.1, University of Nebraska, Lincoln, NE (1995).

23. D. C. Lo, D. H. Allen and C. E. Harris, "Modeling the Progressive Failure of Laminated Composites with Continuum Damage Mechanics," in *Fracture Mechanics: Twenty-Third Symposium*.ASTM STP 1189, Ed. R. Chona, pp. 680-695, ASTM, Philadelphia, PA (1993).

24. P. C. Chou, A. S. D. Wang and H. Miller, "Cumulative Damage Model for Advanced Composite Materials," AFWAL-TR-82-4083, Air Force Wright Aeronautical Laboratories, OH 1982.

25. Morgan Matroc Inc. -- Specification Booklet (1995).

26. Instron Corp. -- Specification 8502 (1994).

27. E. R. Fuller, Jr., Private Communication (1995).

CRACK GROWTH IN FERROELECTRIC CERAMICS
AND ACTUATORS UNDER MECHANICAL
AND ELECTRICAL LOADING

G. A. Schneider, H. Weitzing, and B. Zickgraf *

Technische Universität Hamburg-Harburg, Arbeitsbereich Technische
Keramik, Hamburg, Germany
* Max-Planck-Institut für Metallforschung, Pulvermetallurgisches
Laboratorium, Stuttgart, Germany

ABSTRACT

Fracture mechanical behaviour of ferroelectric components under combined mechanical and electrical loading is investigated in order to improve reliability and understand fatigue mechanisms of multilayer actuator stacks.

Model actuators were cyclically loaded by an electric field above the coercitive field strength. Five different mechanisms of crack growth and failure were detected that depend on the number of cycles the samples were subjected to.

Further, crack growth experiments were carried out using multilayer actuators. Well defined precracks were initiated on 2 sites of the actuator surface using Vickers indentations. Under cyclic electric loading with two different electric driving modes crack growth was measured. Significant crack growth was detected during the polarization process. Under unipolar cyclic electric loading slow crack growth occurs before cracks saturate after some hundred thousand cycles. Bipolar loading leads to crack growth some orders of magnitude faster than unipolar loading and causes a characteristic crack pattern.

A model for crack extension during poling is developed.

INTRODUCTION

Ferroelectric ceramics are widely used in micropositioning systems, as sensors or transducers[1,2]. Since ferroelectric actuators show faster response times and more precise displacement control than electromagnetic devices they are very suitable for replacing electromagnetic components in fuel injection systems[3]. Due to better combustion ceramic injection systems reduce fuel consumption and gas emissions significantly. However, ferroelectric ceramics are very brittle and susceptible to failure. Furthermore they show fatigue behaviour.

In contrast to structural ceramics ferroelectric ceramics are elongated during the application of an electric field. In order to explain the properties of ferroelectric ceramics Bariumtitanate is chosen as a prototype material.

A shift of the titanate ion along the c-axis of the tetragonal unit cell (see Fig. 1) causes the formation of electric dipoles, a process called spontaneous polarization[4]. Domains, i.e. regions of aligned dipoles, are separated by 90° and 180° domain walls as indicated in Fig. 2.

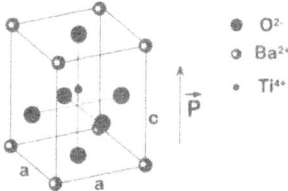

Figure 1. Unit cell of Bariumtitanate.

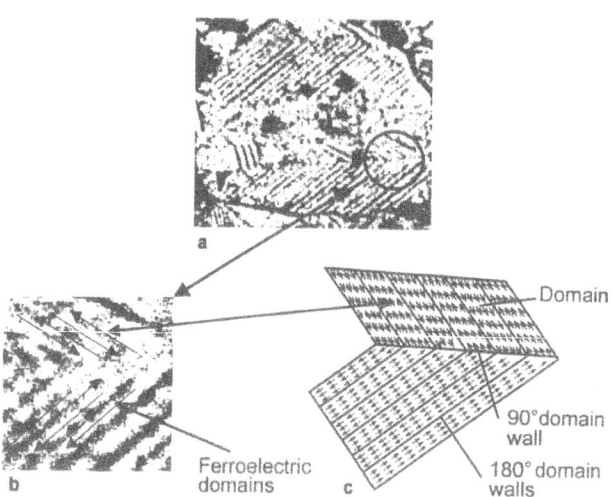

Figure 2. Domain pattern in BaTiO$_3$: (a) Polished and etched surface of unpoled BaTiO$_3$. (b) Ferroelectric domains. (c) Schematic diagram of 90° and 180° domain walls.

Before poling polycrystalline ceramics show no macroscopic displacement effect (see Fig. 3). After applying an electric field above the coercitive field most of the domains are orientated in the direction parallel to the field. Increasing the electric field leads to further elongation. Without applied field the polarization reaches a remanent point. Applying an electric field antiparallel to the poling direction causes 180° switching of the domain orientation and as a consequence a polarization in the opposite direction.

In this study the influence of the applied electric field on crack propagation in ferroelectric ceramics is investigated using multilayer stack actuators.

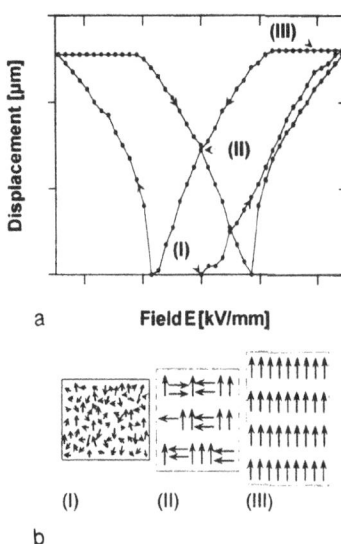

Figure 3. (a) Mechanical displacement of an actuator stack as a function of the electric field. (b) Relation between domain orientation and displacement.

EXPERIMENTS

In order to study the fatigue behaviour of PZT multilayer actuators sintered to a monolithic stack two type of actuator configurations were used both fabricated in the laboratories of the Corporate Research and Development of Siemens AG, Munich, Germany.

(I) Model actuator tests

Cyclic fatigue tests without mechanical loading were performed using model actuators with only two active layers of 200μm thickness, enclosed by two electrically inactive 800μm thick layers. The applied electric field of up to 2.5 kV/mm was above the coercitive field and it was switched bipolar at a frequency of 0.1Hz with a trapezoidal pulse shape. The actuator surface was polished and crack growth was observed using a long distance optical microscope.

(II) Test of Actuator Stacks

Actuator stacks (see Fig. 4) consisting of several hundreds of 80 μm thick ceramic layers separated with internal Ag/Pd-electrodes were polished with 1μm diamond paste and precracked using Vickers indentations of different loads between 3 and 20N. The precracks were initiated either in the PZT ceramic in between two electrodes or directly at the electrode ceramic interface as shown in Fig. 5. Crack lengths were measured using an optical microscope. The actuators were poled at twice the coercitive field. In the following fatigue testing was performed by cyclically loading the actuators employing driving voltages of two different shapes (see Fig. 6).

For unipolar tests a rectangular signal of 150V at 100Hz was applied and for bipolar tests a sinusoidal waveform of 150V at 50Hz was chosen. In addition to that the actuators were mechanically loaded by a spring load installed inside a metal frame. Crack growth due to poling and cyclic fatigue occurred and the influence of the polarization process and of cyclic electric loading on crack growth was investigated.

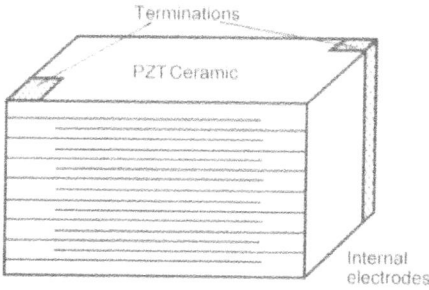

Figure 4. Multilayer actuator stack, fabricated by Corporate Research and Development of Siemens AG, Munich, Germany.

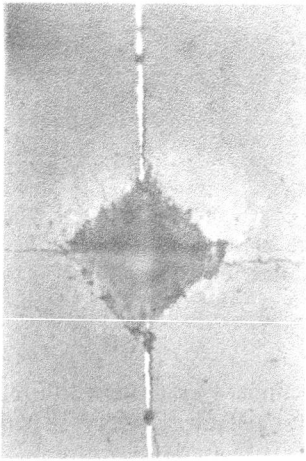 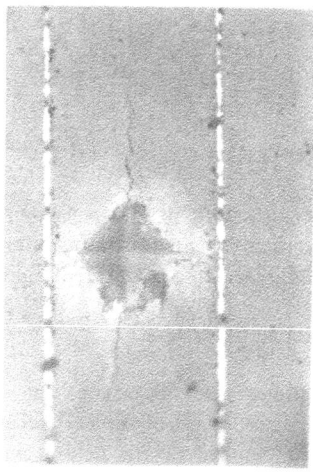

a) Internal Crack   b) Electrode delamination crack
Figure 5. Crack initiation by indentation method.

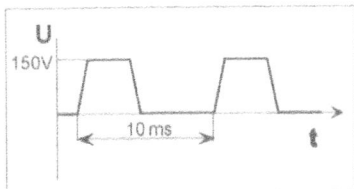 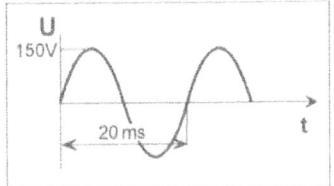

Figure 6. Shape of the driving voltages.

RESULTS

(I) Model actuator tests

When driving the model actuator at the very high field amplitude of 2.5kV/mm five different fracture mechanisms in four different stages of cycling occurred. After only one to ten cycles segmentation cracks developed which are due to contraction in the direction parallel to the electrodes and elongation in the perpendicular direction (see Fig. 7a). From 50 up to 500 cycles cracks propagated from the electrode tips to the opposite electrodes (see Fig. 7b). After 500-2000 cycles subcritical crack growth into the top and bottom cover layers takes place as well as electrode delamination (see Fig. 7c-d). Loaded more than 2000 cycles dielectric failure occurs probably initiated by electromigration on the crack path (see Fig. 7e)[5,6].

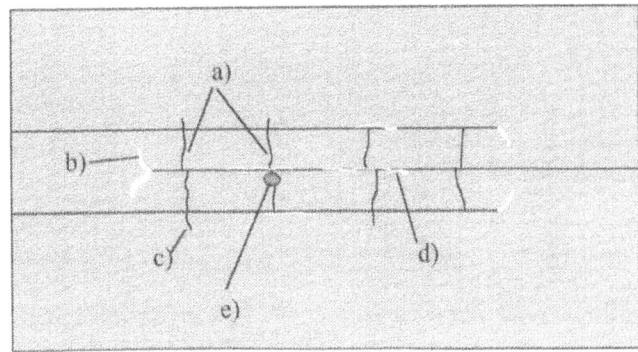

Figure 7. Crack configuration in model actuators:
    (a) Segmentation cracks (1-10 cycles) (b) Electrode tip cracks (50-500 cycles)
    (c) Subcritical crack growth inside the bottom and top layer (500-2000 cycles)
    (d) Electrode delamination (500-2000 cycles) (e) Dielectric failure

(II) Actuator stack experiments

At all sites investigated poling of the actuator stacks caused significant crack growth of 10-20% for the cracks orientated parallel to electrodes. Cracks perpendicular to the electrodes did not grow. Precracks on the electrodes which had caused delamination showed more pronounced crack growth than cracks inside the PZT ceramic.

During unipolar cyclic electric loading further crack growth of the cracks orientated parallel to the electrodes was detected. Crack extension rates were lower than 10^{-9}m/cycle and decreasing with increasing cycle number until a saturation point was reached where crack growth came to arrest (see Fig. 8 and 9). Internal cracks between the electrodes were growing slower than electrode cracks which had caused delamination and their final length was smaller. Furthermore crack extension rates depend on the indent load precracks are initiated with. With decreasing loads extension rates decrease, as well. Precracks initiated by small loads did not grow at all.

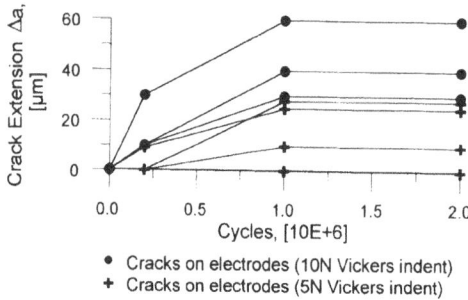

* Cracks on electrodes (10N Vickers indent)
+ Cracks on electrodes (5N Vickers indent)

Figure 8. Crack extension at the electrode ceramic interface (electrode crack) under unipolar cyclic electrical loading.

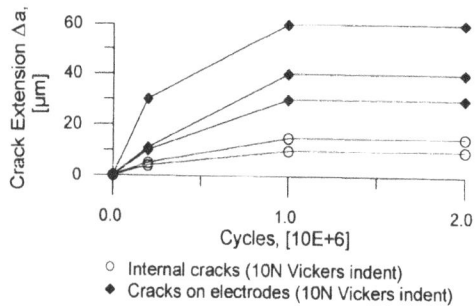

○ Internal cracks (10N Vickers indent)
◆ Cracks on electrodes (10N Vickers indent)

Figure 9. Comparison of crack extension under unipolar cyclic electrical loading between internal and electrode cracks.

Contrary to unipolar loading bipolar electric loading had a much stronger effect on crack growth (Fig. 10) and crack pattern (Fig. 11). Crack growth rates due to bipolar loading were higher than 10^{-7} m/cycle. After a few thousand cycles dielectric failure occurred. Bipolar loading results in a characteristic crack pattern where eight beams grow radially from the indentation.

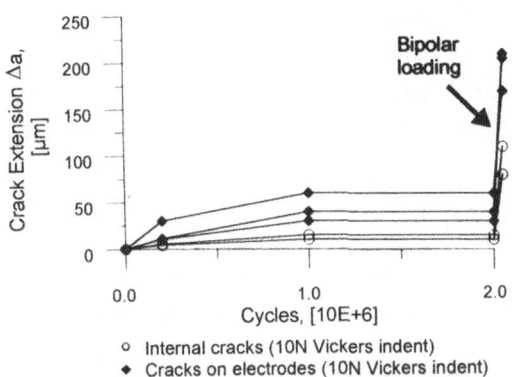

○ Internal cracks (10N Vickers indent)
◆ Cracks on electrodes (10N Vickers indent)

Figure 10. Accelerated crack extension under bipolar cyclic loading subsequent to unipolar loading.

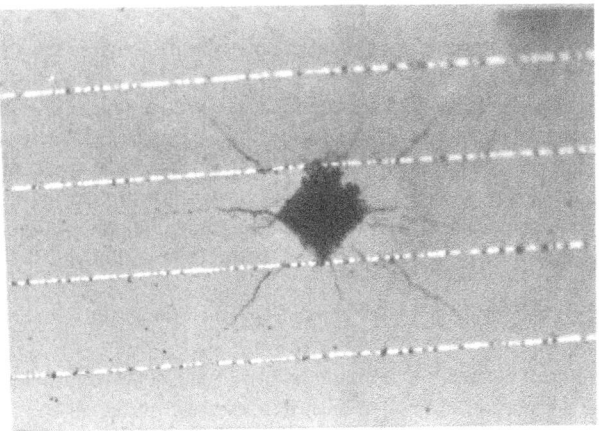

Figure 11. Typical crack pattern after bipolar cyclic electric loading

DISCUSSION

Most of the observed cracks in the model actuator are induced through mechanical constraints [5,6]. The early separation effects are the result of high tensile stresses parallel to the layer structure because of the clamping effects of the upper and lower inactive ceramic parts. These regions not subjected to an electrical field do not follow the contraction of the active layers perpendicular to the electric field direction and therefore high tensile stresses in the active layer arise. These stresses are also responsible for the subcritical crack growth in the top and bottom layer (Fig. 7c). Electrode tip cracks, especially the bifurcation of the crack propagating from the central electrode, have been observed in similar experiments [7,8]. Finite element calculations for a linear piezoelectric material[9] and a geometry similar to the one of the model actuator gave tensile stresses at the electrode tip as high as 100 MPa. For the calculation small signal values for the piezoelectric constants of PZT have been chosen. These calculated tensile stresses at the electrode tip are in the range of the strength values of PZT materials and therefore are high enough to initiate crack growth. A more detailed fracture mechanical calculation of the stress intensity factor for a small crack ahead of the electrode tip is also available[10,11]. This result, calculated under the assumption that the applied field E_{appl} approaches the saturation field E_s, allows to introduce a criterion for the critical ceramic layer thickness h_c [11]:

$$h_c = 8 \left(\frac{K_{Ic} E_s}{Y \gamma_s E_{appl}} \right)^2 \qquad (1)$$

Below h_c the model calculation predicts that no crack growth should occur. E_s denotes the electric field strength were the mechanical displacement and strain γ_s saturates (see Fig. 3a). Taking the Young's modulus Y to be 48 GPa[3] and $E_s = E_{appl}$ the critical layer thickness h_c would be app. 4,8mm for $K_{Ic} = 1$ MPam$^{1/2}$. Therefore no spontaneous stable or unstable crack growth originating from the electrode tips would be expected because the ceramic layer was 200μm thick. Due to the observed slow crack growth of these cracks it is therefore most probable that fatigue crack growth prevails. The occurrence of delamination cracks as the last stage of observed damage before dielectric breakdown might be explained as follows. If there are existing microcracks at the ceramic electrode interface the electric field at the crack tip is intensified leading to strain mismatches which act as the crack driving force.

For the observed crack extension under electric poling in the actuator stacks we propose the following simple model (see Fig. 12). Before poling a random domain orientation exists. An electric field above the coercitive field strength causes a reorientation of domains parallel to the field and thereby an elongation of the ceramic layer. Due to the electric shielding effect of the crack no domain reorientation takes place in the crack environment. This causes a mismatch in the strain distribution around the crack which results in tensile stresses at the crack surface. These stresses lead to crack extension during the polarization process.

A first order approximation of this effect can be performed as follows:

We assume

a) that the electric field is shielded in a spherical volume around the initial crack with radius a_0 as shown in Fig. 12b and

b) isotropic modulus of elasticity Y, Poisson ration ν and isotropic piezoelectric strains γ.

In analogy to the calculation of thermal mismatch stresses[12] in the absence of the crack the tensile stresses σ_{Piezo} inside the circular area due to the piezoelectric mismatch are homogeneous and can be expressed as:

$$\sigma_{Piezo} = \frac{2Y\gamma_s}{3(1-\nu)} \qquad (2)$$

The stress intensity factor K_{Piezo} due to these stresses can be calculated accordingly

$$K_{Piezo} = \frac{2}{\sqrt{\pi a}} \int_0^{a_0} \frac{r\sigma_{Piezo}}{\sqrt{a^2 - r^2}} dr \qquad (3)$$

where a represents half the crack length, r the radial crack tip coordinate.
In addition to this crack driving force the residual stresses of the Vickers indentation must be taken into account[13].

$$K_{Vickers} = \xi (Y/H)^{0.5} Pa^{-1.5} \qquad (4)$$

H : Indentation hardness
P : Indentation load
ξ : Geometric constant

The overall stress intensity factor K_I therefore is:

$$K_I = K_{Piezo} + K_{Vickers} \qquad (5)$$

The calculated stress intensity factors as a function of the crack length are shown in Fig. 13. The following values have been chosen for the calculation:

$\gamma_{coerc} = 10^{-3}$
$E = 48$ GPa
$\nu = 0.3$

In Fig. 13 it can be seen that the original crack extension due to the residual stresses of the Vickers indent produces cracks of a half length of approx. 85 μm. During poling additional stable crack extension of approx. 20μm is predicted in good agreement with the measured values. The K_{Ic} value in the unpoled state was measured by the indentation crack length method.
In our opinion the crack extension during unipolar and bipolar electric loading in the actuator stacks is a fatigue effect because static electrical loading of the actuator during one hour did not show any measurable crack extension. One hour loading corresponds roughly to the loading time of 10^6 cycles during which crack extension was observed. This result also implies that domain switching in the crack wake is not strong enough to produce strain differences which are sufficient for crack propagation.
The result for bipolar electrical loading is in agreement with measurements at PLZT[14]. In this work 20N Vickers indents were used to produce starter cracks in slender beams (2mm x 3mm x 30mm) parallel to the electrodes. The specimen were loaded only electrically with a bipolar electrical field. For an applied electrical field twice of the coercitive field crack extension rates of app. 5×10^{-8} were measured. In addition the crack extension rate was constant for every cycle. It was also found in this work that electric fields below the coercitive field did only result in a minor amount of crack growth of app. 50μm and then no further crack propagation was observed. This result corresponds to our findings during unipolar loading. The authors argue that this crack growth is driven by the residual stress field of the indent and the so called stress intensification of an „impermeable crack". In our opinion this effect should not assist crack growth but hinder crack growth[15].

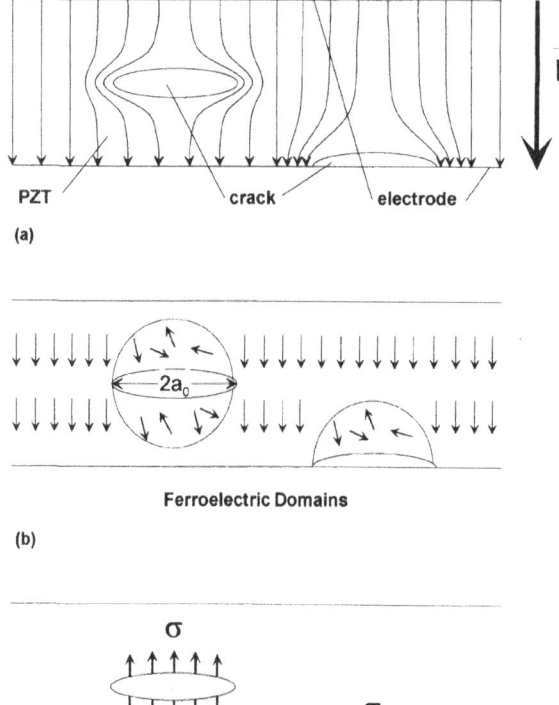

PZT   **crack**   **electrode**

(a)

Ferroelectric Domains

(b)

(c)

Figure 12. Model for crack extension under electric poling
      (a) Electric field distribution
      (b) Domain orientation
      (c) Crack extension forces

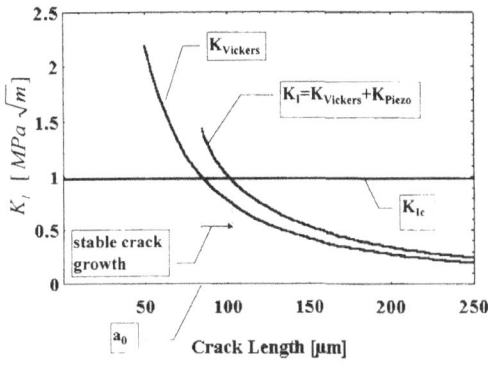

Figure 13. Model calculations for crack extension during poling (for explanation see text).

An explanation of the fatigue effects seems to be possible if we distinguish between
case I)   bipolar loading below the coercitive field strength or unipolar loading and
case II)   bipolar loading above the coercitive field strength.
The difference between the two cases is that in case I only 90° switching of the polar direction of the unit cell is occurring whereas 180° switching happens in case II. A 90° polar rotation results in enormous strains of up to 1% in BaTiO$_3$ ((c-a)/a=0.01, a,c: lattice constants). A 180° polar rotation does not result in any strain. Therefore 90° switching produces high strains and stresses which are most likely responsible for the local fatigue in the ceramic material. That case II loading produces a much more pronounced fatigue crack growth than case I loading is related to the fact that during the complete field reversal the unit cells switch two times by 90° and not at once 180°. Additionally in case II almost all unit cells participate at the switching whereas in case I only a smaller percentage of the unit cells switch. Nevertheless it is not clear to us why the difference in crack growth between the two cases is so tremendous. All the measurements indicate that case I electrical loading must be assisted by other tensile stress fields in order to initiate crack growth. Because the indent stress field declines with a$^{-3/2}$ crack growth stops after a certain extension because K$_{vickers}$ gets too small. This explanation is also supported by the fact that crack growth was not measurable by using smaller indent loads.

CONCLUSIONS

From the results of the present study the following conclusions have been obtained.
- Crack extension of precracks initiated by indentation was observed under unipolar and bipolar loading in PZT actuators.
- Poling of actuators caused significant crack growth.
- Delamination cracks at the electrodes are more critical than cracks inside the PZT material.
- Bipolar loading leads to crack growth some orders of magnitude higher than unipolar loading. Unipolar loading reached a saturation point under the described conditions.
- A threshold of minimum crack length for crack propagation may exist

- The found arresting of crack extension after some 10^6 cycles is an important result which opens up the way to long term endurance of piezoelectric actuator stacks in the most practical application of unipolar driving mode.

ACKNOWLEDGMENT

The authors thank Dr. K Kempter and Dr. K. Lubitz from Corporate Research and Development of Siemens AG, Munich, Germany, for making possible the fabrication of the actuator samples and fruitful discussions.

REFERENCES

1. S.L Swartz, Topics in Electronic Ceramics, IEEE Trans. Electrical Insulation 25 [5], 935 (1990)
2. R.Newnham, G.R. Ruschau, Smart Electroceramics, J. Am. Cer. Soc. 74[3] 463 (1991)
3. A. Wolff, D. Cramer, H. Hellebrand, I. Probst, K. Lubitz, Optical Two Channel Elongation Measurement of PZT Piezoelectric Multilayer Stack Actuators, in: 9th IEEE, International Symposium on the Application of Ferroelectrics, R.K. Pandey, M. Liu, A. Safari, eds., 755 (1994)
4. A.J. Moulson, J.M. Herbert „Electroceramics," Chapman and Hall, London, New York, Tokyo, Melbourne, Madras, 1990
5. G.A. Schneider, A. Rostek, B. Zickgraf, F. Aldinger, Crack Growth in Ferroelectric Ceramics under Mechanical and Electrical Loading, in: Electroceramics IV, Volume II, R. Waser et al., eds., Augustinus Buchhandlung, Aachen, 1211 (1994)
6. B. Zickgraf, Ermüdungsverhalten von Multilayer-Aktoren aus Piezokeramik, Thesis, Universität Stuttgart, Germany, 1995
7. A. Furuta, K Uchino, Dynamic Observation of Crack Propagation in Piezoelectric Multilayer Actuators, J. Am. Cer. Soc. 76: 1615 (1993)
8. H. Aburatani, S. harada, K. Uchino, A. Furuta, Destruction Mechanism of Ceramic Multilayer Actuators, Japanese J. Appl. Phys. 33: 3091 (1994)
9. ᴣ. Takahashi, A. Ochi, M. Yonezawa, T. Yano, T. Hamatsuki, I. Fukui, Internal Electrode Piezoelectric Actuator, Jpn. J. Appl. Phys. 22[Suppl. 22-2]: 157 (1983)
10. T.H. Hao, X. Gong, Z. Suo, Fracture Mechanics for the Design of Ceramic Multilayer Actuators, J. Mech. Phys. Solids. In Press
11. X. Gong, Z. Suo, Reliability of Ceramic Multilayer Actuators: A Nonlinear Finite Element Simulation, Submitted to J. Mech. Phys. Solids
12. J. Selsing, Internal Stresses in Ceramics, J. Am. Cer. Soc. 44: 419 (1961)
13. G.R. Anstis, P. Chantikul, B.R. Lawn, D.B. Marshall, A Critical Evaluation of Indentation Techniques for Measuring Fracture Toughness: I, Direct Crack Measurements, J. Am. Cer. Soc., 64[9]: 533 (1981)
14. H. Cao, A.G. Evans, Electric-Field-Induced Fatigue Crack Growth in Piezoelectrics, J.Am. Cer. Soc. 77[7]: 783 (1994)
15. Z. Suo, Mechanics Concepts for Failure in Ferroelectric Ceramics, in Smart Structures and Materials, Am. Soc. Mech. Engin. 24:1 (1991)

EFFECTS OF RESIDUAL STRESS ON FRACTURE TOUGHNESS AND SUBCRITICAL CRACK GROWTH OF INDENTED CRACKS IN VARIOUS GLASSES

M.Yoda, N.Ogawa and K.Ono

College of Engineering, Nihon University, Koriyama 963, Japan

ABSTRACT

Fracture and subcritical crack growth were investigated using indented cracks in soda-lime glass, aluminosilicate glass and borosilicate glass. The effect of residual stress was experimentally determined by obtaining the fracture toughnesses for cracks of various lengths. Extended cracks of various lengths were produced at the same indentation load. From the results, dimensionless indenter-material constants χ_r were determined for the surface crack and depth crack of semi-elliptical shape. Subcritical crack growth data for the as-indented short cracks show anomalous behavior, i.e. negative dependence of crack velocity on the stress intensity factor during small crack growth. This decrease in the crack velocity for the as-indented short crack can be explained by taking account of the residual stress.

INTRODUCTION

The interest in the as-indented crack is that of elastic/plastic contact, produced in sharp-point indentation, in which an irreversible process plays a major role in the crack evolution. Most significantly the irreversible component of the field gives rise to a residual opening force on the newly developed crack[1]. The determination of fracture toughness was made from direct measurements of the crack traces on the indented surface using the indentation fracture due to the residual stress[2] and toughness values were obtained indirectly from the strength of indented flexural test specimens[3]. Evidence has been provided that as-indented short cracks may behave differently in fatigue and may even have an anomalous or inverse subcritical growth velocity dependence on the applied stress intensity factor[4,5]. These experimental results indicate that the residual component of the indentation stress field is an important factor in fracture toughness measurements and subcritical crack growth. The stress intensity factor due to residual stress decreases with increasing crack growth at the same indentation load.Therefore, the effect of residual stress can be quantitatively determined by obtaining applied fracture toughness of as-intended crack and extended cracks of various sizes at the same indentation load.

In this study, the effects of the residual stress on fracture toughness and subcritical crack growth in soda-lime glass, aluminosilicate glass and borosilicate glass were determined using as-indented short cracks and longer cracks produced under a certain bend stress during the same indentation load.

EXPERIMENTAL PROCEDURE

The materials used in the experiment were soda-lime glass, aluminosilicate glass and borosilicate grass. The specimen geometry was 2 by 10 by 100mm. Vickers indent at the indentation load P=9.8N was placed on the tensile surface of the specimen. The as-indented short cracks were grown by subcritical crack growth under a certain bend stress in water, as shown in Fig.1(b). The specimens with various crack lengths were loaded to unstable fracture in four-point bend at a high loading rate so that little or no subcritical crack growth occurred in air during loading, as shown in Fig.1(a). The fracture surface was observed after fracture and the final surface crack length c and the corresponding depth a were measured from the arrest lines.

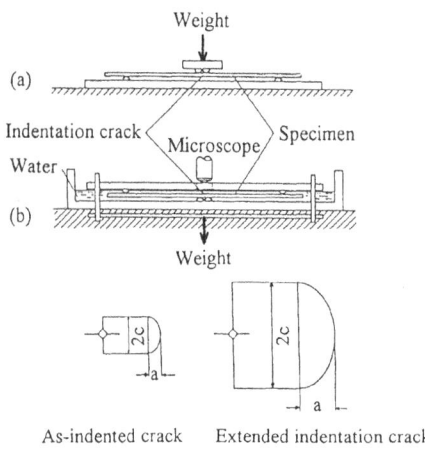

Figure 1. Four-point bending fixture

To determine subcritical crack growth characteristics, the specimens were subjected to sustained bend stress in water using the as-indented short cracks and the extended long cracks. Unloading was periodically repeated so that the arrest lines could appear during crack growth, which marked the crack front in the depth direction. From these arrest lines, the crack depth Δa corresponding to each measurement of surface crack growth Δc was measured after fracture.

EXPERIMENTAL RESULTS

The applied fracture toughnesses were calculated for the surface crack ($\phi = 0$) and at the maximum depth ($\phi = 90°$) of the semi-elliptical crack by[6]

$$K_I = H \sigma \sqrt{(\pi a / Q)} \cdot F(a/h, a/c, a/b, \phi) \tag{1}$$

where σ =bending stress, H=boundary correction factor, h=plate thickness, b=half plate width, Q=shape factor for elliptical crack and ϕ the parametric angle of the ellipse. Figure2 shows applied fracture toughnesses of soda-lime glass for the surface and depth cracks, respectively. The fracture toughnesses at both sides increase with increasing crack length and indicate upper-bound saturation fracture toughness. The reduction in fracture toughness at short cracks is due to the residual stress. Thus, fracture toughness shows little effect of residual stress at longer crack lengths. In addition to the applied stress intensity factor (K_{Iapp}), the stress intensity due to the residual stress (K_{Ires}) should be considered. Thus, the effective stress intensity factor (K_{Ieff}) is in the form

$$K_{Ieff} = K_{Iapp} + K_{Ires} \tag{2}$$

$$K_{Ires} = \chi_r P/(c \text{ or } a)^{3/2} \tag{3}$$

where χ_r is a dimensionless indenter-material constant depending on the surface crack or the depth crack. Therefore, the K_{Ieff} value gives a true fracture toughness in fracture toughness tests. The fracture toughness data for the surface crack and the depth crack are plotted against $c^{-3/2}$ and $a^{-3/2}$ for soda-lime glass, in Fig.3.
The solid lines are the linear least-squares fits of eq.(2) and are used to obtain the true

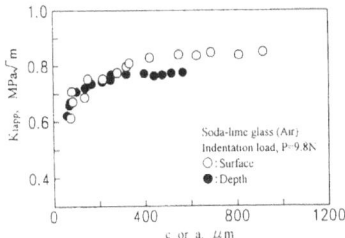

Figure 2. Variation of fracture toughness with c and a for soda-lime glass

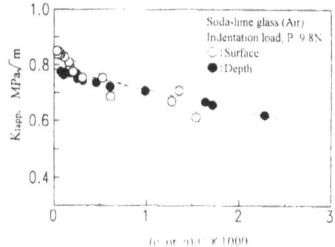

Figure 3. Plot of K_{Iapp} as a function of $c^{-3/2}$ and $a^{-3/2}$ for soda-lime glass

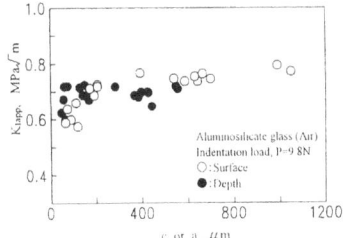

Figure 4. Variation of fracture toughness with c and a for aluminosilicate glass

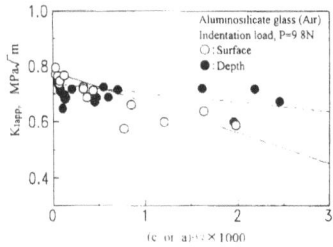

Figure 5. Plot of K_{Iapp} as a function of $c^{-3/2}$ and $a^{-3/2}$ for aluminosilicate glass

Table 1. K_{IC} and χ_r for each material

material	surface		depth	
	χ_r	K_{IC}	χ_r	K_{IC}
soda-lime glass	0.022	0.85	0.014	0.78
aluminosilicate glass	0.011	0.80	0.009	0.75
borosilicate glass	0.028	0.84	0.008	0.82

fracture toughness free from the residual stress and the value of χ_r. Similar experimental data are shown in Figs 4 and 5 for aluminosilicate glass.

Table 1 shows the true fracture toughnesses K_{IC} and the value of χ_r in the surface and the depth directions for each material. The fracture toughnesses free from the residual stress determined from the surface crack and the depth crack are in agreement with those of long cracks measured by conventional methods. This method is one method for evaluating the residual stress effect of the indented cracks. In the short crack regime, the linearity is not good since the lateral crack may prevent the surface crack from growing in the well-developed semi-elliptical shape.

Subcritical growth characteristics of the as-indented short cracks were compared with those of the extended cracks of about 5 times the length of the as-indented short cracks Figures 6 and 7 show the subcritical crack velocity V versus the applied stress intensity factor K_{Iapp} for the as-indented short cracks and the extended long cracks in soda-lime glass, respectively. The crack growth characteristics were similar in the surface direction and the

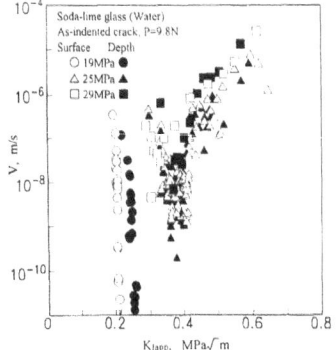

Figure 6. V vs. K_{Iapp} for as-indented crack in soda-lime glass

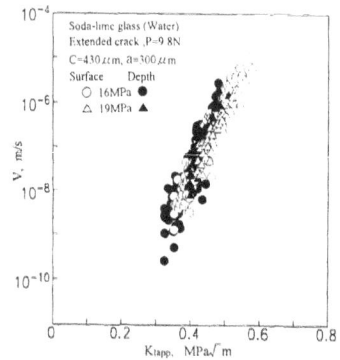

Figure 7. V vs. K_{Iapp} for extended crack in soda-lime glass

depth direction. The V-K_{Iapp} curves for the as-intended short cracks indicate the initial decrease in V with K_{Iapp}, reaching a minimum, and thereafter the increase with K_{Iapp} as in Fig.6, while the crack velocity for the extended long cracks with little residual stress effect has a normal dependence on K_{Iapp}, i.e. it increases with increasing K_{Iapp} from the initial crack growth as in Fig.7. The V-K_{Iapp} curves for the short cracks tend to shift to those for the long cracks as the crack growth increases. Figures 8 and 9 show the V-K_{Iapp} curves for as-indented short cracks and extended long cracks in aluminosilicate glass. The V-K_{Iapp} curves for as-indented and extended cracks have similar tendency to those for soda-lime glass, although the slope of the curves is a little steeper than those for soda-lime glass. The crack growth was sometimes arrested at a small stress for as-indented cracks in Figs 6 and 8 even when the crack growth initially occurred partly due to the effect of residual stress. Initial decrease in V with K_{Iapp} can be explained by taking account of the residual stress as described in fracture toughness measurements.

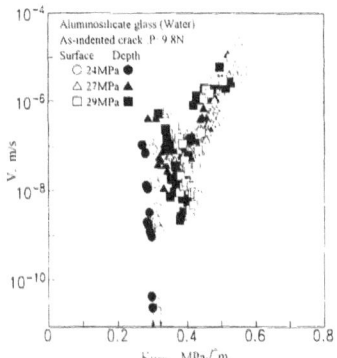

Figure 8. V vs. K_{Iapp} for as-intdented crack in aluminosilicate glass

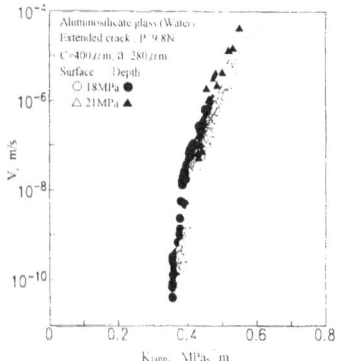

Figure 9. V vs. K_{Iapp} for extended crack in aluminosilicate glass

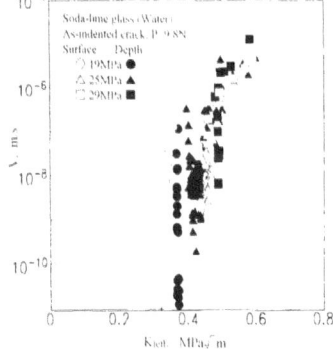

Figure 10. V vs. K_{Ieff} for soda-lime glass

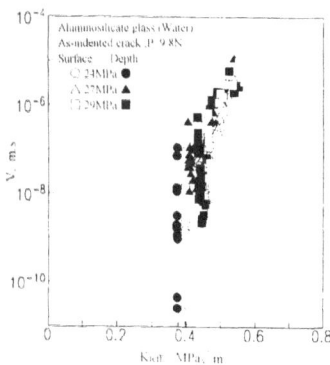

Figure 11. V vs. K_{Ieff} for aluminosilicate glass

The crack velocity V was again plotted against the effective stress intensity factor K_{Ieff} for as-indented cracks in soda-lime glass and aluminosilicate glass, in Figs 10 and 11, respectively. The normal crack growth behavior was observed for as-indented cracks in each material, showing the positive dependence of V. The position and slopes of the V-K_{Ieff} curves for as-indented curves are almost the same as those for extended cracks for each material. The experimental results for borosilicate glass have a similar tendency to those for soda-lime glass and aluminosilicate glass.

CONCLUSIONS

Fracture and subcritical crack growth characteristics were studied using the as-indented short cracks and the extended long cracks in soda-lime glass, aluminosilicate glass and borosilicate glass. From the fracture toughness measurements for various crack lengths at the same indentation load, the component of residual stress was determined experimentally for the surface crack and the depth crack. The subcritical crack growth date indicate that the as-indented short cracks have the negative crack growth dependence on the stress intensity. The anomalous crack growth behavior can be explained using the effective stress intensity factor taking account of the residual stress component determined from the fracture toughness tests.

Acknowledgments - The authors thank A.Kubota, N.Ikeda and T.Higuti for experimental assistance.

REFERENCES

1. D.B.Marshall and B.R.Lawn, Residual stress effects in sharp contact cracking: I.J.Mater.Sci.14(8), 2001-2012

2. D.R.Anstis, P.Chantikul, B.R.Lawn, and D.B.Marshall, A critical evaluation of indentation techniques for measuring fracture toughness:I, direct crack measurements. J.Am.Ceram.Soc.64(9), 533-538(1981)

3. P.Chantikul, G.R.Anstis, B.R.Lawn, and D.B.Marshall, A critical evaluation of indentation techniques for measuring fracture toughness: II , strength method. J.Am.Ceram.Soc.64(9), 539-543(1981)

4. M.Yoda, Subcritical crack growth characteristics on compact type specimens and indentation cracks in glass. J.Engng Mater. Technol.111, 399-403(1989).

5. M.Yoda and M.Inaba, Fracture and subcritical growth characteristics of indented cracks in various glasses, International Conference Ceramic Processing Science and Technology, 1994, Friedrichshafen, Germany.

6. J.C.Newman and I.S.Raju, An empirical stress-intensity factor equation for the surface crack. Engng Fracture Mech.15,185-192(1981)

DETERMINATION OF THRESHOLD STRESS INTENSITY FACTOR FOR SUB-CRITICAL CRACK GROWTH IN CERAMIC MATERIALS BY INTERRUPTED STATIC FATIGUE TEST

Vincenzo M. Sglavo,[1] David J. Green,[2] Steven W. Martz,[2] and Richard E. Tressler[2]

[1] Dipartimento di Ingegneria dei Materiali
Università di Trento, Via Mesiano 77, I-38050 Trento, Italy
[2] Department of Materials Science and Engineering
The Pennsylvania State University, University Park
16802 Pennsylvania

ABSTRACT

The interrupted static fatigue (ISF) test is a useful technique for the evaluation of the threshold for sub-critical crack growth, K_{th}, in ceramic materials. From a theoretical analysis of this technique, different procedures for the evaluation of K_{th} have been suggested. In the first approach the stress intensity factor corresponding to 50% of failures occurring during the stress hold is used as an estimate of the threshold. In the second approach, a stress level is applied in order to obtain some failures during the static hold, which then allows the determination of the "interrupted" strength distribution. The stress intensity factor corresponding to the weakest sample that survives the hold is calculated as the estimate of K_{th}. The value of these estimates is then re-determined as a function of the hold time until it becomes invariant. In some cases, a third approach can be used if the sub-critical crack growth parameters are known. In this approach, the value of the stress intensity factor corresponding to the strength drop-off associated only to the sub-critical growth phenomenon is calculated and compared to the K_{th} estimate.

The first two procedures outlined above were utilized for the experimental determination of K_{th} in soda-lime-silica glass in water environment (room temperature) and a value of ~ 0.15 MPa$\sqrt{m}$ was calculated for holding times of 20 days. ISF tests were also performed on c-axis sapphire fibers at 1200 and 1400°C with a hold time of 10 hours. The third approach was used and values of K_{th} equal to 0.68 and 0.66 MPa$\sqrt{m}$ were obtained at 1200 and 1400°C, respectively.

INTRODUCTION

The phenomenon known as sub-critical crack growth constitutes one of the limiting factors in the use of ceramic materials in structural applications. In fact, this fatigue behaviour is responsible for the strength dependence on time and environmental conditions. Ceramics loaded slowly or forced to support a load for a long time are found relatively weak; conversely, they are stronger if loaded rapidly. Therefore, the design of structural components constituted of materials which exhibit fatigue behaviour must proceed by the estimation of a design stress, which corresponds to a required minimum failure time.

Some studies on sub-critical crack growth in glasses and ceramics have shown that the crack velocity seems to tend to zero for some particular value of the applied stress intensity factor.[1-8] This has been termed fatigue limit or threshold stress intensity factor, K_{th}. From an engineering point of view, the existence of a fatigue limit is extremely desirable, as it allows the definition of an applied stress below which delayed failure does not occur unless a new flaw distribution is introduced during service.

In spite of the importance of the fatigue limit, no comprehensive test methodologies have been presented for its determination. Static fatigue tests and crack propagation measurements have been performed for the evaluation of the threshold for fatigue but extremely long test duration often prevents an accurate determination of K_{th}.[3,5-7,9-13] The interrupted static fatigue (ISF) test has been proposed as an alternative for the measurement of the fatigue limit and has been applied to various ceramics, including silicon carbide, silicon nitride, alumina and glass. Nevertheless, the lack of a solid theoretical background sometimes made the interpretation of the results difficult.[14-19]

In this paper, a previous theoretical analysis[20] of the interrupted static fatigue test is reviewed. This analysis allowed various test methodologies for the ISF test to be proposed[20]. These test procedures have applied previously to the evaluation of the fatigue limit in soda-lime-silica glass in water environment[21] and in sapphire fibers at 1200 and 1400°C,[22] the results of which are summarized in the current paper.

THEORETICAL BACKGROUND

During the ISF test, a sample is subjected to a constant load for a certain time in the active environment. The fracture stress is then measured, unless failure occurs during the static hold. In any case, it is reasonable to imagine that all flaws subjected to tensile stress within the specimen undergo some sub-critical growth. In order to define the strength degradation that occurs during the constant stress hold, the empirical power function of the stress intensity factor, K, is used to describe the sub-critical crack growth behaviour:

$$v = v_0 \left(\frac{K}{K_c} \right)^n \qquad \text{if } K > K_{th} \qquad (1a)$$

$$v = 0 \qquad \text{if } K \le K_{th} \qquad (1b)$$

where v is the crack velocity, K_c is the fracture toughness, v_0 and n are constants which depend of the material and the environment.

When a constant stress, σ_h, is applied for a time t_1 and $K > K_{th}$, the crack size increases from the initial value c_0 to a final value c_1. Correspondingly, the strength decreases from S_0 to S_1. The evaluation of the final strength, S_1, can be performed on the basis of the definition of the stress intensity factor associated with the crack. The typical relationship between K, the applied stress, σ, and the crack length, c, can be considered:

$$K = \sigma \psi \sqrt{c} \qquad (2)$$

where ψ is a geometric factor that includes the effect of flaw size and loading geometry. Substitution of Eq. (2) into Eq. (1), with $\sigma = \sigma_h$, and subsequent integration gives, for $K > K_{th}$:[20]

$$S_1 = \left(S_0^{n-2} - \frac{\sigma_h^n t_1}{B} \right)^{1/n-2} \qquad (3)$$

where $B = \left[2 K_c^2 / \left((n-2) \psi^2 v_0 \right) \right]$. It is clear that, if $K \le K_{th}$, $S_1 - S_0$. Therefore, Eq. (3) defines the strength degradation as a function of the applied stress and time. If the variable is the applied stress and t_1 is a constant, the final strength, S_1, becomes zero for $\sigma_h > \sigma^* = (B S_0^{n-2} / t_1)^{1/n}$.

Equation (3) can also be expressed in terms of the stress intensity factor, K_h, applied at the start of the static hold. As

$$\sigma_h = S_0 \frac{K_h}{K_c} \qquad (4)$$

when $K > K_{th}$, Eq. (3) becomes:[20]

$$S_1 = S_0 \left(1 - \frac{S_0^2 K_h^n t_1}{K_c^n B} \right)^{1/n-2} \qquad (5)$$

In this case, $S_1 = 0$ when $K_h > K^* = K_c [B / (S_0^2 t_1)]^{1/n}$.

Table 1. Fatigue parameters of sapphire fibers at 1200 and 1400°C. The standard deviation is shown in parentheses.

T (°C)	1200	1400
n	13.8 (0.8)	16.0 (1.0)
B (MPa^2 s)	69518 (28402)	8446.3 (4155.6)

The influence of the hold time, t_1, the applied stress, σ_h, and the initial strength, S_0, on the final strength, S_1, can be better understood if Eq. (3) and (5) are represented in graphical form. For illustration purposes, the behavior of sapphire fibers at 1400°C was simulated. The fatigue properties of this material are reported in Table 1, which is discussed in one of the following sections. Figure 1 shows the diagram of the strength after the ISF test, S_1, as a function of the applied stress, σ_h, for three different values of the hold time, as calculated by Eq. (3), in the absence of threshold. Even in absence of a fatigue limit, the strength abruptly decreases over a small range of applied stress values. It can also be observed that the stress value, σ^*, at which "drop-off" occurs, decreases for increasing hold time. If Eq. (5) is considered, the strength can be expressed in terms of the stress intensity factor as shown in Fig. 2. In this case, an arbitrary threshold for sub-critical crack growth equal to 0.55 MPa $\sqrt{m}$ was assumed. A distinct strength drop-off is still evident and K^* decreases for increasing hold times until the threshold is reached. At this point, further increases in hold time do not have any influence on the stress intensity factor corresponding to the strength loss.

The effect of the stress hold on the strength is also evident on the strength distribution.[20] This is schematically represented by a Weibull plot in Fig. 3 where R represents the survival probability. Due to sub-critical crack

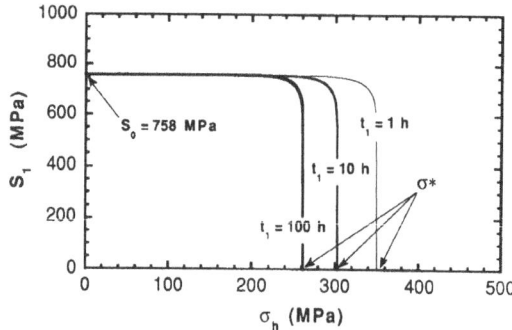

Figure 1. Strength, B, of sapphire fibers after the ISF test as a function of the applied stress, σ_h, for three different holding times, in the absence of a threshold. The initial strength, S_0, and the applied stress, σ^*, corresponding to the strength drop-off are shown.

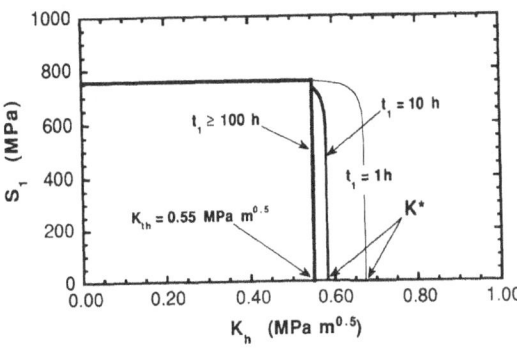

Figure 2. Strength, σ^*, of sapphire fibers after the ISF test as a function of the applied initial stress intensity factor, K_h, for increasing holding times. An arbitrary fatigue limit of 0.55 MPa $\sqrt{m}$ is assumed. The stress intensity factor, K^*, corresponding to the strength drop-off is shown.

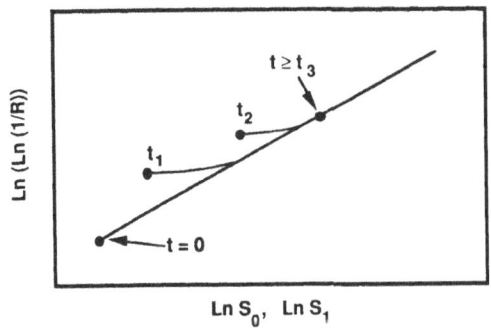

Figure 3. Schematic representation of the strength distributions after the ISF test for increasing hold times, $t_1 < t_2 < t_3$.[20] The weakest element of each distribution is marked with a solid circle.

growth, larger flaws can grow and may lead to failure during the constant stress period. This accounts for the interruption and "twist" in the distribution, though the scatter in the strength values and the finite sampling effects can prevent the "twist" from being observed in real situations. As the holding time is increased, larger number of specimens fail during the stress hold until the threshold conditions are reached. For longer hold times, no further weakening effect is found and the specimens either fail during the static hold or undergo no strength loss.[20]

PROCEDURES FOR K_{th} DETERMINATION BY ISF TEST

On the basis of the results presented in the previous section, two different procedures for a correct determination of the threshold stress intensity factor have been presented.[20] The first approach is related to the number of samples failing during the stress hold and the average strength of the surviving samples. The second one considers the changes in the strength distribution and the strength of the weakest surviving sample.

In the first approach, sample sets are subjected to various stress levels in the active environment. The applied stress, $\sigma_{50\%}$, corresponding to 50% of failures during the stress hold is measured. The corresponding stress intensity factor at the start of the stress hold in the ISF test is determined as:

$$K_{50\%} = K_c \left(\frac{\sigma_{50\%}}{S_{0av}} \right) \tag{6}$$

where S_{0av} is the initial average strength measured in inert environment. The value of $K_{50\%}$ is then re-determined for increasing hold time and when it becomes invariant, $K_{50\%} = K_{th}$.

In the second approach, a stress level is applied in order to obtain some failures during the static hold. This allows the determination of the "interrupted" strength distribution. The stress intensity factor, K_w, at the start of the constant stress phase, corresponding to the weakest sample that survives the hold is calculated as:

$$K_w = K_c \left(\frac{\sigma_h}{S_{0w}} \right) \tag{7}$$

where S_{0w} is the inert strength of the weakest sample in the "interrupted" distribution, calculated on the basis of its rank and the inert strength distribution. K_w is assumed as the estimate of K_{th}. As in the previous case, the value of this estimate is then determined at the same hold stress as a function of the hold time until it becomes invariant.

In some cases, due to limited availability of samples or to experimental costs, re-determination of the threshold estimated for increasing time is difficult. In this situation, a third approach can be used if the sub-critical crack growth parameters are known. The value of the stress intensity factor, K^*, corresponding to the strength "drop-off" is calculated (Eq. (5)) and compared to the K_{th} estimate. If K^* is lower than the fatigue limit estimated by one of the previous procedures, one can conclude that this latter corresponds to the threshold.

FATIGUE LIMIT IN SODA-LIME-SILICA GLASS

The ISF test was used for the determination of the threshold stress intensity factor in soda-lime-silica glass at room temperature in a water environment. Glass rods (0080 Code, Corning Glass Work, USA), with a

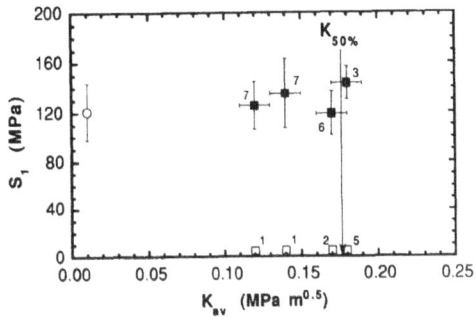

Figure 4. Average strength, S_1, measured in deionized water, of soda-lime silicate glass after the static hold. The initial strength measured in water is shown for comparison at $K_{av} = 0.01$ MPa $\sqrt{m}$. The number of specimens surviving (solid squares) and failing (empty squares) during the stress hold is shown. The error bars represent the standard deviations. The stress intensity factor corresponding to 50% of failures during the ISF is shown.

nominal diameter of 3.2 mm, were used on this purpose.[21] Inert strength was determined in paraffin oil by four-point bend test with inner and outer spans of 20 and 80 mm, respectively. A similar configuration was used to evaluate the strength in deionized water. A cross-head speed of 15 mm/min was used for these tests.

The ISF tests were performed in deionized water and the load was applied in a four point bend configuration with inner and outer spans of 20 and 80 mm, respectively. The constant load was maintained on the specimens for duration ranging from 1 hour and 20 days. Samples which did not fail during the stress hold were broken in four point bending tests.

Figure 4 shows the strength measured in water, after the hold, as a function of the applied stress intensity factor for holding time of 5 days.[21] The number of specimens surviving and failing during the ISF test is shown. Similar diagrams were obtained for hold times ranging from 1 hour to 20 days. The strength of the surviving samples is substantially equivalent to the strength measured initially in water. Following the first procedure described in the previous section, the stress intensity factor, $K_{50\%}$, corresponding to 50% of failure during the stress hold, was evaluated as a function of the holding time. The fracture toughness of the glass was assumed to be 0.75 MPa $\sqrt{m}$.[23]

The strength distributions corresponding to a hold time of 3 days for two different applied stress values are shown in Fig. 5 and compared to the strength distribution measured initially in water.[21] Similar diagrams were obtained for hold times ranging from 1 hour to 20 days. On the basis of the Eq. (7), the stress intensity factor, K_w, applied to the weakest surviving sample was calculated. Different values of the applied stress were used at each hold time. K_w was calculated from two "interrupted" strength distributions in which the survival probability of the weakest sample was close to 50%.[21]

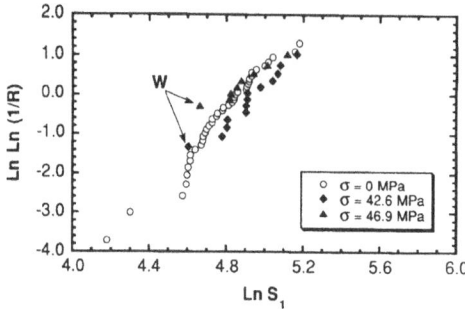

Figure 5. Strength distribution of soda-lime silicate glass after the ISF test, for a holding time of 3 days. The initial strength distribution, measured in water, is shown for comparison. The weakest survivors of the "interrupted" strength distributions are marked with "W".

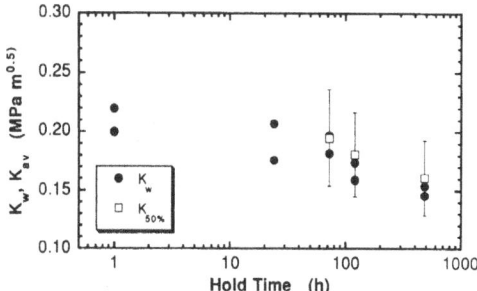

Figure 6. Fatigue limit estimates, $K_{50\%}$ and K_w, as function of the holding time for soda lime silica glass. The error bars represent the standard deviation of $K_{50\%}$.[21]

The results in terms of $K_{50\%}$ and K_w are shown in Fig. 6.[21] The two approaches used to determine the "potential" threshold furnished similar results. Both $K_{50\%}$ and K_w decrease for increased holding times and seem to approach a limiting value around 0.14-0.15 MPa $\sqrt{m}$. Though both threshold estimates do not reach an invariant limit, results calculated for holding times of 20 days are lower than the fatigue limit proposed by other authors, which range from 0.18 to 0.40 MPa $\sqrt{m}$. The measurement of the real value of the threshold for fatigue would require holding times longer than 20 days. The decision to use such long hold times would depend on various factors like experimental costs and importance of the threshold information in component design.

FATIGUE LIMIT IN SAPPHIRE FIBERS

The threshold stress intensity factor for sub-critical crack growth was determined in sapphire fibers at 1200 and 1400°C. Single crystal alumina fibers (Saphikon, Inc., USA) with a nominal diameter of 150 μm and the c-axis aligned 0 to 8° to the principal axis were used. The fatigue parameters n and B were evaluated from the experimental results of a previous work[24] and are given in Table 1. The initial uniaxial tensile strength distribution was determined at 1200 and 1400°C in air by tension using a high cross-head speed (100 mm/min) in order to minimize any sub-critical crack growth process.

The ISF test was performed in uniaxial tension at 1400°C with a constant stress hold time of 10 hours. Fibers which did not fail during the stress hold were immediately broken at the same temperature using a cross head-speed of 100 mm/min.[22]

The "interrupted" strength distributions obtained after the ISF test at 1200 and 1400°C is shown in Fig. 7 together with the initial strength distributions. The weakest sample surviving the ISF test is considered and the corresponding stress intensity factor, K_w, was calculated as $K_w = \sigma_h \left(K_c / S_{0w} \right)$, where σ_h is again the applied stress and S_{0w} is the strength corresponding to the weakest survivor in the initial distribution. A value equal to 1.46 MPa$\sqrt{m}$ was assumed for the fracture toughness.[25] Values of 0.68 and 0.66 MPa$\sqrt{m}$ were calculated for K_w at 1200 and 1400°C, respectively.

In order to demonstrate that the calculated K_w values corresponds to the real threshold for fatigue, the third approach defined in one of the previous section was followed. The stress intensity factor, K^*, corresponding to the strength "drop-off" was calculated. In order to take into account the scatter in the fatigue parameter, the 95% confidence interval of n and B was considered. Values ranging from 0.40 to 0.64 MPa$\sqrt{m}$ were calculated for K^* taking into

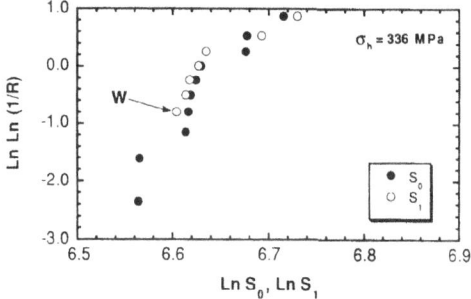

Figure 7. Strength distribution before (S_0) and after (S_1) the static hold at 336 MPa for sapphire fibers at 1400°C. The weakest specimen of the "interrupted" strength distribution is marked with "W".

account each combination of n and B within their 95% confidence interval. Since these values are lower than calculated K_w values, a threshold for fatigue of 0.68 and 0.66 MPa$\sqrt{m}$ can be accepted for sapphire fibers at 1200 and 1400°C, respectively.

CONCLUSIONS

An evaluation of the interrupted static fatigue test for the determination of the threshold stress intensity factor, K_{th} , for sub-critical crack growth has been presented. Various test procedures for the determination of K_{th} have been proposed previously and these were reviewed. In the first approach, the threshold estimate is calculated as the stress intensity factor corresponding to 50% of failures occurring during the stress hold phase of the ISF test. When the second procedure is followed, a constant stress is applied to the material in order to obtain an "interrupted" strength distribution. The stress intensity factor corresponding to the weakest survivor in the censored distribution is considered as an estimate of the fatigue limit. These estimates are then re-calculated for increasing hold time until they become invariant. In the final approach, when the fatigue parameters of the material are known, the estimated K_{th} is compared to the calculated stress intensity factor associated with strength degradation in the absence of a threshold.

The ISF test was utilized for the determination of K_{th} in soda-lime-silica glass in water environment at room temperature and in sapphire fibers at 1200 and 1400°C. In the first case, a threshold around 0.15 MPa $\sqrt{m}$ was calculated for holding times of 20 days. Conversely, values of the fatigue limit of 0.68 and 0.66 MPa$\sqrt{m}$ were obtained for sapphire fibers at 1200 an 1400°C, respectively.

REFERENCES

1. R.E. Mould, and R.D. Southwick, Strength and static fatigue of abraded glass under controlled ambient conditions: II, Effect of various abrasions and the universal fatigue curve, *J. Am. Ceram. Soc.*, 42[12], 582-92 (1959).
2. W.B. Hillig, and R.J. Charles, Surfaces, stresses-dependent surface reactions and strength, in "High Strength Materials", V.F. Zackay, ed., John Wiley and Sons, New York (1965).
3. S.M. Wiederhorn, and L.H. Bolz, Stress corrosion and static fatigue of glass, *J. Am. Ceram. Soc.*, 53[10], 543-48 (1970).
4. J.E. Ritter, and C.L. Sherburne, Dynamic and static fatigue of silicate glasses, *J. Am. Ceram. Soc.*, 54[12], 601-5 (1971) .
5. A.G. Evans, A simple method for evaluating slow crack growth in brittle materials, *Int. J. Fract.*, 9[3], 267-75 (1973).
6. S.M. Wiederhorn, Subcritical crack growth in ceramics, in "Fracture Mechanics of Ceramics", Vol. 2, R.C. Bradt, D.P.H. Hasselman and F.F. Lange, eds., Plenum Press, New York (1974).
7. K. Wan, S. Lathabai, and B.R. Lawn, Crack velocity functions and threshold in brittle solids, *J. Eur. Ceram. Soc.*, 6, 259-68 (1990).

8. K.S. Chan, and R.A. Page, Origin of the creep-crack growth threshold in a glass-ceramic, *J. Am. Ceram. Soc.*, 75[3], 603-12 (1992).

9. E. Gehrke, C. Ullner, and M. Hähnert, Correlation between multistage crack growth and time-dependent strength in commercial silicate glasses. Part 1. Influence of ambient media and types of initial cracks, *Glastech. Ber.*, 60[8], 268-78 (1987).

10. E. Gehrke, C. Ullner, and M. Hähnert, Correlation between multistage crack growth and time-dependent strength in commercial silicate glasses. Part 2. Influence of surface treatment, *Glastech. Ber.*, 60 [10], 340-45 (1987).

11. E. Gehrke, C. Ullner, and M. Hähnert, Effect of corrosive media on crack growth of model glasses and commercial silicate glasses, *Glastech. Ber.*, 63[9], 255-65 (1990).

12. E. Gehrke, C. Ullner, and M. Hähnert, Fatigue limit and crack arrest in alkali-containing silicate glasses, *J. Mater. Sci.*, 26, 5445-55 (1991).

13. T.A. Michalske, The stress corrosion limit: its measurement and implications, *in* "Fracture Mechanics of Ceramics", Vol. 5, R.C. Bradt, D.P.H. Hasselman and F.F. Lange, eds., Plenum Press, New York (1983).

14. B.J.S. Wilkins, and R. Dutton, Static fatigue limit with particular reference to glass. *J. Am. Ceram. Soc.*, 59 [3-4], 108-12 (1975).

15. K. Hayashi, T.E. Easler, and R.C. Bradt, A fracture statistics estimate of the fatigue limit of a borosilicate glass, *J. Eur. Ceram. Soc.*, 12, 487-91 (1993).

16. E.J. Minford, and R.E. Tressler, Determination of threshold stress intensity for crack growth at high temperature in silicon carbide ceramics, *J. Am. Ceram. Soc.*, 66[5], 338-40 (1983).

17. E.J. Minford, D.M. Kupp, and R.E. Tressler, Static fatigue limit for sintered silicon carbide at elevated temperatures, *J. Am. Ceram. Soc.*, 66[11], 769-73 (1983).

18. M.R. Foley, and R.E. Tressler, Threshold stress intensity for crack growth at elevated temperatures in a silicon nitride ceramic, *Adv. Ceram. Mat.*, 34, 383-86 (1988).

19. B.O. Yavuz, and R.E. Tressler, Threshold stress intensity for crack growth in silicon carbide ceramics, *J. Am. Ceram. Soc.*, 76[4], 1017-24 (1993).

20. V.M. Sglavo, and D.J. Green, The interrupted static fatigue test for evaluating threshold stress intensity factor in ceramic materials: A theoretical analysis, to be published on the *J. Eur. Ceram. Soc.*.

21. V.M. Sglavo, and D.J. Green, Threshold stress intensity factor in soda-lime silicate glass by interrupted static fatigue test, submitted for publication to the *J. Eur. Ceram. Soc.*.

22. S.W. Martz, "Evaluation of the Threshold Stress Intensity Factor of c-axis Sapphire Fibers at Elevated Temperature", M.S. Thesis, The Pennsylvania State University (1994).

23. S.M. Wiederhorn, Fracture surface energy of glass, *J. Am. Ceram. Soc.*, 52, 99-105 (1969).

24. S.A. Newcomb, and R.E. Tressler, Slow Crack Growth in Sapphire Fibers at 800°C to 1500°C, *J. Am. Ceram. Soc.*, 76[10], 2505-12 (1993).

25. S.A. Newcomb, and R.E. Tressler, High Temperature Fracture Toughness of Sapphire, *J. Am. Ceram. Soc.*, 77[11], 3030-32 (1994).

FRACTURE TOUGHNESS AND SUBCRITICAL CRACK GROWTH IN AN

ALUMINA/SILICON CARBIDE 'NANOCOMPOSITE'

Mark Hoffman[1], Jürgen Rödel[1], Martin Sternitzke[2] and Richard Brook[2]

[1]FB Materialwissenschaft
Technische Hochschule Darmstadt
64295 Darmstadt, Germany

[2]Department of Materials
University of Oxford
Parks Road, Oxford, UK

INTRODUCTION

Ceramic materials have many attractive properties for use in structural applications especially in environments that involve high temperatures, high abrasion or are corrosive. The breadth of applicability of ceramic materials in engineering situations is severely limited by their low toughness and their brittle failure mode which significantly reduces their reliability. A broad range of toughening processes have been designed 'into' structural ceramics including grain, fibre and whisker bridging, crack and phase transformation toughening (Evans, 1990). These processes essentially all involve crack extension toughening. While strengths are significantly improved by the incorporation of these toughening mechanisms, fracture is still fundamentally controlled by crack propagation from an inherent flaw. Additionally most toughening mechanisms deteriorate with increasing temperature.

A recently developed group of ceramic materials known as 'nanocomposites' involves the incorporation of 5-25% particles of grain size less than 400 nm in a ceramic matrix. Material combinations include Al_2O_3/SiC_p (Niihara, 1991; Otsuka et al, 1994; Zhao et al, 1993), MgO/SiC_p (Niihara & Nakahira, 1990), Al_2O_3/TiN_p (Walker et al, 1994) and Si_3N_4/SiC_p (Pezzotti & Sakai, 1994). With the exception of the latter two, in all cases an improvement in strength relative to the pure matrix material has been observed. In some cases toughness improvements have also been noted.

Increases in strength have been accompanied by a change in fracture mode from intergranular in pure alumina to transgranular in materials containing SiC particles. It is proposed that the fracture behaviour is a result of the compressive radial and tensile hoop residual stresses that surround the particles as a result of thermal mismatch between the matrix and particle phases. The radial stresses exerted by particles near the grain boundaries cause compression at the boundaries raising the stress for grain boundary sliding. Observed rippling of the grain boundaries, as a result of drag by the particle phase, further increases the stress for grain boundary sliding. As fracture energy of the grains is higher than that of the grain boundaries in pure alumina, then the overall fracture properties should be improved by change in the fracture mode from transgranular to intergranular, should grain boundary strengthening be occurring. Theoretical and experimental work by the same workers has found that this should only be the case for SiC contents less than ~5% (Levin et al, 1994; Levin et al, 1995). At greater particle contents the overall average tensile residual stress in the matrix exceeds a critical value which lowers the fracture energy of the composite. This is consistent with experimental data of Niihara & Nakahira (1990) but in conflict with that of Walker et al. (1994) who found that strength increased with increasing particle content up to 25% SiC. In the latter case however it was also noted that this strength increase coincided with a reduction in grain size. In the same theoretical work the authors proposed that the overall tensile stresses in the material could be decreased by reducing the size of the particles consistent with experimental findings of Sawaguchi et al. (1992). It has also been proposed that the tensile stresses surrounding the particles attract the crack into the grain from the grain boundary (Walker et al, 1994). Once the crack has progressed through this small tensile zone it is 'stopped' within the grain. One may ask in this case however, why the crack does not then continue along its original grain boundary path if this is the only effect that the particle has.

Also of note is the reported ease of polishing of nanocomposite materials and improved wear behaviour relative to pure alumina (Walker et al., 1995). It is noted that in both cases the improved grain boundary strength reduces grain pullout leading to improved behaviour. Deterioration of mechanical properties in alumina is attributed usually to grain pullout, therefore it is not surprising that grain boundary strengthening should lead to improvements.

Annealing heat treatments have been found to lead to still further increases in strength in Al_2O_3/SiC_p (Niihara & Nakahira, 1990; Zhao et al, 1993). A study of indentation cracks before and after annealing in pure Al_2O_3 and an Al_2O_3/SiC_p nanocomposite found that the cracks healed in the composite during an annealing heat treatment while they extended in the pure alumina (Thompson et al., 1995). The authors stated that grinding of samples caused compressive residual stresses on the surface which led to observed fracture strength increases. The annealing process relieved these residual stresses but at the same time healed microcracks on the surface which may act as fracture initiation sites. The overall effect is to strengthen the sample.

While there is a recognised strength increase by the inclusion of nano-sized particles in an alumina matrix relative to pure alumina, the presence of a toughness increase is debated in the literature. Some authors have noted a clear increase in fracture toughness (Niihara & Nakahira, 1990) while others have proposed that compressive residual stresses on a ground sample surface would restrict crack propagation in indentation based fracture toughness measurements causing an overestimate of fracture toughness (Thompson et al., 1995). Additionally some authors have proposed that strength increases are due to reduction in the size of inherent flaws in the material (Niihara & Nakahira, 1990). Others have noted, however, that particle agglomeration processing difficulties lead to large flaws yet, despite these flaws, strength is still increased (Walker et al., 1994). This implies some form of toughening in the composite relative to the pure material.

To date all toughness measurements have been made on nanocomposite materials using indentation methods. The main reasons for this have been (i) the simplicity of the test and (ii) that it is supposedly close to the 'small crack' situation as would occur in 'real' situations in these materials. The latter argument is not particularly valid in this case as the crack lengths from indentations (50-300 μm) are significantly longer than inherent facet lengths from either the interparticle spacing (0.2-0.5 μm) or grain size (<3 μm).

Additionally, indentation methods are sensitive to surface preparation, such as residual compressive stresses resulting from grinding, and indentation load.

Despite wide ranging studies there are many conflicting results as to the mechanical performance of nanotoughened materials, especially their toughness and its effect upon strength measurements. The purpose of this study is to qualify toughness for an Al_2O_3/SiC_p composite using compact tension samples which measure through thickness toughness over a wide crack front. Slow crack growth rates are also determined to consider the effects of subcritical crack growth upon toughness measurements.

EXPERIMENTAL WORK

Material and Sample Preparation

The material used was an Al_2O_3 composite containing 5% SiC. Both starting powders had a particle size of less than 300 nm. The SiC[†] powder was ultrasonically dispersed in distilled water. The Al_2O_3[‡] and SiC powders were then attrition milled for two hours at 500 UPM. The mixture was then freeze dried and sieved through a 150 μm mesh. The powder was hot pressed at 1550°C for one hour at 20MPa under Ar using a 50 mm die. Hardness was found to be 18.5±1.8 GPa, Youngs modulus 396 GPa and Poissons ratio 0.23 as measured by ultrasonic resonance (Grindosonic MK4i), and density, as measured by Archimedes method, 3.9 g/cm³. The final average grain size of the composite was 2.5 μm.

From billets bend bars of dimensions 25×2×2.5 mm were cut, polished on the tensile face and the edges bevelled on the tensile face to reduce stress concentrations. A total of 25 bars were made. From the other disc a standard (ASTM E399, 1983) compact tension sample with a width W of 27 mm, and thickness of 3 mm, was machined and polished on both sides to a 1 μm finish. A half chevron notch was cut in the CT sample using a 0.15 mm thick blade on a slow speed saw. A Vickers indentation was placed on the reverse side of the half chevron notch to initiate a sharp crack. The sample was loaded in a purpose built loading device of high stiffness which enabled exceptionally stable crack growth. The crack was grown till it was of equal length of both sides of the sample and well ahead of the half chevron notch. The crack was then cut away till ~20 μm short of the crack tip to eliminate any possible crack wake effects in the toughness measurements.

Indentation fracture toughness tests were made surfaces polished with 3 μm diamond paste.

Experimental Procedure and Results

The bend bars were loaded in a 4-pt bending fixture (outer span 20 mm, inner span 10 mm) and found to have a fracture strength of 595±82 MPa.

Fracture toughness determined from a 5 kg Vickers indentation, according to Anstis et al. (1981), was found to be 3.71±0.5 MPa√m and from a Herzian indentation according to Warren (1995), 3.04±0.39 MPa√m.

The compact tension sample was slowly loaded and the load for crack propagation recorded versus crack length. Stress intensity factors, K, for crack propagation were then calculated using the standard and plotted versus crack extension a seen in Fig.1. A slight increase with crack extension can be seen but essentially no crack resistance curve or R-curve can be seen with the stress intensity factor for crack extension relatively flat at approximately 2.1 MPa√m.

[†] Lonza UF45
[‡] Sumitomo AKP53

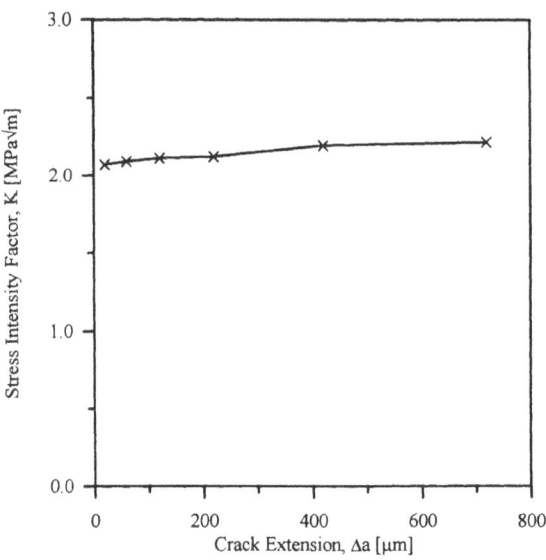

Fig.1 Stress intensity factor, K, for crack propagation as a function of crack extension, Δa, for Al_2O_3/SiC_p nanocomposite

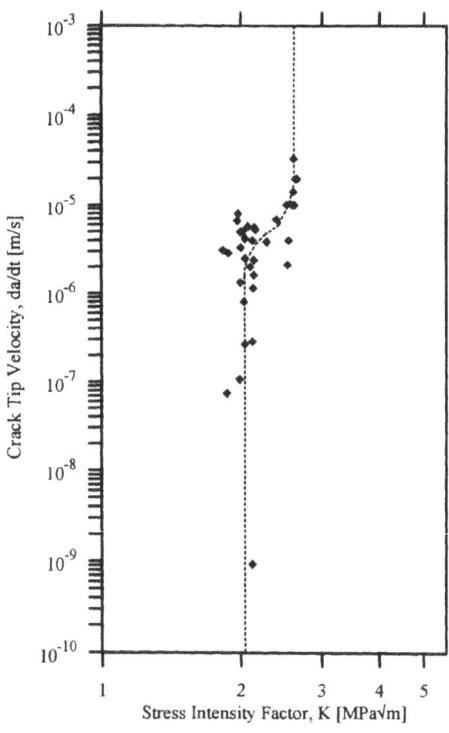

Fig.2 Crack growth rate, da/dt, as a function of stress intensity factor, K, in ambient air.

Following fracture toughness measurements the sample was loaded to a constant displacement at which relatively fast crack extension occurred and crack length, time and load simultaneously recorded. The process was repeated a number of times. Stress intensity factors were again calculated according to the standard and plotted versus crack growth rate as seen in Fig.2. A trimoidal fatigue curve can clearly be seen with a fatigue limit of 2.1 MPa$\sqrt{m}$ and an inert limit of 2.6 MPa$\sqrt{m}$. Both the R-curve testing and slow crack growth rate testing were undertaken consecutively under ambient conditions.

Following the fatigue testing the sample was broken by fast fracture and the fracture surfaces studied in a scanning electron microscope (SEM). Fig.3 compares the fracture surfaces of a relatively fine grained Al_2O_3 (grain size≈ 5 μm) sample with the Al_2O_3/SiC_p composite tested in this work. Fracture in the composite is clearly transgranular and SiC particles can be discerned on the fracture surface. In contrast the Al_2O_3 fractures in a intergranular mode. No clear distinction could be made on the fracture surfaces as a result of the R-curve testing, slow crack growth or fast fracture.

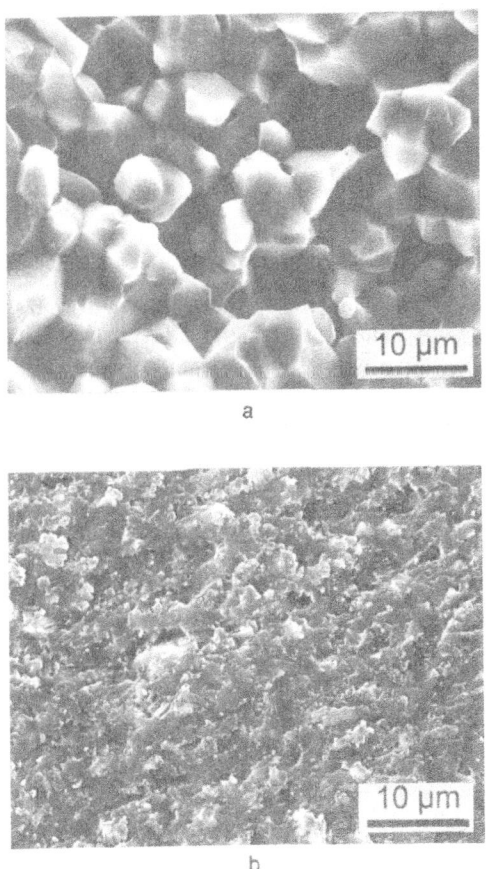

a

b

Fig.3 Fracture surfaces of (a) pure Al_2O_3 and (b) Al_2O_3/SiC_p nanocomposite following fast fracture.

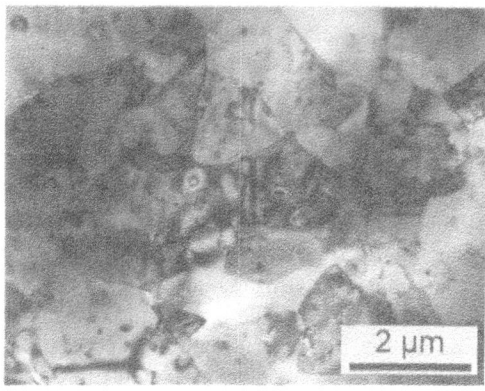

Fig.4 TEM micrograph showing microstructure of Al_2O_3/SiC_p nanocomposite.

A transmission electron microscope (TEM) foil was made to study the microstructure of the material and is shown in Fig.4. SiC particles can be seen at both the grain boundaries and within the grains. There is a wide distribution of both the particle distribution and particle size.

DISCUSSION

It can be seen that the stress intensity for crack propagation is significantly affected by the environment. The stress intensity factor difference between the fatigue limit and inert limit of 0.5 MPa√m is significantly greater than that observed in fine alumina (0.25 MPa√m) over the same crack growth rates (Kishimoto et al., 1994). A closer comparison may be made with slow crack growth in sapphire crystals (Wiederhorn, 1974). This is not an unexpected result as crack growth in the composite is transgranular whereas it occurs along the grain boundaries in alumina. While the fracture toughness of sapphire is orientation sensitive, the inert limit for slow crack growth in our nanocomposite (2.6 MPa√m) compares well with inert fracture toughness values of sapphire along the weaker (1$\bar{1}$02) and (11$\bar{2}$0) orientations of 2.38 MPa√m and 2.43 MPa√m, respectively (Iwasa et al., 1989). It should be noted that the measured strength values of the nanocomposite were higher than those obtained for pure alumina made from the same starting powder: 592 MPa compared with 491 MPa (Sternitzke, 1995). Fracture toughness values of the pure alumina and nanocomposite were similar.

Clearly the fracture toughness testing using the CT sample undertaken in this work took place near the fatigue limit on the left side of the fatigue curve. This is to be expected in testing of this nature where stable crack growth is more easily maintained at low crack growth rates. The fracture toughness values obtained using indentation methods could also be expected to be overestimates due to the low indentation loads used.

Environmental effects could clearly lead to an exacerbation of the difference between strength values obtained before and after annealing. Grinding flaws are open to the environment and cracks originating from them would therefore propagate at lower stress intensity factors than cracks originating from internal flaws. Reduction in the size and volume of surface flaws by a crack healing annealing process would raise the probability of failure from an internal flaw where the stress intensity factor for crack propagation is significantly higher, hence leading to increases in strength.

From prior documented work and this ,yet limited, study, an elucidation of possible nanotoughening mechanisms requires: (i) a study of the distribution of pores for nanotoughened materials as compared to monolithic materials to assess the effects of agglomerate formation, (ii) study of the effect of possible delayed crack initiation at serrated grain boundaries around pores as for example undertaken in alumina by Wakayama & Nishimura (1992) using acoustic emission techniques, and (iii) the influence of slow crack growth on the distribution of fracture initiation sites and thereby fracture strength.

CONCLUSIONS

From this work it may be concluded that:

1. The Al_2O_3/SiC_p composite tested displays no distinct R-curve with low fracture toughness values,

2. Environmental effects significantly affect crack growth rates and the stress intensity factors for crack growth.

Acknowledgments

This work was supported by the Deutsche Forschungsgemeinschaft under contract no. Ro954/6-1. M.S. was supported by EU (BRITE/EURAM 2, Human capital and mobility contract no. ERB BRE2 CP94 3093).

REFERENCES

Anstis, G.R., Chantikul, P., Lawn, B.R. and Marshall, D.B., 1981, A critical evaluation of indentation techniques for measuring fracture toughness: 1, Direct crack measurements, *J. Amer. Cer. Soc.* 64[9]:533.

ASTM E399, 1983, Standard test method for plain-strain fracture toughness of metallic materials, *in*: "1983 ASTM Annual Book or Standards," American Society for Testing and Materials, Philadelphia.

Evans, A.G., 1990, Perspective on the development of high-toughness ceramics, *J. Amer. Cer. Soc.* 73[2]:187.

Iwasa, M., Ueno, T. and Bradt, R.C., 1981, Fracture toughness of quartz and sapphire single crystals at room temperature, *J.Soc.Mat.Sci.Japan* 30:1001.

Kishimoto, H., Ueno, A., Okawara, S. and Kawamoto, H., 1994, Crack propagation behavior of polycrystalline alumina under static and cyclic load, ", *J. Amer. Cer. Soc.* 77[5]:1324.

Levin, I., Kaplan, W. and Brandon, D., 1994, Residual stresses in alumina-SiC nanocomposites, *Acta metall.mater.* 42:1147.

Levin, I., Kaplan, W., Brandon, D. and Layyous, A., 1995, Effect of SiC submicrometer particle size and content on the fracture toughness of alumina-SiC "nanocomposites", *J. Amer. Cer. Soc.* 78[1]:254.

Niihara, K., Nakahira, A., 1990, Particulate strengthened oxide ceramics-nanocomposites *in*: "Advanced Structural Inorganic Composites", ed., P.Vincenzini, Elsevier, Trieste, Italy.

Niihara, K., 1991, New design concept of structural ceramics: ceramic nanocomposites, *J.Cer.Soc.Japan* 99[10]:974.

Otsuka, J., Iio, S., Tajima, Y., Watanabe, M. and Tanaka, K., 1994, Strengthening mechanisms in Al_2O_3/SiC particulate composites, *J.Cer.Soc.Japan* 102[1]:29.

Pezzotti, G. and Sakai, M., 1994, Effect of a silicon carbide "nano-dispersion" on the mechanical properties of silicon nitride, *J. Amer. Cer. Soc.* 77[11]:3039.

Sawaguchi, A., Toda, K. and Niihara, K., 1991, Mechanical and electrical properties of silicon nitride-silicon carbide nanocomposite material, *J. Amer. Cer. Soc.* 74[5]:1142.

Sternitzke, M, 1995, unpublished work.

Thompson, A.M., Chan, H., Harmer, M., 1995, Crack healing and stress relaxation in Al$_2$O$_3$-SiC "nanocomposites", *J. Amer. Cer. Soc.* 78[3]:567.

Wakayama, S. and Nishimura, H., 1992, Critical stress of microcracking in alumina evaluated by acoustic emission, *in*: "Fracture Mechanics of Ceramics, Vol.10", ed. R.C.Bradt, Plenum Press, New York.

Walker, C.N., Borsa, C.E., Todd, R.I., Davidge, R.W. and Brook, R.J., 1994, Fabrication, characterisation and properties of alumina matrix nanocomposites, *British Cer. Proc.* 53:249.

Walker, C., Borsa, C. and Todd, R., 1995, Nanocomposites, *in*: "Ceramic Technology International" I.Birkby,ed., Ampang Press, Malaysia.

Warren, P.D., 1995, Determining the fracture toughness of brittle materials by herzian indentation, ", *J. Eur. Cer. Soc.* 15:201.

Wiederhorn, S., 1974, Subcritical crack growth in ceramics, *in*: "Fracture Mechanics of Ceramics, Vol.2", eds, R.C.Bradt, D.P.H.Hasselman and F.F.Lange, Plenum Publishing, New York.

Zhao, J., Stearns, L., Harmer, M., Chan, H. and Miller, G., 1993, Mechanical behaviour of alumina-silicon carbide "nanocomposites", *J. Amer. Cer. Soc.* 76[2]:503.

HIGH TEMPERATURE CRACK GROWTH IN SILICON NITRIDE WITH TWO DIFFERENT GRAIN SIZES UNDER STATIC AND CYCLIC LOADS

T. Hansson, Y. Miyashita and Y. Mutoh

Department of Mechanical Engineering
Nagaoka University of Technology
940–21 Nagaoka, Japan

ABSTRACT

The crack growth behaviour at 1300°C of two silicon nitrides with different grain sizes was studied and compared with the behaviour at room temperature. It was found that at 1300°C, unlike at room temperature, both materials exhibited a beneficial effect of cyclic loading on the crack growth resistance. From observations of crack paths and fracture surfaces of specimens with cracks grown at elevated temperature but fractured at room temperature it was found that the formation of bridges was promoted by high stress intensities and static loading. In contrast to at room temperature the bridges consisted not only of single grains but also multiple grains. The room temperature fracture toughnesses of testpieces with cracks grown under static and cyclic loads at 1300°C confirmed that more bridges were formed during static loading and at high K–levels. The lower crack growth rates observed under cyclic loading as compared to static loading was suggested to be the result of a higher crack opening displacement rate which increased the load transferred to the bridges even if the amount of bridges were larger under static loading. With increasing K–levels both the crack growth rate and the crack opening displacement rate increased and consequently the effect of bridging increased very rapidly in the case of static loading. For high K–levels the crack growth rates under static loading became similar to those under cyclic loading. The crack growth resistance was higher for the large grain size material both for cyclic and static loading.

1. INTRODUCTION

In recent years the demands for lightweight materials with good high temperature mechanical properties and chemical stability have focused a lot of interest on ceramic materials. Silicon nitride based materials have shown to be main candidates for many high temperature structural applications since they exhibit high strength and creep resistance in combination with good chemical stability at temperatures above 1200°C. High temperature failure of structural ceramics is in most cases determined by creep and/or fatigue. (The term fatigue is throughout this study used strictly for cyclic loading conditions). It is therefore of primary interest, both for the use and for the materials developement, to characterise and to develop an understanding of the creep and fatigue behaviour at elevated temperatures. A number of studies on high temperature creep[1-10] and fatigue[11-19] of silicon nitride have been reported, but a full understanding of the mechanisms for failure has not yet been presented.

For the case of sustained loading, fracture maps have been developed[20-22]. The fracture of silicon nitride under sustained loading can be divided in 3 regimes; fast fracture at high load levels, slow crack growth at intermediate load levels and high temperatures and creep crack growth at low stress levels and high temperatures. For low stresses and low temperatures no fracture occurs. However, these fracture maps give information only about fracture type and do not provide any detailed information about the micro–mechanisms for the different types of fracture.

Traditionally it was believed that most ceramic materials were not sensitive to cyclic loading since they exhibit very limited or no dislocation activity, even at very high temperatures[23,24]. Later it was shown that not only dislocation activity, but also other types of irreversible deformation (i.e. microcracking, martensitic transformation, interfacial sliding and creep) can result in fatigue damage[25].

At room temperature several studies[26-35] have shown that cyclic loading reduces the lifetime of many ceramics and increases the crack propagation rates compared to static loading with a stress intensity equal to the maximum cyclic stress intensity. This has in most cases been attributed to fracture of bridging ligaments and wear of asperity contact points during cyclic loading. At higher temperatures, above the glass transition temperature (T_g) for the amorphous grain boundary phase, the situation is more complicated since the amorphous grain boundary phase becomes viscous which increases the sensitivity to both temperature and loading rate. Most high temperature studies have reported lower crack growth rates during cyclic loading as compared to sustained loading[12,14-16,19] but also higher crack growth rates for cyclic loading has been observed[17-18].

The aim of this work was to study the high temperature crack propagation behaviours of two similar silicon nitrides with different grain sizes under sustained and cyclic loads. The mechanisms for crack growth were studied by observations of crack paths after crack growth and by fractography of testpieces containing cracks grown at high temperature under static and cyclic loads but fractured at room temperature. Fractography of testpieces broken at 1300°C could not reveal any information about the crack growth mechanisms since the fracture surfaces were covered with an amorphous glass phase that formed very rapidly after fracture.

2. MATERIALS AND EXPERIMENTAL PROCEDURES

2.1. Materials

The materials were produced by mixing powder of Si_3N_4 with 5 wt% Y_2O_3 and 3 wt% Al_2O_3 followed by cold isostatic pressing (CIP) at 390 MPa and N_2–gas pressure sintering at 9.3 MPa. The small grain size material (LG) and the large grain size material (SG) were obtained by sintering at 1850°C for 3h and 2000°C for 5h respectively. The resulting microstructures consisted of elongated β–grains surrounded by grain boundary phases, most probably both crystalline and amorphous. The small grain size material had β–grains with a long axis of 2–4 μm and a small axis of 0.3–0.5 μm while the large grain size material had a long axis of 10–20 μm and a small axis of 1.0–1.5 μm. The mechanical properties of both materials are given in Table 1 and the microstructures are shown in Figures. 1a–b. The influence of grain size on the crack growth behaviour could be studied directly since the same starting powders and same processing techniques were employed for both materials.

Table 1. Processing parameters and mechanical data for the materials used.

Material	Grain size (μm)	Processing parameters	Young's modulus, E (GPa)[1]	Poisson's ratio, v[1]	Bending Strength, σ_f (MPa)	Fracture Toughness, K_{IC} (MPam$^{1/2}$)
Si_3N_4	small	1850°C, 3h	332	0.26	1103	6.3 (8.2)[2]
Si_3N_4	large	2000 °C, 5h	309	0.27	739	8.8 (14.6)[2]

[1]Measured from bend tests using strain gauges.
[2]Numbers in brackets are values at 1300°C estimated from specimen fractured during fatigue tests.

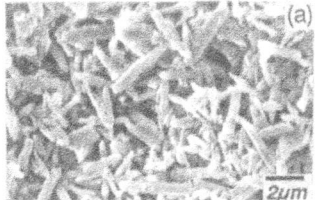 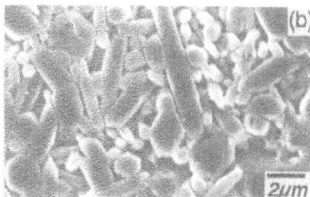

Figure 1. SEM–micrographs showing the microstructure of (a) the small grain size material and (b) the large grain size material.

2.2. Experimental Procedure

The crack growth tests for both cyclic loading and sustained loading were carried out in four point bending at 1300°C using single edge precracked bendbars of dimensions 5 x 10 x 55 mm. Precracking was made by the "Bridge indentation" technique first proposed by Sadahiro and Takatsu[36] and further developed by Warren and Johannesson[37] and Nose and Fuji[38]. First a row of 9 Vickers indentations of 98 N near the edges and 294 N in the mid part were made across the width (5 mm) of the specimen and aligned so that the cracks emanating from each corner linked up to form a shallow defect being the starting point for the final precrack. The testpiece was then placed in the bridge indentation device, as shown in Figure 2. Pre-cracks with a length of 3–4 mm were created by applying a compressive load of 35–50 kN.

The tests were carried out at 1300°C in air using a computer controlled resistance heated electrical furnace with a stability better than ±1°C. Heating and cooling rates were kept below 20°C/min in order to avoid any risks for thermal shock. Before starting the tests, the furnace temperature was kept constant for at least 20 minutes after reaching 1300°C. The bend fixture was made of SiC and had inner and outer spans between loading rollers of 20 and 40 mm respectively.

The cyclic loading was applied in a sinusoidal waveform at a frequency of 5 Hz and a R–ratio (minimum load / maximum load) of 0.1. First a low ΔK was applied for 20 000 cycles and if no crack growth was found ΔK was increased by 0.1–0.2 MPam$^{1/2}$. The crack length was measured in an optical microscope after slowly cooling down the specimen and removing the glassy phase, that was formed by oxidation of the surface at elevated temperature, by gentle polishing with a fine grain diamond suspension. When crack growth was observed, the crack growth rates in the high K_{app} regime was obtained by increasing ΔK in steps of 0.1–0.2 MPam$^{1/2}$, repeatedly heating up, applying a predetermined number of loading cycles and cooling down the specimen for crack length observations. The threshold values were obtained by reducing the applied ΔK, after crack growth was detected, in steps of 0.1 MPam$^{1/2}$ until no crack growth was detected after 100 000 cycles.

For the sustained loading tests a low K was applied for 30–60 minutes and if no crack growth was observed after cooling down and measuring the crack length, the load was increased by 0.1–0.2 MPam$^{1/2}$ until crack growth was detected. Crack propagation rates in the high K_{app} regime were obtained by increasing K_{app} by 0.1–0.2 MPam$^{1/2}$ and measuring the crack length after a predetermined time at load. The threshold values were obtained by decreasing the K_{app} in steps of 0.1 MPam$^{1/2}$ until no crack growth was detected after loading for 5 h. As for the cyclic tests, the glass film formed at elevated temperature had to be removed by gentle polishing with a diamond suspension. The static load was applied for times that were significantly shorter than the estimated characteristic time for the transition between small scale and large scale creep. The estimation of the characteristic transition time and the choice of parameter for crack growth is discussed in section 2.3. If the desired hold time was a large fraction of the transition time, the testpiece was periodically unloaded after a time much shorter than the transition time and reloaded after 5 min or, if the desired hold time was reached, cooled down to room temperature for crack growth measurements.

Some of the tests were interrupted before specimens failed. Information on the high temperature fracture behaviour could be obtained by observing the crack paths and the fracture surfaces after fracturing these testpieces at room temperature. The fracture toughness was measured and fracture surfaces were studied in SEM to reveal the high temperature crack growth mechanisms.

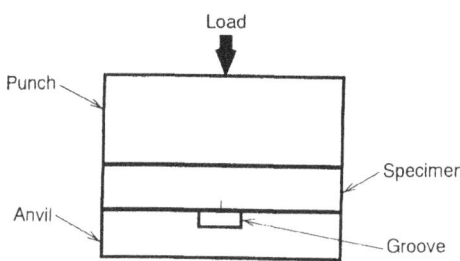

Figure 2. Schematic showing precracking by the bridge indentation technique.

2.3. Choice of Parameter for Crack Growth at High Temperature

At temperatures above T_g for the amorphous grain boundary phase, creep mechanisms become active and if the creep occurring during loading creates a large creep zone around the crack tip, K may not be a valid parameter for characterising the crack growth. As an alternative, Landes and Begley[39] showed that the parameter C^*, a modification of the J integral[40] with strain and displacement replaced with their rates, could be successfully used to correlate the crack growth rates under extensive steady state creep conditions. For bend bars of the present geometry (four point bending) the value of the C^* integral is obtained as[39,41]:

$$C^* = \sigma_0 \dot{\varepsilon}_0 c h_1 \left(\frac{a}{w}, n \right) \left[\frac{M}{M_0} \right]^{n+1} \qquad (1)$$

where σ_0 is the reference stress, $\dot{\varepsilon}_0$ is the reference strain rate, a is the crack length, w is the specimen height, c is the uncracked ligament, M and M_0 are the bending moment and reference bending moment per unit thickness respectively and h_1 is a dimensionless function of a/w and the creep exponent, n. Numerical values for h_1 for single edge notched bend bars subjected to bending were given by Shih and Needleman[41]. These values were calculated for plastic cracks but are also valid for cracks growing under steady state creep conditions. The reference bending moment per unit thickness was also given by Shih and Needleman[41] as:

$$M_0 = 0.364 \sigma_0 c^2 \qquad (2)$$

For steady state creep under the present loading conditions, C^* is given by:

$$C^* = \frac{A c h_1 \left(\frac{a}{w}, n \right) M^{n+1}}{(0.364 c^2)^{n+1}} \qquad (3)$$

Riedel and Rice[42] derived the transition time for small scale creep to extensive creep during creep crack growth:

$$t_T = \frac{K_I^2 (1 - v^2)}{(n+1) E C^*} \qquad (4)$$

where all the parameters are known or can be obtained from creep tests. The transition time for the present material and test conditions was estimated by using tensile creep data at 1300°C for a Si_3N_4 produced by a similar technique using similar composition and amount of the sintering additives[6]. For the present bendbars at K=5 MPam$^{1/2}$ the following parameters were used to calculate t_T: v=0.26, E(1300°C)= 200 GPa[43], n=4.5[6], A=6.17·10^{-43} Pa^{-n6}, c=5 mm, h_1(a/w,n)=1.20[41] and M=533 Nm/m. Using Equations (3) and (4) the transition time was calculated as t_T= 2960 s. Since the frequency used was 5 Hz and the transition time was larger than 2960 s, K can be used as a parameter for characterising the crack growth of the Si_3N_4 under the present conditions. However, as was pointed out by Riedel[44] for the case of rapid unbalanced loading, the rapidly varying component of the near–tip field is determined by K, but the time independent component (= the mean value) cannot be characterised by C^* except for the cases when R is close to 1 or –1.

In the case of sustained loading the time at load was in most cases much shorter than the transition time. If the desired loading time was a significant fraction of the transition time the specimen was unloaded for 5 min before test was continued. Also for the case of sustained loading C^* cannot be used as the parameter to characterise the crack growth since the holding times at load were normally only a small fraction of the transition time. The unloading procedure was used to ensure that K was a valid parameter. However, it is not known to what extent the unloading could recover the creep strains around the crack tip.

Due to the difficulties in using C^* and the lack of tensile creep data for the present materials K was used as the parameter to characterise crack growth throughout this study.

The conditions for linear elastic fracture mechanics according to ASTM E399 are satisfied since width, b and crack length a>2.5 $(K_{IC}/\sigma_y)^2$ (with K_{IC}=6 MPam$^{1/2}$ and σ_y=600 MPa[43]). In this case time dependent deformation is not considered.

3. RESULTS

Figure 3a shows the crack growth rates, da/dN, for different K_{app} under cyclic loading at 1300°C. K_{app} is the maximum stress intensity during one loading cycle for cyclic loading and the applied stress intensity under static loading. Also indicated in Figure 3a are the room temperature crack growth rates[45]. As is seen in Figure 3a, the large grain size material (LG) exhibited a larger crack growth resistance and a higher threshold value for crack growth, K_{th}, compared to the small grain size material (SG) at both room temperature and 1300°C. Further, the crack growth rates at 1300°C were higher for all K_{app}–levels and the threshold values for crack growth were lower at 1300°C compared to at room temperature. It is also interesting to note that the slopes of the crack growth curves at room temperature and 1300°C were similar.

Figure 3b shows the crack propagation rates for different K_{app} for both materials under static loading at 1300° and also indicates the corresponding room temperature crack propagation rates. As in the case of cyclic loading the elevated temperature crack growth rates were generally higher and the threshold values were lower compared to at room temperature. Also under static loading the large grain size material exhibited a higher crack growth resistance and K_{th} compared to the small grain size material at both 1300°C and room temperature. For static loading at 1300°C the crack growth rates showed a relatively mild increase with increasing K_{app} while at room temperature the crack growth curves were nearly vertical.

In order to compare the crack growth rates under static loading, da/dt, and cyclic loading, da/dN, the crack growth rates for cyclic loading da/dN can be converted to da/dt by using the simple relation:

$$\frac{da}{dt} = \frac{da}{dN} \cdot v_c \qquad (5)$$

where v_c is the frequency.

Figures 4a–b show the crack propagation rates as a function of K_{app} at 1300°C and room temperature under both static and cyclic loading for the SG and LG materials respectively. It is interesting to note that the threshold values for both materials are higher for cyclic loading as compared to static loading at 1300°C while at room temperature the threshold values are higher for static loading as compared to cyclic loading. Also for low K_{app}–values the room temperature crack growth rate was higher for cyclic loading as compared to static loading while at 1300°C the crack growth rate was higher under static loading. At high K_{app}–values however, the crack growth rates under static and cyclic loading were similar both at room temperature and 1300°C.

The fracture toughnesses at 1300°C were estimated from fatigue tests using the crack lengths measured on the fracture surfaces and the maximum load under one loading cycle. The high temperature fracture toughness values were 8.2 MPam$^{1/2}$ and 14.6 MPam$^{1/2}$ compared to the room temperature values of 6.3 MPam$^{1/2}$ and 8.8 MPam$^{1/2}$ for the SG and the LG materials respectively.

The results of the fracture toughness tests made at room temperature on specimens containing cracks grown under cyclic and static loads are shown in Table 2. All these testpieces contained stably grown cracks of a length much longer than the bridging zone. As is shown in the table the testpieces containing cracks grown under static loading exhibited higher values than those containing cracks grown under cyclic loading. The LG material exhibited higher fracture toughness values than the SG material. Further, specimens with cracks grown at high K_{app}–levels exhibited larger fracture toughnesses compared to specimens with cracks grown at low K_{app}–levels.

Scanning electron microscopy (SEM) observations of the fracture surfaces of these specimens showed that there was a correlation between the bridging zone size and the fracture toughness values (see Table 2). Further, it was observed that the amount of bridging, at the same distance behind the crack tip, was larger for specimens containing cracks grown under static loading at both high and low K–levels (See Figures 5a–d). The SEM–observations also showed that the majority of the bridges were multi–grain ligaments and single grain bridges and that only very few bridges consisted of a glassy phase only (see Figures 5a–d).

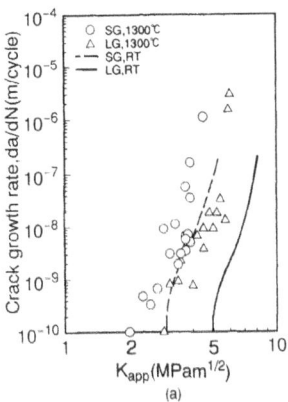 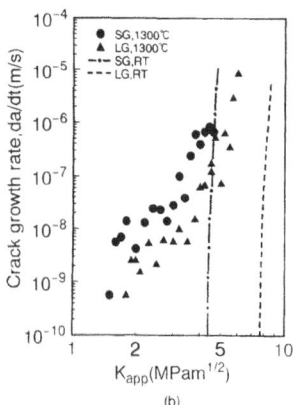

Figure 3. Relationship between applied maximum stress intensity factor and the crack growth rate for both the small and large grain size during (a) cyclic loading and (b) static loading.

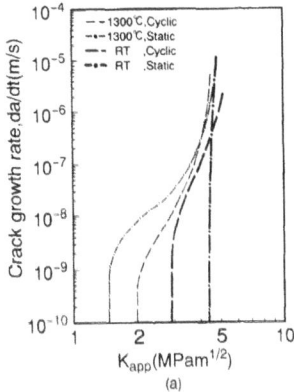 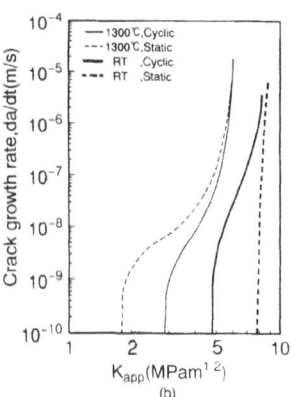

Figure 4. Relationship between applied maximum stress intensity and crack growth rates under static and cyclic loading for (a) the small grain size material and (b) the large grain size material.

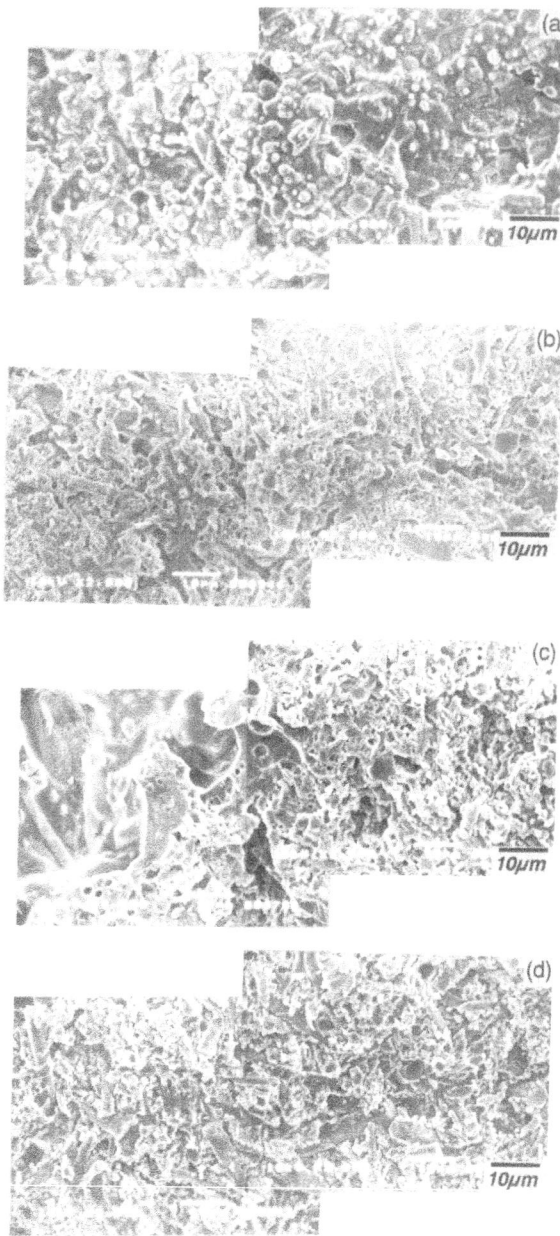

Figure 5. SEM-micrographs showing fracture surfaces of the LG material 700μm behind the crack tip of room temperature fracture toughness specimens with cracks grown under (a) cyclic loading and low K_{app}, (b) cyclic loading and high K_{app}, (c) static loading and low K_{app} and (d) static loading and high K_{app}.

Table 2. Room temperature fracture toughness of specimens containing cracks grown during static and cyclic loading at 1300°C.

Specimen	Crack length (mm)	K_{app} at 1300°C (MPam$^{1/2}$)	K_{IC} measured at room temperature (MPam$^{1/2}$)	Bridging zone size (mm)	Room temperature fracture toughness (MPam$^{1/2}$)
LG, static	6.20	5.7	12.1	0.94	8.8
LG, static	5.40	1.8	9.4	0.42	8.8
LG, cyclic	5.87	6.0	10.6	0.69	8.8
LG, cyclic	4.68	2.9	7.7	0.22	8.8
SG, static	4.91	4.8	9.1	–	6.3
SG, static	4.67	1.5	7.1	–	6.3
SG, cyclic	4.68	2.0	6.6	–	6.3

4. DISCUSSION

The glass transition temperatures, T_g, for the amorphous grain boundary phases in the present materials were believed to be well below 1300°C[46]. However, the mechanical behaviour is determined by the brittle–to–ductile transition temperature, BDTT, rather than by T_g of the amorphous grain boundary phase. The BDTT of a material with an amorphous grain boundary phase depends on the T_g of this phase, but is not constant since it is known to increase with increasing loading rate[47]. However for the present materials it is believed that for all loading conditions used in this study the BDTT is lower than 1300°C. This is supported by the estimation of fracture toughnesses at 1300°C during the highest loading rate condition, cyclic loading at fracture. Both the small and large grain size materials exhibited higher fracture toughnesses under these conditions than at room temperature (see Table 1). The BDTT in ceramics containing an amorphous glass phase is often taken as the temperature when the fracture toughness starts to increase with increasing temperature[48]. Another indication that the BDTT is lower than 1300°C, even for cyclic loading, is that the effect of bridging is larger for high K_{app}–values than for low values (see Table 2). If the BDTT for the present loading rate is higher than 1300°C the effect of bridging is expected to decrease with increasing loading rate (= increasing K_{app}–value).

Crack growth resistance was shown to be promoted by large grain size and aspect ratio since larger grains can bridge the crack at a larger crack opening and they do not fracture as easily as small grains. (Assuming a constant shear stress at the interface, the small grain will fracture more easily since the strength of the grain is directly proportional to r^2 while the load transferred through the interface is directly proportional to r, where r is the grain radius). Further, the deformation resistance of multigrain ligaments in the large grain material was believed to be higher than that of the small grain size material since creep resistance is known to increase with increasing grain size.

It is also worth noting that glass bridging gave only a very minor contribution to bridging. Even if observations of the crack path on the side surface suggested that the crack wake was filled with a glassy phase, fracture surfaces of specimens with cracks grown at high temperature but fractured at room temperature exhibited very few glass bridges (see Figures 5a–d and 6).

4.1. Cyclic Loading

The lower threshold values and the higher crack growth rates observed at 1300°C compared to at room temperature were the result of softening of the amorphous grain boundary phase. This promotes grain boundary sliding and nucleation, growth and coalescence of cavities at grain boundaries which was the dominating crack growth mechanisms at 1300°C (see Figure 6). At room temperature the crack growth occurs by a combination of grain boundary microcracking and transgranular cleavage fracture[28]. The intergranular cavitation and microcracking ahead of the crack tip created ligaments bridging the crack wake. At room temperature mainly single grain bridges formed but at 1300°C also multi–grain ligaments (see Figures 5a–d and 6).

Crack growth direction

Figure 6. SEM–micrograph of crack path of the large grain size material containing a crack grown under cyclic loading at a K_{app} of 6 MPam$^{1/2}$.

The results from the room temperature fracture toughness tests made on specimens containing cracks grown under cyclic loading at high temperature confirmed that the amount of bridging was significant, especially at high K_{app}–levels (see Table 2). At low K–levels the crack growth rate was small and bridging was less effective since many of the bridging ligaments were broken, due to cyclic loading, before new ligaments were formed. For the LG material at K–levels near the threshold value the effect of bridging was less than during a room temperature fracture toughness test (see Table 2).

4.2. Static Loading

The crack propagation behaviour under static loading at 1300°C was significantly different compared to that at room temperature (see Figure 3b). The viscous behaviour of the amorphous grain boundary phase at 1300°C promoted cavitation and consequently crack growth was observed for lower load levels compared to at room temperature. Even if the nucleation, growth and coalescence of cavities resulted in the formation of crack wake bridges bridging was found not to be significant at low K_{app}–levels (see Figure 3b). The amount of bridging ligaments at high and low sustained K_{app}–levels at 1300°C is indicated in Table 2. It shows that the amount of bridging ligaments increases with increasing K_{app}–levels, but a significant amount of bridges were formed also at low K_{app}–levels.

The reason for the small bridging effect at low K_{app}–levels is that the crack opening

196

displacement rate was low and the bridges could only provide a small bridging force since stress relaxation could take place. With increasing K_{app}-level the crack growth rate and also the crack opening displacement rate increased rapidly, which increased the effect of bridging. Consequently, the K_{app} range over which stable crack growth occurred was significantly larger at 1300°C compared to at room temperature. As an example, for the small grain size material at K_{app}-levels near failure the crack growth rates under static loading is lower at 1300°C compared to at room temperature. The effect of crack opening displacement rate on the shielding effect is discussed in sections 4.3 and 4.4.

4.3. Comparison Between Static and Cyclic Loading

The effect of cyclic loading on the crack growth behaviour at 1300°C was the opposite compared to that at room temperature. At room temperature cyclic loading resulted in higher crack growth rates than sustained loading due to fracture of the grain bridges and fretting wear of asperity contacts that were created during crack growth[28]. From the fracture toughness tests made at room temperature on testpieces containing cracks grown under static and cyclic loading at 1300°C it was clear that the amount of bridging ligaments could not explain the beneficial effect of cyclic loading (see Table 2 and Figures 5a–d). The fracture toughness values of specimens with cyclically grown cracks were lower than those for specimens with statically grown cracks, and this gives an indication of the amount of bridges.

The lower crack growth rates observed under cyclic loading as compared to sustained loading, especially for low K_{app}-levels, was believed to be due to the viscous behaviour of the amorphous grain boundary phases at temperatures above T_g. If the grain boundary phase is viscous, the bridging force that can be provided by the bridges increases with increasing pull-out speed of the grains and deformation rate of the multigrain ligaments (see Figure 7). The pull-out speed of grains and deformation rate of multigrain ligaments is mainly determined by the crack opening displacement rate.

In the case of static loading the crack opening displacement rate, $d\delta/dt$, is mainly a function of the crack growth rate. For static K_{app}-values close to K_{th} the crack growth rate and also the crack opening displacement rate were close to zero at 1300°C. Consequently the pull-out speed of single grains and the deformation rate of multi-grain ligaments were very close to zero and therefore the load carrying capacity of the bridging grains and ligaments was very small. The grains could be pulled out slowly and the multigrain ligaments could be deformed without much resistance. This can be understood by replacing the bridges with Maxwell elements where the elastic part of the bridges is represented by the spring and the viscous part is represented by the dash pot (see Figure 7).

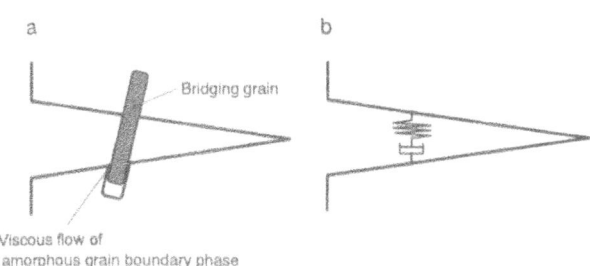

Figure 7. Schematic showing (a) bridging grain and (b) bridging grain replaced by Maxwell element.

In the case of cyclic loading the crack opening displacement rate is mainly determined by the loading rate, dK/dt, which is proportional to the frequency and the load amplitude. Even at K_{app}-values close to the threshold value dK/dt was large enough to provide effective load transfer to the bridging ligaments. It is evident from Table 2 and Figures 5a–d that the amount of bridging grains and ligaments was smaller compared to for sustained loading. The reason for this was that during cyclic loading more load was transferred to the bridging grains and ligaments and some of them were damaged or destroyed. Despite this, bridging in the low K_{app}-level regime was much more effective under cyclic loading than under static loading.

At large K_{app}-values the crack growth rate increased by several orders of magnitude, compared to K_{app}-values close to K_{th}, under both static and cyclic loading. Under static loading also the dδ/dt increased several orders of magnitude since it can be assumed to be proportional to the crack growth rate. For the sustained loading the load carried by the bridging grains and multigrain ligaments increased rapidly with increasing K_{app}-level and consequently the slope in the K_{app}–da/dt diagram was less steep than that for cyclic loading (see Figures 4a–b). Also for the cyclic loading case the crack opening displacement rate increased with increasing K_{app}-levels. However, the increase for cyclic loading was not several orders of magnitude but merely a linear increase with increasing K_{app}. Consequently, the crack growth rate for high K_{app}-levels became similar for both static and cyclic loads. For the small grain size material at very high K_{app}-levels the crack growth resistance was even higher under sustained loading.

The effect of cyclic loading can be more precisely investigated by first assuming that there is no cyclic effect at high temperature. This was suggested in an early study on high temperature crack growth of a Si_3N_4 by Evans et. al.[23]. If this is true it should be possible to predict the crack growth rates under cyclic loading based on the crack growth rates under static loading. The crack growth under static loading for a stress intensity, K, larger than the threshold value can be approximated by:

$$\frac{da}{dt} = A \cdot K^n \qquad (6)$$

where A and n are constants. The values of n and A of the small grain size material were 6.82 and $3.4 \cdot 10^{-11}$ m/(s·(MPam$^{1/2}$)$^{6.82}$) respectively and of the large grain size material 6.69 and $9.71 \cdot 10^{-12}$ m/(s·(MPam$^{1/2}$)$^{6.69}$) respectively. The amount of crack growth under one loading cycle for cyclic loading, which is equal to da/dN, can be predicted by integrating the static crack growth rates over one loading cycle:

$$\frac{da}{dN} = \int_0^{1/v_c} A \cdot K(t)^n dt \qquad (7)$$

where v_c is the frequency of the cyclic loading. In Figures 8a–b the measured crack growth rates under cyclic loading are compared with the crack growth rates predicted from static crack growth rates. For the predicted crack growth rates it was assumed that no crack growth occurred at K-levels below K_{th} for sustained loading. The predicted curves were discontinued at K_{app}-levels when the crack growth under static loading could not be described by equation 6. As is seen in Figures 8a–b the materials in the present study indeed exhibited a beneficial effect of cyclic loading on the crack growth resistance, especially at low K_{app}-levels. As was indicated earlier, this was the effect of the rate dependent deformation resistance of the single and multi–grain bridges formed during high temperature crack growth. Not unexpectedly, the difference in crack growth rates between cyclic and static loading as shown in Figures 8a–b was smaller than that in Figures 4a–b. At the highest K_{app}-levels the predicted crack growth rates were lower than the measured.

4.4. Effect of Temperature and Loading Rate

At temperatures above T_g for the amorphous grain boundary phase the crack growth behaviour under both static and cyclic loading is very sensitive to changes in temperature and loading rates. Earlier studies have shown that both positive and negative effects of cyclic loading on the crack growth resistance are possible. At room temperature silicon nitrides normally exhibit higher crack growth rates under cyclic loading. However, even at temperatures above T_g for the amorphous phase cyclic loading can have a negative effect on the crack growth resistance. Ramamurty et. al[17] showed that cyclic loading had a negative effect on the crack growth resistance of a silicon nitride containing only 4 wt% yttria as sintering additive. In the same study it was also shown that a SiC–whisker–reinforced silicon nitride with both alumina and yttria as sintering additives also showed a negative effect of cyclic loading on crack growth resistance both at 1300 and 1400°C. However, at 1400°C the

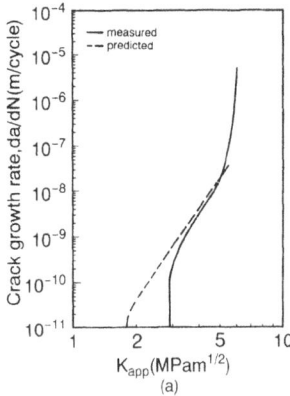
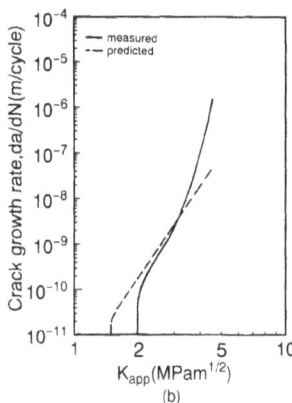

Figure 8. Relationship between K_{app} and measured and predicted crack growth rates for (a) the large grain size material and (b) the small grain size material.

crack growth rates for sustained loading and cyclic loading at a low frequency of 0.1 Hz were similar. Also Ogawa et. al.[18] reported a negative effect of cyclic loading on the crack growth resistance of a silicon nitride at 1200°C. However, in the present study and in a study by Liu et. al.[19] a positive effect of cyclic loading on the crack growth resistance of silicon nitride was reported at 1300 and 1400°C respectively. The reasons for the discrepancy in the results is that silicon nitrides with various amounts and compositions of the amorphous grain boundary phases have been used and that different temperatures and frequencies have been used. As was pointed out by Jacobs and Chen[49] the crack growth behaviour is determined by the balance between crack shielding accumulation and degradation.

At room temperature bridging is a well known phenomena and it is widely accepted that the higher crack growth rates observed under cyclic loading is the effect of degradation of grain bridges[26-35]. At temperatures above T_g for the amorphous grain boundary phase also time dependent deformation processes becomes important, and this promotes the formation of bridging ligaments. Sustained loading consequently results in more bridging ligaments compared to cyclic loading. However, the bridges become effective only if the deformation rate of the bridges is large enough. In the case of sustained loading the deformation rate of the bridging ligaments are only large enough at high crack growth rates, while during cyclic loading the bridging ligaments are effective also at low crack growth rates. If the deformation rate of the ligaments is high enough it will result in a damaging effect of the bridges. The negative effect of cyclic loading reported[17,18] was due to the damaging effect of the bridges (similarly to at room temperature). The positive effect of cyclic loading reported in this study and by Liu et al.[19] was the result of more effective load transfer to the bridges without causing too much damage. The present observations are consistent with other studies[47,50,51] on the effects of loading rate and temperature on the crack growth of ceramics.

5. CONCLUSIONS

At 1300°C the crack growth resistance was shown to be higher under cyclic loading as compared to sustained loading. Fracture toughness tests and fractography of testpieces containing cracks grown at high temperature under static and cyclic loading showed that the beneficial effect of cyclic loading could not be explained by a larger amount of bridging ligaments. Bridges were more frequently observed at high K_{app}–levels and in the testpieces subjected to static loading. Unlike at room temperature, the bridging ligaments consisted both of single and multiple grains. It was concluded that the effect of bridging was determined by the amount of bridges and the crack opening displacement rate. The crack opening displacement rate dependence was the result of the viscous behaviour of the amorphous grain boundary phase. In the case of cyclic loading the crack opening displacement rate is determined by dK/dt and hence bridging was effective also at low K_{app}–levels. In the case of sustained loading the crack opening displacement rate was determined by the crack growth rate and bridging therefore only became effective at high K_{app}–levels. At high K_{app}–levels the crack growth rates were similar under both loading conditions. The crack growth rates were lower for the large grain size material both under static and cyclic loading and at both temperatures.

6. REFERENCES

1. J. A. Todd, Z.-Y. Xu, The high temperature creep deformation of Si_3N_4–$6Y_2O_3$–$2Al_2O_3$, J. Mater. Sci., 24:4443(1989).
2. B. J. Hockey, S. M. Wiederhorn, W. Liu, J. G. Baldoni and S.-T. Buljan, Tensile creep of whisker-reinforced silicon nitride, J. Mater. Sci., 26:3931(1991).
3. S. M. Wiederhorn, B. J. Hockey, D. C. Cranmer and R. Yeckley, Transient creep behaviour of hot isostatically pressed silicon nitride, J. Mater. Sci., 28:445(1993).
4. M. M. Chadwick and D. S. Wilkinsson, Microstructural evolution in annealed and crept silicon nitride, J. Am. Ceram. Soc., 76:376(1993).
5. M. M. Chadwick and D. S. Wilkinson, Creep behavior of a sintered silicon nitride, J. Am. Ceram. Soc., 76:385(1993).
6. T. Ohji and Y. Yamauchi, Tensile creep and creep rupture behavior of monolithic and SiC–whisker-reinforced silicon nitride ceramics, J. Am. Ceram. Soc., 76:3105(1993).
7. M. K. Ferber, M. G. Jenkins, T. A. Nolan and R. L. Yeckley, Comparison of the creep and creep rupture performance of two HIPed silicon nitride ceramics, J. Am. Ceram. Soc., 77:657(1994).
8. J.-L. Ding, K. C. Liu, K. L. More and C. R. Brinkman, Creep and creep rupture of an advanced silicon nitride ceramic, J. Am. Ceram. Soc., 77:867(1994).

9. M. N. Menon, H. T. Fang, D. C. Wu, M. G. Jenkins, M. K. Ferber, K. L. More, C. R. Hubbard and T. A. Nolan, Creep and stress rupture behavior of an advanced silicon nitride part I–III, Experimental observations, Creep rate behavior and Stress rupture and the Monkman–Grant relationship, *J. Am. Ceram. Soc.*, 77:1217(1994).

10. C. J. Gasdaska, Tensile creep in an in-situ reinforced silicon nitride, *J. Am. Ceram. Soc.*, 77:2408(1994).

11. A. G. Evans, L. R. Russel and D. W. Richerson, Slow crack growth in ceramic materials at elevated temperatures, *Met. Trans. A.*, 6A:707(1975).

12. T. Fett, G. Himsolt and D. Munz, Cyclic fatigue of hot–pressed Si_3N_4 at high temperatures, *Adv. Ceram. Mater.*, 1:179(1986).

13. M. Masuda, T, Soma, M. Matsui and I. Oda, Fatigue of ceramics (Part 3)– Cyclic fatigue behavior of sintered Si_3N_4 at high temperatures, *J. Ceram. Soc. Jpn. Inter. Ed.*, 97:601(1989).

14. M. G. Jenkins, M. K. Ferber and C.-K. J. Lin, Apparent enhanced fatigue resistance under cyclic tensile loading for a HIPed silicon nitride, *J. Am. Ceram. Soc.*, 76:788(1993).

15. M. G. Jenkins, M. K. Ferber and C.-K. J. Lin, Beneficial effects of cyclic tensile loading on the fatigue resistance of an Si_3N_4, *J. Mater. Sci.*, 12:1940(1993).

16. C.-K. J. Lin, M. G. Jenkins and M. K. Ferber, Cyclic fatigue of hot isostatically pressed silicon nitride at elevated temperatures, *J. Mater. Sci.*, 29:3517(1994).

17. U. Ramamurty, T. Hansson and S. Suresh, High–temperature crack growth in monolithic and SiC_w–reinforced silicon nitride under static and cyclic loads, *J. Am. Ceram. Soc.*, 77:2985(1994).

18. T. Ogawa, M. Hirose and K. Tokaji, Evaluation of elevated–temperature crack growth in ceramics under static and cyclic loads, *in:* Proc. of Plastic Deformation of Ceramics, Snowbird, Utah, Aug. 7–12, 1994.

19. S.-Y. Liu, I.-W. Chen and T.-Y. Tien, Fatigue crack growth of silicon nitride at 1400°C: A novel fatigue–induced crack–tip bridging phenomenon, *J. Am. Ceram. Soc.*, 77:137(1994).

20. G. D. Quinn, Static fatigue in high–performance ceramics, *in:* "Methods for Assessing the Structural Reliability of Brittle Materials, ASTM STP 844, S. W. Freiman and C. M. Hudson, ed., ASTM, Philadelphia(1984).

21. G. D. Quinn, Fracture mechanism maps for advanced structural ceramics: Part 1. Methodology and hot–pressed silicon nitride results, *J. Mater. Sci.*, 25:4361(1990).

22. G. D. Quinn, Fracture mechanism maps for advanced structural ceramics: Part 2. Sintered silicon nitride, *J. Mater. Sci.*, 25:4377(1990).

23. A. G. Evans and E. R. Fuller, Crack propagation in ceramic materials under cyclic loading, *Met. Trans.*, 5:27(1974).

24. A. G. Evans, L. R. Russel and D. W. Richersson, Slow crack growth in ceramic materials at elevated temperatures, *Met. Trans.*, 6A:707(1975).

25. S. Suresh, "Fatigue of Materials", Cambridge University Press, Cambridge, UK, (1991).

26. S. Horibe and R. Hirahara, Cyclic fatigue of ceramic materials: Influence of crack path and fatigue mechanisms, *Acta Met. Mater.*, 39:1309(1991).

27. R. H. Dauskardt, A frictional–wear mechanism for fatigue–crack growth in grain bridging ceramics, *Acta Met. Mater.*, 41:2765(1993).

28. Y. Mutoh, M. Takahashi and M. Takeuchi, Fatigue crack growth in several ceramic materials, *Fatigue Fract. Engng. Mater. Struct.*, 16:875(1993).

29. B. E. Corneliessen, R. H. Dauskardt, R. O. Ritchie and G. Thomas, Cyclic fatigue and fracture toughness of silicon nitride ceramics sintered with rare–earth oxides, *Acta Met. Mater.*, 42:3055(1994).

30. D. C. Salmon and D. W. Hoeppner, In-situ observation of fatigue crack propagation in NT–154 silicon nitride, *in:* "Cyclic Deformation, Fracture and Nondestructive Evaluation of Advanced Materials:Second Volume, ASTM STP 1184", M. R. Mitchell and O. Buck, ed., ASTM, Philadelphia (1994).

31. G. Choi, S. Horibe and Y. Kawabe, Cyclic fatigue in silicon nitride ceramics, *Acta Met. Mater.*, 42:1407(1994).

32. H. Kishimoto, A. Ueno and S. Okawara, Crack propagation behaviour of polycrystalline alumina under static and cyclic loads, *J. Am. Ceram. Soc.*, 77:1324(1994).

33. D. S. Jacobs and I.-W. Chen, Mechanical and environmental factors in the cyclic and static fatigue of silicon nitride, *J. Am. Ceram. Soc.*, 77:1153(1994).

34. M. Li and F. Guiu, Subcritical fatigue crack growth in alumina–I. Effects of grain size, specimen size and loading mode, *Acta Met. Mater.*, 43:1859(1995).

35. M. Li and F. Guiu, Subcritical crack growth in alumina–II. Crack bridging and cyclic fatigue mechanisms, *Acta Met. Mater.*, 43:1871(1995).

36. T. Sadahiro and S. Takatsu, A new precracking method for fracture toughness testing of cemented carbides, *in:* "Modern Developments in Powder Metallurgy, vol. 14", H. H. Hausner, H. W. Antes and G. D. Smith, ed., Plenum Press, New York (1981).

37. R. Warren and B. Johannesson, Creation of stable cracks in hard metals using 'bridge' indentation, *Powder Metall.*, 27:25(1984).

38. T. Nose and T. J. Fuji, Evaluation of fracture toughness for ceramic materials by a single–edge–precracked–beam method, *J. Am. Ceram. Soc.*, 71:328(1988).

39. J. D. Landes and J. A. Begley, A fracture mechanics approach to creep crack growth, *in*: "ASTM STP 590", American Society for Testing and Materials, (1976).

40. J. R. Rice, A path independent integral and the approximate analysis of strain concentration by notches and cracks, *Trans. ASME– J. Appl. Mech.* 35:379 (1968).

41. C. F. Shih and A. Needleman, Fully plastic crack problems, part 1: Solutions by a penalty method, *Trans. ASME– J. Appl. Mech.* 51:48 (1984).

42. H. Riedel and J. R. Rice, Tensile cracks in creeping solids, *in*: "Fracture Mechanics: Twelfth Conference, ASTM STP 700", American Society for Testing and Materials, (1980).

43. K. Sato, K. Tanaka, Y. Nakano and T. Mori, Temperature dependence of anelastic deformation in polycrystalline silicon nitride, *J. Am. Ceram. Soc.* 76[8]:2042 (1993).

44. H. Riedel, Crack–tip stress fields and crack growth under creep fatigue conditions, *in*: "Elastic–Plastic Fracture:Second Symposium, Volume I–Inelastic crack analysis, ASTM STP 803," C. F. Shih and J. P. Gudas, ed., American Society for Testing and Materials, (1983).

45. Y. Mutoh, Y. Miyashita, T. Hansson and M. Takahashi, Effect of grain size on fatigue crack growth in silicon nitride and alumina, *in:* Proc. of Plastic Deformation of Ceramics, Snowbird, Utah, Aug. 7–12, 1994.

46. A. Lakki, R. Shaller, G. Bernard–Granger and R. Duclos, High temperature anelastic behaviour of silicon nitride studied by mechanical spectroscopy, *Acta Met. Mater.*, 43:419(1995).

47. T. Rouxel and F. Wakai, The brittle to ductile transition temperature in a Si_3N_4/SiC composite with a glassy grain boundary phase, *Acta Met. Mater.*, 41:3203(1993).

48. Y. Mutoh, K. Yamaishi, N. Miyahara and T. Oikawa, Brittle–to–ductile transition in silicon nitride, *in*: "Fracture Mechanics of Ceramics," R. C. Bradt *et al.*, ed., Plenum Press, New York (1992).

49. D. S. Jacobs and I–W. Chen, Cyclic fatigue in ceramics: A balance between crack shielding accumulation and degradation, *J. Am. Ceram. Soc.*, 78:513(1995).

50. H. Peterlik, Viscoelasticity of ceramics at high temperatures, *J. Mat. Sci.*, 29:2401(1994).

51. I.–W. Chen, S.–Y. Liu and D. Jacobs, Effect of temperature, rate and cyclic loading on the strength and toughness of monolithic ceramics, *Acta Met. Mater.*, 43:1439(1995).

SIMULATING NON-EQUILIBRIUM FRACTURE BEHAVIOR

OF CERAMICS IN CONTROLLED BENDING TEST

Sergei P. Kovalev and George G. Pisarenko

Institute for Problems of Strength
National Academy of Sciences of Ukraine
Kiev 252014 Ukraine

INTRODUCTION

Stability of fracture test of ceramics is very important in acquiring valid and representative information using limited amount of material and simple test geometries.[1] It is well-known that displacement-controlled test is more stable, than load-controlled, however, direct quantitative comparisons are not yet made. The latter test will be substantially more stable, if one will use parallel stiffness in the specimen load train.[2] Single-edge notch bend test with parallel stiffness formally allows monitoring nonequilibrium fracture behavior of ceramic material under non-monotonous stress intensity history.[3-5] It is possible to initiate several crack acceleration-retardation runs on a single specimen. The output time-dependence of stress intensity during crack growth in the test is determined by starting notch depth and by the cross-head displacement function, $u_{tm}(t)$. Using computer-aided crack velocity determination by real-time measurement of instantaneous specimen compliance during crack extension makes it possible to receive full description of nonequilibrium fracture behavior in the form of closed crack velocity vs stress intensity curve $(V(K))$ recorded on a single specimen. In practice, it is not easy to choose input $u_{tm}(t)$ to fit desired stress intensity history, especially for material with unknown microstructure contribution to toughness, or in the presence of residual stresses. To overcome these problems, fracture tests could be simulated by numerical solution of a power-law crack-growth differential equation for unknown crack extension.[6,7] Simulating fracture test for a model material response is useful not only for correctly adjusting test parameters and selecting loading history but also in interpretating fracture test data and modeling respective fracture mechanisms. The model material response may incorporate stress intensity of both microstructural crack shielding and macroscopic residual stress along prospective crack path. By comparing controlled fracture measurements and numerical simulations for the same stress intensity history it is possible to determine mechanism of microstructure-related contribution to toughness.

The aim of this work is twofold. At first, various aspects of controlled bend test are discussed to demonstrate quantitative possibilities and limitations of controlled test based on parallel stiffness. After that, measured and simulated nonequilibrium fracture behavior of specific ceramic specimens showing nonlinear crack resistance in controlled test is compared. On the basis of such comparison some conclusions are drawn concerning feasibility of certain toughening models for microcracking and transformation toughening in general case of non-monotonous stress-intensity history.

STABILITY OF CONTROLLED FRACTURE BEND TEST

Stability of fracture in mechanical test is defined by variation in output test parameter, e.g., stress intensity, caused by a given increment of input test parameter to be controlled by operator or computer. Cross-head displacement or load are taken usually as input test parameters. Relation between load and displacement depends on the compliance of the testing arrangement. In a simple fracture test, the compliance is composed of sequentially connected compliances of the specimen and of the testing machine

$$C_{tm} + C_s(x) = u_{tm}(t)/P_{tm}(t) \qquad (1)$$

where P_{tm} is testing machine load, $u_{tm}(t)$ is the function of cross-head displacement, C_{tm} is testing machine compliance, and C_s is the compliance of the specimen load train, specifically, for a three-point single-edge notched bend (3pSENB), Fig. 1, (A).

$$C_s(x) = C_0 + (2L^2/BW^2)(1 - v^2)/E \int_0^x Y^2(x)dx \qquad (2)$$

here C_0 is crack-independent component, which includes the compliance of unnotched specimen, the fixture, and other components of load train contained inside the working space of the machine. These components may include, for example, compliance of the load cell for measuring specimen load, contact compliance, arising from imperfect contacts between the specimen and the fixture, etc. Second term in Eq. (2) corresponds to crack-induced compliance of the 3pSENB specimen, where E is Young's modulus, v is Poisson ratio, L is the lower fixture span, B is specimen thickness, W is specimen width, and $x = a/W$ is relative crack length, Y is a calibration function for 3-point SENB test.[8,9]

In the load-controlled test, stress intensity is of the form

$$K(x, t) = P_{tm}(t)Y(x)(L/WBW^{1/2}), \qquad (3)$$

where $P_{tm}(t)$ is input controlled parameter. Note, that Eq. (3) is independent of compliance of the test arrangement. Respective relation for a displacement-controlled test is derived by combining (3) and (1)

$$K(x, t) = u_{tm}(t) Y(x)(L/WBW^{1/2})/[C_{tm} + C_s(x)], \qquad (4)$$

If a parallel stiffness is introduced in the loading scheme, as shown in Fig.1, (B), redistribution of a load between the specimen and the parallel elastic element (PEL) gives Eq.(4) in the form

$$K(x, t) = u_{tm}(t) \, Y(x)(L/WBW^{1/2})/[C_{tm} + C_s(x)(C_{tm}/C_{pel} + 1)], \qquad (5)$$

where C_{pel} is the PEL compliance and crack-independent part of compliance C_s now includes only those components, which are contained inside the PEL. It is evident, that Eq. (4) represents special case of the Eq. (5) for an infinite value of C_{pel}. Cross-head displacement is controlled parameter in last type of testing.

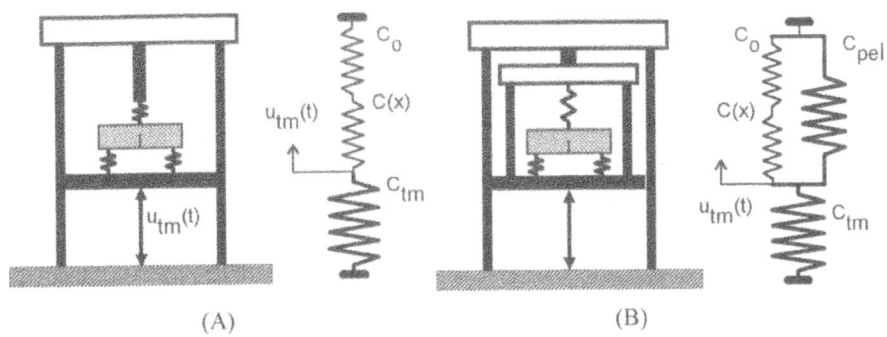

Figure 1. Schematic of compliance components connection in 3-point single-edge notched bend fracture test: (A) simple configuration. (B) configuration with parallel elastic element.

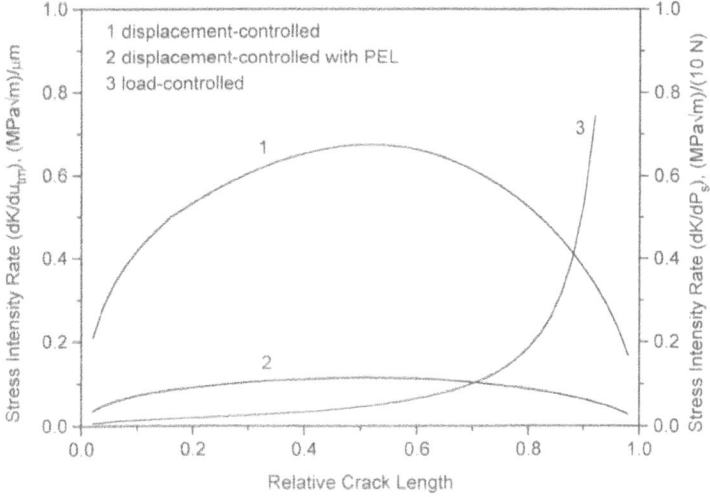

Figure 2. Comparison of fracture test stability. shown as stress intensity increment per 1 micron of cross-head displacement change in displacement-controlled test (per 10 Newtons of specimen load rise in load-controlled test)

Direct comparison of different test control schemes is possible in terms of stress intensity rate per unit of input parameter increment. The lower is the rate the more stable and controllable is the test. Figures 2 shows stability of load-controlled and both kinds of displacement-controlled test, where the unit of input parameter is 10 Newtons and 1 micron, respectively. Another measure of fracture test stability is stress intensity rate per unit crack increment (Fig. 3). It is clear that displacement-controlled test is stable, and load-controlled is unstable. To maintain controlled fracture in the most suitable range of starting notch depths (from $0.4W$ to $0.6W$), very slow controlled loading rates are needed. Figure 2 demonstrates how parallel stiffness could enhance the degree of test stability. Introducing PEL enables to achieve small controllable increments in stress intensity without precise small cross-head displacement, which require special testing machine. As was shown,[3,5] that range of test stability, namely, the range of negative dK/dx values, depends on C_0, where contact compliance components is most difficult to control and estimate. Varying starting notch depth within the region of the rising branch of the calibration curve and selecting an appropriate input function $u_{tm}(t)$ makes it possible to run a test with non-monotonous stress intensity history, which consisted of a combination of crack acceleration and crack retardation stages. Generally, it is possible to initiate several crack acceleration-retardation runs on a single 3pSENB specimen. It should be noted that all conclusions drawn are valid for linear-elastic material.

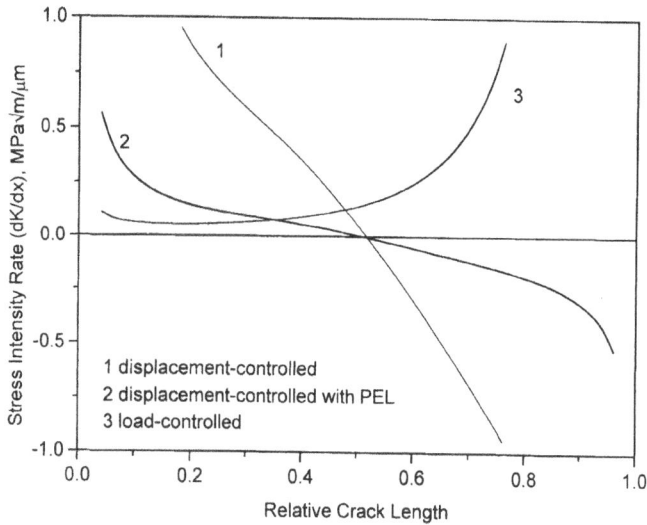

Figure 3. Stress intensity increment per 1 micron of crack advance for different controlled test configurations

EXPERIMENTAL MEASUREMENTS

The tests were conducted under displacement control on the electro-hydraulic testing machine (*ZD-10*, WPM, Dresden, Germany) with the use of parallel stiffness. Sample loading histories are shown in Fig. 4. The function $u_{tm}(t)$ was controlle manually. Instantaneous values of crack length and stress intensity were determined via secant compliance from the real-time digital load-deflection record of the specimen. Respective crack velocities were calculated from crack length vs time using finite differences.[5] Computer-aided fracture data acquisition substantially increases the amount of information obtainable from the individual fracture test. Therefore, experimental data presented for each case of material and type of notch correspond to the test of a single specimen. These results, however, are typical enough to follow general trends of fracture kinetics.

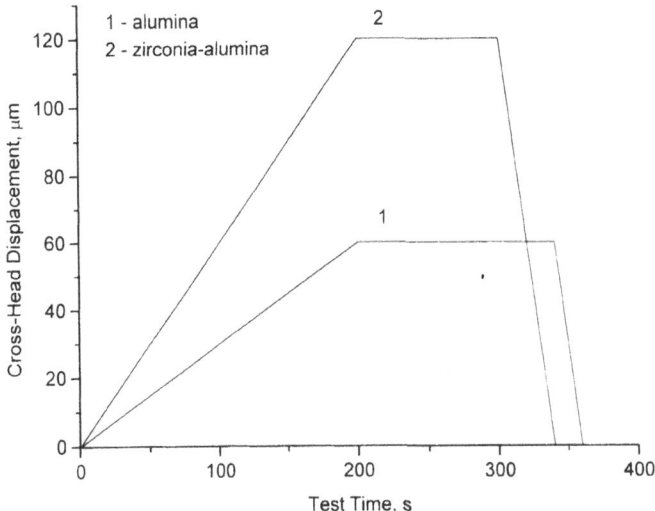

Figure 4. Loading histories for tesing and simulating displacement-controlled fracture test with parallel stiffness.

Table 1. Properties of Materials[1]

Composition	Material notation	Young modulus, GPa	Average grain size, μm	Density, g/cm³
Al_2O_3 + 0.1%MgO	AOT	366	10	3.9
ZrO_2 + 2mol.%Y_2O_3 + 5mol.%Al_2O_3	TZP2Y5A	170	1 (ZrO_2) 2 to 5 (Al_2O_3)	5.1

[1]NPO "Tekhnologhiya", Obninsk, 249020, Russian Federation

Materials and Test Specimens

Experimental results for alumina and tetragonal zirconia-alumina composite are presented (see Table 1), because both exhibit nonlinear fracture behavior presumably resulting from different reasons: microcrack-induced rise in crack-growth resistance in alumina and the transformation toughening in zirconia-alumina composite. Alumina ceramics (AOT) is a high-purity partially translucent material. Yttria-stabilized zirconia-alumina composite (TZP2Y5A) contains 5 mol% of alumina. Mean grain size of the zirconia matrix was estimated as 1 μm, and intergranular inclusions of alumina were from 2 to 5 μm. Ground specimens of dimensions B = 3 mm, W = 6 mm, L = 24 mm were tested. Straight-through and chevron starting notches of 0.08 to 0.09 mm thick were cut using a diamond-contained composite saw. Other details concerning materials, specimens, and testing system can be found elsewhere.[5]

Alumina

Alumina specimens with straight-through starting notches were tested. Starting-notch depth was chosen close to the $dK/dx = 0$. Figure 5 shows the $V(K)$ curve for a single specimen, respective K_r curve could be found in Fig. 6. Crack-acceleration range on the $V(K)$ is of a gradual slope. Crack retardation range shows a more steep slope and it seems that crack growth exponent of crack retardation range depends on the preceding crack extension.[4,5] The observed behavior corresponds well to a "memory" effect, which is attributed to crack-interface grain-bridging as a result of local microcracking and crack-branching during the growth of a macrocrack.[10-12]

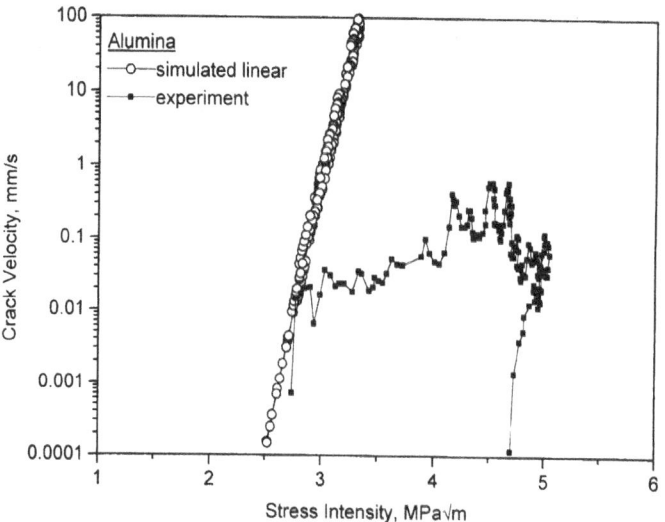

Figure 5. Crack velocity vs stress intensity as measured for alumina (AOT). Linear (environmental) crack growth parameters shown in Table 2 were calculated from the linear fit of the experimental range prior to the supposed start of nonlinearity.

Yttria-Stabilized Zirconia-Alumina

Experimental $V(K)$ curves for zirconia-alumina composite *TZP2Y5A* are shown in Fig. 7. Respective K_r curves are shown in Fig. 8. Both straight-through and chevron starting notches were used. This is because a crack started from the straight-through notch only after attaining high enough level of stress intensity, and crack acceleration range appeared to be obscured by the probable influence of residual transformation stresses in the notch. *Subsequent* stress intensity reduction was accompanied by an insignificant change in the crack velocity, and thus the plateaulike region appears in the $V(K)$ curve. After reducing stress intensity to a certain "threshold" value the crack velocity decreased faster, and the next range of V(K) curve has more steep slope. For transformation-toughened ceramics this plateau could be explained as a result of effective crack retardation by stress-induced phase transformations, which started above a certain level of stress intensity,[13] despite the similarity of this curve with diffusion-limited plateau behavior of environmental slow-crack-growth curve. All reported experiments were conducted in an ambient laboratory environment (20 to 25 C, and a relative humidity of 80% to 90%); therefore, the appearance of diffusion-limited behavior is questionable. Experimental data for chevron notch show something like reversible behavior. That is, low-slope range of the V(K) curve appears in both acceleration and retardation ranges. The appearance of acceleration range resembles that of alumina, however, retardation range was distinctly different. In contrast to alumina, where crack-growth exponent in crack-retardation range becomes higher, and "remember" accumulated crack resistance, in $TZP2Y5A$ ceramics with chevron notch the retardation range follows the trace of acceleration and high stress intensity level at the maximum velocity during retardation reduces to the initial values. The observed behavior is unlikely to be wake-dominated phenomenon, rather, it looks like process-zone dominated non-linear elasticity.

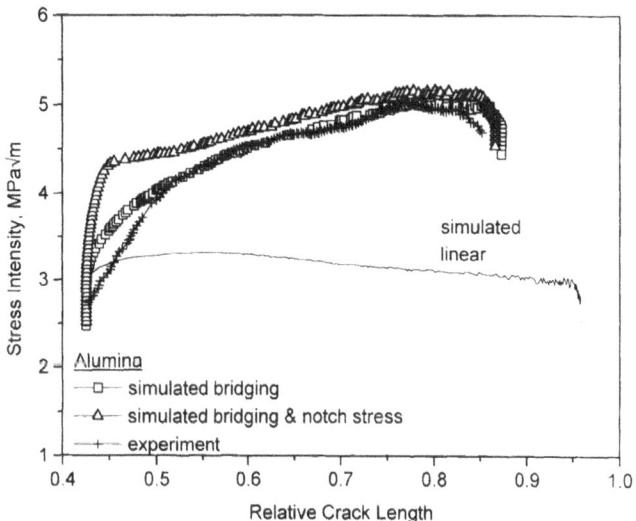

Figure 6. Comparison of simulated and measured stress intensity vs crack increment (K_r curve) for alumina.

SIMULATING PROCEDURE

Presented formulation of controlled test stress intensity calibration via compliance and other parameters of testing system is suitable to simulate non equilibrium fracture behavior of a specimen under a given stress intensity history. Simulating procedure is based on numerical solution of a power-law crack-growth differential equation for unknown crack extension $x=a/W$. The purpose is to simulate fracture behavior of a model material specimen under in the same tetsing environment used in the real experiment discussed in preceding section.

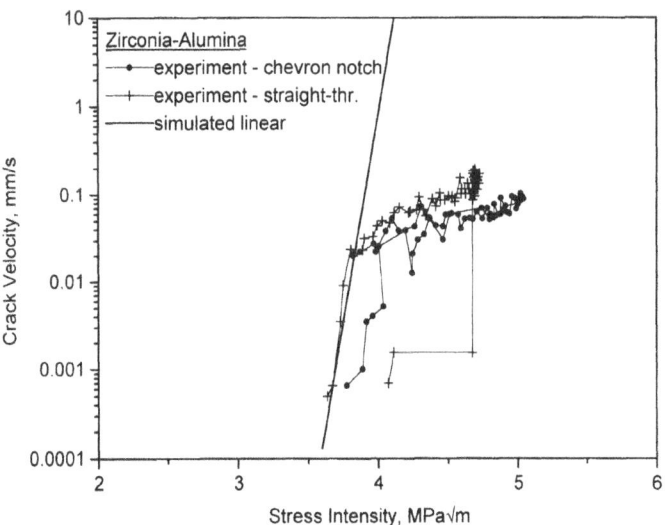

Figure 7. Crack velocity vs stress intensity as measured for zirconia-alumina composite (TZP2Y5A). Linear (environmental) crack growth parameters shown in Table 2 were calculated from the linear fit of the low-K experimental range of the straight-through notched specimen.

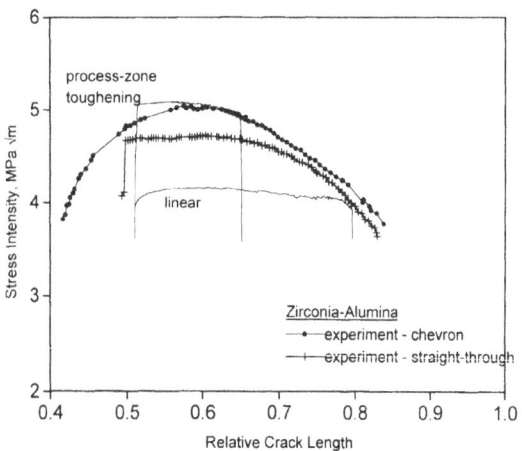

Figure 8. Comparison of simulated and measured stress intensity vs crack increment (K_r curve) for zirconia-alumina for chevron and straight-through notch types.

It is assumed that crack velocity is unambiguously defined by crack-tip stress intensity[7]

$$dx/dt = V_0(k(t,x)/k_0)^n, \qquad (6)$$

Right-hand-side is a function of net crack-tip stress intensity k. Parameters V_0, k_0 and n are considered to be fundamental properties of the given combination of material and environment. These parameters where taken from the least-square-root fit of low-stress intensity ranges of experimental $V(K)$ curves and are shown in Table 2. Explicit form of crack-tip stress intensity can be found from the applied stress intensity function Eq.(5), after introducing certain suggestions concerning microstructure contribution to crack resistance. The most general form of crack-tip stress intensity is

$$k(t,x) = K(t,x) + K_r(x) - K_b(x) - K_p(t,x), \qquad (7)$$

where $K_r(x)$ is stress intensity contribution of residual stress along prospective crack path, $K_b(x)$ and $K_p(t,x)$ are stress intensities of microstructure contributions to toughness arised during crack extension from crack-interface bridging in the wake and frontal process-zone, respectively

Table 2. Simulating Parameters of Environmental Slow-Crack-Growth

Material notation	Slow-crack-growth exponent, n	Benchmark crack velocity V_0, m/s	Benchmark stress intensity k_0, MPa m$^{1/2}$
AOT	47	$0.015 \cdot 10^{-3}$	2.76
TZP2Y5A	83	$0.0287 \cdot 10^{-3}$	3.82

Crack-Interface Bridging in the Wake

As a reasonable approximation function $K_b(x)$ is taken quasistatic or time-independent. Therefore, it is postulated that influence of bridging in the wake of a propagating crack could be separated from the intrinsic resistance of a material to environmental slow crack growth. Function $K_b(x)$ of bridging contribution to toughness is selected in the following form[14,15]

$$K_b(x) = [K(t_s,x_s) - K(t_i,x_i)](1 - \{1-[(x-x_i)/(x_s-x_i)]^{1/2}\}^{m+1}), \quad (x_i < x < x_s) \qquad (8a)$$

$$K_b(x) = [K(t_s,x_s) - K(t_i,x_i)] \qquad (x > x_s) \qquad (8b)$$

where $x_s - x_i = x^*$ is stationary bridging zone length or spatial extent of the K_r curve, x_i and x_s are crack lengths, which correspond to the initial and saturation points of the K_r curve, $K(t_i,x_i)$ and $K(t_s,x_s)$ are respective values of applied stress intensity, m is bridging stress exponent, which varies from 0 (Dugdale model) to 2. Parameters x_i, x_s, $K(t_i,x_i)$, and $K(t_s,x_s)$ are determined from the experimental $V(K)$ curve of a simulated specimen. Crack-interface bridging model was applied to alumina and respective values of the parameters used for simulating are shown in Table 3. One important moment is that saturation point $K(t_s,x_s)$ taken from the experiment has to be corrected for slow-crack-growth occured before t_s. This correction was made using simulated slow-crack growth parameters.

Crack-Tip Process-Zone Toughening

This type of toughening was simulated by a simple model. Toughening increment $K_p(t,x)$ is taken in the form of linear function of *crack-tip* stress intensity. This function is valid within the interval confined by threshold and saturation values of stress intensity. In view of postulated Eq. (6), it means that process-zone toughening depends on crack tip velocity.

$$K_p(t, x) = 0, \qquad\qquad\qquad (k(t, x) < k_i) \qquad (9a)$$

$$K_p(t, x) = [K(t_S,x_S) - k_S] \{[K(t,x) - k_i]/[K(t_S,x_S) - k_i]\}, \qquad (k_i< k(t, x) < k_S) \quad (9b)$$

$$K_p(t, x) = [K(t_S,x_S) - k_S] \qquad\qquad\qquad (k(t, x) < k_S) \qquad (9c)$$

where

$$k_S = k_o(V_S/V_o)^{1/n} \qquad\qquad\qquad (10)$$

$$k_i = k_o; \qquad V_i = V_o; \qquad\qquad\qquad (11)$$

V_i and V_S are initial and saturation velocity limits of nonlinear (low-slope) range of experimental $V(K)$ curve, and $K(t_S,x_S)$ is the saturation stress intensity corresponded to V_S on the same $V(K)$ curve. (Here $K(t_S,x_S)$ differs from that used in the crack-interface bridging model, because the latter value was taken from the K_r curve). Note that $K_p(t, x)$ does not depends on crack advance but solely on the crack-tip stress intensity. Crack-tip process-zone toughening model was applied to zirconia-alumina composite, respective numerical values of simulating parameters are shown in the Table 4.

Table 3. Simulating Parameters of Crack-Interface Bridging

Material notation	Stationary bridging zone length, mm	Bridging stress exponent, m	Starting stress intensity K_i, MPa.m$^{1/2}$	Saturation stress intensity K_S, MPa.m$^{1/2}$
AOT	2.07	0	2.76 ($=k_o$)	4.7

Table 4. Simulating Parameters of Process-Zone Toughening

Material notation	Saturation stress intensity K_S, MPa m$^{1/2}$	Starting crack velocity V_i, m/s	Saturation crack velocity V_S, m/s
TZP2Y5A	3.82 ($=k_o$)	0.0287 10^{-3} ($=V_o$)	0.2 10^{-3}

Simulating Residual Stresses in the Notch

The presence of residual stress can strongly influence fracture behavior, which is reflected by distortions in $V(K)$ curve as well as in K_r curve. Contribution $K_r(x)$ generally is included to account for possible surface or notch root residual stresses. Stress intensity and residual stress distribution were calculated following Tandon and Green,[16] except of applying their model to the notch root instead of specimen surface.

$$K_r(x) = [2.24 \, (xW/\pi)^{1/2}] \int_{x_0}^{x} [\sigma(\xi) / (x^2 - \xi^2)^{-1/2}] d\xi \qquad (12)$$

$$\sigma(x) = \sigma_s\{1 - [(x - x_0)/d]^2\} \qquad (13)$$

where σ_s is the notch surface stress level and d is stressed layer thickness. Equation (13) describes parabolic residual stress distribution with it maximum on the surface.

After considering all contributions, the differential equation (6)-(7) has to be solved numerically for unknown crack advance x. Because crack velocity variation during test can cover several decades, the best method for the solution is predicting-correcting iteration algorithm with automatic adjustable time increment,[12] which is proved to be absolutely and relatively stable. Before simulating, parameter k_0 of linear slow crack growth equation was randomized in the band of $0.02k_0$ as a function of crack length to represent intrinsic scatter of basic crack resistance. Simulated data are presented in the form of crack extension and velocity vs. time. After that applied stress intensity is calculated from Eq. (5).

DISCUSSION OF RESULTS

Alumina and transformation-toughened (TT) zirconia-alumina composite showed distinct effects of nonlinear crack resistance in slow crack growth. Specimens were tested under non-monotonous stress intensity history, which allowed to run crack acceleration-retardation cycle on a single specimen. For alumina a shift of retardation curve to higher stress intensities comparing to acceleration is a clear indication of a "memory" effect probably caused by microcracking. This behavior seems to be reasonably explained by a crack-bridging mechanism.[11,14,15] Simulated K_r and $V(K)$ curves are shown in Fig. 6 and 9, respectively. Simulation gives very close values of applied stress intensity, but crack velocity variation demonstrates only qualitative coincidence with measured values, nevertheless, the validity of crack-interface bridging model is evident. Figure 10 shows the effect of introducing artificial residual stress in the notch root. The presence of notch-root stresses is readily recognized by distinguishing rise of the starting portion of the K_r curve.

The difference in experimental and simulated results seems to reflect the limitations of time-independent bridging stress concept, which could not be compensated by residual stress field, but the approximation could be somewhat better when applying small frontal process zone contribution.

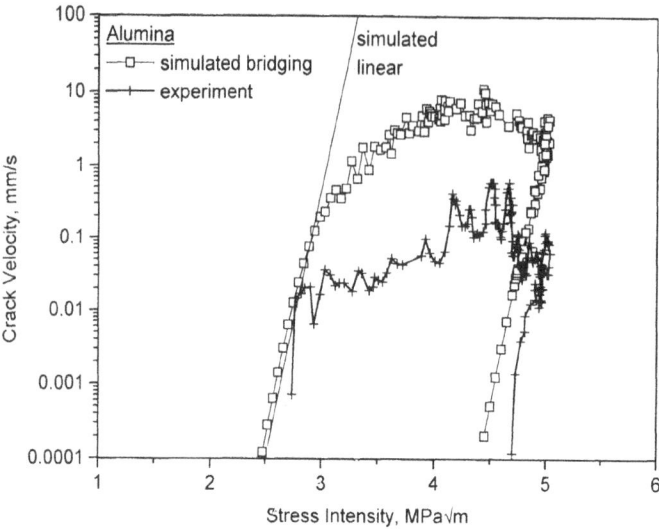

Figure 9. Crack velocity vs stress intensity for simulating crack-interface bridging in alumina (AOT). Parameters of bridging model are shown in Table 3.

Experimental data for zirconia-alumina composite does not show "memory" effect, and a model of process zone toughening was applied in simulating procedure. Fig 11 shows measured and simulated $V(K)$ curves for two types of notches, straight-through and chevron. Simulated results, calculated according to a process-zone toughening concept, display a reasonable fit to experimental $V(K)$ curve. Additional high-toughness steep range appeared on the simulated curve. This feature was observeded in controlled tests for TZP2Y ceramics.[5] At the same time rather poor coincidence of K_r curves and $V(X)$ curve is demonstrated. It seems that adopted models of both crack-interface bridging and process-zone contributions to toughness were based on oversimplified presentation of real non-equilibrium fracture behavior. Slow crack growth curve in nonlinear ceramic material for non-monotonous loading history should be represented as generalized three-dimensional $K(x,V)$ diagram. Figures 12 and 13 compares simulated and experimental data for tested specimens in three dimensions. Present simulating procedure was conducted for real loading history and other actual parameters of testing system. Possible errors were introduced mostly by constructing crack-interface bridging and process-zone toughening using in fact only one projection ($K(x)$ or $V(K)$) of experimental $K(x,V)$ diagram. This situation leads to failure in attemp to get best fit of simulated and measured results. Therefore, simulating procedure for non-equilibrium fracture behavior in case of non-monotonous loading history require more comlicated models of crack-shielding mechanisms.

CONCLUSIONS

Superior stability of displacement-controlled fracture bend test for ceramics with the use of parallel stiffnes is demonstrated, which makes it possible to measure $V(K)$ curves using simple single-edge notched three-point bend specimen under non-monotonous loading history. Test results for two ceramic materials of different fracture behavior are shown. The data for alumina ceramics confirm the "memory" effect, which indicates the presence of wake-dominated type of crack shielding. Zirconia-alumina composite displayed non-equilibrium fracture behavior that could not be related to a wake-dominated type of crack-shielding.

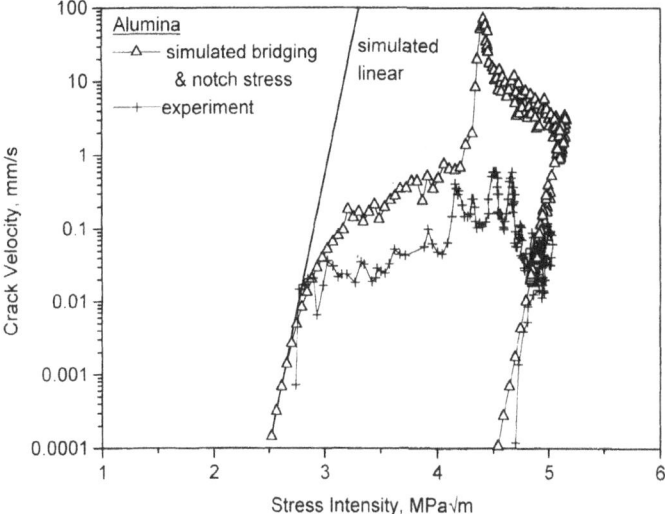

Figure 10. Crack velocity vs stress intensity for simulating crack interface bridging and notch tip residual stress in alumina (AOT). Bridging simulating parameters are the same as in Fig.5, residual stress surface value is equal -50 MPa, the depth of stressed layer is 0.3 mm.

Simulating procedure is considered to compare controlled test experimental data with model non-linear material response. A model of crack-interface bridging was used for alumina, and process-zone toughening model was used to describe behavior od zirconia-alumina composite. Numerical parameters of the models were taken from the real records of controlled testing. Simulated data showed only qualitative coincidence with those measured, especially when comparisons are made on generalized three-dimensional slow-crack-growth diagram, which was introduced to represent generalized picture of fracture behavior of ceramic material under non-monotonous loading history. It is concluded that more complicated models of non-equilibrium fracture are required to achieve better fit of simulated results to experiment. This is of prime importance for valid predictions of lifetime in cyclic and non-monotonic loading histories.

ACKNOWLEDGMENTS

This work was supported in part by the grant U4V000/U4V200 of the International Science Foundation. Authors wish to thank Drs V.M.Chushko and A.I.Kovalev for help with experiments.

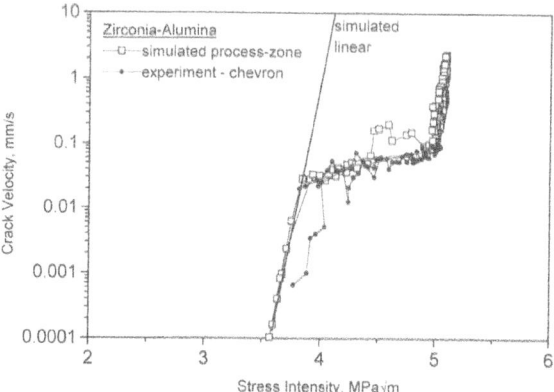

Figure 11. Crack velocity vs stress intensity for simulating crack interface bridging in zirconia-alumina (TZP2Y5A). Parameters of process-zone model are shown in Table 4.

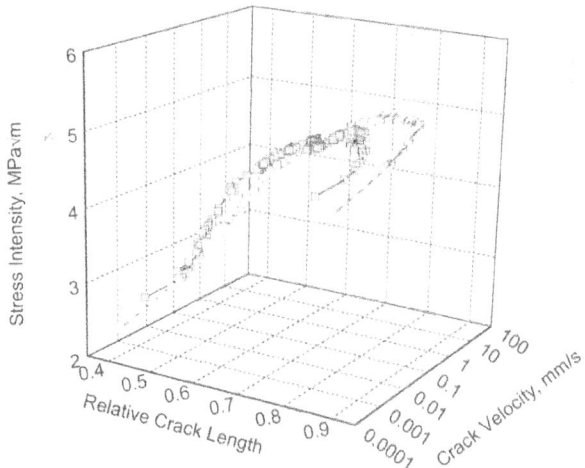

Figure 12. Generalized three-dimensional diagram of non-equilibrium fracture behavior of alumina, experiment is indicated by grayed symbols, simulation - by opened.

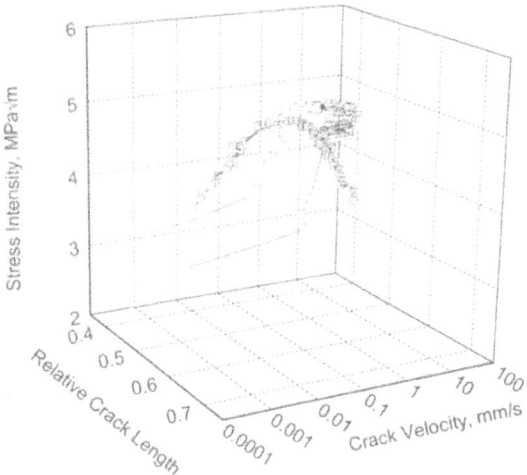

Figure 13. Generalized three-dimensional diagram of non-equilibrium fracture behavior of zirconia-alumina composite, experiment is indicated by grayed symbols, simulation - by opened.

REFERENCES

1. D.Munz, Effect of specimen type on the measured values of fracture toughness of brittle materials, *in:* "Fracture Mechanics of Ceramics," vol.6., R.C.Bradt, D.P.H.Hasselman, F.F.Lange, and A.G.Evans, eds, Plenum Publishing Corp., NY (1983).

2. R.Pabst, Neuere Methoden der Festigkeitsprüfung keramischer Werkstoffe, *Z. Werkstofftechnik*, 6:17 (1975).

3. V.G.Borovik, S.P.Kovalev, and V.M.Chushko, Fracture toughness of ceramic materials with controlled crack growth. 1. Experimental method, *Strength of Mater.* 26:378 (1994).

4. S.P.Kovalev, V.G.Borovik, and V.M.Chushko, Fracture toughness of ceramic materials with controlled crack growth. 2. Influence of microstructure on the kinetics of fracture, *Strength of Mater.* 26:447 (1994).

5. V.G.Borovik, V.M.Chushko, and S.P.Kovalev, Computer-aided single-specimen controlled bending test for fracture-kinetics measurement in ceramics, *J.Am.Ceram.Soc.* 78: (1995).

6. T.B.Troczynski and P.S.Nicholson, Effect of subcritical crack growth on fracture toughness and work-of-fracture test using chevron-notched specimens, *J.Am.Ceram.Soc.* 70:78 (1987).

7. S.Lathabai and B.R.Lawn, Fatigue limits in noncyclic loading of ceramics with crack-resistance curves, *J.Mater.Sci.* 15:4298 (1989).

8. H.Hübner and H.Schuhbauer, Experimental determinations of fracture mechanics stress intensity calibration in four-point bending, *Eng. Fract. Mech.* 9:403 (1977).

9. J.E.Srawley, Wide range stress intensity factor expressions for ASTM E 399 standard fracture toughness specimens, *Int.J.Fract.* 12:475 (1976).

10. R.Knehans and R.Steinbrech, Memory effect of crack resistance during slow crack growth in notched Al_2O_3 bend specimens, *J.Mater.Sci.Lett.* 1:327 (1982).

11. M.Sakai, J.-I. Yoshimura, Y.Goto, and M.Inagaki, R-curve behavior of a polycrystalline graphite: microcracking and grain-bridging in the wake region, *J.Am.Ceram.Soc.* 71:609 (1988).

12. T.Fett and D.Munz, Influence of R-curve effects on lifetime in static tests, *J.Mater. Sci.* 25:1471 (1990).

13. Li-Shing Li and R.F.Pabst, Subcritical crack growth in partially stabilized zirconia (PSZ), *J.Mater.Sci.* 15:2861 (1980).

14. Y.-W. Mai and B.R.Lawn, Crack interface grain bridging as a fracture resistance mechanism in ceramics: II, Theoretical fracture mechanics model, *J.Am.Ceram.Soc.* 70:289 (1987).

15. Xiao-Zhi Hu, E.H.Lutz, and M.V.Swain, Crack-tip-bridging stresses in ceramic materials, *J.Am.Ceram.Soc.* 74:1828 (1991).

16. R.Tandon and D.J.Green, Crack stability and T-curves due to macroscopic residual compressive stress profiles, *J.Am.Ceram.Soc.* 74:1981 (1991).

17. R.W.Hamming. "Numerical Methods for Scientists and Engineers, McGraw-Hill Publishers," NY (1962).

MOLECULAR ORBITAL CALCULATIONS COMPARING WATER

ENHANCED BOND BREAKAGE in SiO$_2$ and Si

G. S. White and W. Wong-Ng

National Institute of Standards and Technolgly
Materials Building 223, A215, Gaithersburgs, MD 20899

INTRODUCTION

Environmentally enhanced crack growth has long been recognized as a problem in most ceramics. In certain environments, cracks will extend in the presence of stresses which are much smaller than those required for crack extension in the absence of the environment; at the same time, in most of these material-environment systems, the cracks will not extend in the absence of a stress. Therefore, both stress and active environment are required to generate environmentally enhanced fracture. Because this behavior results in a time-dependent failure process in service, there have been extensive empirical investigations attempting to clarify the chemical and physical interactions involved in, and required by environmental enhancement of fracture[1,2,3]. This work has culminated with a widely accepted empirically-based set of guidelines for environmentally enhanced fracture in silica based glasses:

1) the environmental molecule must contain a labile proton and a free electron pair and

2) the environmental molecule must be small enough to reach the strained crack tip bond.

To date, every silica based glass which has been investigated has demonstrated environmentally enhanced fracture only for environments which meet these criteria[2,4,5]. In addition, some other materials which have been investigated seem to behave in the same way[6]. However, these criteria do not control environmentally enhanced fracture in all materials[6,7,8].

The most notable of the materials which deviate from the environmentally enhanced fracture criteria listed above is single crystal Si. Because of its tremendous technological importance, a knowledge of which environments might lead to catastrophic mechanical failure is of clear value. While there has been a report of environmentally enhanced crack growth.

The clear distinction in behavior between SiO_2 and Si in environments such as water suggests that these two materials might be used as model systems for an investigation of the basic processes required in environmentally enhanced fracture. An understanding of how the presence of water aids the rupture of strained Si-O bonds in SiO_2 and why it appears to be ineffective in rupturing strained Si-Si bonds would provide the beginnings of an understanding of environmentally enhanced fracture based on knowledge of the chemistry and physics involved, rather than based solely upon empirical evidence. In this manuscript, we report results of an *ab initio* molecular orbital (MO) investigation of water interactions with SiO_2 and Si in an attempt to gain such knowledge.

PROCEDURE

ab initio MO calculations were used for this work. The advantage of using MO techniques is that they can provide information concerning the behavior of individual atoms and electronic orbitals. This information allows the electronic interactions between the strained bond and the environmental molecule to be visualized without *a priori* knowledge of the force laws governing the interaction. The costs of such detail, however, are that 1) the number of atoms which can be included in the calculations is very limited and 2) there is no temperature dependence incorporated into the calculations. In addition, the quantitative values obtained from the calculations depend upon the specific basis set used to model the electrons; i.e., trends can be trusted, but absolute values of numbers should be viewed with a certain degree of caution[11,12,13].

The calculations were made with a commercial software package[a], Gaussian (versions 90 and 92), using the restricted Hartree-Fock technique, and were made on a Silicon Graphics workstation[a] as well as on a Cray Y-MP computer[a]. For the Si system the basis set 6-31g** was used. The basis set used for the reaction and transition state structure calculations for silica was $3\text{-}21g^{**}$. The simpler basis set for the more complicated transition state calculations was necessary due to computer limitations. Since we will be comparing trends in behavior for each system, rather than absolute energy values between systems, the change in basis sets is not expected to be important.

Because the computational expense increases as n^4, where n was the total number of electrons in the system, increasing the number of atoms carried a large price in calculation complexity and computer time. Therefore, SiO_2 was modeled by an $Si_2O_7H_6$ molecule and Si was modeled by an $Si_8H_{17}OH$ molecule (see Figure 1). The $Si_2O_7H_6$ molecule has been investigated[14,15,16] extensively and shown to mimic bond lengths and angles found in bulk SiO_2. Similarly, we have studied[17] the $Si_8H_{17}OH$ molecule and found the optimized structure to reflect angles and bond lengths of single crystal Si. The central Si-O and Si-Si bonds of the two molecules shown in Fig. 1 were designated to be the crack tip bonds. In the Si, one of the two Si atoms bracketing the crack tip was terminated with an OH rather than with an H. This choice was made to incorporate the empirical observation[18] that, on free Si surfaces, water molecules appear to adhere to the Si and then to disassociate to OH⁻ and H⁺. We assumed that water molecules behaved similarly on the crack walls as they do on free surfaces. The two Si atoms bracketing the Si-Si crack tip bond represent nearest neighbor atoms which form part of the crack wall behind the crack tip.

Strain was applied to both molecules in the following manner: 1) the central Si-O or Si-Si bond was extended a chosen amount (i.e., for a nominal bulk strain of 10%, the central bond was extended 10%), 2) the geometry of the remaining atoms was then

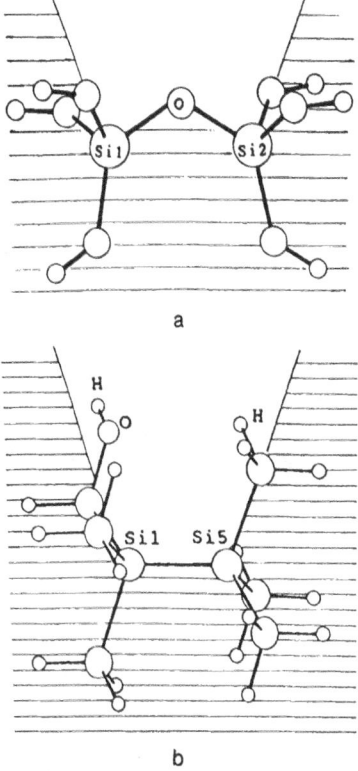

a

b

Figure 1. Schematic drawing of the pyrosilicic acid molecule $H_6Si_2O_7$ as the model for silica and (b) $Si_8H_{17}OH$ as the model for silicon in the presence of water. A sketch of the crack walls in both systems are shown.

optimized by minimizing the energy. While there clearly were many ways in which "strain" could be applied to the molecules, previous work[17,19] has demonstrated that the trends in the electronic behavior as a function of "strain" at the bridging, or crack tip, bond are insensitive to details of the strain on atoms removed from the crack tip.

For the reaction between $H_6Si_2O_7$ and environment, the energy and geometry of the reactants, the transition state structure, and the final products were computed. For the reactants, geometric optimization of the $H_6Si_2O_7$ and environmental molecule was calculated. To obtain the transition state[20], optimization of the combined $H_6Si_2O_7$ and environment system was again obtained with the exception that the "crack-tip" bond was fixed at the desired strain value. As will be discussed below, in the case of $Si_8H_{17}OH$ and water, no transition state was observed.

RESULTS AND DISCUSSION

SiO_2:

Previous molecular orbital investigation of interactions between strained silica and environmental molecules[21] has shown that strain had two effects on the silica/environment system which would enhance reaction rate between environmental molecules and the Si-O crack tip bond. First, strain increased the polarization of the Si and O atoms, thereby increasing the attraction between the crack tip bonds and the environmental molecules. Second, strain reduced the steric hindrances between the environmental molecule and oxygen atoms neighboring the crack tip. Figure 2, from reference 21, shows how a strain of 15% affected the individual molecular orbitals as a water molecule was brought from a

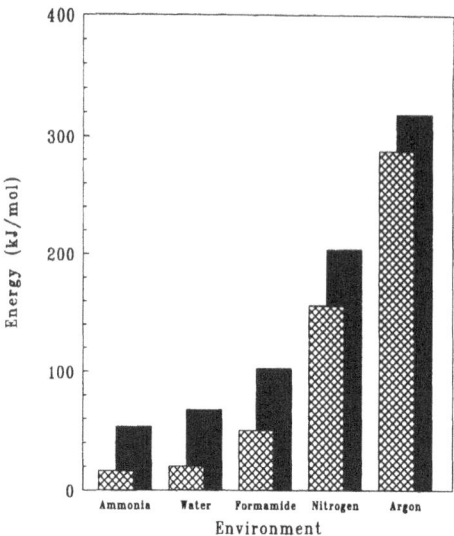

Figure 2. Effect of a 15% strain on the approach energy (δE) of five environmental molecules and the crack tip Si-O bond. Solid bars are for unstrained $H_6Si_2O_7$, cross-hatched bars are for strained $H_6Si_2O_7$.

distance of 2.5 Å from the crack tip to 2.0 Å. The attraction between the H on the water molecule and the O in the Si-O bond increased, as did the attraction between the O in the water and the Si at the crack tip bond. In contrast, steric hindrance forces between the nearest neighbor oxygen atoms and the water molecule were reduced by the strain.

Two questions of particular interest are: 1) what is the reaction path for environmentally enhanced fracture of SiO_2 by water and 2) how does strain of the Si-O bond affect the energy associated with these reactions? We have calculated initial state, transition state, and final state structures and energies (Figure 3) for 3 different strain conditions: 0%, 10%, and 15%.

In the unstrained state (Fig. 3a), there is an energy barrier of ≈ 115 kJ/mol (28 kcal/mol) which must be surmounted before the Si-O bond can be ruptured by the H_2O molecule. For comparison, typical energy barriers for chemical reactions at room temperature are ≈ 60 to 250 kJ/mol[22]. The 115 kJ/mol barrier is consistent with the fact that, in the absence of strain, Si-O bonds are quite stable in the presence of water. It is interesting to note that the final state energy is predicted to be slightly higher than the initial state energy; i.e., the reaction is endothermic. However, the energy barrier for the back reaction,

$$2SiOH \rightarrow SiO + H_2O \qquad (1)$$

remains ≈ 107 kJ/mol (25 kcal/mol). Again, the height of this barrier suggests that the SiOH will be stable at room temperature, despite the fact that the ruptured system is at a slightly higher energy level than the initial state. The net result for the unstrained silica is that, at room temperature, no reaction is expected to occur - either for the Si-O to rupture in the presence of water, or for it to heal.

The energy barrier for reaction of the water with silica for a strain of 10% is shown in Fig. 3b. The barrier height for the forward reaction,

$$H_2O + SiO \rightarrow 2SiOH \qquad (2)$$

has been reduced to ≈ 83kJ/mol (20 kcal/mol). The back reaction (Eqn. 1) remains 102 kJ/mol (24 kcal/mol); since strain in the pre-reacted state would have no effect on the stabilized final state, this result was expected. Finally, as Fig. 3b shows, a strain of 10% shifts the forward reaction (Eqn. 2) from endothermic to exothermic. However, at this strain level, the energy barrier is still too large to expect many reactions to occur between the water and the silica.

Figure 3c shows the effect of a 15% strain on the energy barrier. As expected, the barrier to the reverse direction remains essentially unchanged. However, the forward energy barrier has been reduced from 115 kJ/mol (28 kcal/mol) in the unstrained condition to 49 kJ/mol (12 kcal/mol).

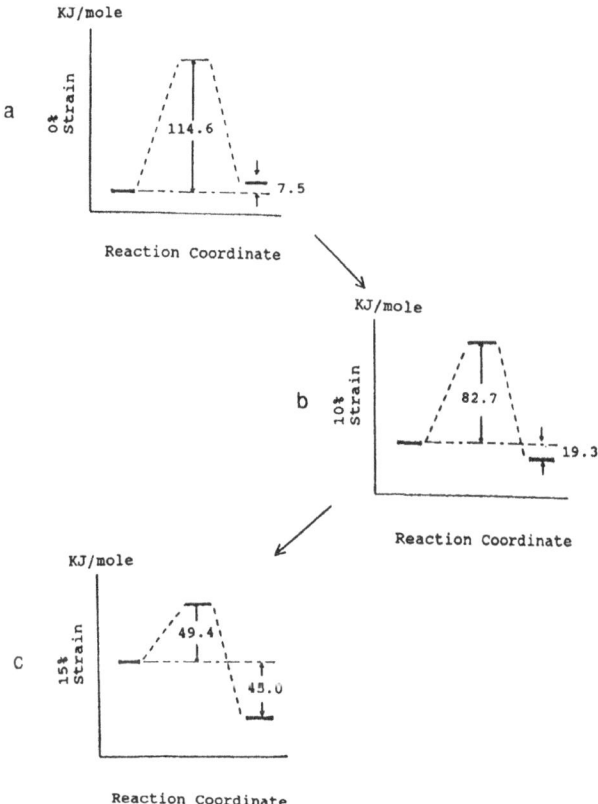

Figure 3. Effect of strain on energy barrier of the reaction between $H_6Si_2O_7 + H_2O$, (a) 0%, (b) 10% and (c) 15% strain on the crack-tip Si-O bond.

Of particular significance in Fig. 3 is the lack of sensitivity of the transition state energy to strain. The decrease in the energy barrier in the forward direction, as a function of strain, results totally from the increased energetic state of the initial configuration rather than from a decrease in the energy of the transition state. Previous work[21] has shown that the energy put into distorting the silica structure results in a decrease in the steric hindrance to the approach of the environment and an increase in the polarization of the Si-O bond which enhances both the attraction and proper orientation of the environmental molecule. The calculations presented here are consistent with the concept of a concerted reaction which was proposed several years ago. The concerted reaction concept proposed that there was little effect of strain on the interaction of the Si-O bond an environmental molecule until a critical strain was achieved. At the critical strain level, the environmental molecule would rupture the Si-O bond, and the crack would advance. This concept was put forward to explain the experimental observations that, in vitreous silica, the slopes of the log(V) vs K_I appear to be independent of environment. The concerted reaction hypothesis postulated that there was no interaction between the environmental molecule and the strained Si-O bond until the critical strain value was reached. The degree of strain required before this interaction occurred was postulated to be environmentally dependent; hence the shift in position of crack growth curves from one active environment to another. The calculations presented here and elsewhere[21] provide a physical interpretation for the empirical observations leading to the concerted reaction model. The lack of sensitivity of

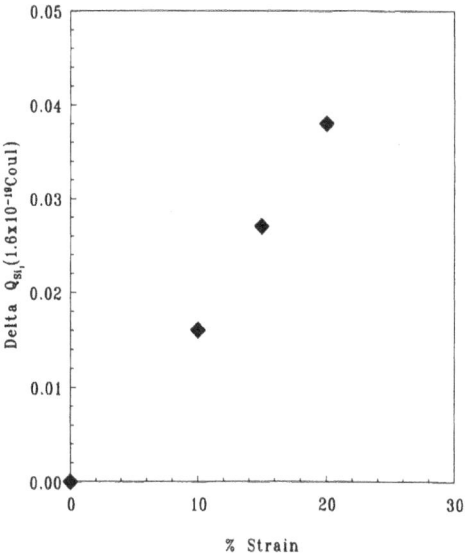

Figure 4. Plot of change in charge, ΔQ, on crack tip $Si1$ as a function of strain. Note that 1.6×10^{-19} Coul is the value of one electronic charge.

the crack tip reaction to the environmental molecule is understood by the observation that, before the environment can interact, it must reach the strained crack tip bonds. Because one of the major effects of strain is to reduce steric hindrance to the approach of the environmental molecule, the reduction of the energy barrier with strain is insensitive to the environment until the environmental molecule approaches closely enough to interact with the Si-O bond. At that point, the identity of the environment becomes important.

   Si:

The behavior of the Si/water system contrasts markedly with that of the silica/water system. Figure 4 is a plot of the calculated charges on the crack tip Si atoms as a function of strain. The charge on each of the atoms remains small and increases with increasing strain. Although the increase of the charge on the crack tip Si atoms, and the concomitant decrease in the charge on the surrounding atoms, may be an artifact of the small number of atoms used (i.e., the presence of the H atoms may affect the charge state of the crack tip atoms), the fact that the charge is the same sign on both crack tip atoms is significant; strain does not polarize the Si-Si bond which would then attract polar environmental molecules. Even the lack of symmetry of the atoms bordering the crack tip, H on one side and OH on the other, does not lift the symmetry enough to cause polarization of the crack tip Si-Si bond.

In addition to the lack of polarization with strain of the Si-Si bond, steric hindrance in Si prevents access to the crack tip bonds even for large strains. A water molecule, with a size of $\approx$ 4Å (taking van der Waals' radii into considerations), cannot pass the H and OH atoms along the crack wall until the material has reached a strain of $\approx$ 20%. The lack of a driving force, in the form of a polarized crack bond, at the crack tip to draw the water

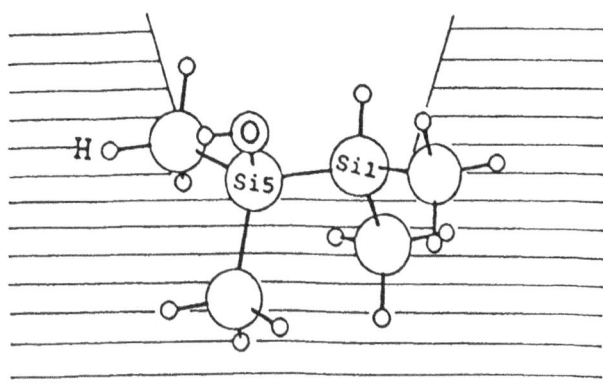

Figure 5. Model molecule $H_{14}Si_6O$ showing 3 Si-Si crack tip bonds.

molecule past the steric barriers along the crack wall means that, even as the strain increases, there is no driving force for environmental interaction with the crack tip.

To investigate this point, we had to widen the crack tip by using a model $Si_6H_{14}O$ in which the crack tip spans of 3 Si-Si bonds as shown in Fig. 5. Even with a wider crack tip space, equilibrium geometry optimization of the system ($Si_6H_{14}O+H_2O$) showed that there is no tendency for the water molecule to be drawn close to the crack tip. Figure 6 shows the condition of a 15% strain applied to the crack-tip Si-si bond. The distances between the crack-tip Si's and water remain large. There is no tendency for reaction to occur.

To summarize the results of the calculations on Si: 1) strain does not polarize the Si-Si bonds as it does for Si-O bonds, 2) steric effects are much more severe in Si than in SiO_2, and 3) even when the water molecule was placed within a reaction distance of the Si-Si bond, it is repulsed rather than attracted.

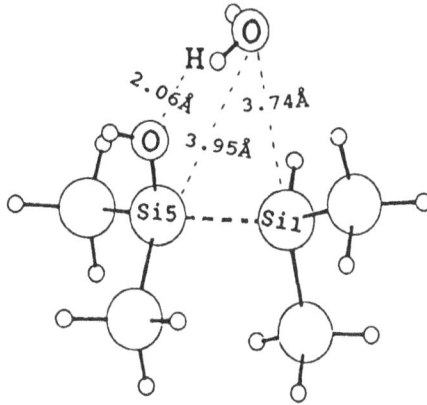

Figure 6. Reaction of $H_{14}Si_6zO + H_2O$ showing large equilibrium distance ($\approx 4\text{Å}$) between water and the crack tip Si-Si bond at a strain level of 15%.

CONCLUSIONS

Based upon *ab initio* MO calculations, we have obtained insight into the atomistic and electronic behavior of water interactions with SiO_2 and Si. This insight has provided explanations both for the susceptibility of SiO_2 and the lack of susceptibility of Si to water enhanced fracture.

Water: Strain reduces the energy barrier to the reaction of water with the crack tip Si-O bond by increasing the initial energy state of the Si-O bond. The distorted initial state of the crack tip region reduces steric hindrance to the approach of the water and enhances the attraction of the water molecule through increased polarization of the Si-O bond. Because the strain has limited effect on the transition state configuration, the reduction of the energy barrier is expected to be insensitive to environmental species.

Si: In contrast to SiO_2, strain has no tendency to polarize the Si-Si bond; therefore, there is no attractive driving force to draw water molecules to the crack tip. In addition, steric hindrances remain a serious obstacle to the approach of water molecules. These results are consistent with our inability to detect water enhanced crack growth in this material.

REFERENCES

1. L. Grenet, Bull. Soc. Enc. Industr. Nat. Paris (ser. 5), **4** 838-48 (1899).

2. S.M. Wiederhorn, J. Am. Ceram. Soc. **50**, 407-14 (1967).

3. T.A. Michalske and S.W. Freiman, J. Am. Ceram. Soc. **66** [4], 288-91 (1983).

4. S.W. Freiman, G.S. White and E.R. Fuller, Jr., J. Am. Ceram. Soc. **68** [3], 108-12 (1985).

5. G.S. White, D.C. Greenspan and S.W. Freiman, J. Am. Ceram. Soc. **69** [1], 38-44 (1986).

6. T.A. Michalske, B.C. Bunker and S.W. Freiman, J. Am. Ceram. Soc. **69** [10], 721-24 (1986).

7. G.S. White, S.W. Freiman and A.M Wilson, J. Am. Ceram. Soc. **72** [11], 2193-95 (1989).

8. G.S. White and L.M. Braun, submitted to J. Am. Ceram. Soc. (1995).

9. B.R. Lawn, J. Mater. Sci. 7 (1980).

10. G.S. White and S.W. Freiman, unpublished data.

11. R.S. Mulliken, J. Chem. Phys. **23**, 1841-46 (1955).

12. J.A. Pople and D.L. Beveridge, **Approximate Molecular Orbital Theory**, McGraw-Hill, New York, NY (1970).

13. A.G. Turner, **Methods in Molecular Orbital Theory**, Prentice-Hall, Englewood Cliff, NJ (1974).

14. G.V. Gibbs, Am. Min. **67**, 421-50 (1982).

15. G.V. Gibbs, E.P. Meagher, M.D. Newton and D.K. Swanson, in **Structure and Bonding in Crystals, Vol. I** (M. O'Keeffe and A. Navrotsky, eds.), Academic Press, New York, NY, pp. 195-225 (1981).

16. M.D. Newton and G.V. Gibbs, Phys. Chem. Min. **6**, 221-46 (1980).

17. W. Wong-Ng, G.S. White, S.W. Freiman and C.G. Lindsay, submitted to J. Am. Ceram. Soc. (1995).

18. P.A. Thiel and T.E. Madey, Surface Sci. Report 7 211-385 (1987).

19. W. Wong-Ng, G.S. White and S.W. Freiman, J. Am. Cer. Soc. **75** [11] 3097-3102 (1992).

20. H. B. Schlegel, J. Comput. Chem. **3** [2], 214 (1982).

21. C.G. Lindsay, S.W. Freiman, G.S. White and W. Wong-Ng, J. Am. Ceram. Soc. **77** [8], 2179-87 (1994).

ª Trade names and companies are identified in order to specify adequately the experimental procedure. In no case does such identification imply that the products are necessarily the best available for the purpose.

INTERFACE MECHANISMS IN CERAMIC-CERAMIC FIBRE COMPOSITES AND THEIR RELATIONS WITH FRACTURE AND FATIGUE BEHAVIOUR

D. Rouby

Groupe d'Etudes de Métallurgie Physique et de Physique des Matériaux
G.E.M.P.P.M. URA CNRS n°341, bâtiment 502
Institut National des Sciences Appliquées de Lyon
69621 Villeurbanne cedex, France

INTRODUCTION

The use of continuous-fibre-reinforced ceramics for aerospace structures such as the space shuttle, high-speed aircraft and hot engines, offers the opportunity to use the same material for resistance to both mechanical loading and thermal aggression. Monolithic ceramic materials present attractive properties such as thermal stability, corrosion and oxydation resistance, and low specific gravity. However, these materials are flaw sensitive and they generally exhibit low damage tolerance and low fracture toughness. An effective way to decrease the brittle character of bulk ceramics is to include a reinforcement phase, such as fibres or whiskers.

In ceramic matrix composites, fibre/matrix interfaces play an important role in the mechanical behaviour[1, 2]. The fibrous reinforcement induce deviation of the matrix crack along the fibres, leading to a damage tolerant material, with pseudo-ductile character. For such improved behaviour, it is necessary that the fibre/matrix interface remains relatively weak; a excessively strong interface prohibits crack deviation and bridging mechanisms, resulting in brittle behaviour. On the other hand, the stress transfer capability, controlled by the interfaces, must be high enough in order to make sure that the effects induced by a given amount of damage remain localized. Hence, the fibre/matrix interface should be relatively strong.

The stress transfer capability is controlled by debonding, and post-debond friction. The friction conditions between fibre and matrix are determined by the the thermal clamping stress and the roughness of the slipping surfaces. This highlights the importance of temperature effects, even at moderate temperatures where neither chemical degradation nor creep occur in the constituents. Finally, chemical or mechanical degradation of the interface, by corrosion or wear phenomena, can lead to delayed failure, as observed, for example, during cyclic fatigue[3].

The aim of this work is to describe some basic features of interface behaviour and their relationship with reinforcement, damage micromechanisms, and the mechanical properties of the material.

The results shown as illustrations or examples in this paper concern two ceramic-ceramic composites having opposite characteristics. Both materials are reinforced with the SiC Nicalon™ NLM 202 fibre from Nippon Carbon, Japan.

The first composite is a SiC(Nicalon)/MAS-L cross-ply laminate processed by sol-gel route by Aérospatiale (Bordeaux, France). Under tensile load, this material exhibits saturated matrix cracking and the coefficient of thermal expansion of the matrix is lower than that of the fibres. Large pull-out lengths (1000 μm) are observed after tensile failure. Indentation tests have been performed on the same composite, but with a unidirectional architecture.

The second material is a 2D woven SiC(Nicalon)/SiC laminate processed by chemical vapour infiltration by the Société Européenne de Propulsion (Bordeaux, France). Here, saturation of matrix cracking is not achieved and $\alpha_m > \alpha_f$. This material exhibits lower pull-out length (100 μm).

The first part of this work deals with debonding characteristics and frictional sliding conditions, measured using a classical microindentation technique. These characteristics are dependent on fibre diameter, and the influence of interface roughness is analyzed.

The second part relates to interfacial parameters directly determined from mechanical tests on standard tensile specimens, by analysis of the area of the load-unload loops, analysis of the pull-out length changes and exploitation of the $p(u)$ curves giving the bridging tractions as a function of crack opening displacement.

BASIC INTERFACE PROPERTIES AND MECHANISMS

In this section, some features about debonding and friction are described. The analyses are based on results of measured interface characteristics obtained by microindentation on the unidirectional SiC/MAS-L composite.

Fibre/matrix debonding

The debonding criterion in mode II, based on an energy balance, can be expressed by the following equation[4]:

$$\left(\sigma_f^D - \sigma_f^{th}\right)^2 = \frac{4\,E_f\,G_c}{R} \tag{1}$$

with σ_f^D: critical longitudinal fibre stress for debonding, σ_f^{th}: longitudinal thermal stress in the fibre, E_f: fibre Young's modulus, R: fibre radius, and G_c: fracture energy of interface. In the case of thermal stress free body ($\sigma_f^{th} = 0$), Eqn. 1 reduces to the early criterion from Outwater and Murphy[5].

Experimentally, the general trend towards a decrease in debond stress as fibre radius increases has been well observed[6]. Fig. 1 shows an example of such relation, measured by the microindentation method on unidirectional SiC/MAS-L composite (details of the measurement procedure are given in reference 6). However, a quantitative analysis of the law shown in Fig. 1 leads to an overestimate of σ_f^{th} with a value of 900 MPa issues Eqn. 1, instead of 300 MPa calculated from the thermal strain misfit between room temperature and processing temperature for this material. Actually, in SiC/MAS-L material the coefficient of thermal expansion of the matrix is lower than that of the fibre, so σ_f^{th} is in tension (about 300 MPa) and in the radial direction the thermal stress is positive too ($\sigma_i^{th} = 60\text{-}80$ MPa). Consequently, fibre/matrix debonding operates in mixed mode (see Fig. 3a). Debonding in mixed mode has been analyzed by Charalambides and Evans[7], and, in a first approximation, the criterion can be written as follows:

$$\frac{1}{R} = \frac{1}{G_c}\left[k_1\, E_f\left(\alpha_m - \alpha_f\right)^2 (T - T_0)^2 + \frac{1}{4\,E_f}\left(\sigma_f^{D^2} - 2\,\sigma_f^{th}\,\sigma_f^D\right)\right] \qquad (2)$$

where k_1: non dimensional constant, determined by finite element analysis[7] (k_1 depends on the elastic properties and geometry of the two phases), T: temperature, T_0: stress free (processing) temperature. If σ_f^{th} is known, G_c can be determined from the linear relation between the term $(\sigma_f^{th})^2 - 2\,\sigma_f^{th}\,\sigma_f^D$ and $1/R$.

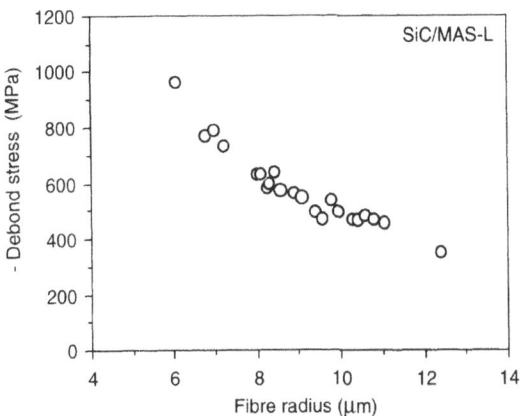

Figure 1. SiC/MAS-L composite (unidirectional). Variation of debond stress, σ_f^D, as a function of fibre radius.

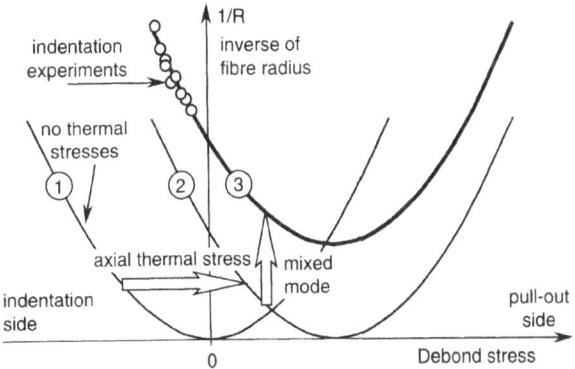

Figure 2. Debonding criteria in the $1/R$ - σ_f^D diagram for pull-out ($\sigma_f^D > 0$) and indentation tests ($\sigma_f^D < 0$). Curve 1: model of Outwater and murphy (mode II, thermal stress free). Curve 2: mode II, effect of axial fibre thermal stress (case of $\sigma_f^{th} > 0$). Curve 3: mixed mode criterion.

A schematic picture of debonding criterion is given in Fig. 2. Curve 1 corresponds to the thermal stress free state, curve 2 is the criterion in pure mode II (σ_f^{th} being in tension) and curve 3 takes into account the existence of some mode I behaviour. Longitudinal thermal stress leads to a horizontal shift (with the same sign) of the parabola. The vertical shift illustrates the effect of mixed mode. For a given fibre radius, mixed mode leads to a reduction of the critical stress σ_f^D measured either by indentation (compressive σ_f^D) or by pull-out (tensile σ_f^D).

In the case of a composite with $\alpha_m > \alpha_f$, the radial thermal stress is compressive and debonding operates in pure mode II, provided that the slipping surfaces are perfectly smooth. When slip occurs between rough surfaces, the surmounting of the asperities (see next section) induces an additional radial displacement, this leads to mixed mode debonding even when the radial thermal stress is compressive. The roughness induces an additional driving force (Fig. 3b) at the debond crack tip.

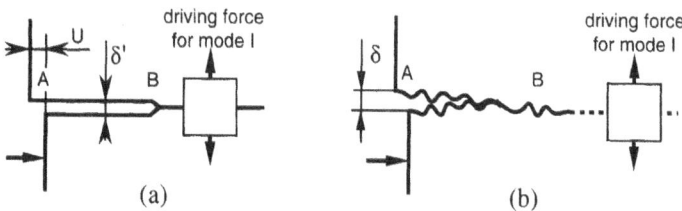

Figure 3. Schematics of radial displacement. (a): the driving force for mode I contribution to debonding is due to the a tensile radial thermal stress leading to the gap δ'. (b): the mode I driving force comes from radial displacement, δ, induced by the roughness. U: slip displacement, A-B: debonded region.

Neglecting the mixed mode nature by using Eqn. 1 for experimental data analysis, finally results in an underestimate of the intrinsic interface fracture energy. Our data, for SiC/MAS-L composite, show $G_c = 16.5$ Jm^{-2}, which is very much higher than the values of 1 Jm^{-2}, or less, reported in the literature for glass-ceramic matrix composites with a thin pyrocarbon layer between the fibre and matrix[4]. This layer is also present in our material. Notice that 16.5 Jm^{-2} is very close to the fracture energy commonly reported in the literature for pyrocarbon, with the crack running in the weak direction.

Due to Poisson's effects, the debonding conditions can be slightly different under indentation or pull-out load. Eqn. 2 does not take into account this effect.

Frictional sliding

The interfacial shear stress related to friction, τ^F has been measured by the classic Marshall's method[6,8] on unidirectional SiC/MAS-L material.

Fig. 4 shows a typical result obtained for SiC/MAS-L composite. Some points stand well away from the main trend and concern fibres of more than 20 μm diameter. These high values of τ^F may be due to more pronounced irregularities in the fibre geometry such as long range changes in diameter or cross-section shape changes along the fibre axis. Fig. 1 exhibits a large amount of scattering, caused by the low accuracy of measurement (about 30%) and especially by the natural variability between fibres. This last feature includes variations in both fibre and interface properties, and differences in the neighbourhood of each tested fibre as well. Nevertheless, the general trend is a linear decrease of τ^F as the inverse of fibre radius decreases. The sliding resistance vanishes at a given fibre radius R*.

The influence of fibre radius on sliding resistance can be explained as follows. In the as-received SiC/MAS-L material, the interface is bonded and a tensile thermal stress takes place in the radial direction (Fig. 5a). During indentation loading, the interface first debonds and the thermal expansion mismatch leads to a gap δ' between the debonded surfaces (Fig. 5a), given by:

$$\delta' = R \left| \alpha_m - \alpha_f \right| \left| T - T_0 \right| \qquad (3)$$

The debonding produces a certain roughness of the two new surfaces. Under these conditions, during post-debonding slip, a radial displacement of the fibre and matrix surfaces takes place. This displacement is controlled by the effective amplitude, δ, of the roughness (see Fig. 3b and c) and generates a compressive radial stress, σ_{rad}, due to the radial stiffness of the system. This stress is given by the following expression:

$$\sigma_{rad} = -\frac{\delta - \delta'}{A\,R} \qquad (4)$$

where A is a radial compliance term, obtained from classical clamping analysis[9, 10]. If δ' is larger than δ then the clamping stress σ_{rad} vanishes.

The interfacial sliding mechanism is assumed to follow a Coulombic law which needs to

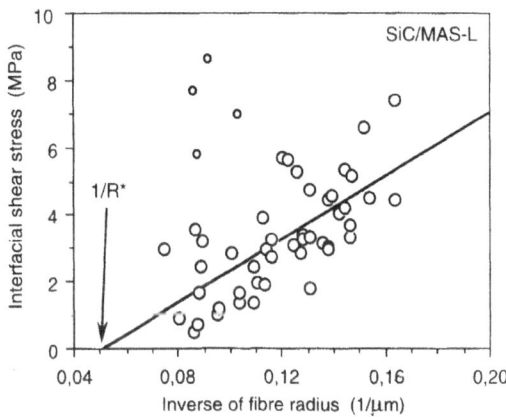

Figure 4. Unidirectional SiC/MAS-L composite. Evolution of sliding resistance, τ^F, as a functi of 1/R.

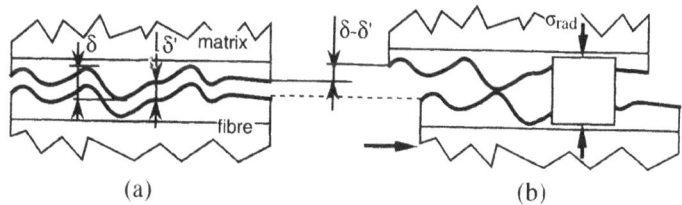

(a)   (b)

Figure 5. Schematics of sliding against roughness. (a): before slip there is a thermal gap δ' and a certain roughness of effective amplitude δ. (b): the asperities induce a radial displacement, δ-δ', leading to a clamping stress σ_{rad}.

233

introduce an effective coefficient of friction, μ:

$$\tau^F = - \mu \, \sigma_{rad} \qquad (5)$$

Finally the sliding resistance can be given by the following expression:

$$\tau^F = \frac{\mu \, \delta}{A \, R} - \frac{\mu}{A}(\alpha_m - \alpha_f)(T - T_0) \qquad (6)$$

This analysis neglegts the effect of Poisson's radial expansion, resulting from fibre compression during indentation testing. Including the Poisson's effect leads to a slightly more complicated expression[11].

Eqn. 6 describes also the influence of temperature, as shown schematically in Fig. 6.

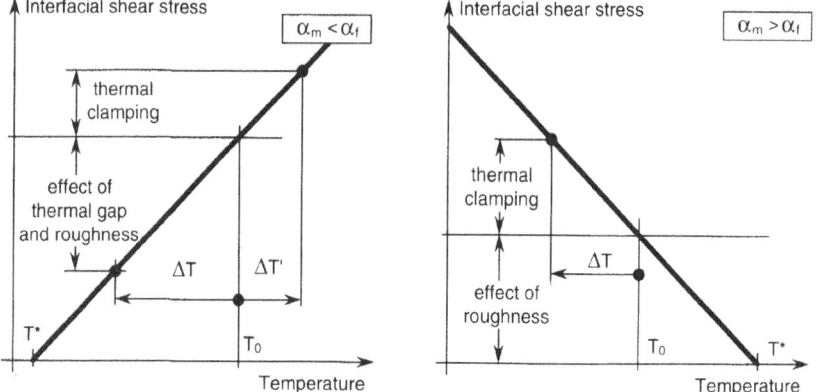

Figure 6. A schematic representation of the variation of sliding resistance as a function of temperature, including the influence of roughness. Left: case of SiC/MAS-L composite. Right: case of SiC/SiC composite.

For a composite having $\alpha_m < \alpha_f$ such as SiC/MAS-L, the sliding resistance increases as temperature increases. At temperature T*, the gap width just equals the amplitude of the roughness and the interfacial shear stress vanishes (below T*, τ^F remains at zero). The increase in τ^F between T* and T_0 is due to the progressive narrowing of the thermal gap. Above T_0, the sliding resistance increases again as a compressive radial stress takes effect and becomes increasingly important.

In the case of the SiC/SiC composite, where $\alpha_m > \alpha_f$, a compressive radial stress appears during cooling after processing. Here, the sliding resistance decreases with increasing temperature. The decrease in τ^F above T_0 is due to the presence of the thermal gap. The SiC matrix is much stiffer than the glass-ceramic, so slight variations in the distance between two adjacent fibres induces large variations in the clamping stress and consequently a large degree of scattering in the sliding resistance measured, for example by microindentation. Therefore, the effect of fibre radius cannot be accurately checked in this kind of composite.

The effect of roughness can also be observed during tensile testing of a unidirectional material[12]. The specimens undergo multiple matrix cracking and the radial displacement induced by the slip between fibre and matrix leads to a transverse expansion of the specimen.

In some cases, the apparent Poisson's ratio of the composite becomes negative.

Several degradation mechanisms can take place at high temperature, either in the interfacial region or in the constituents, such as creep, oxidization, interdiffusion, etc. These problems are not included in the simple analysis presented here, but some basic features shown here can play an important role. For example, in the case of creep under high stress, the matrix is cracked and the fibres are creeping because they are overloaded. Fibres in creep exhibit large Poisson's ratios and the resulting radial displacement is an additional driving force for further debonding, this leads to more fibre overload and to higher a creep rate[13].

In this first section, some basic features of interface behaviour have been described and related with experience. In this field, the results are mainly obtained by using micromechanical tests performed on individual fibres. Such measurements generally give large amount of scatter and to get a reasonable average value, we need a very large number of individual tests. From this point of view, it can be interesting to exploit more conventional mechanical tests for characterizing the interface behaviour. Some examples are shown in the second section below.

INTERFACE ANALYSIS FROM CONVENTIONAL MECHANICAL TESTS

Interface characteristics can be obtained by analysing some results obtained by macroscopic mechanical tests performed on standard specimens. In this second section, we describe three examples of such analysis. The first one concerns the pull-out length measured on fractured tensile specimens or compact-tension (CT) specimens. The second example is focused on the hysteresis of stress-strain loops measured during cyclic fatigue tests in tension-tension. In the last example, emphasis is placed on the bridging mechanism by the analysis of p(u) curves which give access to micromechanical parameters such as sliding resistance, pull-out length and in-situ fibre strength. These curves are also an indication of the fracture crack growth resistance.

Pull-out length analysis

The damage of a unidirectional composite in tension can be described by a serial set of matrix cracks bridged by fibres. Specimen failure occurs when one of these cracks becomes unstable, through an increase in the number of fibres broken. This bridging mechanism will be described in more detail in the later section on p(u) curves. When the fibres undergo overload they have a given probability to fail at a certain distance from the unstable matrix crack. This distance determines the fibre pull-out length after failure of the specimen. Hence, the pull-out length is an indication of the stress transfer profile between fibre and matrix at the moment of failure.

The average pull-out length is given by the following expression[14,15]:

$$<L_p> = \frac{1}{4} F(m) \, (\sigma_0)^{\frac{m}{m+1}} \left(\frac{R}{\tau^F}\right)^{\frac{1}{m+1}} \tag{7}$$

Where m is the fibre Weibull modulus and σ_0 is the scale factor in the Weibull statistics, in turn related to the average fibre strength $<\sigma_f^R>$ (under uniform load and gauge length L) by:

$$<\sigma_f^R> = \frac{\sigma_0}{L^{1/m}} \, \Gamma\left(1+\frac{1}{m}\right) \qquad \Gamma(y): \text{gamma function} \tag{8}$$

The prefactor F(m) depends on the Weibull modulus and on the matrix cracking conditions[14]. In the case of non-saturated matrix cracking (single matrix crack), the fibre failure probability decreases as the distance from the crack increases, and is governed by the fibre overload profile around the crack. In the case of saturated matrix cracking, the stress is more or less uniform along the fibre axis, the fibres undergo multiple failure. Thus, the next fibre failure probability is governed by the location of the former fibre failures. Fig. 7 shows the prefactor variations with m for saturated and non-saturated matrix cracking.

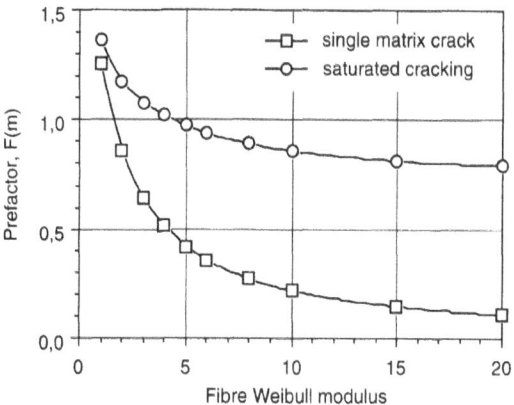

Figure 7. Prefactor, F(m), as a function of Weibull modulus, m, of the fibres.

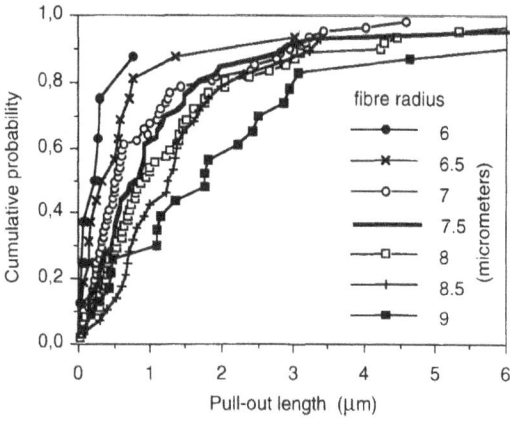

Figure 8. 2D SiC/MAS-L, fatigued in cyclic tension-tension. Cumulative pull-out length distributions measured for different fibre radii.

Eqn. 7 shows that large average pull-out lengths are obtained by low sliding resistance, high fibre strength and low Weibull modulus. The distribution of individual pull-out lengths depends also on the matrix cracking situation[14,15]. The theoretical distributions are only available numerically and they are few depending on Weibull modulus. For further analysis, it is interesting to take an approximate distribution, given simply by an exponential function as shown below:

$$P(L_p) \cdot <L_p> = \exp\left(-\frac{L_p}{<L_p>}\right) \qquad (9)$$

For 2D composites, the pull-out length can be analysed too, provide that one direction of reinforcement is parallel to the loading direction. The pull-out length has been measured on SiC/MAS-L composites as a function of fibre diameter[6]. An example of the cumulative distribution is shown in Fig. 8. Globally, the pull-out length is much larger for higher fibre diameter. Despite the scattering, the distributions can be described by a simple exponential form (Eqn. 9).

Fig. 9 shows the variation of average pull-out length with fibre radius. Cyclic fatigue leads to an overall increase of pull-out length. In the case shown here, the global average pull-out length (all fibres together) increases with fatigue from 850 μm to 1250 μm. This effect is always observed either the specimen is fatigued at a great number of cycles just below the fatigue limit and then fractured by monotonic tension, or failed by fatigue beyond the fatigue limit. The fatigue induced increase in pull-out length is due to a progressive decrease in sliding resistance during fatigue cycling[3]. Cyclic fatigue can also result in a reduction of fibre strength (sub-critical crack growth, friction induced surface cracks, etc.), but this effect leads to a decrease of pull-out length, in these tests this was never observed.

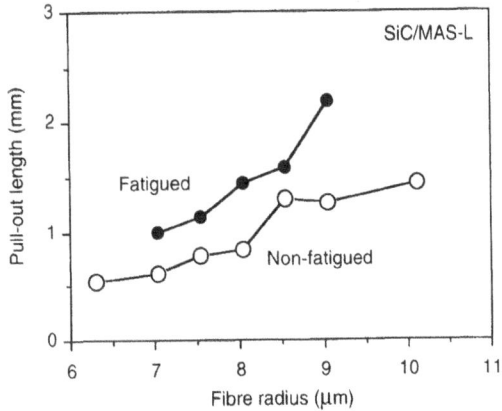

Figure 9. 2D SiC/MAS-L. Variation of the average pull-out length as a function of fibre radius.

By taking a fibre strength of $<\sigma_f{}^R> = 500$ MPa (in-situ value), with a gauge length of 50 mm and a Weibull modulus of $m = 2$, the sliding resistance can be calculated from Eqn. 7 and 8, by considering the saturated matrix cracking situation for SiC/MAS-L composite. The values taken here are confirmed by analysis of the p(u) curves (last section). The results of this calculation are given in Fig. 10. Clearly, the sliding resistance depends linearly on the inverse of the fibre radius as expected from Eqn. 6.

In addition, it is interesting to note that the two best fit lines have the same ordinate at the origin. This seemingly indicates that fatigue cycling reduces the amplitude of the roughness, but the coefficient of friction remains unchanged (see Eqn. 6). More work is needed to confirm this conclusion, because here the accuracy is low and the Weibull modulus is not well defined.

A second example of pull-out length analysis concerns SiC/SiC composite, tested at different temperatures. An increase in average pull-out length was obseved as test temperature

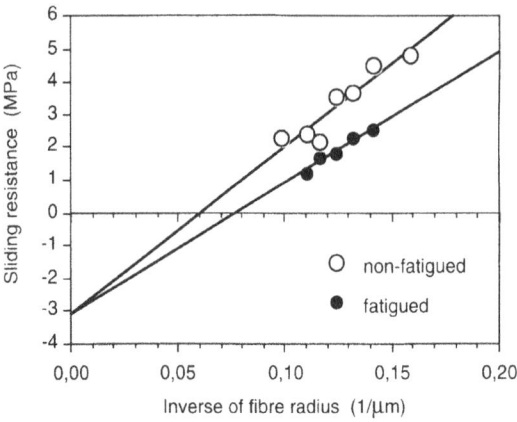

Figure 10. 2D SIC/MAS-L. Interfacial shear stress deduced from pull-out length plotted as a function of the inverse of fibre radius.

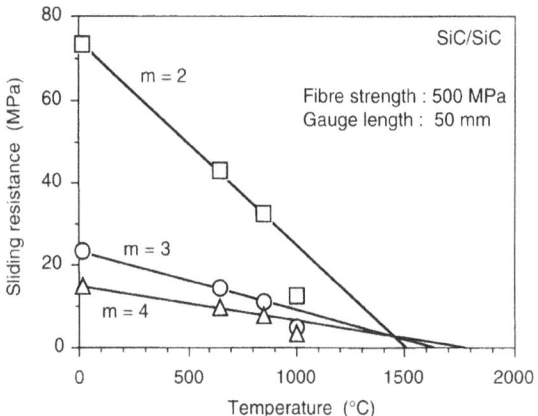

Figure 11. 2D SiC/SiC composite. Interfacial shear stress deduced from pull-out length plotted as a function of test temperature. The tests were performed on CT specimens under argon.

was increased: from 120 μm at room temperature to 170 μm at 650°C, 205 μm at 850°C and 390 μm at 1000°C. The same calculation as before, for the non-saturated matrix cracking situation, gave the results shown in Fig. 11.

The sliding resistance decreases as temperature increases, as depicted in Fig. 6 for the case where $\alpha_m > \alpha_f$. The temperatute T* (where τ^F disappears) is higher than the processing temperature (around 1000°C), the difference comes from the effect of roughness. The slope of the relation betweem τ^F and T depends on the value of Weibull modulus used for the calculation. Indentation tests made at room temperature give a sliding resistance value between 30 and 50 MPa for this material. As shown in Fig. 11, this level of τ^F is reached from pull-out length analysis by taking a Weibull modulus between 2 and 3, perhaps closer to 3.

In conclusion, pull-out length analysis is a powerful tool to get information about interface characteristics. The drawback here is the need to know the in-situ values of fibre strength and Weibull modulus.

Analysis of the load-unload loops

Study of cyclic fatigue in ceramic-ceramic composites has established that the main cause of delayed damage is a degradation of the interfaces by wear phenomena due to the see-saw motion of the sliding surfaces[3, 17, 18]. During the first loading, initial damage is introduced in the material, as multiple matrix cracking and some fibre failure. During the subsequent cycles, the stress transfer capability diminishes, due to interface degradation and, consequently, the crack bridging efficiency decreases as fatigue proceeds to failure.

The hysteresis of the load-unload loops is the result of the inversion of the stress transfer mechanisms as applied load reverses[19, 20]. The changes in interfacial shear stress during fatigue can also be analysed from the changes in the hysteresis of the loops. The relation between sliding resistance and shape of the loops depends on the matrix cracking situation.

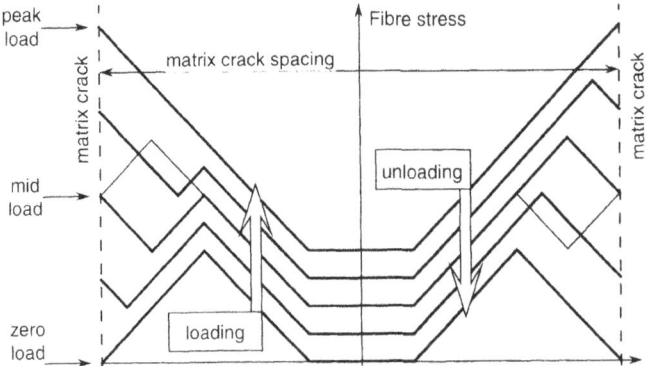

Figure 12. Large matrix spacing. Fibre stress profiles induced by stress transfer mechanism near matrix cracks, at 0, 25, 50, 75 and 100% peak load. Left: loading; right: unloading. The shaded area indicates the difference betwen unloading and loading profile. The shaded area is proportional to loop width at the mid-load.

In the case of large matrix crack spacing (non-saturated matrix cracking), the fibre stress profiles are shown in Fig. 12, for loading and unloading. The profile inversion process continues during the whole stage of loading or unloading, leading to a parabolic relationship between applied stress and strain. Details about the calculations of these relationships are given elsewhere[19, 20]. At mid-load, the unloading strain is larger than loading strain and the mid-load width of the loop is related to the area of the shaded surfaces. The height of these diamond-shaped surfaces is fixed by the applied peak load and the slopes of the stress profile lines are proportional to the sliding resistance. As a result, if τ^F decreases, the loop width increases. This feature is due to a free widening of the stress transfer zone, as long as the present zone does not overlaps the neighbouring zones. The zone overlapping later results in the saturated matrix cracking situation.

For a given crack spacing, D, the saturated situation is reached when τ^F becomes lower than a critical value τ_{sat}.

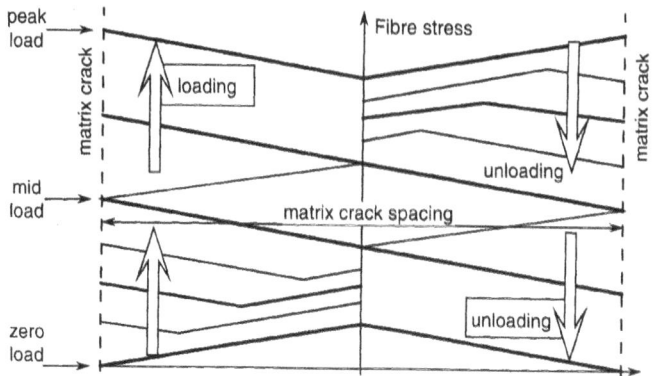

Figure 13. Small matrix spacing. Fibre stress profiles induced by stress transfer mechanism near matrix cracks, at 0, 25, 50, 75 and 100% peak load. Left: loading; right: unloading. The shaded area indicates the difference betwen unloading and loading profile. The shaded area is proportional to the loop width at mid-load.

In the case of small crack spacing (saturated condition), the fibre stress profile reverses in a first step (see Fig. 13, where the profile inversion is finished before mid-load is reached), leading to a parabolic beginning of the loop branch. Thereafter, the stress profile remains unchanged and shifts only vertically up to peak load (by loading) or down to zero load (by unloading). During this second step, the stres-strain relationship is linear, the slope being related to the Young's modulus of the fibres. Here, the mid-load width is proportional to the area of the shaded lozenges again, but the lozenge width is fixed by crack spacing. So, if the slopes (the sliding resistance) decrease, the loop width decreases.

To summarize, Fig. 14 shows the influence of sliding strength (in non-dimentional form: τ^F/τ_{sat}) on loop area (relative to a normalizing parameter ΔW^*).

The relation between sliding resistance and loop width is also explicitly given by the following expressions.

Non saturated matrix cracking ($\tau^F \geq \tau_{sat}$):

$$\frac{\Delta W}{\Delta W^*} = \frac{1}{6}\left(\frac{\tau^F}{\tau_{sat}}\right)^{-1} \tag{10a}$$

Saturated matrix cracking ($\tau^F \leq \tau_{sat}$):

$$\frac{\Delta W}{\Delta W^*} = \frac{1}{2}\frac{\tau^F}{\tau_{sat}}\left(1 - \frac{2}{3}\frac{\tau^F}{\tau_{sat}}\right) \tag{10b}$$

The critical interfacial shear stress τ_{sat} is given by:

$$\tau_{sat} = \frac{R\,E_f\,S}{2\,\eta\,D\,E_x} \tag{11}$$

and the normalizing parameter by:

$$\Delta W^* = \frac{S^2}{\eta\,E_x} \tag{12}$$

with $\eta = E_f V_f / E_m V_m$ (for a non porous unidirectional composite with all fibres intact); E_f, V_f, E_m, V_m: are the fibre and matrix Young's modulus and volume fraction respectively; E_x: composite Young's modulus; S: maximum applied stress during cycling; D: matrix spacing.

Strictly speaking, ΔW^* is not exactly the elastic strain energy exchanged during a cycle, so $\Delta W / \Delta W^*$ is different from the common internal friction parameter.

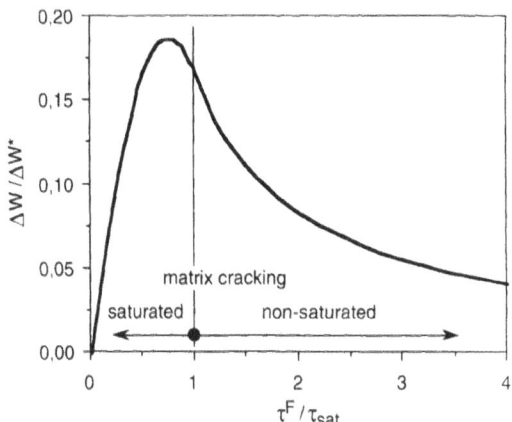

Figure 14. Relation between normalized loop width and relative interfacial shear stress

The behaviour of SiC/SiC material is described by the right hand side of the curve given in Fig. 14, where for large matrix crack spacing the loop area decreases as sliding resistance increases. Conversely, SiC/MAS-L composite corresponds to the left hand side where, for small matrix crack spacing, the loop area increases with sliding resistance.

Experimentally, fatigue cycling has been performed in tension-tension at different temperatures under argon atmosphere on both materials. Peak load was chosen so that the materials were working just below the conventional fatigue limit and was kept constant for all test temperatures, for comparison purposes. Fig. 15 and 16 show the evolution of loop area at different temperatures measured on SiC/SiC and SiC/MAS-L respectively.

For both materials, as cycling temperature increases, the loop area increases. For SiC/SiC, as temperature increases, τ^F decreases (Fig. 6 right) and then, according to Eqn. 10a, the loop area increases. For SiC/MAS-L, the effect of temperature on τF is inversed and from Eqn. 10b, an increasing sliding resistance with temperature leads again to increasing loop area.

Concerning fatigue effect, in both materials τ^F decreases as number of cycles increases. This trend leads to an increases in loop area for SiC/SiC as shown on Fig. 15 (except for the first 10 cycles), and to a decreases in loop area for SiC/MAS-L material as shown in Fig. 16. The accomodation stage observed during the first 10 cycles in the SiC/SiC composite has also been observed by considering the stiffness and tangent moduli changes, it seems be due to delayed damage extension in the matrix[3, 19].

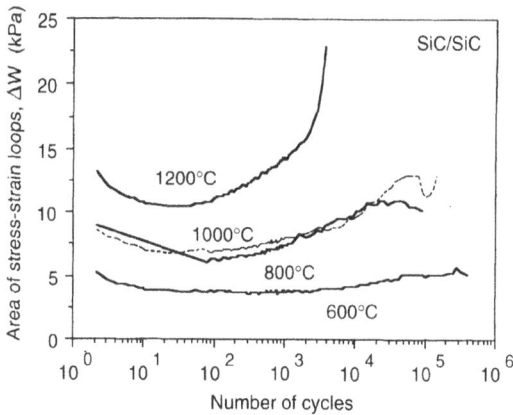

Figure 15. 2D SiC/SiC composite. Load-unload loop area as a function of numer of cycles. Fatigue cycling in tension-tension under argon, frequency : 1 Hz, Peak load : 110 MPa.

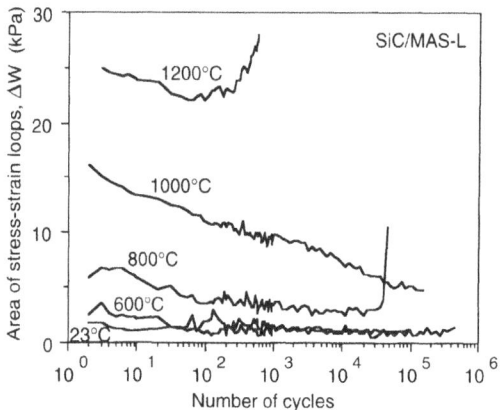

Figure 16. 2D SiC/MAS-L composite. Load-unload loop area as a function of number of cycles. Fatigue cycling in tension-tension under argon, frequency : 1 Hz, Peak load : 130 MPa.

In order to get an approximate for the law of variation of sliding resistance with number of cycles, a very simple analysis can be performed, which does not need a knowledge of material parameters. It is assumed that the fatigue effect can be described relative to the initial sliding resistance, τ_0, which is determined from the loop area of the second cycle, noted ΔW_0. The first cycle is not considered because it exhibits a larger area due to the introduction of damage during initial loading.

For SiC/SiC, Eqn. 10a leads to a relative change of interfacial shear stress, given by:

$$\left(\frac{\tau^F(N)}{\tau_0}\right)_{SiC/SiC} = \frac{\Delta W_0}{\Delta W(N)} \tag{13}$$

ΔW_0 is determined here by extrapolating $\Delta W(N)$ up to $N = 1$.

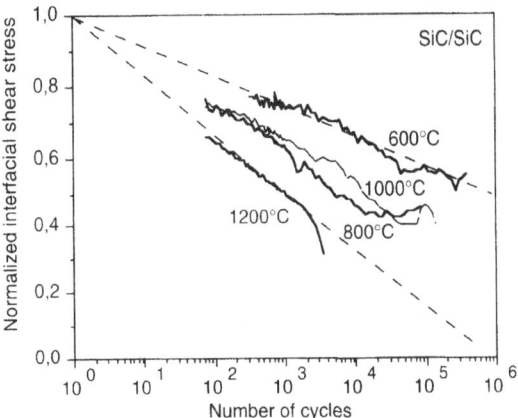

Figure 17. 2D SiC/SiC. Evolution of non-dimensional sliding resistance, τ^F/τ_{sat} with number of fatigue cycles (calculation of τ^F/τ_{sat} made by using results from Fig. 15 and Eqn. 13).

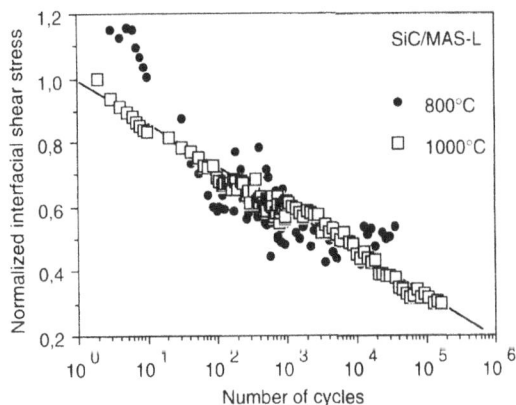

Figure 18. 2D SiC/MAS-L. Evolution of non-dimensional sliding resistance, τ^F/τ_{sat}, with number of fatigue cycles (calculation of τ^F/τ_{sat} is made by using results from Fig. 16 and Eqn. 14).

For SiC/MAS-L, we assume to simplify that τ^F/τ_{sat} is very small compared to 2/3 (this assumption is not very good at high temperature because τ^F is increasing), so that from Eqn. 10b we can write a linear relationship between relative sliding resistance and non-dimentional loop area:

$$\left(\frac{\tau^F(N)}{\tau_0}\right)_{SiC/MAS-L} = \frac{\Delta W(N)}{\Delta W_0} \tag{14}$$

The corresponding results are given in Fig. 17 and 18. To a first approximation, the change of sliding resistance during cyclic fatigue can be described by the following law, for both materials:

$$\tau^F(N) = \tau_0 \left[1 - B \ln(N)\right] \tag{15}$$

243

The interface degradation per cycle, which is given below:

$$\frac{\Delta\tau^F}{\Delta N} = - B \frac{\tau_0}{N} \qquad (16)$$

depends on the fibre clamping stress through the τ_0 parameter: the higher the interfacial pressure (including roughness effect), the more important the effect of wear is. The number of cycles in the denominator indicates that the degradation rate slows as fatigue proceeds, perhaps because the sliding surfaces become increasingly smooth. As shown in Fig. 17, SiC/SiC exhibits a relatively faster degradation at higher temperature. This effect is not yet unterstood, it could be a result of the oversimplyfied assumptions concerning the relations between temperature and clamping stress and between clamping stress and degradation rate; it could also be due to oxidation phenomena in the crushed pyrocarbon between the sliding surfaces since the argon atmosphere is not perfecly clean.

Of course, more work must be done in order to precise the real effect of temperature, the influence of cycling amplitude, etc.. Nevertheless, this example illustrates clearly the basic interest of studying the load-unload hysteresis.

ANALYSIS OF MATRIX CRACK BRIDGING

The increase in toughness of a ceramic matrix reinforced with ceramic fibres is due to the development of a process zone at the crack tip. Several energy dissipating mechanisms take place in the process zone, and in terms of the stress intensity factor, the process zone creates forces that prevent the crack from opening, this causes crack tip shielding[21, 22]. In the composites considered here, matrix crack bridging is the main toughening phenomenon. In fact, the p(u) curve, giving the bridging tractions as a function of crack opening displacement, represents the tensile behaviour of the process zone, and the associated toughening corresponds to the work done by the closing forces acting on the crack surfaces. Therefore, the p(u) curve can be used for characterizing fracture behaviour[23, 24]. In addition, the p(u) curves are linked to the mechanisms operating during matrix crack bridging and should therefore give information on the micromechanical parameters controlling the bridging process, such as fibre strength, pull-out length and sliding resistance.

We will briefly describe below the theoretical aspects of p(u) curves and some results obtained with SiC/SiC and SiC/MAS-L composites.

Theoretical description of p(u) curves

The bridging mechanism has been analysed by several authors[3, 15, 24, 25] and we give here a brief overview of the principal results.

The bridging stress (applied load divided by the area of the bridged region) is due to two additional contributions:

$$p(u) = p_1(u) + p_2(u) \qquad (17)$$

The first contribution, corresponding to the behaviour of the fibre bundle which bridges the crack, can be written as follows:

$$p_1(u) = V_f (C u)^{1/2} \exp\left[- C' (C u)^{(m+1)/2}\right] \qquad (18)$$

with: $C = \dfrac{4\,E_f\,\tau^F\,(1 + \eta')}{\eta'\,R}$, $C' = \dfrac{R}{\sigma_0^m\,\tau^F\,(m + 1)}$ and $\eta' = \dfrac{E_c}{E_f\,V_f\,(1 - q)} - 1$

where R is the fibre radius; V_f the fibre volume fraction; E_f the fibre Young's modulus; E_c the tensile modulus of the composite and τ^F the interfacial shear stress. The proportion of broken fibres, q, is related to the fracture probability P_R of the fibres submitted to overload (Fig. 19) and is given by:

$$q = P_R = 1 - \exp\left[- \frac{R\,T^{m+1}}{\tau\,\sigma_0^m\,(m + 1)} \right] \tag{19}$$

where σ_0 and m are the fibre Weibull's parameters and T is the fibre peak stress. Eqn (18) exhibits a maximum which can be thought of as the instability of the of the fibre bundle bridging the crack. The proportion of broken fibres becomes critical and is then given by:

$$q^* = 1 - \exp\left(- \frac{1}{m+1} \right) \tag{20}$$

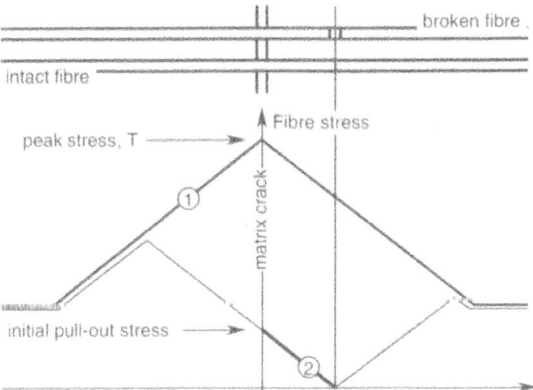

Figure 19. Fibre stress profile around a single matrix crack. Curve 1: intact fibre. Curve 2: broken fibre. The distance between matrix crack and fibre failure defines the pull-out length l_p.

The second contribution due to fibre pull-out increases because the proportion of broken fibres increases and the embedded length remains large; this pull-out contribution then decreases because of a decreasing embedded length. The contribution due to fibre pull-out can be written as follows:

$$p_2(u) = \frac{2\,V_f\,\tau^F\,<l_p>}{R}\ \exp\left[- \frac{2\,u}{<l_p>} \right] \left(1 - \exp\left[- C'\,(C\,u)^{(m+1)/2} \right] \right) \tag{21}$$

where $<l_p>$ is the average pull-out length. This expression is obtained by considering that the pull-out length distribution, $P(l_p)$, given simply by an exponential function (see Eqn. 9).

For a large crack opening displacement, all fibres are broken (q = 1) and Eqn (21) becomes:

$$p_2(u) = \frac{2\, V_f\, \tau^F\, <l_p>}{R}\, \exp\left[-\frac{2\, u}{<l_p>}\right] \qquad (22)$$

When all fibres are broken, the first contribution reduces to zero and $p(u) = p_2(u)$. Thus, Eqn. 22 shows that the tail of the $p(u)$ curve can easily be used for getting information about average pull-out length and sliding resistance.

Experimental p(u) curves and analysis

The $p(u)$ curves are measured by using double edge notched tensile specimens. The specimen is loaded in tension under a carefully controlled strain rate, in order to record the load decreasing stage after instability of the bridging. Details of specimen geometry and test procedure are given elsewhere[23]. The bridging stress, p, is obtained from the applied load and the area of the ligament between the notches. The value of crack opening displacement is obtained from the displacement given by an extensometer, by substracting the linear displacement corresponding to the elastic elongation of the specimen zone situated between the extensometer knives (elastic correction)[23].

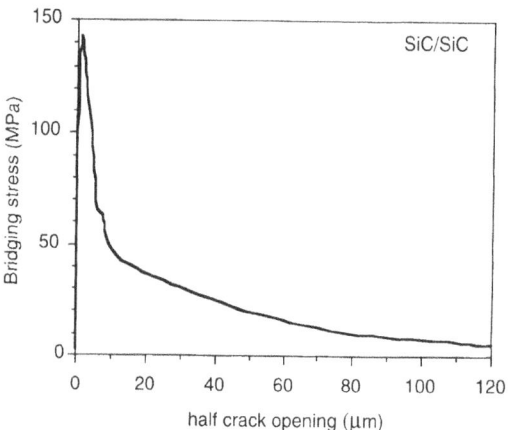

Figure 20. 2D SiC/SIC. p(u) curve. Bridging stress as a function of crack opening displacement

Typical $p(u)$ curves for SiC/SiC and SiC/MAS-L are shown in Fig. 20 and 21 respectively. At the beginning of the test, a steep increase in load is initially observed. During this first stage, one or more matrix cracks initiate and propagation occurs through the matrix between notches. Observations of the cracked zone revealed less than three parallel cracks, some of them do not cross the whole ligament. So we can consider that very quickly a unique effective matrix crack operates and progressively opens as the applied load increases. The proportion of broken fibres increases, following Eqn. 19, up to the critical value, q^* (Eqn. 20), leading to instability of the bridging system and to the maximum of $p_1(u)$ contribution. After the peak is reached, the proportion of surviving fibres quickly decreases and the $p(u)$ curve is then controlled by the pull-out contribution $p_2(u)$.

For SiC/SiC, the peak of load is very sharp and the crack opening displacement is estimated with low accuracy in this region. However, the $p(u)$ curve is well measured in the pull-out region. Concerning SiC/MAS-L composite, the peak of load is not so sharp and the

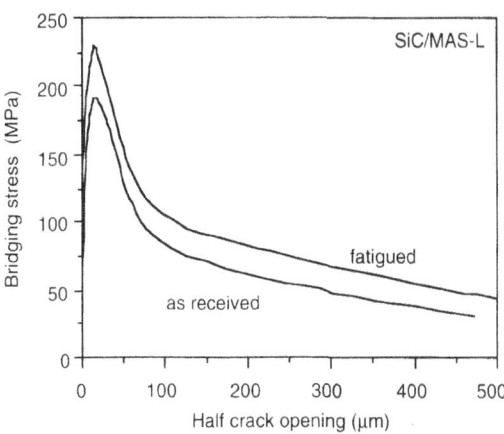

Figure 21. 2D SiC/MAS-L. p(u) curve. Bridging stress as a function of crack opening displacement

whole curve is measured with acceptable accuracy. For a comparison purpose, notches were also machined in fatigue-damaged specimens. Fatigue leads to a more pronounced pull-out phenomenon.

If we consider the tail of the curves, the decrease in pull-out stress as crack opening increases is not linear because the fibres are not of the same length. Eqn. 22 takes into account the existence of a pull-out length distribution, and from this equation, the logarithm of pull-out stress should decrease linearly as crack opening increases. This is well observed for both materials in Fig.22 and the following results are obtained:

SiC/SiC	as received:	$<l_p> =$ 100 μm	$\tau^F =$ 17.5 MPa
SiC/MAS-L	as received:	$<l_p> =$ 790 μm	$\tau^F =$ 2.9 MPa
SiC/MAS-L	fatigued:	$<l_p> =$ 1000 μm	$\tau^F =$ 2.8 MPa

The values of average pull-out length are in close agreement with those measured directly from the fracture surfaces. Analysis of p(u) curves confirms the observation that cyclic fatigue leads to a significant increase in average pull-out length.

In the case of SiC/SiC composite, the interfacial shear stress found using the p(u) curve was rather lower than that found by microindentation (12-59 MPa). This is due to the Poisson's contraction (the fibres are here in tension) effect on the fibre clamping stress and may also be due to wear phenomena, at the interface, because of the relatively long displacements.

In the case of SiC/MAS-L, the values of τ^F from p(u) curve and from indentation are similar. The effect of Poisson contraction for this material is lowered because of a more flexible matrix. In addition, it is well known that Nicalon fibre exhibits changes in diameter and cross-section shape along the fibre axis, over distances in the order of several millimeters. This leads to a wedging mechanism which can compensate for the reduction in τ^F by interfacial wear. The fibre wedging mechanisms arising from macroscopic variations in diameter along the fibre length become important in such long-distance sliding (almost 1mm). This hypothesis is confirmed by identical τ^F values determined from p(u) curves for both undamaged and fatigued samples, where fatigue induces a higher $<l_p>$ due to interface degradation by cyclic sliding between fibre and matrix (see last section).

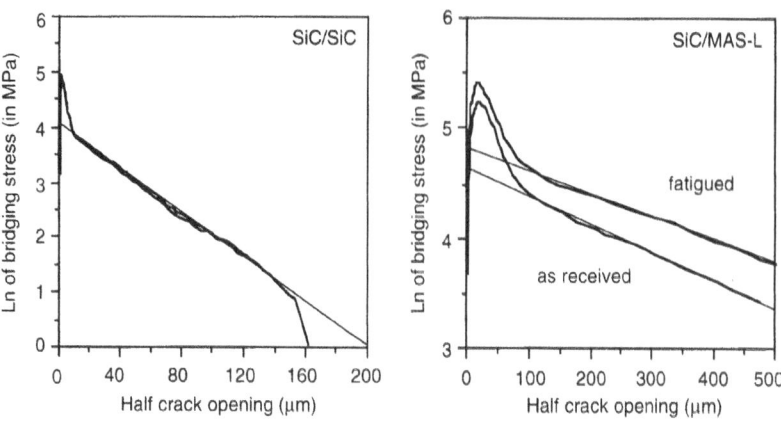

Figure 22. Analysis of the tail of the p(u) curves. Left: SiC/SiC. Right: SiC/MAS-L.

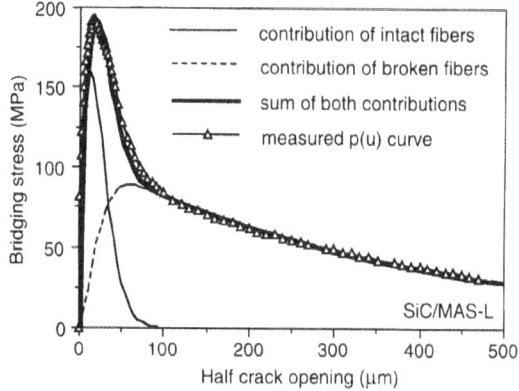

Figure 23. Simulation of the p(u) curve in the case of 2D SiC/MAS-L material. The parameter values taken are given in the text.

Finally, the average pull-out length is an indication of the interfacial characteristics under small slip distances (the slip needed for stress transfer). On the other hand, the tail of the $p(u)$ curve is linked to the interface characteristics under larger slip distances.

Fig. 23 shows a simulation of the $p(u)$ curve in the case of SiC/MAS-L material. It is interesting to note that the $p(u)$ peak is not simply linked to the maximum of $p_1(u)$ intact fibres contribution. Near the $p(u)$ peak, $p_1(u)$ and $p_1(u)$ contributions have nearly the same weight. The best fit with experimental data for as received SiC/MAS-L, was obtained with an average fibre strength $<\sigma_f^R> = 295$ MPa (for a gauge length of $L = 50$ mm, see Eqn. 8) and a Weibull's modulus $m = 2$. This value of fibre strength is very low compared to nominal fibre strength commonly reported (2000 MPa, $L = 50$ mm), but corresponds reasonably with the mean strength as calculated from the mirror fracture on fibre failure surface, where the

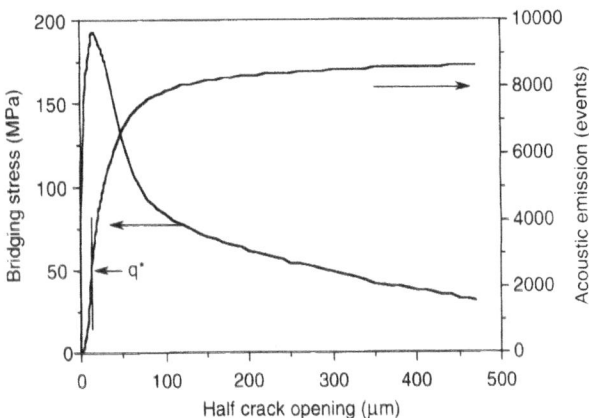

Figure 24. 2D SiC/MAS-L. p(u) curve and number of acoustic emission events recorded. The critical proportion of broken fibres is the number of events recorded at the p_1 peak (see Fig. 23) divided by the total number of events.

equivalent gauge length is twice the average pull-out length (975 MPa, L = 1.6 mm). As a result, composite processing drastically reduces fibre strength.

The present simulation was performed with a quite low value of Weibull's modulus (m = 2) which appears lower than the values commonly reported for Nicalon fibres (m = 3-6). This low value is confirmed by analysis of the acoustic emission recorded during the test. As shown in Fig. 24, the total number of events recorded is very close to the number of fibres present in the ligament cross-section. So each fibre failure leads to one acoustic emission event. By using Eqn. 20, and the number of events recorded up to the peak of $p_1(u)$ contribution, a value close to 2 is found for m. Composite processing induces fibre damage (mainly at larger defects) such that the Weibull's modulus is reduced, hence, m − 2 seems representative of the in-situ fibre strength scattering.

In terms of the characterization of fracture behaviour, the bridging contribution to the work of fracture is given in the following relationship:

$$\Delta G^* = 2 \int_0^{u^*} p(u) \ du \tag{23}$$

where 2u* represents the distance of separation reached by the crack surfaces, thus for complete separation Eqn. 23 becomes:

$$\Delta G_c^* = 2 \int_0^{\infty} p(u) \ du \tag{24}$$

The behaviour of a macrocrack associated with the fracture of the material (initiation, propagation and separation) is usually described by an R-curve, where changes in crack growth resistance, R, are plotted as a function of crack increment, Δa. Above an initiation threshold, R_0, the R-curve starts to increase due to extension of the process zone. A plateau is then reached, this stationary regime arises when the crack tip propagates with nearly constant process zone size. The plateau value, R_p, can then be written as the threshold value plus the increase of thoughness: $R_p = R_0 + \Delta R$. The increase of the crack growth resistance can be estimated by means of Eqn (24), with $\Delta G_c{}^* = \Delta R$ and $R_0 = 0$.

In fact, the uncertainty in the peak position after the elastic correction, does not really affect the integral of the measured p(u) curve. Fig. 25 shows the result in the case of SiC/SiC: integration of the p(u) curve is compared to the R curves (plot of crack growth energy versus crack increment) obtained from mechanical tests on CT specimens with varying geometry and notch size.[16] In order to directly compare the results, the crack increment, Δa, has to be related to crack opening, u. This was done assuming that the effective crack in the CT specimen is triangular in shape (linear variation of u versus position) and that there is a rotation of both arms of the specimen, the rotation axis being located at one third of the remaining ligament legth (see Fig. 25 right).

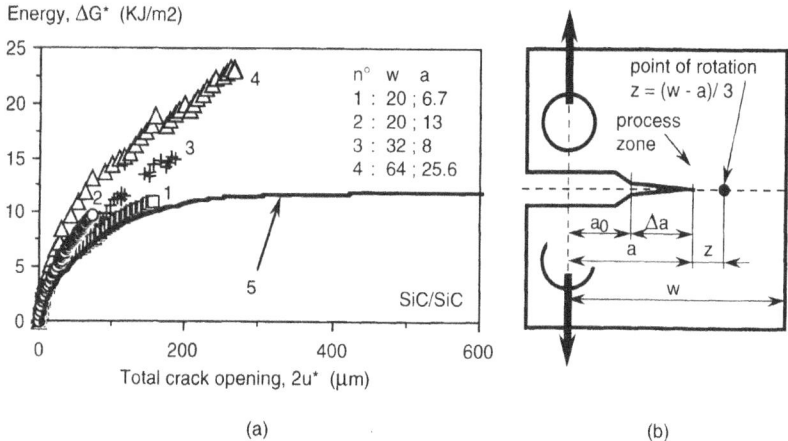

(a)   (b)

Figure 25. 2D SiC/SiC. Exploitation of the p(u) curve in terms of crack growth resistance. (a): as a function of crack opening displacement, comparison of crack growth resistances measured on CT specimens with the integration of the p(u) curve (curve 5). (b): schematic of the CT specimen.

Fig. 25 shows that crack growth energy curves, deduced from R curves, do not reach a plateau and that they are located always above the p(u) integral. This comes from the fact that in the CT specimens, the process zone at the crack tip is always large and interacts then with the compressive zone of the ligament, whereas the crack opening near the loading axis remains small. To avoid these problems, very large CT specimens should be used, where the length of the bridging zone would be much smaller than the crack length. So, the p(u) curve method looks like a promizing alternative to the use of very large CT specimens. Moreover, the equipment needed to measure p(u) curves, is almost similar to that used for tensile tests. It would be easy, then, to modify high temperature tensile tests in order to characterize the high temperature bridging behaviour, and finally, to characterize the high temperature fracture behaviour of a ceramic matrix composite.

In the case of SiC/MAS-L, the use of CT specimens was unsuccessful due to the systematic appearance of a delamination damage between plies in the compressive zone of the ligament. The p(u) curve is, therefore, the only method available to determine the crack growth resistance here. The results for ΔR are 84 kJ/m^2 for the as received material and 140 kJ/m^2 in the fatigued condition. These values can be looked upon as the plateau values because the initial crack growth resistance, R_0, is very small (0.1 to 0.2 kJ/m^2). These high values of ΔR for SiC/MAS-L are caused by the very important contribution of fibre pull-out. A more conservative approach would be to take as the maximum crack opening, the abcissa of the p(u) peak (eqn (23) with $u^* = u_{max}$), leading to $\Delta R = 4.8$ kJ/m^2 (as received case) and 5.4kJ/m^2 (fatigued case).

CONCLUSION

Fibre-matrix debonding is mainly operating in mixed mode because of the post-debonding radial displacement, due in some cases, to thermal strain misfit (when $\alpha_m < \alpha_f$). In most cases, the radial displacement is induced by the roughness of the sliding surfaces.

The interfacial shear stress associated with frictional slip of the debonded surfaces increases as fibre diameter decreases. This behaviour is controlled by the roughness which introduces an additional compressive misfit between the fibres and the matrix. The variation of this interfacial shear stress with temperature depends mainly on the difference in coefficients of thermal expansion between the fibres and the matrix. At temperatures where the phases are not physically or chemically affected, the changes in mechanical properties are due to these variations of thermal misfit.

Pull-out length distributions and average values can be measured directly from fracture surfaces. If fibre strength and Weibull modulus are known, these measurements lead to a good estimation of the interfacial shear stresses operating during stress transfer at specimen failure. Pull-out length measurements performed on hot tested specimens give informations about changes in sliding resistance which are not easy to measure directly, for example by indentation, at high temperature. Clearly, cyclic fatigue leads to increasing pull-out length, because the stress transfer capability is reduced by interfacial degradation phenomena.

The hysteresis of the load-unload loops can easily be measured during cyclic fatigue. This kind of analysis is a powerful in-situ monotoring tool of the interfacial shear stress changes and can potentially give access to the law of interface degradation. However, the link between loop area and frictional stress depends on the amount of multiple matrix cracking.

The p(u) curves giving the bridging tractions as a function of matrix crack opening displacement can be measured. Analysis of these curves leads to information concerning in-situ fibre strength, pull-out length and interfacial shear stress related to large sliding distances. Clearly, composite processing reduces considerably the fibre strength. The p(u) curve represents the constitutive law of the bridging system and can prove fruitfull when used for characterizing the toughnening behaviour.

ACKNOWLEDGMENTS

This work was done within part of the french Scientific Group called: "Thermomechanical behaviour of fibrous ceramic-ceramic composites". The author thanks Aérospatiale and SEP for supplying the materials.

REFERENCES

1. R.J. Kerans, R.S. Ray, N.J. Pagano and T.A. Parthasarathy, The role of the fiber-matrix interface in ceramic composites, *Ceram. Bull.*, 68 : 429-442 (1989).
2. A.G. Evans and F.W. Zok, The physics and mechanics of fibre-reinforced brittle matrix composites, *J. Mater. Sci.*, 29 : 3857-3896 (1994).

3. D. Rouby and P. Reynaud, Fatigue behavior related to interface modification during load cycling in ceramic-matrix fiber composites, *Comp. Sci. Techn.*, 48 : 109-118 (1993).

4. D.B. Marshall, Analysis of fiber debonding and sliding experiments in brittle matrix composites, *Acta Metall. Mater.*, 40 : 427-441 (1992).

5. J.O. Outwater and M.C. Murphy, The fracture energy of unidirectional laminates, Proc. 24th Annual Technical Conf., Society of the Plastic Industry, paper 11C (1969).

6. M. Benoit, Ph.Brenet and D. Rouby, Comportement des interfaces dans des composites céramique-céramique, *Rev. des composites et des matériaux avancés*, 3 : 235-251 (1993).

7. P.G. Charalambides and A.G. Evans, Debonding properties of residual stressed brittle-matrix composites, *J. Am. Ceram. Soc.*, 72 : 746-753 (1989).

8. B. Martin, M. Benoit and D. Rouby, Intefacial sliding resistance in fibre reinforced ceramic matrix composites having tensile radial thermal misfit strain, *Scripta Metall. Mater.*, 28 : 1429-1433 (1993).

9. H.J. Oel and V.D. Frechette, Stress distribution in multiphase systems : II, composite disks with cylindrical interfaces, *J. Am. Ceram. Soc.*, 69 : 342-346 (1986).

10. J.W. Hutchinson and H.M. Jensen, Models of fiber debonding and pullout in brittle composites with friction, *Mech. of Mater.*, 9 : 139-163 (1990).

11. P.D. Jero, R.J. Kerans and T.A. Parthasarathy, Effect of interfacial roughness on the frictional stress measured using pushout tests, *J. Am. Ceram. Soc.*, 74 : 2793-2801 (1991).

12. B.F. Sørensen, Effect of fibre roughness on the overall stress-transverse strain response of ceramic composites, *Scripta Metall. Mater.*, 28 : 435-439 (1993).

13. F. Lamouroux, M. Steen and J.L. Vallés, Uniaxial tensile and creep behaviour of an alumina fibre-reinforced ceramic matrix composite: I. Experimental study, II. Modelling of tertiary creep, *J. Eur. Ceram. Soc.*, 14 : 529-537 and 539-548 (1994).

14. W.A.Curtin, Theory of mechanical properies of ceramic matrix composites, *J. Am. Ceram. Soc.*, 74 : 2837-2845 (1991).

15. M. Sutcu, Weibull statistics applied to fiber failure in ceramic composites and work of fracture, *Acta metall.*, 37 : 651-661 (1989).

16. F. Conchin, "Étude du comportement thermomécanique en relation avec la micro structure de matériaux à renfort fibreux SiC/C/SiC 2D", Thesis, INSA de Lyon (1994).

17. Ph. Brenet, "Étude du comportement mécanique, en relation avec la microstructure, d'un composite céramique-céramique SiC/MAS-L", Thesis, INSA de Lyon (1994)

18. B.F. Sørensen and R. Talreja, Analysis of damage in a ceramic matrix composite, *Int. J. Damage Mech.*, 2 : 246-271 (1993).

19. P. Reynaud, "Étude du comportement en fatigue des matériaux composites à matrice céramique suivi par émission acoustique", Thesis, INSA de Lyon (1992)

20. B.F. Sørensen and J.W. Holmes, Improvement in the fatigue life of fiber-reinforced ceramics by use of interface lubrication, *Scripta Metall. Mater.*, to be published.

21. B.N. Cox, Extrinsic factors in the mechanics of bridged cracks, *Acta Metall. Mater.*, **39** : 1189-1201 (1991).

22. R.N. Stevens and F. Guiu, The application of the *J* integral to problems of crack bridging, *Acta Metall. Mater.*, 42 : 1805-1810, 1994.

23. P. Brenet, F. Conchin, G. Fantozzi, P. Reynaud, D. Rouby and C. Tallaron, Direct measurement of the crack bridging tractions, a new approach of the fracture behavior of brittle matrix composites, *Comp. Sci. Techn.*, to be published.

24. J. Llorca and M. Elices, Fracture resistance of fiber-reinforced ceramic matrix composites, *Acta Metall. Mater.*, 38 : 2485-2492 (1990).

25. M.D. Thouless and A.G. Evans, Effects of pull-out on the mechanical properties of ceramic-matrix composites, *Acta Metall.*, 36 : 517-522 (1988).

FRACTURE CHARACTERISTICS OF SiC MONOFILAMENT /SiAlON CERAMIC COMPOSITE

K.Ueno, T.Inoue and S.Sodeoka

Osaka National Research Institute, AIST
Midorigaoka 1-8-31, Ikeda, 563 Osaka, Japan

INTRODUCTION

Continuous fiber reinforced composite attracts recently much attention. Its high fracture resistance makes the material quite attractive for structural applications in automobile and other industries. Many researchers reported improved properties such as high fracture resistance, reflected by high work-of-fracture(WOF)[1-12]. There are two types of the composites; one is reinforced with multifilament fiber like carbon fiber or Nicalon fiber, and another with monofilaments like CVD-formed SiC fiber or Boron fiber. Remarkable feature of the latter-type composite is its significantly high fracture resistance as compared with former-type composite in spite of low fiber volume percentage[7,12]. It is important to reveal the reason of its high fracture resistance in terms of evaluating technological possibility as engineering composite material. One of other features is that one can arrange the fiber in a uniform array with nearly same fiber-to-fiber distances. Such a composite can be regarded as a simple model of fiber reinforced composite that will supply information on fracture behavior of the composite.

Several composites with filament-type fiber and various ceramic matrix have reported such as SiC fiber/glass or SiC fiber/Si_3N_4[7-12]. β'-SiAlON ($Si_{6-z}Al_zO_zN_{8-z}$, $z \le 4.2$) is a member of silicon nitride family. The material is interesting as a industrial material as it has several good properties such as high oxidation resistance, excellent thermal shock properties and high resistance to corrosive substances like molten metal. Huang et al. reported that SiC fiber/ combustion-synthesized β'-SiAlON composite had a significantly improved fracture toughness and improved ultimate fracture strength[13]. To improve the mechanical properties of the composite, it is necessary to reveal what microfracture process takes place and the role of the matrix/fiber interface.

In this study, SiC monofilament reinforced β'-SiAlON (z=4) matrix composite was fabricated. Its fracture process was monitored by direct observation of crack propagation. Work-of-fracture(WOF) of each microfracture process including fiber breakup, fiber pull-out, and matrix multiple fracture has been analyzed based on measured interfacial frictional stress and fiber pull-out length. The influence of strain-rate on the WOF was also studied. On the

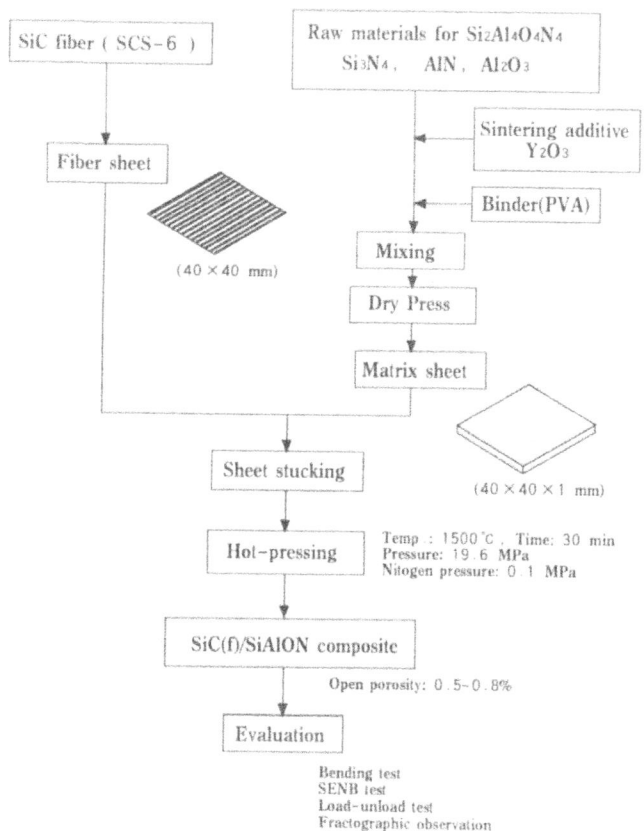

Fig. 1 Schematic illustration of fabrication process of SiC fiber/SiAlON composite

basis of these results, the fracture characteristics was evaluated as a model of a mposite reinforced with uni-directional strong monofilaments with nearly same fiber-to-fiber distance.

EXPERIMENTAL

Materials and fabrication process

As reinforcement, SiC monofilament SCS-6, supplied from Textron Specialty Materials, Lowell,MA, is used. The fiber has a carbon coating layer with the thickness of about 3 μ m. Matrix SiAlON has a chemical composition of $Si_2Al_4O_4N_4$. Powder mixture of Si_3N_4(Herman C. Starck, LC-12), Al_2O_3(Taimei Chemical Industry Co., TM-D), and AlN(Tokuyama Soda Co., G-grade) with 3 mass% Y_2O_3(Nippon Yttrium Co., 99.9%) as a sintering additive are mixed for 24 hours in a polyethylene bottle with Si_3N_4 balls and ethanol solvent.

SiC fiber was unidirectionally aligned with the fiber-to-fiber distance of 0.5mm to make a fiber sheet having the dimensions of 40 x 40 mm. One fiber sheet was placed at the bottom of a metal die, and the matrix powder was poured and die-pressed to form a fiber/matrix green sheet. Five green sheets were stacked in a square(40x40 mm) graphite die and hot-pressed at 1500 °C for 30 min with the pressure of 20 MPa under nitrogen atmosphere. The fabrication process is schematically illustrated in Fig.1. Monolithic SiAlON was also fabricated in the same manner. Sample surface was ground with a diamond wheel to finish the surface layer, and cut into test specimen (3.3x4.1x40 mm, rectangular bar) parallel to the fiber direction.

Fig.2 shows the cross-sectional view of the composite perpendicular to the fiber alignment direction. Fiber/matrix sheets are stacked from downward to upward in the picture. It appears that fiber array has nearly same fiber-to-fiber distance of about 500 μ m. The composite has the fiber content of 4.5 vol%. Table 1 shows fundamental properties of the composite and monolithic SiAlON.

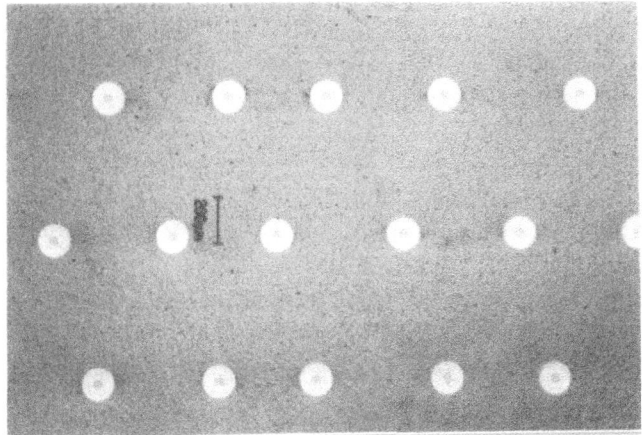

Fig. 2   Cross-sectional view of SiC monofilament/SiAlON composite.

Table 1 Density, Open porosity, Flexural strength, Fracture toughness, and WOF of SiC monofilament/SiAlON and monolithic SiAlON.

	SiC filament/SiAlON	Monolithic SiAlON
Density/ gcm^{-3}	3.15~3.18	3.28
Open porosity/ %	0.5~0.8	0.15
Flexural Strength/MPa	250	300
Fracture toughness/MPam$^{1/2}$	7~8	4

Evaluation of fracture characteristics

Crack extension during the fracture was observed by load-unload method for a sample specimen bent by 3 points mode with the crosshead speed of 0.05 mm/min and the span of 30 mm. Load was applied and released immediately after an microfracture accompanied with abrupt load drop. Cracks in the specimen were then observed by an optical microscope, and was reloaded up to next fracture stage. This procedure was cycled to the final fracture stage.

Flexural strength and the work-of-fracture(WOF) were measured by 4 points bending test with 30 and 10 mm spans with the crosshead speed of 0.1 mm/min. Apparent fracture toughness was measured by SENB method. Straight-through notch with the depth of 1.4 mm (a/W=0.42) was machined by a 0.15 mm thick diamond blade at the center part of the specimen 3.3 mm thick, 4.1 wide, 18 mm long. Three points bending was carried out with the crosshead speed of 0.05 mm/min. Apparent toughness value was calculated from the ultimate stress for 3 specimens.

Influence of strain-rate was examined by 3 points bending with the crosshead speed of 0.05, 0.5 and 5 mm/min. The WOF was measured for three specimen at each crosshead speed. Fiber pull-out length was measured directly by an reading microscope with the accuracy of 1 μ m. Approximate interfacial friction stress was measured by single-fiber push-out method using a Vickers indentor.[14,15] Thin specimens with various thickness from 0.34 to 0.54 mm were sliced perpendicular to fiber axes and polished by 1 μ m diamond. The fiber was indented by a diamond pyramid with various loads and examined whether the filament was pushed out[11]. The interfacial friction stress(τ) was calculated by using following equation, τ =F/2 π rt, where F is the load at which a fiber pushed out, r is the filament radius and t is the specimen thickness. Pushed filament was then push-back from the reverse side. Results obtained by filament push-out and push-back tests gives an interfacial debonding strength and an interfacial frictional stress, respectively.

RESULTS AND DISCUSSION

Observation of Fracture Behavior

Fig.3 shows a load-deflection curve obtained in load-unload test. Along the total fracture, microfractures with small rapid load drops were observed. After each load drop is recorded, stress loading is stopped and released for each fracture stage. In the figure, such fracture stages are numbered. Optical side view of tested specimen after 1st, 2nd, 3rd, 4th and 7th fracture stage are shown in Fig.4 (a), (b), (c), (d) and (e), respectively.

Initial first stage corresponds to elastic deformation of the matrix tensile part. Applied

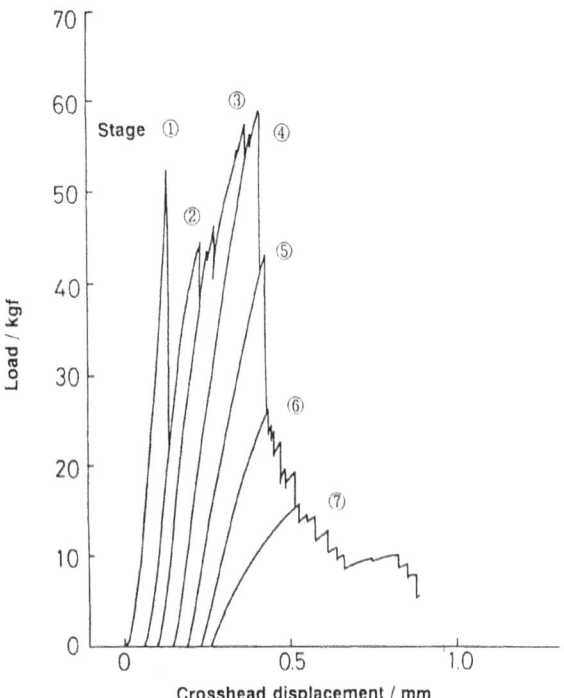

Fig.3 Load-deflection curve obtained in load-unload test.

load increases up to first-matrix-cracking (FMC) stress. Therefore, the WOF in this stage is strain energy. Fig. 4(a) after the 1st stage shows single crack propagating from the tensile surface to compressed part. The length of the crack is about 1.6 mm, corresponding to a/W=0.48 where a and W are crack length and specimen width, respectively. Initiated crack from tensile surface propagates about half of specimen width, and is then arrested by fibers. In 2nd fracture stage after FMC, the load increases again, suggesting that intact fibers can bear the load. At the final fracture of this stage, crack branching was observed as indicated by an arrow in Fig.4 (b). Third and 4th stages exhibit different fracture phase as compared to previous stages(Fig.4 (c) and (d)), that is, matrix multiplex fracture occurs. Two new matrix cracks besides the main crack propagate from the tensile surface. The spacing between these cracks are almost same as shown in Fig.4(c). These three cracks extend and gather in the compressed part of the bar. Such matrix multiplex fracture is a typical fracture manner an an in the 4th stage. In final 7th stage, no new matrix fracture at tensile part takes place, but complicated matrix fracture occurs in compressed side, as shown in Fig.4 (e).

Often distinctive sound of fiber breaking, being different from that of the matrix fracture, has been dictated accompanied with the load drop in each fracture stage. This confirms that the applied load is carried by intact fibers in the wake region after FMC. It suggests that much amount of elastic strain energy is stored in the fiber and is released by its fracture. Complicated matrix fracture at the later fracture stages seems to absorb much energy. The works done by fiber pull-out or the interface debonding should be taken into account although they are not recognized by the optical observation on specimen. These fracture processes raise the fracture energy of the composite. The contribution of each fracture process to the total WOF is studied in next chapter.

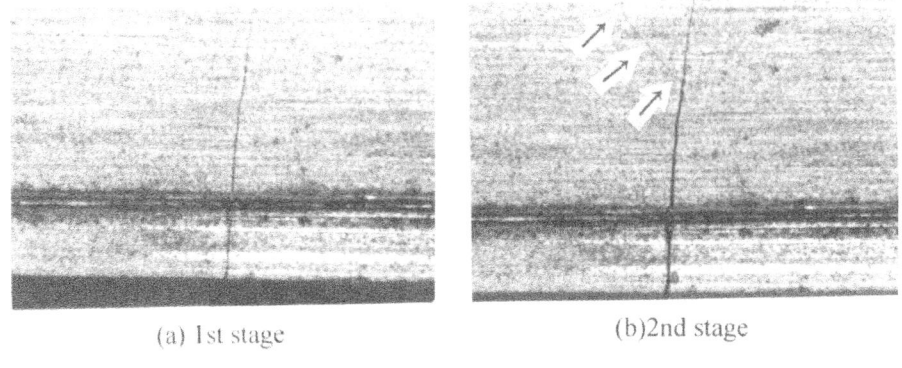

(a) 1st stage   (b)2nd stage

500 μ m

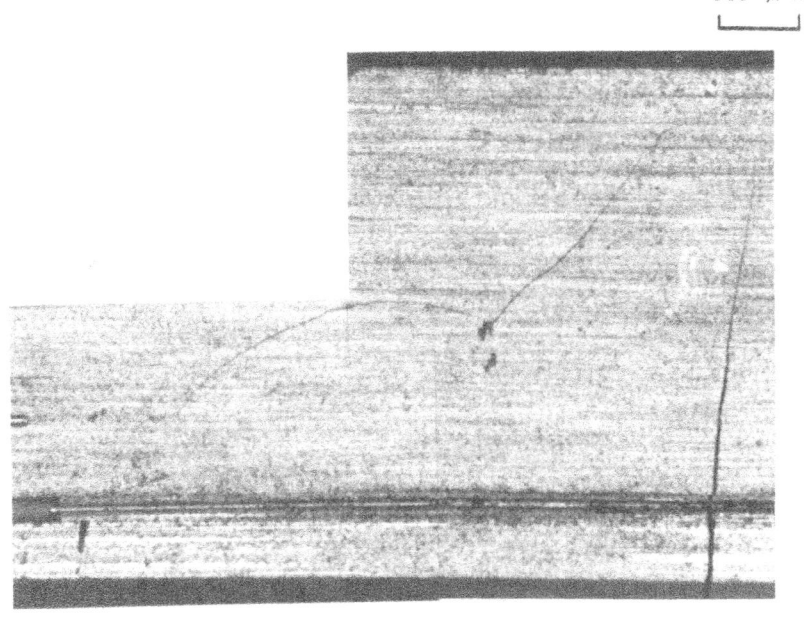

(c)3rd stage

Fig. 4   Optical side view of specimen after   (a) 1st,   (b) 2nd,   (c) 3rd,   (d) 4th and (e) 7th stage in load-unload test.

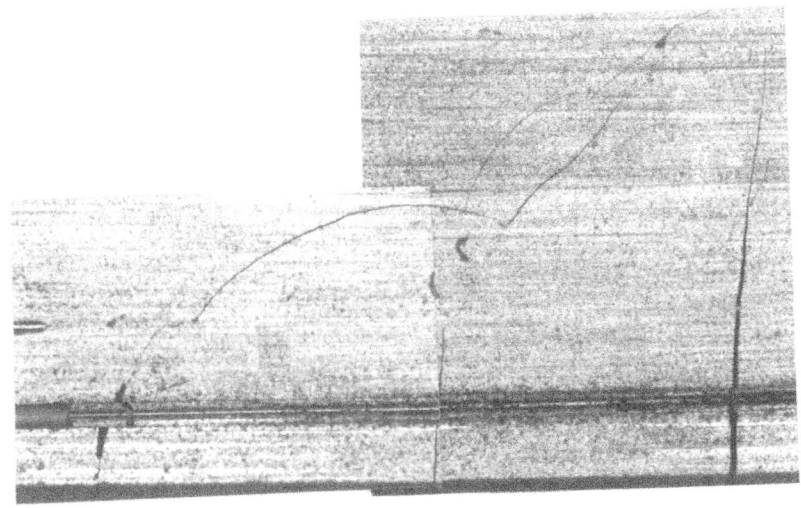

(d)4th stage

500 μ m

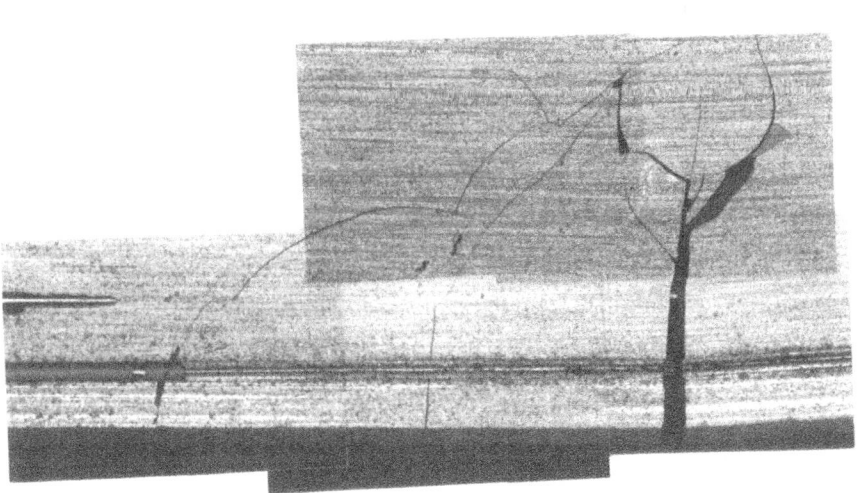

(e)7th stage

Analysis of Fracture Energy

The fracture of SiC monofilament/SiAlON composite includes elastic deformation of tensile matrix and fracture, elastic deformation of intact fibers and fracture, fiber pull-out, and matrix multiplex fracture in tensile and compressed part. The contributions of these process have been analyzed approximately.

The fracture process can be divided into two parts; linear and non-linear part, as shown in Fig.5. The former corresponds to the elastic strain energy of the tensile matrix (including elastic strain energy of the measuring rig in part) during the first fracture stage which ends with FMC. The latter corresponds to the energy absorbed during non-linear behavior of the composite after the FMC. The WOF of linear response can be calculated from the load-deflection curve. Elastic strain energy stored by the intact fibers, and the work done by filament pulling-out were estimated as follows on the basis of the work by Phillips[16].

Elastic strain energy γ_E stored by the intact filaments is approximately calculated on the basis of the equation,

$$\gamma_E \sim (V_f \, \sigma_f l_c / 6E_f) + V_f \, \gamma_f \tag{1}$$

where V_f is filament volume percent, σ_f is fiber strength, l_c is the minimum length of fiber which can be loaded to failure in a composite, and γ_f is the fracture energy of a fiber. l_c is calculated from the average spacing of matrix cracking t, the tensile strength of the matrix $(\sigma_m)_u$, and σ_f.

$$l_c = \sigma_f t \, V_f / (\sigma_m)_u (1-V_f) \tag{2}$$

The calculated l_c is 2.43 ± 0.58 mm. For SiC filament σ_f is 3.79 GPa, and E_f is 400 GPa. SiAlON matrix strength $(\sigma_m)_u$ is 250 MPa. The fracture energy of SiC fiber, γ_f is

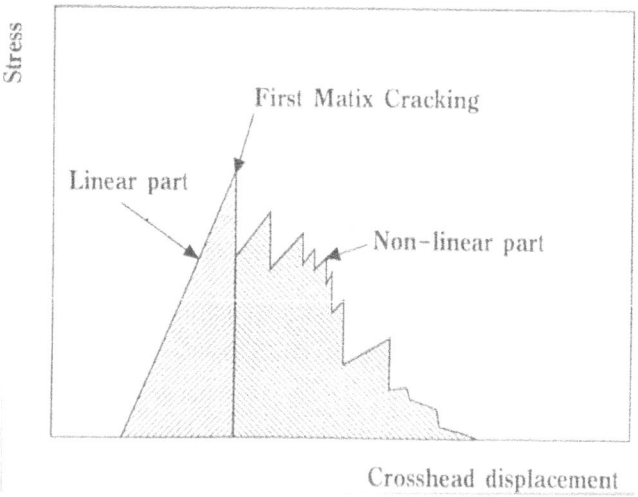

Fig. 5 Schematic illustration of load-deflection curve, divided into linear and non-linear part.

calculated on the basis of the fracture energy of graphite(~ 50 J/m^2) and of SiC(~ 40 J/m^2). These figures are put into eq.(1). Obtained fiber elastic strain energy γ_E is ~ 660 J/m^2.

The work done by fiber pull-out, γ_{PO}, can be derived by an equation,

$$\gamma_{PO} = V_f \; \tau \; \Sigma \; N_i \, l_i^2 \, / \, 2r \; \Sigma \; N_i \tag{3}$$

where N_i is the number of fibers with embedded length of li, r is fiber radius, and τ is interfacial frictional stress. $\Sigma \; N_i \, l_i^2$ has been calculated from the histogram of the filament pull-out length. τ, estimated by single filament push-out and push-back method, is ~ 26 MPa. The estimated γ_{PO} is ~ 165 J/m^2.

The work done during matrix fracture in compressed region is measured by comparing the non-linear WOF of normal flexural specimen and that of the specimen with a straight-through notch in compressed region, to which Mo wire of 0.85 mm diameter is inserted, as illustrated in Fig. 6. The half of the incorporated fiber is not pulled out as the notch depth is half of the specimen height, so that this effect is subtracted from the result. The work done by matrix fracture in compressed region is ~ 80 J/m^2. These results can be summarized in Table 2.

Table 2   Analysis of the work-of-fracture of SiC monofilament/SiAlON composite.

Total WOF	2310 J/m^2
Linear WOF	860 J/m^2
Non-linear WOF	1450 J/m^2
Elastic strain energy stored in filaments γ_E;	~ 660 J/m^2
Work by fiber pull-out γ_{PO};	~ 165 J/m^2
Work by matrix fracture in compressed area;	~ 80 J/m^2
Others;	~ 545 J/m^2

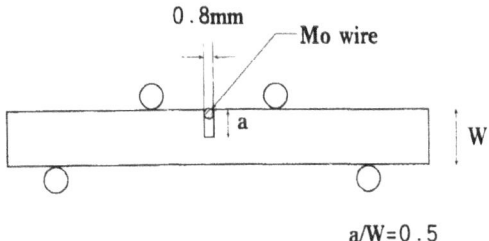

0.8mm

Mo wire

a

W

a/W=0.5

Fig.6 Schematic illustration of a specimen with a notch in compressed region in which molybdenum wire is inserted.

In term "Others", the energy accompanied with multiplex matrix fracture in tensile region, and debonding of fiber/matrix interface, and elastic strain energy stored in measuring rig are included. It is notable that the percentage of filament pull-out in non-linear WOF is not great as compared to a composites reinforced with multifilament fiber, in which the pull-out energy exceeds 60 % to total WOF[16]. While the elastic strain energy stored in intact fibers has much more dominant. The pull-out energy is about one forth of the fiber elastic strain energy, and this ratio is independent of the fiber volume percentage. The reason why the intact filaments govern half of the non-linear part of WOF is that the fiber is so strong that a large amount of strain energy is stored before fiber failure. Even though the matrix fracture is stable, unstable fracture of the fibers are inevitably included and the process absorbs much amount of energy. This is a distinctive feature of the fracture of CVD-filament reinforced ceramic composite.

Influence of Strain Rate on the Fracture Behavior

On an actual application of fiber reinforced composite, stress is applied with various loading conditions. It is doubtful that the composite might be tough under fast strain rate. Therefore, it is necessary to reveal the strain-rate dependence of fracture resistance. SiC monofilament / SiAlON composite was fractured by 3-point bending with the crosshead speed of 0.05, 0.5 and 5 mm/min, and fracture behaviors were compared.

The load-deflection curve does not change as a whole when tested under different strain rate. Fig. 7 shows the change of linear and non-linear part of WOF by crosshead speed. Linear part is independent on the strain-rate. This corresponds to the fact that first-matrix-cracking stress, that is, matrix strength is not influenced by strain rate. On the other hand, non-elastic part of WOF decreased with increasing strain rate. Supposing that the strength of SiC fiber might be independent of strain rate within the range studied here, the processes concerning with fiber/matrix interface such as pull-out and interface debonding should cause the strain-rate dependence of fracture behavior of the composite. Fiber pull-out length was examined.

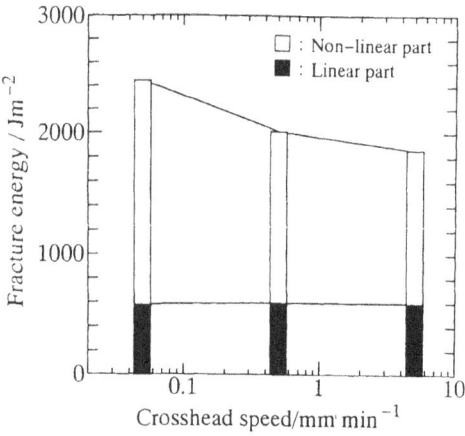

Fig. 7   Change of WOF by crosshead speed.

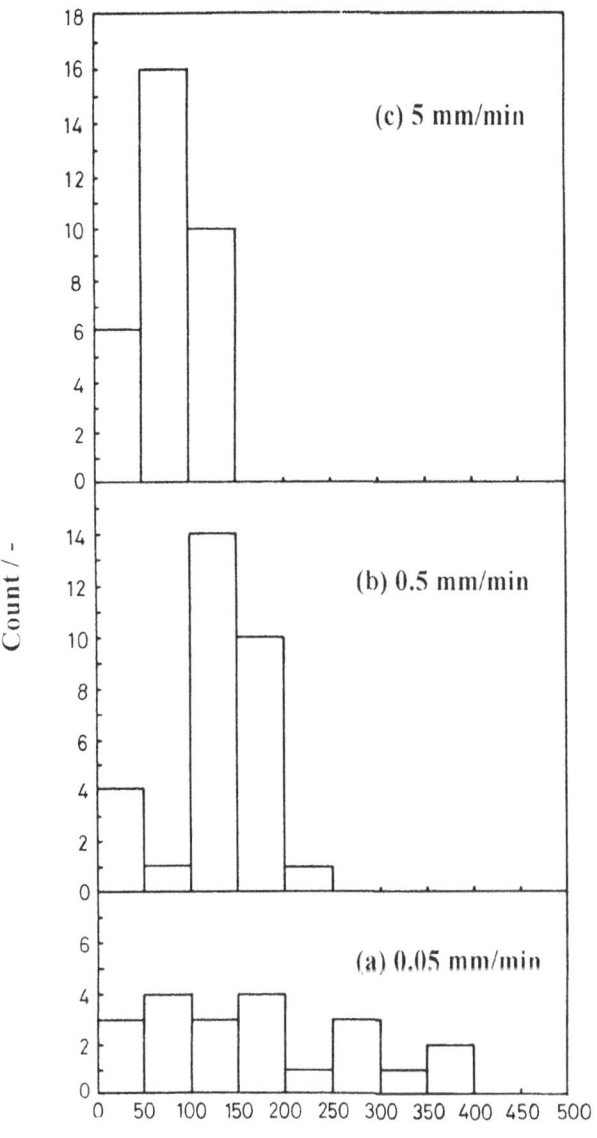

Fig.8　Fiber pull-out length distribution of SiC fiber/SiAlON composite tested under the crosshead speed of (a)0.05 mm/min, (b)0.5 mm/min, and (c)5 mm/min.

Figure 8 shows fiber pull-out length distributions of the composites fractured under different crosshead speeds. With increasing strain-rate, pull-out length decreases and the distribution width also decreases. Figure 9 indicates that average filament pull-out length decreases in proportion to the logarithm of crosshead speed.

The work done by pull-out of a single fiber, W(li), embedded with the length of li is W(li)= π r τ li^2[16]. Total work by fiber pull-out , γ $_{PO}$, is therefore derived on the basis of the pull-out length distribution(equation (3)). It is necessary to consider not an average pull-out length but the parameter (Σ Nili^2 / Σ Ni), as the factor affecting total pull-out energy. The relation between the Σ Nili^2 / Σ Ni and the non-linear part of WOF is shown in Fig.10. Although we cannot estimate quantitatively the work of pull-out because strain-rate dependence of frictional stress is unknown, the contribution of fiber pull-out energy to the non-linear part of WOF increases linearly with decreasing strain-rate.

When there is no contribution of pull-out, that is, Σ Nili^2 / Σ Ni=0, non-linear part of WOF is calculated to be about 1350 J/m^2 by extrapolating the linear relationship in Fig. 10. In the previous chapter, non-linear WOF has been analyzed under the condition of 4-point bending and the crosshead speed of 0.1 mm/min . Non-linear part of WOF is 1450 J/m^2 and γ_{po} is ~ 165 J/m^2. Therefore, the WOF without filament pull-out is ~1285 J/m^2. This value is well coincident to the value obtained from the extrapolation, considering the difference in the fracture mode.

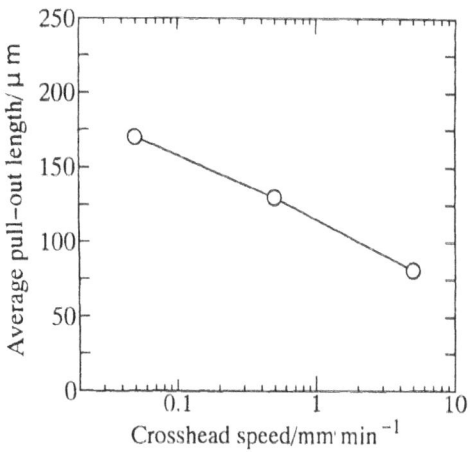

Fig. 9   The relation between average pull-out length and crosshead speed.

The reason of strain-rate dependence of pull-out length is considered as follows. Fracture stage after FMC is supposed. If the strain-to-failureof SiC fiber and crack propagation velocity of interfacial debonding (v) are independent of strain-rate, the time Δ t from the starting of loading to fiber fracture becomes shorter, resulting in smaller debonding length (v Δ t) from matrix crack plane as the strain-rate is faster. Fiber fracture takes place at someplace within $\pm$v Δ t. When the strain-rate is high, the fiber should fracture somewhere within the smaller debonded area, resulting in smaller pull-out length and narrower pull-out length distribution, and vice versa. This explains the result shown in Fig.8 qualitatively. More exact analysis on the phenomenon is required. As fracture resistance by fiber pull-out decreases with increasing strain-rate, it is necessary to take account of the loading condition under which the composite is used in practical application of such type of composite reinforced with monofilaments.

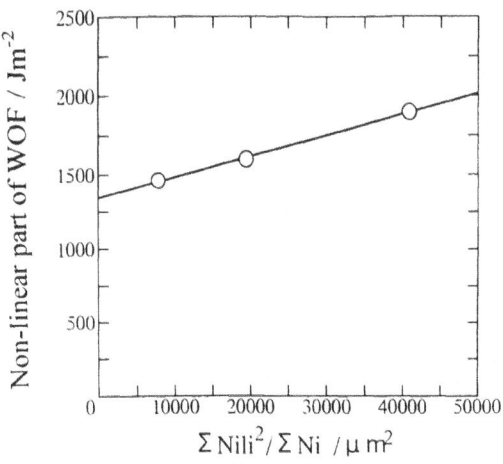

Fig.10 The relation between non-linear part of WOF and $\Sigma \, N_i l_i^2 / \Sigma \, N_i$

SUMMARY AND CONCLUSIONS

Fracture behavior of SiC fiber reinforced SiAlON matrix composite was studied by means of direct observation of fracture stages, fracture energy analysis on various microfracture processes, and the strain-rate dependence of non-linear part of the WOF. Among the processes taking place during the composite fracture, quite large energy was stored by elastic deformation of intact fibers and released by its failure. This energy is not exact fracture energy but is inevitably accompanied with the composite fracture, because it is difficult to carry out stable fracture of fiber incorporated in the composite. The work done by fiber pull-out was estimated to be about one forth of the fiber elastic strain energy, being not main energy absorbing process as like a composite reinforced with multifilament fiber. Large amount of the work was also done by matrix multiplex fracture both in tensile and compressed region of the test specimen. As concerns the influence of strain-rate on the fracture behavior, elastic strain energy of matrix stored before FMC is independent of strain-rate, while non-linear part of WOF after FMC decreases with increasing strain-rate. Fiber pull-out length became shorter, that is, the fracture resistance degraded when the composite was fractured under faster strain-rate.

REFERENCES

1.M.W.Lindley and D.J.Godfrey, Silicon Nitride Ceramic Composites with High Toughness, *Nature*, 229, 192-193 (1971).
2.K.Prewo and J.Brennan, High Strength Silicon Carbide Fiber Reinforced Glass Matrix Composites, *J.Mater.Sci.*, 15[2],463-68(1980).
3.K.Prewo and J.Brennan, Silicon Carbide Yarn Reinforced Glass Matrix Composites, *J.Mater.Sci.*, 17,1201-06(1982).

4. J.Brennan and K.Prewo, Silicon Carbide Fibre Reinforced Glass-ceramic Matrix Composites exhibiting High Strength and Toughness, *J.Mater.Sci.*, 17, 2371-83(1982).

5. J-K.Guo, Z-Q.Mao, C-D.Bao,R-H.Wang, and D-S.Yan, Carbon Fibre-Reinforced Silicon Nitride Composite, *J.Mater.Sci*, 17, 3611-16(1982).

6. R.Lundberg, R.Pompe and R.Carlsson, HIPed Carbon Fiber Reinforced Silicon Nitride Composites, *Ceram.Eng.Sci.Proc.*, 9[7-8],901-05(1988).

7. D.K.Shetty, M.R.Pascucci, B.C.Mutsuddy, and R.R.Wills, SiC Monofilament-Reinforced Si3N4 Matrix Composites, *Ceram.Sci.Eng.Proc.*, 6[7-8],632-45(1985).

8. R.T.Bhatt, Mechanical Properties of SiC Fiber-Reinforced Reaction-Bonded Si3N4, *NASA Technical Report* 85-C-14, (1985).

9. R.T.Bhatt, Mechanical Properties of SiC Fiber-Reinforced Reaction-Bonded Si3N4, Tailoring Multiphase and Composite Ceramics, Edited by R.Tressler, G.L.Messing, C.G.Pantano and R.E.Newnham, Material Science Research Vol.20, pp.675-686, Plenum press, 1985.

10. N.D.Corbin , G.A.Rossetti, Jr., S.D.Hartline, Microstructure/Property Relationships for SiC Filament Reinforced RBSN, *Ceram.Sci.Eng.Proc*, 7,958-68(1986).

11. H.Kodama,H.Sakamoto and T.Miyoshi, Silicon Carbide Monofilament-Reinforced Silicon Nitride or Silicon Carbide Matrix Composites, *J.Am.Cer.Soc.*,72[4]551-58(1989)

12. K.Ueno, S.Kose and M.Kinoshita, Mechanical Properties of SiC Fiber/Si3N4 Composite, Proc.of the 1st International Symposium on the Science of Engineering Ceramics, Edited by S.Kimura and K.Niihara, The Ceram.Soc.Jpn., pp.405-10,1991.

13. C.M.Huang, Y.Xu, D.Zhu and W.M.Kriven, A SiC/Combustion-Synthesized β-SiAlON Composite, *Ceram.Eng.Sci.Proc.*, 15[5],1154-63(1994).

14. D.B.Marshall, An Indentation Method for Measuring Matrix-Fiber Frictional Stresses in Ceramic Composites, *J.Am.Cer.Soc.*, 67[10]C-259-C-260(1984).

15. D.B.Marshall and W.C.Oliver, Measurement of Interfacial Mechanical Properties in Fiber-Reinforced Ceramic Composites, *J.Am.Cer.Soc.*, 70[8]542-48(1987).

16. D.C.Phillips, The Fracture Energy of Carbon-Fiber Reinforced Glass, *J.Mater.Sci.*, 7, 1175-1191(1972).

ONSET OF CUMULATIVE DAMAGE (FIRST MATRIX CRACKING) AND THE EFFECTS OF TEST PARAMETERS ON THE TENSILE BEHAVIOR OF A CONTINUOUS FIBRE-REINFORCED CERAMIC COMPOSITE (CFCC)

Michael G. Jenkins[1], John P. Piccola, Jr.[2], and Edgar Lara-Curzio[3]

[1]Department of Mechanical Engineering
 University of Washington,
 Seattle, Washington, USA 98195-2600
[2]777 Structures
 Boeing Commercial Airplane Group
 Seattle, WA 98124-2207
[3]Metals and Ceramics Division
 Oak Ridge National Laboratory
 Oak Ridge, Tennessee, USA 37830-6064

INTRODUCTION

Continuous-fibre ceramic matrix composites (CFCCs) comprise a recently-introduced subset of ceramic materials and are finding numerous potential applications in the aerospace and energy generation industries. The result of the micromechanical interaction of fibers, fiber coatings, matrix, and their interfaces give CFCCs much greater resistance to catastrophic failure than monolithic ceramics while retaining the high-temperature strengths, corrosion/erosion resistances, high stiffnesses, and low densities characteristic of ceramic materials. As a result, CFCCs have the potential to overcome the inherent brittleness of monolithic advanced ceramics which is often cited as a major limitation preventing these materials from becoming widely accepted in modern designs [1].

Because CFCCs possess greater 'toughness' they exhibit increased reliability and damage tolerance. In this context 'toughness' refers to the ability of the material to absorb energy without catastrophic failure and not necessarily the resistance to the initiation of the fracture process. Macroscopically the area under the stress-strain curve is a measure of 'toughness.' While CFCCs offer greater 'toughness' than monoliths, their strengths are often much less (see Fig. 1), thus necessitating still-evolving and different approaches to mechanical design with CFCCs.

Despite this limitation, numerous industrial uses have been listed for CFCCs including combustor liners, vanes, and nozzles for heat engines [1]. Heat recovery systems could use CFCC tubes and supports. Burners and combustors can employ burners and heat pipes fabricated from CFCCs. Filters, substrates, piping and tanks are some of the applications for separation/filtration and chemical reactor areas. Proposed uses in aerospace applications are leading edges, thermal insulation, and propulsion components.

A conservative approach to design with CFCCs is to exploit only the linear region of the tensile stress-strain curve for the composite as shown in Fig. 1. The maximum stress in the linear region, generally referred to macroscopcially as the proportional limit (PL) stress, the elastic limit stress or the quasi-yield stress, often has been equated (although not always correctly) to the first matrix cracking stress [2-11]. This matrix cracking stress represents the onset of the cumulative damage process which initially causes the fibres and interphase to exposed to the harsh environments of use. Such environments can cause rapid deterioration of the reinforcing materials leading to embrittlement of the CFCC and subsequent loss of the required (and desired) toughness. If a macroscopic property, such as the PL stress, can be linked to a micromechanical event, such as matrix cracking stress, then the upper bound on the design stress can be determined. It is conceivable that a failure theory based on this macroscopic property could then be used by designers for determining the operating loads and component dimensions for exploiting the full capabilities of CFCCs in advanced designs.

In this paper, the PL stress is first briefly described and discussed. Next, experimental results are presented related to determining the PL stress from tensile stress-strain curves and acoustic emission (AE) measurements of the onset of matrix cracking stress. Finally, relations of the theoretical, analytical, and experimental aspects of the macromechanical PL stress, mesomechanical AE measurements, and micromechanical first matrix cracking are discussed.

PROPORTIONAL LIMIT

Mechanical Response

The apparent stress necessary to produce the onset of curvature in the tensile stress-strain relationship is the proportional limit. The PL can be defined as the maximum uniaxial, principal normal stress at which the corresponding normal strain remains directly proportional to stress [12,13] such that:

$$\sigma_{applied} = \sigma_1 = E\varepsilon_1 \tag{1}$$

where $\sigma_{applied}$ is the applied uniaxial tensile stress, σ_1 is the maximum principal normal stress, E is the elastic modulus in the stressing direction, and ε_1 is the normal strain response.

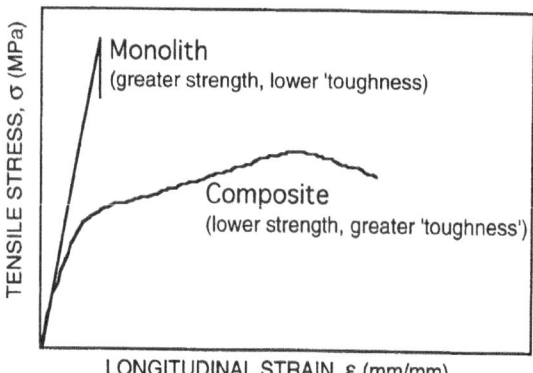

Figure 1 Comparison of stress-strain curves for monolithic and composite ceramics

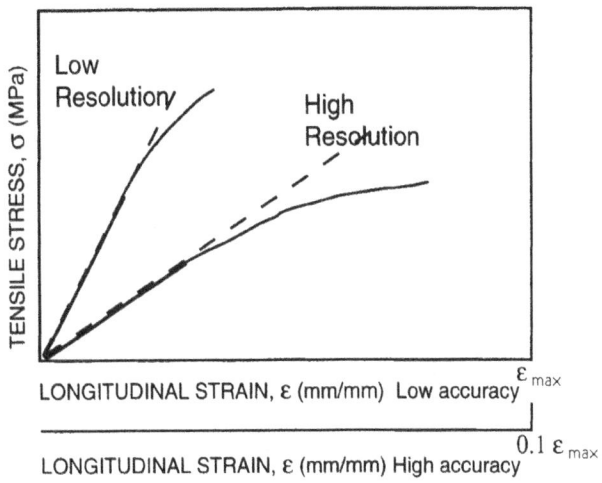

Figure 2 Effect of resolution of strain measurement on the determination of the proportional limit.

Departure from proportionality may be attributed to anelasticity and/or initiation of permanent inelastic deformation. The ability to detect the occurrence of these phenomena during a tensile test is dependent on the accuracy with which stress and strain are measured. The measured value of the PL decreases as the accuracy of the measurement of either stress or strain increases as shown in Fig 2 for strain [12]. Because the measured value of the PL is dependent on test accuracy, for conventional materials such as metals and alloys, the PL is generally not reported as a tensile property. However, the close link of PL and material performance for CFCCs places more emphasis on the determination of PL.

Although conceptually, PL is straight forward, the experimental determination of PL is much less well defined. For example, as shown in Fig. 3 such mechanical properties for CFCCs as elastic modulus, ultimate strength, fracture strength, and even modulus of toughness are easily formulated. In a recently-adopted national (American Society for Testing and Materials, ASTM C1275) test standard [14] for tensile testing CFCCs, single equations for each of these properties are given. However, at least two suggested methods are given for determining PL.

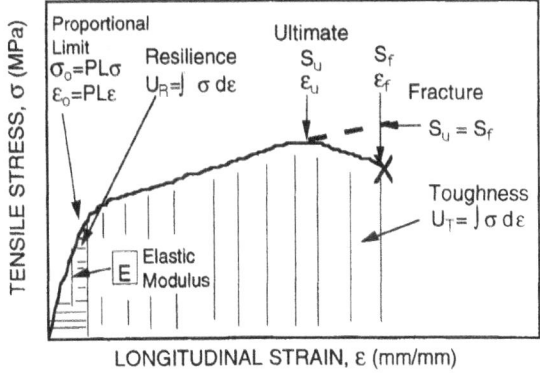

Figure 3 Illustration of various mechanical properties and perfomance of CFCCs [14]

269

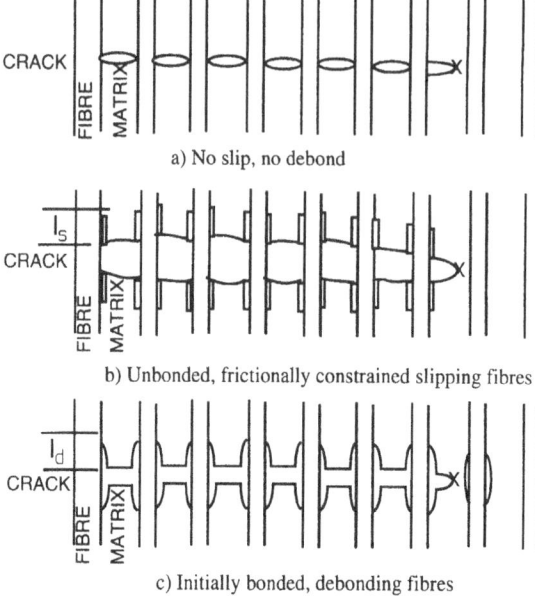

a) No slip, no debond

b) Unbonded, frictionally constrained slipping fibres

c) Initially bonded, debonding fibres

Figure 4 Illustration of various matrix cracking scenarios in CFCCs

In a follow-on study [15,16] of the use of the standard for a bi-directionally reinforced SiC fibre / CVI SiC matrix CFCC, at least five methods (including the two methods of the ASTM standard) for evaluating PL were used to evaluate the same tensile stress-strain curves. The outcome of this study is reported in the Experimental Results section of the paper. A sensitivity analysis of the test methods indicated that the resolution of the strain measurement had the most effect on the calculated results [15], in agreement with the previously-noted study [12] and Fig. 2.

Physical Interpretation

Ceramic composites stressed in the direction of the fibres often exhibit initial damage in the form of a crack extending through the matrix with unbroken fibres bridging the matrix segments (Figs. 4a-4c). This type of damage has been observed in a number of ceramic matrices reinforced with SiC fibres, including glasses, glass-ceramics, and silicon nitride. The well-known, lower-bound theoretical solution for the critical condition in the composite (or matrix) at which the crack first appears in the matrix and extends completely through a section was developed by Aveston, Cooper and Kelly (ACK) [2,7]. The measured PL stress$= \sigma_0$ can be equated to the macroscopic matrix cracking stress (applied stress which causes first matrix cracking), σ_{cr}, which can be linked to a first matrix cracking stress, σ_{mu}, through the relation [2,7]:

$$\sigma_{mu} = (E_c) \sqrt[3]{\frac{12 \, \gamma_m \, \tau \, E_f \, (V_f)^2}{r \, V_m \, E_c \, (E_m)^2}} \tag{2}$$

where τ is the fibre/matrix frictional stress, γ_m is the fracture surface energy of the matrix, V_f and V_m are the volume fractions of the fibre and matrix respectively, r is the fibre radius, and E_c, E_f and E_m are the elastic moduli of the fibre, matrix, and composite, respectively.

Equation (2) represents the expression for critical, large-slip cracking stress [4,7]. A more sophisticated expression [4] has been developed for, σ_{cr}, which takes into account residual stresses in the composite due to conditions such as thermal property mismatches between the constituent materials and the temperature change occurring during cooling from the processing temperature. The expression for σ_{cr} in a composite with unbonded, frictionally-sliding fibres is given as [4]:

$$\sigma_{cr} = \left(E_c\right) \sqrt[3]{\frac{12\,\gamma_m\,\tau\,E_f\,(V_f)^2}{r\,V_m\,E_c\,(E_m)^2}} - \left(\frac{E_c}{E_m}\right)\sigma_{residual} \tag{3}$$

where $\sigma_{residual}$ is the matrix axial residual stress due to the thermal coefficient of expansion, α, mismatch and elastic property differences between the matrix and fibre. In one study [17], $\sigma_{residual}$ was calculated as a tensile stress of ~178 MPa for a braided, primarily-unidirectionally reinforced SiC fibre / CVI SiC matrix CFCC where $\alpha_m = 5 \times 10^{-6}$ /°C, $\alpha_f = 3 \times 10^{-6}$ /°C, $E_m = 350$ GPa and $E_f = 150$ MPa and the difference between the processing temperature, 1000°C, and the test temperature, 25°C, of $\Delta T = -975$°C. Since σ_{mu} was calculated as 273 MPa from Eq. 2, σ_{cr} was calculated as 95 MPa using Eq. 3 and the residual stress result. The measured σ_o was 76±5 MPa for 10 specimens which compares reasonably well with $\sigma_{cr} = 95$ MPa [17] given the single-valued material properties used in the calculations.

Thus, theoretical models for the onset of cumulative damage (first matrix cracking stress) can be linked to the macroscopically measured PL stress, if the complete stress state within the composite is accounted for. Obviously, for composites with no residual stress Eq. 2 may be applied directly. However, Eq. 3 provides a more general expression for the matrix cracking stress.

Previous work [18] has indicated a strong correlation between test rate and the stress-strain response (including PL stress) for one type of CFCC (SiC filament-reinforced Si_3N_4 matrix). Note that neither Eq. 2 or 3 explicitly includes test rate nor do they include other test variables such as the test mode, non uniaxial loading, or volume effects of strengths. Therefore, Eqs. 2 and 3 must be used with caution unless specific information is available regarding these effects.

EXPERIMENTAL RESULTS

In this section, results are presented for a recent study [15,16] of a bi-directionally reinforced SiC fibre / CVI SiC matrix composite in which the effects of test rate (fast vs. slow) were investigated as well as such effects as test mode (load vs. displacement), gripping system (fixed vs. non fixed), specimen geometry (straight-sided vs. contoured), specimen volume (long/thin vs. short/fat), and nonuniform stresses (eccentricity) in uniaxial tension. The results are then presented for a follow-on study of the same material tested in tension with AE measurements conducted to correlate the macroscopically measured PL stress with the theoretical matrix cracking stress and matrix cracking event measured by AE.

Material and Test Setup

The material used in this study was obtained commercially (B.F. Goodrich). The fibre used in all preforms was a ceramic grade (Nicalon™) SiC fibre produced by the pyrolysis of spun polymer-derived ceramic precursor fibres. A two-dimensional reinforced, plain woven cloth was fabricated using ~1800 denier fibre bundles (~500 fibres/bundle) by the fibre producer. The fibre preform was then fabricated by the CFCC manufacturer by stacking 12 plies of cloth ~200 mm square, with the warp of the fibres aligned in the longitudinal direction of the resulting test specimens.

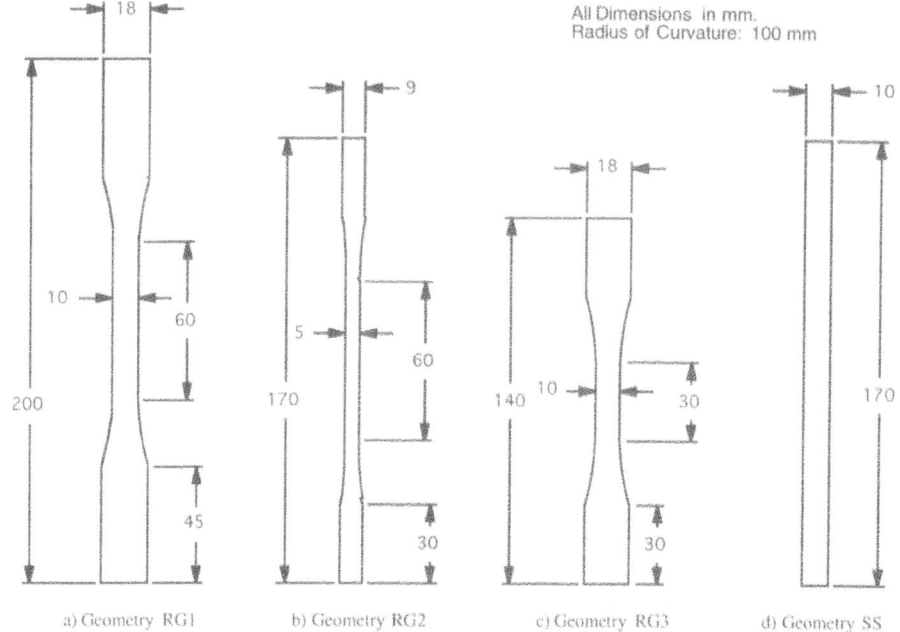

a) Geometry RG1   b) Geometry RG2   c) Geometry RG3   d) Geometry SS

Figure 5 Tensile test specimen geometries a) Geometry reduced-gage (RG) section 1, b) Geometry reduced-gage (RG) section 2, c)Geometry reduced-gage (RG) section 3, d)Geometry straight-sided (SS)

The following steps were included in the processing of the matrix. First a ~0.3 μm interfacial layer of pyrolytic carbon was deposited by chemical vapor infiltration (CVI) onto the fibre preform. The second and final processing step involved the decomposition of methyltrichlorosilane and the infiltration of the preforms by CVI until all the microporosity was filled, thus forming a polycrystalline β-SiC matrix.

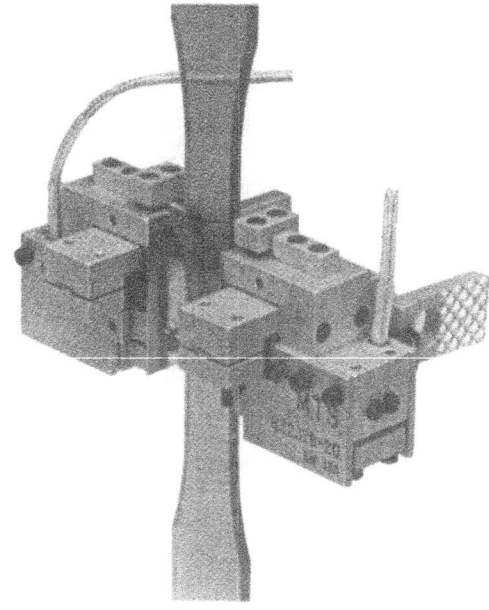

Figure 6 Illustration of dual extensometer in place on a generic tensile specimen

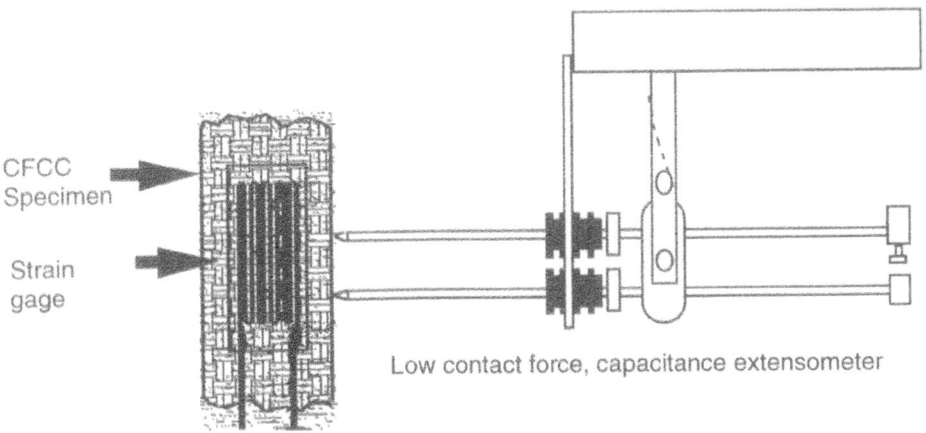

CFCC
Specimen

Strain
gage

Low contact force, capacitance extensometer

Figure 7 Illustration of a strain gage and low contact force capacitance extensometer on a CFCC specimen

Simple, straight-sided flat specimens and reduced-gage section (i.e. dog bone) specimens were fabricated by an undisclosed process (presumably diamond grinding) into four different test specimen geometries designated RG1, RG2, RG3, & SS (Figs. 5a-5d). All specimens were 12 plies thick, equivalent to ~3.7 mm.

All tests were conducted at ambient temperatures and humidity (20-25° C, <65 % RH) on commercial, single-actuator, electro-mechanical materials test systems having load, stroke, strain control capabilities. The test systems included interchangeable, hydraulically-actuated specimen grip systems which could maintain an adjustable grip force without backlash and was independently activated. The digital controller and related software assembled all input and output signals for the test system. Strain was measured using several methods. For the tensile tests a dual-arm, strain-gage based extensometer, axis A and axis B (Fig. 6), was used thereby returning separate but continuous strain values for two opposing sides of the specimen and allowing the calculation of percent bending (PB) (in this case, out of plane bending only). The extensometer operated at a gage length of 25 mm with a range of +5%/-2% strain. For the tensile tests with AE, longitudinal strains were monitored simultaneously using a low-contact force capacitance extensometer (25 mm gage length, 0.5 μm resolution, +0.4%/-0.4% range) and adhesively bonded strain gauges. The surface area covered by the longitudinal strain gauge was several-times larger than the typical cell size of the fabric (Figure 7).

End tabs comprised of an E-glass fibre/epoxy matrix composite were used to protect the specimens from being damaged at the area within the hydraulic grips. Clamping without using end tabs can produce localized damage due to the contact of the grip surface with the specimen. The ratio for the resin and curing agent was so chosen to produce an adhesive with a shear strength greater than ~9.4 MPa, the maximum interfacial shear stress anticipated at the interface of the tab and specimen. If the shear strength was exceeded during testing, delamination at the end tab would result in subsequent damage to the specimen.

As per ASTM C1275 [14] verification of load train alignment was performed with a strain-gaged 'dummy,' reduced-gage section steel alignment specimen prior to and at the completion of testing. The purpose of the verification procedure was to minimize the introduction of bending moments (and non uniform stresses) into the specimen during tensile testing by assuring <5 percent bending (PB) at an average strain of 500 μm/m in the alignment specimen.

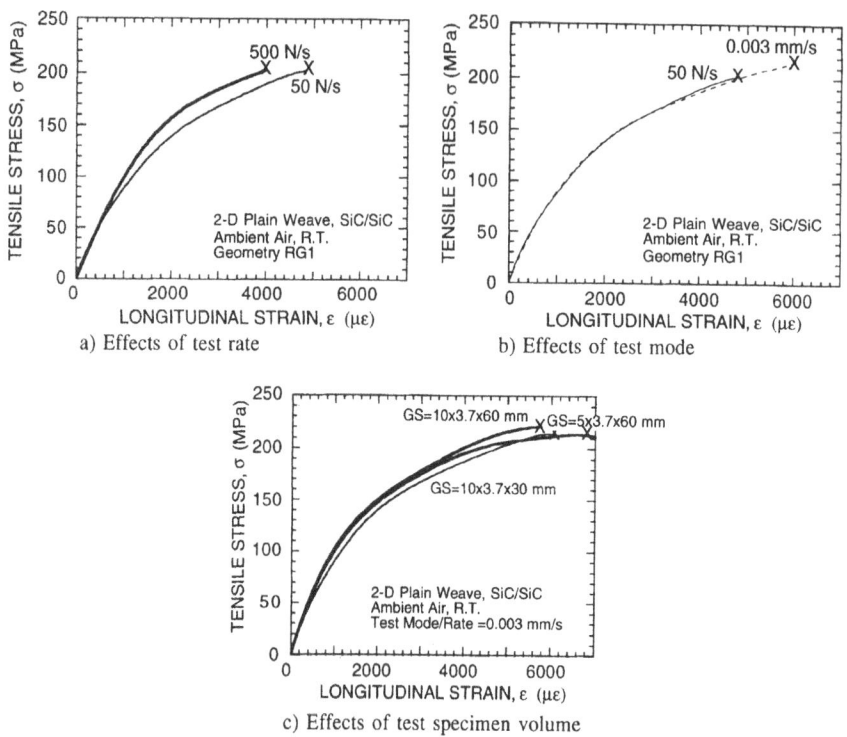

Figure 8 Typical effects of test rate, test mode, and test specimen volume on mechanical behaviour a) effects of test rate, b) effects of test mode, and c) effects of test specimen volume

Tensile Test Measurements

To determine the effect of test rate, if any, an order of magnitude difference was chosen between the minimum and maximum test rates so as to produce failure times of ~20 s and ~200 s. Similar rates were chosen to compare differences in test mode. For displacement control, test rates of 0.003 mm/s and 0.03 mm/s were used, whereas for load control, test rates of 50 and 500 N/s were used.

The stress-strain response is the primary source for analyzing data. In this study, the stress-strain response was reasonably linear up to σ_0 after which the stress increased at a much slower rate to the ultimate tensile strength. Generally, the fracture stress corresponded to the ultimate tensile stress. In addition to the PL stress, the elastic modulus, E, ultimate tensile strength, S_U, fracture strength. S_F and corresponding strain values, (ε_0, ε_U, and ε_F) along with the modulus of toughness, U_T, were extracted from stress-strain curves. Figures 8a-8c show typical effects of test rate, test mode and specimen volume.

Although ASTM C1275 [14] recommends two methods for determining the PL stress (stress at an offset strain and stress at a prescribed strain), three additional methods were employed to determine the 'best' method. These methods are illustrated in Fig. 9 where method A and E represent those recommended in ASTM C1275 with the offset strain and prescribed strain were chosen a priori as 500 μm/m.

For reasons of limited space in this paper, only the PL determined for various test modes and test rates for one geometry (RG1) are shown in Fig. 10. Note that the average PLs determined from method A shows the greatest difference from the average PLs for the other methods. Methods B through E agree well between geometry, test mode, and test rate.

In the following analyses, method C was chosen for determining PL. Methods A and E were discounted because the offset and prescribed strains of 500 μm/m were chosen somewhat arbitrarily. Method B was discounted because of its subjectivity (PL was chosen by the user as the deviation of the stress and strain from linearity). Methods C and D can be shown to be equivalent. However, method D is susceptible to a greater error since strain measurements can introduce greater error than the stress measurements of method C.

For these reasons [15] the PL stress, σ_0, was calculated using the following equation:

$$\sigma_0 = \sigma_i \text{ when } \frac{(\sigma_i - \sigma)}{\sigma} \times 100 \geq 10\% \qquad (4)$$

where σ_i is the stress calculated from the elastic modulus, E, and the corresponding strain at the ith datum, and σ is the actual stress at the ith strain. The PL is the point at which the difference between the actual stress and the calculated stress is equal to 10% (note that 10% is somewhat arbitrary but is consistent with acceptable levels of error for experimental work). Figs. 11a-11b show results of the σ_0 analyses. When specimen geometries are grouped (Fig. 11a), geometry RG1 has the most consistent σ_0 values with geometry SS having the overall greatest values. Comparing test mode/rate (Fig. 11b) shows that displacement control typically results in greater σ_0 values than load control.

For the ultimate tensile strength, S_U, both the specimen geometry and test mode/rate relations show fairly consistent results (Fig. 12). S_U is an important characteristic for materials comparison and may be used intuitively to compare and select a material for design purposes. In many composites, it is thought that at S_U the load is carried almost entirely by the fibres and therefore, often coincides with the fracture strength, S_F, of the composite.

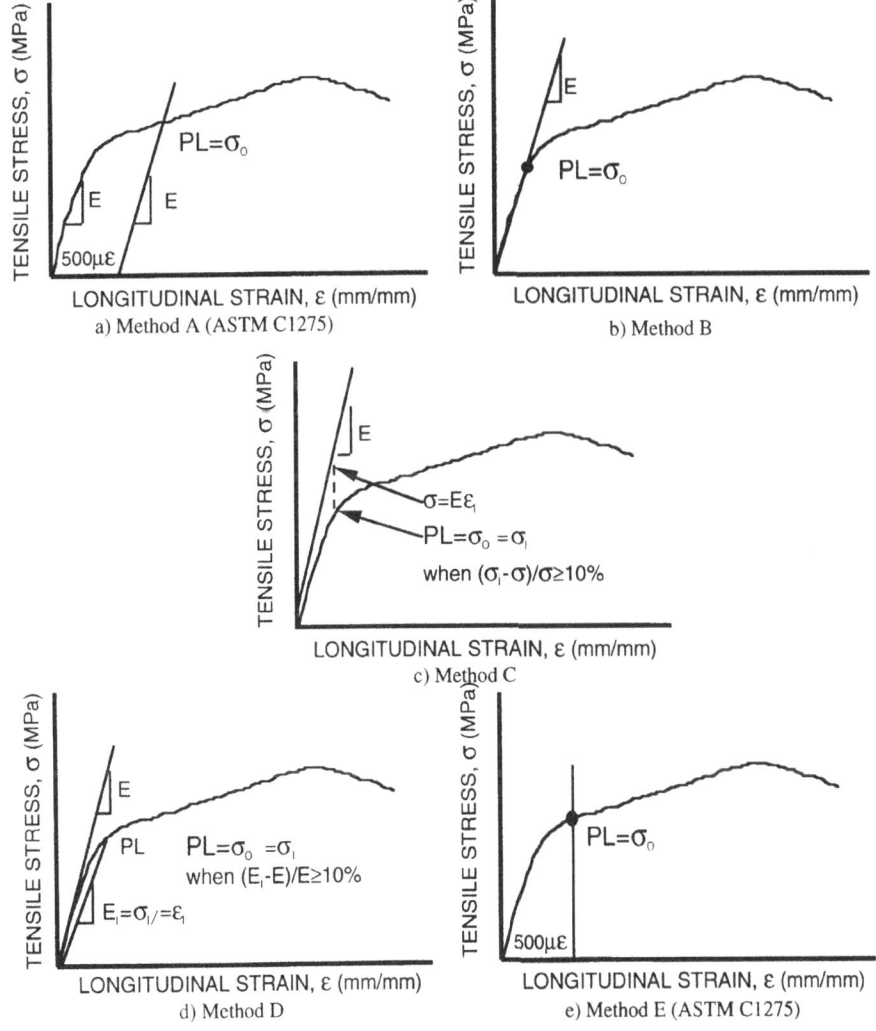

Figure 9 Various methods for determining proportional limit stress [15]

275

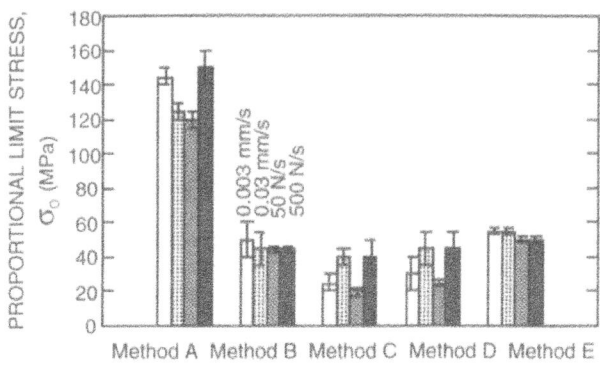

Figure 10 Proportional limit stress for various test modes and test rates for specimen geometry RG1 (error bars represent the upper and lower range of the data)

The use of a dual arm extensometer made it possible to calculate the amount of bending in each specimen. The percent bending (PB) was calculated to compare the uniformity of uniaxial tensile stress on the material properties. Figs. 13a and 13b contain all data points represented as scatter plots to illustrate the effect of PB on σ_0 and S_U. Data points in Figs. 13a and 13b are grouped well around the 2-5 PB range. As expected, σ_0 (~matrix cracking stress), for the fracture stress of a brittle matrix, decreases with increasing percent bending. S_U values in the range of 200-250 MPa show little effect of increasing percent bending, as expected for the fibre-dominated ultimate tensile strength.

A statistical analysis of the data was performed to identify the effects of test mode, test rate, and specimen geometry. Specifically, a two-way ANOVA with replication capable of evaluating more than one hypothesis simultaneously was performed to determine the effect of load rate, displacement rate or load/displacement mode, within a given geometry and between the four test specimen geometries at a 95% significance level [15,16]. In addition, the rate and mode effects for geometry to geometry results for σ_0 and S_U of the tests were analyzed.

In summary, the primary results of the ANOVA show:

1. For σ_0; there was no significance of test rate, there was no significance of test mode, and there was no significance for geometry.

2. For S_U; there was no significance of test rate, there was no significance of test mode, and there <u>was</u> a significance for geometry.

An additional ANOVA was performed to determine the effect of specimen geometry on the fracture location with the following result:

3. For fracture location: there was no significance of test rate, there was no significance of test mode, and there was no significance for geometry.

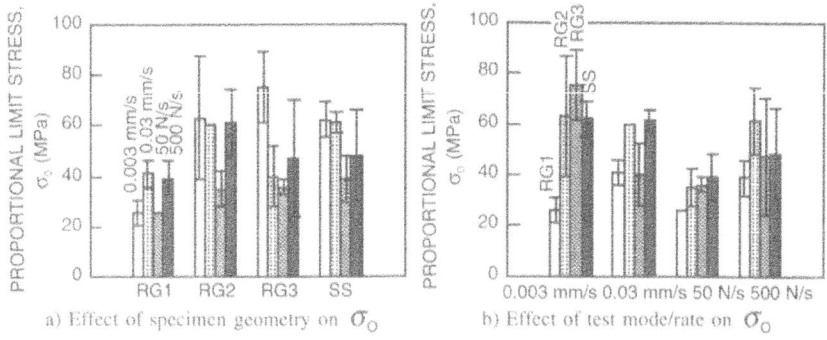

a) Effect of specimen geometry on σ_0

b) Effect of test mode/rate on σ_0

Figure 11 Effect of test parameters on the proportional limit stress a) effect of specimen geometry and b) effect of test mode/rate (error bars represent the upper and lower range of the data)

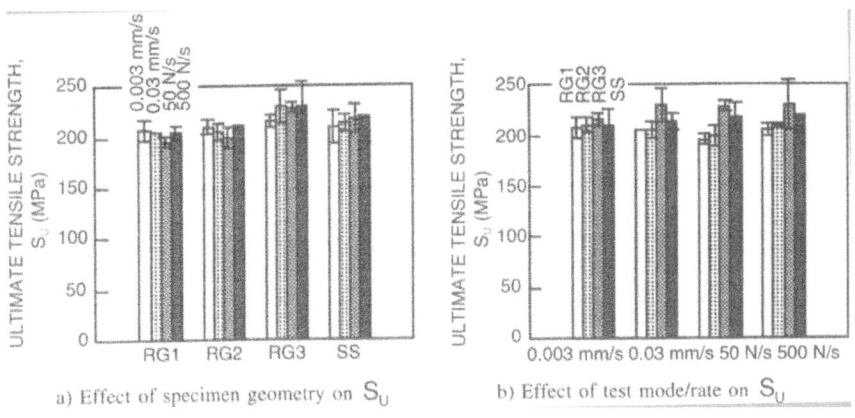

a) Effect of specimen geometry on S_U b) Effect of test mode/rate on S_U

Figure 12 Effect of test parameters on the ultimate tensile strength a) effect of specimen geometry and b) effect of test mode/rate (error bars represent the upper and lower range of the data)

Pooling of the data allows direct comparisons of mean strength values as shown in Fig. 14. Note that in most cases, $S_U \approx S_F$ and that σ_O is only $\sim S_U/5$. Because the strength results could be pooled, the resulting proportional limit stress, σ_O and ultimate tensile strength, S_U distributions for the reduced gage section tensile test specimens were analyzed using Weibull statistics to determine the applicability of a weakest link failure criterion (Weibull statistics are commonly applied to monolithic ceramics). Figures 15a and 15b show the Weibull, probability of failure plots for σ_O and S_U. While both strength distributions indicate the strong possibility of a minimum strength (i.e. a sharp curve downward at the left end of the data), the S_U data appears to be better represented by a Weibull distribution than does σ_O. The value of the Weibull modulus, m, for σ_O is so low as to be nearly random (note that a Weibull modulus of 3.5-4.0 approximates a normal distribution).

A goodness of fit test [19] can be used to test the null hypothesis that the strengths are a two-parameter Weibull distribution. If the null hypothesis is rejected, then other distributions must be considered, including a three-parameter Weibull.

The application of the test is straightforward, requiring the calculation of the test statistic such that [19]:

$$S = \frac{\sum\limits_{i=[r/2]-1}^{r-1} \left[\dfrac{(x_{i+1} - x_i)}{M_i} \right]}{\sum\limits_{i=1}^{r-1} \left[\dfrac{(x_{i+1} - x_i)}{M_i} \right]} \tag{5}$$

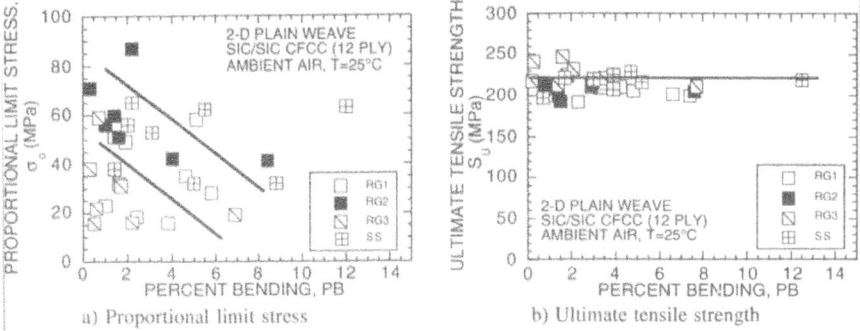

a) Proportional limit stress   b) Ultimate tensile strength

Figure 13 Strength as a function of percent bending a) proportional limit stress and b) ultimate tensile strength

277

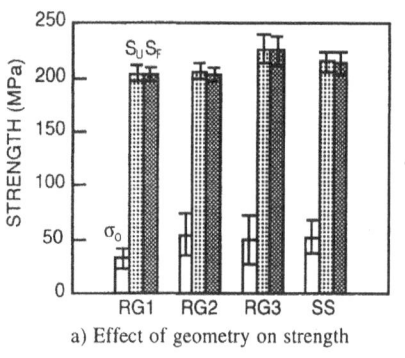 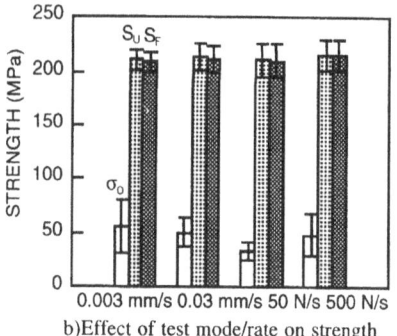

a) Effect of geometry on strength   b)Effect of test mode/rate on strength

Figure 14 Pooled data for strength values a) geometry and b) test mode/rate (error bars represent the range of data)

where r is the number of the first, ordered strengths, x is the natural logarithm of the strength [i.e. x=ln (σ_0 or S_U)], and M is a tabulated statistical parameter. Critical values of S, (i.e. S_C) are tabulated for given levels of significance such that if $S_C \geq S$, the distribution is a two-parameter Weibull [19].

For the 24 tests analyzed, S_C=0.64 at the 95% significance level. The calculated S values are 0.72 and 0.80 for σ_0 and S_U, respectively. Thus, the hypothesis of a two-parameter Weibull must be rejected and other distributions should be considered, such as a three-parameter Weibull.

Three-parameter Weibull analyses were performed as shown in Figs. 15a and 15b. In a three-parameter Weibull analysis, in addition to the shape parameter, m, and scale parameter, σ_θ or S_θ of the two-parameter analysis, a non zero location parameter, δ, is used in the probability density function to fit the data [20]. The values of δ were chosen so as to 'linearize' the data shown in Figs. 15a and 15b [20]. For σ_0, $\delta \approx 15$ MPa and for S_U, $\delta \approx 190$ MPa. In the case of strength data, the physical interpretation of δ is the stress below which σ_0 (matrix cracking) or S_U (ultimate) would not be expected to occur [20].

Acoustic Emission Measurements

During the tensile tests using the AE measurements, at low levels of deformations good correlation was found between the measured longitudinal strain using a low-contact force capacitance extensometer and the strain gage. However, the measurements diverged at greater strains, particularly when substantial cracking of the matrix had occurred.

The mechanical tests were conducted in a cyclic-fashion at constant loading rates (1-3 MPa/s) to increasingly greater stress levels, as shown in Table 1. The purposes of these tests were several: i) to determine the onset of non-linear response in the stress-strain behavior of the material; ii) to determine the onset of hysteresis upon subsequent loading-unloading; ii) to determine AE activity upon repeated loading, and finally, iv) to correlate the stress-strain response to the AE data.

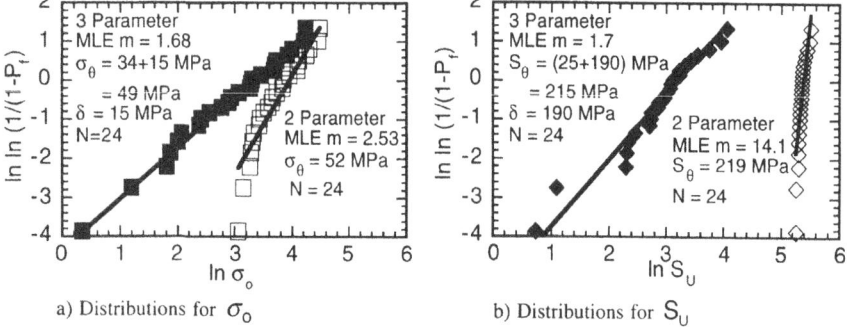

a) Distributions for σ_0 b) Distributions for S_U

Figure 15 Weibull strength distributions a) proportional limit stress and b) ultimate tensile strength. (Maximum likelihood estimators of Weibull modulus and characteristic strength are biased)

Table 1 Peak stress levels and the evolution of Felicity ratio values as a function of applied stress

Peak Stress	Previous Peak Stress	Felicity Ratio
10 MPa	0	-
20 MPa	10 MPa	1
30 MPa	20 MPa	0.95
40 MPa	30 MPa	0.93
50 MPa	40 MPa	0.88
75 MPa	50 MPa	0.84
100 MPa	75 MPa	0.73
150 MPa	100 MPa	0.74
200 MPa	150 MPa	0.70
failure	200 MPa	0.73

Figure 16 shows a composite plot of the tensile stress-strain results. The material was linear elastic (given the resolution of the load and strain measurements) for stresses up to 40 MPa. However, when the material was subjected to stresses greater than 40 MPa, it exhibited hysteretic behavior. The hysteresis loops were not closed and their widths grew larger at greater stress levels as indicated in Figure 16. However, the 'envelope' stress-strain curve coincided with that from a monotonic test.

During the tensile tests a piezoelectric transducer was attached to the test specimen using a thin layer of silicon and a 'C'-clamp. The following parameters were recorded during the tests as a function of the applied stress: counts, amplitude, duration, rise time and measured area under the rectified signal envelope (MARSE). AE amplitudes are directly related to the magnitude of the source event and because they vary over a wide range, they are expressed on a decibel scale. The number of counts are the threshold crossing pulses, in this case 40 dB, and they depend on the acoustic properties and reverberant nature of the specimen and sensor.

Recent test strategies pay much attention to the emission that occurs at loads below the previous maximum during cyclic loading [21]. Regarding this behavior, the Felicity ratio has been defined as the ratio of the load at which emissions begin again divided by the previous maximum load and may be used as a measure of damage in the material [20]. The AE data associated with the test results in Fig. 16 are presented in Fig. 17 in the form of a plot of cumulative counts vs. applied stress.

It was found that the cumulative counts vs. stress curves had three well-defined stages as shown schematically in Figure 18. In the first stage, the cumulative counts started at a stress level, σ_1, and then increased at a small rate with the applied stress. At approximately the previous maximum stress, the second stage started in which the cumulative counts grew much faster with increasing stress. Finally, the third stage coincided with the unloading part of the mechanical cycle in which there were no emissions. It was found that the Felicity ratio decreased as the maximum applied stress increased as indicated by the results in Table 1, a behavior that is typical of most fiber-reinforced materials. For example, systematic decreases in the Felicity ratio as the material approaches failure have been well documented for polymer matrix composites [21].

DISCUSSION AND CONCLUSIONS

In general, in addition to information about the onset of cumulative damage (i.e. first matrix cracking stress) basic information about the tensile mechanical behaviour of a CFCC was also obtained. In particular, as expected [15,16], faster loading rates apparently resulted in greater S_U. Overall, there was apparently a lower fracture strength under displacement control than for load control due to the relaxation of the strain in displacement control, whereas under load control the test machine continued to pull the specimen.

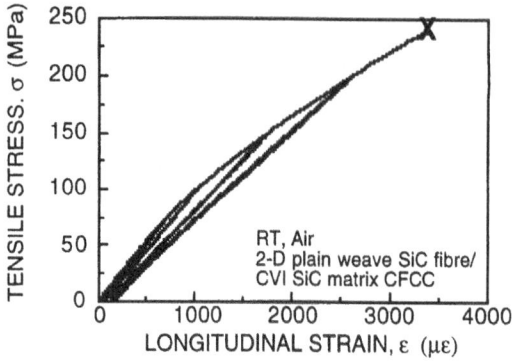

Figure 16 Tensile stress-strain curves conducted under constant loading rate at increasingly greater stress levels as indicated in Table 1

However, as indicated by the ANOVA there was no effect of test rate or test mode at the 95% significance level for either σ_0 or S_U. As for geometry effects, the S_U but not σ_0 seemed to be affected by the volume of the gage section, that is, a smaller volume shows greater S_U values. The variation of S_U with geometry indicates that as expected ultimate tensile strength rested on the strength of the fibres. The ANOVA showed that the data for σ_0 and S_U could be pooled, thus allowing application of Weibull analyses for determining the form of the strength distribution.

Two-parameter Weibull analyses of S_U showed a characteristic strength of $S_\theta \approx 219$ MPa and a reasonably high Weibull modulus of m≈14. Since S_U reflects the strength of the fibres, these statistics are consistent with those expected for uniformly damaged (due to processing of the CFCC) fibres. Weibull analyses of σ_0 showed a characteristic strength of $\sigma_\theta \approx 52$ MPa and a fairly low Weibull modulus of m≈3. Note that σ_0 does not necessarily represent a weakest link fracture, but instead is affected by many random factors such as processing conditions, residual stress, porosity, in-situ formed interphases, etc. The low Weibull modulus (wide dispersion) reflects this randomness.

However, it should be noted that the distributions for both σ_0 and S_U fail goodness-of-fit tests for two-parameter Weibull distributions. Three-parameter Weibull analyses were applied and showed minimum strengths of $\delta \approx 15$ MPa and $\delta \approx 190$ MPa for σ_0 and S_U, respectively.

There has been a controversy for quite some time as to whether or not σ_0 in the monotonic stress-strain curve coincides with matrix cracking. The results presented here indicate that the average σ_0 (~41 MPa) is approximately equal to the stress (~40 MPa) at which hysteretic behavior commenced. However, the Felicity ratio (an indication of the onset of cumulative damage) begins to decrease from a value of unity at stresses between 10 and 20 MPa. In addition, the minimum strength for σ_0 determined from the three-parameter Weibull analysis was ~15 MPa.

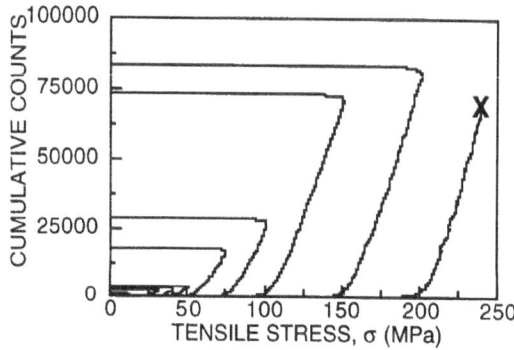

Figure 17 Acoustic emission data for the mechanical tests results in Fig. 16.

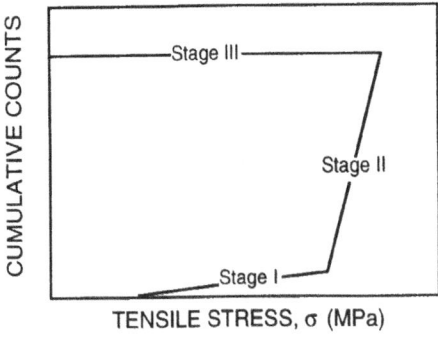

Figure 18 Schematic description of the cumulative counts vs. applied stress curves

From this one could conclude that for this material the stress at which the matrix first cracks is on the order of one half of the macroscopic stress indicating onset of nonlinearity, σ_0. However, it has already been shown that the macroscopic resolution of the PL stress is directly related to the resolution of the stress and strain. Thus, it is conceivable that higher resolution extensometers might be employed to provide a better estimate of the PL stress.

The following conclusive points can be made:

1) For σ_0 at a 95% significance level, there was no significance of test rate, there was no significance of test mode, and there was no significance for geometry. However, for S_u, there was no significance of test rate, there was no significance of test mode, and there <u>was</u> a significance for geometry.

2) Pooling data for σ_0 and S_u resulted in two parameter Weibull strength distributions showing wide dispersion (m≈3) for σ_0 and narrower dispersion (m≈14) for S_u. Three-parameter Weibull analyses were applied and showed minimum strengths of ~15 MPa and ~190 MPa for σ_0 and S_u, respectively.

3) Acoustic emission indicated first matrix cracking in the stress range of 10-20 MPa. However, loading/unloading tensile tests showed onset of hysteresis (cumulative damage) in the 40 MPa range, in good agreement with the average σ_0 of 41 MPa determined from monotonic tensile tests indicating stress and/or strain resolution may play a significant role in determining σ_0 for first matrix cracking.

ACKNOWLEDGMENT

Research sponsored by the U.S. Department of Energy, Assistant Secretary for Conservation and Renewable Energy, Office of Industrial Technologies, as part of the CFCC Program under contract DE-AC05-84OR21400 managed by Lockheed Martin Energy Systems, Inc.

REFERENCES

1. M.A. Karnitz, D.A. Craig, and S.L. Richlen, "Continuous Fibre Ceramic Composite Program," *Ceram. Bull.*, **70** [3] 430-435 (1991)

2. J. Aveston, G.A. Cooper, and A. Kelly, "Single and Multiple Fracture," pp. 15-26 in The Properites of Fibre Composites, National Physical Laboratory, IPC Science and Technology Press, Teddington, U.K. (1971)

3. D.B. Marshall, B.N. Cox and A.G. Evans, "The Mechanics of Matrix Cracking in Brittle Matrix Fibre Composites," *Acta Metall.*, **33** [11] 2013-2021 (1985)

4. B. Budiansky, J.W. Hutchinson, and A.G. Evans, "Matrix Fracture in Fibre-Reinforced Composites," *J. Mech. Phys. Solids*, **34** [2] 167-189 (1986)

5. L.N. McCartney, "Mechanics of Matrix Cracking in Brittle Matrix Fibre Reinforced Composites," *Proc. R. Soc. London, A.*, **409** 329-350 (1987)

6. B.S. Majumdar, G.M. Newaz, and A.R. Rosenfield, "Yielding Behaviour of Ceramic Matrix Composites," pp. 2805-2814 in <u>Advances in Fracture Research</u>, K.Salama, K. Ravi-Chandar, D.M.R. Taplin, and P. Rama Rao, Pegamon Press, Oxford, U.K. (1989).

7. R.W. Davidge and A. Briggs, "The Tensile Failure of Brittle Matrix Composites Reinforced with Unidirectional Continuous Fibres,"*J. Mater. Scien.*, **24**, 2815-2819, (1989).

8. S. Danchaivijit and D.K. Shetty, "Matrix Cracking in Ceramic Matrix Composites," *J. Am. Ceram. Soc.,* **76** [10] 2497-2504 (1993)

9. X.F. Yang and K.M. Knowles, "On the First Matrix-Cracking Stress in Unidirectional Fibre-Reinforced Brittle Materials," *J. Mater. Res.* **8** [2] 371-376 (1993)

10. A.W. Pryce and P.A. Smith, "Matrix Cracking in Unidirectional Ceramic Matrix Composites under Quasi-Static and Cyclic Loading," *Acta. Metall. Mater.,* **41** [4] 1269-1281 (1993)

11. W.A. Curtin, "Multiple Matrix Cracking in Brittle Matrix Composites," *Acta. Metall. Mater.,* **41** [4] 1369-1377 (1993)

12. M.R. Louthan, Jr., "Tensile Testing of Metals and Alloys," pp. 61-104 in <u>Tensile Testing</u>, P. Han, eds. ASM International, Materials Park, Ohio (1992)

13. D. Lewis III, "Tensile Testing of Ceramics and Ceramic Matrix Composites," pp. 147-182 in <u>Tensile Testing</u>, P. Han, eds. ASM International, Materials Park, Ohio (1992)

14. American Society for Testing and Materials, C1275-94, "Standard Test Method for Monotonic Tensile Strength Testing of Continuous Fibre-Reinforced Advanced Ceramics with Solid Rectangular Cross-Sections at ambient Temperatures," ASTM Annual Book of Standards, Vol 15.01, Philadelphia, Penn. (1994)

15. J.P. Piccola, Jr., "Effects of Test Parameters on Tensile Mechanical Behaviour of a Continuous Fibre Ceramic Composite (CFCC)," Master of Science Thesis, University of Washington (1994)

16. J.P. Piccola, Jr. and M.G. Jenkins, "Effects of Test Parameters on Tensile Mechanical Behaviour of a Continuous Fibre Ceramic Composite (CFCC)," *Ceram. Eng. Scien. Proc.,* **16** [4] (1995)

17. M.G. Jenkins and M.D. Mello, "Processing and Characterization of 3-D Braided, Continuous SiC Fibre Reinforced/CVI Matrix Composites," pp. 239-254 in <u>Processing, Design, and Performance of Composite Materials</u>, MD-Vol. 52, American Society of Mechanical Engineers, New York (1994)

18. J.W. Holmes, "A Technique for Tensile Fatigue and Creep Testing of Fibre Reinforced Ceramics," *J. Comp. Mater.,* **26** [6] 916-924 (1992)

19. K.C. Kapur and L.R. Lamberson, pp. 329-331 in "Reliability in Engineering Design," John Wiley and Sons, New York (1977)

20. R.B. Abernethy, J.E. Breneman, C.H. Medlin, and G.L. Reinman, "Weibull Analysis Handbook," AFWAL-TR-83-2079, Wright Patterson AFB, Ohio (1983)

21. A.A. Pollock, "Acoustic Emission Inspection," Physical Acoustic Corporation Technical Report, TR-103-96-12/89

CRACK RESISTANCE OF CERAMIC MATRIX COMPOSITES WITH COULOMB FRICTION CONTROLLED INTERFACIAL PROCESSES

Meinhard Kuntz and Georg Grathwohl

University of Bremen
28359 Bremen
Germany

INTRODUCTION

The mechanical performance of continuous fiber reinforced ceramic matrix composites is generally controlled by the crack bridging effect of the fibers. If the fiber-matrix-bond is not too strong, the interface undergoes debonding during matrix crack propagation. Thus, the crack bridging is first governed by intact fibers. An increase of the stress on the composite leads to accumulated fiber failure. Due to the characteristic flaw distribution of the individual fibers there is a considerable portion of fiber failure events taking place in the matrix along the fiber axis. Consequently, pullout of broken fibers contributes also to the overall bridging effect.

Therefore, it is a fundamental requirement to determine the fiber failure probability at increasing fiber load with respect to the axial stress profile along the fiber axis. This profile $\sigma_z(z)$ is directly obtained from the frictional shear stress τ in the fiber-matrix-interface

$$\frac{d\sigma_z}{dz} = \frac{-2\tau}{R_f} \tag{1}$$

where R_f is the fiber radius. Most studies concerning the mechanical properties of ceramic matrix composites prefer the simplifying assumption of a constant frictional stress in the interface. The resultant linear axial stress profile allows relatively simple analytical solutions.

On the other side, many results from theoretical and experimental micromechanical studies indicate a mismatch between this simplification and realistic stress transfer mechanisms. The most common competitive model is the assumption of Coulomb friction $\tau = \mu \cdot \sigma_n$ where μ is the coefficient of friction and σ_n adds up all contributions to the normal stress at the interface. Such contributions derive from the internal compressive normal stresses due to thermal expansion mismatch or from roughness interactions during fiber sliding which is referred to as the *clamping stress* σ_c. The clamping stress is superposed by the normal stress induced by the Poisson effect during axial loading σ_ν. These contributions result in a not-constant interfacial shear stress and consequently in

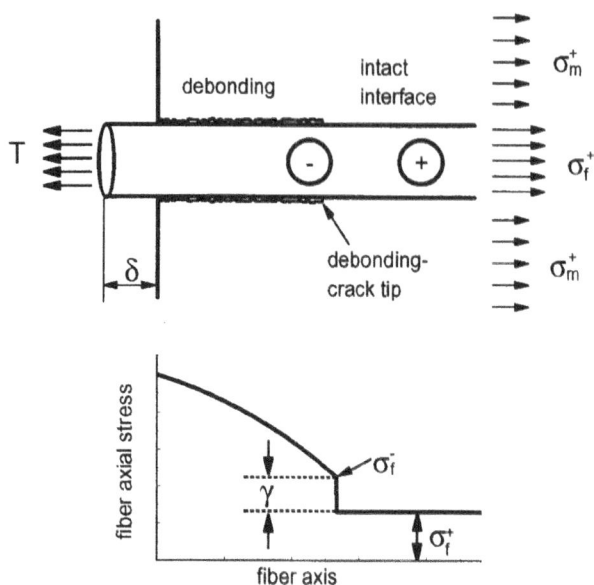

Figure 1. Model for progressive debonding with nonlinear fiber axial stress (factor $b_1 > 0$)

a nonlinear axial stress profile in the fiber. A prominent study based on this principle was presented by Hutchinson and Jensen [1].

Taking into account the Poisson effect it becomes clear that effective bridging can only be provided by selected fiber-matrix combinations characterized by suitable elastic constants of fiber and matrix. Particular combinations of these constants might lead to ineffective fiber bridging.

In this study, the bridging stress law of ceramic matrix composites is derived on the basis of a nonlinear axial stress profile. Main attendance is laid to the influence of the Poisson effect during fiber loading. Some characteristic dependencies are elaborated and demonstrated by variation of the interface parameters.

Furthermore, a fracture mechanical treatment is applied to model the material behavior with respect to the bridging law. Thus, the effects of the bridging stresses can be assessed taking into account the resulting macroscopic material performance.

BASIC MICROMECHANIC RELATIONS

This study is partly based on the analysis of Hutchinson and Jensen [1]. In their study, stress-displacement-relations for intact and broken fibers are calculated. An axisymmetric unit cell is used for the stress-strain-relationship of fiber and matrix. Two types of boundary conditions at the outer matrix radius are introduced for single fiber model experiments (type I) and for aligned fibers in a unidirectional composite (type II). Here only type II is pursued. The matrix is assumed to show isotropic behavior, whereas the fiber can also exhibit transversely isotropic thermoelastic properties. The debond process is treated as a mode II crack propagation along the interface (see figure 1). Radial and axial stresses at the debond crack tip are denoted with a minus (-) sign, in the region of the intact interface a (+) is used.

In this region, axial and radial stresses of the fiber depend on the composite stress

$\bar{\sigma}$ and residual stresses due to thermal expansion mismatch,

$$\sigma_f^+ = a_1\bar{\sigma} - a_2 E_m \epsilon^T \qquad (2)$$

$$\sigma_r^+ = a_3\bar{\sigma} - a_4 E_m \epsilon^T \qquad (3)$$

where a_i are dimensionless constants, representing linear stress-strain-relationship depending on the thermoelastic properties of fiber and matrix, explicitly given in [1]. a_1 can simply be derived from the rule of mixture only if the Poisson ratios of fiber and matrix are identical. E_m is the Young's modulus of the matrix. The terms $a_2 E_m \epsilon^T$ and $a_4 E_m \epsilon^T$ represent axial and radial residual stresses, respectively.

To calculate stresses and strains in the debonded region, the fiber-matrix-interface is implicitly assumed as a perfectly cylindrical body. Thus, the induced radial stress at the interface due to fiber and matrix contraction ("Poisson effect") depends linearly on the axial fiber load[1].

$$\Delta\sigma_r = b_1 \Delta\sigma_f \qquad (4)$$

where $\Delta\ldots$ denotes changes of stress or strain from the debonded to the bonded region. b_1 is an important factor to describe the Poisson effect during fiber loading resulting in an additional normal stress component at the interface. In single fiber pullout experiments (boundary condition I), the Poisson effect always leads to a release of the normal clamping stress, so factor $b_1 > 0$. With boundary conditions type II the normal clamping stress release is partly cancelled by the matrix expansion. In some cases, even an additional interface compression is possible leading to a negative factor b_1. Marshall[3] has performed extensive calculations in order to study the parameters influencing b_1.

The end of the debonded region is treated as a crack tip stressed in mode II. If the debond energy Γ_i is relatively low, its effect on the stress-displacement relationship can be treated as a jump of the fiber axial stress $\gamma = \sigma_f^- - \sigma_f^+$ at the debond crack tip.

$$\gamma = \sigma_f^- - \sigma_f^+ = \frac{1-V_f}{V_f}\frac{1}{c_1 c_3}\left(\frac{E_m \Gamma_i}{R_f}\right)^{1/2} \qquad (5)$$

where c_i are combinations of the constants a_i. V_f is the fiber volume fraction. The radial stress at the debond crack tip can then be given as

$$\sigma_r^- = a_3\bar{\sigma} - a_4 E_m \epsilon^T + b_1\gamma \qquad (6)$$

which is needed as a boundary condition to calculate the axial stress profile in the fiber

$$\sigma_f(z) = \sigma_f^+ + \gamma - \frac{\sigma_r^-}{b_1}\left[1 - exp\left(\frac{2\mu b_1(z - l_d)}{R_f}\right)\right] \qquad (7)$$

where l_d is the debond length. The corresponding crack opening Δ of the composite is calculated from the additional fiber strain in the debonded area with respect to the strain in the bonded area.

$$\Delta = \int_0^{l_d}\left[\epsilon_f(z) - \epsilon^+\right]dz \qquad (8)$$

The axial strain is obtained from the axial stress profile, so that the resulting composite crack opening can be derived for the intact fiber. Similar calculations lead to the pullout-contribution to the bridging stress. To obtain the correlation of bridging stress

---

[1]Recently, the authors[2] of the present work have shown that in the case of roughness interaction during fiber sliding this effect is nonlinear

and crack opening, the fiber failure probability has to be calculated which is shown in the following section.

CALCULATION OF BRIDGING STRESS LAW

The bridging stress in the crack wake of a ceramic matrix composite is provided by the contributions of the intact (σ_i) and the broken (σ_b) fibers.

$$\sigma(\Delta) = \sigma_i(1 - p) + \sigma_b p \tag{9}$$

p represents the portion of the broken fibers which is a priori identical to the fiber failure probability. The stress-displacement relation of intact and broken fibers can be derived from the analysis of Hutchinson and Jensen[1]. Thus, the major task is here to calculate the failure probability of a fiber at a given crack opening Δ.

The analysis of Hutchinson and Jensen has to be extended as the fiber load T_i is no longer directly connected to the average composite stress $\bar{\sigma}$. Once T_i and $\bar{\sigma}$ are known the debond length is calculated as

$$l_d = -\frac{R_f}{2\mu b_1} \cdot \ln\left[1 + \frac{b_1\left(T_i - a_1\bar{\sigma} - a_2 E_m \epsilon^T - \gamma\right)}{\sigma_r^-}\right] \tag{10}$$

and with $\zeta_l = 2\mu b_1 l_d / R_f$ the half crack opening is derived from eq. 8

$$\Delta = b_2\left\{\frac{l_d\gamma}{E_m} - \frac{R_f\sigma_r^-}{2\mu b_1^2 E_m}\left[e^{-\zeta_l} + \zeta_l - 1\right]\right\} \tag{11}$$

b_2 is another dimensionless constant directly connected with b_1. To obtain the fiber load T_i, eqs. 10 and 11 have to be solved numerically and with an iterative procedure as the average composite load $\bar{\sigma}$ is unknown from the beginning.

The fiber strength distribution is usually assumed as a two parameter Weibull distribution which follows from the mathematical treatment of a weakest link statistic. A fiber of unity length l_0 has an average number of flaws causing failure at stresses below σ

$$F(\sigma) = \left(\frac{\sigma}{S_0}\right)^m \tag{12}$$

S_0 and m are the Weibull-parameters.

In brittle matrix composites it is assumed that fibers will only fail in the debonded region. Thus, it has to be taken into account that every point on the fiber has a *starting load* σ_u, the axial stress in the fiber when the debond crack reaches this point. So σ_u is per definitionem the lowest load at which that part of fiber can fail. Consequently, the Weibull expression has to be extended

$$F(\sigma) = \left(\frac{\sigma - \sigma_u}{S_0}\right)^m \tag{13}$$

The calculation of the fiber failure probability is based on the analysis of Oh and Finnie[4], which has been worked out by Thouless and Evans[5]. A probability density function for fiber failure in terms of eq. 13 is expressed as

$$\Phi(T, z) = exp\left(-2\int_0^l \frac{1}{l_0} F(\sigma)dz\right) \frac{\partial}{\partial T}\left[\frac{1}{l_0}F(\sigma)\right] \tag{14}$$

Integration leads finally to fiber failure probability and pullout length

$$p = 2 \int_0^{T_i} \int_0^{l_d} \Phi(T, z) dz dT \tag{15}$$

$$l_p(T_i) = 2 \int_0^{T_i} \int_0^{l_d} z\Phi(T, z) dz dT \tag{16}$$

Note that σ and σ_u are used in the function F as being nonlinear dependent on the fiber axis z so that numerical integration and differentiation is needed.

RESULTS

The objective of this study is to clarify the influence of the fundamental mechanisms responsible for the increase of the crack resistance in ceramic matrix composites. The simplifying model with $\tau = const.$ provides immediately information about the effects of several parameters, e.g. fiber strength distribution, magnitude of sliding resistance or fiber volume fraction. In order to approach systematically the optimized composite there are still some further questions to be solved: what is the source of interfacial friction, how can it be effectively activated? Is the Poisson effect relevant for material properties? If it is relevant, one should expect that particular fiber-matrix combinations are intrinsically unsuitable for effective reinforcement. Consequently, the calculations given below are performed in order to answer these questions.

The following results are elaborated on the basis of the material data of a commercially available fiber reinforced glass (Nicalon/Duran, Schott Glaswerke, Mainz, Germany). The fiber is assumed to be affected by the processing step; the fiber strength is then estimated on the basis of several experimental results. However, it should be noted that these data are not to be transferred to each arbitrary fiber state. The complete data are given in table 1.

Table 1. Material data of the components in the composite Nicalon/Duran

	E [GPa]	ν	α [$10^{-6}K^{-1}$]	d [μm]	V [%]	m	σ_0 [GPa]	l_0 [mm]]
fiber (Nicalon)	200	0.19	3.0	15	40	3.0	2000	1
matrix (Duran)	63	0.24	3.37		60			

The Poisson effect during composite loading is obviously dominated by the Poisson ratio of the fiber. Thus, a hypothetical parameter variation has been performed in order to quantify this phenomenon. The result is given in figure 2.

Figure 2a shows three characteristic bridging stress curves calculated with different Poisson ratios of the fiber. If ν_f is high, the interface radial stresses are partly cancelled due to fiber contraction during loading. This leads to a lower frictional shear stress in the interface and thus to a moderate axial stress gradient in the fiber. Consequently, the average nominal fiber failure stress is lower since the stressed volume of the fiber increases. As can be seen in figure 2a, the maximum bridging stress decreases with a higher Poisson ratio of the fiber whereas the fracture toughness as obtained from the area under the bridging curve increases. These considerations also persist with all parameters that influence the frictional shear stress in the interface, e.g. coefficient of friction and clamping stress.

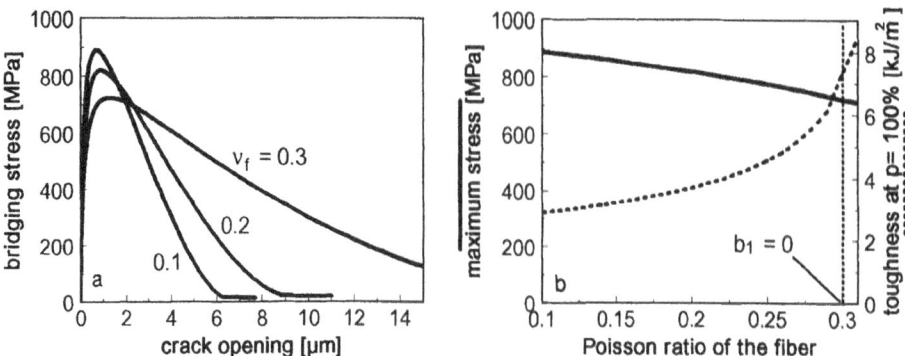

Figure 2. a) Bridging stress functions for various Poisson ratios of the fiber
b) Maximum bridging stress and toughness vs Poisson ratio of the fiber
(Material data for Nicalon/Duran see table 1)

These tendencies are manifested in a more general way by demonstrating the relationship between the maximum bridging stress and toughness versus the parameter under consideration. As a representative value for the toughness, the area under the bridging curve is used when all fibers are broken, i.e. $p = 100$ %. The result can be seen in figure 2b. The maximum bridging stress decreases moderately and nearly linear with increasing fiber Poisson ratio whereas the toughness behaves characteristically different. At low Poisson ratios the increase is moderately, but at higher values there is a sharp increase in toughness. It can be immediately concluded that composite properties would be dramatically improved if this state could be reached.

As pointed out above, the factor b_1 which is a measure for the Poisson effect during fiber loading can be either positive or negative. Even the extreme case $b_1 = 0$ is possible where the shear stress in the interface is in fact constant. For this special case, the extended model presented here is just identical to the model with constant friction. In figure 2b this stage is marked by the doted line, referring to a fiber Poisson ratio of 0.30. This is obviously not a realistic value; for this fiber-matrix-combination such a simplification is then not possible. At lower values of the fiber Poisson ratio the factor b_1 is negative. Thus, it can be concluded that toughening of the composite becomes ineffective with decreasing factor b_1 as already suspected by Hutchinson and Jensen[1].

It is now the further objective of this study to assess the effects of the bridging stresses with varying parameters in terms of macroscopic properties of the composite.

CONSEQUENCES FOR OVERALL CRACK RESISTANCE

From the bridging stress function with varying interface parameters it cannot directly be concluded which curve is particularly advantageous for the macroscopic properties of the composite. Thus, a methodology is needed to obtain information about composite failure behavior from a known bridging stress. In this study, such a methodology is derived for a standard fracture mechanics test (Single Edge Notch Beam, SENB) which has been successfully applied for different ceramic composite materials.

The load-displacement behavior of the SENB-test can be modelled by the method of weight functions which recently has been comprehensively presented by Fett[6]. In this method, the bridging stresses σ_{bridge} are taken into account by transformation into

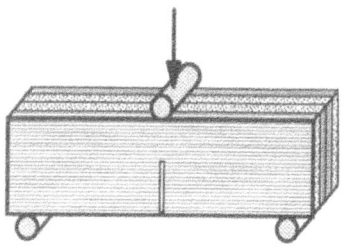

Figure 3. Single edge notch beam test (SENB) applied for a cross-ply fiber reinforced composite

a negative stress intensity factor $K_{I,bridge}$.

$$K_{I,bridge} = \int_0^a \sigma_{bridge}(x)h(a,x)dx \qquad (17)$$

where $h(a,x)$ is a weight function, given in polynom-formulation [7] for explicite specimen geometries. a is the crack length and x the coordinate along the crack.

During stable crack propagation the increasing external stress intensity factor $K_{I,appl}$ is balanced by the bridging stress intensity factor and the internal crack resistance $K_{I,0}$.

$$K_{I,0} = K_{I,appl} + K_{I,bridge} \qquad (18)$$

The external stress intensity factor is derived analogous to $K_{I,bridge}$ using a linear stress profile

$$K_{I,appl} = \sigma^* \int_0^a \left(1 - 2\frac{x}{W}\right) h(a,x)dx \qquad (19)$$

where W is the specimen thickness and σ^* the stress in the outer fiber of an unnotched beam.

The method of weight functions allows the exact determination of the crack opening profile with respect to the bridging stress profile. Once the stress profile normal to the crack surfaces obtained by superimposing bridging stress and applied stress is known the crack opening $\delta(x)$ is derived by

$$\delta(x) = \frac{1}{E^*} \int_0^a \int_{max(x,x')}^a h(a',x)h(a',x')\sigma(x')da'dx' \qquad (20)$$

E^* is the Young's modulus E at plain stress, or $E^* = E/(1-\nu^2)$ at plain strain.

The stress profile itself is a function of the crack opening, so again an iterative algorithm has to be applied. Finally, the loading point deflection is derived from the crack opening profile.

$$\delta_{LP} = \frac{3L}{W^2} \int_0^a \left(1 - 2\frac{x}{W}\right)\delta(x)dx \qquad (21)$$

The complete load-displacement and crack opening curve of a SENB-test is obtained by giving discrete crack lengths and calculating each equilibrium load and loading point deflection.

In figure 4 the crack opening profile and the bridging stress profile are presented at the maximum load of a SENB-test (specimen dimensions $L = 30$ mm, $W = 9$ mm,

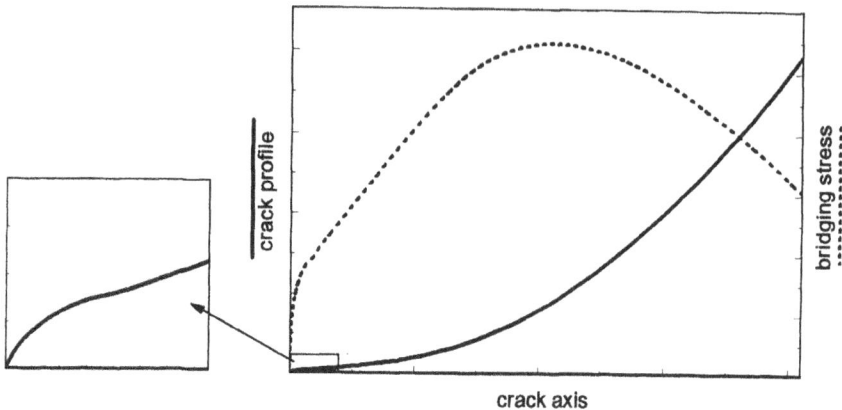

Figure 4. Bridging stress and crack opening profile along the stressed part of the crack surface at a SENB-test

$a_0 = 4$ mm). At the crack tip the profile steeply opens due to the stress intensity in this region. Accordingly the bridging stresses also increase steeply as nearly all fibers remain intact at low crack opening displacements. At increasing distances from the crack tip the crack opening increases relatively slowly resulting in a nearly linear increase of the bridging stresses. The maximum of the bridging stresses is reached approximately in the middle of the bridging zone. Then, the increase of crack opening becomes steeper corresponding to the decreasing crack closing forces.

As mentioned in the preceding chapter it is now possible to assess the influence of bridging stresses on the fracture resistance of the composite. In figure 5a, some calculated bridging stress curves are presented as obtained from the variation of the coefficient of friction in the fiber-matrix interface. The trend is similar to the one discussed for the Poisson ratio of the fiber. A high coefficient of friction leads to a high gradient of the fiber axial stress. Thus, the maximum bridging stress achieves higher values but the toughness decreases.

Spontaneously, it cannot be decided which trend is advantageous for the macroscopic failure behavior. Therefore, for each bridging curve the SENB-calculation has been performed as described above. The results are presented in figure 5b. Obviously, a very high coefficient of friction (0.7) leads to a low load maximum and to a rather brittle failure behavior. The highest load maximum of the SENB-test was achieved with a coefficient of friction of 0.3. With the friction coefficient lower than 0.3, the load maximum decreases again and the load-displacement-behavior is extremely nonlinear. From a practical point of view, the failure behavior assiociated with a coeffcent of friction of 0.3 would probably be preferred, as the structural element keeps a high stiffness up to high loads, and sustains a high nominal load. On the other hand, in some cases an extreme high failure energy might be preferred, so a lower coefficient of friction would be the choice.

CONCLUSIONS

A comprehensive model has been presented to describe the macroscopic failure behavior of ceramic matrix composites in terms of micromechanical data of the composite components, i.e. fiber, matrix and interface. The micromechanical model is based on a universal approach given by Hutchinson and Jensen [1]. The stress that can be trans-

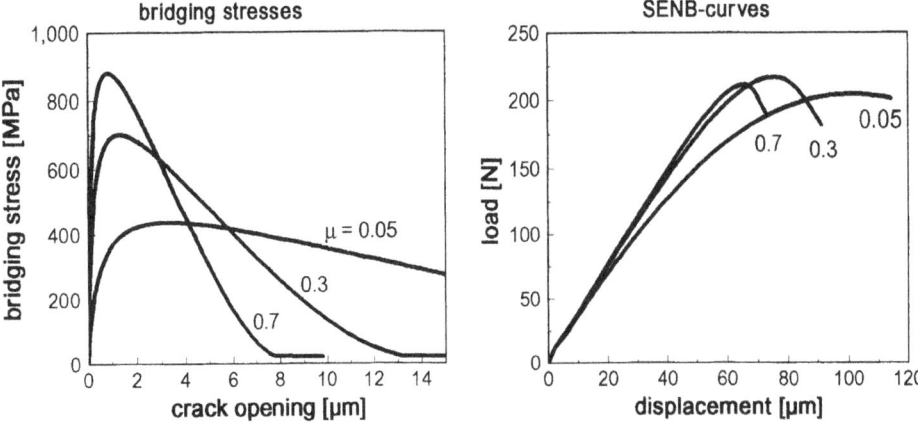

Figure 5. Bridging stresses calculated with different interface coefficients of friction and corresponding load-displacement-curves of the SENB-test

ferred across the surfaces of a macrocrack is calculated using the Weibull distribution of the fiber strength. In this model the simplifying assumption of a constant interfacial shear stress which is usually applied to describe macroscopic phenomena is included as a special case of this solution which is more general due to an additional parameter b_1 quantifying the nonlinearity of the fiber axial stress. The micromechanical model leads to a characteristic function for the bridging stresses versus crack opening for a single crack perpendicular to fiber orientation.

For a selected example, it has been shown that the bridging stress curve can be used to predict macroscopic failure behavior of a specimen with a particular load configuration and geometry. An optimum in macroscopic composite performance is predicted by selecting the proper coefficient of friction. It should be noted that the numerical result is not to be generalized as it depends on a large number of parameters. In the case of other characteristic parameters of fiber, matrix or interface, other trends will be observed. Thus, if this model shall be applied to determine the optimum interface parameters of a given composite, the actual data including all elastic constants, initial strain mismatches between fiber and matrix and the fiber strength distribution have to be taken into account. Also more sophisticated solutions considering the real structure of the interface, e.g. the effect of an interfacial layer between fiber and matrix[8] or the effect of roughness interaction during fiber sliding[2] serve as additional input parameters to the micromechanical model. The semi-analytical solutions for these cases can be explicitly transformed into expressions appropriate to maintain the framework of the Hutchinson-Jensen model.

This exemplary calculation gives some insight in the potential of this comprehensive theoretical treatment. The method of weight functions can be extended to predict load-displacement-crack opening characteristics for any structural element where single crack extension is dominant. The comprehensive methodology should be applied as a tool to obtain a criterion for optimized interface design with respect to the properties of the constitutive components.

References

[1] J.W. Hutchinson & H.M. Jensen. Models of Fiber Debonding and Pullout in Brittle Composites with Friction. *Mechanics of Materials*, 9, 139–163, 1990.

[2] M. Kuntz, K.-H. Schlapschi, B. Meier & G. Grathwohl. Evaluation of interface parameters in push-out and pull-out tests. *Composites*, 25(7), 476–481, 1994.

[3] D.B. Marshall. Analysis of fiber debonding and sliding experiments in brittle matrix composites. *Acta metall.*, 40(3), 427–441, 1992.

[4] H.L. Oh & I. Finnie. On the Location of Fracture in Brittle Solids-I. Due to static loading. *Int. Journ. of Fracture Mech.*, 6(3), 287–300, 1970.

[5] M.D. Thouless & A.G. Evans. Effects of Pull-Out on the Mechanical Properties of Ceramic-Matrix Composites. *Acta metall.*, 36(3), 517–522, 1988.

[6] T. Fett. *Contributions to the R-curve Behaviour of Ceramic Materials, KfK 5291*. Kernforschungszentrum Karlsruhe, 1994.

[7] T. Fett & D. Munz. Evaluation of R-curve effects in ceramics. *J. Mater. Sci.*, 28, 742–752, 1993.

[8] M. Kuntz, B. Meier & G. Grathwohl. Residual Stresses in Fiber-Reinforced Ceramics due to Thermal Expansion Mismatch. *J. Am. Ceram. Soc.*, 76(10), 2607–12, 1993.

MECHANICAL BEHAVIOR OF LARGE SIZE COMPACT TENSION SPECIMENS OF 2D WOVEN SiC-SiC COMPOSITE MATERIALS

Monssef Drissi-Habti

LERMAT URA CNRS n° 1317, ISMRA
6, Bd du Maréchal Juin, 14050 CAEN Cedex, France

ABSTRACT

Due to the heterogeneous structure of continuous ceramic fiber-reinforced ceramic matrix composites (CMCs), traditional design based on R-curves presents several limitations. Particularly, the size effects which lay down the choice of appropriate specimen sizes with regard to the micro-structure. To scale down considerably the damaged zone with respect to the ligament size, very large and very thick specimens are therefore required. For this main reason, the unrealistic of such tests can be understood. In the event of an R-curve not being available for structural part calculations, the aim of the present article is to set out some guidelines which indicate that on 2D woven SiC-SiC composites, significant crack growth resistance values, directly related to the main reinforcement micromechanisms, could be achieved by testing small compact tension (CT) specimens (with a size, $W = 20$ mm) having an appropriate thickness, B.

INTRODUCTION

It is now recognized that engineering techniques based on the linear elastic fracture mechanics (LEFM) have been successful for assessing and predicting strength, operating lifetime and reliability of most monolithic ceramics [1,2]. Nevertheless, limited success has been achieved to date for composite materials, especially in the case of continuous fibers reinforced brittle ceramic matrix composites (CMCs)[3]. Due to the toughening micro-mechanisms developed on purpose in CMCs (such as matrix microcracking, fibers bridging, pull-out,...), crack growth resistance curves (R-curves) based on LEFM formalism had been hinted for a long time as being an useful implement. However, numerous extrinsic factors, involving particularly, fracture mechanics test geometry [4], strongly affects R-curves.

The development of a specimen design based R-curve should be built around the crack length, the ligament length and the thickness required in order that the crack growth resistance curve of the material could be accurately established. However, such a procedure clashes with many difficulties. Pointing out the main limitations, involve considerations of either the extrinsic factors affecting experiments and the limitations of the commonly used theoretical treatments of experimental results.

Regarding the particular structure of CMCs, the test configuration adopted must provide a ligament which can ensure an intrinsic mechanical response of the material. So that, influences of size effects will be particularly canceled[5,6]. To get meaningful crack growth resistance values, the « crack » entity must be defined in an unambiguous way and a straightforward methodology for crack length measurements must be implemented. The latter must be reliable and particularly practicable at elevated temperature. In this respect, in the case of the 2D woven SiC-SiC composite materials, numerous attempts had been made and can be amounted to : (i) crack lengths deriving from the compliance calibration curve obtained from machined notches[7], (ii) the dye penetrant techniques to get insight the height of the crack in the vicinity of the specimens[8,9], (iii) *in-situ* matrix crack lengths measured on a polished surface of the specimens[9,10] and (iv) crack lengths tabulated from the standard elastic compliance formulation of Tada[11,12]. Though the diversity of these methods, there remains the question of their validity. Especially, with respect to experiment based observations which demonstrate that failure does not occur by a dominant mode I crack and even standard elastic compliance formulations underestimate the amount of damage.

Downstream from experimental results, their theoretical treatments remain. In fact, in a CMC (with a sufficiently weak interface), when a crack grows in the matrix initially without breaking fibers, a bridging zone develops behind the crack front, resulting in increasing crack closure tractions as the matrix crack extends. As a consequence, crack growth is dictated by a rising crack growth resistance (R-curve). In general, R-curves of CMCs had been formulated in terms of the potential derived toughness, K_R,[5,12,13,14] ; and the strain elastic energy release rate, G_R,[9,10,14,15,16,17]. Whatever the methodology, design engineers need a R-curve which is representative of the intrinsic resistance of the composite against crack growth. Indeed, correlated to experiment based observations, R-curve provides (with an accurate modelization) an estimate of the amount of the main reinforcement mechanisms to the total crack growth resistance of the composite. However, many extrinsic factors are hindering such a procedure : among them, size effects on the shape and the values of R-curves. Typically, for CMCs endowed with a sufficiently weak interface such that sliding of the fibers are promoted, size effects ensue intimately from fiber-bridging zone size. The problem is that R-curves of CMCs are strongly depending on both the absolute length of the bridging zone and the bridging zone relative to the total crack length and specimen width[5]. Referring to Cox[19], if the length of the bridging zone is comparable to any of specimen dimensions, the bridging contribution to the potential derived toughness, becomes strongly dependent on the specimen size and shape.

Otherwise, in the case of 2D woven SiC-SiC composites, [9] had obtained two shapes of R-curve, depending on the compact tension (CT) specimen sizes. These shapes had been correlated to the propagation of the frontal process zone (FPZ). In this way, before attaining its stationary size, the FPZ interacts with the opposite side of smallest specimens (with a size, W = 20 mm) and this leads to a parabolic increase of K_R values. Whereas for larger CT specimens (W = 40 mm), the FPZ grows, reaches its steady state size along a region where K_R values behaves as $a^{1/2}$. Thereafter, the FPZ interacts with the opposite side of the specimens. Hence, K_R values behave as for small size specimens.

Starting from the precise description of the mechanical behavior of two batches of 2D woven SiC-SiC composites, R-curve prevailing parameters such as, the crack entity, specimen size and test geometry will be analyzed regarding reinforcement micromechanisms. The aim of this part is on one hand, to stand out extrinsic factors such as geometrical effects which affect crack growth resistance values. In the other hand, to set out to convince that on load displacement relationships, over a region ending at the peak load, P_{max}, significant crack growth resistance values could be calculated using small size CT specimens (W = 20 mm) having an appropriate thickness.

MATERIALS AND METHODS

The materials investigated were two types of 2D SiC-SiC composite materials processed by the SEP (Etablissement de Bordeaux, France). These materials are made of a stack of equilibrated woven cloths of SiC Nicalon fibers (NLM 202) with a pyrocarbon interphase. The preform is then densified by a β–SiC matrix deposited by chemical vapor infiltration process. The morphological properties of the 2D SiC-SiC composite materials had been extensively studied in numerous articles [20, 21, 22, 23]. Two types of 2D SiC-SiC composite materials were achieved, namely HUS (**High Ultimate Strain**) and LUS (**Low Ultimate Strain**). The physical characteristics of these composite materials are listed in table I.

Table I. Few physical properties of the 2D SiC-SiC composite materials studied (after [24]).

The matrix	β - SiC (I - CVI)
Elastic modulus	E_m = 360 GPa
Matrix reinforcement	V_m = 40-45%
The reinforcement	SiC Nicalon NLM 202
Elastic modulus	E_f = 200 GPa
Fiber fraction	V_f = 40 % (20 %⊥ ; 20 %‖)
The composite	
Poisson's ratio	ν = 0.12
Density	1.4 - 1.5
Porosity	15 - 20 %
Strain to failure (uniaxial tension)	
• HUS	ε_r > 0.4 %
• LUS	ε_r = 0.2 %
Stress to failure (uniaxial tension)	
• HUS	σ_r = 300 MPa
• LUS	σ_r = 180 MPa

Mechanical tests were carried out on compact tension (CT) specimens according to the ASTM standards E-399-81 specifications[25]. Provided sizes were W = 20, 40, 60 and 80 mm in the case of LUS material and W = 20 and 40 mm in the case of HUS material. Both batches were provided in the tickness, B = 3 mm ; except for the CT size, W = 20mm, where provided ticknesses were B = 3 and 6 mm. Tests were conducted on an electro-mechanical Schenck testing machine (Schenck Treble RMC 100) with a 100 kN load cell. Tests were displacement controlled, at a cross head speed of 50 μm/min. The crack opening (or displacement) was measured by the relative displacement of the loading axis, using a linear variable differential transducer (LVDT).

The matrix crack length measurements were monitored *in-situ* on a polished surface of the specimens using an optical traveling microscope which is connected to both a camera and a complete video recording system. The crack growth resistance curves were worked out using two procedures :

- According to the elastic strain energy release rate concept, G_R, calculated using two ways : (i) the classical one, based on compliance measurement :

$$G_R = (P^2/2B) * (dC/da) \qquad [1]$$

where P is the applied load, B the thickness of the specimen and C the specimen compliance. and (ii) following the composite beam theory proposed by[17], which considers the ligament of the damaged specimen equivalent to two parallel specimens[18]. The first one, microcracked at saturation, corresponding to the matrix microcracked zone around the notch tip (representing the FPZ). The second, exempt of damage, corresponding to the undamaged region of the ligament. Besides, this approach assumes that the macroscopic crack starts propagating at the peak load, P_{max}.
  - Using the potential derived toughness parameter, K_R, which has been defined by LEFM in terms of the applied load, P, and the test specimen geometry, ie width W, thickness B and crack length, a. The K_R parameter has the following analytical form :

$$K_R = (P/BW^{1/2}) * Y(\alpha) \qquad [2]$$

where α (= a / W) is the relative crack length. $Y(\alpha)$, given by expression [3], is the polynomial function, formulated by[26]. It had also been also defined by[2] as being the dimensionless stress intensity factor :

$$Y(\alpha) = ((2 + \alpha) / (1 - \alpha)^{3/2}) * (0.886 + 4.6\,\alpha - 13.32\,\alpha^2 + 14.72\,\alpha^3 - 5.6\,\alpha^4) \qquad [3]$$

RESULTS AND DISCUSSION

Load-displacement relationships and compliance measurements for HUS materials

Based on the proposed results[9, 10, 12], this type of material behaves linear elastic. So, all the specimens were loaded monotonically up to rupture. Figure 1 shows load versus displacement curves for three specimens with a notch depth ratio (NDR), a_0/W = 0.2 ; 0.3 and 0.4. A first noticeable point is related to the outlines of these three P-u curves. Indeed, making abstraction of the maximal load values, the outlines are similar for these three specimens. To get deeper insight the mechanical behavior, the onset of damage is detected in the early stage of the loading in the linear elastic region, at a stress intensity factor, K_c^{ini}, close to 2 MPa√m. The corresponding matrix critical stress intensity factor is K_C^m = 3 and 3.5 MPa√m, when considering the crack propagation analysis of [27, 28], respectively. These comparable values accord with the critical stress intensity factor of the polycrystalline SiC matrix. The end of the linear elastic region (at a load value, P_l) corresponds to a load ratio P_l / P_{max} close to 0.5 for all specimens. Correlating this to observations under microscope show that, up to the proportional load, P_l, an important matrix microcracking mechanism localized around the notch tip takes place. When increasing the load beyond P_l, a macroscopic matrix crack prevails, surrounded by a microcracked zone. This macroscopic crack propagates quickly up to the maximal load, P_{max}, and then slows down considerably, as illustrated in figure 2. Thereafter, the transverse bundles of fibers began to fail sequentially, leading to a non stabilized rupture.

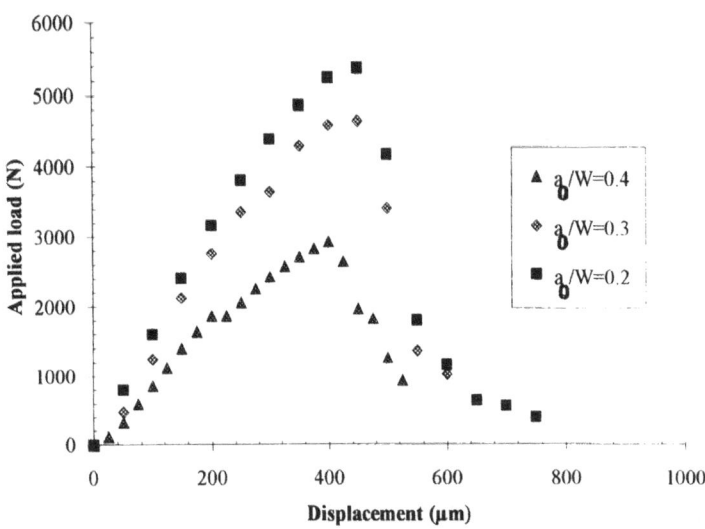

Figure 1.. Typical load displacement relationships for three CT specimens of HUS material.(W = 40 mm and a_0/W = 0.2, 0.3, 0.4)

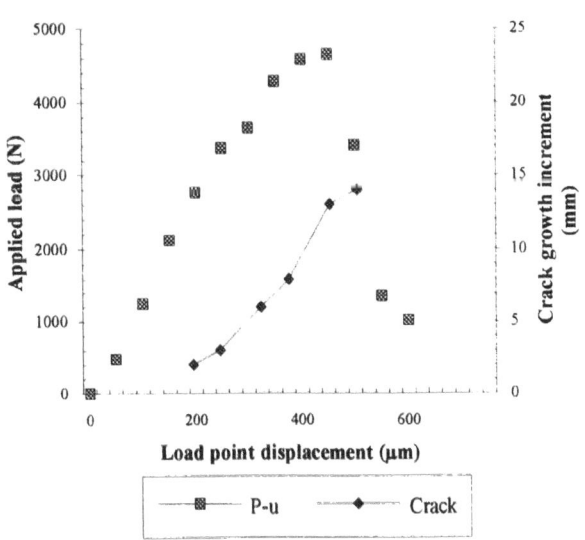

Figure 2.. Load - displacement curve of a HUS CT specimen (W = 40 mm ; a_0/W = 0.3) with corresponding *in-situ* matrix crack lengths.

Compliance values versus either experimental matrix crack lengths and tabulated crack lengths using the Tada expression, are reported in figure 3 for a CT specimen of size W = 40 mm and a NDR, a_0/W = 0.3. As it could be shown, the linear elastic formulation of Tada under estimates the experimental matrix crack lengths. Accordingly, the use of this iterative expression[5, 6] in the case of composite materials appears problematic. Since this expression had been proposed in the case of linear elastic materials, damage micro-mechanisms characterizing the mechanical behavior of CMCs cannot be taken into account. Though a matrix crack length is not wholly reflecting the state of damage in the specimens, it could be considered more significant, due to its experimental character.

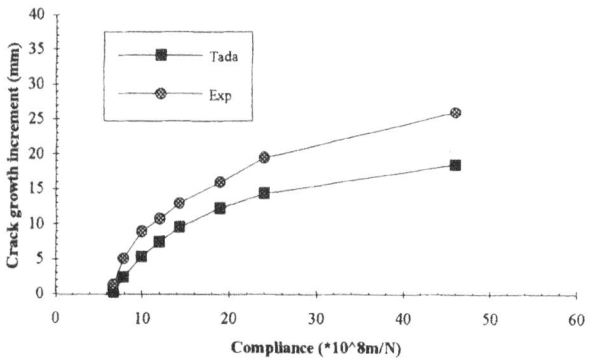

Figure 3.. *In-situ* matrix crack lengths and tabulated ones using the Tada formulation plotted versus compliance values in the case of a HUS CT specimen (W = 40 mm ; a_0/W = 0.3).

Load-displacement curve and compliance measurements for LUS material

As previously proposed[9, 10, 12], the mechanical behavior of this batch of 2D SiC-SiC composite materials is linear non elastic. Consequently, the specimens were unloaded periodically to assess changes in compliance and to measure matrix crack lengths. Figure 4 shows the outlines of load versus displacement curves (P-u) for all specimen sizes, endowed with the same NDR, a_0/W = 0.3. For the smaller CT sizes (W = 20 and 40 mm), the onset of damage, detected by acoustic emission in the proportional region, corresponds to a stress intensity factor, K_C^{ini} close to 3.5 MPa√m. Whereas, for larger sizes (W = 60 and 80 mm), K_C^{ini} values were more elevated (7.5 and 9.5 MPa√m, respectively). Unlike the HUS batch, Observation-based microscope shows a less important matrix microcracking. But an extensive fiber bridging in the wake of the matrix crack is revealed (figure 5a and 5b). The latter seems to be the major reinforcement micro-mechanism in this material. Afterwards the peak load, P_{max}, the specimens present a well controlled rupture, contrasting with the mechanical behavior of HUS material. Once the damage initiated, a "tortuous" major mode I matrix crack is easily prevailing (figure 6a). Its propagation is abrupt in the early stage of the loading and slows down considerably after the peak load. At rupture, high pull-out lengths with a mean size close to 300 μm, are revealed (figure 6b).

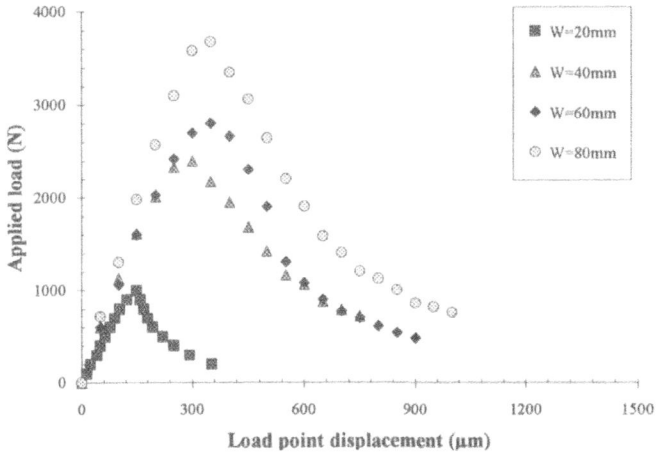

Figure 4. Load displacement relationships for three CT specimens of LUS material. (with W = 20, 40, 60 and 80 mm ; a_0/W = 0.3).

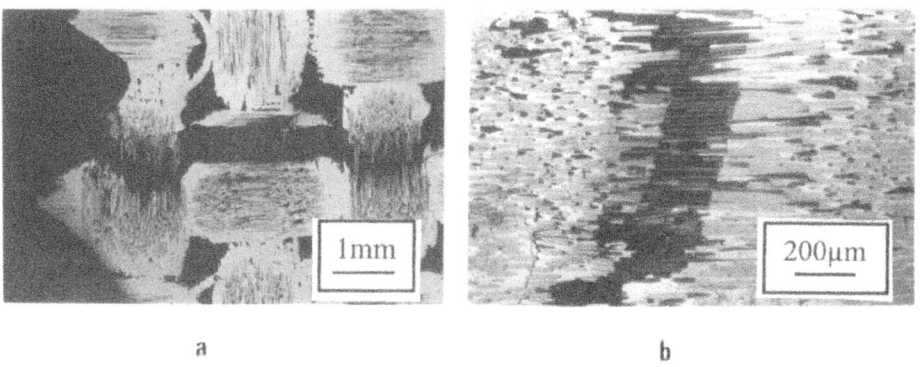

a   b

Figure 5. Micrograph showing the extensive fiber bridging mechanism acting in the wake of the matrix crack in a specimen of LUS batch (a) crack closure by the bundles , (b) crack bridging by the fibers.

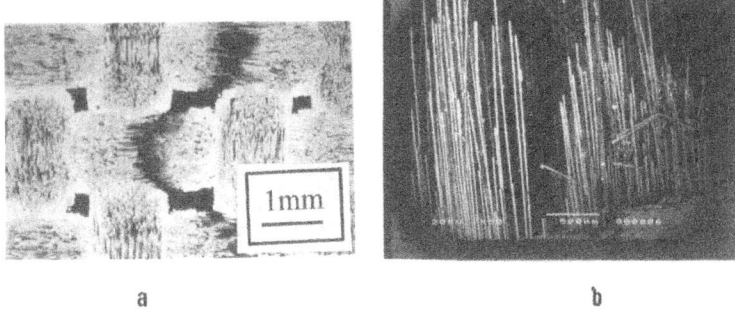

a   b

Figure 6. (a) Micrograph of the crack path in a CT specimen of LUS material, (b) Micrograph of the pull-out in a fractured CT specimen.

Regarding the irreversible deformation observed when loading/unloading this batch, all compliance measurements were made on the reloading loops of the hysteresis. Thenafter, these compliance values were compared to the ones worked out using the Saxena and Hudak expression[29] in the case of a linear elastic material (C_{th}). Two main points should be inferred. The first one is related to the fact that compliance values are constant on over than 80 % of the total ligament in the case of large size compact tension specimens (W = 60 and 80 mm) (figure 7). The second point concerns the results obtained for three specimens of size, W = 40 mm, with NDR, a_0/W = 0.3, 0.4 and 0.5. In fact, for a given matrix crack length, LUS batch is tougher than the linear elastic material having the same dimensions and the same initial elastic characteristics (figure 8).

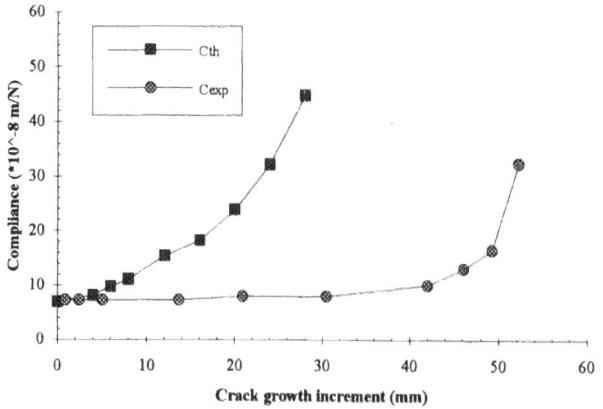

Figure 7. Experimental and linear elastic compliance values versus matrix crack lengths for a CT specimen of size W = 80 mm and a NDR, a_0/W = 0.3).

Besides the above result, it appears that the experimental compliance values could be approximated by straight lines. The slope, p, of the straight line is related to the polynomial factor, Y(a/W), by the following expression :

$$C(a) = C(a_0) + p\int_{a_0}^{a} Y^2 \, da \qquad [4]$$

which leads to the values, Y = 7.5, 8.7 and 6.5 for the NDR, a_0/W = 0.3, 0.4 and 0.5, respectively. Then, these Y values had been reported in the same graph than the polynomial factor, Y(a/W), derived from Saxena and Hudak works[29] (figure 9). The comparison shows that a systematic use of theoretical values of Y(a/W) overestimates the potential derived toughness, K_R, of this batch of 2D SiC-SiC composite materials.

As emphasized above, LUS batch exhibits an efficient fiber shielding of the matrix crack in the wake zone. Starting all from the well known Dugdale-Barenblatt model for crack closure, relevant attempts had been made towards the estimate of either the bridging zone length and the bridging stress[5, 14, 30]. Since the Dugdale-Barenblatt model assumes a uniform crack closure stress which resists the applied load in the crack tip of a linear elastic material, the implement of this model on CMCs becomes very crude when considering that the bridged

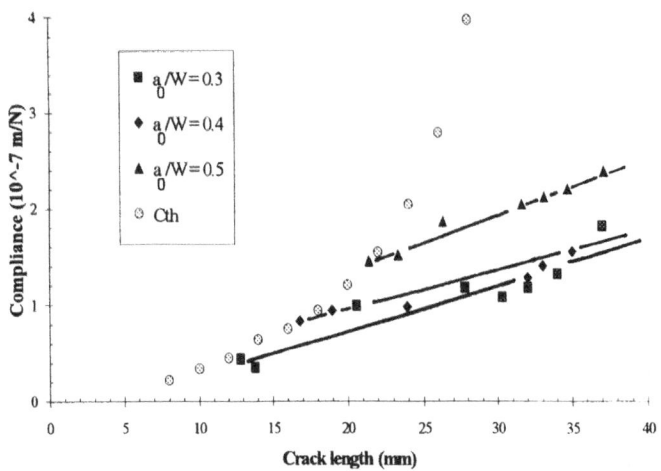

Figure 8. Experimental and linear elastic compliance values plotted as a function of the matrix crack lengths for three LUS CT specimens of size W = 40 mm endowed with NDR, a_0/W = 0.3, 0.4 and 0.5.

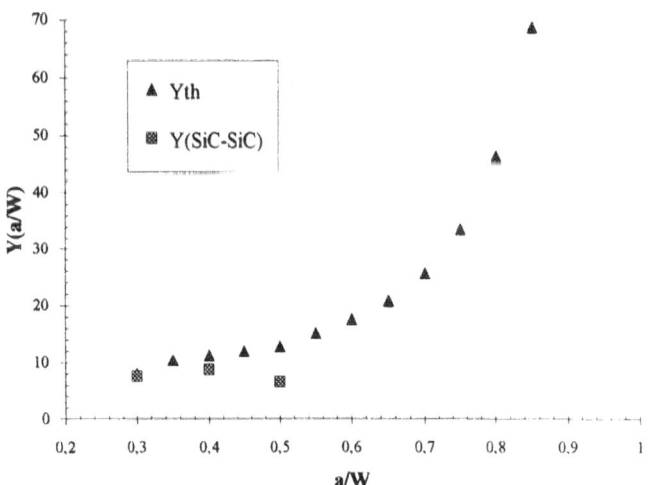

Figure 9. Calculated values of the polynomial factor (using expression [4]) and tabulated ones using the Saxena and Hudak formulation (Yth).

zone length is in the same order of magnitude than provided ligaments. For lack of an accurate methodology for estimating the bridging stress, it would be interesting to assess the length of the bridged zone. Such an estimate could be carried out using compliance values (figure 8). Indeed, the reasoning consists of considering, on one hand, that the increase in the theoretical compliance is merely due to the loss of the SiC matrix stiffness. On the other hand, the experimental compliance, integrating irreversible deformations, is mainly due to the fiber bridging mechanism. According to these points, the related gap to the linear elastic behavior, expressed by the term $(C_{th} - C)/C$, could be related to the amplitude of anelastic micromechanisms. Upon that, the fictitious crack length having the same effect than the bridged zone length is X_B, given by the following expression :

$$X_B = (W - a) * ((C_{th} - C) / C) \qquad [5]$$

The results are plotted in figure 10 as a function of the matrix crack length in the case of three specimens of size W = 40 mm, and NDR, $a_0/W = 0.3$, 0.4 and 0.5. In the main, calculations coupled to observations under microscope stand out the following points : (i) X_B values calculated using expression [5] accord with microscope based measurements, (ii) even after a crack growth, $\Delta a = 20$ mm, a stationary bridged zone length is not observed, (iii) the maximum of the bridged zone length corresponds to the maximal load, P_{max}, and (iv) the decrease of X_B values coincides with the beginning of the bundles breakage (starting at the peak load). Correlating these points to the evolution of the potential derived toughness values for the same material (figure 11 ; 31), shows that the increase as $a^{1/2}$ of K_R values corresponds to the increasing part of X_B values. Thereafter, the change in K_R curve shapes (ie the increase as a^n (with n > 1)) is recorded when X_B values start decreasing. Such a result would indicate a close relation between K_R curves and the evolution of the bridging zone in LUS batch. It would be therefore interesting to get on this main assumption. In other words, the remaining question is : can we get a steady state propagation of the bridging zone when testing larger size CT specimens (W = 60 and 80 mm) ? Such a behavior would be clearly reflected by K_R curves presenting a plateau like value.

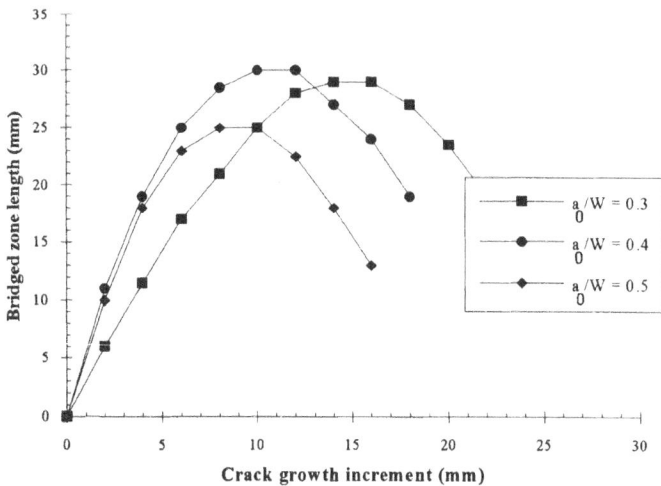

Figure 10. Bridged zone size plotted as a function of the crack growth for three LUS specimens (W = 40 ; a_0/W = 0.3, 0.4, 0.5).

302

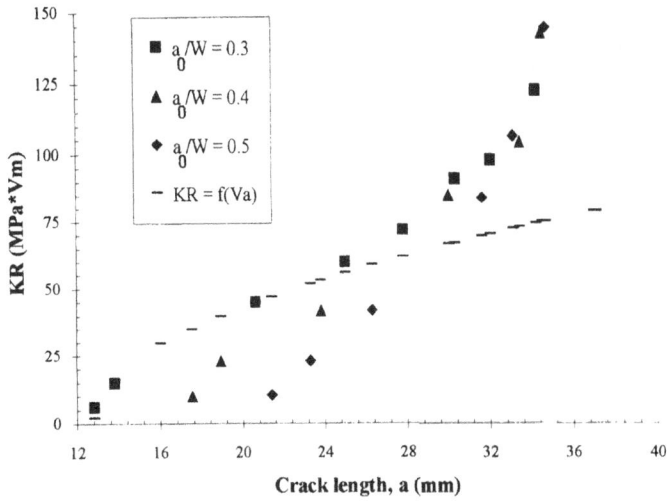

Figure 11. K_R values plotted as a function of matrix crack lengths for three LUS CT specimens with a size, W = 40 mm and NDR, a_0/W = 0.3. 0.4 and 0.5.

Crack growth resistance curves for HUS material

As stated above, crack growth resistance values were tabulated using the strain elastic energy release rate, G_R, and the potential derived toughness parameter, K_R, in the case of CT specimens of size, W = 40 mm, with NDR, a_0/W = 0.2, 0.3 and 0.4. In this way, G_R values are plotted in figure 12 as a function of matrix crack lengths. While G_T values, calculated using the composite beam theory[17], are reported versus the applied load, P, in figure 13.

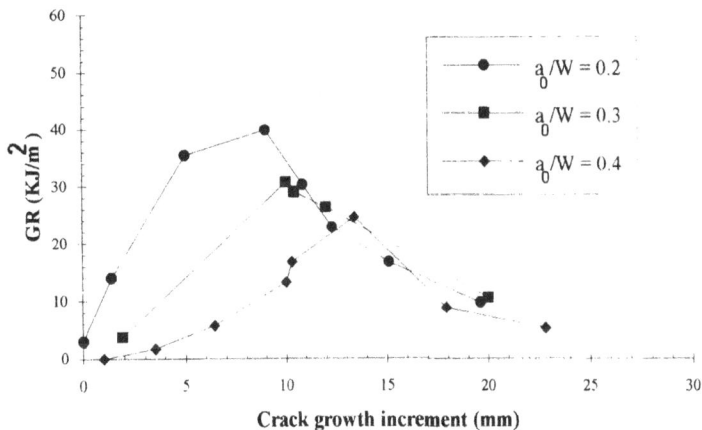

Figure 12. Crack growth resistance values plotted as a function of the crack growth for three specimens of HUS material with a size, W = 40 mm and a notch depth ratio (NDR), a_0/W = 0.2, 0.3 and 0.4.

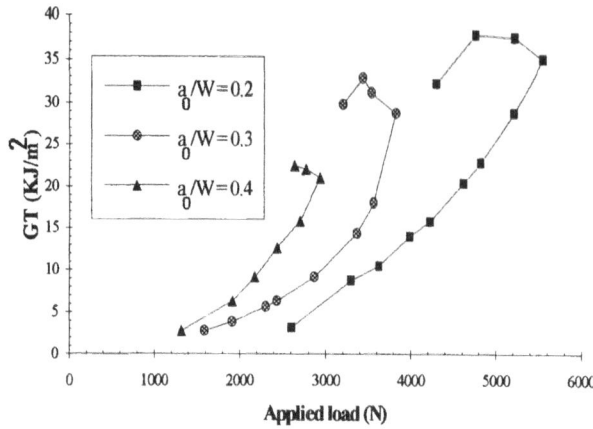

Figure 13. Crack growth resistance values, G_T, calculated using the composite beam theory proposed by[17] plotted as a function of the applied load, P.

Using both procedures, the general features are : (i) crack growth resistance values increase up to a maximum and then decrease, (ii) the values obtained are similar, particularly at the maximums, G_{Rmax} and G_{Tmax}, (iii) small is the saw cut length, elevated are the maximum energy values. The observations underlined above could be explained when considering that G_R is the rate of energy provided for creating matrix microcracks. Then, as the saw cut length is smaller, as more matrix microcracks will be created, leading to a more elevated energy values, G_R. However, the main inferring information is that, these decreasing crack growth resistance values as the saw cut length increases cannot be considered as a material characteristic.

Let us now analyze the evolution of the potential derived toughness parameter, K_R, versus matrix crack lengths for provided CT specimens of HUS material (W = 20 and 40 mm). K_R values plotted versus crack lengths for three specimens (with the same size, W = 40 mm and different NDR, a_0/W = 0.2, 0.3 and 0.4), are presented in figure 14. As it could be shown, initiation values of the crack growth resistance ($\approx$ 3 MPa√m) are not affected by the initial saw cut length. Moreover, values apart, R-curve shapes are identical to the ones of LUS material (figure 11). At a first step, K_R values are identical and increase as a function of √a over a length, Δa = 12, 8 and 4 mm for NDR a_0/W = 0.2, 0.3 and 0.4, respectively. When the matrix crack equals half the total ligament (a/W = 0.5) (figure 15), the shape of the slope changes and a parabolic increase of the crack growth resistance values is recorded.

The comparison between provided CT specimens sizes (W = 20 and 40 mm) leads to the following interesting points. As shown in figure 16, presenting K_R values versus crack growth increments, resistance values are the same over a length, Δa = 4 mm. Whereupon, an abnormal increase of K_R values is recorded for the smallest CT size. Otherwise, when plotting K_R values versus relative crack lengths, a/W (figure 17), one have to face facts that a unique R-curve shape is obtained with specific features : an increase as a function of √(a/W) up to a value, a/W = 0.5 (corresponding to a crack growth increment, Δa = 4 and 8 mm, for W = 20 and 40 mm, respectively). Whereupon, the R-curve shape changes to get a parabolic increase. It is also noticeable that resistance values are slightly lower for the size W = 20 mm. Explanations of this R-curve behavior will be detailed in the following paragraphs.

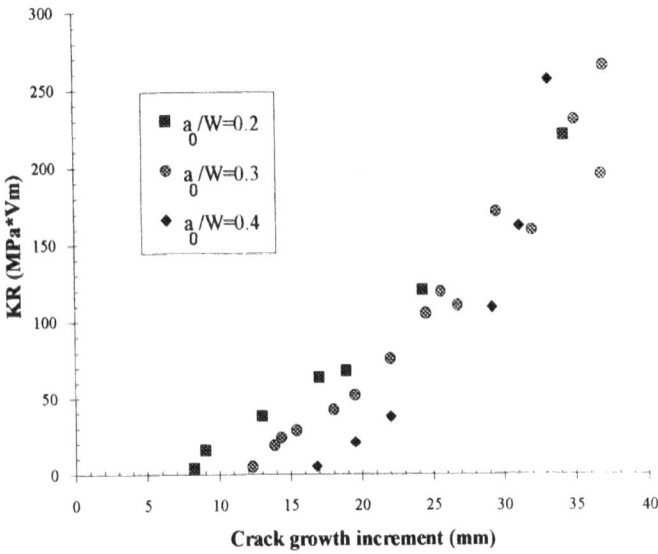

Figure 14. K_R values plotted as a function of the matrix crack lengths for HUS specimens (W = 40mm ; a_0/W = 0.2, 0.3 and 0.4).

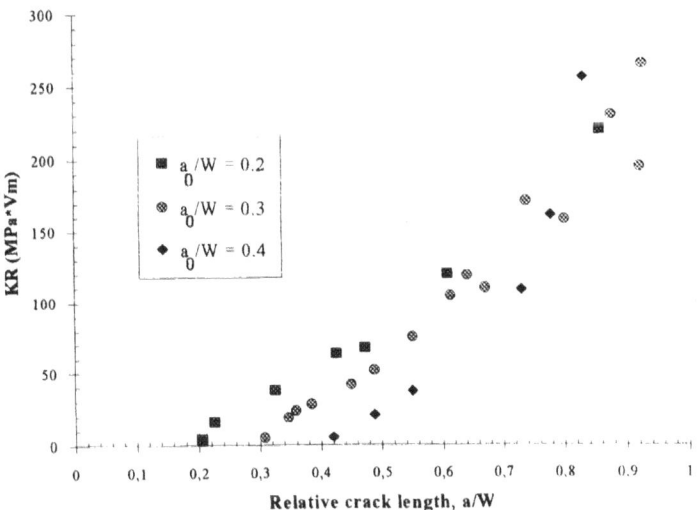

Figure 15. K_R values plotted as a function of the relative crack lengths for HUS specimens (W = 40 mm ; a_0/W = 0.2, 0.3 and 0.4).

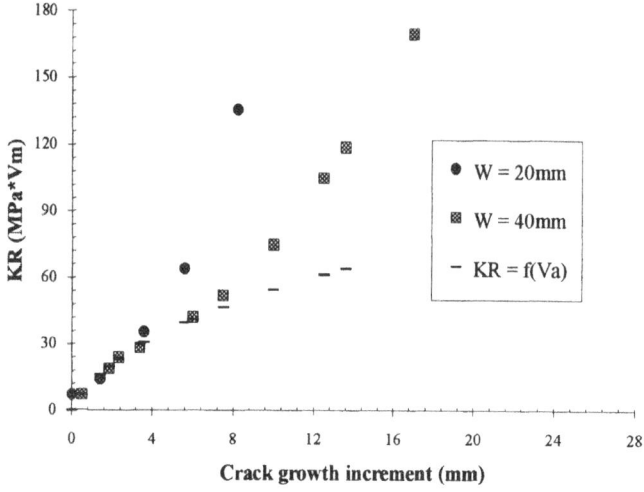

Figure 16. K_R values plotted as a function of matrix crack lengths for HUS specimens (W = 20 and 40 mm ; $a_0/W = 0.3$).

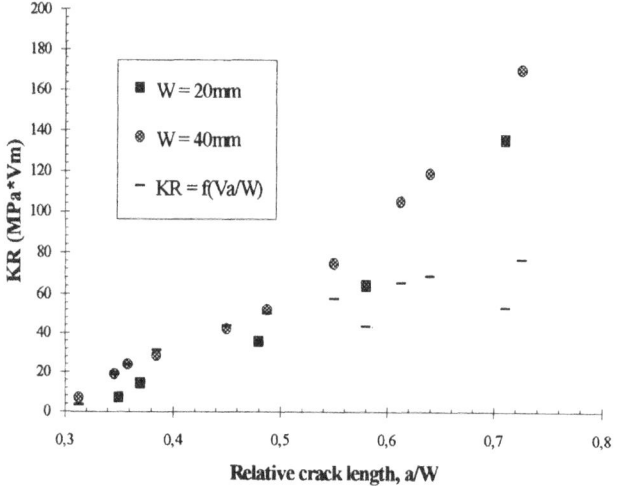

Figure 17. K_R values plotted as a function of relative crack lengths, a/W, for HUS specimens (W=20 and 40mm ; $a_0/W = 0.3$).

Crack growth resistance curves for LUS material

K_R values plotted as a function of matrix crack lengths in the case of CT specimens of size, W = 80 mm (with a NDR, $a_0/W = 0.3$) and W = 60 mm ($a_0/W = 0.3$ and 0.4) are presented in figures 18 and 19. A first noticeable point is related to the energy initiation values (K_C^{ini}). Indeed, for smaller CT size (W =20 and 40 mm), $K_C^{ini} = 3.5$ MPa√m, which is consistent with the toughness of the SiC matrix. Whereas, K_C^{ini} equal 9.5, 7.5 and 6 MPa√m for the CT size W = 80 mm and 60 mm, respectively. These values are higher than the toughness of the polycrystalline SiC matrix. This discrepancy could be the consequence of the low thickness (B = 3mm) of larger size specimens with regards to their ligament, W. Making a crude analogy with steel materials, solicitation under conditions close to plane stress ones could be involved.

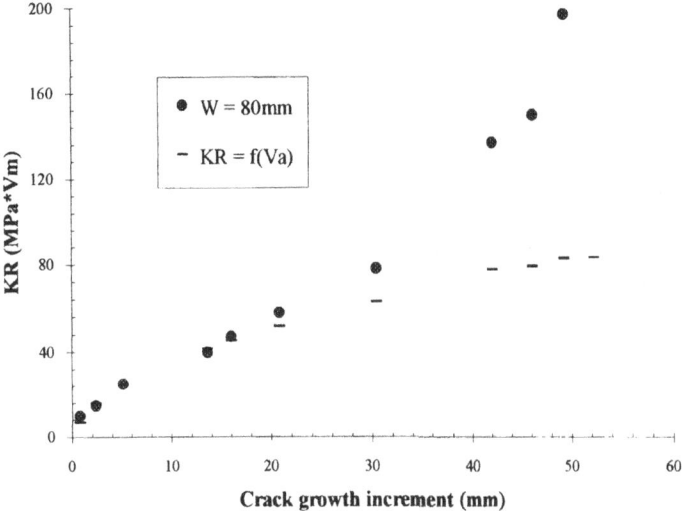

Figure 18. K_R curve for a LUS CT specimen of size, W = 80 mm (a_0/W = 0.3).

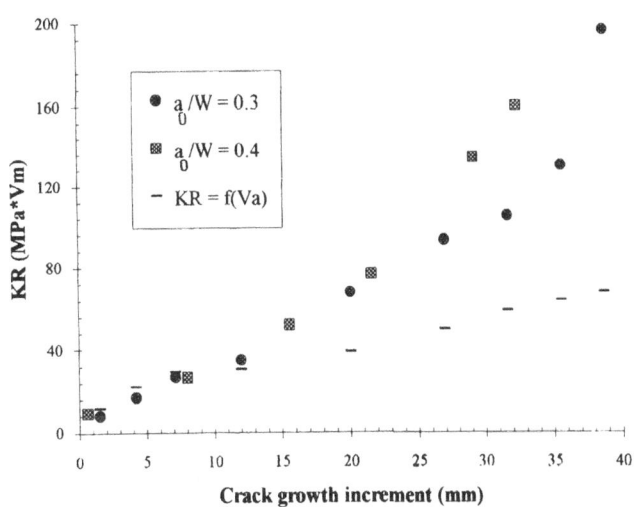

Figure 19. K_R values versus crack growth increments in the case of LUS CT specimens of size W = 60 mm (with a NDR, a_0/W = 0.3 and 0.4).

The second feature related to R-curve behavior is, though tested ligaments are larger than $W = 20$ mm, no R-curve plateau-like value is revealed. R-curve behavior is the same than reported above. The domain of behavior as $a^{1/2}$ is along an increment, $\Delta a = 8$ and 12 mm (corresponding to a crack length, $a = a_0 + \Delta a = 30$ mm $= W/2$) in the case of CT specimens of size $W = 60$ mm with NDR $a_0/W = 0.3$ and 0.4, respectively ; and along $\Delta a = 16$ mm (corresponding to a crack length, $a = a_0 + \Delta a = 40$ mm $= W/2$) for the CT size, $W = 80$ mm.

Getting deeper into the analysis, let us compare the R-curve behavior of all specimen sizes. It is ensuing from figure 20, the existence of a region along $\Delta a = 4$ mm, where equal resistance values are calculated using all CT sizes. Beyond this value, the constraint to the movement of the crack resulting from the interaction with the compressive region of the smallest size specimen ($W = 20$ mm) (side effect) leads to the virtual increase of the fracture resistance values and so forth at $\Delta a = 8$, 12 and 16 mm for the CT sizes, $W = 40$, 60 and 80 mm, respectively. However, when K_R values are plotted versus relative matrix crack lengths (a/W), a unique R-curve shape is obtained whatever the CT size (figure 21). This K_R-curve increases as $\sqrt{(a/W)}$ over half the total ligament ($W/2$). Thereafter, crack growth resistance values get a parabolic increase. Nevertheless, as previously emphasized on HUS specimens, the resistance values calculated using the smallest CT size are slightly lower that those calculated using larger size specimens.

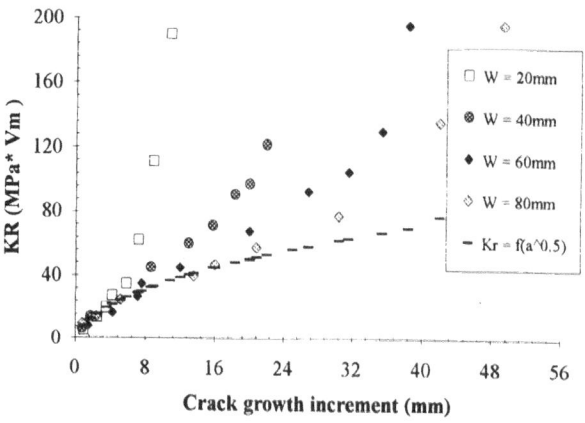

Figure 20. K_R values plotted as a function of the crack growth increment, Δa, for LUS CT sizes, $W = 20$, 40, 60 and 80 mm (with a NDR $a_0/W = 0.3$).

From crack growth resistance calculations summarized above for both batches, several points must be clarified, namely : (i) the unique K_R-curve shape obtained, when plotting crack growth resistance values versus relative matrix crack lengths for CT specimens of LUS material ($W \geq 20$) and HUS specimens ($W = 20$ and 40 mm), (ii) the R-curve behavior regarding damage micromechanisms (the matrix cracking and the fiber breakage) and CT sizes, (iii) the slightly lower crack growth resistance calculated using smallest size specimens (figures 17 and 21) and R-curves with no evidence of a plateau like value, despite the large size of provided ligaments. To answer these questions, one must involves, on one hand, analysis of load versus displacement relationships, in conjunction with corresponding matrix crack lengths. On the other hand, load point displacement versus matrix crack lengths curves.

(i) Concerning the first point related to the unique K_R - a/W shape. Explanations come when plotting relative matrix crack lengths at the peak load, $(a/W)_{Pmax}$, versus specimen sizes endowed with the same NDR (figure 22). As it could be shown, the maximal load, P_{max}, which shows the beginning of the bundles breakage, is always occurring when the relative matrix crack length is half the total ligament $((a/W)_{Pmax} = W/2)$. An obvious explanation could be inferred : there exists an homothetic effect which governs P_{max} and its corresponding matrix crack length relationship. In other words, whatever the CT size, up to the peak load (P_{max}), the matrix crack is entirely bridged by the bundles of fibers. This totally bridged regions (which equals W/2 at P_{max} for all the specimens) corresponds to the domain where K_R-a/W curves behave as $\sqrt{(a/W)}$. Such a result would say that on 2D woven SiC-SiC composites, significant pieces of information about the bundles bridging mechanism could be extracted from the « useful » part of K_R-a/W curves. The « useful » part of K_R curves, being the region of behavior as $\sqrt{(a/W)}$ (figure 21). Nevertheless, the result summarized above (figure 21 and 22) demonstrates also that geometrical effects governs load displacement relationships, thus affecting crack growth resistance shapes and values.

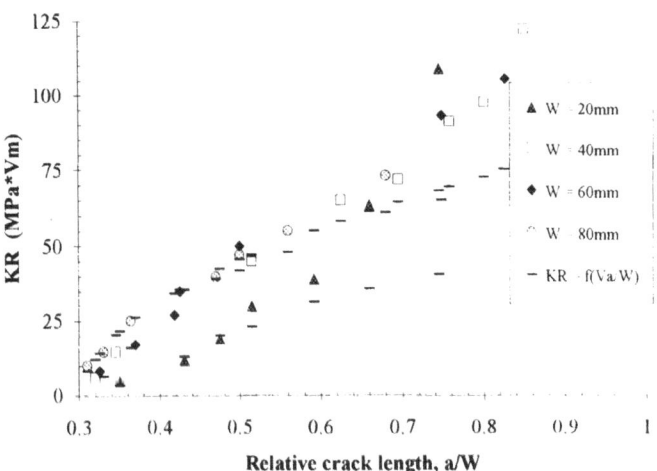

Figure 21. K_R values plotted as a function of the relative matrix crack lengths, a/W, for CT sizes, W = 20, 40, 60 and 80 mm (a_0/W = 0.3).

(ii) Analyzing R-curves with respect to the mechanical behavior and damage micromechanisms (figure 4, 20), shows that, as long as the matrix crack is wholly bridged by the bundles of fibers (up to the peak load, P_{max} ; figure 4), crack growth resistance values are the same for all CT sizes. But, beyond the peak load, when the bundles of fibers began to fail, the resistance values calculated using the smallest specimen start progressively moving apart (with a parabolic shape) from the crack growth resistance values calculated using larger sizes and so on.

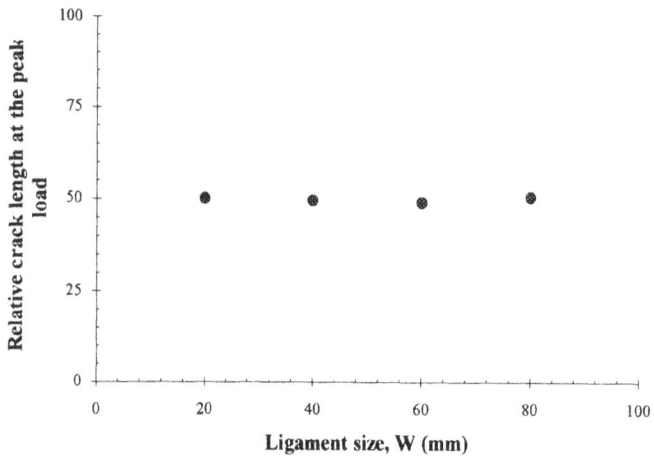

Figure 22 . Relative matrix crack lengths at the peak load versus the CT size, for CT sizes, W = 20, 40, 60 and 80 mm (with the same NDR, $a_0/W = 0.3$).

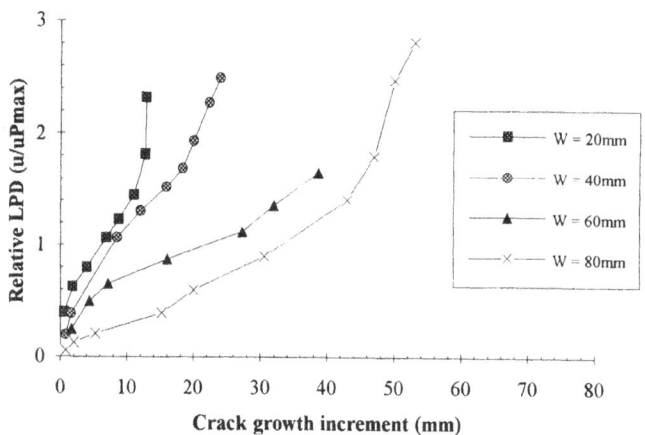

Figure 23 . Relative load point displacement, (u/u_{Pmax}), plotted as a function of the crack growth increment, for all LUS CT sizes $(a_0/W = 0.3)$.

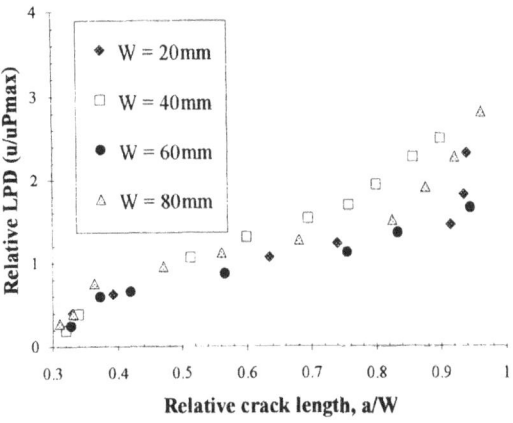

Figure 24. Relative load point displacement versus relative crack lengths, for all LUS CT specimen sizes with a saw cut length, $a_0/W = 0.3$.

Load point displacement (LPD) versus crack lengths relationships (figure 23, 24, and 25) furnish numerous parallels with the K_R - curve behavior. This could be noticeable when plotting relative LPD values (u/u_{Pmax}) versus matrix crack growths (Δa) and relative matrix crack lengths (a/W), respectively (figures 23 and 24). Indeed, along the entirely bridged region (*ie* up to the peak load), LPD values behave as $\sqrt{(a/W)}$ (figure 25) and get a parabolic increase beyond, due to the failing of the bundles of fibers.

On the basis of LPD results, it would be possible, by analogy with K_R-a/W curves, to define the « useful » part of LPD-a/W curves as being the region of behavior as $(a/W)^{1/2}$ (figure 24, 26). This significant region could be then coupled to damage micro-mechanisms, especially to the fiber bridging mechanism in 2D woven SiC-SiC composites. Arguments pleading for this assumption come when referring to the works of [32] on steel fibers reinforced mortar. In such materials, the crack starts propagating regardless of its length when its opening angle (COA) reaches a critical value, called critical crack opening angle (CCOA). In 2D woven SiC-SiC composites, the evolution of LPD values (or COA) as a function of the matrix crack is not a linear relationship (figure 23, 24 and 25), but behaves as $\sqrt{a}$ along the wholly bridged region (ending at P_{max}). Considering the fact that a linear shape is realistic only if the crack is allowed to propagate without constraint[32], it is therefore understandable that the micro-mechanisms acting in the wake of the matrix crack such as fiber-matrix debonding and the extensive sliding/friction along the debonded regions are responsible of the curved shape (especially since in this composite, the interface is radially under residual compression). When at a first approximation, all damage micromechanisms are put into the bundles bridging contribution, several pieces of information could then be gathered by coupling LPD-a/W and K_R-a/W results.

Let us now analyze the mechanical behavior of the composite with regards to the compact tension specimen geometry. In this way, it appears that up to the peak load, CT specimens behave as notched plates subject to uniaxial tensile loading. When the yarns began to fail (at the peak load), a significant increase in the crack opening angle is recorded. Accordingly, beyond P_{max}, CT specimens are put into bending. Starting from this reasoning, the rotation point seems to be located at half the ligament ($W/2$). Its occurrence is dictated by the beginning of the failure of the yarns. The stress-strain field at the front of the matrix crack is thus perturbed by the bundles breakage in the wake of the matrix crack, leading to a virtual increase of K_R and LPD values. Such a behavior is illustrated by the parabolic increase of K_R and LPD values (figure 26). In fact, this result could be foreseeable, considering that the stress intensity factor, which describes the stress-strain field in the front of the matrix crack cannot takes into account damage micromechanisms acting in the wake of the matrix crack, such as yarns failure.

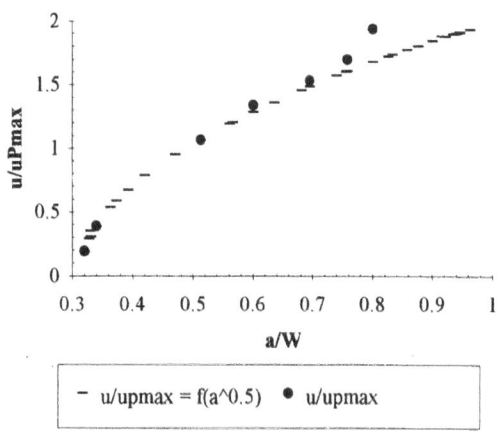

Figure 25. Experimental and theoretical relative load point displacement plotted as a function of a/W for a CT size, W = 40 mm.

(iii) Since LEFM has been proposed to study the crack growth resistance in 2D woven SiC-SiC composite materials, one must find a theoretical issue of the absence of a plateau like value in K_R - curves and the slightly lower crack growth resistance values obtained on smallest size specimens. A rational one comes when considering, at a first approximation, the damaged zone in CMCs analogous to the plastic zone in steel specimens. Namely, this crude assumption involves consideration of plane strain or plane stress loading conditions in these composites. It emerges from the fundamental works on the subject, that the meaningful use of LEFM is based on the size of the damaged zone, associated with a discontinuity (or a crack), should be small, so that the linear elastic stress intensity factor could reflect the conditions of strain and stress in the vicinity around the crack. It is therefore important that the constraint relieving influence on the free faces of the specimen should be minimized and the plane strain conditions should exist across the major portion of the crack tip. These conditions seem to be satisfied when analyzing R-curves obtained by[33] on DCB specimens of polycrystalline alumina (with a mean grain size of 20 μm and a thickness, B = 3 mm). Indeed, R-curves established on these specimens show the existence of a plateau-like value. What's about in CMCs ? R-curve with a plateau-like value in CMCs with a sufficiently weak interface to promote fiber-sliding rather than fiber breakage, means that conditions close to plane strain ones should exist and that a rate of physical displacement of a stationary bridging zone size is achieved. This is visibly not the case with the thickness B of 3 mm provided by these CT specimens. On the basis of ASTM standards E-399-81 specifications, a ratio $W/B = 2$ is recommended for establishing meaningful R-curves. Accordingly, considering that on provided smallest size specimens, the ratio W/B (which equals 6.6 and 3.3 for the thickness $B = 3$ and 6 mm, respectively), ties up to ASTM specifications more than for larger size specimens, this leads to the assumption that crack growth resistance values obtained on the smallest specimen size are more significant than the ones calculated using larger CT sizes. Two main points plead in favor of this assumption, namely :

• K_C^{ini} values show considerable variation when calculated using larger CT sizes (about three times the value calculated on smaller sizes). This variation could be related to plane stress loading conditions, as shown by [34].

• K_R curves of the HUS batch of 2D SiC-SiC composites seem to be influenced by the specimen thickness. This could be noticeable when plotting crack growth resistance values calculated using HUS CT specimens having the same size, $W = 20$ mm, different thickness ($B = 3$ and 6 mm) and notched at a NDR, $a_0/W = 0.4$ (figure 27).

To get deeper on this assumption, further efforts have to be made towards testing small CT specimens ($W = 20$ mm) with increasing thickness values, (up to $B = 10$ mm). If a significant variation on crack growth resistance values is noticed along the region of behavior as $a^{1/2}$ of K_R curves, the diagram showing large scale bridging effects on the toughness enhancement proposed by [5] would be reestablished by adding a third dimension : namely, the W/B ratio. Effects of large scale bridging would then be analyzed on surfaces combining : X_B / a, a / W and W / B ratios.

On the basis of the above results on 2D woven SiC-SiC composites, three main points must be underlined : (i) LEFM use is meaningful only along the region of P-u curves ending at the peak load, due mainly to the low gap to the linear elastic behavior and the absence of the bundles breakage in the wake of the matrix crack, (ii) K_R - curves of 2D SiC-SiC composites, calculated using CT specimens, must be regarded as a function of the relative crack length, a/W, rather than the crack length only (due to geometrical effects) and (iii) the most prevailing and limiting parameter when establishing R-curves seems to be the ratio : size to thickness (W/B), rather than the ligament alone.

CONCLUSIONS

The mechanical behavior of 2D woven SiC-SiC composites had been assessed using compact tension specimens of different sizes. The main points deriving from the present study are :

The measured matrix crack lengths on specimen surfaces need polishing, which affect partially external yarns and makes crack path tortuous rather than straight-line in the volume as demonstrated by [8]. However, similar results are obtained from both measures[9]. This special point concerning the definition of a meaningful crack entity in CMCs had been much debated, due mainly to the fact that failure is not due to a dominant mode I crack propagation[35], but to a number of damage micromechanisms acting simultaneously. Nevertheless, this seems to be a non-problem in the case of 2D SiC-SiC composites, considering that on polycrystalline ceramics, crack paths are also sinuous due to interlocking grains[36]. So, when scaling up in the particular case of 2D woven SiC-SiC composites, mode I matrix crack propagation could be a meaningful assumption, though matrix microcracking and fiber breakage micromechanisms. The main limitation comes from the difficulty to transpose this direct crack measure based microscope to high temperature tests because the uneasiness estimate. An alternative to direct

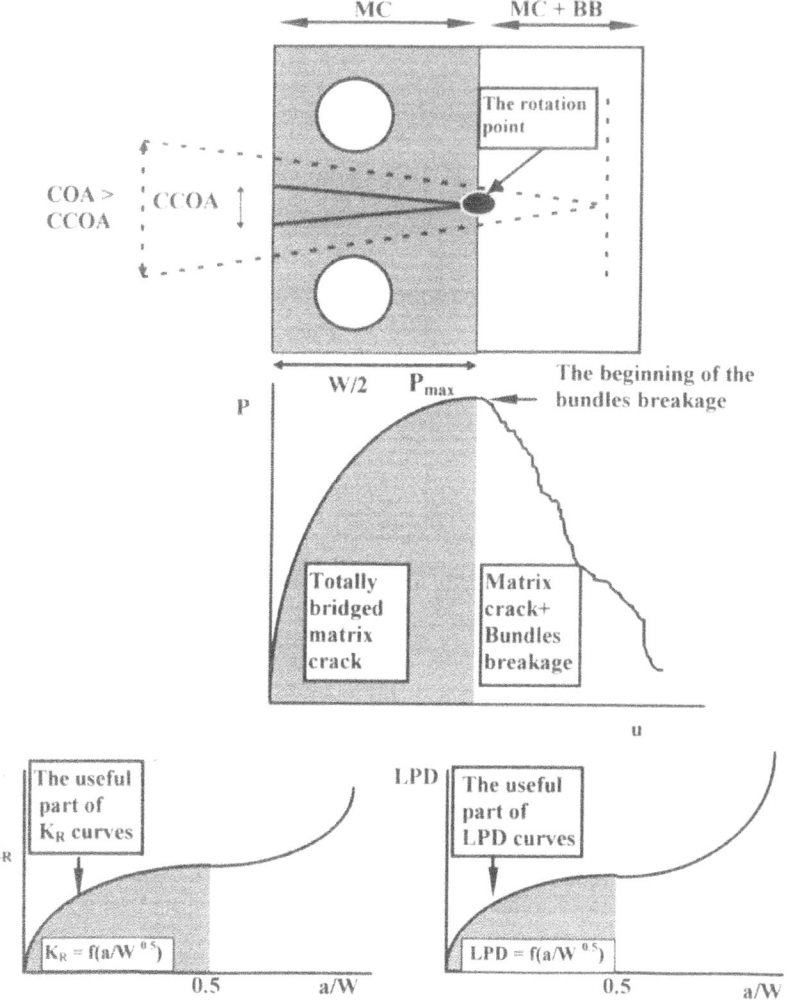

Figure 26. Illustration of the evolution of the crack opening angle (COA) as the matrix crack propagates. Up to the peak load, only matrix cracking (MC) is recorded, the matrix crack which equals W/2 is totally bridged by the bundles. The corresponding part of K_s and LPD values behave as $a/W^{0.5}$ could be directly related to the bundle bridging mechanism. Beyond the peak load, when the bundles began to break (BB), a parabolic increase of K_s and LPD values is recorded.

measurement could be provided by establishing compliance calibration curves, but there will remain the sampling problem.

R-curves calculated using CT specimens of 2D SiC-SiC composites, rather reflecting the intrinsic crack growth resistance, are test geometry dependents. This seems to be resulting from the specimens low thickness, which lead to solicitations under conditions analogous to plane stress ones in the case of steel specimens. Nevertheless, two main points should be emphasized : (i) crack energy initiation values obtained on smallest CT size could be of significant use and (ii) present results would indicate that meaningful crack growth resistance values could be calculated using smallest size specimens (W = 20 mm) having thickness values,

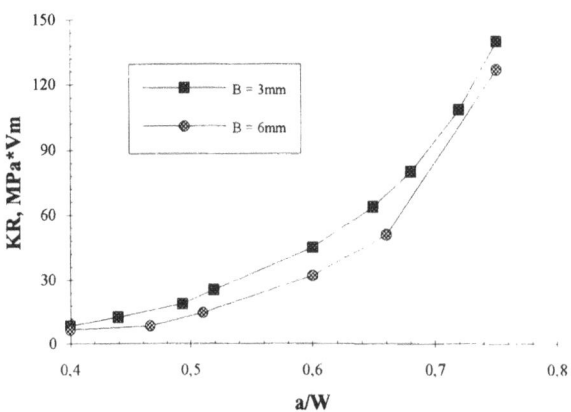

Figure 27.   Influence of the specimen tickness on the R curve behavior of two CT specimens of HUS material
(W = 20mm ; with a tickness, B = 3 and 6 mm and the same saw cut length, a_0/W = 0.4).

$B \geq 3$ mm. Such an assumption must be verified. It seems useless to test larger size specimens (W > 20 mm), due to the fact that, ensuring plane strain specimen geometry is equivalent to test much thicker specimens. At a laboratory   scale, the unrealistic of such tests is easily understandable.

The main question now is, how to devoid geometrical effects and characterize the resistance of such composite materials against crack propagation ? The philosophy of the consistent answer is, a methodology which is, on one hand, explicitly and closely related to CMCs constituents (matrix, fiber and interface). On the other hand, since CMCs are hinted to be use in aeronautical structural parts in low thickness, it is therefor time to focus on testing specimens endowed with flaws reflecting "realistic" and probable ones (as holes). The results inferring from such a methodology would provide a practical and a useful implement to the designers. Recently, two attempts had been proposed and are based on testing, under uniaxial tensile loading, samples with machined holes[37] or double-notched specimens[38,39]. The former, with all reserve that side effects are devoid, had the advantage being reflecting the most probable occurring flaws in machined CMC structures : rivets. Since the latter, based on considering the fiber-matrix tractions as a composite characteristic and despite the numerous simplifications, provides crack growth resistance values in agreement with those calculated using CT specimens[39]. The main difficulty resides on preserving the specimen symmetry by developing matrix cracks of equal lengths from the two lateral notches.

ACKNOWLEDGMENTS

This work is a part of the PhD thesis of Monssef DRISSI-HABTI, defended on the 13th October 1994. It had been performed within the joint French program GS4C (Comportement thermomécanique des composites céramiques-céramiques à fibres), supported by CNES, CNRS, DRET, MRE and Aérospatiale, SEP and SNECMA companies. We thank mainly the CNRS for the financial support and the SEP for delivery of the specimens. The author thanks also Pr. D. Rouby for fruitful discussions.

REFERENCES

(1) Hertzberg R. W., Deformation and Fracture Mechanics of Engineering Materials (Wiley, New York, 1983), Chaps. 7-9.

(2) Broek D., Elementary Engineering Fracture Mechanics (Nijhoff, The Hague, 1982), Chap. 5.

(3) Myajima T., Sakaï M.,"Fiber bridging of a carbon fiber-reinforced carbon matrix lamina composite", J. Mater. Res., vol. 6, n°3, 1991.

(4) Marshall D. B., Evans A.G., J. Amer. Ceram.Soc., 68, 225, (1985).

(5) Zok F.W., Hom L.M., "Large scale bridging in brittle matrix composites", Acta metall., 38, 1895-1904 (1990).

(6) Zok F.W., Hom L.M., "Mode I fracture resistance of a laminated fiber-reinforced ceramic", J. Amer. Ceram. Soc., 74 [1] 187-93 (1991).

(7) Bouquet M., Thèse de l'Université de Bordeaux, 1989.

(8) Navarre G., "Etude des mécanismes d'endommagement et de rupture des matériaux composites à fibres et à matrice céramique", Thèse de l'INSA de Lyon, April 1990.

(9) Rouillon M.H.,"Résistance à la propagation de fissure de matériaux composites composites céramiques SiC/C/SiC 2D", Thèse de l'Université de Caen, February 1993.

(10) Drissi-Habti M., Thèse de l'Université de Caen, October 1994.

(11) Tada H., Paris P. C., Irwin G. R., The Stress Analysis of Cracks Handbook. Del Research Corp., St Louis, Mo, 1985.

(12) Conchin F., Thèse de l'INSA de Lyon, February, 1994.

(13) R'mili M.,"Application de la mécanique de la rupture au composite carbone-carbone bidirectionnel", Thèse de l'INSA de Lyon, 1987.

(14) Sakaï M., Myajima T., Comp. Sci. Tech., 40, 231, (1991).

(15) Sakaï M., Bradt R.C., Fracture Mechanics of Ceramics, edited by R. C. Bradt, A. G. Evans, D. P.H. Hasselman, and F.F. Lange (Plenum Press, New York, 1986), vol. 7, p. 127.

(16) Gomina M., Rouillon M.H., "The crack growth resistance of SiC-SiC ceramic-ceramic composite materials", Frac. Mech. Ceram., vol. 9, edited by Bradt et al., New York, 1992.

(17) Droillard C., Voisard P., Heibst C., Lamon J., "Matrix cracking as a toughening mechanism in ceramic matrix composites", HT-CMC 1, pp. 473-481, Bordeaux, France, September 1993.

(18) Inghels E., Thèse de Docteur-Ingénieur, Ecole des Mines de Paris, 1987.

(19) Cox N.B., "Extrinsic factors in the mechanics of bridged cracks", Acta metall. mater., vol. 39, N° 6, pp. 1189-1201, 1991.

(20) Bernhart G., Lamicq P., Mace J., "Reliability of Ceramic-Ceramic Composites", Industr. Ceram., 790, p 51, 1985.

(21) Lamicq P., Bernhart G., Dauchier M., Macé J. G., "SiC-SiC Composite Ceramic", Am. Ceram. Soc. Bull., 65 [2], p 336-338, 1986.

(22) Abbé F., Chermant L., Coster M., Gomina M., Chermant J. L., "Morphological characterization of ceramic-ceramic composites by image analysis", Comp. Sci. Tech., 34, 37, p 109, 1989.

(23) Cojean D., Monthioux M., Oberlin A., in Fantozzi G. and Fleishman P. (Eds.), JNC-7, Comptes rendus des septièmes Journées sur les Matériaux Composites, Lyon, November 6-8 1990, AMAC Publications, Paris, p. 381, 1990.

(24) Aubard X., Lamon, J., Allix O., « Model of Non Linear Mechanical Behavior of 2D SiC-SiC Chemical Vapor Infiltration Composites », J. Amer. Ceram. Soc., 77 [8] 2118-26 (1994).

(25) ASTM Standard E-399-81, 1983, Annual Book of ASTM standards, part 10, American Society for Testing and Materials, Philadelphia, PA.

(26) Srawley J. E., "Wide range Stress Intensity Factor Expressions for ASTM E399 standard Fracture Toughness Specimens", Int. J. Fract., 12, 475-476, 1976.

(27) Marshall D. B., Cox B.N., Evans A.G., "The mechanics of matrix cracking in brittle matrix fiber composites", Acta metall. mater., 33, 2013, 1985.

(28) McCartney L. N., Proc. R. Soc. London A-409, 329 (1987).

(29) Saxena A. Hudak S. J., Jr., Int. J. Fract. 14 (5), 453 (1978).

(30) Hsueh C.H., Becher P.F., "Evaluation of bridging stress from R-curve behavior for nontransforming ceramics", J.Amer. Ceram. Soc., 71 [5] C-234-C-237 (1988).

(31) Drissi-Habti M., "R-curve behavior of compact tension SiC-SiC composite materials", Topic on the 3rd Euro-Ceramics Conference, Madrid Spain, September 1993.

(32) Visalvanish K., Naaman A., « Fracture Model for Fiber Reinforced Concrete », Technical paper, ACI Journal, Title n° 80-14, p. 128-138, Marsh-April 1983.

(33) Steinbrech R.W., Reichl A., Schaarwächter W., "R-curve behavior of long cracks in alumina", J. Amer. Ceram. Soc., 73 [7] 2009-2015 (1990).

(34) Andrianopoulos N.P., Boulougouris V.C., « On a intrinsic relationship between plane stress and plane strain critical stress intensity factor », Int. Journ. of Fracture, 67 : R9 - R12, 1994.

(35) Evans A.G., "Ceramic matrix composites : challenges and opportunities, AGARD Workshop "Introduction of Ceramics into Aerospace Structural Composites", Antalya, Turkey, April 1993.

(36) Drissi-Habti M., M. Gomina M., "Crack growth resistance values calculated from natural crack lengths", Journal of Alloys and Compounds, 188, 259-63, 1992.

(37) Heredia F.E., Spearing S.M., He M.Y., Mackin T.J., Bronsted P.A., Evans A.G. and Mosher P., J. Amer. Ceram. Soc., in press.

(38) Cady C. Evans A.G., « Silicon carbide/Calcium aluminosilicate : a notch insensitive ceramic-matrix composite », J. Amer. Ceram. Soc., 78 (1) 77-82 (1995).

(39) Brenet P., Conchin F., Rouby D., "Direct measurement of the bridging tractions versus crack opening displacement in fibre reinforced ceramic matrix composites. A new approach of fracture behaviour and microstructural characteristics », CCC III, Mons, Belgium, 1994, to be published in Silicates Industrielles.

317

HIGH TEMPERATURE FRACTURE RESISTANCE
AND STRENGTH OF CERAMIC COMPOSITES
AFTER THERMAL EXPOSURE AT 1200 °C

J. A. Celemín, J. Y. Pastor, J. LLorca, M. Elices, and A. Martín

Department of Materials Science
Polytechnic University of Madrid
E. T. S. de Ingenieros de Caminos. 28040 - Madrid, Spain

INTRODUCTION

The reinforcement of ceramic matrices with high strength ceramic fibers is aimed at manufacturing composite materials which combine the excellent properties of ceramics (low density, chemical stability, high wear and thermal shock resistance as well as high melting point) with a flaw insensitive, ductile behaviour. However, fiber-reinforced ceramics (FRC) cannot compete with standard metallic alloys in structural elements at ambient temperature for obvious reasons. Their applications have to be found at very high temperatures and in aggressive environments, where they exhibit evident advantages over their metallic counterparts. Industria de Turbo Propulsores (ITP), a company involved in the design and manufacture of engine components for the aircraft industry, has promoted and financed a research project to study the use of FRC under such conditions. This paper presents some results obtained within the framework of this project, which is aimed at evaluating the potential of the FRC as materials for very high temperature structural components in jet engines.

The critical factor to obtain a tough composite is the nature of the fiber/matrix bonding. If the fibers are strongly bonded to the matrix, a crack nucleated in the matrix breaks the fibers as it propagates, and the composite fails catastrophically. On the contrary, when the fiber/matrix interface is weak the cracks propagate upon loading through the matrix without breaking the fibers. The non-broken fibers in the crack wake slide with respect to the matrix, and the elastic energy stored in the fibers, together with the energy employed in pulling-out the fibers which are eventually broken within the matrix, reduce the energy available to propagate the crack. These toughening mechanisms have been extensively documented at ambient temperature, where fiber-reinforced ceramics can reach fracture toughness similar to that of high strength Al alloys (Hillig, 1987; Curtin, 1991; LLorca and Singh, 1991). Different experimental studies have demonstrated, however, that the strength of FRC was significantly reduced above 1000°C and that this change was accompanied by a ductile to brittle transition in the failure mode (Mah et al., 1987; Singh, 1993; Woodford et al., 1993; Xu et al., 1995a).

The reduction in the mechanical properties of FRC at elevated temperatures has usually been attributed to the degradation of the fiber/matrix interface by oxidation (Gomina *et al.*, 1992; Woodford *et al.*, 1993). This hypothesis was supported by experimental observations which showed that the fiber coating deposited to aid fiber/matrix decohesion disappeared by oxidation during high temperature exposure (Cojean and Monthioux, 1992) and that the interfacial strength increased with temperature (Heuer *et al.*, 1990). However, other authors postulated that the drop in mechanical properties was also due to degradation of the ceramic fibers or even that fiber degradation was more important than interface oxidation (Mah *et al.*, 1987; Singh, 1993; Fareed, 1993; Fareed and Schiroky, 1994). In this respect, there is experimental evidence of the strength reduction in Al_2O_3 and SiC fibers above 1000^0C (Simon and Bunsell, 1984; Clark *et al.*, 1986; Di Carlo, 1991).

It should also be noticed that the amount of experimental data on the high temperature mechanical properties of FRC is limited, mainly due to the intrinsic problems of high temperature testing. This is specially true for the fracture toughness and fracture resistance, which have been determined above 1000^0C in only a few studies (Gomina *et al.*, 1992; Nair and Wang, 1992; Fareed *et al.*, 1993; Xu *et al.*, 1995b). Thus the effects of high temperature exposure on the composite microstructure and failure mechanisms are not well understood for many composite systems, even though this information is critical in the manufacture of new materials with good damage tolerance at high temperatures. In addition, the design of structural components requires a knowledge of the influence of temperature and exposure time on the strength and fracture resistance of these composites. The influence of the temperature (1200^0C) and exposure time (1 and 100 hours) on the flexure strength and fracture resistance in two SiC-fiber-reinforced ceramic-matrix composites is studied in this paper. The results of the mechanical tests, together with detailed analyses of the failure mechanisms by means of quantitative microscopy, were used to elucidate the relationships between the macroscopic behaviour and the microstructural features.

MATERIALS

The first material studied was a SiC matrix bidirectionally (0^0-90^0) reinforced with 40 vol. % Nicalon SiC fibers. The preform was manufactured by stacking together several layers of Nicalon plain satin weave fabric. A very thin layer ($\approx$ 0.1 μm) of pirolitic C was deposited on the fiber surface, and the SiC matrix was introduced into the preform by the CVI process. The composite was received in the form of prismatic bars of 10 mm x 3.5 mm cross-section. Porosity was around 10-12%. The second material was an Al_2O_3 matrix bidirectionally (0^0-90^0) reinforced with 37 vol. % Nicalon SiC fibers. The preform was manufactured by stacking together several layers of Nicalon harness satin weave fabric. The fibers were coated by CVI with a very thin ($\approx$ 0.4 μm) layer of BN and afterwards with a thicker layer of SiC (in the range 3-4 μm) onto the BN. The preform was then brought into contact with molten Al in air. The Al reacted with the oxygen to form a matrix of Al_2O_3 which grows into the preform. Finally, the residual Al was removed from the Al_2O_3 matrix (Fareed, 1993). The composite was received in the form of prismatic bars of 10 mm x 3 mm cross-section. Porosity was around 7-8%. The polished cross-sections of both composites can be observed in Fig. 1.

EXPERIMENTAL TECHNIQUES

Flexure and fracture tests were carried out on a ceramic three-point bend testing fixture with 50 mm loading span and with the specimen edgewise on the fixture (*W*

= 10 mm). The specimen and the fixture, placed in the high temperature furnace, were loaded through two alumina rods connected to the actuator and to the load cell, respectively, of a servo-mechanical testing machine. The external ends of the rods were water-cooled to avoid overheating the actuator and the load cell.

Flexure and fracture tests were carried out on each material in three different conditions: ambient temperature and 1200^0C after 1 hour and 100 hours exposure at this temperature. The specimen temperature was controlled by two B-type thermo-couples in contact with the specimen. The heating rate was 12^0C per minute and, as indicated above, the specimen was held at the test temperature (1200^0C) during 1 or 100 hours prior to testing. All the tests were performed under stroke control, with a cross-head speed of 50 μm per minute.

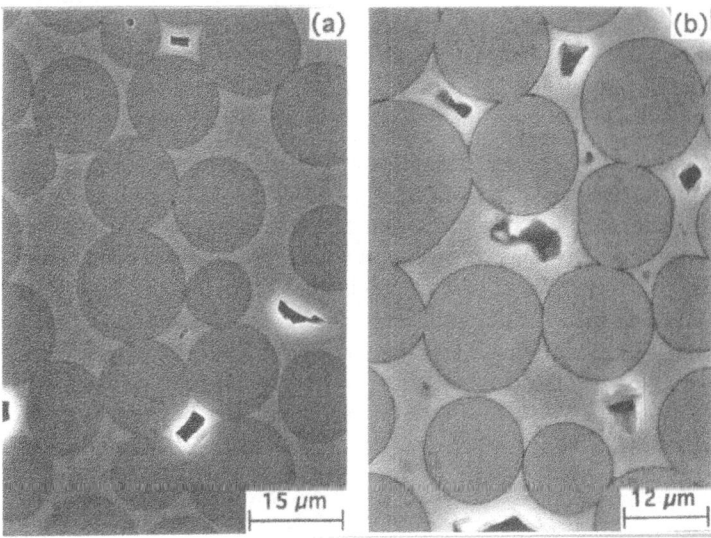

Figure 1. Polished cross-sections of the composites in the as-received condition. (a) SiC/SiC. (b) Al_2O_3/SiC.

The load (P) and the midspan deflection (δ) of the prismatic bar with respect to the supporting rollers were continuously monitored during the flexure tests, the latter through a laser extensometer which can be used at high temperature. This extensometer was a low power (< 1 mW) He-Ne laser emitter, which sends a scanning laser beam through two silica windows across the furnace walls. The laser beam is limited by the specimen and the SiC fixture and the resulting intensity distribution is measured by a detector placed on the opposite side of the furnace. A microcomputer processes the intensity distribution to determine the width of the laser beam and thus the deflection of the specimen during the test.

Although the theoretical resolution of the system is very high, the accuracy was limited by two factors which lead to undesired fluctuations of the optical properties of air: the air seepage around the loading rods and the thermal gradients along the optical path. Preliminary calibrations of the extensometer indicated that of the two air seepage was dominant in our system, perhaps because the large size of the furnace led to smaller temperature gradients along the laser path (Pastor et al., 1993 and 1995a). Thus special care was taken to reduce the chimney effect as much as possible by closing the gaps between the loading rods and the furnace with high-temperature frictionless seals made of heat-resistant alumina tissue. The good thermal insulation of the furnace provided by the 15 cm-thick insulating walls allowed the laser source and detector to be located very close to the silica glass without danger of overheating, reducing the fluctuations produced outside the furnace. In addition to all these precautions, the measurements were averaged over 150 scans for each displacement to minimize the random noise.

A notch of around 2 mm in length and 150 μm in thickness was cut in the prismatic bars with a very thin diamond wire for the fracture tests. Three magnitudes were measured during the tests, namely the load (P), the crack mouth opening displacement ($CMOD$), and the cross-head displacement of the testing machine with respect to the frame (v). The $CMOD$ was determined by two alumina pins glued symmetrically to the notch mouth on the tensile surface of the specimen. The distance between the pins was obtained through the laser extensometer from the shadow projected by the pins on the detector. The cross-head displacement, v, was monitored by means of a linear-variable differential transducer placed outside the furnace.

Once broken, the fracture surfaces were examined in the scanning electron microscope to determine the dominant fracture mechanisms for each material and temperature. In addition, the specimens were sliced far away from the fracture surface perpendicularly to their longest dimension with a low speed diamond saw. The surfaces were polished successively on diamond cloths of 40, 9, 3 and 1 μm grain size and finally on alumina with 0.3 μm grain size. They were cleaned during 30 minutes by ultrasound in acetone to remove the alumina from polishing and were then observed in the scanning electron microscope. The polished surfaces of the Al_2O_3-based composites were sputtered with Au-Pd during three minutes before being introduced in the microscope.

RESULTS

Mechanical Properties

Representative load-deflection ($P - \delta$) curves measured during the flexure tests are shown in Figs. 2a and 2b for the SiC/SiC and the Al_2O_3/SiC composites, respectively. The flexure strength, σ_u, was calculated from the maximum load in these curves following the Strength of Materials theory for an elastic beam. The magnitude of σ_u in all the tests is plotted in Fig. 3.

The results in these Figs. show that the SiC/SiC material exhibited excellent properties at ambient temperature. However, the behaviour at 1200 ^{0}C was disappointing and the flexure strength was reduced from 450 MPa to less than 200 MPa after 1 hour of exposure at 1200 ^{0}C. The degradation of the mechanical properties was mainly dependent on the temperature and not on the exposure time: the mechanical behaviour after 1 and 100 hours of exposure was very similar within the experimental scatter. The Al_2O_3-based composite also presented remarkable properties at ambient temperature which were partially maintained at 1200 ^{0}C, and the flexure strength at 1200 ^{0}C (between 300 and 350 MPa) was significantly higher than in the SiC-matrix material. Exposure during 100 hours to 1200 ^{0}C led to a small reduction in strength but the effect of exposure time was less.

Three load-crack mouth opening displacement ($P - CMOD$) curves obtained

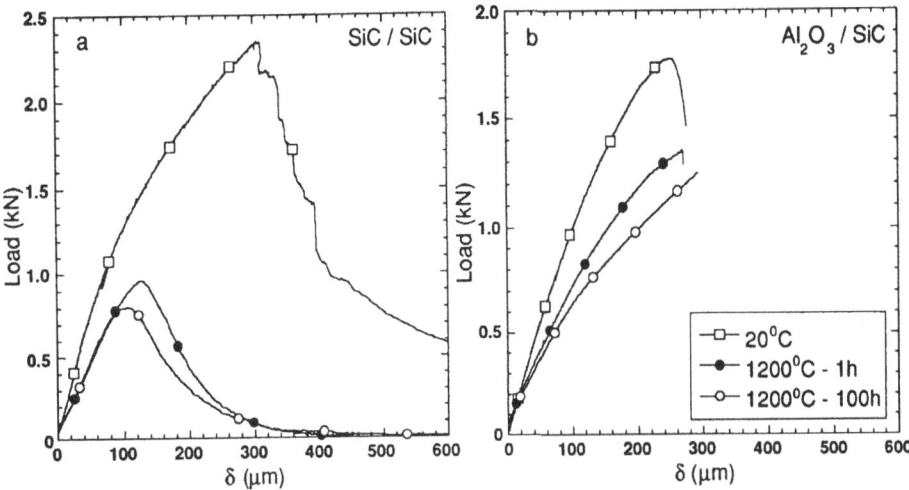

Figure 2. Load-deflection curves $(P - \delta)$ during the flexure tests. (a) SiC/SiC. (b) Al$_2$O$_3$/SiC.

from the fracture tests at 20 °C and 1200 °C on the SiC/SiC composite are shown in Fig. 4a. Similar curves for the Al$_2$O$_3$/SiC composite are plotted in Fig. 4b. These tests confirmed the results obtained in the flexure tests. Both materials presented an excellent behaviour at ambient temperature, and the $P - CMOD$ curves exhibited a non-linear zone before the maximum load, which is found in ductile, damage tolerant materials. In addition, the load drop after the maximum load was gradual and the complete failure of the specimens took place for cross-head displacements over 2 mm. The properties of both materials dropped at 1200 °C, but the reduction in the Al$_2$O$_3$/SiC composite was much less marked than in the SiC/SiC material. Exposure during 100 hours at 1200 °C again reduced slightly the properties of the Al$_2$O$_3$-matrix composite, while the SiC-matrix material was not further degraded by the heat treatment.

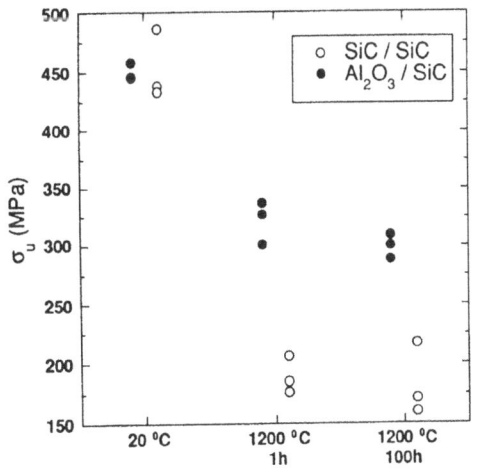

Figure 3. Influence of the temperature and exposure time on the flexure strength of both composites.

The nominal fracture toughness, K_Q, was calculated from the maximum load in the fracture tests, P_u, and the initial notch length, a_0, according to (Pastor *et al.*, 1995b):

$$K_Q = \frac{3\,P_u\,L\,\sqrt{\alpha}\left[1.9179 - 1.2795\alpha + 3.3532\alpha^2 - 3.2260\alpha^3 + 1.2235\alpha^4\right]}{2B\,W^{3/2}\,(1-\alpha)^{3/2}(1+2\alpha)} \tag{1}$$

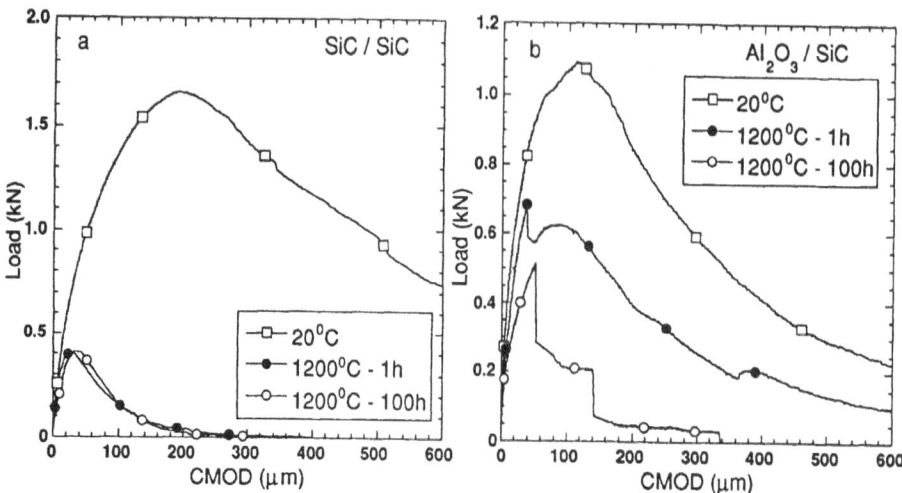

Figure 4. Load-crack mouth opening displacement curves $(P - CMOD)$ during the fracture tests. (a) SiC/SiC. (b) Al_2O_3/SiC.

where L is the span, B the specimen thickness, and $\alpha = a_0/W$. This expression is valid for $0 < \alpha < 1$ when $L/W = 5$. It should be noted that K_Q cannot be considered, properly speaking, a material property (the fracture toughness) because the fracture process zone, where fiber bridging and pull-out take place, was not small enough as compared to the characteristic dimensions of the specimen, mainly a_0 and W. Under such conditions, K_Q is dependent on the specimen size and notch length (Llorca and Elices, 1993) and very large specimens would be needed to obtain *true* fracture toughness values independent of the specimen geometry and size. Even so, K_Q can still be used to compare the resistance to crack initiation of samples whose size is similar to the specimens used in this investigation.

The specific fracture energy, G_F, was also calculated from the fracture tests as the energy spent to create a unit area of free surface. Mathematically,

$$G_F = \frac{1}{B\,(W - a_0)} \int P\,dv \tag{2}$$

where the integral stands for the area under the load *vs.* cross-head displacement curve, e.g. the energy supplied to break the specimen. While K_Q stands for the resistance to crack initiation, G_F provides an average value of the resistance to crack propagation during the whole fracture process. Thus both magnitudes, K_Q and G_F, characterize the fracture behaviour of the composites: they are plotted in Figs. 5a and 5b respectively. At ambient temperature, both materials presented an outstanding resistance to crack initiation for a ceramic with K_Q values around 25 MPa$\sqrt{\text{m}}$. The specific fracture energy was also very good, specially for the SiC/SiC composite. The properties of both materials dropped at 1200 ^{0}C, and the reduction in the Al_2O_3/SiC composite was again much less marked than in the SiC/SiC material. In fact, K_Q and G_F for the Al_2O_3-matrix composite at 1200 ^{0}C were still much higher than those measured for any monolithic ceramic at this temperature. On the contrary, the resistance to crack initiation in the SiC/SiC composite at 1200 ^{0}C (around 6.5 MPa$\sqrt{\text{m}}$) was similar to the values reported in the literature for monolithic ceramics with elevated fracture resistance. The lack of ductility in the SiC/SiC composite at 1200 ^{0}C was also observed in the deflection under maximum load (δ_u) during the

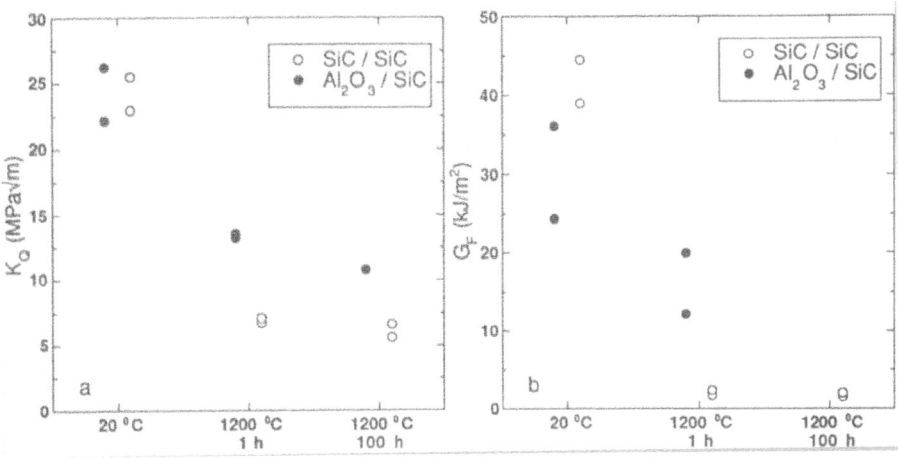

Figure 5. Influence of the temperature and exposure time on the (a) fracture toughness, K_Q. (b) fracture energy, G_F. The magnitude of G_F for the Al_2O_3-matrix composite at 1200 ^{0}C - 100 h was not obtained because the tests were not stable.

flexure tests. δ_u was almost constant in the Al_2O_3-matrix material but fell from 300 μm to 100 μm at 1200 ^{0}C in the SiC/SiC composite (Figs. 2a and 2b).

Failure Mechanisms

As was expected, the fracture surfaces of the SiC/SiC composite at 20 ^{0}C and 1200 ^{0}C were quite different (Fig. 6). Fibers pulled out from the matrix were observed in the former, while fiber fracture normally took place in the crack plane at 1200 ^{0}C. The mechanisms of fiber fracture also changed with temperature. At ambient temperature, the pulled-out fibers in the SiC/SiC composite failed in tension from surface defects (up to 1 μm), and the fiber fracture surfaces were abrupt (Fig. 7a). On the contrary, the fibers broken in the crack plane at elevated temperature presented a specular fracture surface with no trace of defects on the fiber surface (Fig. 7b), which indicates that the fiber was fractured by the propagation of the matrix crack into the fiber.

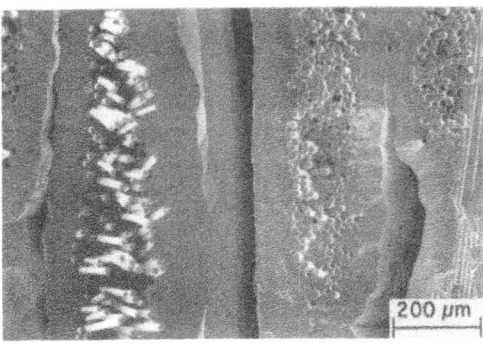

Figure 6. Fracture surfaces of the SiC/SiC composite tested at 20 ^{0}C (left) and 1200 ^{0}C (right).

Fiber pull-out was also observed on the ambient temperature fracture surfaces of the Al_2O_3/SiC composite (Fig. 8a), although the average pull-out length in this material was shorter than in the SiC-based composite. It should be noted that the defect size on the fiber surface was smaller owing to the shorter pull-out length. The smaller defect size led to the formation of the mirror-mist-hackle features, which are often found in Nicalon fibers broken in tension (Curtin, 1991; Fareed, 1993). Fibers pulled out from the matrix and broken with this mirror-mist-hackle morphology were also seen on the fracture surfaces created at 1200 °C, together with fibers broken from very large defects (Fig. 8b). These fracture mechanisms were predominant at elevated temperature although regions without pull-out and where the fiber fracture surface was specular were observed near the specimen surfaces and the notch tip. It should also be noticed that the BN coating was seen on the fiber surface after 100 hours of exposure at 1200 °C and that the BN layer was often debonded from the SiC outer layer (Fig. 8c).

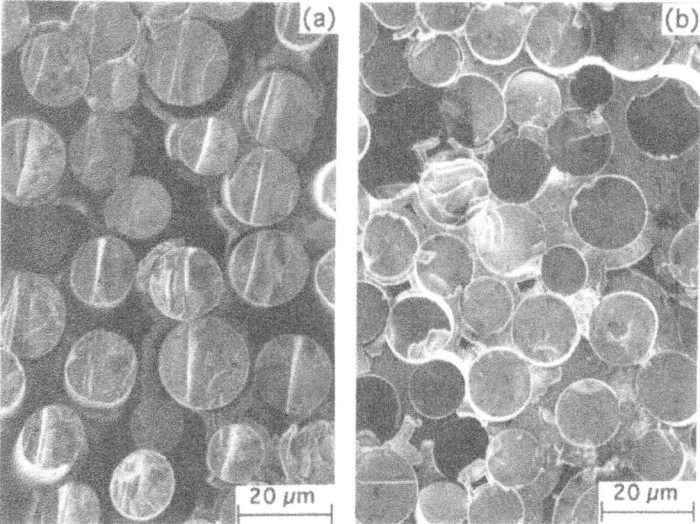

Figure 7. Fiber fracture in the SiC/SiC composite. (a) 20 °C. (b) 1200 °C.

The very large defects (up to 4.6 μm) found in the fibers after high temperature testing seemed to indicate that fiber damage had taken place during high temperature exposure of the Al_2O_3-matrix composite. These defects were never seen in the specimens tested at 20 °C. To characterize the extent of fiber damage, metallographic samples were prepared from the specimens tested at 1200 °C, and they are shown in Fig. 8d. These micrographs, when compared with Fig. 1b, clearly demonstrate that fiber damage occurred in the Al_2O_3-based composite during high temperature exposure. It should also be mentioned that the ceramic fibers of the SiC-matrix specimens tested at 1200 °C were analyzed as well, and they did not exhibit extensive damage after high temperature exposure.

The size and shape of the fiber defects in the Al_2O_3-matrix composite was studied by quantitative microscopy. Micrographs were taken randomly from the polished cross-sections and the depth (a) and the surface length ($2c$) of longest defect in each

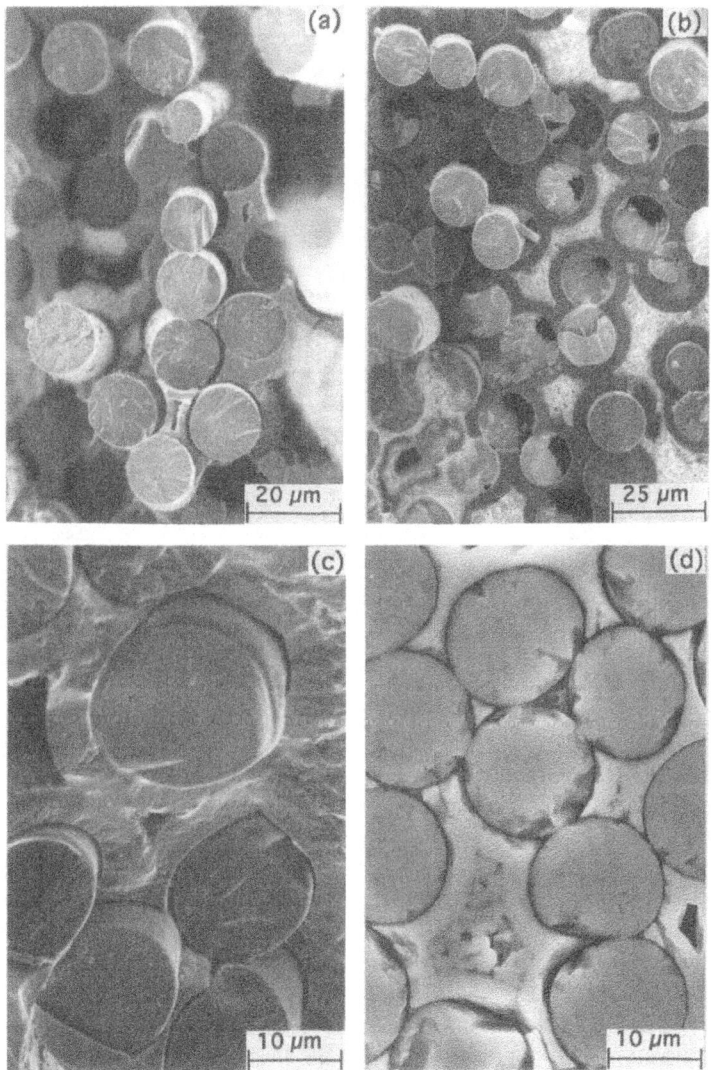

Figure 8. Fracture surfaces of the Al_2O_3/SiC composite. (a) 20 ^{0}C. (b) and (c) 1200 ^{0}C. (d) Polished cross-section after 100 hours of exposure at 1200 ^{0}C.

fiber was measured. About 100 fibers were studied for each testing condition and the cumulative probability of a fiber having a defect whose depth was shorter than a is plotted in Fig. 9a. The results show that forty percent of the fibers in the as-received material presented defects shorter than 0.5 μm. This was reduced to 10 percent after high temperature exposure. In addition, the fraction of the fibers exhibiting defects over 1 μm increased from 20 to 40 percent after the thermal treatment, and the depth of the longest defect also grew from 2 up to 4.6 μm. It should also be noted that the defect size distributions were very similar after 1 and 100 hours of exposure at 1200 ^{0}C.

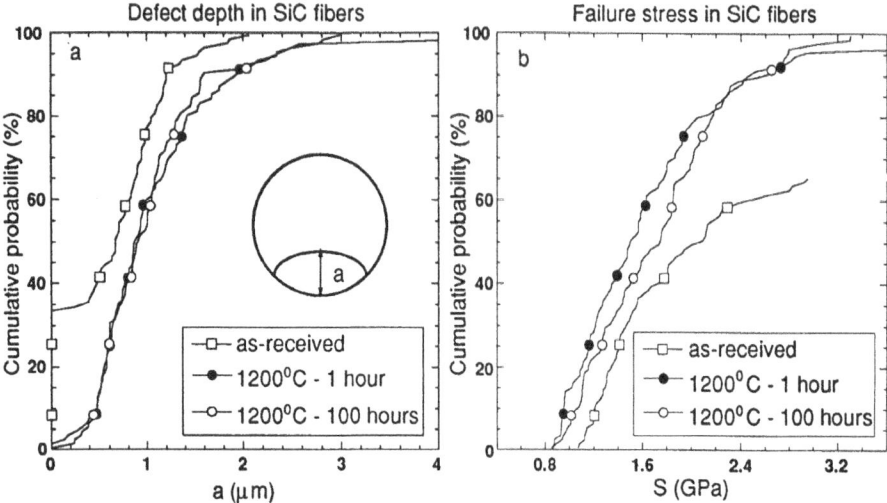

Figure 9. Cumulative probability of a fiber: (a) having a defect whose depth is shorter than a. (b) failing under an applied tensile stress S.

DISCUSSION

The mechanical tests at high temperature demonstrated that the excellent properties of these FRC were significantly reduced at 1200 ^{0}C. The extent of this degradation and the underlying mechanisms were, however, very different in the two materials. The SiC/SiC composite behaviour at 1200 ^{0}C was very poor, with a fracture toughness barely higher than that of monolithic ceramics. Other key mechanical properties, such as the flexure strength and the deflection under maximum load were also dramatically reduced at 1200 ^{0}C. Obviously, the fracture surfaces at ambient and elevated temperature reflected these results. While fiber pull-out was seen throughout the ambient temperature fracture surfaces, the specimens tested at 1200 ^{0}C presented flat fracture surfaces. The absence of fiber pull-out at 1200 ^{0}C suggests interface oxidation as the culprit for the degradation in the mechanical properties. In fact, the fibers broken at elevated temperature presented a specular fracture surface with no trace of defects on the fiber surface, indicating that no fiber/matrix decohesion took place and that fibers were fractured as the matrix crack propagated into them.

The hypothesis of interface oxidation in the SiC/SiC composite is supported by other studies (Frety and Boussuge, 1990; Cojean and Monthioux, 1992; Gomina *et al.*, 1992) that showed that the pyrolitic C layer, which provides a weak interface, was replaced by silica when the composite was exposed to temperatures above 700 ^{0}C in air. This silica layer prevented fiber/matrix debonding, leading to the embrittlement of the composite (Gomina *et al.*, 1992). The silica layer was not observed after high temperature exposure in inert atmosphere and the ambient temperature properties of specimens aged under such conditions were not modified by the thermal treatment (Frety and Boussuge, 1990; Cojean and Monthioux, 1992).

On the contrary, the reduction in the mechanical properties of the Al_2O_3/SiC composite at 1200 ^{0}C cannot be attributed mainly to interface oxidation. Fiber pull-out was still present (although the average pull-out length was shorter) at elevated temperature, indicating that fiber/matrix decohesion was possible. In fact, decohesion between the BN/SiC coatings was observed on the high temperature fracture surfaces (Fig. 8c). Previous studies in this material attributed to fiber degradation the reduction in the mechanical properties at elevated temperature (Fareed et al., 1993), but no attempt was made to quantify this hypothesis.

Recently, a theoretical model was proposed by Curtin (1991) to estimate the strength and fracture energy of FRC as functions of the fiber, matrix and interface properties. By assuming that the fibers fracture independently and that global load redistribution occurs upon fiber fracture, the successive fragmentation of the fibers in the composite was analyzed. The results indicated that the composite strength, σ_u is given by,

$$\sigma_u = \lambda_\sigma(m)\, f \left[\sigma_0^m \, L_0 \right]^{\frac{1}{m+1}} \left[\frac{\tau}{R} \right]^{\frac{1}{m+1}} \qquad (3)$$

and the fracture energy, G_F, can be obtained as

$$G_F = \lambda_G(m)\, f \left[\sigma_0^m \, L_0 \right]^{-\frac{2}{m+1}} \left[\frac{R}{\tau} \right]^{\frac{m-1}{m+1}} \qquad (4)$$

where f stands for the volume fraction of fibers oriented perpendicular to the crack plane, and λ_σ and λ_G are non-dimensional functions of the Weibull modulus of the fibers, m. τ and R stand for the fiber/matrix frictional stress and the fiber radius respectively, and σ_0 and L_0 are two constants with dimensions of stress and length, respectively, which characterize the fiber mechanical strength in such a way that the fracture probability of fiber of length L_0 subjected to a tensile stress σ_0 is $1 - 1/e$.

These expressions indicate that the strength and fracture energy of FRC depend on the fiber strength according to $\sigma_0^{\frac{m}{m+1}}$ and $\sigma_0^{\frac{2m}{m+1}}$ respectively. Assuming that the fiber/matrix frictional stress did not change significantly with temperature, the ratio between ambient to elevated temperature mechanical properties could be obtained from the corresponding values of σ_0 and m. These latter parameters can be estimated from the quantitative microscopy analyses assuming that fiber strength is dictated by the largest defect on the fiber. Thus the strength of an individual fiber, S, can be expressed as

$$S = \frac{K_c}{\sqrt{\pi a}} \frac{1}{F(a/c,\, a/R)} \qquad (5)$$

where K_c is the fracture toughness of the fiber, which is close to 2 MPa$\sqrt{\mathrm{m}}$ (Sawyer et al., 1987) and it was assumed constant within the temperature range studied. The shape function F was evaluated from the numerical results of Astiz (1986) for a

329

semielliptical surface crack in a circular bar when $a < c$ while the expressions given by Newman and Raju (1983) for an embedded elliptical crack were used for $a > c$. The fracture probability, P, for a fiber subjected to a stress S is plotted in Fig. 9b for the three testing conditions. The parameter σ_0 can be obtained from this plot as the stress which gives a fracture probability equal to $1 - 1/e$. The values of σ_0 are given in Table 1. In addition, the fiber fracture probability according to the Weibull statistics is given by

$$P = 1 - \exp\left\{-\left[\frac{S}{\sigma_0}\right]^m\right\}$$ (6)

and m can be obtained by the least squares fitting of (6) to the experimental results plotted in Fig. 9b. The values of m are also presented in Table 1 and it is worth noting that they were not affected by the high temperature exposure, in agreement with previous observations (Simon and Bunsell, 1984).

Once σ_0 and m are known, it is possible to determine the ratio between the elevated and ambient temperature strength and fracture energy in the Al_2O_3/SiC composite. These ratios are shown in Table 2, together with the experimental ones, calculated from the average values of σ_u and G_F plotted in Figs. 3 and 5b. There is reasonable agreement between the experimental and the theoretical results, which supports the hypothesis that the dominant degradation mechanism at 1200 ^{0}C in the Al_2O_3/SiC composite was the development of defects in the SiC fibers.

Table 1. Temperature effect on the fiber strength in the Al_2O_3/SiC composite.

Testing condition	σ_0 (MPa)	m
20 ^{0}C	2810	3.3
1200 ^{0}C - 1 hour	1782	3.2
1200 ^{0}C - 100 hours	1860	3.4

Table 2. Temperature effect on σ_u and G_F in the Al_2O_3/SiC composite.

	$\dfrac{\sigma_u\,(1200^0C-1h)}{\sigma_u\,(20^0C)}$	$\dfrac{\sigma_u\,(1200^0C-100h)}{\sigma_u\,(20^0C)}$	$\dfrac{G_F\,(1200^0C-1h)}{G_F\,(20^0C)}$
Theoretical results	0.676	0.758	0.458
Experimental results	0.715	0.665	0.532

CONCLUDING REMARKS

The strength and fracture resistance of two FRC were measured at 20 ^{0}C and 1200 ^{0}C. The mechanical properties of both materials were degraded at high temperature, although the extent and the causes of this degradation were different. Fiber pull-out disappeared at 1200 ^{0}C in the SiC/SiC, very likely due to the oxidation of the C layer which provided a weak fiber/matrix interface. As a consequence, the material presented a brittle behaviour and the fracture toughness and flexure strength were reduced to values unsuitable for structural applications.

Oxygen can diffuse to the interface through matrix microcracks and pores or via diffusion along the fiber/matrix interfacial region from mechanized surfaces. With sufficient time, the affected zone extends deeper below the surface, leading to the embrittlement of the composite. Thus to prevent the access of oxygen to the C layer, it would be neccesary to use an oxygen-resistant seal coating on the specimen surfaces and to operate at loading levels below the matrix cracking stress. As these solutions present evident problems, other approaches were attempted to achieve interfaces with the desired strength and oxidation resistance. One of them involved the coating of the fibers with a double layer of BN and SiC prior to the composite fabrication (Naslain et al., 1991; Sun et al., 1994). The rationale of this selection is complex and will not be detailed here. It can be indicated, however, that oxygen ingress reaching the BN layer would lead to the local formation of boria, a glassy phase, which would act as a sealant. The SiC outer coating was applied to prevent the massive oxidation of the BN during composite processing and elevated temperature exposure. SiC was chosen for several reasons, including its oxidation resistance and good thermochemical stability in relation to BN and Al_2O_3.

The use of the dual BN/SiC was successful in the Al_2O_3/SiC composite, which presented fiber pull-out at elevated temperature. The material also presented, however, a reduction in the mechanical properties at 1200 ^{0}C although less significant than in the SiC/SiC composite. This degradation was attributed to the development of large defects in the SiC fibers during high temperature exposure. A theoretical estimation of the composite degradation due to this mechanism was in good agreement with the experimental results, the fiber strength at 20 ^{0}C and 1200 ^{0}C being inferred from quantitative microscopy studies of the fiber defects. The origin of the defects in the fibers tested at high temperature (which did not appear in the SiC/SiC material) was not clear. Chemical reaction of the fibers with the Al_2O_3 matrix, or even with residual Al from the manufacturing process, was proposed, but definite conclusions could not be reached at this point.

Acknowledgments

This investigation was supported by Industria de Turbo Propulsores, S. A. The discussions with E. Erauzkin, J. Estevas, M. Gutierrez y A. Odriozola in the course of this research are gratefully acknowledged. The experimental results reported in this paper are property of ITP, S. A. and cannot be used without permission. Three of the authors (JAC, AM and JLL) also acknowledge the support from CICYT, Spain, through grant MAT 95-787.

REFERENCES

Astiz, M. A., 1986, An incompatible singular elastic element for two- and three-dimensional crack problems, *Int. J. Fracture*, 31: 105.

Clark, T. J., Jaffe, M., Rabe, J., Langley, N. R., 1986, Thermal stability characterization of SiC ceramic fibers: I, mechanical property and chemical structure effects, *Ceram. Engng. Sci. Proc.*, 7: 901.

Cojean, D., and Monthioux, M., 1992, Unexpected behaviour of interfacial carbon in SiC/SiC composites during oxidation, *Br. Ceram. Trans.*, 91: 188.

Curtin, W. A., 1991, Theory of mechanical properties of ceramic-matrix composites, *J. Am. Ceram. Soc.*, 74: 2837.

DiCarlo, J., 1991, *High Temperature Structural Fibers: Status and Needs*, NASA Technical Memorandum 105174, NASA Lewis Research Center, Cleveland.

Fareed, A. S., Schiroky, G. H., and Kennedy, C. R., 1993, Development of BN/SiC duplex fiber coatings for fiber-reinforced alumina matrix composites fabricated by direct metal oxidation, *Ceram. Engng. Sci. Proc.*, 14: 794.

Fareed, A. S., and Schiroky, G. H., 1994, Microstructure and properties of Nextel 610 fiber reinforced ceramic and metal matrix composites, *Ceram. Engng. Sci. Proc.*, 15: 344.

Frety, N., and Boussuge, M., 1990, Relationship between high-temperature development of fibre-matrix interfaces and the mechanical behaviour of SiC-SiC composites, *Comp. Sci. Techno.*, 37: 177.

Gomina, M., Chermant, J. L., and Fourvel, P., 1992, Effect of temperature and oxidation on the mechanical behaviour of uncoated SiC-SiC composite materials, in *Fracture Mechanics of Ceramics, Vol. 9*, R. C. Bradt *et al.*, eds., Plenum Press, New York.

Heuer, A. H., Morscher, G., and Pirouz, P., 1990, Temperature dependence of interfacial strength in SiC-fiber-reinforced reaction-bonded Si_3N_4, *J. Am. Ceram. Soc.*, 73: 713.

Hillig, W. B., 1987, Strength and toughness of ceramic matrix composites, *Ann. Rev. Mater. Sci.*, 17: 341.

LLorca, J., and Singh, R.N., 1991, Influence of fiber and interfacial properties on fracture behaviour of fiber-reinforced ceramics, *J. Am. Ceram. Soc.*, 74: 2882.

LLorca, J., and Elices, M., 1993, Influence of specimen geometry and size on fracture of fiber-reinforced ceramic-matrix composites, *Engng. Fract. Mech.*, 44: 341.

Mah, T.-I., Mendirata, M. G., Katz, A. P., and Mazdiyasni, S., 1987, Recent developments in fibre-reinforced high temperature ceramic composites, *Ceram. Bull.*, 66: 304.

Nair, S. V., and Wang, Y. L., 1992, Failure behaviour of a 2-D woven SiC fiber/SiC matrix composite at ambient and elevated temperature, *Ceram. Engng. Sci. Proc.*, 13: 433.

Naslain, R., Dugne, O., Guette, A., Sevely, J., Brosse, C. R., Rocher, J.-P., and Cotteret, J., 1991, Boron nitride interphase in ceramic-matrix composites, *J. Am. Ceram. Soc.*, 74: 2482.

Newman, J. C., and Raju, I. S., 1983, Stress intensity factor equations for cracks in three-dimensional finite bodies, in *Fracture Mechanics, 14th Symposium, STP 791*, J. C. Lewis and G. Sines, eds., ASTM, Philadelphia.

Pastor, J. Y., LLorca, J., Planas, J., and Elices, M., 1993, Stable crack growth in ceramics at ambient and elevated temperatures, *J. Engng. Mater. Techno.*, 115: 281.

Pastor, J. Y., Planas, J., and Elices, M., 1995a, A new technique for fracture characterization of ceramics at room and high temperature, *J. Test. Eval.*, 23: 209.

Pastor, J. Y., Guinea, G., Planas, J., and Elices, M., 1995b, Nueva expresión del factor de intensidad de tensiones para probetas de flexión en tres puntos, *Anal. Mec. Fract.*, 12: 85.

Sawyer, L. C., Jamieson, M., Brikowski, D., Haider, M. I., and Chen, R. T., 1987, Strength, structure and fracture properties of ceramic fibers produced from polymeric precursors: I, base line studies, *J. Am. Ceram. Soc.*, 70: 798.

Simon, G., and Bunsell, A. R., 1984, Mechanical and structural characterization of the Nicalon silicon carbide fibre, *J. Mater. Sci.*, 19: 3649.

Singh, R. N., 1993, Interfacial properties and high-temperature mechanical behaviour of fiber-reinforced ceramic composites, *Mater. Sci. Engng.*, A166: 185.

Sun, E. Y., Nutt, S. R., and Brennan, J. J., 1994, Interfacial microstructure and chemistry of SiC/BN dual-coated nicalon-fiber-reinforced glass-ceramic matrix compostes, *J. Am. Ceram. Soc.*, 77: 1329.

Woodford, D. A., Van Steele, D. R., Brehm, J. A., Timms, L. A., and Palko, J. E., 1993, Testing the tensile properties of ceramic-matrix composites, *J. of Metals*, May: 57.

Xu, H. H. K., Braun, L. M., Ostertag, C. P., Krause, R. F., and Lloyd, I. K., 1995a, Failure modes of SiC fiber/Si_3N_4-matrix composites at elevated temperatures, *J. Am. Ceram. Soc.*, 78: 388.

Xu, H. H. K., Ostertag, C. P., Fuller, E. R., Braun, L. M., and Lloyd, I. K., 1995b, Fracture resistance of SiC-fiber-reinforced Si_3N_4 composites at ambient and elevated temperatures, *J. Am. Ceram. Soc.*, 78: 698.

MICROCRACKING IN SILICON CARBIDE-TITANIUM DIBORIDE PARTICULATE COMPOSITES

David J. Green, Ming-Jen Pan and John R. Hellmann

Department of Materials Science and Engineering
The Pennsylvania State University
University Park, PA 16802

ABSTRACT

Elastic constant measurements were made from room temperature to 1200°C for silicon carbide/15 vol% titanium diboride particulate composites. Anomalously low values were obtained and this behavior was ascribed to microcracking. The discrepancy was established from a comparison of the elastic constants with the constitutive equations. Before this comparison could be performed, it was first critical to determine the accuracy of existing elastic constant data for the single phase constituents.

Models were established to quantify the influence of interfacial microcracking on the elastic properties, using a combination of theoretical and numerical techniques. The approach was based on energy principles and combined finite element analysis with a differential effective medium technique. The results allowed the bulk modulus to be determined as a function microcrack density and size. For limiting cases, the models were found to agree with existing theories at both dilute and finite concentrations of particles. From microstructural measurements on the silicon carbide/titanium diboride composites, it was also possible to estimate the critical particle size for microcracking in this system.

One approach to controlling microcrack formation is to utilize coated second phase particles. The presence of such a coating on a second phase particle will influence the nature of the residual stress field. This structure was modelled using linear elasticity and the residual stress field could be expressed analytically. The nature of this solution will be discussed with respect to microstructural design.

INTRODUCTION

Microcracking is a well-known phenomenon in brittle composites and is often the result of thermal expansion mismatch between the components. It is known that microcracking in particulate composites can be suppressed if the size of the second component is kept below a critical size[1-4]. One of the major challenges in this area is the ability to detect microcracks and to quantify structural information on these features. One approach to this problem is to

Table 1. Selected physical properties of Hexoloy ST

Density	3.30 g/cm^3
Flexural Strength	
Four point bend (RT)	448 MPa
Three point bend (RT)	520 MPa
Modulus of Elasticity	427 GPa
Poisson's ratio	0.15
Weibull modulus (2 parameter)	12
Fracture toughness	
Double torsion and SENB	8.0 MPa√m
Coefficient of thermal expansion	4.02 ppm K^{-1}

Form A-12.032 Standard Oil Engineered Materials, Niagara Falls, NY

measure properties of the composite that are sensitive to the presence of microcracks. For example, it is known the elastic constants of a material are reduced by the presence of microcracks[5] and such reductions have been used extensively as indirect confirmation of microcracking in brittle materials. Elastic constants can be measured to a high degree of accuracy and thus, could be used to obtain quantitative information on the microcracking, if suitable methodologies can be developed. The current paper summarizes recent studies, in which this approach was used to study microcracking in silicon carbide-titanium diboride particulate composites[6-8]. The emphasis here, will be on one particular composition, SiC 15 vol% TiB$_2$ (Hexoloy ST, Carborundum Co., NY). This material has been shown to undergo stress-induced microcracking[9] but in this paper, the focus will be on microcracks present in the as-received material (spontaneous microcracks).

MATERIAL CHARACTERIZATION

The physical properties of Hexoloy ST are summarized in Table 1 and the microstructure is shown in Fig. 1. A polished section contains both porosity and pull-out, in addition to the second phase TiB$_2$ in the SiC matrix (primarily 6H polytype). The mean grain sizes of the SiC and TiB$_2$ were found to be 6.6 μm and 4.7 μm respectively[6]. The particle size distribution of the TiB$_2$ component is shown in Fig. 2. Transmission electron microscopy (TEM) was used to confirm the presence of microcracks in the as-received material[6].

ELASTIC CONSTANT MEASUREMENTS

The elastic constants were measured from room temperature to 1200°C for Hexoloy ST, as well as polycrystalline SiC and TiB$_2$, using dynamic resonance[6] and the results are shown in Fig. 3. For the theoretical analysis, the Young's modulus (E) data were corrected for porosity effects to values for theoretically dense materials (E$_0$) using[10,11]

$$E_0 = \frac{E}{(1 - P)^2}$$

(1)

where P is the fractional porosity.

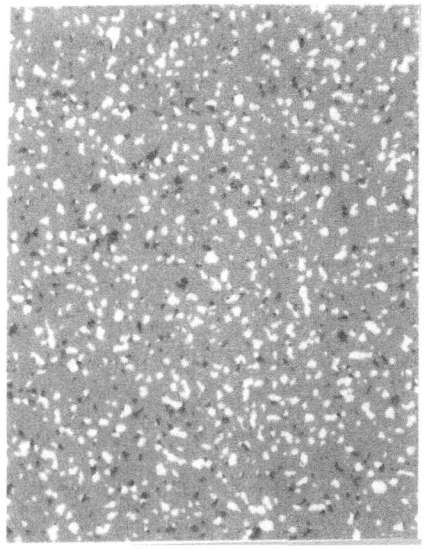

Figure 1. Microstructure of SiC/15 vol. % TiB$_2$ particulate composite (Hexoloy ST).

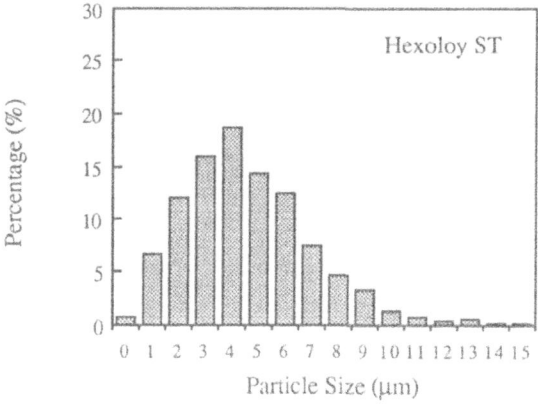

Figure 2. Size distribution of TiB$_2$ particles in the as-received material.

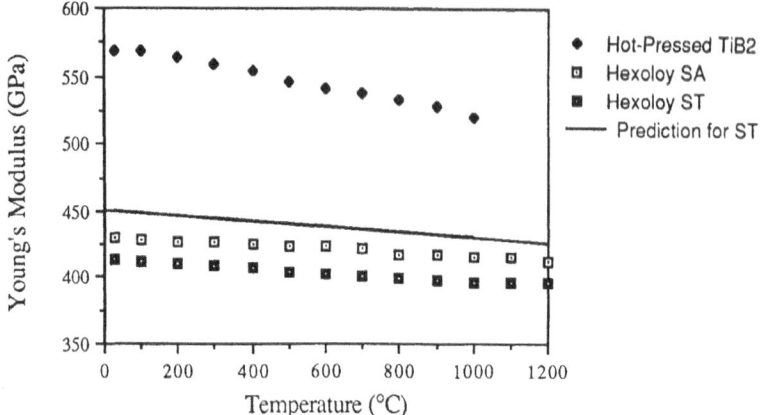

Figure 3. Young's modulus as a function of temperature for Hexoloy ST, polycrystalline α-SiC and TiB$_2$. The composite modulus is also compared to the theoretical value obtained from the self-consistent approach (spherical particles).

Table 2. Single crystal elastic stiffness constants of α-SiC and TiB$_2$ (GPa)

Material	c$_{11}$	c$_{12}$	c$_{13}$	c$_{33}$	c$_{44}$
α-SiC[13]	500	92	--	564	168
α-SiC[13]	504	98	--	566	170
α-SiC[14]	479	98	56	521	148
TiB$_2$[6]	660	48	93	432	260

Table 3. Comparison of experimental and calculated values of elastic constants (GPa) of Hexoloy ST, polycrystalline α-SiC and TiB$_2$.

Material	Young's modulus		Shear modulus	
	Measured	Predicted	Measured	Predicted
α-SiC	430	419-420[1]	191	179.2-179.5[1]
TiB$_2$	569	578-580[1]	259	262-263[1]
Hexoloy ST	414	449[2]	186	200[2]

[1]Hashin-Shtrikman bounds[15].
[2]Self-consistent solution from measured values[16,17]

There are numerous theoretical approaches to determine the elastic constants of a composite from the properties of its constituents[12]. For this work, the self-consistent scheme was used, though the Hashin-Shtrikman bounds are rather narrow and could also have been utilized. One of the difficulties of comparing the data in Figure 3 to a theoretical analysis, is confirmation of the accuracy of the polycrystalline data. For SiC, this can be easily accomplished by comparing the data with the numerous previous studies[6]. In cases when such extensive data is not available, however, it is useful to compare the experimental values with those calculated from the single crystal elastic constants (e.g., Hashin-Shtrikman bounds). This approach confirmed the values for the elastic constants of SiC but not for TiB_2[6]. It was suspected the single crystal elastic constants for TiB_2 were in error and thus, these constants were measured using dynamic resonance[6]. The single crystal elastic constants of SiC and TiB_2 are given in Table 2 and the comparison between the experimental and calculated polycrystalline data is given in Table 3. With the revised single crystal data, agreement between the theoretical and experimental values was obtained for TiB_2. Returning to the composite material, one can now compare the experimental data on Hexoloy ST to the self-consistent scheme for spherical particles[16,17]. As shown in Fig. 3 and Table 3, the elastic constants are significantly lower than expected from the theoretical calculations, confirming the presence of microcracks in the as-received materials.

EFFECT OF INTERFACIAL MICROCRACKS ON THE ELASTIC CONSTANTS

For SiC/TiB_2 particulate composites, TEM observations have shown microcracks to be present at the interface between the TiB_2 particles and the SiC matrix[6,9]. In the post-sintering cooling, cracks form under the action of the residual stress because the particles possess a higher thermal expansion coefficient than the SiC matrix (approximately 7.5 and 4.5 ppm/K, respectively). In order to model the interfacial microcracking, a numerical scheme was devised in which cracked composite spheres were added to an effective medium, Fig. 4. By applying hydrostatic stresses to the medium, the bulk modulus, K, can be determined from the strain energy difference caused by the microcracking. This energy difference was obtained using a finite element analysis [6,7]. In order to model a finite concentration of microcracked particles, the differential scheme was utilized[6,7] and the data were fitted for each increment to an equation of the form

$$K^* = K_m + f(A - BK_m) \qquad (2)$$

where A and B are numerical constants, f is the volume fraction of microcracked composite spheres and K_m is bulk modulus of the effective medium. The effective bulk modulus can be expressed in the following form[6,7],

$$K^*(f) = \frac{A}{B} + \left(K_0 - \frac{A}{B} \right)(1 - f)^B \qquad (3)$$

The results of the analysis are shown in Fig. 5 and the constants are given in Table 4. A test of the analysis was comparison of the bulk modulus of completely cracked particles ($\phi = 180^0$) with the constitutive relationships for porosity. The results of the numerical analysis were <1% below the Hashin-Shtrikman upper bound and ~2% higher than the self-consistent solution (spherical pores). The measured bulk modulus of Hexoloy ST was 177.5 GPa and comparison with Fig. 5 shows this is reasonable but information on the microcrack size would be needed before a fraction of microcracked particles could be established. For a crack size coressponding to $\phi = 90^0$, approximately 63 vol% of the TiB_2 particles would need to be microcracked. Converting this value to a numerical percentage (19%) leads to a critical size for microcracking of ~7 μm to be obtained from Fig. 2.

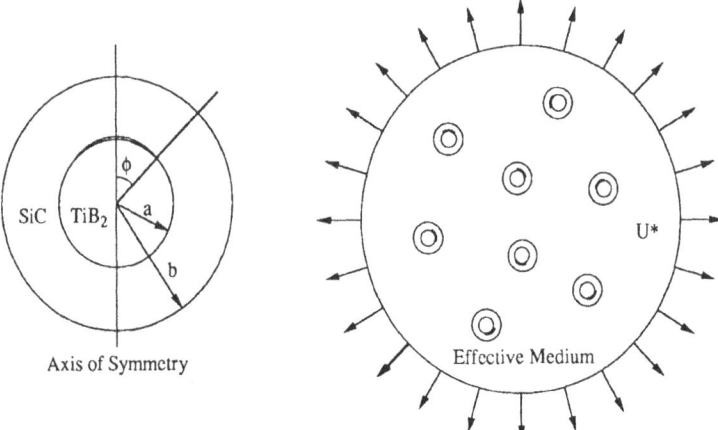

Figure 4. Microcracked composite spheres, as shown on left, were added to an effective medium to model the elastic behavior. The bulk modulus was obtained using a strain energy difference calculation with the model in a state of hydrostatic stress.

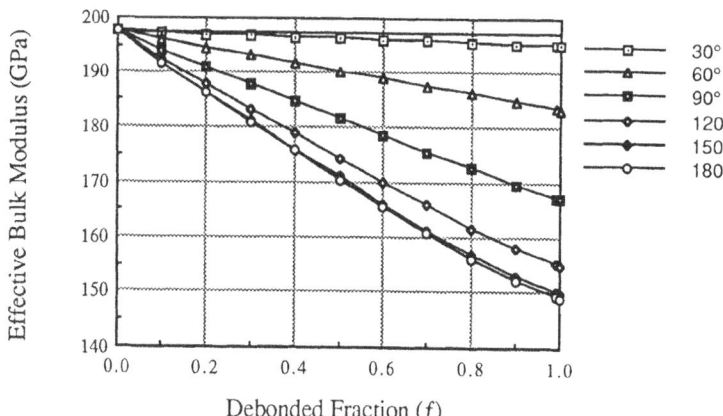

Figure 5. The effective bulk modulus of an interfacially microcracked composite as a function of the volume fraction of microcracked composite spheres.

Table 4. Numerical constants determined from the finite element analysis.

Crack Size	A (GPa)	B
30°	186.6	0.9560
60°	183.7	1.003
90°	180.8	1.082
120°	178.2	1.149
150°	177.3	1.183
180°	176.7	1.185

EFFECT OF PARTICLE COATING ON RESIDUAL STRESSES

In the study of other silicon carbide/titanium diboride composites, microcracking was found to be absent and this was considered a result of the presence of a glassy phase[6]. In order to understand the effect of a surface coating on the particulate phase, modelling was performed using the spherical geometry shown in Fig. 6. For this geometry, the residual stress field can be expressed in analytical form for all volume fractions[6,8]. For example, at low volume fractions, the radial stresses at interfaces 1 and 2 are given by

$$\sigma_1 = \frac{12\,K_p[\,3\,K_c(G_c - G_m)\,\varepsilon_1 a^3 - G_c(4\,G_m + 3\,K_c)\,\varepsilon_1 b^3 - G_m(4\,G_c + 3\,K_c)\,\varepsilon_2 b^3\,]}{12\,(K_c - K_p)(G_c - G_m)\,a^3 - (4\,G_c + 3\,K_p)(4\,G_m + 3\,K_c)\,b^3} \tag{4}$$

$$\sigma_2 = \frac{12\,K_p[\,-K_p(4\,G_c + 3\,K_c)\,\varepsilon_1 a^3 + 4\,G_c(K_c - K_p)\,\varepsilon_2 a^3 - K_c(4\,G_c + 3\,K_p)\,\varepsilon_2 b^3\,]}{12\,(K_c - K_p)(G_c - G_m)\,a^3 - (4\,G_c + 3\,K_p)(4\,G_m + 3\,K_c)\,b^3} \tag{5}$$

where K and G are the bulk and shear moduli and the subscripts p, m and c indicate particle, matrix and coating respectively. The misfit strains are given by $\varepsilon_1 = (\alpha_p - \alpha_c)\,\Delta T$ and $\varepsilon_2 = (\alpha_c - \alpha_m)\,\Delta T$, where α_i are the thermal expansion coefficients and ΔT is the temperature difference over which the residual stresses arise. The complete set of equations are presented elsewhere[8]. One interesting result that arose from this analysis was that a simple expression exists for the residual stresses in uncoated particles, i.e.,

$$\frac{12\,K_p G_m K_m(\alpha_p - \alpha_m)\,\Delta T\,(1 - V_p)}{3\,K_p K_m + 4\,G_m K_m + 4\,G_m(K_p - K_m)\,V_p} \tag{6}$$

An understanding of the effect of particle coating on the residual stress field is important in developing new particulate composites. For example, if one wishes to design a composite system without microcracking, equations such as 4 and 5 could be used to select particle coatings that would act to reduce the residual stresses in a chosen particulate composite system[8]. Figure 7 shows an example of the coating properties that could be used to reduce the radial stresses associated with a TiB$_2$ particles in a SiC matrix (Poisson's ratio of coating 0.2, b/a = 1.2).

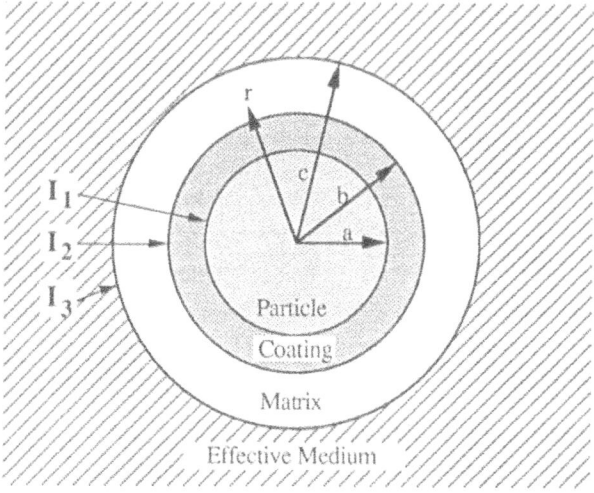

Figure 6. Finite concentration model that considers the volume fractions of constituents. The radii a, b, and c are related to the volume fractions.

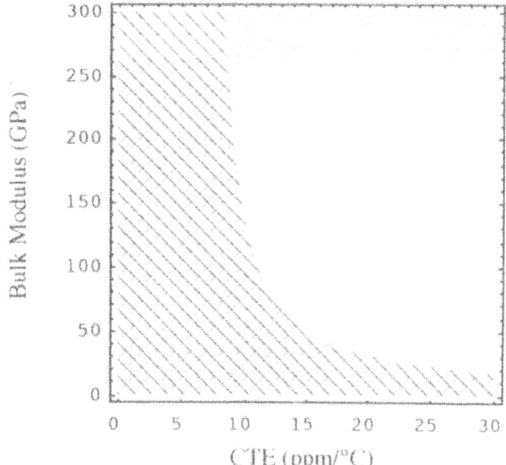

Figure 7. For the SiC/TiB$_2$ system, coating properties that would reduce the interfacial radial stresses can be identified (cross-hatched region).

CONCLUSIONS

The elastic constants of a silicon carbide/15 vol% titanium diboride particulate composite were shown to be less than values obtained from the constitutive equations and this discrepancy was ascribed to microcracking. In order to establish the accuracy of the data, the theoretical equations were applied to both the single phase components and the composite. The influence of interfacial microcracking on the elastic properties was modelled using a combination of theoretical and numerical techniques, which allowed the bulk modulus to be determined as a function microcrack density and size. For the limiting case of completely debonded particles, the models were found to agree with existing theoretical expressions. The critical particle size for microcracking was estimated to be ~7μm in this system. Coatings can be used to control microcracking in particulate composites and for spherical coated particles, analytical expressions for the residual stress field are available.

REFERENCES

1. R.W. Davidge and T. J. Green, The strength of two-phase ceramic/glass materials, *J. Mater. Sci.*, **3**:629 (1968).
2. F.F. Lange, Criteria for crack extension and arrest in residual localized stress fields associated with second phase particles, *in::* "Fracture Mechanics of Ceramics, Vol 2," R.C. Bradt et al., eds., Plenum Press, New York, 1974.
3. A.G. Evans, The role of inclusions in the fracture of ceramic materials, *J. Mater. Sci.*, **9**:1145 (1974).
4. D.J. Green, Microcracking mechanisms in ceramics, *in::* "Fracture Mechanics of Ceramics, Vol 5," R.C. Bradt et al., eds., Plenum Press, New York, 1983.
5. B. Budiansky and R. J. O'Connell, Elastic moduli of a cracked solid, *Int. J. Solids Structures*, **12**:81 (1976).
6. M-J. Pan, "Microcracking Behavior of Particulate Titanium Diboride-Silicon Carbide Composites and Its Influence on Elastic Properties," Ph.D thesis, The Pennsylvania State University, December 1994.
7. M-J. Pan, D. J. Green and J. R. Hellmann, Influence of interfacial microcracks on the elastic properties of composites, *J. Mater. Sci.*, submitted June 1995.
8. M-J. Pan, D. J. Green and J. R. Hellmann, Residual stresses in coated particulate composites, *J. Comp. Mater.*, submitted June 1995.
9. W.H. Gu, K. T. Faber, and R. W. Steinbrech, Microcracking and R-curve behavior in SiC-TiB$_2$ composites,*Acta Metall. Mater.*, **40**:3121 (1992).
10. J. K. Mackenzie, The elastic constants of a solid containing spherical holes," *Proc. Phys. Soc.*, **63**:2 (1950).
11. L. J. Gibson and M. F. Ashby, The mechanics of three-dimensional cellular materials, *Proc. R. Soc. London, Ser. A*, **382**:43 (1982).
12. J.P. Watt, G. F. Davies, and R. J. O'Connell, The elastic properties of composite materials, *Rev. Geophys. Space Phys.*, **14**:541 (1976).
13. G. Arlt and G. R. Schodder, Some elastic constants of silicon carbide, *J. Acoust. Soc. Am.*, **37**:384 (1965).
14. Z. Li and R. C. Bradt, The single crystal elastic constants of hexagonal SiC to 1000^{0}C, *Int. J. High Technology Ceramics*, **4**:1 (1988).
15. L. Peselnick and R. Meister, Variational method of determining effective moduli of polycrystals, *J. Appl. Phys.*, **36**:2879 (1965).
16. B. Budiansky, On the elastic moduli of some heterogeneous materials, *J. Mech. Phys. Solids*, **13**:223 (1965).
17. R. Hill, A self-consistent mechanics of composite materials, *J. Mech. Phys. Solids*, **13**:213 (1965).

DETERMINATION OF THE CRITICAL STRESS FOR MICROCRACKING

IN ALUMINA AND ALUMINA MATRIX COMPOSITES

Shuichi Wakayama, Hidefumi Naito and Byung-Nam Kim

Department of Mechanical Engineering, Faculty of Engineering
Tokyo Metropolitan University
1-1 Minami-Ohsawa, Hachioji-shi, Tokyo 192-03, Japan

INTRODUCTION

Whisker reinforced ceramic composites have been investigated because of the excessive toughness, contributed by the toughening mechanisms due to the crack deflection[1], crack bowing[2] and especially crack bridging[3] by whiskers. But those toughening mechanisms contribute to the enhancement of the resistance to crack growth but not crack initiation. From the view point of structural design of ceramics, it is necessary to enhance the resistance to crack initiation as well as crack growth.

Toughening mechanisms, including the mechanisms responsible for the whisker reinforced ceramics, have been also investigated[1-6]. But the further development may be restricted by the fact that the parameters obtained experimentally are macroscopic and the parameters included in theoretical models are microscopic. Therefore, it is required to establish an experimental technique to evaluate the microscopic parameters.

On the other hand, the authors have studied the toughening mechanisms and fracture behavior in ceramic materials using acoustic emission technique[6-8], because of its ability to detect the nucleation times and locations of microcracks in the material. Especially, an experimental technique to evaluate the critical stress for maincrack formation have been developed[7,8], using acoustic emission monitoring and a fluorescent dye penetrant observation of fracture process during bending test of the monolithic alumina.

In this study, two kinds of alumina matrix composites reinforced with different whisker (silicon carbide and alumina) were fabricated by hot pressing. Therefore, the residual stress around the whisker due to thermal mismatch and the bonding strength at the interface between whisker and matrix of both materials were quite different. The microfracture process during four point bending tests were evaluated by acoustic emission technique. Especially, the critical stress of microcracking in matrix of the composites was determined experimentally. Consequently, the influences of the whisker arrangement and the combination of whisker/matrix materials on the fracture behavior and the critical stress were investigated.

PREPARATION OF MATERIALS

In this study, three types of materials were prepared, i.e. monolithic alumina and alumina reinforced with silicon carbide or alumina whisker. They were sintered by the hot pressing in Ar gas with different sintering conditions for each materials.

Monolithic Alumina

Monolithic alumina was sintered from pure alumina powder with a purity of 99.99 % and average grain size of 0.23 μm, without any additives. Sintering was done at a temperature of 1650 °C with a holding time of 3 hours and a pressure of 20 MPa. The relative density and the average grain size of sintered material were 99.7 % and 7 μm, respectively.

Silicon Carbide Whisker Reinforced Alumina

Silicon carbide whisker was used as received from manufacturer and with an average diameter of 0.49 μm, average length of 34.3 μm, tensile strength of 21 GPa and elastic modulus of 481 GPa. Matrix alumina was made from the same powder as the monolithic material. The whisker and powder were mixed by ball milling for 24 hours. After the drying and screening, the composites were sintered by hot pressing. Since sufficient density of the composite was not obtained under the same condition as monolithic materials, the sintering temperature and holding time were 1800 °C and 12 hours, respectively. The volume fraction of whisker was 10 and 20 %, and the relative density of the fabricated materials was greater than 99 %.

Alumina Whisker Reinforced Alumina

Alumina whisker used in this investigation was obtained by the crystallographic change of $9Al_2O_3 \cdot 2B_2O_3$ whisker caused by the thermal decomposition of material at 1700 °C for 72 hours[9]. Therefore, sintering temperature must be below 1700 °C. In this study, the chemical composition of matrices in composites were desired to be as equivalent as possible in order to investigate the effect of whiskers on the microcrackings in matrix . However, it was difficult to obtain dense composites at such low sintering temperature and pressure. Therefore, 3 percents by volume of Y_2O_3 was added. Sintering temperature, holding time and pressure were 1600 °C, 8 hours and 40 MPa, respectively. Consequently, volume fraction of whisker was 10 and 20 %, and relative density was greater than 99.5%.

TESTING PROCEDURES

The fracture toughness was measured by the indentation fracture (IF) method in this study. Four point bending tests were also carried out where the microfracture process in composites were investigated by AE technique.

Bending Tests

Specimens were cut from sintered discs by diamond saw to the dimensions of $3 \times 4 \times 40$ mm. Each specimens were chamfered to 0.1 mm and polished with diamond powder of 1 μm. In order to avoid the influence of corrosion by water on the microfracture process, specimens were dried in a vacuum of 10^{-4} Torr at 150 °C for 60 min. Bending tests were carried out using the Instron-type tensile testing machine in air (temperature ≈ 20 °C. relative humidity ≈ 60 %) with upper and lower span of 10 and 30 mm.

Cutting direction of samples was selected so that hot press axis is parallel (P direction) or normal (N direction) to the surface where tensile stress was applied during bending tests. The whiskers in the P direction samples were parallel to the tensile axis, where the toughening mechanism due to crack bridging caused by whisker is effective, while the orientations of those in the N samples were 2-dimensionally random. Both P and N direction samples were available for silicon carbide whisker reinforced alumina ($SiC/Al_2O_3[P]$ and $SiC/Al_2O_3[N]$), but other materials were only P direction samples (Al_2O_3 and $Al_2O_3/Al_2O_3[P]$).

AE Measurement

The AE measuring system used in this study is shown in Figure 1, schematically. Two piezo-electric elements were used directly as AE sensors and attached on both ends of the specimen. In this study, AE source locations were calculated from the difference of arrival times between two sensors and used for the removal of mechanical noise as mentioned later. Since the accuracy of AE source location strongly depends on the equilibrium of

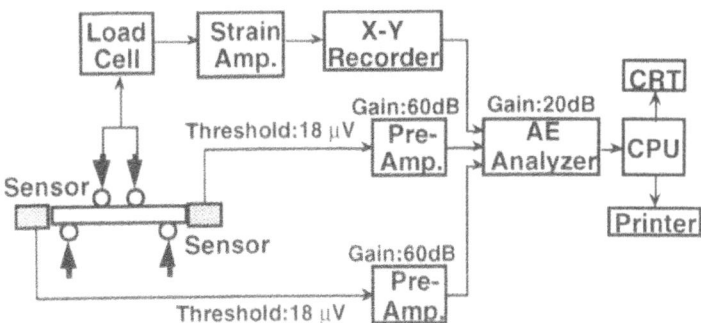

Figure 1. AE measuring system used in this study. AE sensors are direct piezo-electric elements and the connections between sensors and pre-amplifiers were modified the same way as differential type transducers, in which the electromagnetic noise can be automatically canceled between the opposite phase signal cables.

345

sensitivities of sensors, the sensitivity of two sensors, especially those equilibrium, were calibrated carefully using pencil lead breaking as a simulated source, before each testing.

Since it is well known that the AE activity of ceramic materials is quite low, the minimization of noise level of the system is indispensable for the AE measurement of such materials[6-8]. In this study, noise-filter-transformers were used and the connections between sensors and pre-amplifiers were modified the same way as differential type transducers, in which the electromagnetic noise can be automatically canceled between the opposite phase signal cables. Consequently, the noise level at the input terminal of pre-amplifier was decreased to 14 µV, and then the threshold level was selected as 18 µV. AE signals were measured by the AE analyzer with load signals and sent to a personal computer through the RS-232C interface to be analyzed.

RESULTS

Fracture Toughness and Vickers Hardness

Fracture toughness was measured by the indentation fracture (IF) method on the polished surface of the bending test samples. Therefore, for the P direction samples, in which whisker orientations are parallel to the length, the difference in orthogonal crack lengths was not negligible. But the effect of toughness, especially the resistance to crack growth (R-curve behavior), on the fracture process after the maincrack formation during bending tests of whisker reinforced ceramics is discussed in this paper. Since the fracture

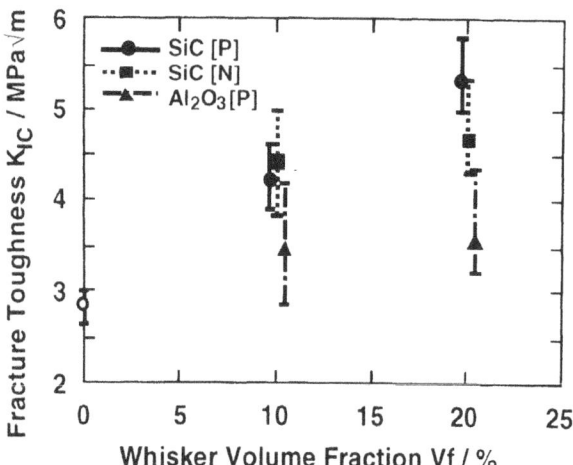

Figure 2. Fracture toughness of materials measured by the indentation fracture (IF) method. Symbols indicate the average values; "O" for monolithic alumina, "●" and "■" for SiC whisker reinforced Al_2O_3 for P direction (SiC/Al_2O_3[P]) and N direction (SiC/Al_2O_3[N]), and "▲" for Al_2O_3 whisker reinforced Al_2O_3 for P direction (Al_2O_3/Al_2O_3[P]), respectively.

toughness values obtained by IF method represent the R-curve behavior of materials, they might be able to used for the discussion on the strengthening behavior of composites in this study.

Figure 2 shows the fracture toughness, K_{IC}, of materials investigated in this study. For silicon carbide reinforced alumina, P direction samples ($SiC/Al_2O_3[P]$) have larger toughness than N direction samples ($SiC/Al_2O_3[N]$) at 20 % volume fraction, Vf, of whisker, while, on the contrary, the toughness of $SiC/Al_2O_3[N]$ is a little larger than that of $SiC/Al_2O_3[P]$ at a Vf of 10 %. The toughness of $SiC/Al_2O_3[P]$ with 20 % Vf is 5.3 MPa√m and almost twice of the monolithic alumina (2.8 MPa√m). But the toughness values of $Al_2O_3/Al_2O_3[P]$ with 10 and 20 % Vf are almost same as 3.6 MPa√m, and they are not so enhanced by whisker as SiC/Al_2O_3. These behaviors can be related with the results of SEM observation of crack path, i.e. crack bridging by whisker in $Al_2O_3/Al_2O_3[P]$ is not so remarkable comparing with SiC/Al_2O_3.

Bending Strength

Figure 3 shows the bending strength, σ_B, of materials. It appears that the strengthening by whisker in silicon carbide whisker reinforced alumina for both P and N directions ($SiC/Al_2O_3[P]$ and $SiC/Al_2O_3[N]$) is much clearer than that of fracture toughness; the increase in σ_B of 10 % $SiC/Al_2O_3[P]$ is 230 MPa, 20 % $SiC/Al_2O_3[P]$ is 290 MPa, 10 % $SiC/Al_2O_3[N]$ is 180 MPa and 20 % $SiC/Al_2O_3[N]$ is 260 MPa. Furthermore, the increase in strength of $SiC/Al_2O_3[P]$ is larger than that of N direction $SiC/Al_2O_3[N]$: the effect of whisker arrangement on bending strength is much larger than that on fracture toughness measured by IF method.

On the other hand, the increases in the bending strength of alumina whisker reinforced alumina ($Al_2O_3/Al_2O_3[P]$) are 110 MPa at a Vf of 10 % and 180 MPa at 20 %, and smaller than those in $SiC/Al_2O_3[P]$ and $SiC/Al_2O_3[N]$. These results can be explained by the fact that pull out of whiskers was rarely observed on the fracture surface of $Al_2O_3/Al_2O_3[P]$.

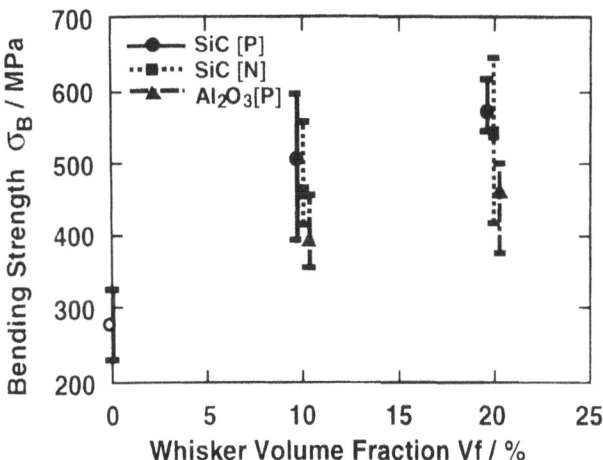

Figure 3. Bending strength of materials. Symbols are the same as in Figure 2.

AE Generation Behavior

In this study, electromagnetic noise was minimized as mentioned above. While, mechanical noise was removed by using Teflon sheets between the loading rods and specimen, and then the negligibility was ascertained by the pre-loading using the specimen with twice height and same width. Furthermore, when AE generation pattern was analyzed, the AE source location was also used for the removal of mechanical noise, i.e. the events located out of the upper span were neglected. Therefore it can be concluded that the sources of those AE events are microcrackings.

The typical results of bending tests and AE generation patterns during bending tests are shown in Figure 4, for (a) monolithic alumina, (b) 20 % Vf silicon carbide whisker reinforced alumina ($SiC/Al_2O_3[P]$) and 20 % Vf alumina whisker reinforced alumina ($Al_2O_3/Al_2O_3[P]$). It can be observed in Figure 4 (a) that cumulative AE events increases rapidly at about 350 s. It has been already made clear by the authors[7,8] that the apparent bending stress, σ_C, at that point corresponds to the critical stress for maincrack formation, investigating the dependence of the critical stress on AE threshold level, AE source locations and the observation of fracture process using fluorescent dye penetrant method. Those points can be also determined in the AE generation patterns of $SiC/Al_2O_3[P]$ at about 250 s (Figure 4 (b)) and $Al_2O_3/Al_2O_3[P]$ at about 570 s (Figure 4 (c)). The apparent stresses at those points were determined to be the critical stress for maincrack formation in alumina matrix by the same procedure of the previous paper[7] except for the dye penetrant observation.

The critical stress, σ_C, of monolithic alumina (Figure 4 (a)) is 230 MPa and close to the bending strength, σ_B. σ_C of $SiC/Al_2O_3[P]$ (Figure 4 (b)) is 240 MPa and less than 50 % of σ_B. It is important that the σ_C of $SiC/Al_2O_3[P]$ is equivalent to that of monolithic material. On the contrary, the σ_C of $Al_2O_3/Al_2O_3[P]$ (Figure 4 (c)) is 410 MPa and approximately 80 % of σ_B. It is much larger than those of monolithic alumina and SiC reinforced composite.

DISCUSSIONS

Microcracking in Matrix of Whisker Reinforced Composites

In this study, microcrackings, especially maincrack formation, in matrix of silicon carbide and alumina whisker reinforced alumina composites were detected and the critical stress, σ_C, was determined by acoustic emission technique. From the view points of both structural and material design of ceramic composites, it is important to investigate their characteristics.

The relationships of σ_C and volume fraction of whisker are shown in Figure 5. Figure 5 (a), where materials are monolithic alumina and silicon carbide reinforced alumina for P ($SiC/Al_2O_3[P]$) and N ($SiC/Al_2O_3[N]$) direction, represents the effect of whisker arrangement on σ_C. At a Vf of 10 %, σ_C is 280 MPa for P direction and 300 MPa for N direction, and at 20 %, σ_C is 350 MPa for P direction and 360 MPa for N direction. It is understood that the value of σ_C is hardly influenced by the arrangement of whisker.

On the other hand, Figure 5 (b), where materials are monolithic alumina and silicon carbide ($SiC/Al_2O_3[P]$) and alumina whisker ($Al_2O_3/Al_2O_3[P]$) reinforced alumina for P direction, describes the effect of whisker material on σ_C. The value of σ_C for

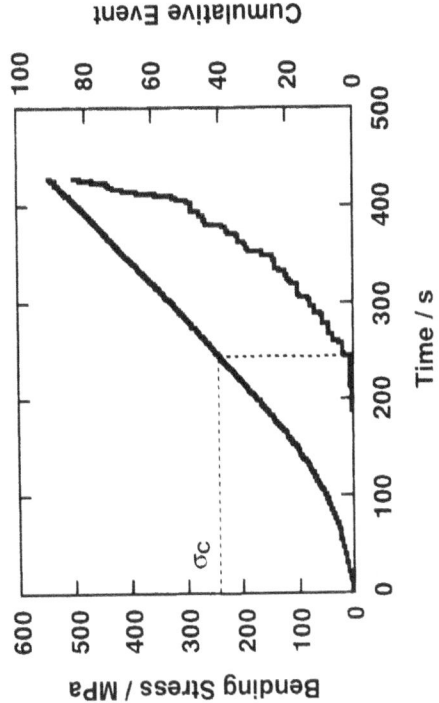

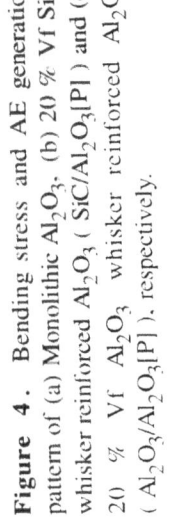

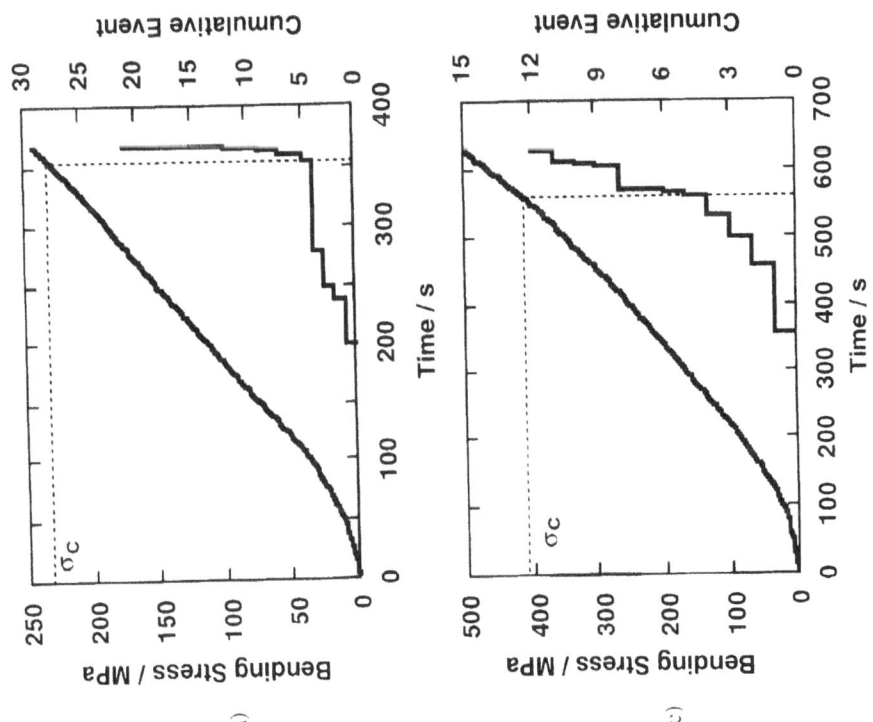

Figure 4. Bending stress and AE generation pattern of (a) Monolithic Al_2O_3. (b) 20 % Vf SiC whisker reinforced Al_2O_3 ($SiC/Al_2O_3[P]$) and (c) 20 % Vf Al_2O_3 whisker reinforced Al_2O_3 ($Al_2O_3/Al_2O_3[P]$). respectively.

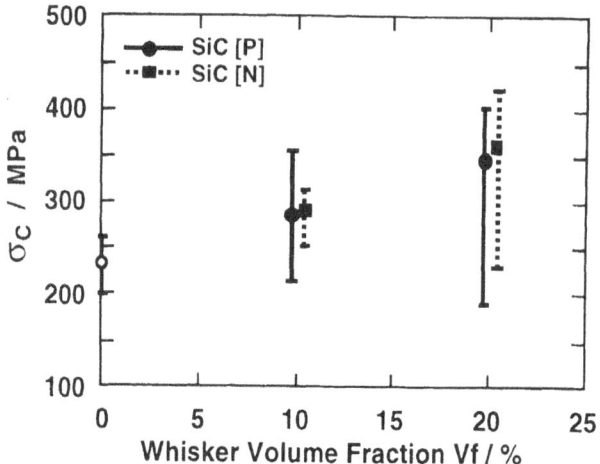

(a) Effect of whisker arrangement on σ_C.

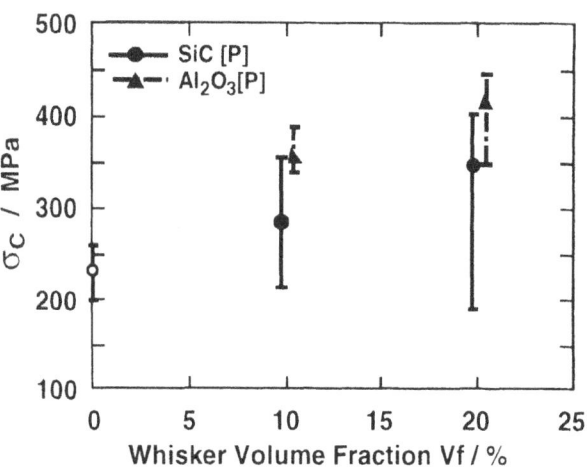

(b) Effect of whisker material on σ_C.

Figure 5. Increase in σ_C with the increase in Vf with regard to (a) whisker arrangement and (b) whisker material. Materials are monolithic alumina ("O"), SiC whisker reinforced Al_2O_3 for P direction (SiC/Al_2O_3[P] ("●")) and for N direction (SiC/Al_2O_3[N] ("■")), and Al_2O_3 whisker reinforced Al_2O_3 for P direction (Al_2O_3/Al_2O_3[P] ("▲"))

$Al_2O_3/Al_2O_3[P]$ is larger than $SiC/Al_2O_3[P]$. The differences are 80 MPa at a Vf of 10 % and 70 MPa at 20 %, respectively, and they are much larger than the differences between $SiC/Al_2O_3[P]$ and $SiC/Al_2O_3[N]$.

The value of σ_C measured in this study is the macroscopic critical stress for maincrack formation. The microscopic criteria for microcrack initiation was introduced by Laws et al.[10] as

$$l_c \cdot (\sigma_{yy}^2 + \sigma_{xy}^2) = b \cdot K_c^2 \qquad (1)$$

where l_c is the critical crack length, K_c is the microscopic fracture toughness, b is a constant and σ_{yy} and σ_{xy} are local tensile and shear stress, respectively. Since the monolithic alumina and the matrices of $SiC/Al_2O_3[P]$ and $SiC/Al_2O_3[N]$ were sintered from pure alumina powder without any additives, the differences of l_c and K_c between the materials are quite small. Consequently, it is concluded that the criteria for microcracking as well as maincrack formation is dominantly controlled by local stress, σ_{yy} and σ_{xy}, which are superposition of the applied stress and the residual stress due to the thermal mismatch between whisker and matrix. Because the thermal residual stress of $SiC/Al_2O_3[P]$ and $SiC/Al_2O_3[N]$ is considered as equivalent, the effect of whisker arrangement on σ_C was quite small (Figure 5 (a)). And the value of σ_C for $Al_2O_3/Al_2O_3[P]$ was larger than $SiC/Al_2O_3[P]$.(Figure 5 (b)), because the residual stress in $Al_2O_3/Al_2O_3[P]$ is smaller than $SiC/Al_2O_3[P]$.

Effect of Volume Fraction of Whisker on Critical Stress, σ_C

Considering the unit square with an area of λ^2 containing one whisker, the volume fraction of whisker, Vf, can be represented as

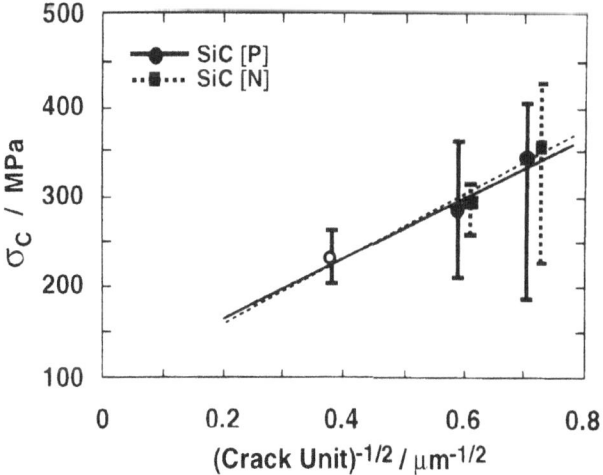

Figure 6. Dependence of σ_C on crack units. Materials are monolithic alumina (**O**), SiC whisker reinforced Al_2O_3 for P direction ($SiC/Al_2O_3[P]$ ("●")) and for N direction ($SiC/Al_2O_3[N]$ ("■")).

$$Vf = \frac{\pi \cdot r^2}{\lambda^2} \qquad (2)$$

where r is the radius of whisker. Assuming the homogeneous and one dimensional arrangement of whiskers, λ can be considered as the average distance between whiskers. It was shown by the authors[8] that the critical stress, σ_C, as well as bending strength, σ_B, of high purity monolithic alumina increases with the decrease in grain size, according to Petch's law. Therefore, it is considered that the maincrack formation yields the fracture mechanical criterion (K - criterion), which is consist with Equation (1), and the length of maincrack as initiated strongly depends on the length of crack unit relating to grain boundary facets. In whisker reinforced ceramic composites, the crack unit is considered as the distance between whiskers.

Figure 6 shows the dependence of the critical stress, σ_C, on the inverse square root of crack unit for SiC whisker reinforced Al_2O_3 for P direction ($SiC/Al_2O_3[P]$) and for N direction ($SiC/Al_2O_3[N]$). The fitting lines show good agreement with data; it is understood from this result that the criterion for maincrack formation in whisker reinforced composites is also fracture mechanical and the increase in volume fraction of whisker is useful to avoid the matrix cracking by this way.

Microfracture Process in Whisker Reinforced Composites

Figure 7 shows the relationships between the bending strength, σ_B, and the critical stress for maincrack formation, σ_C, for (a) SiC whisker reinforced Al_2O_3 for P direction ($SiC/Al_2O_3[P]$) and (b) N direction ($SiC/Al_2O_3[N]$), and (c) Al_2O_3 whisker reinforced Al_2O_3 for P direction ($Al_2O_3/Al_2O_3[P]$), respectively. As shown in the figure, the differences between σ_B and σ_C ($\sigma_B - \sigma_C$) strongly depends on material system; 40 MPa for monolithic alumina, 230 MPa for $SiC/Al_2O_3[P]$, 180 MPa for $SiC/Al_2O_3[N]$ and 40 MPa for $Al_2O_3/Al_2O_3[P]$. The differences of fracture toughness measured by IF method between composites and monolithic material (Figure 2) are 1.4 - 2.5 MPa$\sqrt{m}$ for $SiC/Al_2O_3[P]$, 1.7 - 1.8 MPa$\sqrt{m}$ for $SiC/Al_2O_3[N]$, and 0.7 - 0.8 MPa$\sqrt{m}$ for $Al_2O_3/Al_2O_3[P]$. On the other hand, the increases in $\sigma_B - \sigma_C$ are 190 MPa for $SiC/Al_2O_3[P]$, 140 MPa for $SiC/Al_2O_3[N]$, and 0 MPa for $Al_2O_3/Al_2O_3[P]$, respectively. It is understood from these results that $\sigma_B - \sigma_C$ strongly depends on the increase in fracture toughness. The toughening of ceramic composites resulting from crack bridging of whisker (a rise in R-curve), dK^w, was estimated by Becher et al.[3] as

$$dK^w = \sigma_f^w \left(\frac{Vf \cdot r}{6(1 - v^2)} \cdot \frac{E^c}{E^w} \cdot \frac{G^m}{G^i} \right)^{1/2} \qquad (3)$$

where, σ_f^w, r and E^w are the strength, radius and elastic modulus of whisker, v and E^c are Poisson's ratio and elastic modulus of the composite, and G^m and G^i are the strain energy release rates of matrix and interface, respectively. The value of G^i of Al_2O_3/Al_2O_3 is considered larger than SiC/Al_2O_3, because the materials of whisker and matrix are same. Therefore, the toughening of Al_2O_3/Al_2O_3 is poorer than that of SiC/Al_2O_3. Although precise measurement of R - curve of the materials is needed, it can be concluded that the fracture process after σ_C is controlled by the R - curve behavior of the material.

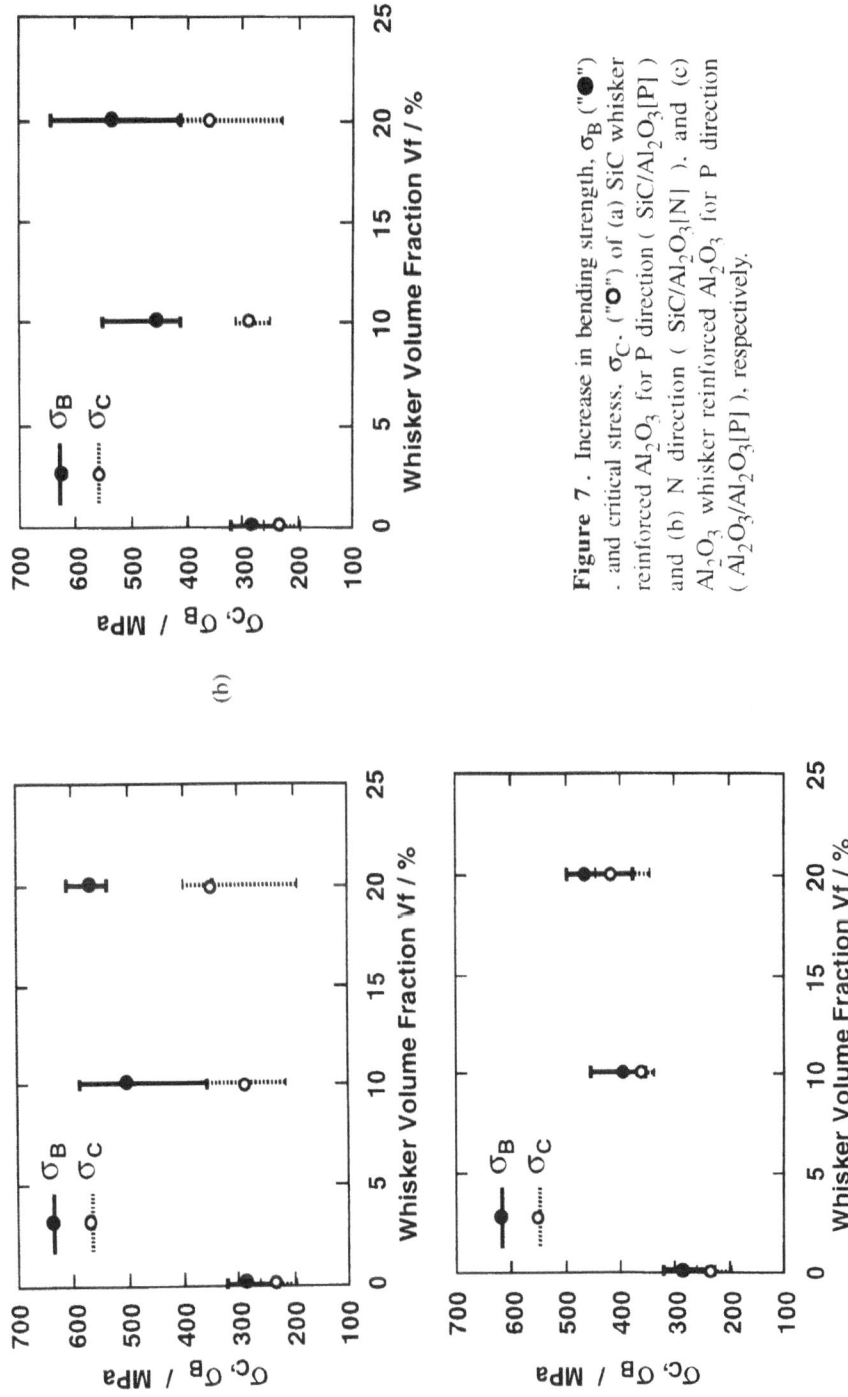

Figure 7. Increase in bending strength, σ_B ("●") and critical stress, σ_C ("○") of (a) SiC whisker reinforced Al_2O_3 for P direction ($SiC/Al_2O_3[P]$) and (b) N direction ($SiC/Al_2O_3[N]$), and (c) Al_2O_3 whisker reinforced Al_2O_3 for P direction ($Al_2O_3/Al_2O_3[P]$), respectively.

It is understood from the Figure 7 that the lifetime after the maincrack formation in silicon carbide whisker composites are much larger than those in alumina whisker composites. Consequently, it can be concluded that reinforcing silicon carbide whisker is effective for the enhancement of crack growth resistance while alumina whisker contributes the increase in crack initiation resistance.

CONCLUSIONS

In this study, monolithic alumina, silicon carbide reinforced alumina and alumina whisker reinforced alumina were sintered by hot pressing, and fracture toughness and bending strength of the materials were measured. Furthermore, microfracture process during four point bending tests was evaluated by AE technique.

The critical stress, σ_C, for microcracking, especially the maincrack formation, in matrix was determined from AE generation behavior. The value of σ_C is hardly influenced by the arrangement of whiskers in SiC whisker reinforced Al_2O_3. But it strongly depends on the combination of whisker and matrix materials; σ_C of Al_2O_3 whisker reinforced Al_2O_3 was larger than that of SiC whisker reinforced Al_2O_3. These results suggest that the microcracking in the matrix of whisker reinforced composites is controlled by the residual stress due to the thermal mismatch between whisker and matrix.

The fracture process after the maincrack formation during bending test yields the R-curve behavior of the composites. Consequently, the contributions of whisker to the strengthening and toughening were understood.

ACKNOWLEDGMENT

Present study was supported by Grant-in-Aid for Co-operative Research (A) (Project No. 05302049), the Ministry of Education, Science and Culture.

REFERENCES

1. K. T. Faber and A. G. Evans, Acta Metall., 31:565(1983).
2. A.G. Evans, Phil. Mag., 26:1327(1972).
3. P.R. Becher, C.H. Hsueh, P. Angelini and T.N. Tiegs, J. Am. Ceram. Soc.,71:1050 (1988).
4. Y. W. May and B. R. Lawn, J. Am. Ceram. Soc., 70:289(1987).
5. A. G. Evans and K. T. Faber, J. Am. Ceram. Soc., 67:255(1984).
6. T. Kishi, S. Wakayama and S. Kohara, Fracture Mechanics of Ceramics, Vol. 8, R. C. Bradt, A. G. Evans, D. P. H. Hasselman and F. F. Lange, ed., Plenum Press, New York, 85(1985).
7. S. Wakayama and H. Nishimura, Fracture Mechanics of Ceramics, Vol. 10, R. C. Bradt, D. P. H. Hasselman, M. Sakai and V. Ya. Shevchenco, ed., Plenum Press, New York, 59(1985).
8. S. Wakayama and M. Kawahara, CERAMICS adding the value, Vol. 2, M.J.Bannister, ed.,CSIRO, Melbourne, 938(1992).
9. K. Okada, M. Mutoh, N. Otsuka and T. Yano, J. Mat. Sci. Let., 10:588(1991).
10. N. Laws and J.C. Lee, J. Mech. Phys. Solids, 37:603(1989).

APPLICATION OF A MULTIPARTICLE MODEL TO ESTIMATE THE BRITTLE STRENGTH OF A PARTICLE REINFORCED COMPOSITE

V.I. Kushch and V.T. Golovchan

Institute for Superhard Materials
National Academy of Sciences
Kiev, 254074 Ukraine

ABSTRACT

The theoretical model for estimation of the strength of particle composites is proposed; it is based on a rigorous analytical solution of the elasticity boundary-value problem for a multi-particle model. The essence of the method is the representation of the displacement vector by a series of partial vectorial solutions of Lame's equation. By strictly satisfying all interfacial conditions the primary boundary-value problem is reduced to an infinite set of linear algebraic equations. The method allows the detailed analysis of stress distribution in each phase of the composite. To estimate brittle strength of a composite, this solution is combined with the statistical theory of brittle fracture under multiaxial loading. The model developed takes into account the statistical (Weibull's) brittle strength parameters of phase materials and internal stress concentration caused by thermoelastic mismatch and by the interaction of neighbouring particles of the disperse phase. It permits to predict the most probable fracture mode (matrix, inclusions or interfaces), the fracture probability of a composite specimen under complex loading and its mean brittle strength. The numerical results are presented; they demonstrate the possibilities of an approach to calculate stress microfields and to estimate the strength of real composite materials.

INTRODUCTION

The modern advanced composites and ceramics are, as a rule, the strongly heterogeneous materials with high volume contents of disperse phases. It is known that their properties are greatly influenced by the microstructure and internal microstresses caused by interaction of disperse phase particles. As a result, the one-particle models traditionally used to investigate properties of composites (self-consistens scheme, three-phase model, etc.) are insufficient for a reliable prediction of the properties of these materials. To take into account more completely particle-particle interaction, the multi-particle models consisting of a homogeneous medium with a finite or infinite number of inclusions must be considered. The theoretical analysis of these models calls for a complex mathematics to be applied. At the same time the multi-particle model reflects more adequately the microstructure of a composite and produces results which are in better agreement with experimental data. The most widely used variant of the multi-particle model is the lattice model with disperse particles arranged in some periodic array. The model was used successfully to investigate thermoelastic properties of composites with spherical inclusions[1-5]. As it was shown in these publications and in studies on the conductivity of periodic particle composites, the combination of this structural model with a rigorous method of analysis gives the satisfactory results even for highly heterogeneous materials with near-to-close packing of disperse phase particles. This model can be generalised also to describe the spatial distribution of particles in non-ordered composites[6].

Phase interaction becomes even more relevant when brittle strength of composite materials is considered because the fracture is governed, most likely, by the maximum value stresses rather than by their average values which determine mainly the effective elastic properties. So between hard particles considerable stress concentration can arise which initiates fracture. It is obvious that a reliable prediction of the composite strength is possible only on the basis of a detailed analysis of microstresses in structural models taking into account the essential peculiarities of specific composites and using reliable fracture criteria of phase materials in a complex stressed state. In the present study the attempt is made to elaborate the micromechanical brittle strength theory of particle composites satisfying these requirements.

THE PROBLEM AND METHOD OF SOLUTION

Let us consider the elastic isotropic medium containing a number N of elastic spherical inclusions with centres in points O_p, $p = 1, 2, ..., N$. We introduce N equally oriented Cartesian coordinate systems $O_p x_p y_p z_p$ with origins in centres of particles and corresponding spherical coordinates (r_p, Θ_p, ϕ_p). The mutual positions of particles with numbers p and q is determined by the vector $\mathbf{R}_{pq} = \mathbf{r}_p - \mathbf{r}_q$, where $\mathbf{r}_p$ is the radius-vector of the p-th local coordinate system. The heterogeneous medium is assumed to be loaded by constant remote stresses. The displacement vector u ($\mathbf{u} = \mathbf{u}^{(0)}$ in the matrix, $\mathbf{u} = \mathbf{u}^{(p)}$ in the p-th inclusion) in all points of the domain satisfies the Lame's equation

$$\frac{2(1-\nu)}{1-2\nu}\nabla(\nabla \cdot \mathbf{u}) - \nabla \times \nabla \times \mathbf{u} = 0, \tag{1}$$

where ν is the Poisson's ratio. On interfaces the conditions of perfect mechanical contact are supposed to be satisfied

$$\left[\mathbf{u}^{(0)} - \mathbf{u}^{(p)}\right]_{r_p = R_p} = 0; \quad \left[T_r(\mathbf{u}^{(0)}) - T_r(\mathbf{u}^{(p)})\right]_{r_p = R_p} = 0; \quad p = 1, 2, ..., N; \tag{2}$$

where R_p is the radius of the p-th particle $T_r(\mathbf{u}) = \sigma_r \mathbf{e}_r + \tau_{r\Theta}\mathbf{e}_\Theta + \tau_{r\phi} \cdot \mathbf{e}_\phi$ is the normal stress vector on the surface r = const.

To solve this problem analytically we use the method[1,2] first proposed by Golovchan[7]. The method consists in representation of the displacement vector in each phase of the composite by a series of partial vectorial solutions of Lame's equation in spherical coordinates (Appendix A). So, the series expansion of the inclusion vector $\mathbf{u}^{(p)}$ contains solutions $\mathbf{u}^{(i)}_{ts}$ (A1) only:

$$\mathbf{u}^{(p)} = \sum_{i=1}^{3} \sum_{t=0}^{\infty} \sum_{s=-t}^{t} D_{ts}^{(i)(p)} \mathbf{u}_{ts}^{(i)}(\mathbf{r}_p), \tag{3}$$

where $D_{ts}^{(i)(p)}$ are the constants found from the boundary conditions (2). We represent the displacement vector in the matrix $\mathbf{u}^{(0)}$ as a sum

$$\mathbf{u}^{(0)} = \mathbf{u}_l + \mathbf{u}_d, \tag{4}$$

where $\mathbf{u}_e = E \cdot \mathbf{r}_1$ describes the deformation of the homogeneous medium without inclusions, $\mathbf{u}_d$ is the disturbance caused by the presence of inhomogeneities. The matrix E has a sense of deformation tensor for $r \to \infty$. According to the superposition principle[8],

$$\mathbf{u}_d = \sum_{p=1}^{N} \sum_{i=1}^{3} \sum_{t=0}^{\infty} \sum_{s=-t}^{t} A_{ts}^{(i)(p)} U_{ts}^{(i)}(\mathbf{r}_p), \tag{5}$$

where $A_{ts}^{(i)(p)}$ are the unknown coefficients, $U^{(i)}_{ts}$ are the external partial solutions of equation (1) in spherical coordinates defined by (A2).

The set of algebraic equations for determination of the unknown coefficients in (2) and (5) can be obtained by substitution of these expressions into conditions (2) and using the vectorial harmonics (A3). We consider, for example, fulfillment of the boundary conditions (2) for particle with the number p. So,

$u^{(p)}$ is written already in variables of the p-th local basis and does not require any additional transformations. The linear part of (4) has the form

$$\mathbf{u}_l = \mathbf{E} \cdot (\mathbf{r}_p + \mathbf{R}_{pl}) = \mathbf{u}_l(\mathbf{r}_p) + \mathbf{u}_l(\mathbf{R}_{pl}) \tag{6}$$

The second term in (6) is the constant vector determining the motion of the whole solid; it does not influence the stress tensor. The expression of the first term in a spherical basis is

$$\mathbf{u}_l(\mathbf{r}_p) = \frac{\mathbf{u}_{00}^{(3)}(\mathbf{r}_p)}{3\gamma_0(\nu_0)}(E_{11} + E_{22} + E_{33}) + \frac{\mathbf{u}_{20}^{(1)}(\mathbf{r}_p)}{3}(2E_{33} - E_{11} - E_{22}) +$$

$$2\mathrm{Re}\left[\mathbf{u}_{20}^{(1)}(\mathbf{r}_p)(E_{13} - iE_{23}) + \mathbf{u}_{20}^{(1)}(\mathbf{r}_p)(E_{11} - E_{22} - 2iE_{12})\right] = \tag{7}$$

$$\sum_{i=1}^{3} \sum_{t=0}^{\infty} \sum_{s=-t}^{t} e_{ts}^{(i)}\, \mathbf{u}_{ts}^{(i)}(\mathbf{r}_p), \qquad e_{ts}^{(i)} \equiv 0 \quad \text{for} \quad t > 2.$$

The representation of u_d in the p-th local basis is based on addition theorems used for the partial solutions of Lame's equation (Appendix B). Substitution of (B1) into (5) gives us the next expression for u_d:

$$\mathbf{u}_d(\mathbf{r}_p) = \sum_{i=1}^{3} \sum_{t=0}^{\infty} \sum_{s=-t}^{t} \left[A_{ts}^{(i)(p)}\, \mathbf{U}_{ts}^{(i)}(\mathbf{r}_p) + a_{ts}^{(i)(p)}\, \mathbf{u}_{ts}^{(i)}(\mathbf{r}_p) \right], \tag{8}$$

where

$$a_{ts}^{(i)(p)} = \sum_{\substack{q=1 \\ q \neq p}}^{N} \sum_{j=1}^{3} \sum_{k=0}^{\infty} \sum_{l=-k}^{k} A_{kl}^{(j)(q)}\, \eta_{klts}^{(j)(i)}(\mathbf{R}_{qp}), \tag{9}$$

The matrix form of (9) is

$$\mathbf{a}_{ts}^{(p)} = \sum_{\substack{q=1 \\ q \neq p}}^{N} \sum_{k=0}^{\infty} \sum_{l=-k}^{k} \left[\eta_{klts}(\mathbf{R}_{qp}) \right]^{\mathrm{T}} \cdot \mathbf{A}_{kl}^{(q)}, \tag{10}$$

where

$$\mathbf{A}_{ts}^{(p)} = \left\| A_{ts}^{(i)(p)} \right\|^{\mathrm{T}}; \quad \mathbf{a}_{ts}^{(p)} = \left\| a_{ts}^{(i)(p)} \right\|^{\mathrm{T}}; \quad \eta_{tskl} = \left\| \eta_{tskl}^{(i)(j)} \right\|.$$

Substitution of (3) and (4) into the first condition of (2) taking into account (7) and (8) and using the orthogonality of the vectorial harmonics (A3) gives us the set of algebraic equations

$$(t-s)!(t+s)!\,UG_t^{(0)}(R_p)\cdot A_{ts}^{(p)} + UM_t^{(0)}(R_p)\cdot(a_{ts}^{(p)} + e_{ts}) = UM_t^{(p)}(R_p)\cdot D_{ts}^{(p)},$$

(11)

$$t=1,2,\ldots, \quad |s|\le t; \quad p=1,2,\ldots,N.$$

In (11) $UG_t^{(p)}$ and $UM_t^{(p)}$ are the matrices

$$UG_t^{(p)}(r)=r^{-(t+2)}\begin{pmatrix} 1 & 0 & r^2\beta_{-(t+1)}(v_p) \\ 0 & r/t & 0 \\ -t-1 & 0 & r^2\gamma_{-(t+1)}(v_p) \end{pmatrix}; \quad UM_t^{(p)}(r)=r^{t-1}\begin{pmatrix} 1 & 0 & r^2\beta_t(v_p) \\ 0 & -r/(t+1) & 0 \\ t & 0 & r^2\gamma_t(v_p) \end{pmatrix}.$$

Fulfillment of the second condition in (2) is quite similar, but in this case one must use additionally the expression (A4) of the normal stress vector for the partial solutions (A1) and (A2). After transformation we have

$$(t-s)!(t+s)!\,TG_t^{(0)}(R_p)\cdot A_{ts}^{(p)} + TM_t^{(0)}(R_p)\cdot(a_{ts}^{(p)} + e_{ts}) = \frac{\mu_p}{\mu_0} TM_t^{(p)}(R_p)\cdot D_{ts}^{(p)},$$

(12)

$$t=1,2,\ldots, \quad |s|\le t; \quad p=1,2,\ldots,N.$$

Here the matrices $TG_t^{(p)}$ and $TM_t^{(p)}$ have the form

$$TG_t^{(p)}(r)=r^{-(t+3)}\begin{pmatrix} -(t+2) & 0 & r^2 b_{-(t+1)}(v_p) \\ 0 & \dfrac{t+2}{2t}r & 0 \\ (t+1)(t+2) & 0 & r^2 a_{-(t+1)}(v_p) \end{pmatrix},$$

$$TM_t^{(p)}(r)=r^{t-2}\begin{pmatrix} t-1 & 0 & r^2 b_t(v_p) \\ 0 & -r\dfrac{t-1}{2(t+1)} & 0 \\ (t-1)t & 0 & r^2 a_t(v_p) \end{pmatrix},$$

and μ_p is the shear modulus of the p-th phase material.

Relations (11) and (12) together form a closed infinite set of linear algebraic equations. It contains as parameters the components of the matrix E determining the kind and intensity of external loading. By exclusion of $D^{(p)}_{ts}$ the number of unknowns is reduced by the factor 2. It was shown[1] that the algebraic set obtained has the normal determinant. Hence, its solution can be found by the reduction method or by the method of successive approximations. The latter is preferable when the number of inclusions becomes large enough. The application of an iterative procedure to multi-particle problems is described in detail by Kushch[9].

It is easy to see that the method outlined can be applied to analyse a model consisting of a medium with an infinite number of equal particles with centres lying in poles of some spatial lattice. For this purpose, it is sufficient to replace in all formulae above the finite sum by sum over all lattice poles. In this case, the problem even simplifies because for uniform load the solution has the same periodicity features as the structure. Hence, $A^{(p)}_{ts} = A_{ts}$, $a^{(p)}_{ts} = a_{ts}$ and $D^{(p)}_{ts} = D_{ts}$ for all values of p, fulfillment of conditions (2) for one trial particle (say, with p = 0) means their fulfillment for the rest of particles in the lattice. The expression of a_{ts} becomes

$$\textbf{a}_{ts} = \sum_{k=0}^{\infty} \sum_{l=-k}^{k} \left[\overset{\bullet}{\eta}_{klts} \right]^{T} \cdot A^{(q)}_{kl},$$

(13)

where

$$\overset{\bullet}{\eta}_{tskl} = \sum_{q \neq l} \eta_{tskl}(\mathbf{R}_{ql})$$

are the lattice sums. For their evaluation the effective numerical procedures are available[1].

The solution obtained provides the analysis of spatial stress distribution in each phase of composites[2,10] and computation of the effective tensor of elastic moduli[1,3].

ESTIMATE OF BRITTLE STRENGTH OF THE COMPOSITE

The solution indicated can be used as a background to elaborate various micromechanics theories of brittle fracture of particle composites. For this the deterministic as well as the statistical criteria of strength may be examined. The latter are obviously more appropriate due to the statistics of fracture occurring in brittle materials. It is known that the strength of homogeneous materials with microdefects is described satisfactorily by Weibull's theory. This theory describes, in part, as an essential feature of brittle materials the size-dependence of strength. Recently, the series of generalisations of this theory has been proposed[11-14] for the complex stressed state of specimen.

So, Matsuo[12] has proposed a multiaxial distribution function to be applied to arbitrarily selected fracture criteria. This function for fracture caused by internal, randomly oriented microcracks may be represented by the equation

$$P_f = 1 - \exp\left[-\left(\frac{\sigma_{1\max}}{\sigma_0} \right)^m V_{eff} \right],$$

(14)

where V_{eff} is the effective volume

$$V_{eff} = \frac{2}{\pi \Omega} \int_V \int_0^{\pi/2} \int_0^{\pi/2} \left(\frac{Z}{\sigma_{1\max}} \right)^m \sin\theta \, d\theta \, d\varphi \, dV,$$

(15)

m and σ_0 are the Weibull's parameters. In a general case $Z = Z(\sigma_n, \tau_n)$, where σ_n is the normal stress vertical to the crack surface and τ_n is the shear stress parallel to the crack surface. The expression of Z can vary depending on the specific material considered[13,14]. We will restrict our consideration now to only the multiaxial Weibull's theory with $Z = \sigma_n$. The mean (expected) strength value of the sample is

$$\sigma_f = E(\sigma_{1\max}) = \sigma_0 \left(V_{eff} \right)^{-1/m} \Gamma(1 + 1/m). \tag{16}$$

where $\Gamma(z)$ is the Gamma-function.

The formulae (14) - (16) describe the fracture of brittle material under uniformity assumption on the distribution and orientation of microcracks in the volume. The more general case of non-uniform crack distribution may also be considered in this way[11]. When fracture is caused by surface defects on the external boundary or on interfaces, its fracture probability is also described by formula (14) where the effective surface S_{eff} should be considered instead of the effective volume. The value of S_{eff} is defined by formula (15) where the volume integral should be replaced by the surface integral.

Now we use the theory given to estimate brittle strength of a model composite with a simple cubic lattice of spherical inclusions. Due to the periodicity of the structure, the elementary structure cell (cube with centrally placed spherical inclusion) is also the representative volume of the composite. We suggest that the mean crack size is much smaller than the typical size of inhomogeneities. Therefore, the strength of this structural model may serve as strength characteristics of the composite material. The fracture probability of the cell includes risk of fracture of the matrix, inclusion and interface

$$P_f = 1 - \exp \left[- \left(\frac{\sigma_{1\max}}{\sigma_0^{(0)}} \right)^{m^{(0)}} V_{eff}^{(0)} - \left(\frac{\sigma_{1\max}}{\sigma_0^{(1)}} \right)^{m^{(1)}} V_{eff}^{(1)} - \left(\frac{\sigma_{1\max}}{\sigma_0^{(2)}} \right)^{m^{(2)}} S_{eff} \right], \tag{17}$$

where Weibull's parameters $m^{(i)}$ and $\sigma_0^{(i)}$ for each term in general are different.

NUMERICAL RESULTS

To verify the validity of the approach proposed, a comparison with experimental data is needed. For three materials chosen for comparison the effect on strength of the disperse phase volume content was carefully studied[15-18]. These materials have the following common features:
a) Matrix phases are the glasses with similar properties.
b) Inclusions are spherical with small deviation from size and uniform spatial distribution.
c) Deformations of composite phases are linearly-elastic up to fracture.
d) Fracture arises in the matrix phase.
Properties of these materials, denoted as M1, M2, and M3, are given in Table 1.

Table 1. Properties of materials under consideration

Material	M1[15]	M2[15]	M3[18]
matrix	70 SiO$_2$ 14Na$_2$0 16B$_2$0$_3$ glass	70 SiO$_2$ 14Na$_2$0 16B$_2$0$_3$ glass	80 SiO$_2$ 15Na$_2$0 5Ca0 glass
filler	alumina	cavities	silica
particle size, μm	60	60	26
v_0	0.197	0.197	0.2
v_1	0.257	-	0.2
μ_1 / μ_0	5.0	0.0	1.2
$\alpha_0 - \alpha_1$, K^{-1}	0.0	-	8.0·10^{-6}

Here α_i is the linear coefficient of thermal expansion of the i-th phase. M2 is the porous material, M1 is the composite with identical phase thermal expansion coefficients. Hence, the residual thermal stresses in these materials are practically absent and do not influence the strength of materials. By contrast, in M3 the elastic properties of phase materials are similar and the main source of stress concentration is the difference of thermal expansion.

We present at the beginning some results of stress concentration studies on these materials. So, in Fig. 1 the stress $\sigma^{(0)}_{33}$ distribution in the matrix near the contact surface is presented for some values of the volume content of the disperse phase c. The stressed state is induced by uniaxial

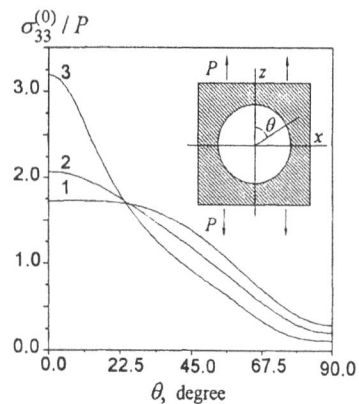

Figure 1. Stress $\sigma^{(0)}_{33}$ distribution near the contact surface due to uniaxial tension of composite M1: line 1 - c=0.1; line 2 - c=0.3; line 3 - c=0.5.

tensile load P in z-direction. The curve 1 (c = 0.1) is similar to the solution for a single inclusion in an infinite medium, where the stress concentration increases with increasing c. This effect is the result of particle interaction. It cannot be described by known approximate theories[19,20] predicting the decline of stresses when c increases.

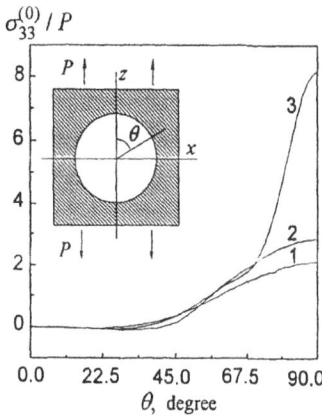

$$\sigma_{33}^{(0)} / P$$

Figure 2. Stress $\sigma_{33}^{(0)}$ distribution near the cavity surface due to uniaxial tension of porous material M2: line 1 - c=0.1; line 2 - c=0.3; line 3 - c=0.5

The analogous data for the porous material M2 are presented in Fig. 2. The stress concentration in cavities is greater than in case of hard inclusions, especially for high porosity. Hence, the lower level of destructive load can be predicted for this material.

The curves in Fig. 3 represent the distribution of thermal stresses in material M3. For small values of c the interface stress slightly decreases and then grows rapidly for $c > 0.3$.

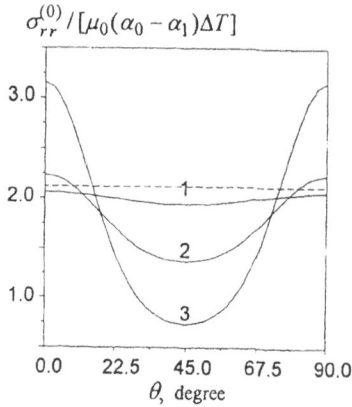

$$\sigma_{rr}^{(0)} / [\mu_0(\alpha_0 - \alpha_1)\Delta T]$$

Figure 3. Thermal stress distribution on the interface of composite M3: line 1 - c=0.1; line 2 - c=0.3; line 3 - c=0.5.

The dashed line in this diagramme represents the value

$$\beta = (\alpha_0 - \alpha_1)\Delta T \left(\frac{1 + \nu_0}{2E_0} + \frac{1 - 2\nu_1}{E_1} \right)^{-1}, \tag{18}$$

used[18] to estimate interfacial thermal stress. It is seen from the plot that formula (18) is valid for very small values of c. With c increased, the difference between the accurate and approximate solutions becomes more and more significant.

Now we estimate the brittle strength of these materials using the statistical criterion mentioned before. We suppose that the weakest link is the matrix. Hence, it is sufficient to carry out the integration in (17) over the matrix volume within the elementary structure cell only. With relation

$$\sigma_f(c_1)/\sigma_f(c_2) = \left[V_{eff}(c_1)/V_{eff}(c_2)\right]^{-1/m}, \tag{19}$$

taken into account which follows directly from (16), the dimensionless value $\sigma_f(c)/\sigma_f(0)$ does not contain the parameter σ_0 and depends on m only. In Fig. 4 the experimental data[15] and calculated concentration dependencies are plotted for composite M1. Curve 1 is calculated for m = 5 and curve 2 is calculated for m = 10. The dashed curve corresponds to formula[15] $\sigma_f(c)/\sigma_f(0) = (1 - c)^{-1/2}$. All data in this and the following figures are related to the strength value of the same volume of the matrix material. Unfortunately, the exact value of m for a given material is unknown, but for the majority of glasses m $\approx$ 5. As can be seen from the figure, curve 1 (m = 5) is a good enough approximation of empirical data.

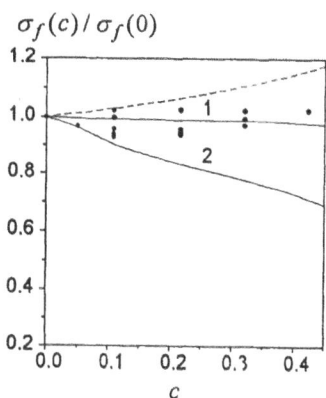

Figure 4. Comparison with experiment for composite material M1.

The analogous results for porous material M2 are depicted in Fig. 5 where black points represent the experiment[15]. The behaviour of the other available data[16] is similar. In this case the effect of Weibull's modulus of strength is less significant. Curve 1 (m = 5) as well as curve 2 (m = 10) describe well the decline of strength with increasing porosity.

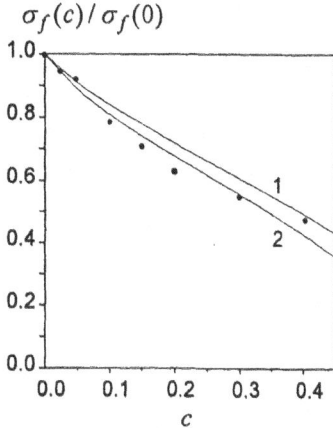

Figure 5. Comparison with experiment for porous material M2.

The thermal stresses may affect essentially the strength of a composite. The points in Fig. 6 represent experiment data[18] for material M3 composed of phases with different thermal expansion coefficient. Line 1 is calculated under the assumption that thermal stresses are absent ($\beta = 0$) and predicts a slightly growth of strength when the fraction of the disperse phase increases. Note that this curve and the dashed line in Fig. 4 are close enough to each other. This means that the theory[15] is valid mainly for weakly heterogeneous materials where no significant stress concentrations arise.

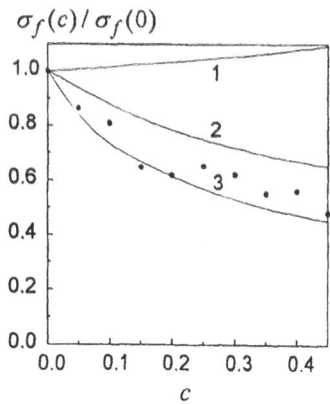

Figure 6. Comparison with experiment for composite material M3.

In fact, the force and thermal load are applied simultaneously and fracture is the result of their common action. The value of the effective volume in this case depends additionally on the ratio ß/P. The comparison with experiment[18] is rather difficult because the matrix solidification temperature is not reported there and the absolute value of thermal stress is unknown. Curve 2 in Fig. 6 is calculated for ß/P = 1.0 and curve 3 is calculated for ß/P = 2.0. The last value is obviously more reasonable for glass (it gives $\Delta T \approx 450$ K for P = 100 MPa). It is seen that internal thermal stresses change radically the σ_f dependence on c. As in previous cases, the correlation is satisfactory of calculated and measured values. Note, however, that in a real experiment the ratio ß/P depends on c. Taking into account this fact may change (not significantly) the course of theoretical curves but, to carry out of these calculations additional information is needed about the material under consideration.

CONCLUSION

The results of the comparison above show that the approach has some potential in predicting the strength characteristics of brittle composite materials. Even for the model of the simplest structure the theoretical and experimental data are in satisfactory agreement. The further development of this approach requires at first the geometrical model be developed taking into account the microstructural parameters of the specific material (size and spatial distribution rule of particles, their shape, presence in the interfacial zones of a third phase, etc.). Second, the experimentally obtained Weibull's parameters of each phase and interfaces of the composites considered are needed. Third, the fracture criterion must be chosen which predicts best the fracture of phase materials. Meeting of these conditions will permit to create the model providing the more reliable theoretical estimation of brittle strength of particle reinforced composites.

REFERENCES

[1]   V.T. Golovchan, "Anisotropy of Physical and Mechanical Properties of Composite Materials", Naukova dumka, Kiev (1987) (in Russian).

[2]   V.T. Golovchan, A.N. Guz, Yu. V. Kohanenko, and V.I. Kushch, "Mechanics of Composites (in 12 V.) Vol. 1. Statistics of materials", Naukova dumka, Kiev (1993) (in Russian).

[3]   V.I. Kushch, "Computation of effective elastic moduli of a granular composite material of regular structure", Soviet Applied Mechanics 23:362 (1987).

[4]   C.K. Nunan and J.B. Keller, "Effective elasticity tensor of a periodic composite, J. Mech. Phys. Solids. 32:253 (1984).

[5]   A.S. Sangani and W. Lu, "Elastic coefficients of composites containing spherical inclusions in a periodic array", J. Mech. Phys. Solids. 35:1 (1987).

[6]   A.S. Sangani and C. Yao, "Bulk thermal conductivity of composites with spherical inclusions", J. Appl. Phys. 65:1334 (1988).

[7]   V.T. Golovchan, " To solution of elasticity boundary-value problems of solids bounded by spherical surfaces", Dopovidi AN of Ukrainian SSR. N°1. 64 (1974) (in Russian).

[8]   M.G. Slobodyansky, "The general forms of solution of elasticity equations for single-connected and multiply-connected domains expressed through the harmonic functions", Prikladnaya matematika i mehanika. 18:55 (1954) (in Russian).

[9]   V.I. Kushch, " Elastic equilibrium of a medium containing finite number of aligned spheroidal inclusions", Int. J. Solids Structures, in press (1995).

[10]   V.I. Kushch, " Elastic equilibrium of a medium containing periodic spherical inclusions", Soviet Applied Mechanics, 21:435 (1985).

[11]   V.T. Golovchan, " On one statistical theory of brittle fracture", Soviet Applied Mechanics, 25:101 (1989).

[12]   Y. Matsuo, Transactions of Japanese Society of Mechanical Engineers. A, 46:605 (1980).

[13]   Y. Nakasuji, N. Yamada, H. Tsuruta, M. Masuda, and M. Matsui, "Fracture behaviour of non-oxide ceramics under biaxial stresses" - in: "Fracture Mechanics of Ceramics", Vol. 10, R.C. Bradt et al., eds., 211, Plenum Press, New York (1992).

[14] S. Hayashi and A. Suzuki, "Bending fracture strength of silicon nitride disks with shoulder fillet at room temperature", ibid., 247.

[15] D.P.H. Hasselman and R.M. Fulrath, "Proposed fracture theory of dispersion strengthened glass matrix", J. Amer. Ceram. Soc. 49:68 (1966).

[16] R.L. Bertolotti and R.M. Fulrath, "Effect of micromechanical stress concentration on strength of porous glass", J. Amer. Ceram. Soc. 50:558 (1967).

[15] D.P.H. Hasselman and R.M. Fulrath, "Micromechanical stress concentration in two-phase brittle-matrix ceramic composites", J. Amer. Ceram. Soc. 50:399 (1967).

[18] N. Miyata, S. Akada, H. Omura, and H. Jinno, "Microcrack toughening mechanism in brittle matrix composites" - in: "Fracture Mechanics of Ceramics", Vol. 9, R.C. Bradt et al., eds., 339, Plenum Press, New York (1992).

[19] V.M. Levin, "On stress concentrations on inclusions in composite materials", - Prikladnaya matematika i mehanika, 41:735 (1977) (in Russian).

[20] G.P. Tandon and G.J. Weng, "Stress distribution in and around spheroidal inclusions and voids at finite concentration", Trans. ASME. J. Appl. Mech. 53:511 (1986).

[21] A.F. Ultiko, "The Method of Eigenfunctions in the Spatial Problems of Elasticity", Naukova dumka, Kiev (1979) (in Russian).

APPENDIX A

Partial vectorial solutions of Lame's equation in a spherical basis[1,2]

Internal solutions (constrained at $r = 0$) are

$$\mathbf{u}_{ts}^{(1)} = \frac{r^{t-1}}{(t+s)!}\left(\mathbf{B}_{ts}^{(1)} + t\mathbf{B}_{ts}^{(3)}\right), \quad \mathbf{u}_{ts}^{(2)} = -\frac{i}{(t+1)}\frac{r^t}{(t+s)!}\mathbf{B}_{ts}^{(2)},$$

$$\mathbf{u}_{ts}^{(3)} = \frac{r^{t+1}}{(t+s)!}\left(\beta_t \mathbf{B}_{ts}^{(1)} + \gamma_t \mathbf{B}_{ts}^{(3)}\right), \tag{A1}$$

external solutions (disappearing at infinity) are

$$U_{ts}^{(1)} = \frac{(t-s)!}{r^{t+2}}\left[B_{ts}^{(1)} - (t+1)B_{ts}^{(3)}\right], \quad U_{ts}^{(2)} = \frac{i}{t}\frac{(t-s)!}{r^{t+1}}B_{ts}^{(2)},$$

$$U_{ts}^{(3)} = \frac{(t-s)!}{r^t}\left[\beta_{-(t+1)}B_{ts}^{(1)} + \gamma_{-(t+1)}B_{ts}^{(3)}\right],$$

$$(A2)$$

where

$$\beta_t = \frac{t+5-4\nu}{(t+1)(2t+3)}, \quad \gamma_t = \frac{t-2+4\nu}{2t+3};$$

$$B_{ts}^{(1)} = e_\theta \frac{\partial}{\partial\theta}\chi_t^s + \frac{e_\varphi}{\sin\theta}\frac{\partial}{\partial\varphi}\chi_t^s, \qquad B_{ts}^{(3)} = e_r\chi_t^s,$$

$$(A3)$$

$$B_{ts}^{(2)} = \frac{e_\theta}{\sin\theta}\frac{\partial}{\partial\varphi}\chi_t^s - e_\varphi\frac{\partial}{\partial\theta}\chi_t^s, \qquad t = 0,1,\ldots; \ |s|\le t;$$

is the full and orthogonal set of vectorial harmonics in spherical coordinates χ^s_t $(\theta, \varphi) = Ps_t (\cos\theta) \exp (is \varphi)$, Ps_t are the associated Legendre's polynomials. The normal stress vector T_r (u) for partial solutions (A1) and (A2) has the form

$$\frac{1}{2\mu}T_r(u_{ts}^{(1)}) = \frac{(t-1)}{r}u_{ts}^{(1)}, \qquad \frac{1}{2\mu}T_r(u_{ts}^{(2)}) = \frac{(t-1)}{2r}u_{ts}^{(2)},$$

$$\frac{1}{2\mu}T_r(u_{ts}^{(3)}) = \frac{r^t}{(t+s)!}\left(b_t B_{ts}^{(1)} + a_t B_{ts}^{(3)}\right),$$

$$(A4)$$

$$\frac{1}{2\mu}T_r(U_{ts}^{(1)}) = -\frac{(t+2)}{r}U_{ts}^{(1)}, \quad \frac{1}{2\mu}T_r(U_{ts}^{(2)}) = -\frac{(t+2)}{2r}U_{ts}^{(2)},$$

$$\frac{1}{2\mu}T_r(U_{ts}^{(3)}) = \frac{(t-s)!}{r^{t+1}}\left[b_{-(t+1)}B_{ts}^{(1)} + a_{-(t+1)}B_{ts}^{(3)}\right],$$

$$a_t = (t+1)\gamma_t - 2\nu, \qquad b_t = (t+1)\beta_t - 2(1-\nu)/(t+1).$$

With use of these formulae the conditions (2) can be satisfied without any difficulties.

APPENDIX B

Addition theorems for external solutions of Lames's equation in spherical basis

The re-expansion formulae of partial solutions can be expressed by[7]

$$\mathbf{U}_{ts}^{(i)}(\mathbf{r}+\mathbf{R}) = \sum_{j=1}^{3}\sum_{k=0}^{\infty}\sum_{l=-k}^{k} (-1)^{k+l}\, \eta_{tskl}^{(i)(j)}(\mathbf{R})u_{kl}^{(j)}(\mathbf{r}), \quad \|\mathbf{r}\| < \|\mathbf{R}\|; \tag{B1}$$

where

$$\eta_{tskl}^{(i)(j)} = 0, \quad j > i; \qquad \eta_{tskl}^{(i)(i)} = Y_{t+k}^{s-l}, \quad i=1,\,2,\,3;$$

$$\eta_{tskl}^{(2)(1)} = -\left(\frac{l}{k}+\frac{s}{t}\right)Y_{t+k-1}^{s-l}, \qquad \eta_{tskl}^{(3)(2)} = 4(1-v)\eta_{tskl}^{(2)(1)}, \quad k\geq 1;$$

$$\eta_{tskl}^{(3)(1)} = -Z_{t+k}^{s-l} - Y_{t+k}^{s-l}\left[\frac{(t+k-1)^2-(s-l)^2}{2t+2k-1} + C_{-(t+1),s} + C_{k-2,l} - 4(1-v)\frac{l}{k}\left(\frac{s}{t}+\frac{l}{k-1}\right)\right],$$

$$|l| \neq k;$$

$$\eta_{tsnn}^{(3)(1)} = -Z_{t+n}^{s-n} - Y_{t+n}^{s-n}\left[\frac{(t+n-1)^2-(s-n)^2}{2t+2n-1} - \frac{t^2-(s-1)^2}{2t+1} - (t+s)(t+s-1)\varepsilon_{-(t+1)}\right],$$

$$\tag{B2}$$

$$\eta_{tsn,-n}^{(3)(1)} = -Z_{t+n}^{s+n} - Y_{t+n}^{s+n}\left[\frac{(t+n-1)^2-(s+n)^2}{2t+2n-1} - \frac{t^2-(s+1)^2}{2t+1} - (t-s)(t-s-1)\varepsilon_{-(t+1)}\right],$$

$$C_{ts} = \beta_t\left[(t+1)^2 - s^2\right], \quad \varepsilon_t = \frac{3-4v-(2t+1)^{-1}}{(t+1)(2t+3)}.$$

In (B2) $Y_t^s = \dfrac{(t-s)!}{r^{t+1}}\,\chi_t^s(\theta,\phi)$ are the scalar harmonics, $Z_t^s = \dfrac{r^2}{(2t-1)}\,Y_t^s$ are the

scalar biharmonics in spherical basis.

FRACTURE BEHAVIOUR AND TOUGHENING OF ALUMINA-BASED COMPOSITES FABRICATED BY MICROSTRUCTURAL CONTROL

Byung-Koog Jang*(**), Manabu Enoki*,
Teruo Kishi*, Sang-Ho Lee** and Hee-Kap Oh**

*Research Center for Advanced Science and Technology, The University of Tokyo, 4-6-1, Komaba, Meguro-ku, Tokyo 153, Japan
**Ssangyong Research Center, Ssangyong Cement Industrial Co., Ltd. P.O. Box 12, Yuseong, DaeJeon 305-345, Korea

ABSTRACT

Methods for improving the mechanical properties of Al_2O_3 ceramics were investigated using Al_2O_3/5vol% SiC composites fabricated by hot-pressing. Nano-sized SiC particulates were dispersed uniformly into Al_2O_3 matrix to create a nearly intragranular type of reinforcement. Flexural strength increased because sub-grain boundaries that formed through distribution of the SiC particulates reduced the critical flaw size. This decrease in critical flaw size could be attributed to the grain-boundary pinning effect of the SiC particulates, which effectively restrained grain growth in the Al_2O_3/5vol%SiC composites. The flexural strength was inversely proportional to the square root of the matrix grain size. The composites also showed higher fracture toughness than did monolithic Al_2O_3, and the fracture toughness values tended to increase with increases in the hot-pressing temperature. An investigation by SEM of the deflection behavior of propagated cracks induced by Vickers indentation showed a larger crack-path deflection angle for composites than for monolithic Al_2O_3, resulting in improved toughness. Transmission electron microscopic observation indicated that propagating cracks were deflected by the dispersed SiC particulates. The toughening of the Al_2O_3/SiC composites in the present study was attributed to crack deflection by the SiC particulates and to residual stress caused by a thermal expansion mismatch between Al_2O_3 and SiC.

INTRODUCTION

Alumina ceramics possess numerous excellent properties such as high hardness, low electrical conductivity, oxidation resistance, and chemical stability. The mechanical properties of strength and fracture toughness for Al_2O_3 ceramics, however, are lower than for the other structural ceramics such as Si_3N_4 and ZrO_2.[1,2] Lately, increased interest has focused on strengthening and toughening polycrystalline ceramics through the incorporation of dispersed particulates or second phases. Recently, attempts have been made to improve the mechanical properties of Al_2O_3-based ceramic composites by incorporating second phases and whiskers and platelets as, for example, ZrO_2 particles,[3-5] SiC whiskers,[6,7] and SiC platelets,[8] respectively. Micro-structural control is imperative for Al_2O_3 ceramics because the mechanical properties of these ceramics are closely related to the microstructure. A better approach toward enhancing mechanical properties through the suppression of grain growth thus may be achieved by incorporating minor amounts of second phases into the matrix. Alumina ceramics also can be strengthened and toughened by adding some reinforced second phases such as SiC or ZrO_2 and by adding SiC particulates. For example, second phase particulate dispersion in Al_2O_3 ceramics has been reported to effectively enhance toughness.[9-11] Recently, high-temperature structural applications have been investigated for nano-composites containing dispersed SiC[12,13] or Si_3N_4[12] particulates in an Al_2O_3 matrix. The mechanism by which the second phase improves the mechanical properties of such composites, however, remains unclear. The present work represents an attempt to fabricate composites by the simultaneous consolidation of nano-sized SiC particulates into an Al_2O_3 matrix, to investigate propagating-crack behavior, and to evaluate the effect of the second phase on the toughening mechanism and the residual stresses in composites vs monolithic Al_2O_3.

EXPERIMENTAL PROCEDURE

MATERIALS

The Al_2O_3 powder used as a matrix had average particle size of 0.2 μm, and the β-SiC powder used as reinforcing material had particle sizes averaging 0.3 μm, with the total particle distribution in the range under 1 μm. The composite studied consisted of Al_2O_3 powder and 5 vol% β-SiC. These powders were mixed for 24 h in a methanol solution using a polyethylene mill and Al_2O_3 balls. After the powders had been ball-milled, the resulting slurry was dried and sieved through 200 mesh. The mixed powders then were uniaxially hot-pressed in graphite dies coated with BN spray in the temperature range 1000°C to 1800°C for 2h under 30 MPa of pressure in an Ar atmosphere.

EVALUATION OF PROPERTIES

The hot-pressed specimens were nominally cut into 3 by 3 by 40 mm strips using diamond blades. The strips then were polished with 3 μm diamond paste and, finally, with 1 μm diamond paste to achieve a mirror surface. The grain structure was revealed by performing thermal etching at 1500°C in an Ar atmosphere. The grain size of the thermally etched specimens was measured by the liner intercept method. The flexural strength of the polished specimens was determined using 4-point bending apparatus. Fracture toughness was estimated by the Vickers indentation microfracture method.[14] To observe crack behavior, a Vickers diamond pyramid was used to induce controlled crack generation. A load of 98 N then was applied for 15 S using a Vickers indentor on the polished surfaces. Crack propagation on the surfaces of the polished and thermally etched specimens was observed by scanning electron microscopy (SEM) and transmission electron microscopy (TEM) with an X-ray energy dispersive spectroscopy (EDS) analyzer attachment. The TEM samples were prepared by a conventional thinning procedure that involved mechanical slicing and polishing to give a thickness of 200 μm. After the specimens had been polished, crack were introduced by Vickers indentation to allow study of crack propagation. Specimens measuring 3 mm in diameter were continuously prepared using an ultrasonic cutter. The back of each indented plane was polished to ~30 nm thick by dimple grinding and successively ion milling.

RESULTS AND DISCUSSION

MECHANICAL PROPERTIES AND FABRICATION OF COMPOSITES

Figure 1 shows transmission electron micrographs of Al_2O_3/5vol%SiC composites hot-pressed at 1600°C. These composites reached approximately full density beyond 1600°C. As shown in Fig. 1, very fine SiC particulates exist preferentially in an intragranular form

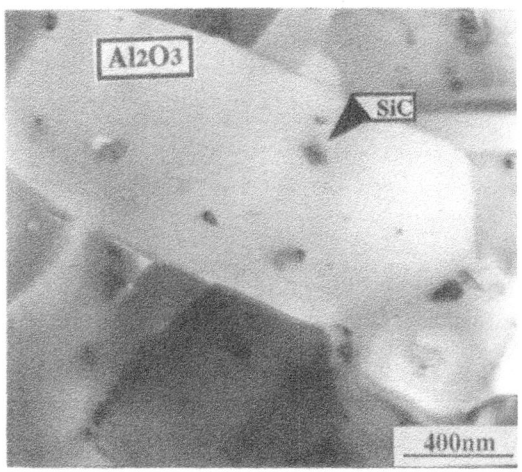

Figure 1. TEM Photograph of Al₂O₃/5vol%SiC composites hot-pressed at 1600°C.

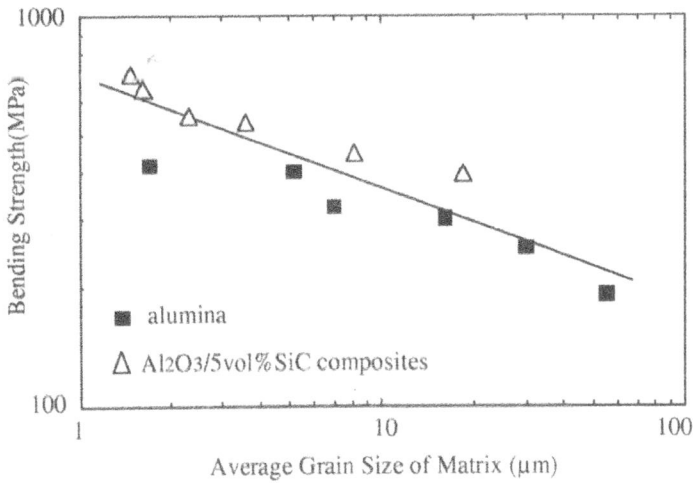

Figure 2. Flexural strength of hot-pressed monolithic alumina and Al₂O₃/5vol%SiC composites as function of grain size of alumina matrix.

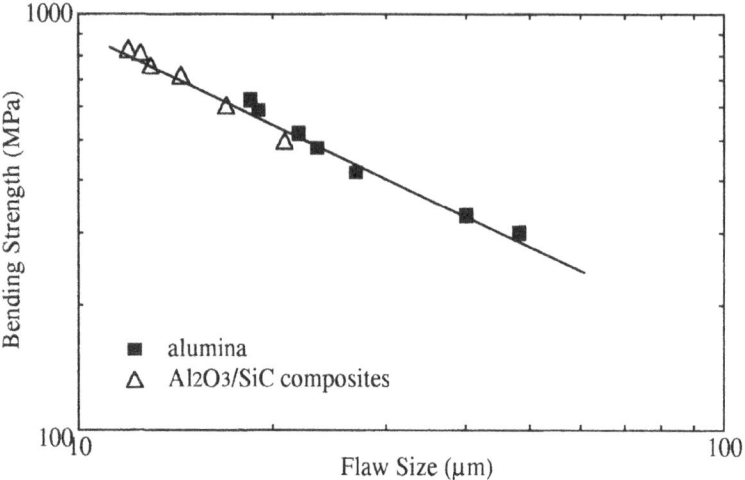

Figure 3. The relation ship of flaw size and flexural strength for monolithic alumina and Al₂O₃/5vol%SiC composites hot-pressed at 1600°C.

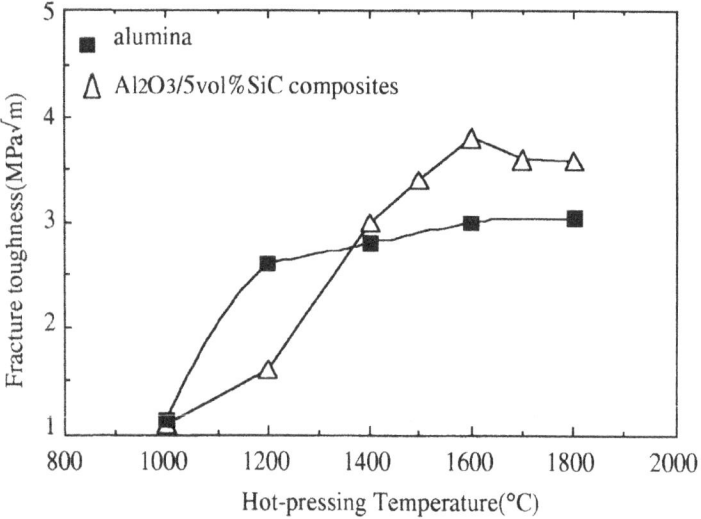

Figure 4. Fracture toughness of hot-pressed monolithic alumina and Al₂O₃/5vol%SiC composites as function of hot-pressing temperature.

within the Al$_2$O$_3$ grains because of the large difference in size between the Al$_2$O$_3$ and the SiC grains. Since the Al$_2$O$_3$ grains are much larger than the fine SiC grains, intragranular, nano-sized SiC grains are trapped during the later coarsening stage of sintering; some larger SiC particulates also were observed at the grain boundaries. The intragranular SiC is attributed to a process similar to that causing the formation of trapped intragranular pores in the sintered bodies. Lange[15] reported that small ZrO$_2$ grains also occurred within large Al$_2$O$_3$ grains, acting as trapped pores, in Al$_2$O$_3$/ZrO$_2$ composites. The size of the intragranular SiC particulates depends entirely on the Al$_2$O$_3$ grain size. In turn, the size of the Al$_2$O$_3$ grains depends on grain-boundary mobility determined by the effect of hot-pressing on the curvature of the Al$_2$O$_3$ grain. The flexural strength of hot-pressed bodies is plotted as a function of matrix grain size in Fig. 2. The flexural strength was inversely proportional to the square root of matrix grain size in the present study. From the Hall Petch equation,[16] the relationship between flexural strength (σ_c) and grain size of the matrix (D) can be expressed as

$$\sigma_c = \sigma_o + KD^{-1/2} \tag{1}$$

where flexural strength (σ_o) is 329 MPa and K a proportional constant with a value of 506 MP·m$^{1/2}$, as determined from the results of Fig. 2. In addition, the flexural strength of the composites were higher than that of monolithic Al$_2$O$_3$. The improved flexural strength of the composites is attributed to a reduction in the critical flaw size resulting from the suppression of grain growth by the nano-sized SiC particulates. The significant decrease in critical flaw size of the composites thus probably is caused by the finer, more uniform microstructure created through disturbance of the moving grain boundaries of the Al$_2$O$_3$ grains. The critical flaw size of the composites was ~10 μm, one-half that of monolithic Al$_2$O$_3$. Figure 3 represents the relationship between flexural strength and flaw size in the interior of fractured surface in materials. The flaw sizes of the nano-composites were lower than those of monolithic Al$_2$O$_3$, and strength decreased with increasing flaw size, regardless of material. Control of grain-size distribution thus apparently affected the critical flaw size in the matrix. Accordingly, the decreased flaw size in the composites can be attributed to suppression of grain growth by the SiC particulates. Since the flaw size of the nano-composites were less than one-half that of monolithic Al$_2$O$_3$, SiC particulates in composites can serve as grain-growth inhibitors, resulting in a stabilizing effect that produces a fine, uniform microstructure and leads to improved strength and toughness. Figure 4 shows the results between fracture toughness and hot-pressing temperature. The fracture toughness of Al$_2$O$_3$/5vol% SiC composites increased with hot-pressing temperature, reaching a maximum at ~1600°C. Below 1600°C, insufficient densification seem to yield more porous composites, with a resulting decrease in fracture toughness. The fracture toughness for all of the composites was higher than for monolithic Al$_2$O$_3$. Mechanical properties therefore seem to be related to microstructure.

INFLUENCE OF SIC PARTICULATES ON CRACK PROPAGATION

Figure 5 shows the behavior of crack propagation introduced by Vickers indentation at the corners of thermally etched specimens hot-pressed at 1600°C. Propagating cracks tended to deflect along the grains and significantly more deflection was exhibited by the composites than by monolithic Al$_2$O$_3$. For monolithic Al$_2$O$_3$, the overall path of crack propagation was comparatively straight, with few observable deflection effects. Mostly intergranular crack propagation was observed in the small grains, whereas intragranular crack propagation occurred in the large grains. In contrast, grains of the Al$_2$O$_3$ matrix in Al$_2$O$_3$/5vol% composites were smaller than those of monolithic Al$_2$O$_3$, and mainly intergranular crack propagation along the Al$_2$O$_3$ grain boundaries was observed. The extent of crack propagation could be measured from the deflection angles along the crack paths. The averaged values of deflection angles is 22° for the composites were larger than 16° for monolithic Al$_2$O$_3$, resulting in improved fracture toughness, as in Fig. 4. These deflection-angle results seem to indicate the existence of a toughening mechanism in the Al$_2$O$_3$/5vol% composites. The different results for the deflection angles are related to microstructural properties such as decreased average grain size and incorporation of the nano-sized second phase. A more thorough investigation of crack deflection was conducted by TEM observation, and Fig. 6 show transmission electron micrographs of crack propagation. In monolithic Al$_2$O$_3$, the

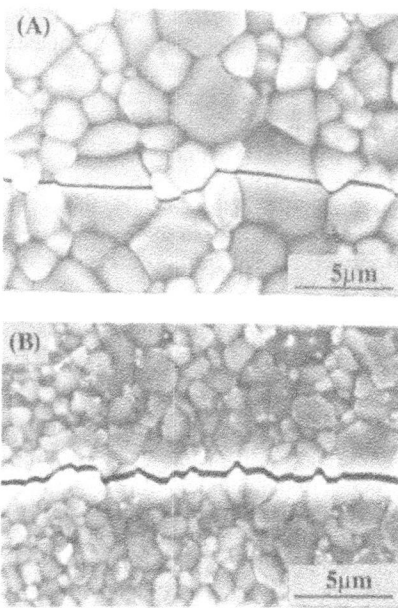

Figure 5. SEM photographs of crack deflection induced at the corner of Vickers indentation
for thermally etched hot-pressed, (A) monolithic alumina
(B) $Al_2O_3/5vol\%SiC$ composites.

fracture mode was linear and representative of intragranular fracture. Consequently, the propagated cracks had a tendency toward a low angle of deflection because of the absence of obstacles such as a second phase. On the other hand, the fracture mode of the $Al_2O_3/5vol\%$ composites was intergranular and deflected at the SiC particulates. In the composites, the crack paths were deflected by the nano-sized SiC particulates, resulting in a somewhat complicated crack propagation and a higher fracture energy than in monolithic Al_2O_3. Since the fracture toughness (K_{IC}) was proportional to the square root of fracture energy (γ_f), according to equation

$$K_{IC} \propto \gamma_f^{1/2} \tag{2}$$

the crack deflection behavior led to enhanced toughening in the composites. Consequently, crack propagation in the $Al_2O_3/5vol\%SiC$ composites tended toward relatively many deflected cracks and resulted in large deflection angles. These results are consistent with those for the distribution of the deflected angles. Crack deflection thus might be generated easily because the SiC particulates dispersed in the Al_2O_3 act as a barrier, disturbing crack propagation. It has been reported that fracture toughness improves because of toughening by crack deflection at a second phase in particulate-dispersed composites.[17] Such crack-deflection behavior accounts for the improved toughness of the composites. In addition, a theoretical analysis by Faber and Evans[18] postulated that fracture toughness is increased by the crack deflection. Improved fracture toughness in the $Al_2O_3/5vol\%SiC$ composites also may be attributable to reasons other than crack deflection. It is possible that fracture toughness improves simultaneously with increasing residual stress. The thermal expansion mismatch between the matrix and the dispersed particulates can lead to residual stresses. It has been reported[19] that local residual stresses induced by the mismatch of thermal expansion coefficients in Al_2O_3/SiC and SiC/Al_2O_3 composites can increase the fracture toughness of particulate-dispersed composites. The critical stress intensity factor increases proportionally to the residual stress, resulting in improved fracture toughness. Such residual stresses also

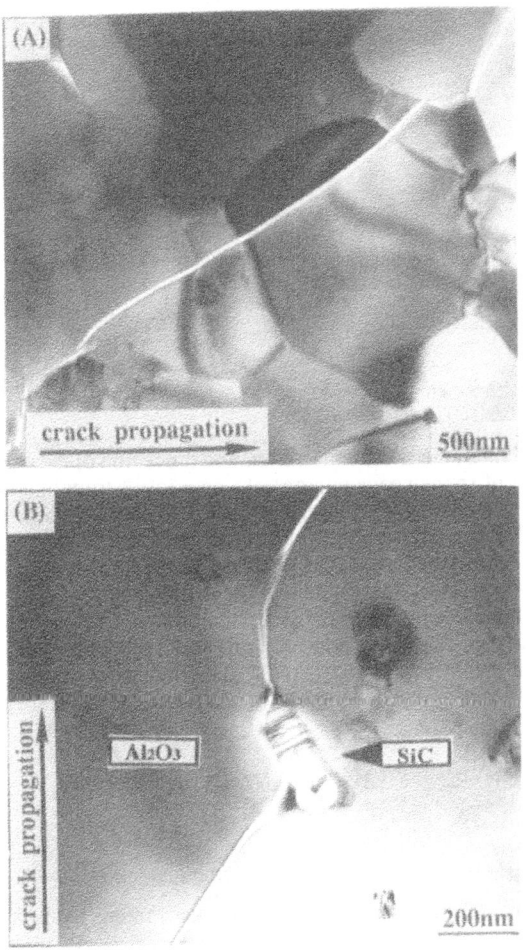

Figure 6. TEM photographs of propagating crack induced at the corner of Vickers indentation, (A) monolithic alumina (B) Al2O3/5vol%SiC composites.

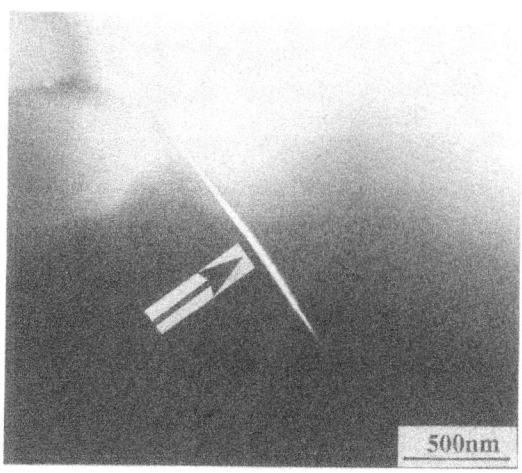

Figure 7. TEM photographs of microcrack in
Al_2O_3/5vol%SiC composites.

may lead to a relaxation of the stress concentration in the region of the crack tip. On the other hand, the generation of spontaneous microcracks may result from internal stresses between the grains caused by incompatible thermal expansions between the matrix and a second phase. Such spontaneous microcracks would decrease fracture toughness. In the present work, however, fracture toughness did increase, and so the absence of spontaneous microcracks was expected. A critical grain size or particulate size (d_c) for the generation of spontaneous microcracks is known to exist, and under those conditions spontaneous microcracks could occur. The generation of microcracks has been attributed to the thermal expansion mismatch between the matrix and the second phase in composites. The critical grain size for spontaneous microcracking can be expressed by the following equation.

$$d_c = \Psi \gamma_f / E_m (\Delta \alpha \Delta T)^2 \qquad (3)$$

Here, Ψ is a constant ranging from 10 to 70, γ_f is the fracture energy, $\Delta\alpha$ is the thermal expansion mismatch, ΔT is the difference between room temperature and sintering temperature, and E_m is the Young's modulus of the matrix. In the case of Al_2O_3/SiC composites, the critical grain size (d_c) of the SiC platelet is ~24 μm for the generation of spontaneous microcracks.[19] The present work revealed no probability for the generation of spontaneous microcracks because the average grain size of the SiC particulates was 0.3 μm, below the critical grain size. Nevertheless, microcracks were observed around the propagating cracks in the composites studied. Figure 7 shows a microcrack around the propagating crack in a composite. This microcrack, which was generated during crack propagation and seemed to be caused by residual stress, lead to relaxation of the stress concentration at the crack tip and a consequent improvement in fracture toughness. No such microcracks occurred in monolithic Al_2O_3, however, because of the absence of residual stresses. Taya et al.[20] reported that a thermal expansion mismatch between TiB_2 and SiC in

SiC/TiB$_2$ composites led to the generation of internal residual stresses. In the SiC/TiB$_2$ composites, the thermal expansion coefficient of the matrix was lower than that of the TiB$_2$ particulates and, according to these researchers, the difference led to an average compressive stress in the matrix and a corresponding increase in fracture toughness. The existence of residual stresses has been reported in other composites such as Al$_2$O$_3$/mullite[21] and Al$_2$O$_3$/SiC.[22]

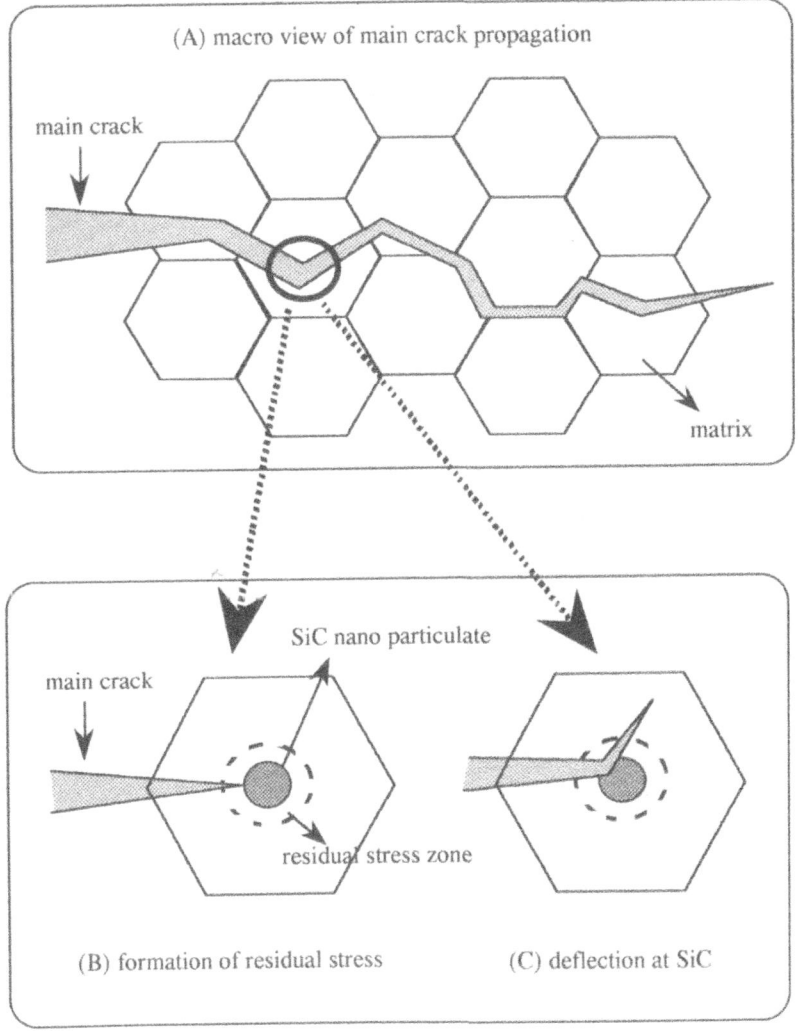

(A) macro view of main crack propagation

main crack

matrix

SiC nano particulate

main crack

residual stress zone

(B) formation of residual stress

(C) deflection at SiC

Figure 8. Schematic illustrating model of crack deflection due to existence of SiC particulates for Al$_2$O$_3$/5vol%SiC composites.

TOUGHENING MECHANISM

Figure 8 schematically describes deflection behavior in terms of the existence of nano-sized SiC particulates in Al_2O_3/SiC composites, supporting the theory that such toughening effects as crack deflection and residual stresses are induced by these particulates. The proposed behavior of the crack path during the propagation of a main crack is represented by Fig. 8 (A). The nano-sized SiC particulates, which exist randomly in the Al_2O_3 matrix of the composites, may serve as obstacles to the typical crack propagation, as shown in Fig. 8 (A). Cracks can be deflected at the insides of the Al_2O_3 grains (circled area in Fig. 8 (A)), as well as at the grain boundaries. For monolithic Al_2O_3, no crack deflection was observed inside the Al_2O_3 grains. The crack deflection in the composites seemed to result from residual stress caused by the existence of SiC particulates because of the differing thermal expansion coefficients between the Al_2O_3 and the SiC. Because of the thermal expansion mismatch between the Al_2O_3 and the SiC, a radial residual stress field may be induced around the dispersed SiC particulates during cooling of the composite after sintering, as shown in the microview of Fig. 8 (B). The average elastic properties of Al_2O_3 and SiC are remarkably similar in absolute magnitude, but the thermal expansion coefficients of the two materials are very different. A residual stress field can occur around second phase particulates because of either elastic or thermal expansion mismatch between the matrix and the dispersed

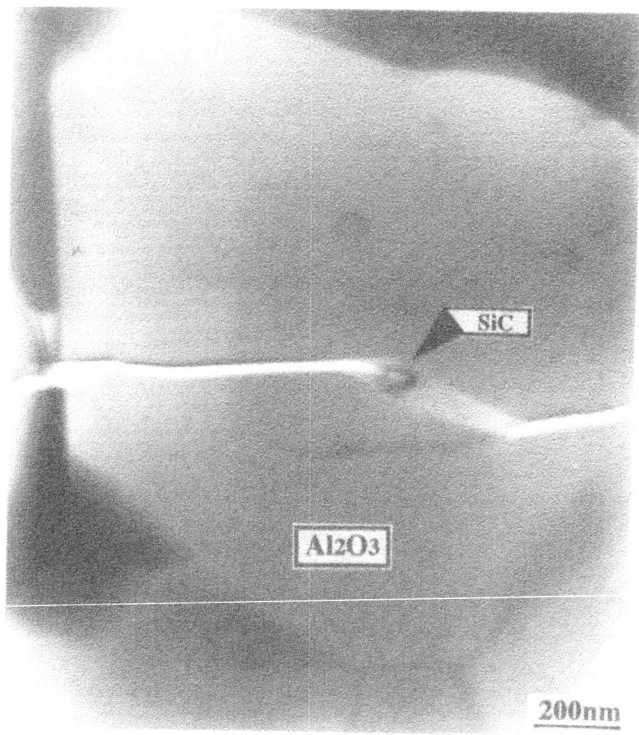

Figure 9. TEM photographs of deflected crack by existence of SiC particulates for $Al_2O_3/5vol\%SiC$ composites.

particulates. Since the thermal expansion coefficient of Al2O3 is ~8.0×10^{-6}/°C and that SiC ~4.5×10^{-6}/°C, residual stress will accumulate around the SiC particulates when the hot-pressed bodies are cooled. As a result, a residual stress zone such as that shown in Fig. 8 (B) would be expected. When propagating cracks collide with a dispersed SiC particulate, then, each crack is deflected at the particulate, resulting in a residual stress field such as that shown in Fig. 8 (C). Consequently, the stress concentration at the crack tip may relax, causing a toughening effect. Apparently, the crack paths in the composites are occupied by second phase particulate, and this path limitation acts to toughen the composites more than the monolithic Al_2O_3. Overall, the Al_2O_3/SiC composites are much tougher than monolithic Al_2O_3, as shown in Fig. 4. Figure 9 shows transmission electron micrographs of the deflected crack at a nano-sized SiC particulate dispersed in an Al_2O_3 grain. This micrograph attests to the toughening model represented in Fig. 8. The crack deflection occurring because of nano-sized SiC particulates in the Al_2O_3 grains is an important characteristic of the toughening observed in the present study. Propagating cracks can be understood to have been deflected at the nano-sized particulates and to have exhibited general deflection behavior. It has been reported that improved toughness by crack deflection is achieved through the diminished stress intensity at the crack tip, that is, by reduction of the driving force of the crack. Improving the toughness of ceramic composites thus is possible through a toughening mechanism such as crack deflection at a dispersed particulate. The present work has led to the conclusion that crack deflection and residual stresses increase fracture toughness while finer grains promote higher strength in ceramic composite.

CONCLUSIONS

Alumina-based composites containing dispersed nano-sized SiC particulates were fabricated by hot-pressing in the present study. The influence of SiC particulates on the toughening of the composites was investigated, and the following conclusions were derived from those results.
(1) Al_2O_3/SiC composites were successfully obtained by dispersing nano-sized SiC particulates intragranularly within the Al_2O_3 matrix grains and larger particulates at the grain boundaries. The fracture mode of the Al_2O_3/SiC composites was predominantly intergranular, whereas monolithic Al_2O_3 exhibited a mixed fracture mode consisting of intergranular fracture within small grains and intragranular fracture in the large grains.
(2) Strength and fracture toughness were effectively increased by the dispersion of nano-sized particulates within the composites. Strength improvement was attributed to the diminution of critical flaw size resulting from suppressed grain growth, that is, to grain boundary pinning and the generation of sub-grain boundaries by the distribution of the nano-sized particulates.
(3) The fracture toughness of the composites was higher than that of the monolithic Al_2O_3 and increased with increases in hot-pressing temperature. The composites exhibited outstanding crack deflection, which was caused by the disturbance of crack propagation at the SiC particulates. Apparently, the important mechanisms for toughening and strengthening composites are crack deflection, due to the dispersion of SiC particulates, and residual stress. Nano-sized SiC particulates occurring as a strong second phase in Al_2O_3-based ceramic composites act as reinforcement, by presenting a barrier to crack propagation and improving the fracture toughness of the ceramics.

ACKNOWLEDGMENT

The authors would like to thank Mr. Kim J. Ostreicher of GTE Laboratories Inc. for his helpful advice on this present work.

REFERENCES

1. J. L. Shi B. S. Li, and T. S. Yen, "Mechanical Properties of Al_2O_3 Particle - Y-TZP Matrix Composite and its Toughening Mechanism," J. Mater. Sci., 28: 4019 (1993).

2. P. F. Becher, "Microstructural Design of Toughened Ceramics" J. Am. Ceram. Soc., 74: 255 (1991).
3. L.A. Xue, K. Meyer, and I-W. Chen,"Control of Grain-Boundary Pinning in Al_2O_3/ZrO_2 Composite with Ce^{3+}/Ce^{4+} Doping," J.Am.Ceram.Soc.,75:822 (1992)
4. D.J. Green, "Critical Microstructures for Microcracking in Al_2O_3-ZrO_2 Composites," J.Am.Ceram.Soc., 65:610 (1982).
5. M. Rühle, N.Claussen, and A.H. Heuer,"Transformation and Microcrack Toughing as Complementary Processes in ZrO_2 -Toughened Al_2O_3," J.Am.Ceram.Soc., 69:195 (1986).
6. J.Homeny, W.L. Vaugh, and M.K. Ferber,"Processing and Mechanical Properties of SiC Whisker-Al_2O_3 Matrix Composites," Am.Ceram.Soc.Bull., 65:333 (1986).
7. T.N. Tiegs and P.F. Becker,"Sintered Al_2O_3-SiC-Whisker Composites," Am.Ceram.Soc.Bull., 66:339 (1987).
8. Y. S. Chou and D. J. Green, "Silicon Carbide Platelet/Alumina Composites: I. Effect of Forming Technique on Platelet Orientation," J.Am.Ceram.Soc., 75:3346 (1992).
9. G. C. Wei and P. F. Becher, "Improvements in Mechanical Properties in SiC by the Addition of TiC Particles," J.Am.Ceram.Soc., 67:571 (1984).
10. D. W. Shin and K. K. Orr, "Microstructure-Mechanical Property Relationships in Hot Isostatically Pressed Alumina and Zirconia-Toughened Alumina," J.Am.Ceram.Soc.,73: 1181 (1990).
11. A. G. Evans and R. M. McMeeking, "On the Toughening of Ceramics by Strong Reinforcements," Acta Metall., 34:2435 (1986).
12. K. Niihara, "New Design Concept of Structural Ceramics - Ceramic Nanocomposites-," J.Ceram.Soc.Jpn.,99:974 (1991).
13. M. P. Harmer, H. M. Chan, and G. A. Miller, "Unique Opportunities for Microstructural Engineering with Duplex and Laminar Ceramic Composites," J.Am.Ceram.Soc.,75:1715 (1992).
14. G. R. Anstis, P. Chantikul, B. R. Lawn and D. B. Marshall, "A Critical Evaluation of Indentation Techiques for Measuring Fracture Toughness: I, Direct Crack Measurements," J.Am.Ceram.Soc., 64:533 (1981).
15. F.F. Lange and M.M. Hirlinger,"Hindrance of Grain Growth in Al_2O_3 by ZrO_2 Inclusions,"J.Am.Ceram.S31.,oc.,67:164 (1984).
16. J. H. Bucher, J. D. Grozier, and J. F. Enrietto, "Strength and Toughness of Hot-Rolled Ferrite-Pearlite" ; pp. 253-54 in Fracture, Vol. IV. Edited by H. Liebowitz. Academic Press, New York, 1969.
17. D. H. Kim and C. H. Kim, "Toughening Behavior of Silicon Carbide with Additions of Yttria and Alumina," J. Am. Ceram. Soc., 73:1431 (1990).
18. K. T. Faber and A. G. Evans, "Crack Deflection Process-I. Theory," Acta Metall. 31: 565 (1983).
19. Y. S. Chou and D. J. Green, "Silicon Carbide Platelet/Alumina composites: II, Mechanical Properties," J. Am. Ceram. Soc., 76:1452 (1993).
20. M. Taya, S. Hayashi, A. S. Kobayashi and H. S. Yoon, "Toughening of a Particulate-Reinforced Ceramic-Matrix Composite by Thermal Residual Stress," J. Am. Ceram. Soc., 73:1382 (1990).
21. J. H. Root and J. D. Sullivan, "Residual Stress in Alumina-Mullite Composites," J. Am. Ceram. Soc., 74: 579 (1991).
22. A. Abuhasan, C. Balasingh, and P. Predecki, "Residual Stresses in Alumina/Silicon Carbide(Whisker)" J. Am. Ceram. Soc., 73:2474 (1990).

FRACTURE CHARACTERIZATION OF SILICON NITRIDE BASED LAYERED COMPOSITES

J.Dusza[1], P.Šajgalík[2], E.Rudnayová[1], P.Hvizdoš[1], and Z.Lenčéš[2]

[1]Institute of Materials Research, Slovak Academy of Sciences, Košice, Slovakia
[2]Institute of Inorganic Chemistry, Slovak Academy of Sciences, Bratislava, Slovakia

INTRODUCTION

Silicon nitride based structural ceramics are a family of advanced materials that exhibit a combination of high hardness, high strength, good corrosion and erosion behaviour, high elastic modulus and dimensional stability. Major application of these ceramics includes wear components, cutting tools and parts of engines (turbochargers, bearings, etc.). Their wide application is, however, still limited mainly due to their brittleness, low flaw tolerance and low reliability[1,2]. In recent years nitride based ceramics have been very intensively investigated all over the world with the aim to improve their mechanical properties and make them suitable for structural applications. The main ways of improving the room temperature mechanical properties of silicon nitride based ceramics can be summarized as follows:

- improving the strength level and reducing the strength values scatter, i.e., enhancing the reliability by reduction of the critical defect size (improved properties of powders, clean room manufacturing, etc.) - the flaw diminution approach[3,4];
- promoting the localized bridging behind the crack tip (in the form of frictional and mechanical interlocking, or pull out) by which the flaw tolerance of the material can be improved - the flaw tolerance approach[5-7];
- improving the strength values by incorporating into the matrix the nano-sized, second-phase particles with different expansion coefficients - the nano-particle dispersion strengthening[8];
- improving the structural reliability by designing novel laminar composites with a promoted crack deflection at the interlayer boundaries and utilizing the compressive residual stresses arisen during cooling down from the sintering temperature because of the differences in the thermal expansions between the layers which have different compositions - the laminar structure approach[9-12].

Up to now, laminar ceramic composites were developed mainly on the basis of oxide ceramics. Russo et al.[13] used tape casting to produce a three layered composite with a surface layer consisting of a homogeneous alumina - 20 % aluminum titanate (A-HT20) and a flaw tolerant inner layer made of inhomogeneous H-AT20. According to their results at an optimum surface layer thickness (approx. 100 µm) and small flaw sizes the high strength of the outer homogeneous layer was dominant, in case of high flaw sizes the flaw tolerant inner layer. Marshall et al.[10] fabricated laminar composites containing alternating layers of Ce-TZP and a mixture of Al_2O_3 and $Ce-ZrO_2$ using a colloidal technique. They found out that layers interacted with the transformation zones surrounding the cracks; it led to the spreading of the zones and to enhanced fracture toughness. These laminar composites exhibit R-curve behaviour for cracks oriented in normal direction to the layers with a K_{IC} value up to 17.5 $MPa.mm^{1/2}$.

The aim of the present contribution is to show that the laminar silicon nitride based composites also provide advantage for tailoring the properties by stacking layers of different composition (microstructure) in a suitable sequence.

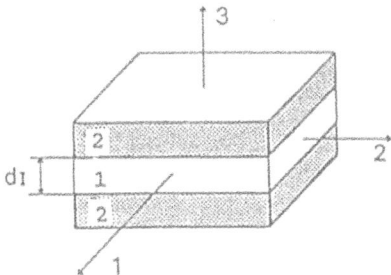

Figure 1. In-plane stress state with stress normal to the surface, $\sigma_{33}=0$ and $\sigma_{11}=\sigma_{22}$ in a three layer composite

THEORETICAL CONSIDERATIONS

Internal stresses appear inside the individual layers after processing the laminar ceramic composites with layers of different chemical composition and/or different microstructure. These stresses can arise due to the different thermal expansion between the layers or due to the difference in their sintering behaviour. For laminar composites with a symmetrical stacking and a low thickness compared with the area of the layer, the stress value normal to the layers can be taken as zero, Fig. 1.

In case of a good bonding between the layers the total deformation is the same for all layers[14]:

$$\epsilon_{11}^{I} = \frac{1-\nu_I}{E_I} \cdot \sigma_{11}^{1} + \alpha_1 \cdot \Delta T \tag{1}$$

and $\Delta\sigma_y = \sigma_{y2} - \sigma_{y1}$.

As it was described by the authors[16], the effect of interfacial decohesion between the brittle/brittle laminates can
influence their mechanical properties in a very strong way.
Using a mathematical simulation and a spring/network model they
found that the brittle/brittle laminates with a weak interface
can be 50% stronger and 300% tougher. The enhancement of the
normalized fracture energy was approximated as follows:

$$W = W_0 - A \cdot \gamma_i / \gamma_f \quad \text{for } \gamma_i / \gamma_f < 0.6 \text{ and}$$
$$W = 1 \qquad\qquad \text{for } \gamma_i / \gamma_f > 0.6,$$
where $W_0 = 2.9$ and $A = 3.1$.

As it was demonstrated, the final mechanical properties of the
layered ceramics/ceramics composites can be influenced by the
stress conditions in the neighbouring layers and by the
fracture energy level of the interface.

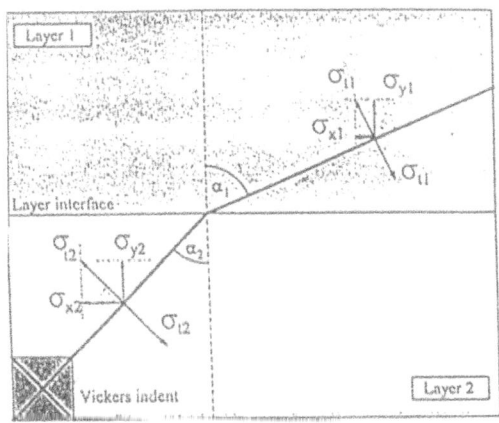

Figure 2. Schematic representation of the
crack deflection crossing the layer boundary and the stress anisotrophy in the
neighbouring layers

EXPERIMENTAL MATERIALS AND PROCEDURE

The experimental materials investigated in this work have the
compositions described in Tab. I.
Starting mixtures were prepared by attrition milling of Si_3N_4
powder with sintering aids in dry isopropanol for 4 h. Si_3N_4-
whiskers were added to the starting mixture during mixing and
simultaneous ultrasonic treatment, followed by steering and
drying. In the second case, SiC platelets were added to the
starting mixture during 90 h homogenization in isopropanol in
plastic bottles with Si_3N_4 balls.
Green samples were formed by cold isostatic pressing (CIP) at
a pressure of 200 MPa. Before CIP individual layers with
different composition were formed by strowing the powders into
the rubber form. Thereafter the samples were gas pressure

where $\Delta T = T_{sint.} - T_{room}$, E_I and σ_I are Young modulus and Poisson ratio for layer I and α_1 is the thermal expansion coefficient.

It is easy to show that for symmetrical laminar composites with $2n + 1$ alternate layers (type 1 and 2):

$$(n+1)\,\sigma_{11}^1 + n\sigma_{11}^2\,d_2 = 0 \tag{2}$$

From equations (1) and (2) follows:

$$\sigma_{11}^1 = \frac{nE_1E_2.d_2\,(\alpha_2 - \alpha_1).\Delta T}{n(1-\nu_I)\,E_2\,d_2 + (n+1)\,(1-\nu_2)\,E_I d_I} \tag{3}$$

and

$$\sigma_{11}^2 = \frac{(n+1)\,E_1E_2\,d_1\,(\alpha_2 - \alpha_1)\,\Delta T}{n(1-\nu_1)\,E_2\,d_2 + (n+1)\,(1-\nu_2)\,E_1\,d_1} \tag{4}$$

As it is visible, the residual stresses are determined by thermal expansion mismatch, by Young modulus of layers and by their relative thickness.

Brittle materials as, e.g., structural ceramics are extremely sensitive to the surface flaws. Therefore, the composite structure has to be tailored in order to develop compressive residual stresses in the outer layer with a maximal tensile stress. For the composite illustrated in Fig. 1 this can be achieved by choosing the composition of layers in such a way that the thermal expansion of the layer 1 should be lower than that of layer 2.

For the simplest case (three layers' composite, n=1) the compressive and tensile stresses can be expressed as a function of the layer thickness ratio in the form:

$$\sigma_{11}^1\,(x) = \frac{E_1E_2\,(\alpha_2 - \alpha_1)\,\Delta T}{(1-\nu)} \cdot \frac{X}{2\,E_1 + E_2\,X} \tag{5}$$

$$\sigma_{11}^2\,(x) = -2\,\frac{\sigma_{11}^1\,(x)}{x} \tag{6}$$

with $\nu = \nu_1 = \nu_2$

The authors[15] showed that the direction of the crack propagation can be influenced by the boundaries between layers, as it is schematically illustrated in Fig. 2. The deflection angle of the crack crossing the boundary between the layers can be expressed by the following equation:

$$\sin\Delta\alpha = \sin(\alpha_1 - \alpha_2) = \frac{\sigma_{y1}.\sigma_{x2} - \sigma_{x1}.\sigma_{y2}}{\sigma_{t1}.\sigma_{t2}} \tag{7}$$

where $\Delta\alpha$ has a positive or negative value according to the stress condition in the neighboring layers. If $\Delta\alpha > 0$, the crack path crossing the layers is longer, which can have a positive influence on the mechanical properties. This condition is fulfilled in the case when $\Delta\sigma_x > \Delta\sigma_y$, where $\Delta\sigma_x = \sigma_{x2} - \sigma_{x1}$

sintered (GPS) with a heating rate 20 $^{\circ}$C up to 1700 $^{\circ}$C and 10 $^{\circ}$C up to 1900 $^{\circ}$C. The sintering was realized at 1900 $^{\circ}$C for 3 hours under nitrogen pressure of 10 MPa.

Table I. Composition of starting powders

	Si_3N_4 [wt.%]	Sintering Aids [wt.%]		Powders used for reinforcing [wt.%]	
		Al_2O_3	Y_2O_3	β-Si_3N_4- whiskers	SiC- plate- lets
C-SN[1]	92	5.0	3.0		
F-SN[2]	92	5.0	3.0		
SN-wh.	72	5.0	3.0	20[3]	
SN-SiC	72	5.0	3.0		20[4]

Starting powders:
[1] Si_3N_4, UBE SN 03, lot No.B44203 (coarse grained)
[2] Si_3N_4, UBE SN E10, lot No. A14801 (fine grained)
     Y_2O_3, H.C.Starck, grade Fine, Lot No.1/91
     Al_2O_3, ALCOA, A16
[3] Si_3N_4 -whiskers, Chernogolovka, Inst.of Macrokinetics, SHS-
     process
[4] SiC   platelets, C-Axis Technology, Grade SF 11 μm, Lot
     No.79
     BN spray, Sintek Keramik GmbH

Billets with dimensions of 10x20x45 mm were cut and ground to bars (3x4x45 mm) with the tensile face having 15 μm finish.
After processing we have got symmetrically layered composites with 2n+1 alternated layers of types 1 and 2. In Tab. 1 the composition of layers can be seen. The n = 1 (three layer composite) was chosen in the case of a strong interlayer connection and that of n = 2 (five layer composite) in the case when a BN layer was used to weak the interlayer connections, see Fig. 3. The coarse grain sized silicon nitride C-SN layer was present in all investigated composites.
The bending strength was measured using a four-point bending fixture with the inner/outer span 20/40 mm and a crosshead rate of 0.5 mm/min.
The fracture toughness was measured using IS (Indentation Strength) and IF (Indentation Fracture) methods[2]. Using IS method the indents (20 kg) on the tensile surface of specimens were placed either perpendicularly or parallel to the layers' plane ($\equiv\uparrow$, or $\equiv\leftarrow$). The IF method (10 and 20 kg) was used for measuring the fracture toughness in individual layers in parallel and perpendicular directions to the layers. Shetty's formula[17] was used for calculation of K_{IC} values. The IF method was used also for predicting the magnitude of residual stresses from the different lengths of indentation cracks in the presence of residual stresses when compared with the stress-free state[18]. The critical stress intensity in the presence of

residual stresses is:

$$K_{IC} = K_{ind.} + K_{res.} \qquad \text{(8)}$$

with

$$K_{res.} = 2\,\sigma_{res.}\,(\frac{c}{\pi})^{1/2} \qquad \text{(9)}$$

where 2c is the length of the surface crack[19].
Structure of the layered composites on the macroscopic level, microstructure of the individual layers, fracture origins and reinforcing mechanisms at the layer/layer boundaries and inside the layers were studied using light microscopy and SEM.

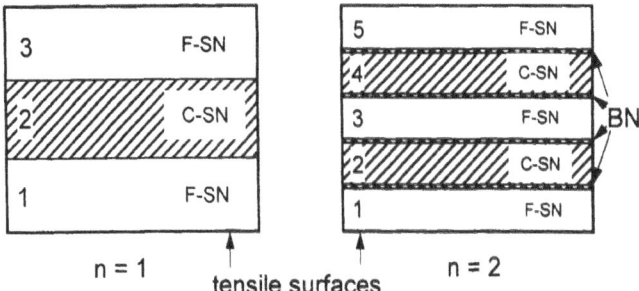

Figure 3. Schematical representation of layers in the studied composites with an example for layers arrangement, see composites No.1 and 3 in Table II.

RESULTS AND DISCUSSION

Microstructure characteristics of the layers

Characteristic microstructure of the individual layers and layer/layer boundaries are shown in Figs. 4a-c and Fig. 5a,b, respectively. The F-SN layer consists of a fine, equiaxial Si_3N_4 grains with the grain size about 0.6 μm and of whisker shaped grains with the main diameter approx. 1.0 μm (aspect ratio 4), Fig. 4.a. The C-SN layer consists of similar Si_3N_4 grains (approx. 0.8 μm in diameter) and coarser whisker shaped grains (approx. 2.0 μm wide and with an aspect ratio 5), Fig. 4b. The SN-wh. layers have a coarser microstructure, with coarse β-Si_3N_4 (up to 5 μm wide) whiskers and an aspect ratio from 5 to 6. This layer contains a relatively high percentage of porosity, Fig. 4c. The matrix of the SN-SiC layer was very similar to F-SN with homogeneously dispersed SiC platelets (diameter from 5 to 10 μm, approx.), Fig. 4d.

Figure 4. Characteristic microstructure of individual layers in studied composites;
a) fine grain sized Si_3N_4 (F-SN), b) coarse grain sized Si_3N_4 (C-SN), c) $Si_3N_4+Si_3N_4$-wh. (SN-wh.), d) Si_3N_4+SiC platelets (SN-SiC).

The boundary between F-SN|C-SN and C-SN|SN-SiC layers with BN interlayers (F-SN|BN|C-SN and C-SN|BN|SN-SiC) is illustrated on Fig. 5a,b. As it is visible, the BN layer is not thin enough and its width is not regular (varies from 5 to 20 μm). The "strong" boundaries without BN layers were clean, without any defects.

Mechanical Properties

Mechanical properties of studied layered composites are summarized in Tab. II.

Table II. Mechanical properties of the studied layered composites

		$K_{IC\ IS}$ $\parallel$	$K_{IC\ IS}$ $\perp$	$K_{IC\ IF}$ $\parallel$	$K_{IC\ IF}$ $\perp$	σ_{4-bend} [MPa]
1.	F-SN|C-SN n=1	6.89		7.55/8.1	7.6/7.8	862
2.	C-SN|F-SN n=1		6.75/6.7	8.1/7.55	7.8/7.6	870
3.	F-SN|BN|C-SN n=2	7.37		7.65/7.9	7.6/7.9	971[*]
4.	C-SN|BN|F-SN n=2			7.95/7.65	7.9/7.6	955[*]
5.	C-SN|SN-wh. n=1	6.7		7.7/6.4	7.55/7.2	721
6.	SN-wh.|C-SN n=1			6.4/7.7	7.2/7.55	420
7.	C-SN|BN|SN-wh. n=2	7.43		7.95/6.3	7.9/6.4	485
8.	SN-wh.|BN|C-SN n=2			6.3/7.95	6.4/7.9	435
9.	C-SN|SN-SiC n=1	7.55	8.2/6.25	6.69/8.1	8.25/8.2	794
10.	SN-SiC|C-SN n=1			8.1/6.7	8.2/8.25	520
11.	C-SN|BN|SN-SiC n=2	7.62		7.4/8.05	8.3/8.1	524
12.	SN-SiC|BN|C-SN n=2			8.05/7.4	8.1/8.3	520

[*]In these mean values are not included those values at which the failure was originated on BN clusters;

$K_{IC\ IS}$ $\parallel$:IS fracture toughness with the indentation load paralell to the layers

$K_{IC\ IS}$ $\perp$:IS fracture toughness with the indentation load perpendicular to the layers

$K_{IC\ IF}$ $\parallel$:IF fracture toughness calculated from the length of the crack paralell to the layer boundary

$K_{IC\ IF}$ $\perp$:IF fracture toughness calculated from the length of the crack perpendicular to the layer boundary

Fracture toughness: The IS fracture toughness measurement of F-SN\|C-SN and C-SN\|F-SN composites with indents perpendicular to the layer's plane show values 6.7 and 6.75 MPa.mm$^{1/2}$. These values are very similar and probably are representative for the monolithic F-SN and C-SN materials. The toughness of these composites are determined by the reinforcing mechanisms acting

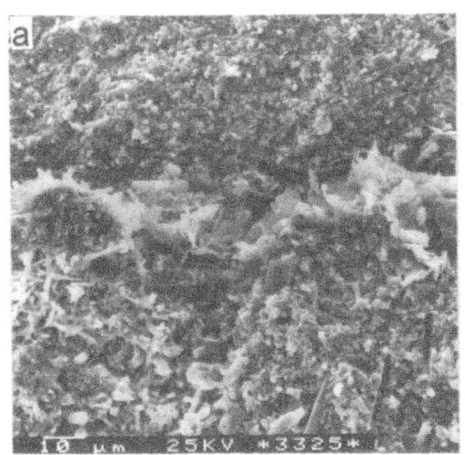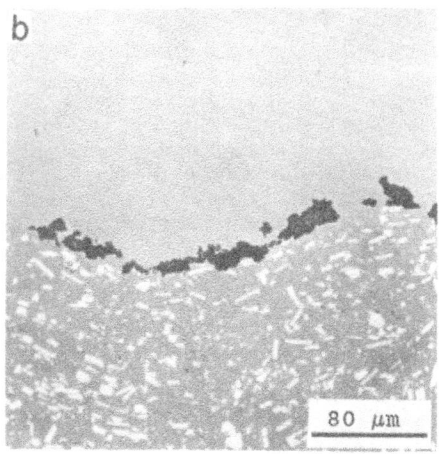

Figure 5. Weak layer boundaries between: a) F-SN|BN|C-SN - SEM, b) C-SN|BN|SN-SiC - light microscopy

Table III. Young modulus, Poisson ratio and thermal expansion coefficient of coarse grain sized silicon nitride and Si_3N_4+SiC platelets layers

	E [GPa]	$\alpha.\Delta T$	ν
C-SN	310	$- 73.10^{-4}$	0.24
SN-SiC	332	$- 79.10^{-4}$	0.24

mainly at microscopic level (crack deflection and crack bridging) and is not significantly influenced by mechanisms taking place at the macroscopic level or by residual stresses. However, some anisotropy in fracture toughness values (IF) was found in this composite, too. More visible is the difference in the fracture toughness values of C-SN|SN-SiC composite measured in samples with C-SN or SN-SiC layers on the tensile surface (8.2 and 6.75 MPa.mm$^{1/2}$). Using relations (5) and (6) and data in Table III. it is easy to show that in C-SN|SN-SiC composite the compressive stresses in the C-SN layer are from 80 to 130 MPa and the tensile stresses in SN-SiC layer vary from 100 to 140 MPa (considering the ratio of SN-SiC to C-SN layer thickness from 1.1 to 2.1, which is the case of this work). The change of the compressive and tensile residual stresses in this composite are in relation to the layer thickness ratio, as it is illustrated in Fig. 6. The anisotropy in IF-fracture toughness values in this composite is in some cases very high, Fig. 7.

Using eqs. (9) and (10) there exists an other way to estimate the level of the residual compressive stress in the silicon nitride part of C-SN/SN-SiC composite. With the assumption that the stress intensity factor for a residual stress free C-SN layer is 6.75 MPa.mm$_{1/2}$, using eqs. (9) and (10) the value of approx. 135 MPa was calculated for the residual compressive stress in the vicinity of the two layer boundary. This value is in a very good agreement with the residual stress level achieved from the theory presented by eqs. (5) and (6), (see Fig. 6).

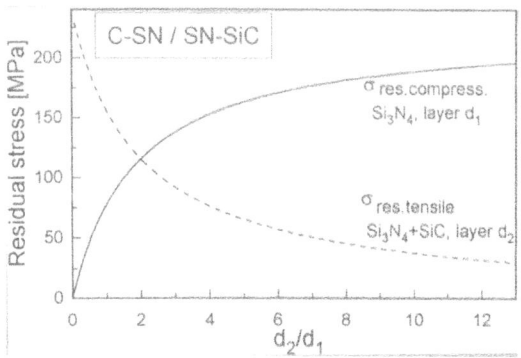

Figure 6. Influence of layer thickness ratio on the compressive and tensile residual stresses developed in C-SN and SN-SiC layers

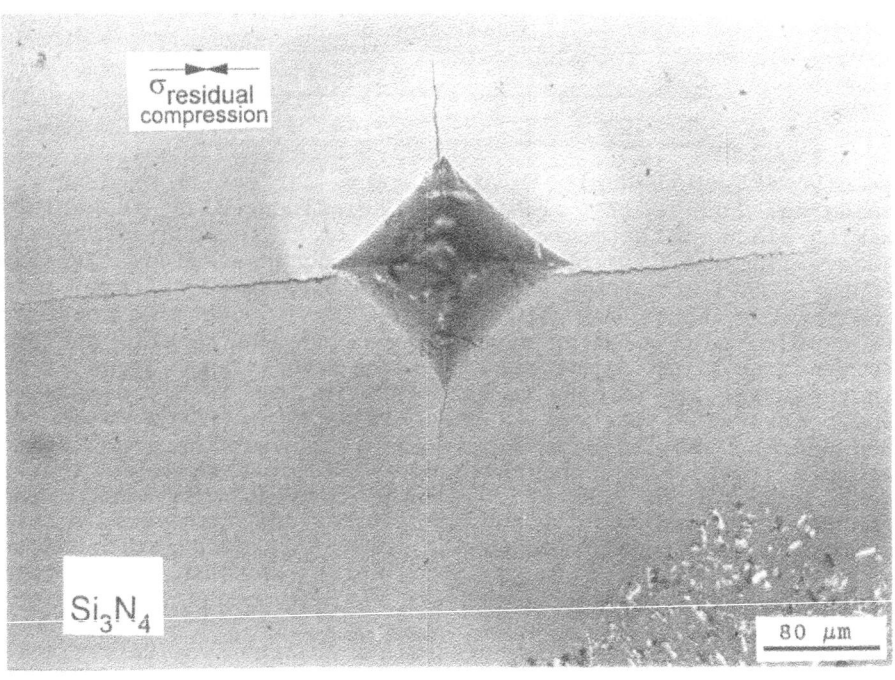

Figure 7. Anisotrophy in IF-fracture toughness values in C-SN|SN-SiC composite

An exact evidence for presence of residual stresses (compression and tensile as well) in layers with different composition and/or microstructure was found by the study of indentation cracks propagation on the survival cracks on the tensile surfaces of specimens after loading them in four-point bending mode. This process was most apparent in the composite C-SN/SN-SiC as it is schematically illustrated on Fig. 8a and on the ceramographic section (tensile surface) in Fig. 8b.

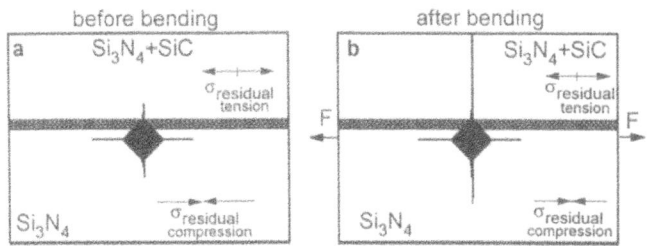

Figure 8a. Schematic illustration of the indentation crack propagation on the tensile surface of the C-SN|BN|SN-SiC composite during loading in the four-point bending mode (survival crack)

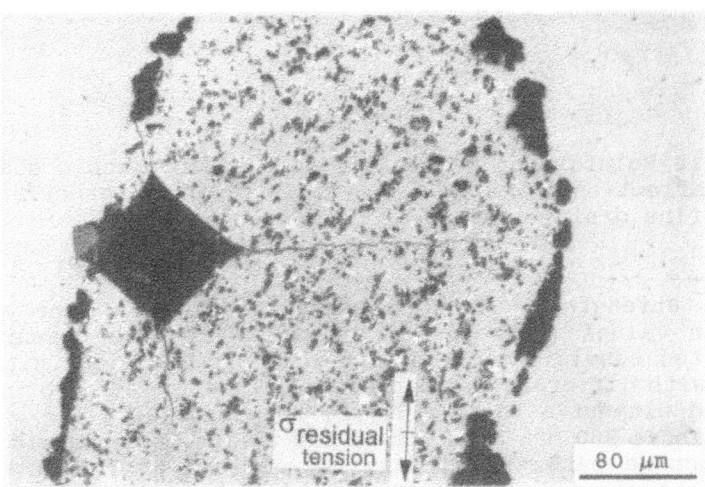

Figure 8b. As Fig. 8a but on the ceramographic section

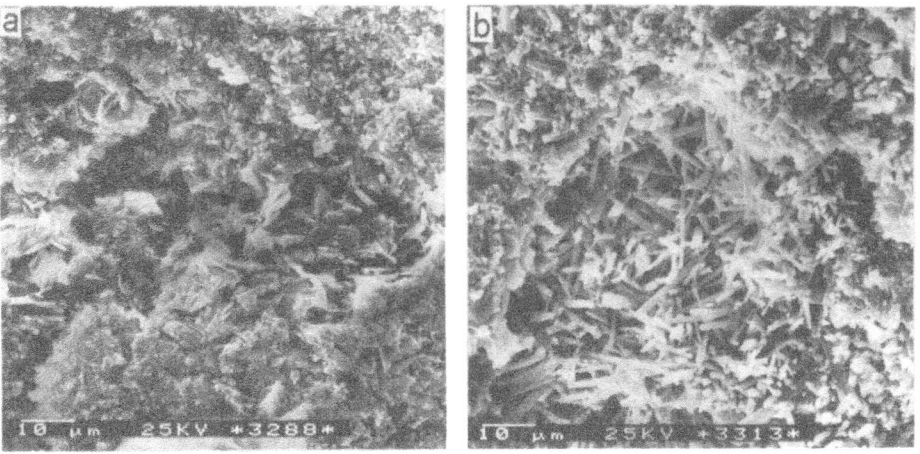

Figure 9. Fracture origins: a) a cluster of BN particles in F-SN|BN|C-SN composite, b) a pore filled with whisker shaped Si_3N_4 grains in the C-SN layer

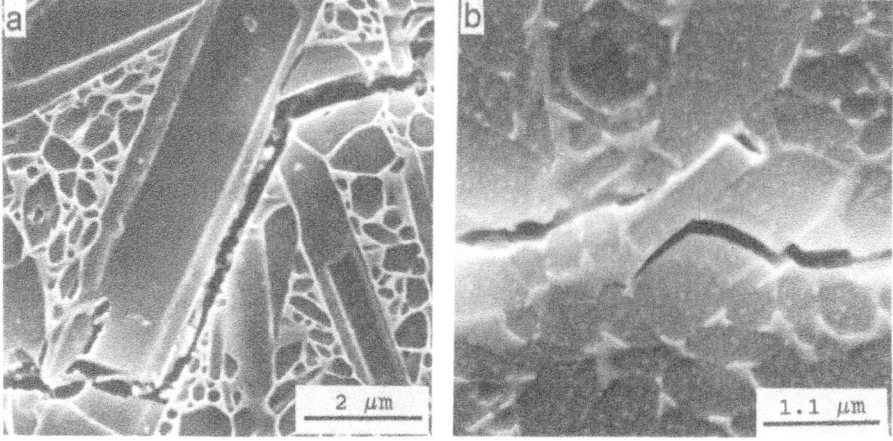

Figure 10. Reinforcing mechanisms at the microscopic scale: a) crack deflection in $Si_3N_4+Si_3N_4$-wh. layer and b) crack bridging in the fine grain sized Si_3N_4

Flexure Strength - Strength Degrading Defects: The highest strength values (between 862 and 955 MPa) were measured for composites consisting of in-situ reinforced silicon nitride layers with different microstructure. This is valid for systems with and without a BN interlayer. Specimens with the strength values above 900 MPa exploded at the failure to many pieces and the fracture path showed many deflections at the interlayer boundaries. Some specimens from the F-SN|BN|C-SN and C-SN|BN|F-SN (these were not included in the mean value) fractured at a lower stress level. In these cases, evident failure origins in the form of clusters of BN particles were found by fractographic examination, Fig. 9. Beside the clusters of BN only one type of fracture origin was found, i.e. a pore filled in by whisker shaped Si_3N_4 grains in the C-SN layer. The strength values of other composites were also lower in comparison with those based on the in-situ reinforced silicon nitride. This strength degradation can be, mainly in the case of systems containing BN interlayers, explained by a lower homogeneity of these materials. In the C-SN|SN-SiC system with compressive residual stresses in the outer layer the failure in some cases was originated at defects located in the second layer (from the sample surface) with tensile residual stresses (at a low thickness ratio of SN-SiC|C-SN layers). This can be probably the reason of slightly lower strength values of this composite (794 MPa) when compared with the in-situ reinforced Si_3N_4 composites.

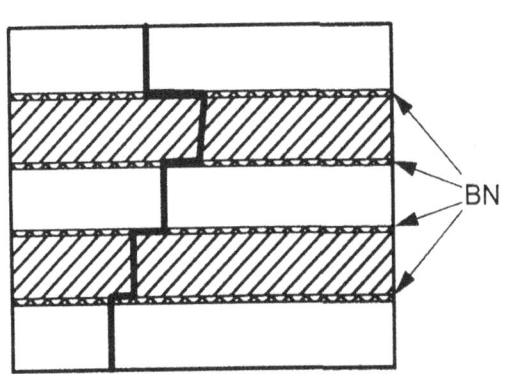

Figure 11. Schematic representation of the crack jumping

Reinforcing Mechanisms: Reinforcing mechanisms at microscopic level occur very often in each layer. These are mainly crack deflection and crack bridging in the in-situ reinforced Si_3N_4 based ceramics, crack deflection at SiC platelets in Si_3N_4+ SiC-pl. layer, and crack deflection, and bridging in Si_3N_4+ Si_3N_4wh. layer, Fig. 10a,b. These mechanisms were studied in detail and their influence on fracture toughness was recently analyzed by many authors[5,6,19]. But there exist also other reinforcing mechanisms working on a macroscopic scale, when the crack is crossing the layer boundaries. The main reinforcing mechanisms working at this scale are: crack deflection at the layer/layer boundary, crack jump when the crack crosses the boundary between layers, see Fig. 11. Small scale jump arising at propagation of an indentation crack, Fig. 12a, or a large scale jump arising during propagation of the main crack causing the specimens' failure at bending strength tests or fracture toughness tests, Fig. 12b,c. Crack branching at the entrance of the crack into the $Si_3N_4+\beta$-Si_3N_4wh. is also frequently occurring reinforcing mechanism, Fig. 12d.

The crack jumping was characteristic for composites with a weaker layer/layer boundary. However, the relative interfacial cohesion energy with respect to the matrix, γ_i/γ_f, is too high (γ_i/γ_f is probably higher than 0.6), which is probably caused

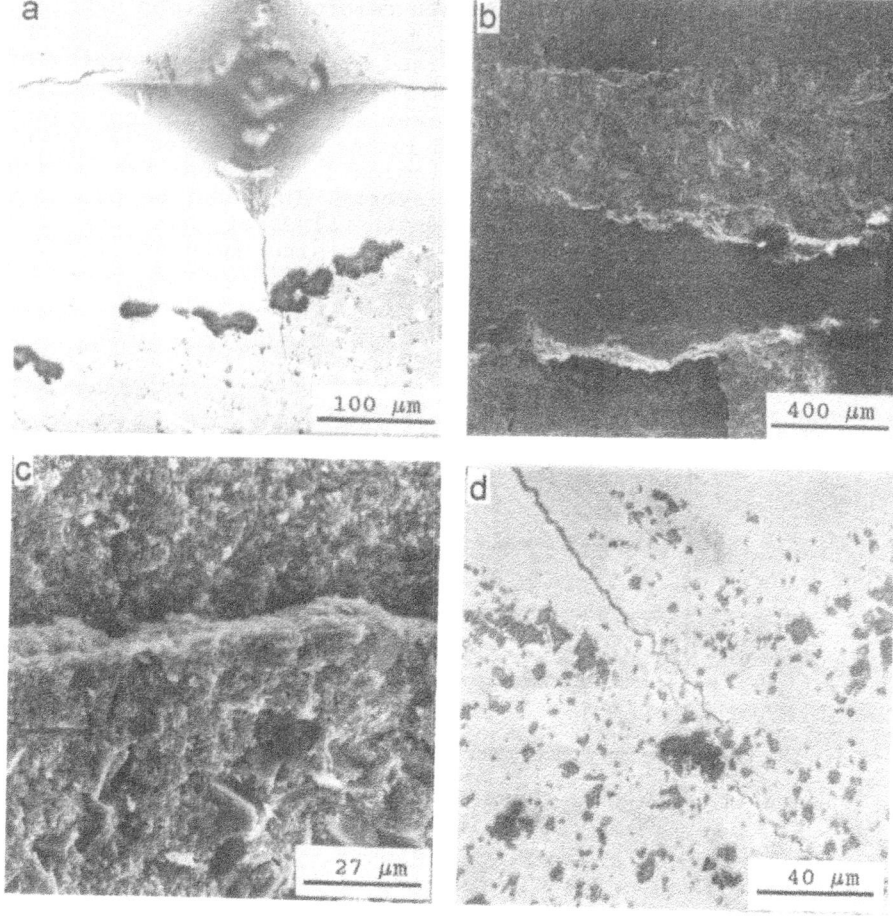

Figure 12. Crack jumping (a,b,c) and crack branching (d) in layered composites

by dimension of the BN layers and their inhomogeneity, therefore the fracture energy of the composite cannot be significantly improved by the incorporation of layers.

CONCLUSIONS

Silicon nitride based layered composites consisting of alternate layers with different microstructure, reinforced by whisker shape grains, whiskers and platelets, have been fabricated by gas pressure sintering. As it was shown their mechanical properties are influenced beside the toughening mechanisms acting at the microstructural scale also by mechanisms acting due to the character of the layered structure (physical properties, width of layers, etc.) and due to the presence of an interface between layers.

The main results can be concluded as follows:
- different sintering rates of Si_3N_4 powders did not cause significant residual stresses in layers. Such stresses were found only in Si_3N_4/Si_3N_4+SiC platelets layered composite due to the differences in Young modulus and thermal expansion coefficients in both layers. Compressive residual stresses improved fracture toughness values and flaw tolerance in this composite.
- The BN interlayer, used in order to weaken the interlayer boundaries, be often made the fracture origin during bending strength test and decreased the strength value. On the other side, the weaker boundaries were places where the crack jumping was observed which certainly can positively influence mechanical properties of these composites.
- The further large scale reinforcing mechanisms observed in these materials were: crack deflection, crack stop and crack branching at the interlayer boundaries.
- In order to achieve the maximum effect of layered composite layers without defects with different Young modulus, thermal expansion coefficients and optimal interface bounding should be used.

The achieved results illustrate the potentials for layered silicon nitride composites as interesting materials for structural applications. In order to obtain the best material having the required complex final properties for respective application, layers with different (wear resistent, creep resistent, damage tolerated etc.) properties have to be combined.

Acknowledgment

The authors (J. Dusza and P.Šajgalík) thank the Alexander von Humboldt Stiftung for financial support.

REFERENCES

1. R.W.Davidge,"Mechanical behaviour of ceramics", Cambridge Press, Cambridge, U.K., 1979.
2. D.Munz and T.Fett,"Mechanisches Verhalten keramischer Werkstoffe", Springer Verlag, Berlin/Heidelberg 1989.
3. A.G.Evans, Structural reliability, a processing-dependent phenomena, J.Am.Ceram.Soc.,73, (1982), p.127.
4. G.D.Quinn and R.Morell, Design data for engineering ceramics: a review of the flexure test, J.Am.Ceram.Soc., 74, (1991), p.2037.
5. P.F.Becher, Microstructural design of toughened ceramics, J.Am.Ceram.Soc., 74 (1991), p.255.
6. G.H. Campbell, M. Rühle, B.J.Dalgleish and A.G.Evans, Whisker toughening: a comparison between alumina oxide and silicon nitride toughened with silicon carbide, J. Am. Ceram.Soc., 73, (1990), p.521.
7. T.Fett, D.Munz, Influence of R-curve effects on lifetimes for specimens with natural cracks, in: Proc."Fracture processes in concrete, rock and ceramics", Ed. J.G.M.van Mier, J.G.Rots and A.Bakker, (1991), p.365.
8. K.Niihara and A.Nakahire, Strengthening of oxide ceramics by SiC and Si_3N_4 dispersions, on:"Ceramic materials and

components for engines" Edited by V.J.Tennery, American Ceram.Soc., Westerville, OH, 1989, p.919.

9. W.J.Clegg, K.Kendall, N.McN.Alford, T.W.Button and J.D. Birchall, A simple way to make tough ceramics, Nature (London), 347, p.455, (1990).

10. D.B.Marshall, J.J.Ratto and F.F.Lange, Enhanced fracture toughness in layered microcomposites of Ce-ZrO$_2$ and Al$_2$O$_3$, J.Am.Ceram.Soc., 74, 12, (1991), p.2979.

11. M.P.Hammer, H.M.Chan and G.A.Miller, Unique opportunities for microstructural engineering with duplex and laminar ceramic composites, J.Am.Ceram.Soc., 75,7, (1992), p. 1715.

12. P.Šajgalík, J.Dusza and Z.Lenčéš, Layered Si$_3$N$_4$ based composites, in Proc.:"Int.conf. on ceramic processing, science and technology, Sept. 11-14, 1994, Friedrichshafen, Germany, to be published.

13. C.J.Russo, M.P.Harmer, H.M.Charz and G.A.Miller, Design of a laminated ceramic composite for improved strength and toughness. 93rd Annual Meeting of the Amer.Ceram. Soc., Cincinnati, 1991 (Ceramic Matrix Composites Symp., Paper No.110-SVI-91).

14. T.Chartier, D.Merle and J.L.Bessou, Laminar ceramic composites, J.Europ.Ceram.Soc., 15, (1995), p.101.

15. P.Šajgalík, J.Dusza, Z.Lenccéš, Layered Si$_3$N$_4$ composites with enhanced room temperature properties, J.Mater.Sci., submitted.

16. S.P.Chen, Modelling of brittle/brittle laminates: The effect of the interfacial cohesion, Scripta Met. et Mater., 31 (1994), 10, p.1437.

17. D.K.Shetty , I.G.Wright, P.M.Mincer, A.H.Claver, Indentation fracture of WC-Co cermets, J.Mater.Sci., 20, (1985), p.1873.

18. G.Dreier, G.Elssner, S.Schmauder, T.Suga, Determination of residual stresses in bimaterials, J.Mater.Sci.,29, (1994), p.1441.

19. J.Dusza, P.Šajgalík, Fracture toughness and strength testing of ceramic composites,in:"Handbook of Advanced materials testing", Ed.N.P.Cheremisinoff and P.N.Cheremisinoff, Marced Dekker Inc., New York-Basel-Hong Kong, 1995, pp.399-436.

MODELLING STRESS DISTRIBUTIONS IN
BRITTLE PARTICULATE - BRITTLE MATRIX COMPOSITES

Ranjan Biswas, J. Leslie Henshall, and Richard J. Wakeman

School of Engineering, Exeter University
Exeter, Devon, U.K.

ABSTRACT

Finite element analysis has been used to investigate the stress concentrations which occur in bonded brittle particle - brittle matrix composites. The modelling was performed using a commercial package, ANSYS, with plain strain plate elements restricted to uniform loading in one direction and only elastic stresses. The parameters varied in this study were particle geometry (square, hexagonal or circular cross-section), and particle:matrix elastic modulus ratio in the range 1:1 to 100:1. The maximum principal and shear stresses were found to occur at, or in close proximity to, the particle:matrix interface. The stress concentration effect in the principal stress increased as the elastic modulus mismatch increased and was greatest for the square cross-section (maximum stress concentration > 3) and least for the circular cross-section (maximum stress concentration ~ 1.18).

INTRODUCTION

Fibre composite systems have been the subject of much research and development and their mechanics have been well elucidated, *e.g.* Piggott, 1980; Vinson and Sierakowski, 1986; Kelly, 1989. In contrast, the fundamental mechanics of particulate composite systems has received less consideration, despite their commercial significance. It is possible to analytically predict the elastic moduli of two phase particulate composites, *e.g.* Eshelby, 1957; Hashin and Shtrikman, 1962 and 1963; Davy and Guild, 1988; Guild and Young, 1989; Kwon and Dharan, 1995, and also phenomenologically correlate the hardness of the composite with hardnesses of the individual phases and their respective volume fractions, *e.g.* Lee and Gurland, 1978; Hooper et al., 1991; John 1992. However, the prediction of the strength properties of particulate reinforced composites from a knowledge of the component properties is still imprecise, *e.g.* Hasselman and Fulrath, 1966; Lange, 1970; Swearengen et al., 1978; Faber and Evans, 1983; Ashby, 1993; Lutz, 1994.

Practically, the manufacture of fibre reinforced composites of the high strength advanced ceramics, *e.g.* silicon carbide, silcon nitride, high purity alumina and toughened zirconia, is difficult. It is also not clear that fibre reinforcement is actually beneficial in these sytems, Piller et al. (1993). Consequently there is currently a growing interest in particulate composites of engineering ceramics, *e.g.* zirconia toughened alumina, nanosize particle composites, Niihara (1991) and silicon carbide or titanium nitride reinforced alumina, Walker et al. (1994). However the development of new ceramic composite materials remains primarily a matter of informed trial-and-error, with the economic penalties that this approach entails. There is a clear requirement to be able to provide a framework and analytical basis to guide the development of these materials. In addition, there is also current interest in analysing the micromechanics of the strength behaviour of such materials as concrete, *e.g.* Fanella (1990), and two-phase minerals.

This study is based upon using the finite element method to investigate the effects of particle shape, square, circular or hexagonal cross-section, and particle:matrix elastic moduli ratios in the range 1:1 to 100:1, on the local stresses developed in uniaxially loaded matrix-bonded particle composites.

FINITE ELEMENT MODEL

The basis of the model has been to examine at the mesoscopic level the stresses generated within a defined cell, concentrating on the stress distributions at the region of elastic modulus mismatch. A notional square 4 x 4 unit cell (where 1 unit has been specified as 100 µm within ANSYS) has been used with plane strain conditions. A single particle has been placed at the centre of this cell (it should be noted that since the analyses have been performed using plain strain plate elements the "particle" is in effect a rod extending through the thickness), and the analyses performed for one quadrant of the cell since the remainder will be symmetrical.

Figure 1 shows the geometries of the particles and meshes. Figure 1(a) represents a square cross-section particle of side length two units; Figure 1(b) represents a regular hexagonal cross-section of side length one unit; and Figure 1(c) has a circular cross-section of radius one unit. The meshes comprise a mixture of six-noded triangular and eight-noded quadrilateral plane strain plate elements. The basic geometry of the system was initially defined and then the meshes generated automatically. The situation being modelled assumed "perfect" bonding across the particle-matrix interface, *i.e.* no contact or other interfacial elements were inserted along the particle-matrix boundaries. The total numbers of elements and nodes were in the ranges 2241 to 3343 and 5626 to 8118, respectively. The boundary conditions specified were planes of symmetry along the vertical and lower horizontal interfaces and free along the top boundary.

The particle:matrix elastic moduli ratios examined were 1:1, 2.5:1, 10:1, 26.8:1 and 100:1. In all cases, except 26.8:1, Poisson's ratio was taken as 0.4 in both regions. The 26.8:1 results were generated to compare with the results of an experimental investigation into the mechanical properties of quartz:perspex composites, Biswas et al., (1995), and hence the Poisson's ratios of quartz and perspex, *i.e.* 0.16 and 0.4, were used. This experimental study (Biswas et al., 1995) has shown that particle size had a greater effect on compressive and Brazilian tensile strength than volume fraction, for volume fractions in the range 10% - 40%. Hence this study has maintained the primary dimension of the particles similar for comparison purposes, rather than adjusting this dimension to give similar volume fractions.

Two methods of loading the system were compared, *i.e.* fixed displacement of the upper surface and constant pressure. When the upper surface was displaced uniformly by a fixed amount there appeared to be slight inconsistencies in the numerical solutions since two regions of high localised stress were generated, *cf* the stress countours in the top part

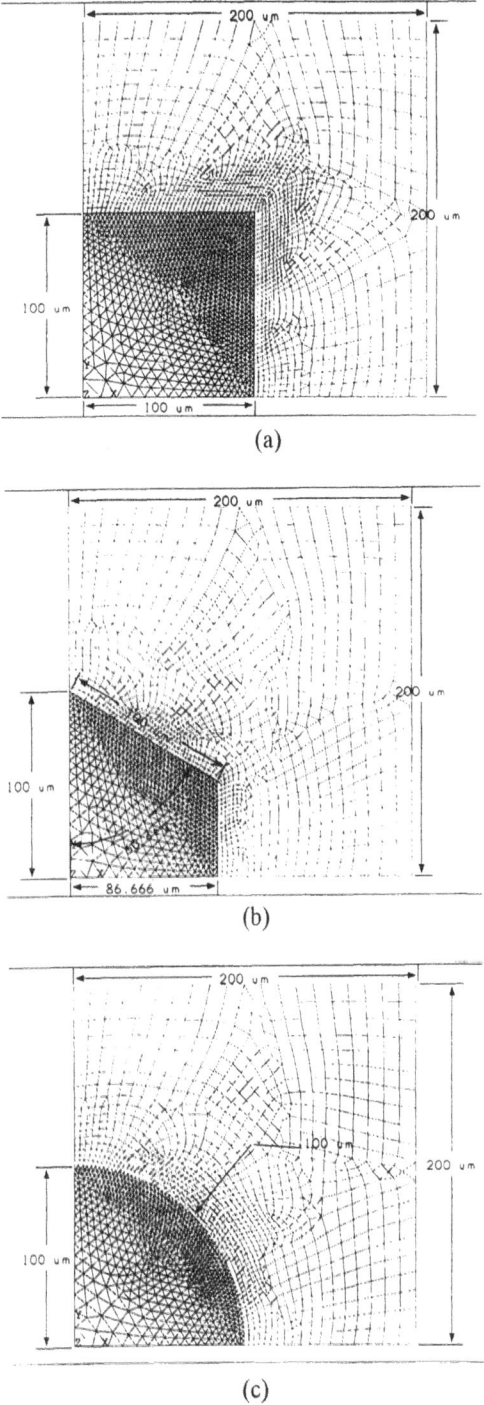

Figure 1. Finite element geometries and meshes for (a) square cross-section, (b) hexagonal cross-section, and (c) circular cross-section particles.

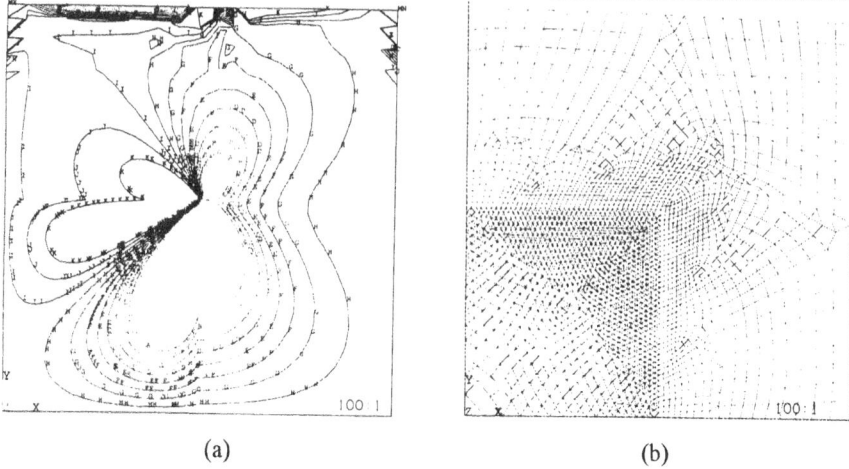

| (a) | (b) |

Figure 2. (a) map of stress contours for square cross-section particle, 100:1 modulus ratio, subjected to a constant displacement along the top surface. The stress contours along the top surface are condidered to be artificial. (b) deformed mesh for square cross-section particle, 100:1 modulus ratio, subjected to uniform pressure along the top surface. The initial mesh boundaries are shown by the dashed lines.

of Figure 2(a), which is for a square section particle, 100:1 elastic modulus ratio. For comparison purposes it was necessary to estimate the effective elastic modulus of the composite by applying a small displacement to the upper surface and calculating the total applied force along the upper surface. From this it is then possible to ratio the displacements for the different conditions to ensure that the value of the total surface loading is the same in all cases. However, it was not certain that, given these small high-stress regions, this procedure would be correct. With a constant applied pressure to the top surface a non-uniform displacement of the top surface occurred, *cf* comparison of the initial boundaries with the deformed mesh, Figure 2(b). This effect was more pronounced for the higher modulus particles. It was decided however that an applied constant pressure along the top surface would allow a more consistent comparison of particle shape and elastic modulus effects.

RESULTS

Analyses were performed for the three meshes shown in Figure 1 with the particle:matrix elastic modulus ratio of 1:1, subject to a nominal uniform 10 MPa equivalent compressive loading on the top surface. It was found that in all cases a uniform stress, equal in magnitude to the applied loading, was obtained. The results presented in this section all apply to a nominal uniform 10 MPa compressive loading, except Figure 10. The notation used for the principal stresses is $S_1 > S_2 > S_3$, and hence in compression the numerically largest stress value is S_3.

Square Cross-Section

Figure 3 shows the S_3 stress contours for the square-cross section particle along the interface with the normal in the X-direction. It can be seen that as the elastic modulus ratio increases, the number of countours increases, and their density along/adjacent to the

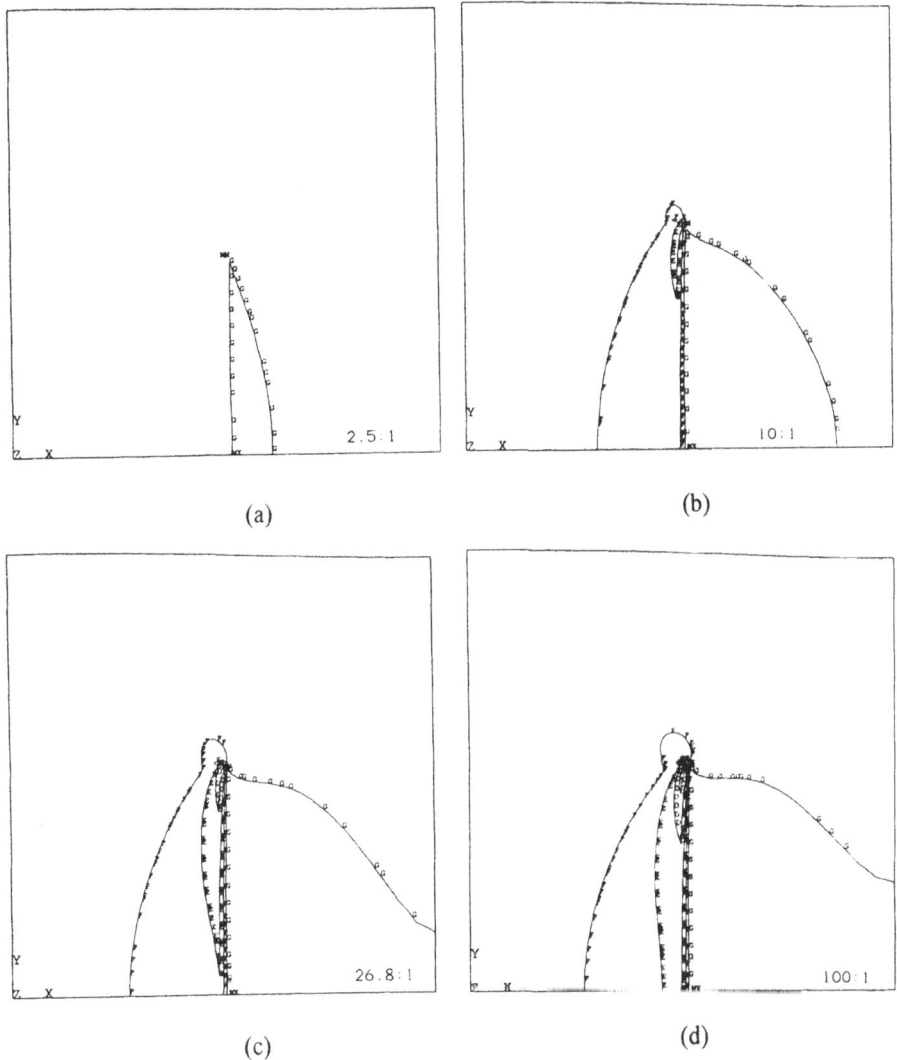

Figure 3. Stress contour maps for square cross-section particles with moduli ratios of (a) 2.5:1, (b) 10:1, (c) 26.8:1, and (d) 100:1. An equivalent uniform pressure of 10 MPa is applied along the top surface. The corresponding stress contour values are: A = -37 MPa, B = -32.2 MPa, C = -27.4 MPa, D = -22.6 MPa, E = -17.8 MPa, F = -13 MPa, G = -8.2 MPa, H = -3.4 MPa, I = 1.4 MPa, J = 6.2 MPa, and K = 11 MPa.

interface, particularly near to the corner, increases. The non-uniformity of the various components of the stress state was greatest in all cases for the 100:1 modulus ratio, as shown in Figure 4 for S_x, S_{xy}, and the ANSYS defined parameter $S_{int} = 2 \times$ the maximum shear stress. The maximum stresses, and stress gradients, are all in the vicinity of the corner of the square, as would be expected.

The values of S_3 and the maximum shear stress along the interface perpendicular to the X direction have been selected from the data output for the different elastic moduli ratios and are shown in Figures 5(a) and (b). The axes have been nondimensionalised with respect to the applied pressure, and are thus effectively stress concentration factors, and distance along the interface. These curves show more quantitatively than Figure 3 the degree of stress concentration at the corner of the particle, and the stress gradients. Figures

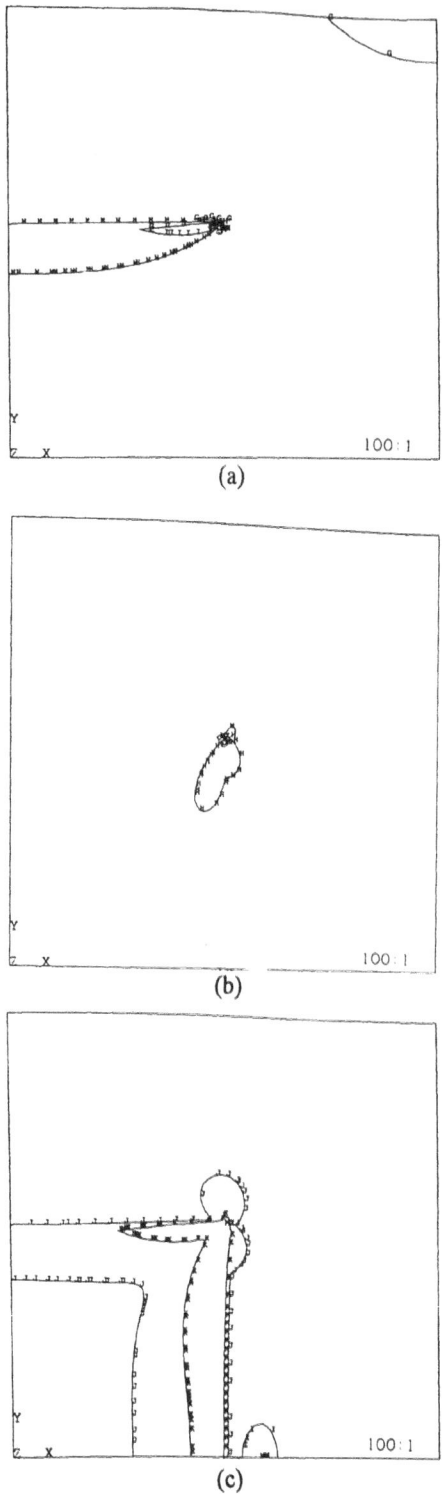

Figure 4. Stress contours of (a) S_x, (b) S_{xy}, and (c) maximum shear stress, for a square cross-section particle with 100:1 modulus ratio. The letters correspond to the same numerical values as in Figure 3.

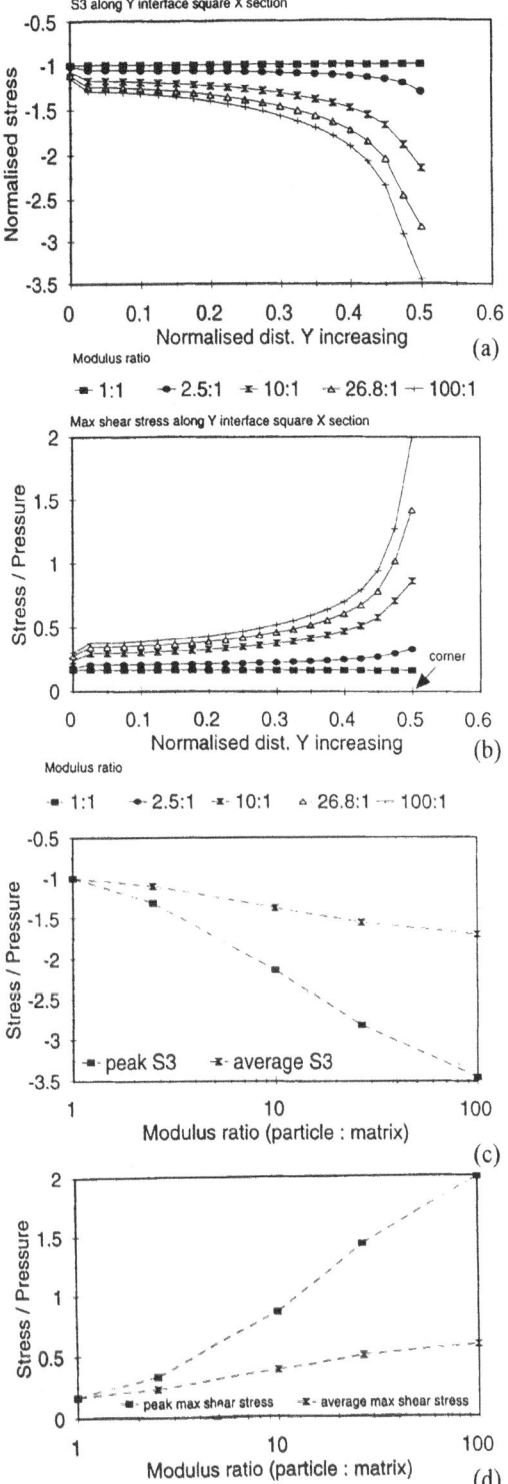

Figure 5. (a) Variation of S_3, nondimensionalised with respect to the applied pressure, along the interface perpendicular to X-axis, *i.e.* vertical face, for the square cross-section particle for the elastic moduli ratios shown, (b) similar to (a), but showing the maximum shear stress, (c) the variation of peak S_3 and average along the interface S_3, values with elastic modulus ratio, and (d) similar to (c) but for maximum shear stress.

5 (c) and (d) depict the variation of the maximum direct and shear stress, and the average values of these stresses along the interface, with elastic modulus ratio. These parameters have been selected, since the peak stresses would probably be expected to control crack nucleation, and the total force along the interface might well be associated with the driving force for crack extension, *e.g.* Fanella, (1990). The maximum stresses increase markedly with increasing modulus ratio, *i.e.* S_3 by a factor of 3.45 and maximum shear stress by 11.9 at the 100:1 ratio. The average values of S_3 and maximum shear stress along the interface increase by 70% and 250%, respectively, as the modulus ratio changes from 1:1 to 100:1.

Hexagonal Cross-Section

The contour maps of S_3, and the variation with elastic modulus ratio, for the hexagonal cross-section particles are shown in Figure 6. The regions of maximum stress are along the interface perpendicular to the X-axis and near the apex on the Y-axis. The more

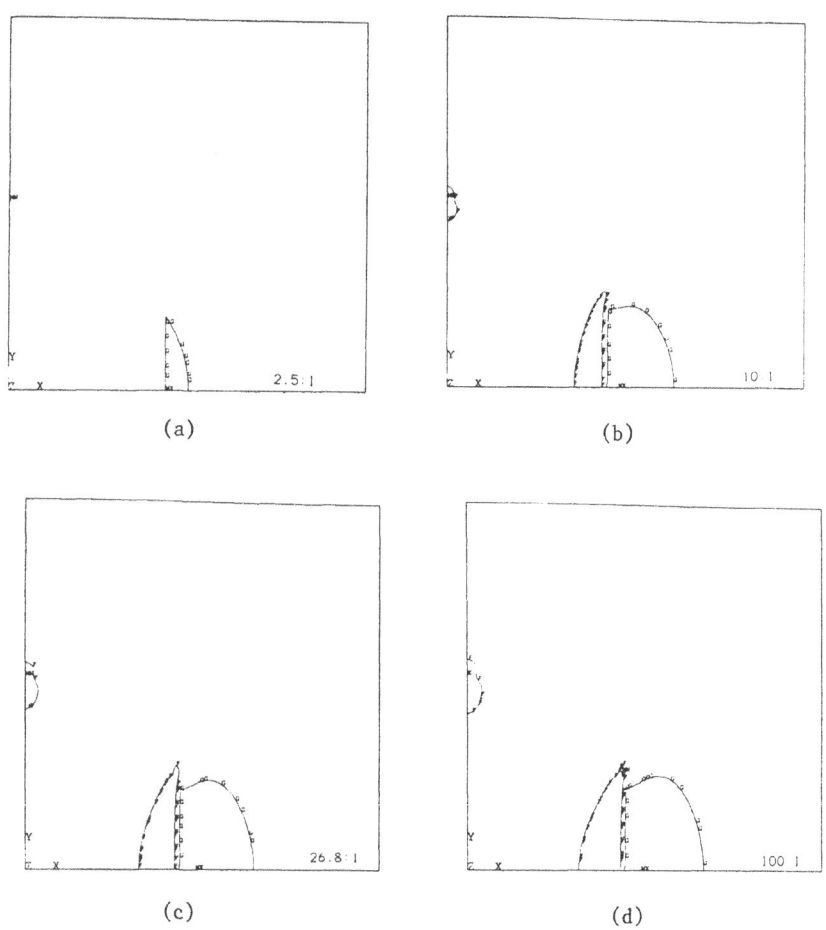

(a)

(b)

(c)

(d)

Figure 6. Stress contour maps for hexagonal cross-section particles with moduli ratios of (a) 2.5:1, (b) 10:1, (c) 26.8:1, and (d) 100:1. An equivalent uniform pressure of 10 MPa is applied along the top surface. The stress contours correspond to those in Figures 3 and 4, *i.e.* E = -17.8 MPa, F = -13 MPa, and G = -8.2 MPa.

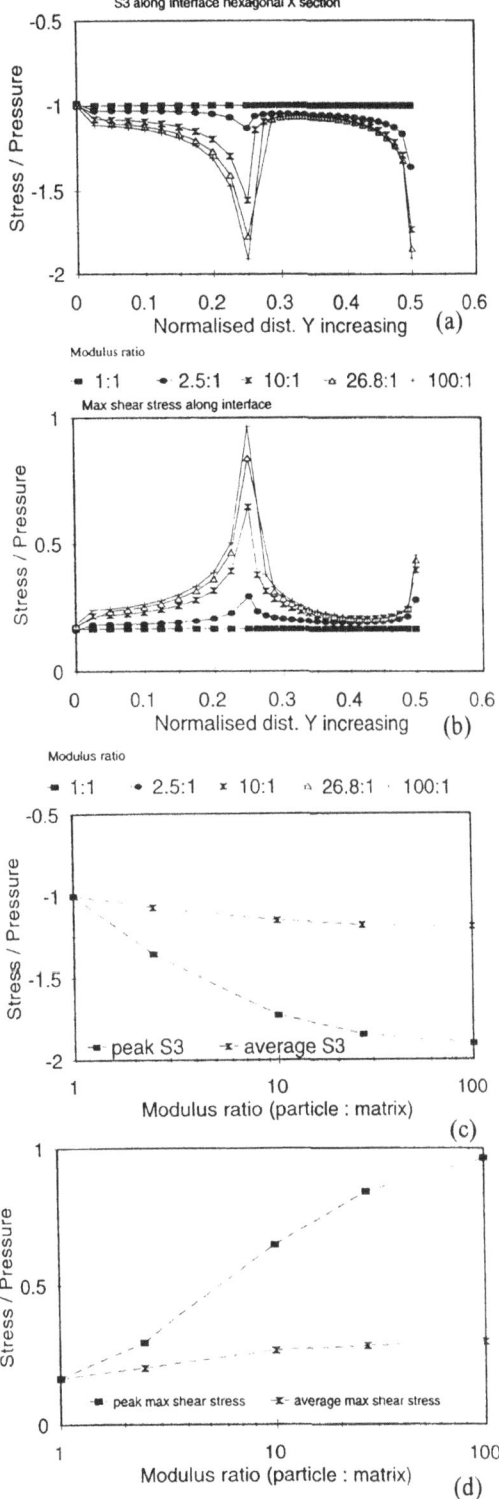

Figure 7. (a) Variation of S_3, nondimensionalised with respect to the applied pressure, along the interface hexagonal cross-section particle for the elastic moduli ratios shown, (b) similar to (a), but showing the maximum shear stress, (c) the variation of peak S_3 and average along the interface S_3, values with elastic modulus ratio, and (d) similar to (c) but for maximum shear stress.

highly stressed regions increase in extent with increasing modulus ratio, as does the stress concentration factor. Figures 7(a) and (b) give the variation of S_3 and maximum shear stress, respectively, along the interface. The two corners correspond to the two stress concentration points. For the lower elastic modulus ratios, S_3 is significantly higher at the top apex, *i.e.* on the Y-axis, but the difference in the stress concentrations at the corners decreases with increasing elastic modulus ratio. The situation with regard to the maximum shear stress is almost the converse of this, *i.e.* the stress concentration effects of the two corners are similar with the low modulus ratio, and as this ratio increases the stress at the lower apex increases more rapidly. Figures 7(c) and (d) show the variation of the peak and average-along-the-interface S_3 and maximum shear stresses, respectively, with elastic modulus ratio. The values have again been nondimensionalised with respect to the applied load. The peak values of S_3 and maximum shear stress increase by factors of 1.90 and 5.78, and the averaged values along the interface by 16% and 76%, respectively as the modulus ratio changes from 1:1 to 100:1.

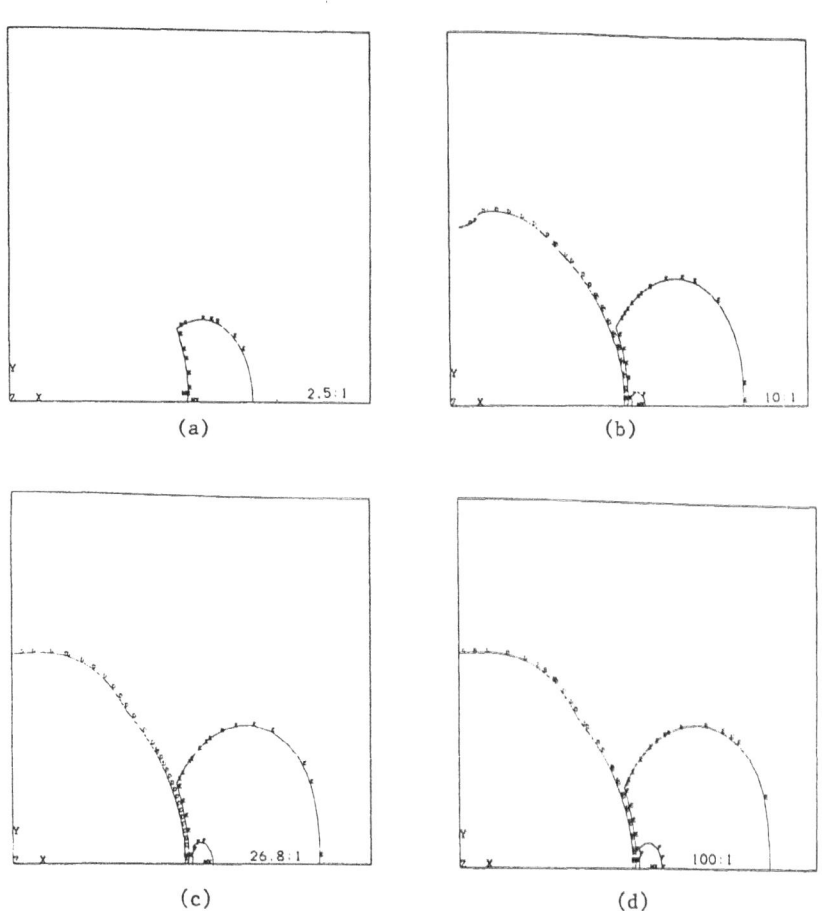

(a)

(b)

(c)

(d)

Figure 8. Stress contour maps for circular cross-section particles with moduli ratios of (a) 2.5:1, (b) 10:1, (c) 26.8:1, and (d) 100:1. An equivalent uniform pressure of 10 MPa is applied along the top surface. The stress contours correspond to 0.5 x the values in Figures 3, 4 and 5, *i.e.* D = -11.3 MPa, E = -8.9 MPa, and F = - 0.65 MPa.

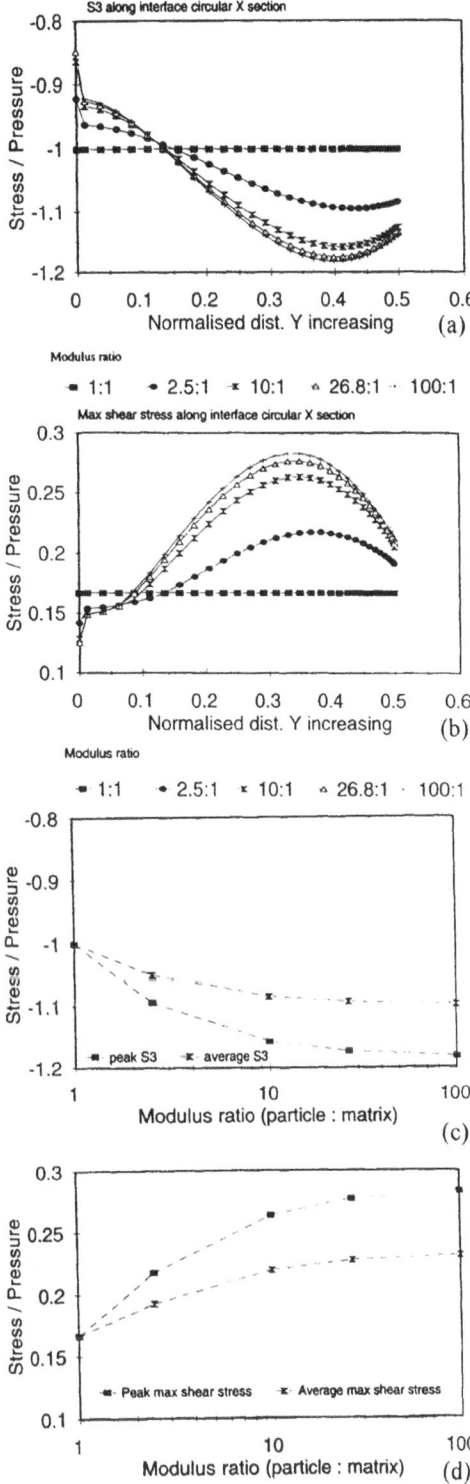

Figure 9. (a) Variation of S_3, nondimensionalised with respect to the applied pressure, along the interface for a circular cross-section particle for the elastic moduli ratios shown, (b) similar to (a), but showing the maximum shear stress, (c) the variation of peak S_3 and average along the interface S_3, values with elastic modulus ratio, and (d) similar to (c) but for maximum shear stress.

Circular Cross-Section

Figure 8 displays the variation of the contour lines of S_3 for the circular cross-section particles for the different moduli ratios (note that the stress value associated with a given letter in this case is one-half the value for the same letter for the square and hexagonal cross-sections: this enables several contours to be depicted). In these cases, the difference in contours between 10:1 and 100:1 is small.

The variations of S_3 and maximum shear stress along the interface are shown in Figures 9(a) and (b). The maximum stress concentrations in these cases are 1.18 and 1.70, respectively, close to the top of the particle. These figures also reinforce the observation that there is little change in stresses as the modulus ratio changes from 10:1 to 100:1. The average increase in the values of S_3 and maximum shear stress are 10.5% and 38% respectively, Figures 9 (c) and (d).

DISCUSSION

The results in the previous section have been presented in terms of compressive loading. However, since the system is linear changing the sign of the applied loading to tension would only change the sign of the results, and not the magnitudes. An additional confirmation of the self-consistency of ANSYS was performed by repeating some of the analyses for tension. For example for the circular cross-section, Figure 10 can be seen to be the same as Figure 9(a), apart from the change in sign.

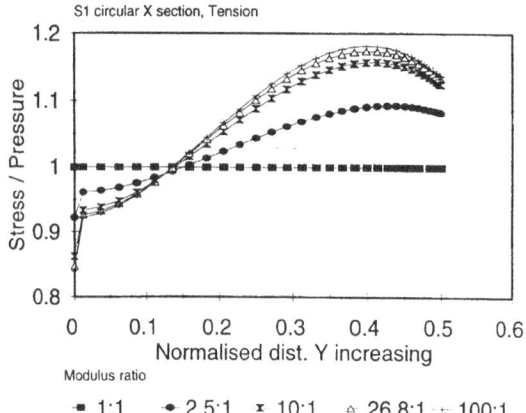

Figure 10. Nondimensionalised maximmum principal stress, S_1, for circular cross-section particles subjected to a uniform tensile stress on the top surface.

Particles that are added to form composites are conventionally classified as angular or rounded. The results of this modelling would indicate that, for bonded particles, the interfacial stress concentrations are markedly higher for angular, *i.e.* square or hexagonal, than for "rounded", *i.e.* circular. There is also a clear difference in the results for square and hexagonal cross-sections, which would suggest that the broad category of "angular" should really be further sub-divided in terms of the geometry, presumably based upon the typical number of faces of the particles. The effect of mismatch in the elastic moduli on the interfacial stresses is also very marked for the square and hexagonal cross-sections, but less so for the circular.

The results for the circular cross-section particles are similar to those obtained by Guild and Young (1989), who used an axisymmetric finite element analysis to investigate the effects of volume fraction of spherical particles on the elastic moduli, but also included some results on stress contours.

CONCLUSIONS

The main conclusions that can be drawn from this analytical modelling of bonded brittle particle - brittle matrix composites are:

(a) the effect of elastic modulus mismatch is to generate stress concentrations along the particle - matrix interface, the greater the modulus mismatch, the greater the effect, and

(b) there is a significant effect of particle shape on the maximum and average direct and shear stresses along the particle-matrix interface, e.g. with a modulus ratio of 100:1, the peak stress concentrations for a square cross-section are 3.45 and 11.9 (for direct and shear stress respectively), whereas the equivalent values for the circular cross-section are 1.18 and 1.70.

REFERENCES

Ashby, M.F., 1993, Criteria for selecting the components of composites, *Acta Metall. Mater.*, 41:1313.
Biswas, R., Henshall, J.L., and Wakeman, R.J., 1995, Deformation and fracture in a model brittle particulate composite system: quartz-perspex, in: Extended Abstracts 7th International Conference on the Mechanical Behaviour of Materials, The Hague, Netherlands.
Davy, P.J., and Guild, F.J., 1988, The distribution of interparticle distance and its application in finite-element modelling of composite materials, *Proc. Roy. Soc. London A*, A418:95.
Eshelby, J.D., 1957, The determination of the elastic field of an ellipsoidal inclusion, and related problems, *Proc. Roy. Soc. London A*, A241:376
Faber, K.T., and Evans, A.G., 1983, Crack deflection processes - I theory, *Acta Metall.*, 31:565.
Fanella, D.A., 1990, Fracture and failure of concrete in uniaxial and biaxial loading, *J. Eng. Mech.*, 116:2341.
Guild, F.J., and Young, R.J., 1989, A predictive model for particulate-filled composite materials, part 1, hard particles, *J. Mater. Sci.*, 24:298.
Hashin, Z., and Shtrikman, S, 1962, On some variational principles in anisotropic and nonhomogeneous elasticity, *J. Mech. Phys. Solids*, 10:335.
Hashin, Z., and Shtrikman, S. 1963, A variational approach to the theory of the elastic behaviour of multiphase materials, *J. Mech. Phys. Solids*, 11:127.
Hasselman, D.P.H., and Fulrath, R.M., 1966, Proposed fracture theory of a dispersion-strengthened glass matrix, *J. Amer. Ceram. Soc.*, 49:69.
John, V., 1992, "Introduction to Engineering Materials", MacMillan, London.
Hooper, R.M., Guillou, M-O., and Henshall, J.L., 1991, Indentation studies of cBN-TiC composites, *J. Hard Materials*, 2:223.
Kelly, A., 1989, "Concise Encyclopedia of Composite Materials", Pergamon, Oxford.
Kwon, P., and Dharan, C.K.H., 1995, Effective moduli of high volume fraction particulate composites, *Acta Metall. Mater.*, 43:1141.
Lange, F.F., 1970, The interaction of a crack front with a second phase dispersion, *Philos. Mag.*, 22:983.
Lee, H.C., and Gurland, J., 1978, Hardness and deformation of cemented tungsten carbide, *Mat. Sci. Eng.*, 33:122.
Lutz, E.H., 1994, Predictability of the mechanical properties of inclusion-containing ceramics, *J. Amer. Ceram. Soc.*, 77:1901.

Niihara, K., 1991, New design concept of structural ceramics: ceramic nanocomposites, *The Centennial Memorial Issue of the Ceramic Society of Japan*, 99:974.

Pigott, M.R., 1980, "Load Bearing Fibre Composites", Pergamon, Oxford.

Piller, R.C., Friend, S.J., Davidge, R.W., Sleurs, J., and Gilissen, R., 1993, Hot isostatic pressing of silicon nitride and silicon carbide using glass encapsulation, *in*: "Engineering Ceramics: Fabrication Science and Technology", D.P. Thompson, ed., Institute of Materials, London.

Swearengen, J.C., Beauchamp, E.K., and Eagan, R.J., 1978, Fracture toughness of reinforced glasses, *in*: Fracture Mechanics of Ceramics, Vol. 4, R.C. Bradt, D.P.H. Hasselman and F.F. Lange, eds., Plenum, New York.

Vinson J.R. and Sierakowski, R.L., 1986, "The Behavior of Structures Composed of Composite Materials", Dordrecht, Nijhoff.

Walker, C.N., Borsa, C.E., Todd, R.I., Davidge, R.W., and Brook, R.J., 1994, Fabrication, characterisation and properties of alumina matrix nanocomposites, *in*: "Novel Synthesis and Processing of Ceramics", F.R. Sale, ed., Institute of Materials, London.

IDENTIFICATION OF DAMAGE VARIABLE IN CERAMIC MATRIX COMPOSITE WITH DIFFERENT BEHAVIOUR IN TENSION AND COMPRESSION

Alexander Zolochevsky

Physical Engineering Department
Kharkov Technical University
Eidemana 15-104, Kharkov, 310118, Ukraine

INTRODUCTION

In the past two decades, in particular, Continuum Damage Mechanics (CDM) has been applied to creep damage, brittle/elastic damage, ductile plastic damage, fatigue damage. The majority of the approaches in CDM are development and generalization of the Kachanov-Rabotnov concept about a material damage parameter.

On the other hand, a new separate branch in mechanics of solid materials has recently been formed, i.e. mechanics of materials with different behaviour in tension and compression. Intense development of it is connected with considerable engineering applications since light alloys, gray cast irons, graphite, concrete, polymers as well as ceramics and ceramic composites, whose mechanical characteristics depend on the type of loading, are used extensively in various fields of mechanical and civil engineering.

Damage in ceramics and ceramic composites is not directly accessible to measurement. There are different methods of nondirect measurement of damage variable from continuum mechanics point of view. Most of these methods are related with the coupling between deformation and damage, i.e., with the degradation of the mechanical properties caused by damage.

Many ceramic materials show in basic experiments the following properties (Ladeveze et.al.,1994; Munz and Fett, 1989; Shan et.al.,1994):
- different rupture limits in tension and compression,
- different elastic behaviour in tension and compression (for example, nonlinear behaviour in tension and linear behaviour in compression),
- the initial anisotropy,
- the unilateral nature of the damage (the active/passive damage effects), when the microcracks are open or closed, respectively,
- different damage in tension and compression,
- damage induced anisotropy under non-proportional loading, due to the directionality of the microcracks.

One of the modern difficulties in the CDM is connected with the simultaneous description of the anisotropic nature of damage and the difference between the damaging processes under tensile and compressive loading types (Chaboche, 1992). The review of various models of the CDM, using scalars, vectors, second-order and fourth-order tensors as a damage variables has been made by Chaboche (1992) in the case of the elastic damage. It is established that existing models do not describe simultaneously the damage induced anisotropy and the unilateral character of the damage. Also note that recently proposed theory by Chaboche (1993) does not take into account the initial anisotropy of materials. The damage theory developed by Ladeveze et al. (1994) and Shan et al (1994) bases on the information about the decrease of the elasticity modulus of the damaged material and does not describe the damage of ceramic composites for compression loading. Furthermore, it is not clear how the last approach may be used in the case when the principal stresses $\sigma_1, \sigma_2, \sigma_3$ have the opposite signs.

Thus, even in the elastic case, up to now there is no theory for ceramic composites which is able to reproduce simultaneously the initial anisotropy, the damage induced anisotropy and the different damage in tension and compression. The absence of this damage theory makes it impossible to perform the identification of damage variable in ceramic materials. The aim of this paper is to solve such an important problem within the framework of the phenomenological approach of CDM. We shall consider small elastic strains in isothermal and time-independent processes.

ISOTROPIC DAMAGE MODEL FOR CERAMICS

We start from the consideration of elastic deformation and isotropic damage in ceramics.

Stress-Strain Relation

In order to construct a connection between components of symmetrical tensors for the elastic infinitesimal strain $\underline{\varepsilon}$ and the CAUCHY stress $\underline{\sigma}$ in ceramics we use the assumption of the existence of a potential

$$F = \frac{1}{2}\sigma_e^2 \tag{1}$$

Then

$$\underline{\varepsilon} = \frac{\partial F}{\partial \underline{\sigma}} \tag{2}$$

Here σ_e is the equivalent stress.

We consider ceramics as initially isotropic materials with elastic damage. In this case the equivalent stress σ_e, which establishes equivalence of uniaxial and complex stressed states, should be a function of stress tensor invariants and damage. Let us introduce into the consideration the first

$$J_1(\underline{\sigma}) = \underline{\sigma} \cdot \cdot I = tr\underline{\sigma} = \sigma_{kk} \tag{3}$$

and the second

$$J_2(\underline{\sigma}) = \underline{\sigma} \cdot \cdot \underline{\sigma} = tr\underline{\sigma}^2 = \sigma_{kn}\sigma_{kn} \tag{4}$$

414

invariants of the stress tensor, where $\underline{I}$ is the second order unit tensor, "." denotes the scalar product operation and "tr" denotes the trace of a second order tensor. The influence of the third invariant of the stress tensor is neglected. Then taking into account that the equivalent stress must be a uniform function of arguments, the equivalent stress can be written in a form

$$\sigma_e = \sigma_2 + \alpha\sigma_1 \tag{5}$$

Here

$$\sigma_1 = B\,J_1(\underline{\sigma}) \tag{6a}$$

$$\sigma_2^2 = A\,J_1^2(\underline{\sigma}) + C\,J_2(\underline{\sigma}) \tag{6b}$$

are some scalar functions of stresses and the damage; A, B are some scalar functions of the damage; α is numerical coefficient which takes account of the specific weight for the function σ_1 in the representation for σ_e. The proposed expression (5) for the equivalent stress is quite general. It is possible to analyse a number special cases resulting from (5), (6a,b) and having more simple structure.

Equivalent Stress in The Classical Linear Elastic Potential. By placing in (5), (6b)

$$\alpha = 0, \quad A = -\frac{\nu}{E}, \quad C = \frac{1+\nu}{E} \tag{7}$$

we arrive at the equivalent stress

$$\sigma_e = \sqrt{\frac{1+\nu}{E}J_2 - \frac{\nu}{E}J_1^2} \tag{8}$$

Equivalent Stress in the Potential of von Mises. If we suppose in (5), (6b)

$$\alpha = 0, C = -3A, A = -\frac{1}{2} \tag{9}$$

we obtain the equivalent stress

$$\sigma_e = \sigma_i \tag{10}$$

in the well-known potential of von Mises. Here

$$\sigma_i = \sqrt{\frac{3}{2}\underline{s}\cdot\cdot\underline{s}} \tag{11}$$

is the stress intensity,

$$\underline{s} = \underline{\sigma} - \frac{1}{3}J_1\underline{I} \tag{12}$$

is the stress deviator.

415

Equivalent Stress in the Potential of Drucker and Prager. In the case

$$C = -3A, A = -\frac{1}{2}, \alpha B = \bar{\alpha} \tag{13}$$

the relation (5) can be rewritten in the following form

$$\sigma_e = \sigma_1 + \bar{\alpha} J_1 \tag{14}$$

where $\bar{\alpha}$ is some constant material parameter. Thus, we arrive at the expression for the equivalent stress in the potential of Drucker and Prager.

Equivalent Stress in the Potential of Green. If we take

$$\alpha = 0 \tag{15}$$

in the representation (5), we obtain the relation

$$\sigma_e = \sigma_2 \tag{16}$$

used for porous undamaged materials in the potential of Green.

Let us return to the determination of the elastic strain tensor according to the equation (2). Then using expressions (5), (6a,b) and relations

$$\frac{\partial F}{\partial \underline{\sigma}} = \frac{\partial F}{\partial \sigma_1} \frac{\partial \sigma_1}{\partial \underline{\sigma}} + \frac{\partial F}{\partial \sigma_2} \frac{\partial \sigma_2}{\partial \underline{\sigma}} \tag{17a}$$

$$\frac{\partial F}{\partial \sigma_1} = \alpha \sigma_e, \quad \frac{\partial F}{\partial \sigma_2} = \sigma_e, \quad \frac{\partial \sigma_1}{\partial \underline{\sigma}} = B\underline{I}, \quad \frac{\partial \sigma_2}{\partial \underline{\sigma}} = \frac{AJ_1 I + C\sigma}{\sigma_2} \tag{17b,c}$$

one obtains the constitutive equation

$$\underline{\varepsilon} = \sigma_e \left(\frac{AJ_1 I + C\underline{\sigma}}{\sigma_2} + \alpha B I \right) \tag{18}$$

of elastic deformation in ceramics.

Thermodynamic Consideration

First note that elastic potential (1) is a homogeneous positively definite quadratic function. Therefore, according to the Euler's theorem, we have

$$\underline{\sigma} \cdot \cdot \frac{\partial F}{\partial \underline{\sigma}} = 2F \tag{19}$$

Let

$$W = \frac{1}{2}\underline{\sigma} \cdot \underline{\varepsilon} \qquad (20)$$

be the density of the elastic strain energy. Then using relation (2), we obtain from equations (19), (20)

$$W = F \qquad (21)$$

Thus, the potential (1) is equal to the density of the elastic strain energy.

On the other hand, elastic potential (1) can be interpreted as a thermodynamic potential which contains all the information about the influence of the damage on the behaviour of ceramics. State variables in this thermodynamic potential are the stress tensor and scalar damage functions A, B, C. In the present case, three damage energy release rates, corresponding to A, B, C, appear as scalars (Chaboche, 1992)

$$Y_A = \frac{\partial F}{\partial A}, \quad Y_B = \frac{\partial F}{\partial B}, \quad Y_C = \frac{\partial F}{\partial C} \qquad (22a,b,c)$$

It is easily to find from (1), (5), (6a,b) that

$$\frac{\partial F}{\partial A} = \sigma_e \frac{\partial \sigma_2}{\partial A}, \quad \frac{\partial F}{\partial B} = \alpha\sigma_e \frac{\partial \sigma_1}{\partial B}, \quad \frac{\partial F}{\partial C} = \sigma_e \frac{\partial \sigma_2}{\partial C}$$

$$\qquad (23a,b,c,d,e,f)$$

$$\frac{\partial \sigma_2}{\partial A} = \frac{J_1^2}{2\sigma_2}, \quad \frac{\partial \sigma_1}{\partial B} = J_1, \quad \frac{\partial \sigma_3}{\partial C} = \frac{J_2}{2\sigma_2}$$

Therefore, relations (22a,b,c) can be rewritten

$$Y_A = \sigma_e \frac{J_1^2}{2\sigma_2}, \quad Y_B = \alpha\sigma_e J_1, \quad Y_C = \sigma_e \frac{J_2}{2\sigma_2} \qquad (24a,b,c)$$

Let us define such a damage criterion

$$g(z,r) \equiv z(Y_A, Y_B, Y_C, A, B, C) - r \leq 0 \qquad (25)$$

Here z is the suitable scalar homogeneous function, r is the damage threshold at current time t depending on the history loading. If the value z is equal to the damage threshold at current time, the damage is increasing. Then let us introduce the following function

$$z = \frac{1}{2}(BY_B + 2AY_A + 2CY_C) \qquad (26)$$

It is not difficult to see on the basis of relations (1),(5),(6a,b),(21),(24a,b,c) that a function z in a case (26) has a property of the density of the elastic strain energy, i.e.

$$z = W \qquad (27)$$

Evolution of damage is defined with the following equations

$$\dot{A} = \dot{\lambda}\frac{\partial z}{\partial Y_A}, \quad \dot{B} = \dot{\lambda}\frac{\partial z}{\partial Y_B}, \quad \dot{C} = \dot{\lambda}\frac{\partial z}{\partial Y_C} \qquad (28a,b,c)$$

which leads to:

$$\dot{A} = \dot{\lambda}A, \quad \dot{B} = \dot{\lambda}B, \quad \dot{C} = \dot{\lambda}C \qquad (29a,b,c)$$

Here the dot above the symbol denotes differentiation with respect to current time, $\dot{\lambda}$ is a certain scalar multiplier.

Using the principle of maximum damage dissipation, one can show (Ju, 1989) that $\dot{\lambda}$ satisfy the Kuhn-Tucker relations

$$\dot{\lambda} > 0, \quad g(z,r) \le 0, \quad \dot{\lambda}g(z,r) = 0 \qquad (30a,b,c)$$

For example, if the damage in ceramics is increasing we have $\dot{\lambda} > 0$. Then, by condition (30c) it follows: $g(z,r) = 0$. If, on the other hand, the damage criterion is not satisfied and $g(z,r) < 0$, the condition (30c) implies that $\dot{\lambda} = 0$.

The second thermodynamic principle has the form:

$$\phi = Y_A\dot{A} + Y_B\dot{B} + Y_C\dot{C} \qquad (31)$$

Substituting the values (24a,b,c),(29a,b,c) into (31), we obtain

$$\phi = \frac{1}{2}\dot{\lambda}\sigma_e^2 \qquad (32)$$

Since $\dot{\lambda} \ge 0$ from (30a), we see that $\phi \ge 0$. Therefore the second thermodynamic principle for presented model of ceramics is always valid.

Basic Experiments

Isotropic damage model does not necessary imply a model with single scalar damage variable. In this connection note, that isotropic model under the consideration is related with three scalar damage functions A, B, C. These functions are independent in the common case. Therefore we must use three data of the basic experiments on standard specimens for the determining the functions A, B, C. Let us consider these basic experiments.

Uniaxial Tension: In the case of uniaxial tension ($\sigma_{11} > 0$) one has the following strains in the direction of loading

$$\varepsilon_{11} = \frac{\sigma_{11}}{E^+} \qquad (33)$$

and in transverse direction

$$\varepsilon_{22} = -v^+ \varepsilon_{11} \tag{34}$$

Here E^+ is the secant elastic modulus of ceramics in tension, and v^+ is the ratio of transverse strain in tension.

On the other hand, it is easily to see on the basis of (3)-(5), (6a,b), (18) that

$$\varepsilon_{11} = \left(\sqrt{A+C} + \alpha B\right)^2 \sigma_{11}, \quad \varepsilon_{22} = \left(\sqrt{A+C} + \alpha B\right)\left(\frac{A}{\sqrt{A+C}} + \alpha B\right)\sigma_{11} \tag{35a,b}$$

Uniaxial compression: In the case of uniaxial compression $\left(\sigma_{11} < 0\right)$ we obtain:

$$\varepsilon_{11} = -\frac{|\sigma_{11}|}{E^-} \tag{36}$$

where E^- is the secant elastic modulus of ceramics in compression.

On the other hand, using equations (3)-(5), (6a,b), (18) it has been found that

$$\varepsilon_{11} = -\left(\sqrt{A+C} - \alpha B\right)^2 |\sigma_{11}| \tag{37}$$

On the basis of data of these basic experiments, equating relations (33) and (35a), (34) and (35b), (36) and (37), we find the damage functions in constitutive equation (18):

$$\alpha B = \left[\left(E_-\right)^{-1/2} - \left(E_+\right)^{-1/2}\right]/2, \quad \sqrt{A+C} = \left[\left(E_+\right)^{-1/2} + \left(E_-\right)^{-1/2}\right]/2 \tag{38a,b}$$

$$A = -\sqrt{A+C}\left[v_+\left(E_+\right)^{-1/2} + \alpha B\right] \tag{38c}$$

Material parameters E_+, E_-, v_+ in (38a,b,c) must be assigned from stress-strain diagrams.

Identification of Damage Variable in Ceramics

The character of the rupture of solids bodies is different depending on the conditions of loading. Therefore, the problem of the rupture of ceramics has many aspects. In this connection we note that the conclusion about no damage of ceramic materials for compression loading in the damage modelling approach of Ladeveze et al.(1994) and Shan et al.(1994) is rather uncertain and inaccurate. Actually, the rupture under compressive loading type occurs as a result of the accumulation of the damage. It is well known indeed that the process of deformation of ceramics is accompanied by the formation of microscopic cracks and a change of the structural state of ceramics. Phenomenologically one can say in this case that damage-accumulation occurs. Obviously, that this inaccurate conclusion of Ladeveze et al.(1994) and Shan et al.(1994) is related with the representation of damage variable on the basis of the elasticity modulus of the material. Thus, we must use another damage measure for ceramics. It is evident that elastic deformation and damage accumulation occur in parallel with each other and they have a reciprocal effect. In order to describe these phenomena, we take the density of the elastic strain energy (20) as the damage variable

$$\omega = W \tag{39}$$

Then we assume that the damage functions A, B, C are depended on the damage variable ω, i.e. $A=A(\omega)$, $B=B(\omega)$ and $C=C(\omega)$. Thus, we introduce in a formal way the scalar parameter (39) which we denote the damage variable. This variable $\omega \in [0, \omega^*]$ characterizes the damaged state of ceramics. Initial value $\omega = 0$ corresponds to the nondamaged state, critical value $\omega = \omega^*$ corresponds to the fracture of ceramics. Last value depends on the type of loading. In the case of fracture a damage criterion (25), (27) has a form

$$\omega^* = r^* \tag{40}$$

where critical value of the damage threshold r^* depends on the type of loading. We suppose

$$r^* = a\sigma_i + bJ_1 \tag{41}$$

Here a and b are some material parameters which can be found on the basis of experiments.

Uniaxial Tension: In the case of uniaxial tension ($\sigma_{11}^* = \sigma_b, \varepsilon_{11}^* = \varepsilon_b$) we obtain

$$\sigma_b \varepsilon_b = \sigma_b (a + b) \tag{42}$$

Therefore

$$a+b=\varepsilon_b \tag{43}$$

Here σ_b is the ultimate tensile strength, ε_b is the limit strain in tension.

Uniaxial Compression: In the case of uniaxial compression ($\sigma_{11}^* = -\sigma_c, \varepsilon_{11}^* = -\varepsilon_c$)

$$\sigma_c \varepsilon_c = \sigma_c (a - b) \tag{44}$$

where σ_c is the ultimate compressive strength, ε_c is the limit strain in compression. Therefore

$$a-b=\varepsilon_c \tag{45}$$

Now it is not difficult to obtain from (43), (45) that

$$a = \frac{1}{2}(\varepsilon_b + \varepsilon_c), \quad b = \frac{1}{2}(\varepsilon_b - \varepsilon_c) \tag{46a,b}$$

ANISOTROPIC DAMAGE MODEL FOR CERAMIC COMPOSITES

Stress-Strain Relation

We consider a unidirectionally fiber-reinforced ceramic matrix composite as initially transversely isotropic material, in which the fibre direction may be characterized by a unit vector **d**. It is assumed that the damage in ceramic composite is connected with the degradation of the elastic properties due to parallel surface-like microcracks. The orientation of an array of parallel microcracks may be characterized by a unit vector $\mathbf{n}$. Let $\Omega = \omega\mathbf{n}$ be a damage vector and ω be a scalar damage variable. Note that elasticity theory for fiber composites with same behaviour in tension and compression, gradually degrading by the evolution of surface-like microcracks, is discussed by Matzenmiller and Sackman (1994).

Constitutive equation of elastic deformation is based on the assumption of the existence of a potential in a form (1). In our case the equivalent stress σ_e is a function (Spencer, 1984; Matzenmiller and Sackman, 1994) of the CAUCHY stress tensor $\underline{\sigma}$, the dyadic product $\Omega \otimes \mathbf{n} = \omega\mathbf{n} \otimes \mathbf{n}$ and the symmetric second-order tensor $\mathbf{d} \otimes \mathbf{d}$, i.e.

$$\sigma_e = \sigma_e(\underline{\sigma}, \omega\mathbf{n} \otimes \mathbf{n}, \mathbf{d} \otimes \mathbf{d}) \tag{47}$$

Then the integrity basis, formed by three symmetric second-order tensors considered here, consists of ten following invariants (Spencer, 1984; Matzenmiller and Sackman, 1994):

$$I_1 = \text{tr}\,\underline{\sigma} = \sigma_{kk}, \quad I_2 = \text{tr}\,\underline{\sigma}^2 = \sigma_{km}\sigma_{mk}, \quad I_3 = \text{tr}\,\underline{\sigma}^3 = \sigma_{ki}\sigma_{im}\sigma_{mk} \tag{48a,b,c}$$

$$I_4 = \text{tr}(\omega\mathbf{n} \otimes \mathbf{n}) = \omega, \quad I_5 = \text{tr}\left[(\omega\mathbf{n} \otimes \mathbf{n})\underline{\sigma}\right] = \omega\mathbf{n} \cdot \underline{\sigma} \cdot \mathbf{n} = \omega n_k \sigma_{km} n_m \tag{48d,e}$$

$$I_6 = \text{tr}\left[(\omega\mathbf{n} \otimes \mathbf{n})\underline{\sigma}^2\right] = \omega\mathbf{n} \cdot \underline{\sigma}^2 \cdot \mathbf{n} = \omega n_k \sigma_{km}\sigma_{ml} n_l \tag{48f}$$

$$I_7 = \text{tr}(\mathbf{d} \otimes \mathbf{d}\underline{\sigma}) = \mathbf{d} \cdot \underline{\sigma} \cdot \mathbf{d} = d_k \sigma_{km} d_m \tag{48g}$$

$$I_8 = \text{tr}(\mathbf{d} \otimes \mathbf{d}\underline{\sigma}^2) = \mathbf{d} \cdot \underline{\sigma}^2 \cdot \mathbf{d} = d_k \sigma_{kl}\sigma_{lm} d_m \tag{48h}$$

$$I_9 = \text{tr}\left[\underline{\sigma}(\omega\mathbf{n} \otimes \mathbf{n})(\mathbf{d} \otimes \mathbf{d})\right] = \omega(\mathbf{d} \cdot \mathbf{n})\mathbf{d} \cdot \underline{\sigma} \cdot \mathbf{n} = \omega(d_k n_k)(d_m \sigma_{ml} n_l) \tag{48i}$$

$$I_{10} = \text{tr}\left[(\omega\mathbf{n} \otimes \mathbf{n})(\mathbf{d} \otimes \mathbf{d})\right] = \omega(\mathbf{d} \cdot \mathbf{n})^2 = \omega(d_k n_k)^2 \tag{48j}$$

Let us form the linear

$$\sigma_1 = B_1(\omega)I_7 + \overline{B}_2(\omega)I_5 \tag{49}$$

and quadratic

$$\sigma_2^2 = A_1(\omega)I_1^2 + A_2(\omega)I_2 + A_3(\omega)I_7^2 + A_4(\omega)I_1 I_7 + A_5(\omega)I_8 \tag{50}$$

scalar invariant functions, where $B_1(\omega), \overline{B}_2(\omega), A_1(\omega), \ldots, A_5(\omega)$ are some functions of the damage variable. We take the expression for the equivalent stress in a form (5). Note that if $\alpha = 0$ in (5) we obtain the equivalent stress

$$\sigma_e = \sigma_2 \tag{51}$$

for ceramic composites with same behaviour in tension and compression. Now using equations (1), (2), (5), (17a,b), (48a,b,e,g,h), (49), (50) and relations

$$\frac{\partial \sigma_1}{\partial \underline{\sigma}} = B_1(\omega)\mathbf{d} \otimes \mathbf{d} + \bar{B}_2(\omega)\omega\,\mathbf{n} \otimes \mathbf{n}, \quad \frac{\partial \sigma_2}{\partial \underline{\sigma}} = \frac{A_1(\omega)I_1\mathbf{I} + A_2(\omega)\sigma}{\sigma_2} +$$

$$\frac{2A_3(\omega)I_1\mathbf{d} \otimes \mathbf{d} + A_4(\omega)(I_1\mathbf{d} \otimes \mathbf{d} + I_7\mathbf{I}) + A_5(\omega)(\mathbf{d} \otimes \mathbf{d} \cdot \underline{\sigma} + \underline{\sigma} \cdot \mathbf{d} \otimes \mathbf{d})}{2\sigma_2} \tag{52a,b}$$

we arrive at the constitutive equation of elastic deformation

$$\underline{\varepsilon} = \sigma_e\Big[\alpha B_1(\omega)\mathbf{d} \otimes \mathbf{d} + \alpha B_2(\omega)\,\mathbf{n} \otimes \mathbf{n} + \frac{A_1(\omega)I_1\mathbf{I} + A_2(\omega)\sigma}{\sigma_2} +$$

$$\frac{2A_3(\omega)I_1\mathbf{d} \otimes \mathbf{d} + A_4(\omega)(I_1\mathbf{d} \otimes \mathbf{d} + I_7\mathbf{I}) + A_5(\omega)(\mathbf{d} \otimes \mathbf{d} \cdot \underline{\sigma} + \underline{\sigma} \cdot \mathbf{d} \otimes \mathbf{d})}{2\sigma_2} \Big] \tag{53}$$

for unidirectionally fiber-reinforced ceramic matrix composites. Here the damage dependent function $B_2(\omega)$ is substituted for $\omega\bar{B}_2(\omega)$.

Thermodynamic Consideration

In the case of ceramic composites we have also that elastic potential (1), (5), (49), (50) is equal to the density of the elastic strain energy (20). On the other hand, let us consider this elastic potential as a thermodynamic potential, in which state variables are the stress tensor and the scalar damage functions $B_1, \bar{B}_2, A_1, \ldots, A_5$. Then from (1), (5), (49), (50) we can define the damage energy release rates, corresponding to $B_1, \bar{B}_2, A_1, \ldots, A_5$

$$y_1 = \frac{\partial F}{\partial B_1}, \ y_2 = \frac{\partial F}{\partial \bar{B}_2}, \ Y_1 = \frac{\partial F}{\partial A_1}, \ Y_2 = \frac{\partial F}{\partial A_2}, \ Y_3 = \frac{\partial F}{\partial A_3}, \ Y_4 = \frac{\partial F}{\partial A_4}, \ Y_5 = \frac{\partial F}{\partial A_5} \tag{54}$$

Therefore, one obtains

$$\frac{\partial F}{\partial B_1} = \alpha\sigma_e\frac{\partial \sigma_1}{\partial B_1}, \quad \frac{\partial F}{\partial \bar{B}_2} = \alpha\sigma_e\frac{\partial \sigma_1}{\partial \bar{B}_2}, \quad \frac{\partial F}{\partial A_1} = \sigma_e\frac{\partial \sigma_2}{\partial A_1}, \quad \frac{\partial F}{\partial A_2} = \sigma_e\frac{\partial \sigma_2}{\partial A_2}$$

$$\frac{\partial F}{\partial A_3} = \sigma_e\frac{\partial \sigma_2}{\partial A_3}, \quad \frac{\partial F}{\partial A_4} = \sigma_e\frac{\partial \sigma_2}{\partial A_4}, \quad \frac{\partial F}{\partial A_5} = \sigma_e\frac{\partial \sigma_2}{\partial A_5}, \quad \frac{\partial \sigma_1}{\partial B_1} = I_7, \quad \frac{\partial \sigma_1}{\partial \bar{B}_2} = I_5 \tag{55}$$

$$\frac{\partial \sigma_2}{\partial A_1} = \frac{I_1^2}{2\sigma_2}, \quad \frac{\partial \sigma_2}{\partial A_2} = \frac{I_2}{2\sigma_2}, \quad \frac{\partial \sigma_2}{\partial A_3} = \frac{I_7^2}{2\sigma_2}, \quad \frac{\partial \sigma_2}{\partial A_4} = \frac{I_1 I_7}{2\sigma_2}, \quad \frac{\partial \sigma_2}{\partial A_5} = \frac{I_8}{2\sigma_2}$$

Thus, we have

$$y_1 = \alpha\sigma_e I_7, \quad y_2 = \alpha\sigma_e I_5, \quad Y_1 = \sigma_e\frac{I_1^2}{2\sigma_2}, \quad Y_2 = \sigma_e\frac{I_2}{2\sigma_2} \tag{56a,b,c,d}$$

$$Y_3 = \sigma_e\frac{I_7^2}{2\sigma_2}, \quad Y_4 = \sigma_e\frac{I_1 I_7}{2\sigma_2}, \quad Y_5 = \sigma_e\frac{I_8}{2\sigma_2} \tag{56e,f,g}$$

By analogy with relation (25) the damage loading surface is expressed as a criterion

$$g(z,r) = z\left(y_1, y_2, Y_1, \ldots, Y_5, B_1, \overline{B}_2, A_1, \ldots, A_5\right) - r \leq 0 \qquad (57)$$

where r depends on the loading history. We assume that the function z has a form

$$z = \frac{1}{2}\left(B_1 y_1 + \overline{B}_2 y_2 + 2A_1 Y_1 + 2A_2 Y_2 + 2A_3 Y_3 + 2A_4 Y_4 + 2A_5 Y_5\right) \qquad (58)$$

Then using equations (1), (5), (20), (21), (49), (50), we can say that the function z has a property of the density of the elastic strain energy as well as in a case of ceramics, i.e.

$$z = W \qquad (59)$$

The damage process is characterized by the following equations of evolution

$$\dot{B}_1 = \dot{\lambda}\frac{\partial z}{\partial y_1}, \quad \dot{\overline{B}}_2 = \dot{\lambda}\frac{\partial z}{\partial y_2}, \quad \dot{A}_1 = \dot{\lambda}\frac{\partial z}{\partial Y_1}, \quad \dot{A}_2 = \dot{\lambda}\frac{\partial z}{\partial Y_2} \qquad (60a,b,c,d)$$

$$\dot{A}_3 = \dot{\lambda}\frac{\partial z}{\partial Y_3}, \quad \dot{A}_4 = \dot{\lambda}\frac{\partial z}{\partial Y_4}, \quad \dot{A}_5 = \dot{\lambda}\frac{\partial z}{\partial Y_5} \qquad (60e,f,g)$$

Therefore, we obtain

$$\dot{B}_1 = \frac{1}{2}\dot{\lambda}B_1, \quad \dot{\overline{B}}_2 = \frac{1}{2}\dot{\lambda}\overline{B}_2, \quad \dot{A}_1 = \dot{\lambda}A_1, \quad \dot{A}_2 = \dot{\lambda}A_2 \qquad (61a,b,c,d)$$

$$\dot{A}_3 = \dot{\lambda}A_3, \quad \dot{A}_4 = \dot{\lambda}A_4, \quad \dot{A}_5 = \dot{\lambda}A_5 \qquad (61e,f,g)$$

Note also that by analogy with ceramics, we have in our case conditions (30a,b,c).
Within this framework, the second thermodynamic principle has the following form:

$$\phi - y_1\dot{B}_1 + y_2\dot{\overline{B}}_2 + Y_1\dot{A}_1 + Y_2\dot{A}_2 + Y_3\dot{A}_3 + Y_4\dot{A}_4 + Y_5\dot{A}_5 \qquad (62)$$

Then, substituting values (56a,b,c,d,e,f,g), (61a,b,c,d,e,f,g) into relation (62) gives

$$\phi = \frac{1}{2}\dot{\lambda}\sigma_e^2 \qquad (63)$$

Since $\dot{\lambda} \geq 0$ from (30a), we obtain that $\phi \geq 0$. Thus, the second thermodynamic principle for anisotropic damage model under consideration is always valid.

Basic Experiments

Let the material is reinforced with a single family of fibres in the direction 1. Let us consider a method for the determining the scalar damage functions $B_1(\omega), B_2(\omega), A_1(\omega), \ldots, A_5(\omega)$ in the constitutive equation (53) on the basis of data of basic experiments. We shall assume that microcracks in ceramic composites are always orthogonal to the direction of the maximum principle stress. This assumption follows from the direct experimental data for different ceramic composites, in which brittle fracturing takes place.

Uniaxial Tension in the Direction 1. In the case of uniaxial tension in the direction of fibres ($\sigma_{11} > 0$) we obtain from this experiment the following strains in the direction of loading and in transverse direction 2 :

$$\varepsilon_{11} = \frac{\sigma_{11}}{E_1^+}, \quad \varepsilon_{22} = -v_{21}^+ \frac{\sigma_{11}}{E_1^+} \tag{64a,b}$$

Here $E_1^+ = E_1^+(\omega)$ is the secant elastic modulus in tension in the direction of fibres and $v_{21}^+ = v_{21}^+(\omega)$ is the ratio of transverse strain under tension in the direction 1.

On the other hand, we shall use the constitutive equation (53).First of all, note that in this case of loading the majority of microcracks run perpendicular to the axis of loading. Thus, we have following vectors for fibre and damage directions $\mathbf{d} = [1,0,0]^T ; \mathbf{n} = [1,0,0]^T$. Then we obtain from (53)

$$\varepsilon_{11} = \left(\sqrt{A_1 + A_2 + A_3 + A_4 + A_5} + \alpha B_1 + \alpha B_2 \right)^2 \sigma_{11}$$

$$\varepsilon_{22} = \left(\sqrt{A_1 + A_2 + A_3 + A_4 + A_5} + \alpha B_1 + \alpha B_2 \right) \left(\frac{A_1 + \frac{1}{2} A_4}{\sqrt{A_1 + A_2 + A_3 + A_4 + A_5}} \right) \sigma_{11} \tag{65a,b}$$

Equating (64a) and (65a), (64b) and (65b), one has

$$\left(\sqrt{A_1 + A_2 + A_3 + A_4 + A_5} + \alpha B_1 + \alpha B_2 \right) = \frac{1}{\sqrt{E_1^+}} \tag{66a}$$

$$\left(\frac{A_1 + \frac{1}{2} A_4}{\sqrt{A_1 + A_2 + A_3 + A_4 + A_5}} \right) = -\frac{v_{21}^+}{\sqrt{E_1^+}} \tag{66b}$$

Uniaxial Compression in the Direction 1. Under uniaxial compression in the direction of fibres $(\sigma_{11} < 0)$ the following relation can be written

$$\varepsilon_{11} = -\frac{|\sigma_{11}|}{E_1^-} \tag{67}$$

where $E_1^- = E_1^-(\omega)$ is the secant elastic modulus in compression in the direction of fibres.

Note that in this case microcracks run parallel to the axis of compressive loading. In this connection we have $\mathbf{d} = [1,0,0]^T ; \mathbf{n} = [0,1,0]^T$. Therefore, we obtain from (53) the relation

$$\varepsilon_{11} = -\left(\sqrt{A_1 + A_2 + A_3 + A_4 + A_5} - \alpha B_1 \right)^2 |\sigma_{11}| \tag{68}$$

424

Equating (67) and (68), we have

$$\left(\sqrt{A_1 + A_2 + A_3 + A_4 + A_5} - \alpha B_1\right) = \frac{1}{\sqrt{E_1^-}} \tag{69}$$

Uniaxial Tension in Transverse Direction 2. In the case of uniaxial tension in transverse direction 2 ($\sigma_{22} > 0$) we obtain from this experiment the following strains in the direction of loading and in transverse direction 3 :

$$\varepsilon_{22} = \frac{\sigma_{22}}{E_2^+}, \quad \varepsilon_{33} = -v_{32}^+ \frac{\sigma_{22}}{E_2^+} \tag{70a,b}$$

Here $E_2^+ = E_2^+(\omega)$ is the secant elastic modulus in tension in the direction 2 and $v_{32}^+ = v_{32}^+(\omega)$ is the ratio of transverse strain under tension in the direction 2.

On the other hand, in this case we have $\mathbf{d} = [1,0,0]^T; \mathbf{n} = [0,1,0]^T$ and obtain relations

$$\varepsilon_{22} = \left(\sqrt{A_1 + A_2} + \alpha B_2\right)^2 \sigma_{22} \tag{71a}$$

$$\varepsilon_{33} = \left(\sqrt{A_1 + A_2} + \alpha B_2\right)\left(\frac{A_1}{\sqrt{A_1 + A_2}}\right)\sigma_{22} \tag{71b}$$

Equating (70a) and (71a), (70b) and (71b), one has

$$\left(\sqrt{A_1 + A_2} + \alpha B_2\right) = \frac{1}{\sqrt{E_2^+}} \tag{72a}$$

$$\left(\frac{A_1}{\sqrt{A_1 + A_2}}\right) = -\frac{v_{32}^+}{\sqrt{E_2^+}} \tag{72b}$$

Uniaxial Compression in Transverse Direction 2. Under uniaxial compression in transverse direction 2 $\left(\sigma_{22} < 0\right)$ the following relation can be obtained

$$\varepsilon_{22} = -\frac{|\sigma_{22}|}{E_2^-} \tag{73}$$

where $E_2^- = E_2^-(\omega)$ is the secant elastic modulus under compression in the direction 2.

On the other hand, we have $\mathbf{d} = [1,0,0]^T; \mathbf{n} = [1,0,0]^T$. Therefore, we obtain from (53)

$$\varepsilon_{22} = -\left(\sqrt{A_1 + A_2}\right)^2 |\sigma_{22}| \tag{74}$$

Equating (73) and (74), one has

$$\left(\sqrt{A_1 + A_2}\right) = \frac{1}{\sqrt{E_2^-}} \tag{75}$$

Pure Torsion in the Plane 1-2. In the case of pure torsion in the plane 1-2 $\left(\sigma_{12} \neq 0\right)$ we have the following relation

$$\varepsilon_{12} = \frac{\sigma_{12}}{2G_{12}} \tag{76}$$

where $G_{12} = G_{12}(\omega)$ is the secant shear modulus.

Note that under pure torsion microcracks run by the angle $\frac{\pi}{4}$ to axes 1,2. Therefore, we obtain the following vectors for fibre and damage directions $\mathbf{d} = [1,0,0]^T; \mathbf{n} = \left[\frac{1}{\sqrt{2}}, \frac{1}{\sqrt{2}}, 0\right]^T$.

Constitutive equation (53) is then transformed into the relation

$$\varepsilon_{12} = \frac{1}{2}\left(\sqrt{2A_2 + \frac{1}{2}A_5} + \alpha B_2\right)^2 \sigma_{12} \tag{77}$$

Equating (76) and (77), we have

$$\left(\sqrt{2A_2 + \frac{1}{2}A_5} + \alpha B_2\right) = \frac{1}{\sqrt{G_{12}}} \tag{78}$$

Now it is easy to find from (66a,b), (69), (72a,b), (75), (78) the damage functions

$$A_1 = -\frac{v_{32}^+}{\sqrt{E_2^+ E_2^-}}, \quad A_2 = \frac{1}{E_2^-} - A_1, \quad \alpha B_2 = \frac{1}{\sqrt{E_2^+}} - \frac{1}{\sqrt{E_2^-}} \tag{79a,b,c}$$

$$A_5 = 2\left(\frac{1}{\sqrt{G_{12}}} - \alpha B_2\right)^2 - 4A_2, \quad \alpha B_1 = \frac{1}{2}\left(\frac{1}{\sqrt{E_1^+}} - \frac{1}{\sqrt{E_1^-}}\right) - \frac{1}{2}\alpha B_2 \tag{79d,e}$$

$$A_4 = -2\frac{v_{21}^+}{\sqrt{E_1^+}}\left(\frac{1}{\sqrt{E_1^-}} + \alpha B_1\right) - 2A_1, \quad A_3 = \left(\frac{1}{\sqrt{E_1^-}} + \alpha B_1\right)^2 - \frac{1}{E_2^-} - A_4 - A_5 \tag{79f,g}$$

Material parameters $E_1^+(\omega), E_2^+(\omega), E_1^-(\omega), E_2^-(\omega), v_{21}^+(\omega), v_{32}^+(\omega), G_{12}^+(\omega)$ may be determined from stress-strain diagrams in tension, compression and torsion.

Identification of Damage Variable in Ceramic Composites

As well as in the case of ceramics we introduce for ceramic composites considered above the damage variable as the density of the elastic strain energy (20), i.e.

$$\omega = W \tag{80}$$

where $\omega \in [0, \omega^*]$. Now phenomenologically we can easy take into account on the bases of (53), (80) the influence of matrix microcracking, wake debonding, fiber failure, crack front debonding on the elastic deformation of ceramic composites. Critical value of the damage variable ω^* can be found from a damage criterion (57), (59) which in the case of fracture has such a form

$$\omega^* = r^* \tag{81}$$

We assume that critical value of the damage threshold is defined as

$$r^* = b_1 I_7 + b_2 \mathbf{n} \cdot \underline{\sigma} \cdot \mathbf{n} + \sqrt{a_1 I_1^2 + a_2 I_2 + a_3 I_7^2 + a_4 I_1 I_7 + a_5 I_8} \tag{82}$$

Here $b_1, b_2, a_1, \ldots, a_5$ are some material parameters which can be found on the basis of rupture data from the following basic experiments.

Uniaxial Tension in the Direction 1. In the case of uniaxial tension in the direction of fibres we obtain from equations (81),(82) the following limit longitudinal strain

$$\varepsilon_{11}^+ = b_1 + b_2 + \sqrt{a_1 + a_2 + a_3 + a_4 + a_5} \tag{83}$$

Uniaxial Compression in the Direction 1. Under uniaxial compression in the direction of fibres one has the following limit longitudinal strain

$$\varepsilon_{11}^- = -b_1 + \sqrt{a_1 + a_2 + a_3 + a_4 + a_5} \tag{84}$$

Uniaxial Tension in Transverse Direction 2. Under uniaxial tension in transverse direction 2 we obtain the following limit strain

$$\varepsilon_{22}^+ = b_2 + \sqrt{a_1 + a_2} \tag{85}$$

Uniaxial Compression in Transverse Direction 2. Under uniaxial compression in transverse direction 2 the following limit strain in the direction of loading can be obtained

$$\varepsilon_{22}^- = \sqrt{a_1 + a_2} \tag{86}$$

Pure Torsion in the Plane 1-2. In the case of pure torsion in the plane 1-2 one has from equations (81), (82) the following limit angular strain

$$\gamma_{12}^* = b_2 + \sqrt{2a_2 + \frac{1}{2}a_5} \tag{87}$$

Pure Torsion in the Plane 2-3. Under pure torsion in the plane 2-3 we obtain the following vectors $\mathbf{d} = [1,0,0]^T$; $\mathbf{n} = \left[0, \dfrac{1}{\sqrt{2}}, \dfrac{1}{\sqrt{2}}\right]^T$ and the following limit angular strain

$$\gamma_{23}^* = b_2 + \sqrt{2a_2} \tag{88}$$

Loading of thinwalled tube by inner pressure: In the case of loading by inner pressure of thinwalled tube with the orientation in the direction of fibres we have $\mathbf{n} = [0,1,0]^T$ and

$$\varepsilon_{11}^{*} + 2\varepsilon_{22}^{*} = b_1 + 2b_2 + \sqrt{9a_1 + 5a_2 + a_3 + 3a_4 + a_5} \tag{89}$$

where $\varepsilon_{11}^{*}, \varepsilon_{22}^{*}$ are limit strains in this basic experiment.

The solution of the system of equations (83)-(89) is

$$b_2 = \varepsilon_{22}^{+} - \varepsilon_{22}^{-}, \ b_1 = \frac{1}{2}\left(\varepsilon_{11}^{+} - \varepsilon_{11}^{-} - b_2\right), \ a_2 = \frac{1}{2}\left(\gamma_{23}^{+} - b_2\right)^2, \ a_1 = \left(\varepsilon_{22}^{-}\right)^2 - a_2 \tag{90a,b,c,d}$$

$$a_5 = 2\left(\gamma_{12}^{+} - b_2\right)^2 - 4a_2, \ a_4 = \frac{1}{2}\left(\varepsilon_{11}^{*} + 2\varepsilon_{22}^{*} - b_1 - 2b_2\right)^2 - \frac{1}{2}\left(\varepsilon_{11}^{-} - b_1\right)^2 - 4a_1 - 2a_2$$

$$a_3 - \left(\varepsilon_{11}^{-} + b_1\right)^2 - \left(\varepsilon_{11}^{+}\right)^2 - a_4 - a_5 \tag{90e,f,g}$$

Now it is possible to investigate in ceramic composites on the basis of (80)-(82) the damage and the rupture as a result of the accumulation of the damage.

CONCLUSION

Isotropic damage model for ceramics and anisotropic damage model for ceramic matrix composites were proposed. Tensor-linear equations of elastic deformation of considered materials were obtained. These constitutive equations take into account different behaviour of materials in tension and compression. The structure of equations is quite simple and equations are written in convenient tensor-invariant form.

ACKNOWLEDGMENT

A.Zolochevsky gratefully acknowledges the financial support of the Deutsche Forschungsgemeinschaft in Bonn.

REFERENCES

Chaboche, J.L. , 1992, Damage induced anisotropy: on the difficulties associated with the active/passive unilateral condition, Int. J. Damage Mech. 1: 148.

Chaboche, J.L., 1993, Development of continuum damage mechanics for elastic solids sustaining anisotropic unilateral damage, Int. J. Damage Mech. 2: 311.

Ju, J.W., 1989, On energy-based coupled elastoplastic damage theories: constitutive modeling and computational aspects, Int. J. Solids and Struct. 25: 803.

Ladeveze,P., Gasser,A., and Allix, O.,1994, Damage mechanisms modeling for ceramic composites, Trans. ASME. J. Eng. Mat. Tech. 116: 331.

Matzenmiller, A., and Sackman ,J.L., 1994, On damage induced anisotropy for fiber composites, Int. J. Damage Mech. 3: 71.

Munz, D., and Fett, T., 1989, "Mechanisches Verhalten Keramischer Werkstoffe,"Springer-Verlag, Berlin, Heidelberg, New York.

Shan, H.-Z.,Pluvinage, P.,Parvizi-Majidi, A.,and Chou, T.-W., 1994, Damage mechanics of two-dimensional woven SIC/SIC composites. Trans. ASME. J. Eng. Mat. Tech. 116: 403.

Spencer, A.J.M.,1984, Constitutive theory for strongly anisotropic solids, in: "Continuum Theory of Mechanics of Fibre-Reinforced Composites," A.J.M.Spencer, ed., Springer-Verlag, Wien, New York.

THERMAL PROPERTIES AND THERMAL STRESSES IN
DELAMINATED CERAMIC MATRIX COMPOSITES

Yangsheng Lu, Kimberly Y. Donaldson, and D. P. H. Hasselman

Thermophysical Research Laboratory
Department of Materials Science and Engineering
Virginia Polytechnic Institute and State University
Blacksburg, VA 24061-0237

ABSTRACT

Using a simple model it is shown that, analogous to crack formation in single phase ceramics, delaminations can significantly decrease the effective transverse thermal conductivity, coefficient of thermal expansion, and, depending on the external constraints, the magnitude of maximum thermal stress and/or thermal deformation in laminated fiber-reinforced ceramic matrix composites subjected to transverse heat flow.

INTRODUCTION

Structural materials considered for high-temperature applications typically are highly brittle and consequently prone to catastrophic fracture by thermal shock.[1-3] However, thermal stress resistance parameters can be used to select the material with maximum thermal shock resistance.[1,3] It has been shown theoretically[3-6] and demonstrated experimentally[4,5] that the introduction of microcracks into a brittle matrix can improve its damage resistance under thermal shock conditions so severe that the onset of crack propagation cannot be avoided. The presence of the microcracks generally causes a greater decrease in effective Young's modulus than in strength, resulting in a significant increase in the strain-at-fracture.[6-9] The accompanying decrease in thermal conductivity due to the cracks is particularly important for those designs requiring high thermal shock resistance coupled with high thermal insulating ability.

Delaminations in laminated fiber-reinforced composites represent a special case of internal crack formation which, from a mechanical load-bearing perspective, generally is not desirable. However, this study presents an analysis showing that the delaminations could be useful in reducing thermal conductivity, the coefficient of thermal expansion and the magnitude of thermal stress as well as accompanying thermal deformations if the direction of heat flow is perpendicular to the delaminations. Although the literature on the delamination mechanics of fiber-reinforced composites is extensive and yields models that quickly become quite complex,[11-16] to illustrate basic principles, the present analysis is based on a very simple model using the most simplifying assumptions.

ANALYSIS AND DISCUSSION

Consider an unconstrained, initially flat, infinite composite plate made up of N plies per unit thickness, t, each ply having a thickness d such that d = 1/N. The plate is located within the x-y plane and is subjected to steady-state heat flow in the transverse or z-direction due to a mean temperature gradient ∇T between its two surfaces. The thermal conductivity in the transverse direction is taken to be equal for each ply and independent of position and temperature, thereby resulting in a linear temperature distribution through the thickness. Young's modulus and the coefficient of thermal expansion are assumed to be isotropic within the plane of the plate and independent of temperature.

Now assume delaminations occur between each ply and extend over the total area of the plate, resulting in a total of (N - 1) evenly spaced delaminations per unit thickness. Heat transfer between the plies is characterized by an interdelamination thermal conductance, h_d, taken to be equal for each delamination and independent of position. The degree of delamination is assumed to approach the limiting case of no mechanical interaction between plies.

The thermal resistance per unit thickness of this delaminated plate is:

$$R_{th} = \frac{1}{K_p} + \frac{(N - 1)}{h_d} \qquad (1)$$

where K_p is the thermal conductivity of each ply in the z-direction.

The heat flux, q (per unit area and unit time), through the delaminated plate is:

$$q = \nabla T \left[\frac{1}{K_p} + \frac{(N - 1)}{h_d} \right]^{-1} \qquad (2)$$

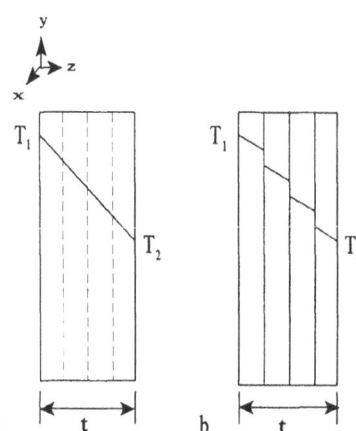

Figure 1. Temperature profiles in [a] delamination-free (infinite h_d) and [b] delaminated (finite h_d) composite plate.

430

From equation 2, the effective transverse thermal conductivity of the plate can be derived to be:

$$K_{eff} = K_p \left(1 + \frac{(N - 1)K_p}{h_d} \right)^{-1} \qquad (3)$$

As h_d approaches 0, K_{eff} approaches 0, suggesting that the presence of the delaminations can reduce the transverse thermal conductivity significantly.

The heat flux of equation 2 creates a discontinuity in temperature ΔT_d at each delamination:

$$\Delta T_d = \frac{\nabla T \left[\dfrac{1}{K_p} + \dfrac{(N - 1)}{h_d} \right]^{-1}}{h_d} \qquad (4)$$

The temperature gradient within the plies due to the heat flux of equation 2 becomes:

$$\nabla T_p = \nabla T \left[1 + \frac{(N - 1)K_p}{h_d} \right]^{-1} \qquad (5)$$

Equation 5 indicates that the presence of the delaminations and associated thermal resistance serves to decrease the temperature gradients within the plies, i. e., for any non-zero value of $(N-1)K_p/h_d$, $\nabla T_p < \nabla T$. Figure 1 compares the temperature profiles for the delamination-free and delaminated plates.

In general, for deflections small in comparison to its thickness, a plate without external constraints subjected to uniform linear heat flow will deform to exhibit curvature with radius:[17,18]

$$R - \frac{1}{\alpha \nabla T} \qquad (6)$$

where α is the coefficient of thermal expansion within the plane of the plate and ∇T is the temperature gradient.

For the delaminated plate of this study, the radius of curvature in the plate is simply the radius of curvature of a single ply, R:

$$R = \frac{\left[1 + \dfrac{(N - 1)K_p}{h_d} \right]}{\alpha \nabla T_p} \qquad (7)$$

Comparison of equation 7 with equation 6 shows that the plate exhibits an effective coefficient of thermal expansion in bending, α_{eff}:

$$\alpha_{eff} = \alpha \left[1 + \frac{(N - 1)K_p}{h_d} \right]^{-1} \qquad (8)$$

which indicates that, from the perspective of bending due to a transverse temperature gradient, the delaminations cause a decrease in the effective coefficient of thermal expansion. For a delamination-free plate, i. e., $N = 1$, $\alpha_{eff} = \alpha$, as expected. Similarly, as $N \to \infty$, $\alpha_{eff} \to 0$.

For small deflections and no mechanical interaction between plies, the magnitude of maximum thermal stress in the plate can be found by solving for the stress resulting from the application of a biaxial bending moment to just overcome the curvature in a single ply resulting from the imposed temperature gradient.[19] In this manner, the maximum stress in each ply can be derived to be:

$$\sigma_{max} = \frac{\alpha E (\nabla T_p)}{2(1 - v)N\left(1 + \dfrac{(N - 1)K_p}{h_d}\right)} \tag{9}$$

In the absence of delaminations (i. e., $N = 1$):

$$\sigma_{max} = \frac{\alpha E (\nabla T)}{2(1 - v)} \tag{10a}$$

in agreement with Timoshenko.[17]

When $N \to \infty$:

$$\sigma_{max} \to 0 \tag{10b}$$

CONCLUSIONS

The above results show that the magnitude of thermal stress resulting from the delamination is reduced by the combination of the mechanical debonding and the reduction of the temperature gradient in each individual ply. These factors could be useful in designs requiring high thermal shock resistance coupled with thermal insulating ability. Although the above results are strictly valid only for a flat plate, they should be generally valid for most other geometries. However, for heat transfer conditions involving convective heating and cooling at the plate surfaces, although the magnitude of maximum thermal stress is reduced, the increase in the plate's thermal resistance due to the presence of the delaminations will cause an increase in its hot side temperature.[20] Under these latter conditions and in service environments that involve thermally activated mechanisms of degradation, such as corrosion, creep and fatigue, the degree of delamination should be kept to a minimum. The tradeoffs for delamination should be analyzed considering specific design requirements.

REFERENCES

1. W. D. Kingery, Factors affecting thermal stress resistance of ceramic materials, *J. Am. Ceram. Soc.* 1:3 (1955).
2. D. P. H. Hasselman, Thermal stress resistance parameters for brittle refractory ceramics: a compendium, *Am. Ceram. Soc. Bull.* 49:1933 (1970).
3. D. P. H. Hasselman, Unified theory of thermal shock fracture initiation and crack propagation of structural ceramics, *J. Am. Ceram. Soc.* 52:600 (1969).
4. D. P. H. Hasselman, Strength behavior of polycrystalline alumina subjected to thermal shock, *J. Am. Ceram. Soc.* 53:490 (1970).
5. R. C. Rossi, Thermal-shock-resistant ceramic composites, *Am. Ceram. Soc. Bull.* 48:736 (1969).
6. D. P. H. Hasselman, Analysis of the strain at fracture of brittle solids with high densities of microcracks, *J. Am. Ceram. Soc.* 52:458 (1969).

7. E. A. Bush and F. A. Hummel, High-temperature mechanical properties of ceramic materials, *J. Am. Ceram. Soc.* 41:189 (1958).
8. E. A. Bush and F. A. Hummel, High-temperature mechanical properties of ceramic materials, *J. Am. Ceram. Soc.* 42:388 (1959).
9. D. P. H. Hasselman and J. P. Singh, Analysis of the thermal stress resistance of microcracked materials, *Am. Ceram. Soc. Bull.* 58:856 (1979).
10. D. P. H. Hasselman, Effect of cracks on thermal conductivity, *J. Comp. Mat.* 12:403 (1978).
11. N. J. Pagano, "Interlaminar Response of Composite Materials," Composite Materials Series Vol. 5, Elsevier, Amsterdam, The Netherlands (1989).
12. E. J. Barbero and J. N. Reddy, Modeling of delamination in composite laminates using a layer-wise plate theory, *Int. J. Solids Struct.* 28:373 (1991).
13. O. Allix and P. Ladevèze, Interlaminar interface modelling for the prediction of delamination, *Composite Struct.* 22:235 (1992).
14. J. A. Nairn and S. Hu, The initiation and growth of delaminations induced by matrix microcracks in laminated composites, *Int. J. Fracture* 57:1 (1992).
15. M. Qingchun and Z. Xing, Analytical-generalized variational method of solution for delaminations of laminates, *Engr. Frac. Mech*, 46:797 (1993).
16. A. Corigliano, Formulation, identification and use of interface models in the numerical analysis of composite delamination, *Int. J. Solids Struct.* 30:2779 (1993).
17. S. Timoshenko. "Theory of Plates and Shells," McGraw-Hill Book Company, Inc. New York, NY (1940).
18. B. A. Boley and J. H. Weiner. "Theory of Thermal Stresses," John Wiley, New York(1960).
19. S. Timoshenko and J. N. Goodier. Chapter 14 in "Theory of Elasticity," McGraw-Hill, Inc., New York, NY(1951).
20. R. Van Stone, "Design Issues for Advanced Engines," Presented at University Research Initiative Winter Study Group, University of California, Santa Barbara, CA. January 4, 1994.

INTERFACIAL GLASS STRUCTURE AFFECTING MICROMECHANISM OF FRACTURE IN A FLUORINE-DOPED Si$_3$N$_4$-SiC COMPOSITE

Hans-Joachim Kleebe[1] and Giuseppe Pezzotti[2]

[1]University of Bayreuth, Institute of Materials Research, Ludwig-Thoma-Str. 36 B, D-95447 Bayreuth, Germany
[2]Toyohashi University of Technology, Department of Materials Science Hibarigaoka, Tempaku-cho 1-1, Toyohachi 441, Japan

INTRODUCTION

In the field of structural ceramics, ceramic-ceramic composites such as Si$_3$N$_4$-SiC components have gained wide interest owing to their potential application at elevated temperatures. The incoporation of ceramic reinforcements, e.g., SiC whiskers or platelets, into a ceramic matrix can improve mechanical properties, however, densification is rendered difficult. Due to the high covalent bonding character of Si$_3$N$_4$, liquid-assisted sintering is required for complete densification even of monolithic materials [1,2]. The addition of metal oxides or transition metal oxides, which react with the SiO$_2$ present on the Si$_3$N$_4$-particle surface to form a silica-rich liquid at high sintering temperatures, enables liquid-phase sintering. The remains of this liquid are commonly present at triple-grain junctions and along grain boundaries as a secondary glass. Post-densification heat treatment can partially crystallize these glass pockets [3]. One of the aims to further improve high-temperature performance of ceramic composites is to drastically reduce the amount of secondary phases. A high volume fraction of an amorphous phase strongly decreases the mechanical properties, because the glass softens at relatively low service temperatures, i.e., about 900°C. Therefore, materials were prepared without the addition of further sintering aids [4,5]. Hereby, liquid-assisted densification is achieved by the SiO$_2$ present in the Si$_3$N$_4$ starting powder. Owing to the high melting temperature of pure silica glass, hot-isostatic pressing (HIPing) was utilized as the densification technique. These materials, formed by HIPing without the addition of sintering aids, could be fully densified and, moreover, showed superior creep behavior, when compared with commercial materials [6]. It should be emphasized that such Si_3N_4 materials with pure silica present at triple pockets and along grain boundaries, revealed a rather low fracture resistance, which is due to the predominantly transgranular mode of fracture [7,8].

In this paper we report on Si_3N_4-SiC composites, HIPed without the addition of sintering aids. The silica glass structure was deliberately changed by the addition of fluorine to the starting powder. The micromechanics of such a fluorine-doped composite, compared to a pure SiO_2-containing material, is described with respect to the correlation between secondary-phase glass structure and resulting mechanical properties.

EXPERIMENTAL PROCEDURE

High purity α-Si_3N_4 starting powder (E-10, Ube Ind. Ltd., Ube, Japan) was doped with fluorine by ball milling with powderized teflon (Teflon, E.I. du Pont de Nemours, Wilmington, DE). This powder blend was pre-heated in vacuum at 1200°C in order to depolymerize the teflon structure to tetrafluorethylene C_2F_4. It is thought that the incorporation of fluorine into the SiO_2-glass structure is achieved at temperatures above 1000°C and that the reaction is accompanied by the formation of gaseous CO. The pre-heated specimens were encapsulated in an evacuated boron-silicate glass tube to enable complete densification as well as to avoid the evaporation of fluorine. Isostatically pressed powder compacts were densified by hot-isostatic pressing (HIPing) at 1900°C for 2 h at 180 MPa Ar-gas pressure. Full density (>99.5%) was achieved under the applied processing conditions. As a reference material, an undoped Si_3N_4/SiC-platelet composite was fabricated under nearly identical processing conditions. However, the HIPing temperature was about 50°C higher than for the F-doped material, in order to adjust for the difference in grain size, which was found when heat treated at the same temperature (diffusion controlled grain growth).

Grain-boundary structure and interface chemistry of the Si_3N_4 material, with fluorine as the only sintering aid, were studied by transmission electron microscopy (TEM). TEM-foil preparation was performed by standard techniques, which involve diamond-blade cutting, mechanical grinding and dimpling, and Ar-ion beam thinning to perforation, followed by a light carbon coating to minimize electrostatic charging under the electron beam. TEM and analytical electron microscopy (AEM), were performed using a Philips CM20FEG (field emission gun) operating at 200 kV with a point resolution of 0.24 nm. The microscope was fitted with an energy-dispersive x-ray Ge detector (Tracor Voyager 2100) and an electron energy-loss spectrometer with parallel detection (PEELS, Gatan 666). To accurately determine the thickness of intergranular films at two-grain junctions, high-resolution electron microscopy (HREM) studies were performed using a JEOL 4000EX (top entry) microscope with a point resolution of 0.18 nm when operated at 400 kV.

436

RESULTS AND DISCUSSION

A Si_3N_4-SiC-platelet containing composite was densified via hot-isostatic pressing employing the glass encapsulation technique. In general, the composite was processed without the addition of commonly used sintering aids, however, a small amount of fluorine was added to the system. The incorporation of fluorine into the material was deliberately chosen, in order to modify the residual glass structure and, therefore, to weaken the interfaces. Subsequently, the micromechanical response of the composite compared to the undoped material was studied.

The overall microstructure of the F-doped material consisted of relatively fine grained, equiaxed ß-Si_3N_4 grains, which were surrounded by a small fraction of amorphous secondary phase, as shown in Figure 1.

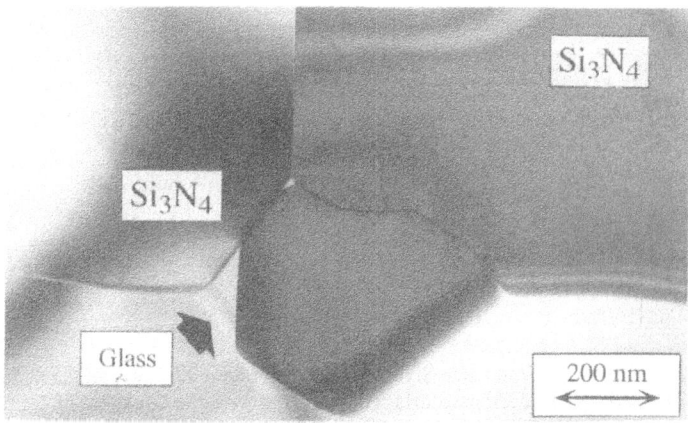

Fig. 1.   TEM-micrograph of the overall microstructure observed in the F-doped material. Note that this microstructure is indistinguishable from the microstructure found in the undoped reference material, which was processed at a slightly higher HIPing temperature to enhance diffusion controlled grain growth.

No crystalline phases other than ß-Si_3N_4 were observed. The formation of Si_2N_2O could be excluded, since the processing temperature was higher than the decomposition temperature of Si_2N_2O. ß-Si_3N_4 grains were about 0.2-1 μm in diameter, with only a small amount of elongated ß-grains observed by TEM. It should be noted that the overall microstructure of this F-doped material was indistinguishable from the undoped reference material. The diameter of the triple-grain pockets, which contained the glass phase that formed at elevated temperatures, was on the order of 10-200 nm in diameter. Since a careful TEM inspection of such triple pockets revealed no crystalline phases, it was concluded that no devitrification of the secondary glass phase had occurred upon cooling. The multi-grain glass pockets, which contained most of the amorphous residue, are interconnected by a three-dimensional network of thin grain-boundary films. HREM investigations confirmed the presence of an amorphous interlayer along all ß-Si_3N_4 grain boundaries studied. In order to precisely determine the width of such intergranular films, the HREM imaging technique was utilized [9]. This technique proved to be most accurate (±0.1 nm), when compared to the diffuse dark field (DDF) imaging and the defocus Fresnel fringe imaging techniques, as reported elsewhere [10]. Figure 2 shows an amorphous intergranular film, present at a grain boundary (i.e., a Si_3N_4/Si_3N_4 interface) in the F-doped material.

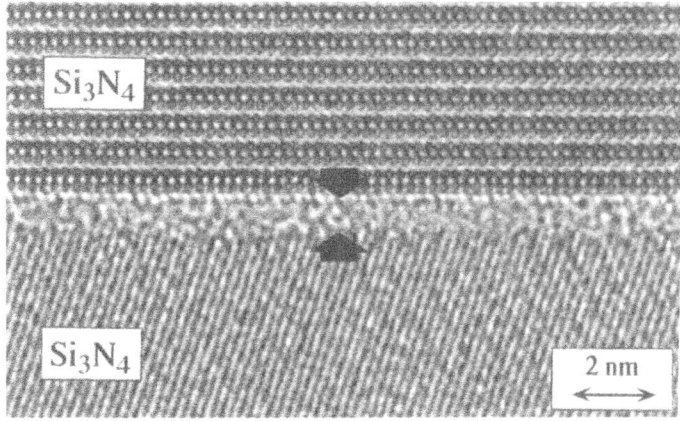

Fig. 2.   HREM-micrograph of a Si_3N_4/Si_3N_4 interface. The two adjacent Si_3N_4 grains are separated by a thin interfacial glass film with the thickness of 1.1 nm.

In Si_3N_4 materials, such amorphous intergranular films typically reveal a characteristic film thickness, which depends on the local chemistry. This observation is consistent with earlier studies, which indicate that the grain-boundary film thickness is governed by the interface chemistry. The major parameters affecting the intergranular film thickness are the chemistry of the adjacent grains and the chemical composition of the amorphous phase itself [11]. Quantitative measurements of grain-boundary films in this HIPed, fluorine-doped material resulted in an equilibrium film thickness of 1.1 nm (compare Figure 2). In comparison to the undoped reference material, which was fabricated with pure SiO_2 as the only sintering aid and which revealed a film width of 1.0 nm [12], a small increase in grain-boundary film thickness of 0.1 nm was observed.

Apart from Si_3N_4/Si_3N_4 interfaces, the phase boundaries were also studied. As observed for the grain boundaries, a thin amorphous film was also detected along the

Si$_3$N$_4$/SiC-platelet phase boundaries. Commonly, the film thickness of phase boundaries is about 2-5 times wider in comparison to grain boundaries [13]. Thus, at these interfaces, a film thickness of about 5 nm was determined. It is important to note that a rather high number of phase boundaries were pre-cracked. In some cases, the microcrack circumsphered the SiC-platelet. This microstructural observation seemingly suggests the presence of intrinsic residual stresses along the phase boundaries. Upon cooling, owing to the thermal mismatch between SiC and Si$_3$N$_4$, relatively high local stresses can form. It should be born in mind that such microcracks, as shown in Figure 3, were not observed in the undoped reference material. Hence, pre-cracking of interfaces was not due to the mechanical grinding procedure involved during TEM-foil preparation, rather than a result of the weakening of the interfaces, i.e., the modification of the glass structure.

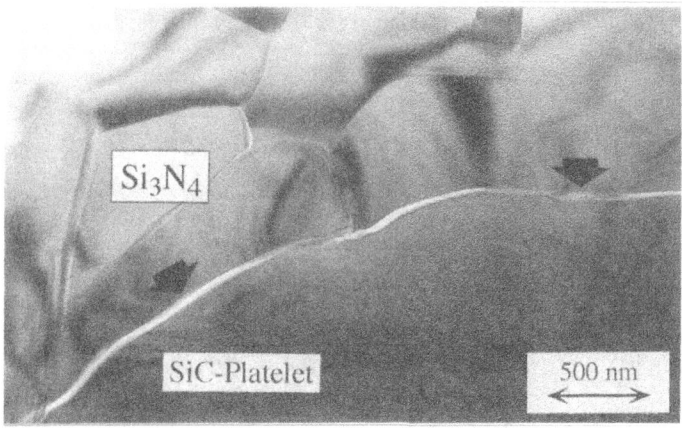

Fig. 3.   TEM-micrograph of a pre-cracked Si$_3$N$_4$/SiC-platelet phase boundary. The microcrack runs within the amorphous film that is about 5 nm wide and separates the two adjacent grains.

Electron energy-loss spectroscopy revealed the presence of fluorine both at triple-grain junctions and two-grain boundaries. A representative spectrum taken at an interface is given in Figure 4. Moreover, fluorine was also detected at Si$_3$N$_4$/SiC-platelet phase boundaries. Owing to the incorporation of fluorine into the SiO$_2$-glass structure, a small increase in grain-boundary film thickness was expected. A replacement of oxygen with fluorine, which both have almost idential ionic radii with 0.132 nm and 0.133 nm for O^{2-} and F$^-$, respectively, requires two fluorine anions in order to allow for charge neutrality. Therefore, two F$^-$-ions are incorporated into the glass structure replacing one O^{2-}-ion. This leads to the observed slight increase of 0.1 nm in intergranular film thickness, when compared to the pure SiO$_2$-glass structure.

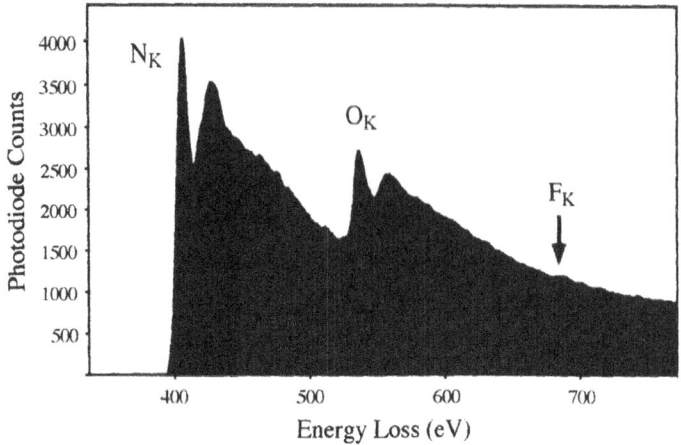

Fig. 4.   EELS spectrum the amorphous phase observed at a Si₃N₄/SiC-platelet phase
boundary (compare Figure 3). As detected for triple pockets and grain
boundaries, fluorine was also present at phase boundaries which is consistent
with the observed pre-cracking of such interfaces.

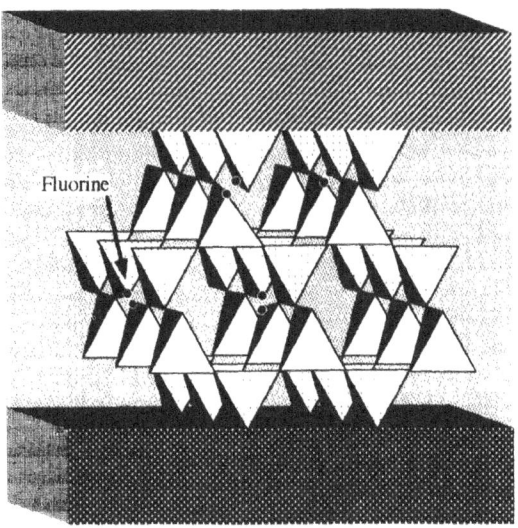

Fig. 5.   Simplified interfacial glass-structure model (segregation of fluorine at the
interface), which is based on both TEM observations and mechanical testing.

Based on TEM observations and EELS analysis in conjunction with the mechanical
testing results, a simplified structure model of the interfacial glass phase can be proposed, as
shown in Figure 5.

Changes in interface structure and chemistry, due to the presence of fluorine, affect
the mechanical response of the composite. The cohesive strength of internal interfaces is
remarkably lowered by segregation of fluorine at grain boundaries. This phenomenon leads
to an increased fraction of intergranular fracture during crack propagation. Moreover, the

high-temperature mechanical behavior of the composite was characterized. Fluorine-doped samples revealed markedly higher creep rates as well as a shift in the damping temperature curves, recorded during internal friction measurements, to lower temperatures, when compared to the undoped composite [14]. This latter data was consistent with the lower HIPing temperature needed in the F-doped system to achieve full densification as compared to the undoped system (i.e., 1900°C versus 1950°C). The replacement of bridging O^{2-} anions by two non-bridging F^- anions reduces the connecting sites within the glass structure and hence reduces the intrinsic viscosity of the glass. This modified glass structure also enhances intergranular fracture because the cohesive strength of the interface is weakened, which promotes debonding ahead of the crack tip. The pre-cracked phase boundaries, Si_3N_4/SiC-platelet interfaces, observed in the F-doped composites, underline the result of a weakened interface. The changes in mechanical response of the F-doped samples compared to the undoped materials can be directly attributed to the variation in glass structure and/or chemistry.

CONCLUSIONS

The incorporation of fluorine into the structure of the amorphous residue, which formed upon cooling of the HIPed Si_3N_4 material without sintering aids and which was present at triple pockets and along grain boundaries, markedly altered the micromechanical behavior of the bulk material. The interconnected network of interfacial glass dominated the performance of the material at room and at elevated temperatures. Compared to an undoped reference material, the F-doped material showed a creep rate several orders of magnitude higher and, in addition, a damping curve of the internal friction measurement that was shifted to lower temperature. This implies a strong reduction in glass viscosity. Therefore, the intrinsic glass structure, which was the only parameter changed during processing, affected the mechanical testing results. Since two fluorine ions replaced one oxygen ion within the glass structure, in order to keep charge balance, a weakened or broken glass network formed. The replacement of bridging oxygen anions by two non-bridging fluorine anions reduces the connecting sites within the glass structure and, hence, lowers the cohesive interface strength. Crack propagation studies revealed a pronounced intergranular fracture when doped with F, compared to the transgranular fracture commonly observed in the undoped samples. Pre-cracked phase boundaries observed in the F-doped composites only underline the given results. Materials performance is dominated by the residual glass structure.

Acknowledgments: We are indebted to Prof. M. Rühle for his support during the HREM investigations performed at the MPI in Stuttgart. The authors would also like to thank Prof. G. Ziegler and Prof. T. Nishida for their support throughout this work. C. Kunert is acknowledged for her assistance during TEM-foil preparation.

References

1. G. Ziegler, J. Heinrich, G. Wötting, "Review: Relationships Between Processing, Microstructure and Properties of Dense and Reaction-Bonded Silicon Nitride," *J. Mater. Sci.*, **22** 3041-86 (1987).

2. N. Hirosaki, A. Okada, K. Matoba, "Sintering of Si_3N_4 With the Addition of Rare-Earth Oxides," *J. Am. Ceram. Soc.*, **71** [3] C-144-47 (1988).

3. A. Tsuge, K. Nishida, M. Komatsu, "Effect of Crystallizing the Grain-Boundary Glass Phase on the High-Temperature Strength of Hot-Pressed Si_3N_4 Containing Y_2O_3," *J. Am. Ceram. Soc.*, **58** [7-8] 323-326 (1975).

4. K. Homma, H. Okada, T. Fujikawa, T. Tatuno, "HIP Sintering of Silicon Nitride Without Additives," *Yogyo Kyokaishi*, **95** [2] 229-34 (1987).

5. I. Tanaka, G. Pezzotti, T. Okamoto, Y. Miyamoto, M. Koizumi, "Hot Isostatic Press Sintering and Properties of Silicon Nitride Without Additives," *J. Am. Ceram. Soc.*, **72** [2] 1656-60 (1989).

6. G. Pezzotti, K. Matsuchita, H.-J. Keebe, Y. Okamoto, T. Nishida, "Viscous Behavior of Grain and Phase Boundaries in Fluorine Doped Si_3N_4-SiC Composites," *Acta metall.*, (1995), in press.

7. G. Pezzotti, I. Tanaka, T. Nishida, "Intrinsic Fracture Energy of Polycrystalline Silicon Nitride," *Phil. Mag. Letters*, **67** [2] 95-100 (1993).

8. G. Pezzotti, "Si_3N_4/SiC-Platelet Composite Without Sintering Aids: A Candidate for Gas Turbine Engines," *J. Am. Ceram. Soc.*, **76** [5] 1313-20 (1993).

9. H.-J. Kleebe, M.K. Cinibulk, M. Rühle, "Statistical Analysis of the Intergranular Film Thickness in Silicon Nitride Ceramics," *J. Am. Ceram. Soc.*, **76** [8] 1969-77 (1993).

10. M.K. Cinibulk, H.-J. Kleebe, M. Rühle, "Quantitative Comparison of TEM Techniques for Determining Amorphous Intergranular Film Thickness," *J. Am. Ceram. Soc.*, **76** [2] 426-32 (1993).

11. D.R. Clarke, "On the Equilibrium Thickness of Intergranular Glass Phases in Ceramic Materials," *J. Am. Ceram. Soc.*, **70** [1] 15-22 (1987).

12. H.-J. Kleebe, M.K. Cinibulk, I. Tanaka, J. Bruley, R.M. Cannon, D.R. Clarke, M. Rühle, "High-Resolution Electron Microscopy Observations of Grain-Boundary Films in Silicon Nitride Ceramics," *Mat. Res. Soc. Symp. Proc.*, **287** (1993), 65-78.

13. H.-J. Kleebe, M.J. Hoffmann, M. Rühle, "Influence of Secondary Phase Chemistry on Grain-Boundary Film Thickness in Silicon Nitride," *Z. Metallkd.*, **83** [8] 610-17 (1992).

14. I. Tanaka, K. Igashira, H.-J. Kleebe, M. Rühle, "High-Temperature Strength of Fluorine-Doped Silicon Nitride," *J. Am. Ceram. Soc.*, **77** [1] 275-77 (1994).

INTERFACIAL FRACTURE IN THE PRESENCE OF RESIDUAL STRESSES

S. Schmauder

Staatliche Materialprüfungsanstalt (MPA), University of Stuttgart
Pfaffenwaldring 32, D-70569 Stuttgart, Germany

SUMMARY

At the tip of interface cracks in bimaterials, residual stresses provide strong mode II contributions while four-point-bending loading leads to dominant mode I components of applied stress intensity factors (ASIF). Thermal stress intensity factors (TSIFs) primarily depend on crack length ratio a/w and Dundurs' first parameter α. ASIFs due to bending are mainly dependent on crack length whereas Dundurs' second parameter β plays a minor role in both, thermal and mechanical loading cases. In this paper, effective SIFs for combined loading situations are derived and discussed.

INTRODUCTION

Differences in thermal expansion coefficients of different constituents often lead to interfacial failure in composites and bimaterials. Especially, in the case of brittle components thermal residual stresses should not be neglected. They may strongly contribute to the driving energy of interface cracks[1]. Normal as well as shear stresses are acting simultaneously at the tip of interface cracks. Thus, interface cracks often tend to kink out of their initial plane, even if pure mode I loading is applied externally[1,2]. Whether or not interface cracks remain at the interface strongly depends on the mixed mode loading conditions at the crack tip[2]. For these reasons it is necessary to quantify the local mode mixity at the crack tip as a function of elastic properties, crack length ratios as well as external and internal loading conditions. In this paper, results of a systematic study on the influence of thermal and applied stresses on interface cracks are presented and briefly discussed.

BIMATERIALS

Since their discovery, Dundurs' parameters α and β have often proven their usefulness in characterizing elastic properties of bimaterials and other material joints[1-3]. Elastic solutions for interface crack problems depend only on α and β[4] which are contractions of the four elastic constants (Young's moduli E_i and Poisson's ratios v_i) according to (plane strain)[5]

$$\alpha = \frac{E_2/(1-v_2{}^2) - E_1/(1-v_1{}^2)}{E_2/(1-v_2{}^2) + E_1/(1-v_1{}^2)} \tag{1}$$

$$\beta = \frac{1}{4}\left[\left(\frac{1-2v_1}{1-v_1} + \frac{1-2v_2}{1-v_2}\right)\alpha + \left(\frac{1-2v_1}{1-v_1} - \frac{1-2v_2}{1-v_2}\right)\right] \tag{2}$$

These parameters are limited within the following parallelogram, $-1.0 \leq \alpha \leq 1.0$ and $-0.25+\alpha/4 \leq \beta \leq 0.25+\alpha/4$. The relationship $\beta=\alpha/4$ is derived from eqn. (2) when $v_1=v_2=1/3$. As the sign of α and β is reverted when materials are interchanged it is sufficient to examine one half of the valid α-β-regime, e.g., $\alpha \geq 0$. For many material combinations: $|\alpha| \leq 0.6$ and $\beta \leq \alpha/4$ (Fig. 1). In several studies, Dundurs' second parameter was even set to zero arguing that β is sufficiently small[6].

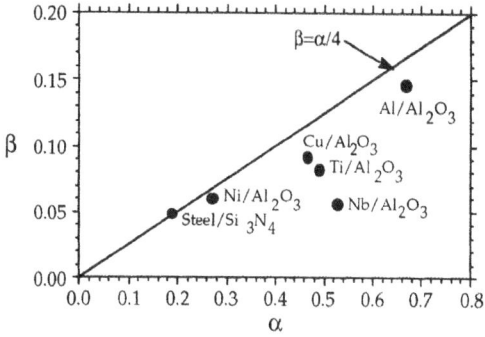

Figure 1. Dundurs' parameters for a number of metal/ceramic combinations.

MODEL

In the following, we consider a bimaterial under external mode I (bending) or thermal loading containing an interfacial edge crack with crack tip coordinates as shown in Fig. 2. Thermal expansion coefficients are chosen such that $\alpha_1 \geq \alpha_2$. A 'cut and paste'-technique[7] is used to calculate TSIFs and a virtual crack extension method based on Rice's J-integral concept[8] is applied for derivingSIFs from applied loads in case of interface cracks.

THEORY

The local stress field at the tip of an interface crack in real material joints is a consequence of residual stresses as well as applied stresses. In the vicinity of the crack tip the local stress field along the interface $\theta=0$) is given by

$$(\sigma_{\theta\theta} + i\tau_{r\theta})_{\theta=0} = \frac{Kr^{i\varepsilon}}{\sqrt{2\pi r}} \tag{3}$$

The complex stress intensity factor K ineqn. (3) has the generic form

$$K = K_1 + iK_2 = Y P \sqrt{a} \; a^{-i\varepsilon} e^{i\psi} \tag{4}$$

with crack length a, and representative stress amplitude P. By definition, ψ is the phase of $Ka^{i\varepsilon}$ where ψ can be interpreted as the phase of the tractions at r=a, assuming that eqn. (4) still holds at this distance ahead of the crack tip. Y is a dimensionless geometric function of material properties, loading conditions and crack length ratios.

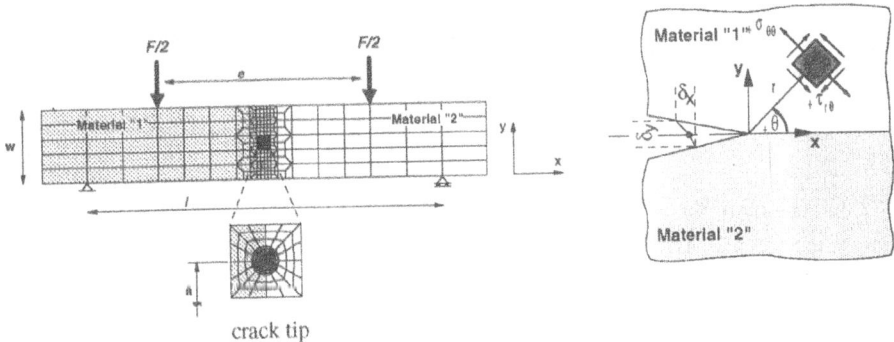

Figure 2. FE model of a four-point bending specimen (thickness t) with an interfacial edge crack and crack tip coordinates.

According to Rice[9], a global SIF for interface cracks may, therefore, be defined in the usual manner, if the radial distance from the crack tip, r, is chosen as a fixed length quantity, $r=\hat{r}$. Due to this substitution we can rearrange eqn. (4) to

$$K \hat{r}^{i\varepsilon} = K_I(\hat{r}) + iK_{II}(\hat{r}) = Y P \sqrt{a} \left(\frac{\hat{r}}{a}\right)^{i\varepsilon} e^{i\psi} \tag{5}$$

where

$$K_I(\hat{r}) = K_1 \cos(\varepsilon\ln(\hat{r})) - K_2 \sin(\varepsilon\ln(\hat{r})) \tag{6a}$$

$$K_{II}(\hat{r}) = K_1 \sin(\varepsilon \ln(\hat{r})) + K_2 \cos(\varepsilon \ln(\hat{r})) \qquad (6b)$$

These global SIFs $K_i(\hat{r})$ (i=I,II) have the usual dimensions (MPa$\sqrt{m}$) and may be interpreted in the conventional manner according to eqn. (6). A detailed discussion of stress oscillation and the effect of length quantities can be found in[9]. For our subsequent treatment of interfacial stress intensity factors we incorporate at this point $K_i(\hat{r}) \equiv K_i$ (i=I,II) and $\hat{r}$=a. The fact that $\hat{r}$=a lies obviously outside the zone of K-dominance is of no consequence as long as $\hat{r}$ is recorded along with the results of $\Psi_{\hat{r}} = \Psi_a$ and as long as one is familiar with the ψ-transformation from distance a to r given through

$$\psi_r = \psi_a + \varepsilon \ln\left(\frac{a}{r}\right) \qquad (7)$$

Interface toughness values K_c are usually calibrated by tabulated Y^{app}-functions, which are given for a wide range of test geometries. As these kinds of calibrations do not take into account inherent stresses their application will lead to different K_c values for bimaterials with identical elastic properties and crack length ratios but different residual stress states. However, by definition K_c must be taken as a material characterizing parameter.

A correct K calibration can be performed by superimposing single mode SIFs due to applied loads (app) and residual stresses (res) according to

$$K_i = K_i^{app} + K_i^{res} \qquad (i = 1, 2) \qquad (8)$$

In analogy to the SIF definition in the case of applied loads, eqn. (4), we may now express the complex TSIF through the nominal residual stresses, $\sigma^{res} = E^* \Delta\alpha\Delta T$, using the relation

$$K^{res} = Y^{res} \sigma^{res} \sqrt{a} \; a^{-i\varepsilon} e^{i\psi^{res}} \qquad (9)$$

where Y^{res} is the dimensionless geometric function of Dundurs' parameters α and β as well as normalized crack length a/w. The difference in thermal expansion coefficients in plane strain is obtained from Hooke's law as $\Delta\alpha = (1+v_1)\alpha_1 - (1+v_2)\alpha_2$. The cooling interval is determined by the difference between room temperature T_0 and processing temperature T_p, $\Delta T = T_0 - T_p$ while the mean Young's modulus is given by $1/E^* = ((1-v_1)/E_1 + (1-v_2)/E_2)/2$. According to eqn. (9), the thermal calibration of a bimaterial interface crack geometry is thus reduced to determining Y^{res} and ψ^{res} for the interesting range of material combinations and crack length ratios.

The main objective of the following section is, to combine thermal and mechanical correction functions as well as phase angles to *effective correction functions and effective phase angles (eff)* in order to discuss the influence of thermal stresses on the effective toughness values for interfacial failure in bimaterials

$$K^{eff}_{a} i^{\varepsilon} = Y^{eff} \sigma^{app} \sqrt{a} \ e i \psi^{eff} \tag{10}$$

RESULTS AND DISCUSSION

Calculated thermal and mechanical correction functions as well as phase angles are shown in Figs. 3 and 4 for different α-values and crack lengths for bimaterials with $\beta=0$ (a significant β-dependence of results has not been found in this study).

Thermal stress intensity factors of interface cracks

In the following, results are presented for the geometry function of thermal residual stress intensity factors Y^{res} and the corresponding mode mixity ψ^{res} for use in eqn. (9). The dependence of geometry function Y^{res} on α is shown in Fig. 3. Obviously, Y^{res} is a decreasing function of crack length but increases with increasing α and β values. For short cracks (a/w $\approx$ 0.1) there is a distinct effect of the elastic mismatch α, β on Y^{res}, whereas long interface cracks (a/w $\approx$ 0.5) produce a modest change in this geometry function.

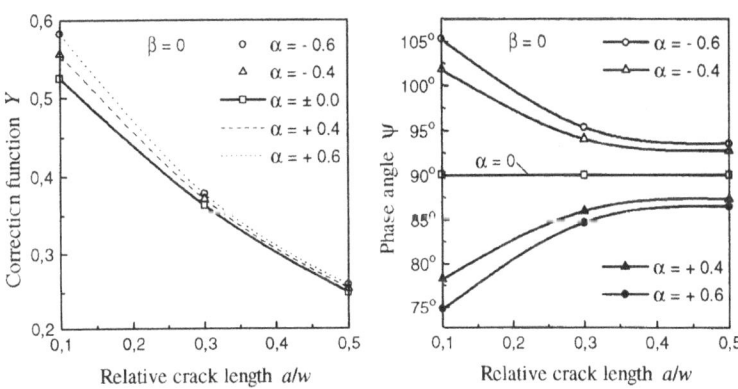

Figure 3. Correction functions Y^{res} and phase angles ψ^{res} for bimaterials under mode I loading.

Similarly, the local phase angle ψ^{res} develops significant mode I fractions for short cracks and increasing α and β values in the range of moderate α values ($\alpha \leq 0.5$, Fig. 1) and typical notch lengths of $0.2 \leq$ a/w ≤ 0.5. Therefore, residual thermal stresses result in a nearly pure shear loading at the tip of the interface crack (Fig. 3).

For engineering purposes the following approximations may be used for $0.1 \leq$ a/w ≤ 0.5 and $\beta = \alpha/4$ [10]

$$Y^{res} = \{0.265 + 0.104\alpha^2\} - \{0.249 + 0.195\alpha^2\} \frac{a}{w} \qquad (11a)$$

$$\psi^{res} = 90 - \{69.2 - 424.9 \frac{a}{w} + 1006.1 \left(\frac{a}{w}\right)^2 - 844.3 \left(\frac{a}{w}\right)^3\} \alpha \qquad (11b)$$

It is worth mentioning that the mode mixity is fully determined by the elastic mismatch and the crack length: for non-zero $\Delta\alpha$ values, the phase angle ψ^{res} is independent of the thermal mismatch, $\Delta\alpha$, and the cooling interval, ΔT. Only the absolute amount of K^{res} changes with $\sigma^{res} = E^*\Delta\alpha\Delta T$ following eqn. (9). In the presence of applied stresses, e.g. when testing the toughness of real bimaterials, TSIFs can significantly increase the mode mixity and the effective SIF at the tip of interface cracks.

Applied stress intensity factors of interface cracks

SIFs from bending loading are found to be identical for all bimaterials and, therefore, Y^{app} may be represented by the usual correction function for homogeneous materials (Fig. 4). In contrast, TSIFs show some dependence on α, decrease with crack length and are smaller by a factor of 3 to 11 depending on crack length compared to SIFs from bending. This fact manifests a neglectable influence of Y^{res} on effective correction functions Y^{eff} over a wide range of loading ratios ($0 \leq \xi < 0.4$) in Fig. 5, where $\xi = \sigma^{res}/(\sigma^{res}+\sigma^{app})$.

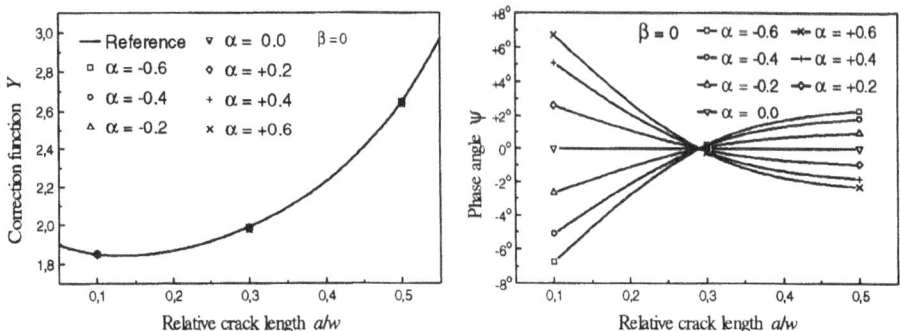

Figure 4. Correction functions Y^{app} and phase angles ψ^{app} for bimaterials under four-point bending loading.

Effective stress intensity factors of interface cracks

Effective Y^{eff}- and ψ^{eff}-values for combined loading conditions and one crack length are depicted in Fig. 5. Normalization of Y^{eff} is done with respect to Y_0, the correction function for the case of a straight interface crack in a homogeneous material under pure bending.

For all phase angles, bending loading provides small phase angles, $\psi^{app} \approx 0°$, which are even identical for a/w = 0.3 (Fig. 4). In opposite, $\psi^{res} \approx 90°$ for homogeneous bimaterials, but deviations above and below this value are observed for elastically inhomogeneous materials. Thus, effective phase angles reflect a dependence on α for large thermal stress contributions, $\xi > 0.4$. It should be emphasized that the sign of ψ^{res} changes when the mismatch is inverted to $\alpha_1 \leq \alpha_2$.

But the influence of Dundurs' first parameter α on the effective correction function is seen to be limited. Moreover, effective SIFs are notably higher compared to pure mechanical SIFs, only for large values of nominal residual stresses ($\xi>0.5$). Thus, apparent SIFs of bimaterials obtained by neglecting TSIFs can be underestimates of more than 25%.

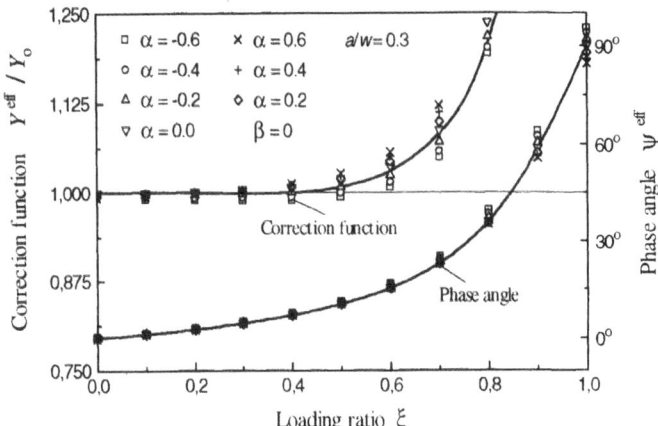

Figure 5. Y^{eff} and ψ^{eff} for bimaterials with interface cracks under combined thermal and four-point bending loading, a/w=0.3 ($\xi=\sigma^{res}/(\sigma^{res}+\sigma^{app})$).

ACKNOWLEDGMENT

Financial support by the DFG (project El 53/13-1) is gratefully acknowledged.

REFERENCES

1. G. Dreier, S. Schmauder, G. Elssner, Propagation and Facture Energy of Interface Cracks in Elastically Similar Brittle Materials under Mixed Mode Loading Conditions, in: "Mixed Mode Fracture and Fatigue", Eds.: H.P. Rossmanith, K.J. Miller, European Structural Integrity Society, Salisbury, U.K. 185 (1993).
2. M. Meyer, S. Schmauder, G. Elssner, Mixed-Mode Fracture Investigations of Interface Cracks in Dissimilar Media, in: "Mixed Mode Fracture and Fatigue", Eds.: H.P. Rossmanith, K.J. Miller, European Structural Integrity Society, Salisbury, U.K. 303 (1993).
3. D.B. Bogy, "On the Problem of Edge-Bonded Elastic Quarter-Planes Loaded at the Boundary", Int. J. Sol. Struct. 6:1287 (1979).
4. J.W. Dundurs, Discussion, Edge-Bonded Dissimilar Orthogonal Elastic Wedges Under Normal and Shear Loading, J. Appl. Mech. 36:650 (1969).

5. S. Schmauder, M. Meyer, Correlation Between Dundurs' Parameters and Elastic Constants, Z. Metallkd. 83: 524 (1992).

6. J.W. Hutchinson, Mixed Mode Fracture Mechanics of Interfaces, in: "Metal-Ceramic Interfaces", Acta Scripta Metallurgica Proceedings, Series 4, Eds.: M. Rühle, A.G. Evans, M.F. Ashby, J.P. Hirth, Pergamon Press, Oxford, England, 295 (1990).

7. N.P. O'Dowd, C.F. Shih and M.G. Stout, Test Geometries for Measuring Interfacial Fracture Toughness, Int. J. Solids Structures 29: 571 (1992).

8. J.R. Rice, A Path-Independent and Approximate Analysis of Strain Concentration by Notches and Cracks, J. Appl. Mech. 35: 379 (1968).

9. J.R. Rice, Elastic Fracture Mechanics Concepts for Interfacial Cracks, J.Appl. Mech. 55: 98 (1988).

10. M. Meyer, Einfluß überlagerter Spannungen auf den Grenzflächenbruch von Bimaterialien, Fortschrittsberichte, VDI-Reihe 18, Nr. 136, VDI-Verlag, Düsseldorf (1993).

FRACTURE BEHAVIOR OF INTERPHASE-MODIFIED METAL/CERAMIC INTERFACES

Gang Liu and J. K. Shang

Department of Materials Science and Engineering
University of Illinois, Urbana, IL 61801

ABSTRACT

The use of interfacial precipitation to control the fracture resistance of metal/ceramic interfaces was explored in the Al_2O_3/Al-Cu system. The bi-material interfaces were formed by bonding high-purity Al_2O_3 with molten Al-5%Cu alloy under pressure. The specimens were then heat-treated so that the Al-Cu alloy reached peak-aged and extended-overaged conditions. The fracture resistance curve was measured for the two interfaces using flexural peel specimens. While the steady-state fracture toughness of the interface scaled with the yield strength of the metal, the initiation toughness was related to the size and spacing of interfacial precipitates. Implication of the weakening effect of the interfacial precipitation to the fracture behavior of metal/ceramic laminates is demonstrated.

INTRODUCTION

Interface control is essential to the performance of a variety of engineering materials and structures. In metal matrix composites, a strong interface is considered critical for an effective load transfer from the metal matrix to the strong ceramic reinforcement. Conversely, in ceramic matrix composites, the interface has to be considerably weaker to ensure the debonding at the interface, which imparts the needed toughening to the ceramic. For ceramic/metal multi-layered structures[1-11], interfaces too strong can cause cracking of the brittle layer whereas too weak an interface can lead to easy delamination between the layers.

Modifying the interfacial chemistry to induce chemical interaction at the interface has been one of the major routes to achieving the desired interface control. In metal/ceramic systems[12-15], this is often accomplished by adding selected alloying elements in the metal or introducing an interfacial transition layer between the metal and ceramic. A wide range of techniques, liquid, solid state and vapor phase, have been used to obtain the required chemistry.

While most studies have focused on the initial bond formation at the interface, interfacial reaction does occur in many systems after the interfacial bonding is completed[6,9,12-15]. This is especially true when the interface serves as the site for heterogeneous nucleation of new phases. While the thermodynamics of the phase formation has been well established, the effect of the interfacial phase (interphase) on the fracture behavior of the interface has not been understood. In this paper, the fracture behavior of Al/Al_2O_3 interfaces modified by interfacial precipitates was examined. The effect of interfacial precipitates in weakening the interface was measured by conducting crack growth experiments on flexural peel specimens, and analyzed in terms of both the steady-state peak toughness and the initiation toughness. The application of the interfacial weakening by precipitation to the crack arrest in metal/laminates is demonstrated.

MATERIALS AND PROCEDURE

Model interfaces for this study were made from a high-purity polycrystalline Al_2O_3 ceramic and an Al-Cu alloy. They were chosen because of their potential applications in Al-base metal matrix composites and in ductile phase toughened ceramic matrix composites. Furthermore, previous work on Al_2O_3/Al interface has shown that the interface between pure Al and Al_2O_3 is very strong that the fracture follows either the softer Al or brittle ceramic, but rarely through the interface[1,2]. An Al-Cu alloy was used in place of Al to explore the potential weakening effect of the interfacial precipitates, which, if substantiated, would be used to control the debonding at the interface in metal/ceramic laminates.

The ceramic was a high-purity polycrystalline alumina (Al_2O_3 content greater than 99.5%), produced by the Coors Ceramic Company* (Coors designation AD995). The grain-size distribution was bimodal and the average grain size determined by the line intercept method was 18 μm. The room temperature flexural strength of the ceramic was 379 MPa and the fracture toughness, 45 J/m^2. The surface of the ceramic was lapped with 45 μm diamond paste and polished to 3 μm diamond surface finish prior to bonding. The Al-Cu alloy contained 5% Cu as the major alloy element, and was made by Alcoa#(Alcoa designation 2519). The alloy was cold rolled to the required thickness for bonding.

The bonding of the ceramic and metal was formed by a molten metal process, where a sandwich of ceramic outerlayers and a metal interlayer was heated above the melting temperature of the metal in an argon atmosphere (Fig. 1). The thickness of the metal layer was controlled at 0.3 mm by selecting the initial thickness of the metal sheet and the weight, P, on top of the sandwich. The sample was kept at the bonding temperature for ~1 hr before it was cooled down to room temperature by furnace cooling. As shown in Fig. 2, a uniform bonding was produced between the ceramic and metal.

The sandwiched specimen was sliced into rectangular beams, 3 mm x 6 mm x 50 mm (Fig. 1), and heat treated. The heat treatment consisted of solutionizing at 535 °C for 4 hours, oil-quenching and artificially aging at 177 °C for 12 hr to peak-aged condition (PA), or at 300 °C for 150 hr to an extended overaged condition (OA). The microstructure of the metal interlayer near and away from the interface was examined in a Philips 420 transmission electron microscope. As shown in Fig. 3, the interface of the peak-aged specimen was covered with a distribution of fine Al-Cu precipitates with an average size of 30 nm. In comparison, in the overaged specimen, the interfacial precipitates were much coarser and averaged 600 nm. The side-by-side spacings between the precipitates were 200 nm and 300 nm respectively. Far away from the interface, the precipitate microstructure was similar to the microstructure of the bulk alloy. Microhardneses of the metal layer were measured using Vickers microhardness tester to be 52 Kg/mm^2 and 33 Kg/mm^2 for the peak-aged and overaged conditions respectively. These values are very close to those of the bulk alloy heat treated to the same conditions. For the bulk alloy, the room temperature yield strengths were 435 MPa and 221 MPa for the peak-aged and overaged conditions.

The sandwich beam specimens were tested along both the crack delamination and arrest orientations shown in Fig. 4. The purpose of the delamination tests was to characterize the interfacial crack growth behavior. To do this, the sandwich beam specimens were first converted into flexural peel specimens shown in Fig. 5. A precrack of 2 mm was introduced by loading a notched beam under three-point bending as shown in Fig. 5b). The precracked specimen was loaded under displacement control at a loading rate of 0.16 N/sec, while both the applied load and crack length were recorded. From the applied load and instantaneous crack length, the strain energy release rate, G, was calculated from the interface-fracture mechanics solutions for the flexural peel specimen[16]. For both types of the interfaces, the elastic properties, external loading, and geometry were the same. Therefore, no attempt was made to separate the total strain energy release rate into Mode-I and Mode-II components

---

* Coors Technical Ceramic Company, Oak-Ridge, TN.
# Aluminum Company of America, Pittsburg, PA.

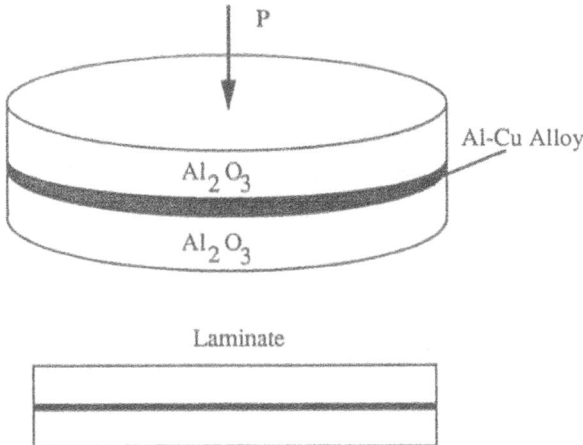

Figure 1. Schematic of the specimen fabrication process.

Figure 2. Morphology of the Al$_2$O$_3$/Al-Cu interface.

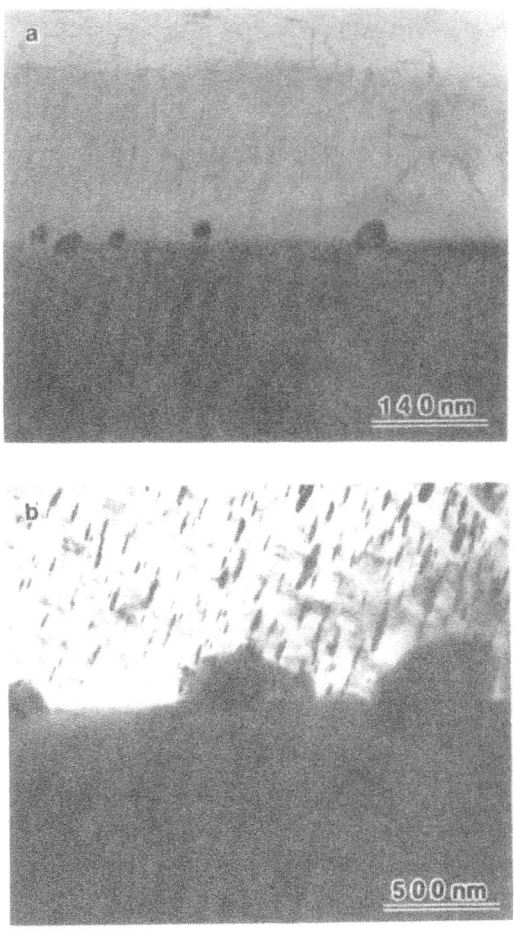

Figure 3. Interfacial precipitates in a) peak-aged and b) overaged conditions.

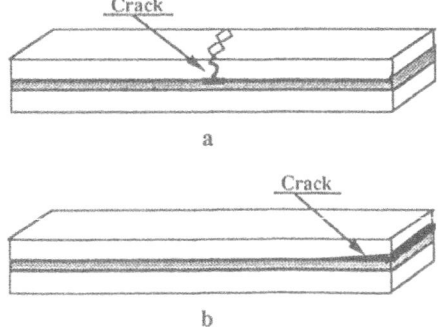

Figure 4. Schematic of the crack paths in sandwich specimen, a) crack arrest and b) delamination

although the estimated phase angle of the specimen was about 40°[16]. The results of the crack delamination tests are presented as fracture resistance curves or R-curves in terms of G as a function of crack extension, Δa.

Crack arrest tests were performed to determine whether the weakened interface was sufficient to deflect the crack in metal/ceramic laminates. As shown in Fig. 4a), the top ceramic surface of the specimen was first pre-cracked by placing a series of Knoop indentations. The pre-cracked specimens were then loaded in three-point bending mode until the specimen was broken, while the load vs. displacement response of the specimen was recorded.

Following both delamination and crack arrest experiments, broken surfaces were examined in the scanning electron microscope (SEM) to determine whether the failure was interfacial or cohesive. On select specimens, crack paths were photographed by interrupting the crack growth experiment after different crack extensions.

RESULTS AND DISCUSSION

Fracture Resistance Curves

The fracture resistance curves are given in Fig. 6 for the two interfaces. The fracture energy ranged from a few J/m^2 to less than 100 J/m^2, orders of magnitude smaller than the fracture energy reported previously on molten Al bonded Al_2O_3 specimens[1]. This difference is related to the difference in the fracture mechanism. In pure Al bonded Al_2O_3 specimens, the crack did not follow the interface but rather propagated through the ductile Al layer[1,2]. However, as shown in Fig. 7, the crack path in the aged Al_2O_3/Al-Cu interfaces was interfacial. On the side of the specimen, there was no visible stretching bands, which were also reported in the pure Al/Al_2O_3 specimens[1]. The fracture surface of the Al-Cu/Al_2O_3 interface is shown in Fig. 8, where the metal side of the broken surface looked like a replica of the grain structure of the polycrystalline Al_2O_3, as opposed to resembling the ductile dimples observed in the pure Al/Al_2O_3 specimens and in bulk Al-alloys. While these differences in the fracture mechanism preclude a direct quantitative comparison between the Al/Al_2O_3 interface and the aged Al-Cu/Al_2O_3 interfaces, in qualitative terms, the aged Al-Cu/Al_2O_3 interfaces are considerably weaker.

Between the two Al-Cu/Al_2O_3 interfaces, the overaged interface was weaker in terms of both the initiation toughness and the steady-state peak toughness. When the R-curve data are normalized by the yield strength of the metal, as in Fig. 9, the difference in the peak portion of the curves essentially disappears. The peak-toughness therefore scales with the yield strength of the metal, consistent with the plasticity model by Reimanis et al.[17], where the steady-state plastic work was shown to vary in proportion to the product of the yield strength and strain of the metal. However, a rather large difference still remains in the initial portion of the curves after the normalization.

Fracture Initiation Toughness

The fracture initiation toughness, indicated by the onset of the R-curve, differed by a factor of eight between the peak-aged and overaged interfaces. The difference can be explained by a fracture initiation model, which assumes that the interface fracture will initiate if the stress is sufficient to cause the debonding of an interfacial precipitate located at some characteristic distance, ξ, ahead of the crack tip, as schematically shown in Fig. 10. The condition of the precipitate debonding can be analyzed from the energy consideration, which leads to the following expression for the fracture initiation toughness[18]:

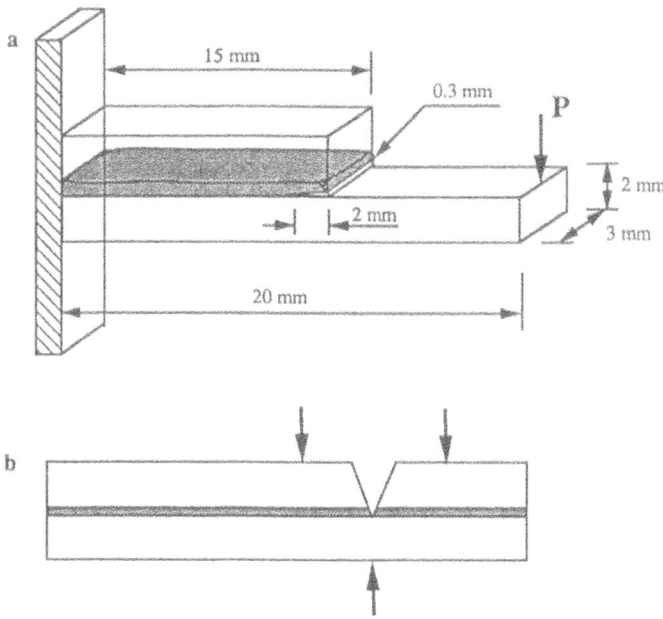

Figure 5. a) Flexural peel specimen used for the crack growth study. All dimensions are in mm. b) Method used to introduce precracks for interface crack growth studies.

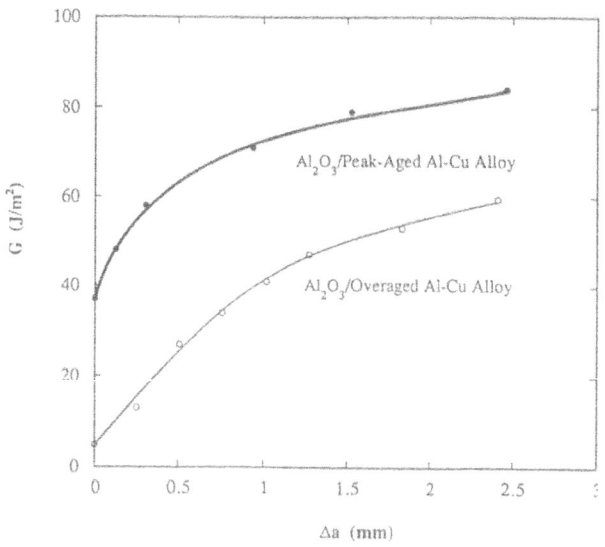

Figure 6. Fracture resistance curves (R-curves) for the peak-aged and overaged interfaces.

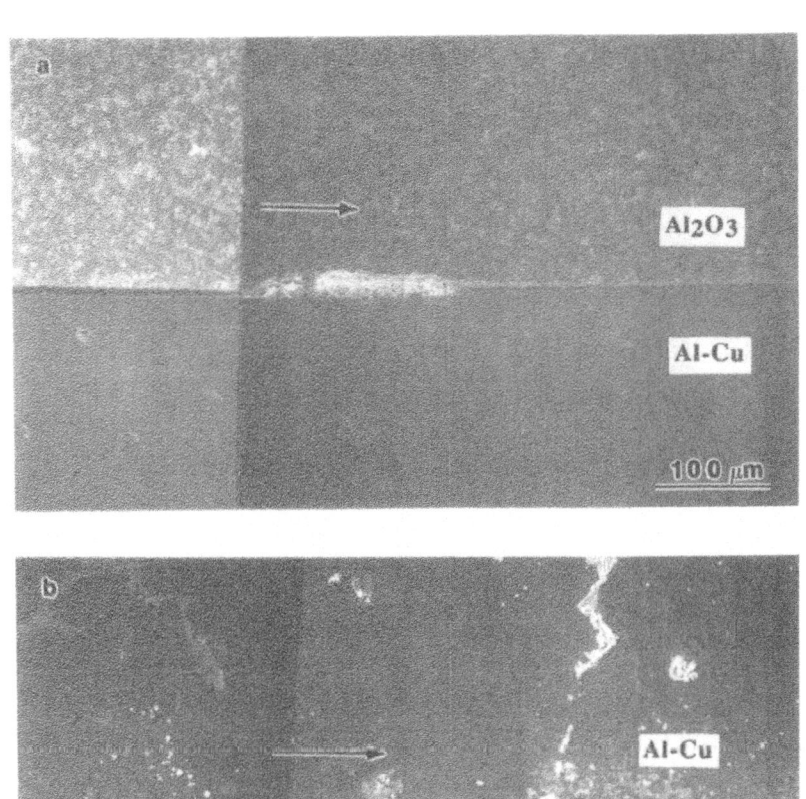

Figure 7. Crack paths in a) peak-aged and b) overaged interface specimens.

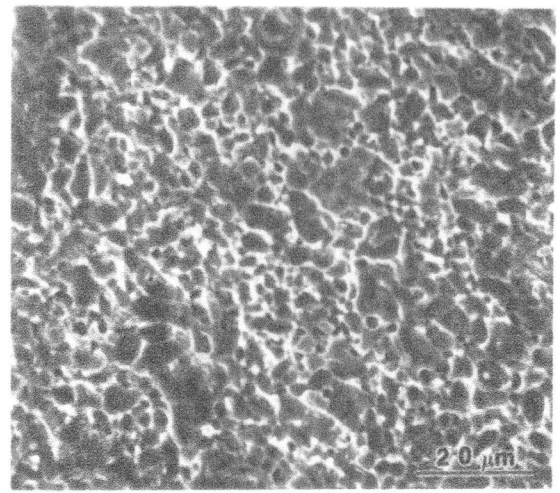

Figure 8. Fracture surface of the flexural peel specimen.

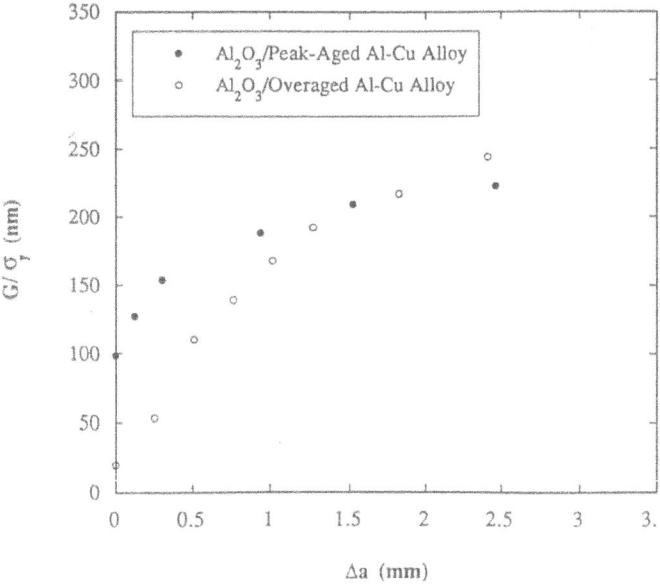

Figure 9. Normalization of the R-curve by the yield strength of the metal, σ_y.

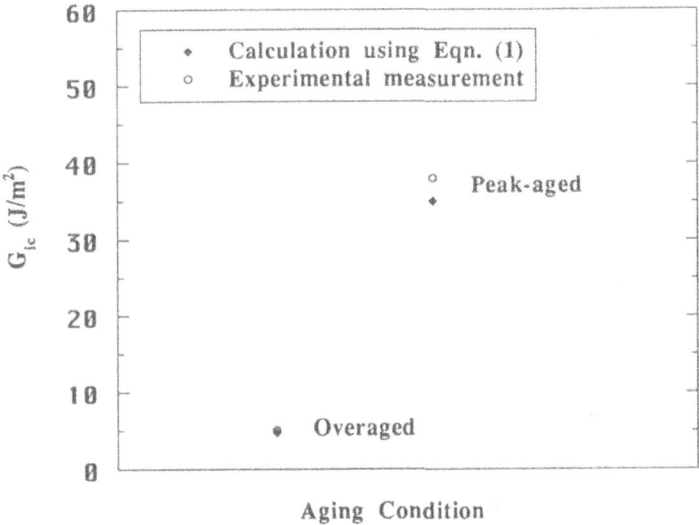

Figure 10. Comparison of the fracture initiation toughness between the experimental measurement vales and the calculated values at two aging conditions

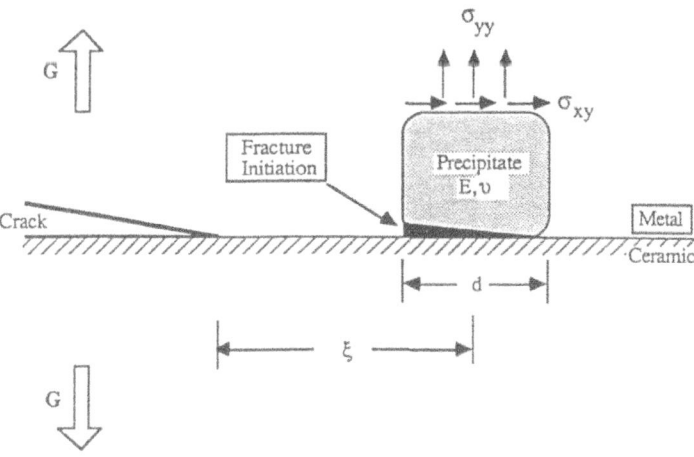

Figure 11. Interface fracture model for estimating the fracture initiation toughness, G_{ic}.

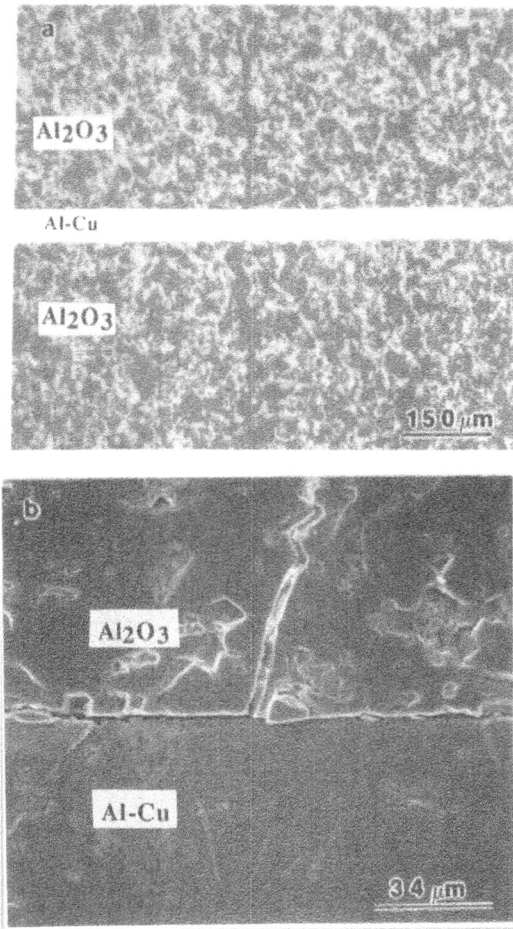

Figure 12. Crack paths in a) peak-aged and b) overaged specimens.

$$G_{ic} = \frac{4\pi\alpha(1 + \lambda^2)E\Gamma\xi}{\{[\cos(\varepsilon\ln\frac{\xi}{h}) - \lambda\sin(\varepsilon\ln\frac{\xi}{h})]^2 + 2(1 + \upsilon)[\lambda\cos(\varepsilon\ln\frac{\xi}{h}) + \sin(\varepsilon\ln\frac{\xi}{h})]^2\}d} \quad (1),$$

where Γ is the interface energy, d, the precipitate size, E and υ, the modulus and Poisson's ratio of the precipitate, h, the thickness of the metal interlayer, α, ε and λ, the bi-material constants. The term inside the large bracket in Eqn. (1) is relatively insensitive to ξ, and therefore the initiation toughness is inversely proportional to the precipitate size.

For the aged Al-Cu/Al$_2$O$_3$ interfaces, ε, α, λ, and h are all known. If ξ is taken as the center-to-center spacing between the interfacial precipitates, the only other quantities needed for calculating G_{ic} from Eqn. (1) are the values of E, Γ and υ for the precipitate, which are not available. The values for the Al are therefore used instead. The interface energy, Γ, between the precipitate and Al$_2$O$_3$ is taken to be the same as the work of adhesion[19] for Al and Al$_2$O$_3$, which is about 1 J/m^2. Since the actual value of Γ is lower than 1 J/m^2, the substitution tends to overestimate G_{ic}. On the other hand, the modulus of the intermetallic phase (precipitate) is higher than the modulus of Al. The use of Al data for E underestimates G_{ic}. With these approximations, the fracture initiation toughness is calculated for the peak-aged and overaged condition using Eqn. (1) and the results are compared to the experimental measurements in Fig. 11. The predicted values are slightly smaller. However, considering there is no adjustable parameter in Eqn. (1), the agreement is reasonable.

Crack Arrest in Metal/Ceramic Laminates

The results of interfacial studies so far have shown that the Al/Al$_2$O$_3$ interface is weakened by interfacial precipitation and that the extent of the weakening depends on the yield strength of the metal and on the interfacial microstructure. When applied to ceramic/metal composites, these findings suggest that the aged Al-Cu/Al$_2$O$_3$ interface may be used to deflect the crack in the brittle ceramic matrix. Accordingly, the sandwich specimens were tested in the crack arrest orientation. The failure patterns are shown in Fig. 12 for both Al-Cu/Al$_2$O$_3$ interfaces. In the peak-aged condition, where the fracture initiation toughness of the interface is comparable to the toughness of the ceramic, the matrix crack ran through the tri-layers and the fracture toughness of the laminate was 40 J/m^2, about the same as the toughness of the alumina. In contrast, the crack in the overaged specimen was clearly deflected when it first hit the interface. Although the initiation toughness of the interface is only one ninth of the fracture toughness of the alumina, the toughness of the laminate is 305 J/m^2.

CONCLUSIONS

Based on our investigation of the effect of interfacial precipitation on the fracture behavior in the Al$_2$O$_3$/Al-Cu system, the following conclusions have been reached:

1. Crack growth at the precipitation-modified Al$_2$O$_3$/Al-Cu interface was mostly interfacial and showed strong R-curve behavior.

2. Interface toughness depended on the microstructures of the interface and metal. While the peak-toughness of the interface scaled with the yield strength of the metal, the initiation toughness of the interface was inversely proportional to the size of interfacial precipitates.

3. Fracture toughness of metal-ceramic laminates depended strongly on the ratio of interface toughness vs. matrix toughness. Crack arrest was observed at the overaged interface but not at the peak-aged interface.

REFERENCES

1. B.J. Dalgleish, K.P. Trumble, and A.G. Evans, "The strength and fracture of alumina bonded with aluminum alloys", *Acta Metall.*, 37:1923 (1989).
2. R.M. Cannon, B.J. Dalgleish, R.H. Dauskardt, T.S. Oh, and R.O. Ritchie, "Cyclic fatigue-crack propagation along ceramic/metal interfaces", *Acta metall.*, 39:2145 (1991).
3. A.G. Evans, B.J. Dalgleish, M. He, and J.W. Hutchinson, "On crack path selection and the interface fracture energy in bimaterial system", *Acta Metall.*, 37:3249 (1989).
4. A.G. Evans, M. Ruhle, B.J. Dalgleish, and P.G. Charalambides, "The fracture energy of bimaterial interfaces", *Metall. trans.*, 21A:2419 (1990).
5. A.G. Evans, A. Bartlett, J.B. Davies, B.D. Flinn, M. Turner, and I.E. Reimanis, "The fracture resistance of metal/ceramic/intermetallic interfaces", *Scr. Metall.*, 25:1003 (1991).
6. H.F. Wang and W.W. Gerberich, "Fracture mechanics of Ti/Al$_2$O$_3$ interfaces", *Acta Metall. Mater.*, 41:2425 (1993).
7. A. G. Evans, M. C. Lu, S. Schmauder, and M. Ruhle, "Some aspects of the mechanical strength of ceramic/metal bonded systems", *Acta Metall.*, 34:1643 (1986).
8. V. Tvergaard, "Failure by ductile cavity growth at a metal-ceramic interface", *Acta Metall.*, 39:419 (1991).
9. A. Bartlett and A.G. Evans, "The effect of reaction products on the fracture resistance of a metal ceramic interface", *Acta Metall. Mater*, 41:497 (1993).
10. A.G. Evans and B.J. Dalgleish, "The fracture resistance of metal-ceramic interfaces", *Acta Metall.*, 40:295 (1992).
11. H.C. Cao, M.D. Thouless, and A.G. Evans, "Residual stresses and cracking in brittle solids bonded with a thin ductile layer", *Acta Metall.*, 36:2317 (1988).
12. C.G. Levi, G.J. Abbaschian, and R. Mehrabian, "Interface interactions during fabrication of aluminum alloy-alumina fiber composites", *Metall. Mater. Trans.* 9A:697 (1978).
13. B.F. Quigley, G.J. Abbaschian, R. Wunderlin, and R. Mehrabian, "A method for fabrication of aluminum-alumina composites", *Metall. Trans.* 13A:93 (1982)
14. A.R. Champion, W.H. Krueger, H.S. Hartmann, and A.K. Dhingra, "Fiber FP reinforced metal matrix composites", *in::* Proc. 1978 Int. Conf. on composite materials (ICCM/2), eds. T B. Noton et al., TMS-AIME, New York, 883 (1978).
15. Z. Zhou, Y.Y. Xu, W.D. Wei, and Z.Q. Hu, "Effect of matrix microstructure on mechanical properties of 2124 aluminum alloy-SiC particle composite", *Mater. Sci. Tech.* 7:592 (1991).
16. Z. Zhang, and J.K. Shang, "Subcritical crack growth at bi-material interfaces: Part I. Flexural peel technique", *Acta metall. mater.*, (in press).
17. I.E. Reimanis, B.J. Dalgleish, M. Brahy, M. Ruhle, and A.G. Evans, "Effects of plasticity on the crack propagation resistance of metal/ceramic interface", *Acta metall. mater.*, 38:2645 (1990).
18. G. Liu and J. K. Shang, Subcritical crack growth at bi-material interfaces: Part II. Microstructural effects on fracture resistance of metal/ceramic interfaces", *Acta metall. mater.*, (in press)
19. V. Laurent, D. Chatain, C. Chatillon, and N. Eustathopoulos, "Wettability of monocrystalline alumina by aluminum between its melting point and 1273 K", *Acta Metall.*, 36:1797 (1988).

STRESSES AND FRACTURE IN CERAMIC-METAL JOINTS

A. Brückner-Foit[1], D. Munz[1,2], M. Tilscher[2], and Y.Y. Yang[1]

[1]Universität Karlsruhe
Institut für Zuverlässigkeit und Schadenskunde im Maschinenbau
[2]Forschungszentrum Karlsruhe
Institut für Materialforschung II
D-76021 Karlsruhe
Germany

INTRODUCTION

Ceramic-metal joints are applied in mechanical engineering constructions in order to profit from the high-temperature strength, the low specific weight or the resistance to wear or corrosion of ceramic materials. In practice, however, it has turned out that large residual thermal stresses due to the differences in thermal expansion may cause failure of the ceramic components during or after a change in temperature. Singular stresses occur at the edge of the interface which leads to unstable growth of flaws located in this highly stressed zone. To investigate the fracture behaviour of ceramic-metal joints, first the singular stress field in the un-cracked structure is calculated. In a second step, stress intensity factors of flaws located in the singular stress field are determined by use of the weight function method. The flaws are assumed to be planar and parallel to the interface.

Ceramic materials contain many natural flaws induced by manufacturing and surface flaws induced by surface finishing such as grinding or polishing. The actual number of these flaws in a ceramic component as well as their locations and sizes are assumed to be randomly distributed. The failure probability caused by the singular stress field in the ceramic component is derived with the scatter of these variables taken into account. Various thermally loaded ceramic-metal joints are investigated and the failure probabilities of the ceramic components are discussed.

STRESS SINGULARITIES

The exact solution of the boundary value problem of a thermally loaded rectangular bimaterial joint in two dimensional linear elasticity leads, in general, to a singular stress field at the free edge of the interface (Figure 1). The components of the stress tensor σ_{ij} can be described as a function of the polar coordinates r and θ

$$\sigma_{ij}(r, \theta) = \frac{K_L}{(r/L)^\omega} f_{ij}(\theta) + \sigma_0 f_{ij0}(\theta). \tag{1}$$

The stress exponent ω is obtained by solving a transcendental equation, which together with the equations for the angular functions f_{ij}, f_{ij0} and the quantity σ_0 is given in [1]. The stress intensity factor K_L is the only parameter which cannot be evaluated analytically. Empirical functions for the evaluation of K_L are given in [2] and [3].

The singular stress field caused by a homogeneous change in temperature is calculated for several ceramic-metal joints. The material properties of the components are listed in Table 1. For plane strain conditions and a thermal load of $\Delta T = -100K$ the parameters of Eq.(1) describing the singular stress field are given in Table 2.

Table 1. Material properties of the investigated ceramic-metal joints.

Materials	$E\,(GPa)$	ν	$\alpha\,(\times 10^{-6}K^{-1})$
Al_2O_3	375	0.27	8.2
Si_3N_4	314	0.28	2.7
ZrO_2	240	0.3	11
Al-alloy	71	0.33	22.5
Steel	215	0.28	11.5

Table 2. Parameters of Eq.(1) for the investigated ceramic-metal joints for plane strain conditions and a thermal load of $\Delta T = -100K$.

Ceramic-metal joint	K_L (MPa)	ω	$f_y(90°)$	$f_{xy}(0°)$	σ_0 (MPa)
Al_2O_3/Al-alloy	257.3	0.1572	1.719	-0.2197	-370.4
Si_3N_4/Al-alloy	379.7	0.1375	1.636	-0.2035	-525.1
ZrO_2/Al-alloy	263.8	0.1053	1.503	-0.1744	-342.9
Si_3N_4/Steel	2053	0.01092	1.093	-0.04642	-2143

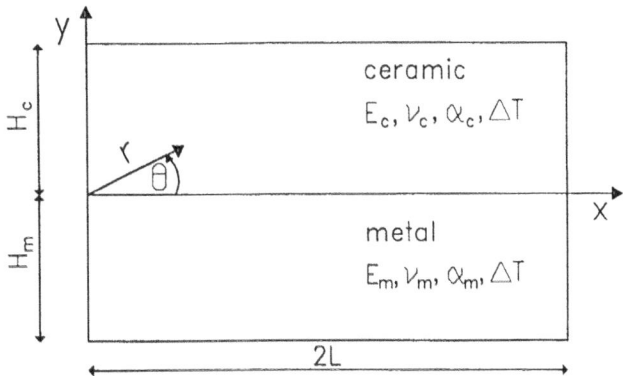

Figure 1. Thermally loaded ceramic-metal joint.

Moreover, the following relations hold: $f_y(0°) = 1$, $f_{y0} = 1$ and $f_{x0} = f_{xy0} = 0$. The results of Eq.(1) as well as the stress values obtained by the finite element method are plotted in Figure 2 for the Al_2O_3/Al-alloy joint. It can be seen that there is good agreement between both methods in the vicinity of the singularity. At great distance from it, the influence of the singularity vanishes and Eq.(1) is no longer valid.

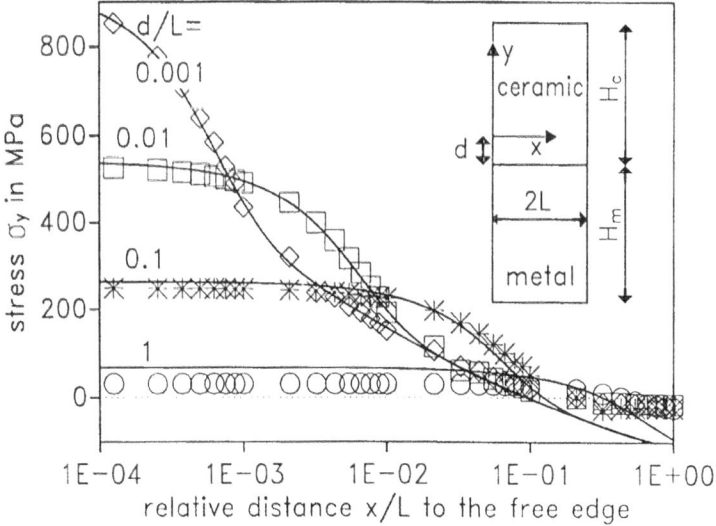

Figure 2a. Normal stresses σ_y in the ceramic component of an Al_2O_3/Al-alloy joint after a change in temperature of $\Delta T = -100K$, obtained by the finite element method (symbols) and with Eq.(1) (solid lines).

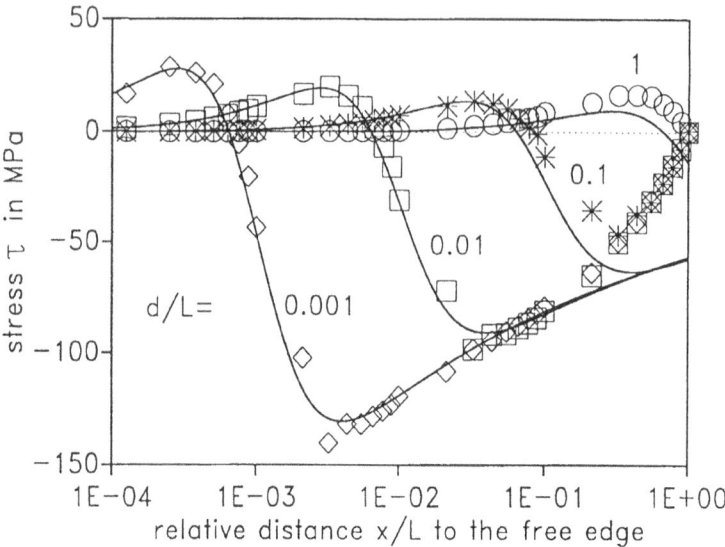

Figure 2b. Shear stresses τ in the ceramic component of an Al_2O_3/Al-alloy joint after a change in temperature of $\Delta T = -100K$, obtained by the finite element method (symbols) and with Eq.(1) (solid lines).

STRESS INTENSITY FACTORS

Ceramic materials contain external as well as internal flaws. These flaws are idealised as planar cracks parallel to the interface. An external crack of the length a and the distance d to the interface is shown in Figure 3. The weight function method developed by Bückner [4] is used to determine its stress intensity factor. The mode I stress intensity factor for an external crack in a bimaterial joint can be written as [5]

$$K_I^e = \int_0^a \left(h_I^{(\sigma)}(x,a)\, \sigma_y(x) + h_I^{(\tau)}(x,a)\, \tau(x) \right) dx. \qquad (2)$$

The quantities $\sigma_y(x)$ and $\tau(x)$ are the stresses in the uncracked component at the location of the prospective crack. The weight functions $h_I^{(\sigma)}$ and $h_I^{(\tau)}$ are given in [6].

An internal crack of the length $2a$ and the distance b to the surface is shown in Figure 4. The stress intensity factor for the crack tip facing the free edge becomes infinite if the relative closeness to the surface $\varepsilon = a/b$ reaches unity. This means that the crack tip of internal cracks located in the vicinity of the surface may become unstable and burst open. The criterion of bursting open is $\varepsilon \geq \varepsilon_c$. The quantity ε_c is the critical closeness of an internal crack to the free edge. Cracks fulfilling the bursting open criterion are considered as external cracks. Using the normal stresses σ_y in the uncracked structure, the mode I stress intensity factor for an internal crack can be estimated as

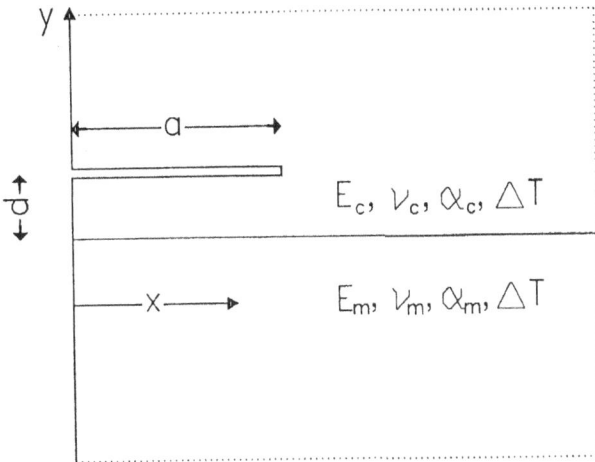

Figure 3. External crack in the ceramic component of a ceramic-metal joint.

$$K_I^i \approx g_i \, g_s \int_{-a}^{a} h(x,a) \, \sigma_y(x,y) \, da. \tag{3}$$

The quantity g_i takes the effect of the inhomogeneous material into account and is a function of the elastic constants of the joint. The quantity g_s is related to the influence of the surface and depends on the choice of ε_c. The weight function $h(x,a)$ is that for an internal crack in a homogeneous infinite plate. Table 3a shows the values of g_i for the investigated ceramic-metal joints and Table 3b those of g_s for several values of ε_c [7].

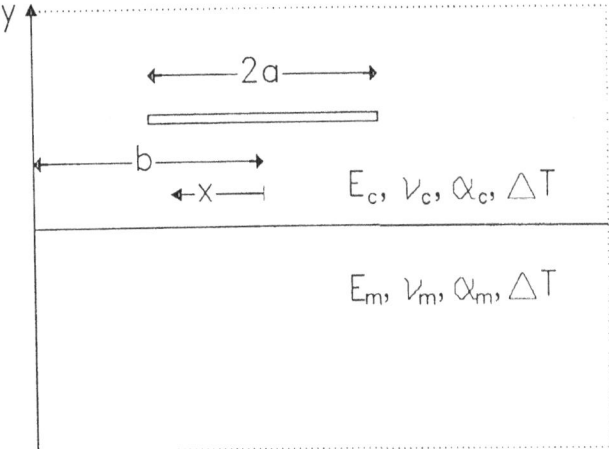

Figure 4. Internal crack in the ceramic component of a ceramic-metal joint.

Table 3a. The quantity g_i (Eq.(3)) for the investigated joints.

Table 3b. The quantity g_s (Eq.(3)) for several values of ε_c.

Ceramic-metal joint	g_i
Al_2O_3/Al-alloy	1.55
Si_3N_4/Al-alloy	1.49
ZrO_2/Al-alloy	1.47
Si_3N_4/Steel	1.24

ε_c	g_s
0.1	1.005
0.5	1.088
0.9	1.801
0.99	4.243

FAILURE PROBABILITY

Two groups of crack populations are considered: natural flaws induced by manufacturing (manufacturing cracks) and surface flaws induced by finishing such as polishing or grinding (finishing cracks). In fracture mechanics a distinction is made between internal or volume cracks inside of the material and external or surface cracks which are open at the surface. Each finishing crack is an external crack, but a manufacturing crack can be internal as well as external. It becomes external if it is cut through during finishing or if it is located so close to the surface that the crack tip near the surface becomes unstable and bursts open. All cracks are assumed to be planar and parallel to the interface. Figure 5 shows different kinds of cracks. There are:

a) Cut through manufacturing cracks (external).
b) Burst open manufacturing cracks (external).
c) Manufacturing cracks (internal).
d) Finishing cracks (external).

The size a of manufacturing and finishing cracks is assumed to be randomly distributed and described by the probability density functions $f_a^m(a)$ and $f_a^f(a)$, respectively. The corresponding cumulative distribution functions are $F_a^m(a)$ and $F_a^f(a)$. No crack in the joint can be larger than the joint itself. Therefore, the length a of manufacturing cracks is limited to L and that of finishing cracks to $2L$ (Figure 1). The probability density function for the crack length a of cut through manufacturing cracks can be written as [7]

$$f_a^{c,m}(a) = \frac{1}{L}\left(1 - F_a^m\left(\frac{a}{2}\right)\right).\tag{4}$$

The probability density function for the crack length a of burst open manufacturing cracks with $\varepsilon \geq \varepsilon_c$ is determined by

$$f_a^{b,m}(a) = \frac{1}{L}\left(F_a^m\left(\frac{a}{2}\right) - F_a^m\left(\frac{\varepsilon_c}{1 + \varepsilon_c}a\right)\right).\tag{5}$$

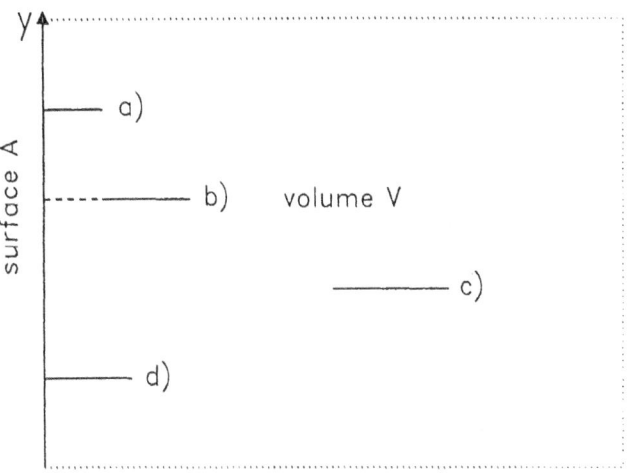

Figure 5. Different kinds of cracks.

The probability density function of external cracks caused through manufacturing cracks is given by

$$f_a^{e,m}(a) = f_a^{c,m}(a) + f_a^{b,m}(a) = \frac{1}{L}\left(1 - F_a^m\left(\frac{\varepsilon_c}{1 + \varepsilon_c}a\right)\right). \tag{6}$$

The probability that a given external manufacturing crack causes failure is given by the probability that its random crack size, whose scatter is described by the probability density function $f_a^{e,m}(a)$, exceeds the critical crack length $a_c^e(y)$ for external cracks

$$P_1 = \int_{a_c^e(y)}^{2L} f_a^{e,m}(a)\, da. \tag{7}$$

The critical crack length for external cracks follows from

$$K_I^e(a = a_c^e) = K_{Ic} \tag{8}$$

with K_I^e from Eq.(2) and the fracture toughness K_{Ic}. If a uniform distribution describes the scatter of crack location, the infinitesimal probability of finding an external crack at a specific point is given by

$$dP_2 = \frac{1}{A}\, dA, \tag{9}$$

where A is the surface area of the component. Multiplying P_1 by dP_2 and summing up over all possible crack locations yield the probability that failure of a component containing exactly one manufacturing crack is caused by an external crack

$$Q_1^{e,m} = \frac{1}{A} \int_A \int_{a_c^e(y)}^{2L} f_a^{e,m}(a) \, da \, dA. \quad \cdot \tag{10}$$

The scatter of the crack length of internal manufacturing cracks with $\varepsilon < \varepsilon_c$ is described by the probability density function

$$f_a^{i,m}(a, \vec{x}) = f_a^m(a) \, \theta(\varepsilon_c - \varepsilon) \tag{11}$$

with the step function

$$\theta(x) = \begin{cases} 1 \text{ for } x > 0 \,, \\ 0 \text{ for } x \leq 0 \,, \end{cases} \tag{12}$$

the relative closeness to the surface $\varepsilon = a/b$, and the critical closeness to the surface ε_c. The probability that a given internal crack causes failure is given by the probability that its random crack size, whose scatter is described by the probability density function $f_a^{i,m}(a,\vec{x})$, exceeds the critical crack length $a_c^i(\vec{x})$ for internal cracks

$$P_3 = \int_{a_c^i(\vec{x})}^{L} f_a^{i,m}(a, \vec{x}) \, da. \tag{13}$$

The critical crack length for internal cracks follows from

$$K_I^i(a = a_c^i) = K_{Ic} \tag{14}$$

with K_I^i from Eq.(3). The infinitesimal probability of finding an internal crack at a certain point is given by

$$dP_4 = \frac{1}{V} \, dV. \tag{15}$$

Multiplying P_3 by dP_4 and summing up over all possible crack locations yield the probability that failure of a component containing exactly one manufacturing crack is caused by an internal crack

$$Q_1^{i,m} = \frac{1}{V} \int_V \int_{a_c^i(\vec{x})}^{L} f_a^{i,m}(a, \vec{x}) \, da \, dV. \tag{16}$$

The probability that a structure with exactly one manufacturing crack fails is

$$Q_1^m = Q_1^{e,m} + Q_1^{i,m}. \tag{17}$$

The survival probability R_1^m of a component containing exactly one manufacturing flaw is given by

$$R_1^m = 1 - Q_1^m. \tag{18}$$

For k manufacturing flaws being present, the total survival probability R_k^m is equal to the product of the survival probability of each individual manufacturing flaw

$$R_k^m = (1 - Q_1^m)^k. \tag{19}$$

Let M^m be the average number of manufacturing cracks in the considered volume. The actual number k of manufacturing cracks in the component is a Poisson distributed random variable, if the volume elements dV are stochastically independent of each other. The probability of having k manufacturing cracks in a component of the volume V is given by

$$P_k^m = \frac{(M^m)^k e^{-M^m}}{k!}. \tag{20}$$

The survival probability of a component containing an arbitrary number of manufacturing flaws follows from multiplying the probability of having k cracks (Eq.(20)) by the corresponding survival probability (Eq.(19)) and summing up over all possible numbers of flaws

$$P_s^m = \sum_{k=1}^{\infty} P_k^m R_k^m. \tag{21}$$

The following expression is obtained by inserting Eqs.(19) and (20) into Eq.(21) and using the definition of the exponential as an infinite series

$$P_s^m = \exp(-M^m Q_1^m). \tag{22}$$

Hence, the failure probability is given by

$$P_f^m = 1 - P_s^m = 1 - \exp(-M^m Q_1^m). \tag{23}$$

By analogy, the probability that a component with exactly one finishing crack fails is

$$Q_1^f = \frac{1}{A} \int_A \int_{a_c^e(y)}^{2L} f_a^f(a) \, da \, dA. \tag{24}$$

If M^f is the average number of finishing cracks in the considered surface and the actual number of finishing cracks is described as a Poisson distributed random variable, the probability that a component with an arbitrary number of finishing cracks fails is given by

$$P_f^f = 1 - \exp(-M^f Q_1^f). \tag{25}$$

The failure probability of a component containing manufacturing and finishing cracks is

$$P_f = 1 - \exp\left(-(M^m Q_1^m + M^f Q_1^f)\right). \tag{26}$$

DESCRIPTION OF THE CRACK POPULATION

According to the results of extreme value theory [9], the crack size distribution function can be described by a power law (a^{-r}) for large values of a. Considering the lower limit a_0 and the upper limit L for the crack size a, the probability density function of manufacturing flaws $f_a^m(a)$ is written as

$$f_a^m(a) = \begin{cases} \dfrac{(r-1)a_0^{(r-1)}}{1 - (a_0/L)^{(r-1)}} \dfrac{1}{a^r} & , \quad a_0 < a < L. \\[4mm] 0 & , \quad else. \end{cases} \tag{27}$$

The corresponding cumulative distribution function is

$$F_a^m(a) = \begin{cases} 0 & , \quad a < a_0. \\[4mm] \dfrac{1}{1 - (a_0/L)^{r-1}} \left(1 - \left(\dfrac{a_0}{a}\right)^{r-1}\right) & , \quad a_0 \le a \le L. \end{cases} \tag{28}$$

By analogy, the probability density function of finishing flaws with the lower limit a_0 and the upper limit $2L$ can be written as

$$f_a^f(a) = \begin{cases} \dfrac{(r-1)a_0^{(r-1)}}{1 - (a_0/2L)^{(r-1)}} \dfrac{1}{a^r} & , \quad a_0 < a < 2L. \\[4mm] 0 & , \quad else. \end{cases} \tag{29}$$

with the corresponding cumulative distribution function

$$F_a^f(a) = \begin{cases} 0 & , a < a_0. \\[2ex] \dfrac{1}{1 - (a_0/2L)^{r-1}} \left(1 - \left(\dfrac{a_0}{a}\right)^{r-1}\right) & , a_0 \le a \le 2L. \end{cases} \tag{30}$$

EXAMPLE

The failure probabilities of some ceramic-metal joints with the geometry $H_c = H_m = 2L = 20mm$ are calculated. Only the vicinity of the singularity, which is given by $0 \le x/L, y/L \le 0.01L$, is taken into account (Figure 1). A thickness of unity and a state of plane strain are assumed. The elastic constants and the thermal expansion coefficients of the joint materials are given in Table 1. The singular stress field results from a homogeneous change in temperature of $\Delta T = -100K$. Equation (1) with its parameters given in Table 2 describes the stress distribution of the singular stress field. The chosen fracture toughness of the ceramic materials is $K_{Ic} = 6MPa\sqrt{m}$. The Weibull parameter is set $m = 5$. The lower limit of the size a of manufacturing flaws is $a_0 = 10^{-4}$ and the average number of manufacturing flaws in the unit volume V_0 is $M_0^m = 1.8 \times 10^6$. This data set corresponds to a ceramic with a Weibull parameter $b = 530MPa$, obtained in a four-point bending test. The bursting open criterion is chosen to be $\varepsilon_c = 0.9$. For the sake of simplicity, the statistical properties of the finishing flaw population are asumed to be the same as those of manufacturing flaws $(a_0 = 10^{-4}, M_0^f = 1.8 \times 10^6)$. The quantities $Q_1^{e,m}$, $Q_1^{i,m}$ and Q_1^f are listed in Table 4. The results for the failure probabilities are shown in Table 5.

Table 4. The quantity Q of the investigated ceramic-metal joints.

Ceramic-metal joint	$Q_1^{e,m}$	$Q_1^{i,m}$	Q_1^f
Al$_2$O$_3$/Al-alloy	1.47×10^{-11}	2.75×10^{-9}	2.09×10^{-9}
Si$_3$N$_4$/Al-alloy	4.49×10^{-11}	5.12×10^{-9}	1.16×10^{-8}
ZrO$_2$/Al-alloy	≈ 0	7.14×10^{-11}	≈ 0
Si$_3$N$_4$/Steel	≈ 0	1.65×10^{-11}	≈ 0

Table 5. Failure probability caused by the singular stress field in the ceramic components ($\Delta T = -100K$).

Ceramic-metal joint	P_f^m	P_f^f	P_f
Al$_2$O$_3$/Al-alloy	8.66×10^{-1}	7.31×10^{-2}	8.76×10^{-1}
Si$_3$N$_4$/Al-alloy	9.77×10^{-1}	3.45×10^{-1}	9.85×10^{-1}
ZrO$_2$/Al-alloy	5.07×10^{-2}	≈ 0	5.07×10^{-2}
Si$_3$N$_4$/Steel	1.20×10^{-3}	≈ 0	1.20×10^{-3}

There is no obvious correlation between the parameters of the singular stress field (Eq.(1), Table 2) and the failure probability. The ceramic component of the Si_3N_4/Steel joint with the highest parameter K_L has the lowest failure probability. The ceramic component of the Si_3N_4/Al-alloy joint with the highest failure probability has a relatively large stress exponent ω combined with a relatively large stress intensity factor K_L.

CONCLUSION

The stresses in thermally loaded ceramic-metal joints are calculated analytically in the vicinity of the singularity and compared to the thermal stress field obtained by the finite element method. The method of weight functions is applied to determine the stress intensity factors for external as well as internal cracks located in the singular stress field. The failure probability caused by the singular stress field in the ceramic component of a thermally loaded ceramic-metal joint is derived with the different populations of manufacturing and finishing cracks taken into account. The statistical properties of the flaw populations and the fracture toughness of the ceramics are chosen in such a way that realistic Weibull parameters of a four-point-bending test are obtained. Four ceramic-metal joints are investigated and the correlation between the parameters of the singular stress field and the failure probability is discussed.

REFERENCES

[1] K.Mizuno, K.Miyazawa, T.Suga, J. of the Faculty of Eng., University of Tokyo (B) Vol.39, No.4, 401 (1988).
[2] D.Munz, Y.Y.Yang, J. of Appl. Mech., 59, 857 (1992).
[3] M.Tilscher, D.Munz, Y.Y.Yang, Int. J. of Adhesion, 49, 1-21 (1995).
[4] H.Bückner, Z. Angew. Math. Mech. 50, 529 (1970).
[5] T.Fett, D.Munz, M.Tilscher, Int. J. Solids Structures (to be published).
[6] T.Fett, M.Tilscher, D.Munz, Eng. Fract. Mech. (to be published).
[7] M.Tilscher, Internal Report (in German), Forschungszentrum Karlsruhe (1995).
[8] T.Thiemeier, A.Brückner-Foit, H.Kölker, J. Am. Ceram. Soc., 74, 48 (1991).
[9] E.J. Gumbel, Statistics of Extremes, Columbia University Press, New York (1958).

INDENTATION CRACKING OF BRITTLE THIN FILMS
ON BRITTLE SUBSTRATES

E. Weppelmann [1], M. Wittling [2], M. V. Swain [3,4] and D. Munz [2,5]

[1] Fraunhofer-Institute of Material Mechanics
79108 Freiburg, Germany
[2] IZSM, University of Karlsruhe
Kaiserstr. 12, 76131 Karlsruhe, Germany
[3] CSIRO - Division of Applied Physics
Lindfield NSW 2070, Australia
[4] Department of Mechanical & Mechatronic Engineering
University of Sydney, NSW 2006, Australia
[5] IMF-II, Research Center Karlsruhe
Po.-Box 3640, 76021 Karlsruhe, Germany

ABSTRACT

Observations are presented of cracking produced by spherical pointed indentations made onto brittle films on brittle substrates. The major emphasis is with TiN films on substrates of silicon and sapphire. A range of film thicknesses and two deposition temperatures are considered resulting in different microstructures and residual stresses. The observations with small spherical tipped indenters are found to be the most amenable to interpretation. Fracture mechanics is invoked to explain the basis of through film cracking and interface delamination cracking.

INTRODUCTION

Thin films are now widely used as a means of providing or improving functionality to cheaper bulk substrates. The range of areas where such films are applied continues to increase being driven by economics and abilities to deposit large homogeneous films on a wide variety of substrates. Other driving forces include the high valued microelectronics industry with its quests for yet higher density of devices available for integrated circuits and information storage media. However, with all these thin film coatings the question of adhesion and mechanical property characterization and monitoring are vital.

Numerous approaches have been developed and explored to investigate the adhesion of thin films on substrates [1]. Probably the most popular technique is that of the scratch test with a spherical tipped indenter, typically a Rockwell diamond indenter with a nominal radius of ~ 200 μm. This method pioneered by Benjamin and Weaver [2] has been developed and standardized by over the last few decades to the point where a critical load for the onset of acoustic emission events during scratching is deemed as characterizing the adhesion. Whilst the simplicity of such an empirical approach has certain appeal to industry, the influences of film thickness, substrate, residual stress, surface texture etc. on such measurements is not only extremely complex from an analytical perspective but problematic for predicting industrial performance. A number of studies have investigated the gross deformation and failure mechanism [3] of the film/substrate system during such scratching but the basic mechanics issues pose daunting problems for those considering finite element evaluations.

Most emphasis on thin film mechanical property characterization today is currently focused on nano-indentation. This approach pioneered by Pethica et al. [4] and subsequently developed by a number of groups has utilized very sharp triangular Berkovich diamond indenters to probe the resistance to penetration in the sub-micron region of surfaces. Such contacts leads to virtually unobservable impressions in materials even with high resolution SEM and AFM systems thereby placing great emphasis on the force-displacement data to determine estimates of hardness and modulus of the film/substrate system. However, for many materials, particularly brittle materials, the physical meaning or interpretation of hardness is a problem and is invariably associated with cracking [5].

An alternative approach pioneered by Field and colleagues [6-8] has replaced nominally sharp indenters with very small spherical tipped indenters for nano-indentation testing. In this manner the elastic, elastic/plastic and elastic/brittle response of materials including thin films may be investigated. The approach provides a better analytical basis for interpreting the measured force-displacement response and enables estimates of the nominal stress-strain behavior to be identified. Considerable success has been had in the interpretation of the phase transformations in silicon and that of complex materials such as glassy carbons with this approach [8]. Furthermore, Field & Swain [7] have developed a novel loading procedure that enables a near continuous determination of mechanical properties and a method of accounting for pile-up and sinking about an indenter.

The aim of this paper is to report on observations of the various forms of cracking about indentations in thin films on brittle substrates. Observations of both cracking about such films with pointed and spherical indenters will be given along with some appreciation of the influence of film thickness, substrate material and residual stress. Various simple means of observations and analysis will be compared together with preliminary finite element approaches to this problem.

MATERIALS AND METHODS

Measurements of the force-displacement response during indentation were made with an ultra micro-indentation system (UMIS-2000). Indenters were diamond triangular pyramids (Berkovich) and small spherical tipped diamond cones with radii from 2 to 20 μm.

In this study the films were of TiN deposited by a magnetically ducted filter attached to an arc evaporation system. Two different films were deposited, one series at 350°C on silicon and another series at around room temperature on silicon and sapphire. The thickness of the films ranged from 0.2 to 4.5 μm with the thicker films deposited at 350°C and the thinner at room temperature. The residual stress in the films was estimated from disk curvature measurements before and after deposition.

Acoustic emission was utilized to monitor events during the course of indenting. This system is shown schematically in Figure 1 and was connected to the PC controlling the UMIS so that acoustic emission events were recorded along with force and displacement data. More details of this system have been presented elsewhere [9].

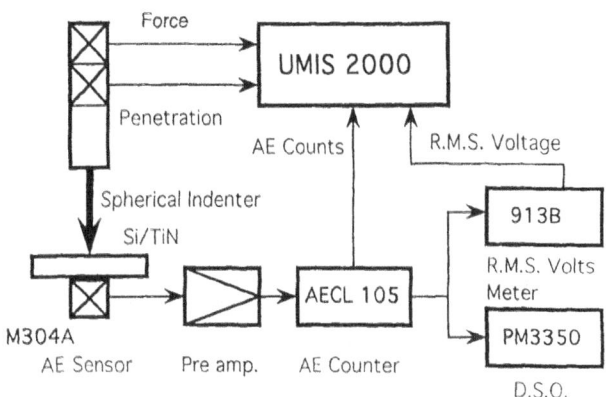

Figure 1. Schematic arrangement of the UMIS and acoustic emission system.

OBSERVATIONS

Typical force-displacement data for pointed indentations on TiN films on silicon and sapphire substrates are shown in Figure 2. In both instances the influence of film thickness on the force-displacement response is observed. At very low loads the initial displacements superimpose and at heavier loads they begin to diverge. As discussed by [10] differentiation of these curves tends to show two linear portions which measure the resistance to penetration of the film followed by that of the substrate. Acoustic emission data shows that a near continuous emission of events occurs throughout the course of the test [11].

More definitive information could be gleaned from spherical indentations made on the two different film substrates combinations. Typical observations of the force-displacement response are shown in Figure 3. These results were obtained with a 5 μm radius indenter. The initial response is elastic in all instances unless the surface roughness of the film is

a) TiN on silicon

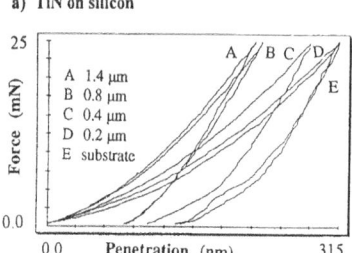

b) TiN on sapphire

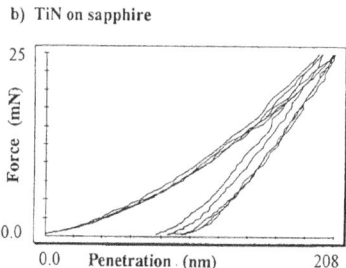

Figure 2. Typical force-displacement curves with a pointed Berkovich indenter into TiN films of different thickness on silicon and sapphire substrates.

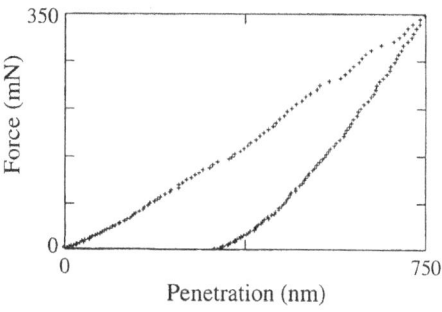

Figure 3. Typical force-displacement curve of an indentation test on a TiN film on silicon using a 5 μm radiused indenter.

significant. The limit of the elastic range is often noted by a small discontinuity although for softer or more deformable substrates the onset of permanent set is virtually imperceptible initially but is often followed by a more significant discontinuity. Acoustic emission monitoring indicated that the acoustic emission events were uniquely associated with the initial discontinuity in the force-displacement curves, as shown in Figure 4. At heavier loads acoustic emission events were more continuous.

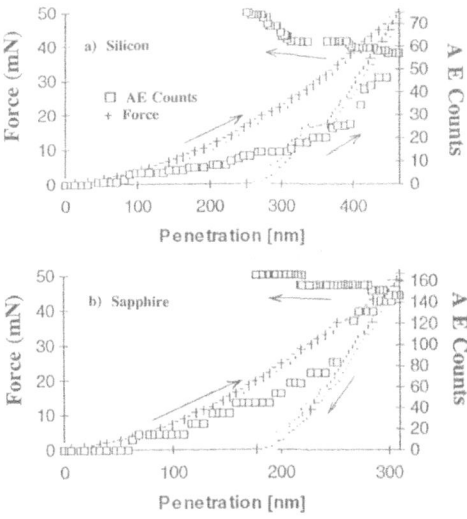

Figure 4. Superimposed force-displacement and acoustic emission events during indentation with a small spherical tipped indenter.

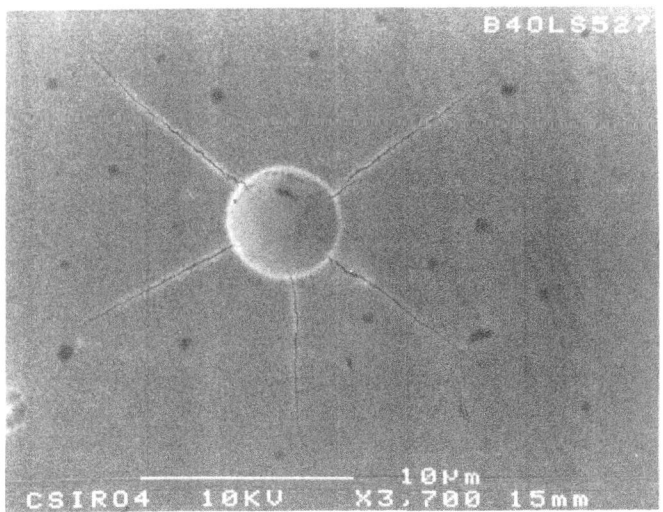

Figure 5. Scanning electron micrograph images of the cracking about spherical indentations into a TiN film on Silicon.

The observations of spherical indentation tests on TiN films on sapphire revealed some interesting differences from that for the films on silicon. In particular, it was noted in sapphire that the onset of deformation took place at a critical load with a major 'pop-in'. With a TiN film on the sapphire the same behavior was observed but at higher loads. Acoustic emission data showed no activity prior to the onset of the major discontinuity and thereafter almost continuous acoustic emission development.

A SEM observation of an indentation on a TiN film made with a 5 μm radiused indenter is shown in Figure 5. These observations, particularly high resolution images, show well defined radial cracks. Circumferential cracks also occurred within the contact area. Those cracks are not visible on the SEM picture in Figure 5 but could be observed on SEM observations of spherical indentations carried out with bigger radiused indenters. However, at heavier loads, particularly with thicker films deposited with high compressive residual stresses, delamination spalling often occurs as shown in Figure 6.

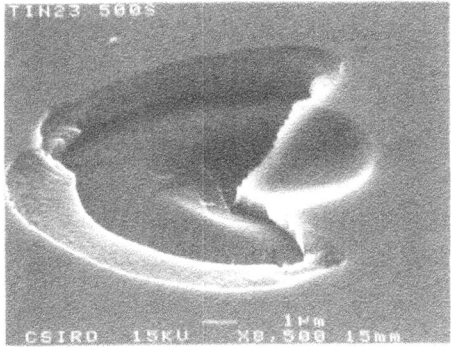

Figure 6. Scanning electron micrograph of the delamination cracking about an spherical indentation into a TiN film with high compressive residual stresses.

More informative SEM observations of the deformation of the film substrate system may be had from cross-sections of the impressions. A cross-sectional observation of a spherical indentation impression on a TiN film carried out with a 5 μm radiused indenter is shown in Figure 7. These observations clearly show the nature of the circumferential cracking within the films and the sub-surface deformation of the underlying silicon. Also evident is the interfacial delamination between the film and substrate which extends some 5 to 10 times the contact diameter. The size of this delaminated region scaled with the load applied on the indenter for loads in excess of the first discontinuity.

Alternative means of investigating the deformation and particularly the delamination is possible using an optical interferometric technique and acoustic microscopy. The former displays the contours about the impression due to pile-up or delamination. A typical example is shown in Figure 8 for the uplift about a TiN film on silicon. Acoustic microscopy of the same region is shown in Figure 9 for 400 MHz ultrasonic waves reflected from the delaminated TiN interface. The results of the SEM cross-sections, optical interferometry and acoustic microscopy show excellent agreement in the size of the delaminated zone about the impression.

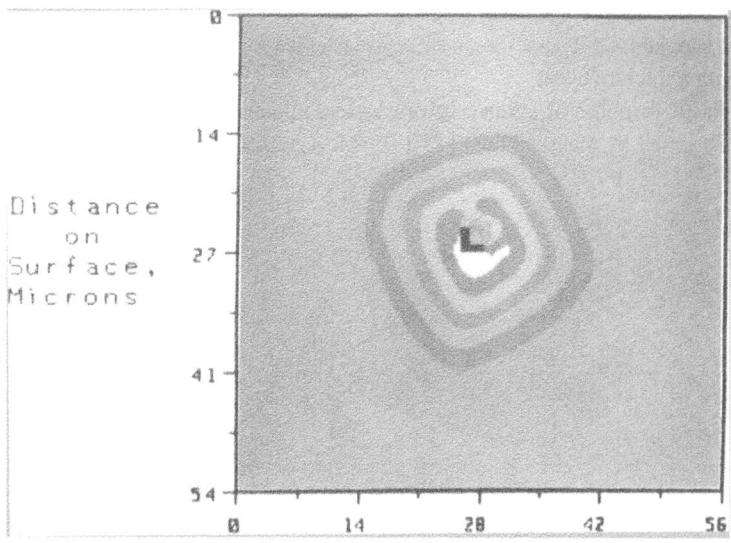

Figure 7. Cross-sectional scanning electron micrographs of the cracking and deformation of a TiN film on silicon indented with a spherical indenter.

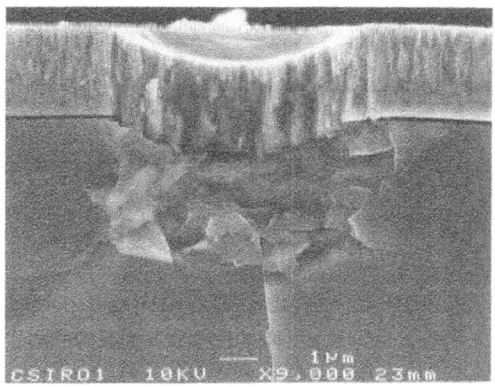

Figure 8. Optical contouring of the uplift about a spherical indentation into a TiN film on silicon.

DISCUSSION

The observations shown above indicate that the spherical indenters lend themselves more readily to analysis. The initial response is elastic and in these observations for both the silicon and sapphire substrates the onset of inelastic response or discontinuity appeared to coincide with a fracture event.

The elastic response of a film/substrate system to indentation with a spherical indenter has been discussed by many authors [12-17]. Most authors have pursued a finite element

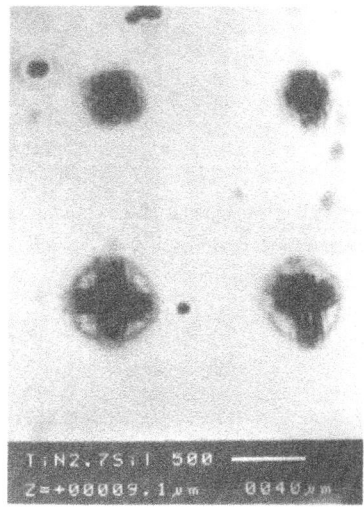

Figure 9. Acoustic microscopy showing the region of delamination about an indentation into a TiN film on silicon.

formulation to solve this difficult problem although a recent study by Schwarzer and Richter [13] has developed an analytical solution. A useful basis for assessment of the film-substrate response has been developed by Swain and Weppelmann [14] using a load/partial unloading procedure. Using this method one usually finds in the case of a TiN film on silicon that the contact pressure increases with load whereas the estimated modulus decreases. The latter arises because the contact area continues to expand, but at a slower rate, with load and the influence of substrate on the overall system stiffness becomes more significant. A simple asymptotic analysis of this problem has been provided by Gao et al. [15] that enables one to deconvolute the modulus of the films from that of the substrate provided the film thickness and contact diameter is known.

Two aspects of the above SEM observations naturally lead to fracture mechanics considerations namely that of the crack propagation behavior through the film and along the interface. The through thickness cracking of the film occurs during loading whereas the cracking along the interface often takes place upon unloading. Some features of the fracture mechanics of both of these issues will now be addressed.

Recent FEM analysis of the spherical contact of thin films on substrates of different properties reveal complex varying stress fields in the vicinity of the contact circle where the ring type cracks are seen to develop [16,17]. The SEM observations (see Figure 7) show that the crack propagation is normal to the film surface and is strongly influenced by the weak boundaries between the very small columnar grains. A finite element simulation of the indentation process was carried out by the authors in order to calculate the stresses along the predicted crack path [18]. The gradients of the radial tensile stresses and the shear stresses through the film are highly dependent upon the ratio of the contact dimension to the thickness of the film. The authors found that for the cases where the films were thin compared to the contact dimensions fracture can initiate at a pre-existing flaw and develops into a fully circumferential crack by high radial tensile stresses. For the cases where the films were thicker a bi-axial stress field of radial tensile stresses and shear stresses exists that can initiate a crack but such a crack will develop into a circumferential crack under high shear stresses only. In this study a fracture mechanics analysis is presented in order to calculate the stress intensity factors K_I and K_{II}. It was shown that the values of the stress intensity factors calculated were sufficiently high to initiate a crack even from very small pre-existing flaws they assumed in their analysis. An additional consideration for calculation of the stress intensity factor is the influence of the residual stresses, in this instance ~ 2.5 GPa bi-axial tension.

Delamination is clearly visible about the indentation site of the present materials as shown in Figures 6 to 9. There have been various proposals to rationalize the extent of cracking and to enable an estimation of interfacial energy between film and substrate. Marshall et al. [19] used a Vickers indenter to estimate the interfacial toughness of ZnO films on glass and subsequently incorporated the role of residual stress in the film. Ashcroft and Derby [20] attempted to use these approaches to measure the adhesion of glass on copper but calculated unacceptably high values for the adhesion energy. The driving force for the interfacial cracks in the latter system was very different from the lateral cracking model proposed by Marshall et al. [19].

Loubet et al. [21] proposed a simple, more realistic model for systems where the driving force was not that due to residual stress caused by plastic flow in the film. Weppelmann et al. [22] also investigated this behavior and proposed a fracture mechanics solution. This latter analysis relied upon film/indenter interference and did not include the influence of residual stress. More recently Drory and Hutchinson [23] have proposed a very simple relationship for the influence of residual stress on the estimated the interfacial toughness. These authors considered the delamination crack-size large in comparison with the contact dimensions and normalized their data in terms of the energy release G_0, for spontaneous delamination conditions namely

$$G_0 = (1 - v^2) \sigma_r^2 \frac{h}{2} E_f$$

where σ_r is the bi-axial residual compressive stress, h the film thickness and E_f the film modulus. A simplified relationship for interfacial toughness is then shown graphically in Figure 10. The data generated by Wittling [24] will now be considered in light of this figure.

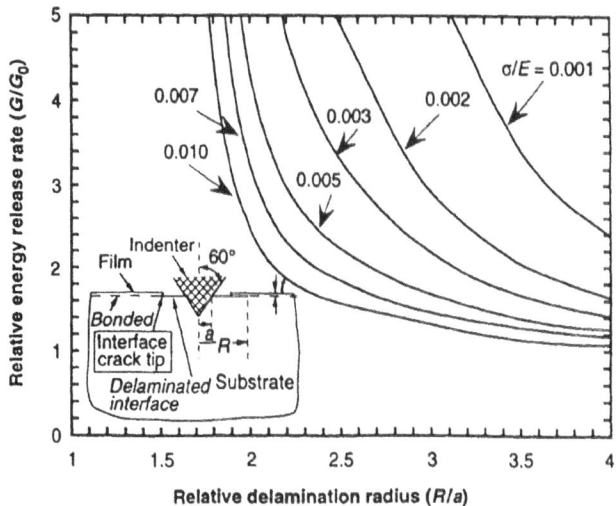

Figure 10. The relationship between the normalized interface toughness G_c/G_0 versus normalized crack radius for films under various values of residual stress.

Table 1. Interfacial Toughness for TiN films on Silicon

Film thickness h (μm)	Residual Stress σ_r (GPa)	Crack Size c (μm)	Interfacial Toughness G_c (Jm^{-2})
0.2	8.5	11	16
0.4	6.4	11	17
0.8	6.0	10	37
1.4	4.4	10	38

Table 1 compares the delamination crack radius, film thickness, residual stress for a number of TiN films deposited under similar circumstances. The estimated interfacial toughness from these values and figure 10 are also included in the table. The values for the interfacial toughness are in the range 20 to 30 Jm^{-2} which is very high for brittle-brittle interfaces but is similar to what Drory and Hutchinson [23] found for diamond films on titanium.

The approach outlined above is only appropriate for films under compressive stresses whereas a previous study by the authors (Weppelmann et al. [22]) noted that delamination cracks also occurred about indentations into thicker TiN films on silicon that were existing in a state of tension. An estimate of the interfacial energy can be obtained from the height of the uplift of the circular crack about the impression. A slight modification of an expression given by Loubet et al. [21] leads to the relationship

$$G = 4/3 \frac{Eh^3 d^2}{(1-v^2)c^4}$$

where d is the uplift height, h the film thickness and c the radius of the interfacial crack. Measurements of the uplift, crack size for a range of loads with 5 and 10 μm radius indenters on the 4.5 μm thick film gave values of 5.6 to 11 Jm^{-2}.

These values for interface measurement and the differences depending upon evaluation approach suggest that a more critical appraisal is required. There are a number of issues that are clear, namely that high residual compressive stresses are conducive to significant delamination, is also evident that uplift of substrate material within and about the impression also contributes to delamination. The two approaches, particularly for very sharp indenters as used by Drory and Hutchinson [23], may be significant and cannot be ignored. A finite element analysis that considers both residual stress within the film as well as substrate response such as pile up, sinking-in on pop-out is required. A major difficulty is knowing the appropriate constitutive response for the film and substrate in order to conduct such investigations.

CONCLUSIONS

The present investigation has highlighted the nature of the contact and deformation of brittle TiN films on softer and comparable hardness substrates. The major emphasis has been with small spherical tipped indenters whereby the elastic response is followed by elastic/brittle film/plastic substrate behavior. Interpretation of the force-displacement response was assisted by simultaneous measurement of acoustic emission activity. Small steps or discontinuities in the loading response corresponded to bursts of acoustic emission, particularly before the onset of significant inelastic behavior. Cross-sectional SEM observations revealed virtually no plastic deformation of the film only cracks through the film and significant deformation of both the silicon and sapphire substrates. An FEM analysis of the through thickness film cracking behavior is presented.

Observations also showed significant delamination occurred about the impression. The driving force for delamination varied from very high residual stresses for the low temperature deposited TiN films (4 to 8 GPa) whereas for the thick TiN films on silicon which were under bi-axial tension the pop-out of the silicon upon unloading was proposed as the driving mechanism. Two expressions available in the literature were used to estimate the interfacial toughness, namely that by Drory & Hutchison [23] as well as a modification of a relationship proposed by Loubet et al. [21]. The former gave rather high values for the interfacial energy (20 to 30 Jm^{-2}) whereas the latter were in the range 5 to 12 Jm^{-2}.

REFERENCES

[1] K.L.Mittal, *J. Adhesion Sci. Technol.*, 1 (1987) 247.

[2] P.Benjamin and C.Weaver, *Proc. Roy. Soc., London*, A2554 (1960) 163.

[3] S.J.Bull and D.R.Rickerby, *Surface Coating Technol.*, 42 (1990) 149.

[4] J.B.Pethica and R.Hutchins and W.C.Oliver, *Phil. Mag.*, A48 (1983) 593.

[5] S.V.Hainsworth, T.Bartlett and T.F.Page, *Thin Solid Films*, 236 (1993) 214-218.

[6] J.S.Field and M.V.Swain, *J. Materials Res.*, 8 (1993) 297-306.

[7] J.S.Field and M.V.Swain, *J. Materials Res.*, 10 (1995) 101.

[8] E.Weppelmann, J.S.Field and M.V.Swain, *J. Materials Res.*, 8 (1993) 830-841.

[9] M.Shiwa, E.Weppelmann, A.Bendeli, M.V.Swain, D.Munz and T.Kishi, *Surface Coatings and Technol.*, 68/69 (1994) 598.

[10] B.Rother and D.A.Dietrich, *Phys. Stat. Sol.*, (a) 142 (1994) 389.

[11] M.Wittling, A.Bendavid, P.J.Martin and M.V.Swain, *Thin Solid Films*, (in press 1995).

[12] K.Komvopoulos, *J. Tribology (Trans ASME)*, 111 (1989) 430-439.

[13] N.Schwarzer and F.Richter, *J. Adhesion Sci. and Tech.*, (in press).

[14] M.V.Swain and E.Weppelmann, *MRS Symp. Proc.*, 308 (1993) 289-294.

[15] H.Gao, C-H.Chiu and J.Lee, *Int. J. Solids & Struct.*, 29 (1992) 2471.

[16] H.Djabella and R.D.Arnell, *Thin Solid Films*, 213 (1992) 205-219.

[17] H.Djabella and R.D.Arnell, *Thin Solid Films*, 213 (1993) 98-108.

[18] E.Weppelmann, *Thin Solid Films*, to be published.

[19] D.B.Marshall and A.G.Evans, *J. Applied Physics*, 58 (1984) 2632.

[20] I.Ashcroft and B.Derby, *J. Materials Science*, 28 (1993) 2989.

[21] J.L.Loubet, J.M.Georges and P.L.Kapsa, *Mechanics of Coatings, Proc. 16th Leeds-Lyon Symp. on Tribology, Lyon Sept. 1989*, Elsevier (1990), Paper XIX, p. 429.

[22] E.Weppelmann, X-Z.Hu and M.V.Swain, *J. Adhesion Sci. and Techn.*, 8 (1994) 611-624.

[23] M.D.Drory and J.W.Hutchinson, *Science*, 263 (1994) 1753.

[24] M.Wittling, *Diploma thesis at the IZSM*, University of Karlruhe, 1994.

INFLUENCE OF A FUNCTIONALLY GRADIENT SURFACE
ON CRACKING IN WC/Co HARDMETALS

J. Rohde[1,2] and S. Schmauder[2]

[1]Max-Planck-Institut für Metallforschung
Institut für Werkstoffwissenschaft
Seestrasse 92, D-70174 Stuttgart, Germany
[2]Staatliche Materialprüfungsanstalt (MPA), Universität Stuttgart
Pfaffenwaldring 32, D-70569 Stuttgart, Germany

INTRODUCTION

Tungsten carbide-cobalt hardmetals (WC/Co) are the most important cutting tool materials in modern technology. High hardness and wear resistance on the one hand side as well as high strength and chipping resistance on the other hand make them superior to high speed steels and ceramics[1,2].

Coating the inserts with thin hard layers (TiN or TiC) by chemical vapor deposition yields a longer edge life in comparison to uncoated inserts. However, residual stresses appear in the tool as a result of cooling after coating process which may cause initial cracks in the coating[3-7]. A tough surface zone underneath the coating with a high crack resistance prevents crack growth into the tool. Especially, cobalt enriched functionally graded surface zones provide an improved crack resistance, Figure 1. Therefore, cracks nucleated in the coating are frequently arrested in the ductile binder[8]. Additional cobalt striations in the gradient close to the surface ensure that cracks arrest before their length becomes critical.

Real structure modelling is much too expensive and time consuming for WC/Co hardmetals. On the other hand, assuming the gradient zone as a homogeneous layer will not reflect the real material behavior. Therefore, a mesomechanical model has been developed and is applied to simulate different gradient materials of different geometries.

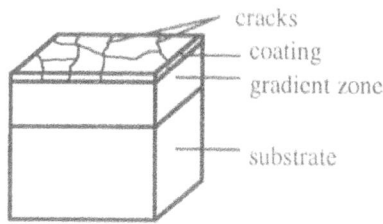

Figure 1. Model of coated hardmetal tool with gradient zone.

MESOSCOPIC MODEL

The linear elastic model presented in this section includes the variation of material properties in the coating, gradient zone and substrate. Isotropic material behaviour can be assumed for the coating and the substrate. In the gradient zone the material data are considered as functions of the distance from surface.

This feature can be approximated by a *multilayer model* where constant material data are assumed for layers of finite thickness. In the case of finite element modelling the finite element mesh has to be generated with respect to such layers. Each layer represents a material with constant properties. All problems of discontinuity at layer interfaces are introduced artificially by such a model.

In this work, a new approch is adopted: *functionally gradient elements* have been developed which allow to assign different material properties to each Gaussian integration point. Thus, the gradient bahaviour is approximated in each element which prevents from restrictions with respect to mesh generation. A high accuracy of results can be achieved even with coarse meshes. The mechanical behaviour of the gradient is better represented by a functionally gradient model compared to the multilayer model (Figure 2).

(a)

(b)

Figure 2. (a) Multilayer model and (b) functional gradient model (using functionally gradient elements).

Three hardmetal grades with various gradient zones have been investigated[6], such as:

- The grade $\gamma Free$ contains a Co enriched zone free of cubic carbides. The material properties are considered as cubic functions of the distance from the surface.

- The grade *CoStri* contains a Co enriched zone with additional cobalt striations. Firstly the thermo-elastic data of this grade are considered as continuous linear functions, in a second model the striations are taken into account where, the gradient is not described by a continous function. The striations are modelled by layers of $1 - 2\mu m$ thickness with the material data of cobalt.

• The grade *Conv* is a conventional hardmetal without modified surface zone which is investigated for comparison reasons.

The functions which have been used for the Young's modulus are shown in Figure 3.

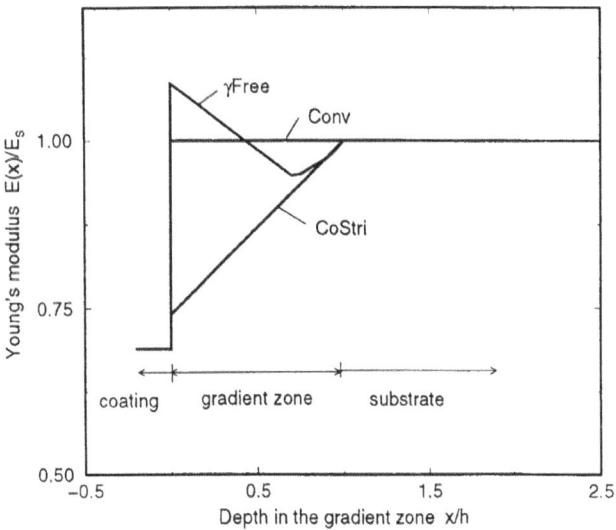

Figure 3. Relative Young's modulus as function of the depth in the gradient zone for the three grades.

The residual stresses have been calculated and the crack/gradient interaction has been studied for these grades.

RESIDUAL STRESSES

Cooling after coating process causes residual stresses due to the changing thermo-elastic material behaviour. A one–dimensional analytical model is introduced which yields a simple formula to approximate the residual stresses as a function from the distance from the surface. Additionally, the stress distribution in the tool is calculated numerically with the finite element method in an axisymmetric two dimensional model.

One-Dimensional Analytical Model

A simple model of an infinite plate with varying material data in the thickness direction is adopted to calculate the residual stresses in a material surface with a functional gradient. The principal stress components which have the same magnitude by symmetry reasons are located in the plane of the plate. The stress component in thickness direction is neglected. The tool is assumed to be symmetric with regard to the center plane.

Now, a cylinder of radius $R \gg h$ is considered, where $2h$ indicates the thickness of the cylinder (Figure 4). The non-zero stress components are the radial and the circumferential stresses which are identical.

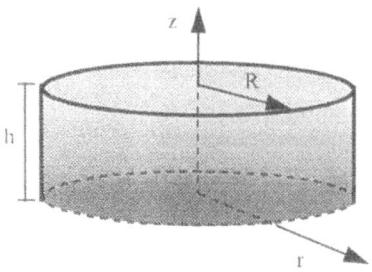

Figure 4. Cylindric model with gradient properties for calculating the residual stresses analytically.

The radial stress is varying in thickness direction and must fulfill the conditions of equilibrium

$$\int_0^h \sigma\, dx = 0 \tag{1}$$

The radial strain in the plate is constant

$$\varepsilon(z) = \varepsilon(\zeta) = const \qquad z, \zeta = 0\ldots, h \tag{2}$$

In the plane with the thickness coordinate z the conditions of plane stress are assumed

$$\varepsilon(z) = \frac{1-\nu(z)}{E(z)}\,\sigma(z) + \alpha(z)\Delta T \tag{3}$$

where ΔT is the cooling interval, $E(z)$ the Young's modulus, $\alpha(z)$ the thermal expansion coefficient and $\nu(z)$ the Poisson's ratio.

Combining equations (2) and (3) results in the local stress at the depth ζ

$$\sigma(\zeta) = \frac{E(\zeta)}{1-\nu(\zeta)}\left[\frac{1-\nu(z)}{E(z)}\sigma(z) + (\alpha(z) - \alpha(\zeta))\Delta T\right] \tag{4}$$

$\sigma(\zeta)$ must satisfy the equilibrium conditions (1):

$$\int \frac{E(\zeta)}{1-\nu(\zeta)}\,d\zeta\,\frac{1-\nu(z)}{E(z)}\sigma(z) + \int \frac{E(\zeta)(\alpha(z) - \alpha(\zeta))\Delta T}{1-\nu(\zeta)}\,d\zeta = 0 \tag{5}$$

This equation can be solved for $\sigma(z)$

$$\sigma(z) = \frac{E(z)}{1-\nu(z)}\,(\bar{\alpha} - \alpha(z))\Delta T \tag{6}$$

where

$$\bar{\alpha} = \frac{\int \frac{E(y)\alpha(y)}{1-\nu(y)}\,dy}{\int \frac{E(y)}{1-\nu(y)}\,dy} \tag{7}$$

Assuming constant but different material data in the coating, gradient zone and substrate, the residual stresses are constant in the different layers. For continuously varying elastic data $\sigma(z)$ is a continuous function.

Two-dimensional numerical model

Numerically, a two dimensional axisymmetric model is considered. For a cylinder with the diameter of $20mm$ and the height of $5mm$ a finite element mesh was created. The model comprises 1200 axisymmetric eight noded elements which accounts to a total of 3749 nodes. The coating with a thickness of $5\mu m$ is modelled with four element layers, while ten layers are used for the gradient zone with thickness $25\mu m$. The gradient zone is modelled with functionally gradient elements, which are shown in Figure 5. Thus any kind of gradient is easily approximated.

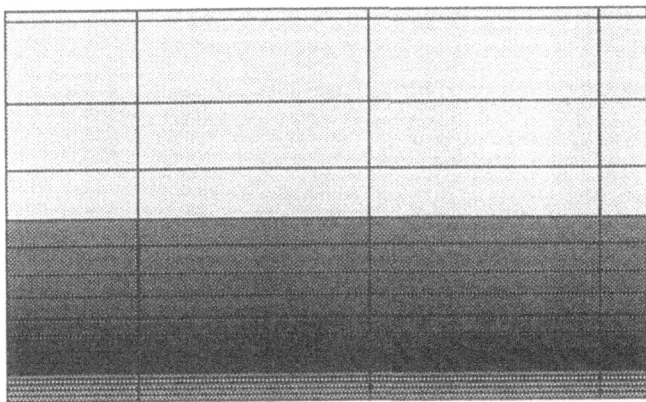

Figure 5. Part of the finite element mesh with functionally gradient elements.

Results and discussion

In the substrate, the average stresses resulting from cooling after the coating process are small compared to the stresses in the gradient zone and the coating layer. The residual stresses in the gradient zone depend on the composition within the gradient. Large differences of the material data between the gradient zone and the substrate and large variations of the material data in the gradient zone result in respective variations of the stresses.

Figure 6 shows the residual stresses which were calculated using formula (6) and the numerical model for three kinds of gradients: constant, linear varying and quadratic varying material data in the gradient zone.

Good agreement between the analytical and the numerical calculations are observed. Thus, the simple formula (6) can be used for a first approximation of residual stresses in a multi-layer or a gradient material.

In the coating, high tensile stresses are present which can lead to initial cracks through the coating. In the following, the crack/gradient interaction is studied with respect to different crack spacings, different grades and addititonal Co striations under mechanical and thermal loading.

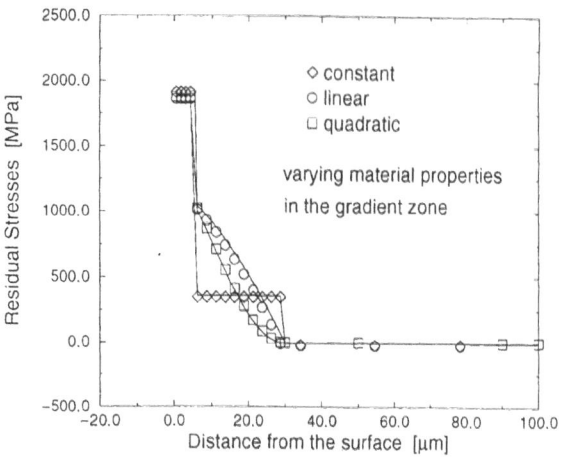

Figure 6. Residual stresses in the material, numerical and analytical calculation for different grades.

CRACK/GRADIENT INTERACTION

A mesomechanical model (Figure 7) with equally spaced multiple cracks of length a and the distance $2L$ is considered. The thickness of the coating is t, the thickness of the gradient zone is designated as h and that of the substrate as H.

Mechanical loading

The model is loaded due to a remotely applied uniform strain ε_0 perpendicular to the cracks. The energy release rate $\mathcal{G}$ is calculated for systematically varied crack

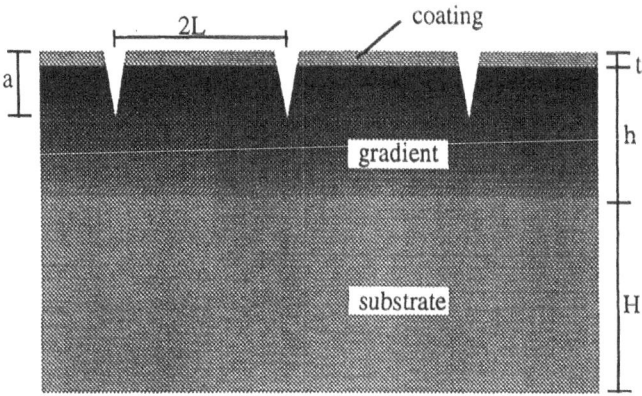

Figure 7. Mesomechanical model.

lengths between t and $t + h$. The average stress in the uncracked model is given as

$$\sigma = \frac{\varepsilon_0}{t + h + H} \int_{-t}^{h+H} \frac{E(z)}{1 - \nu(z)^2} \, dz \tag{8}$$

The energy release rate $\mathcal{G}$, normalized by the square of the average stress σ and the Young's modulus of the substrate E_S as well as the thickness of the gradient zone h results in a dimensionless function

$$\Psi\left(\frac{L}{h}, \frac{t}{h}, \frac{E_C}{E_S}, \frac{E(a)}{E_s}\right) = \frac{\mathcal{G}E_S}{\sigma^2 h} \tag{9}$$

of geometric and material data relations.

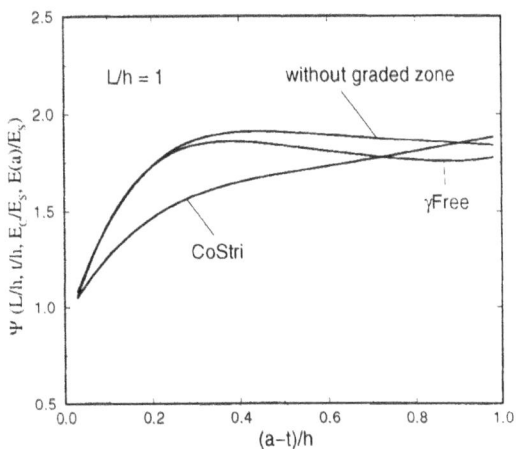

Figure 8. Mechanical loading – Comparison of different grades.

Figure 8 shows a comparison of Ψ for various gradient zones. Young's modulus which is a function of the distance from the surface is decreasing in the grade γFree and increasing in grade CoStri. Therefore, the energy release rate in grade CoStri is much lower than in the conventional grade in the first half of the gradient zone beneath the coating while minor differences between γFree and Conv are found for longer cracks. The differences in energy release rates are directly related to the elastic properties in each grade.

The energy release rate was found to be an increasing function of the crack spacing L/h (Figure 9). This crack driving force converges to an asymptotic maximum value, $\Psi_{L/h=10}$, for large crack spacings ($L/h \geq 10$). In the case of thin coatings only few pre-cracks are expected. Thus, the energy release rate of these surface cracks under mechanical loading is simply described by $\Psi_{L/h=10}$.

Additional cobalt striations in the gradient zone underneath the coating may strongly influence the fracture behaviour of the materials under consideration. The normalized energy release rate increases near the striation and decreases rapidly inside the cobalt, Figure 10. Since the value of Ψ reaches a pronounced minimum inside the striations, cracks will probably arrest inside the Co striations. Moreover, plasticity effects

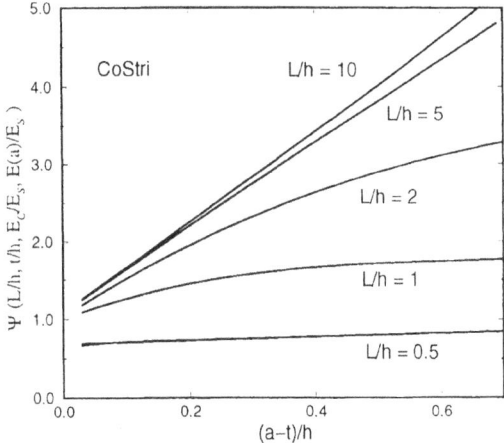

Figure 9. Mechanical loading – Comparison of different crack spacings.

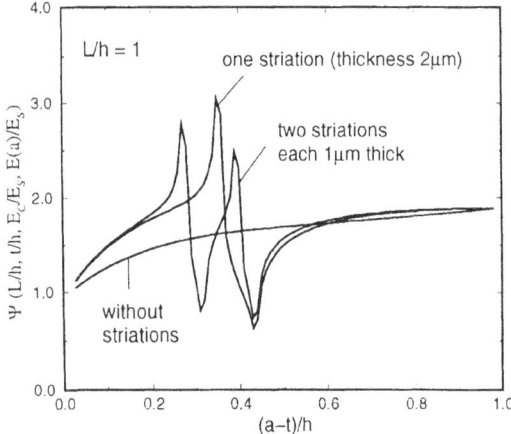

Figure 10. Mechanical loading – Influence of striations.

may consume additional fracture energy and thus result in further material improvement.

Thermal loading

Thermal loading is realized by applying a constant temperature change ΔT at above described mesomechanical model. In this case, the energy release rate is normalized by the square of the thermal mismatch between coating and hardmetal

$$\sigma_T = \frac{E_S}{1 - \nu_S}(\alpha_c - \alpha_S)\Delta T \tag{10}$$

The resulting dimensionless crack driving force is abbreviated as

$$\Phi(\frac{L}{h}, \frac{t}{h}, \frac{E_C}{E_S}, \frac{E(a)}{E_S}) = \frac{\mathcal{G}E_S}{\sigma_T^2 h}. \tag{11}$$

In these thermal calculations, the gradient zones (γFree and CoStri) are assumed to possess linear varying material properties. Young's modulus is a decreasing function in γFree and increasing in CoStri. The thermal expansion coefficient increases in γFree and decreases in CoStri. For Conv, the material properties in the hardmetal are constant. The influence of these parameters is reflected in the energy release rates of the considered cracks, Figure 11. Significant differences exist between strain energy release rates for cracks in CoStri and Conv. However, small differences are visible between γFree and Conv. This effect occurs as a result of the different ranges within which the material constants vary in the gradient zone. The influence of the Co striations is considerable also in these thermal calculations. Similarly as in the calculations with pure mechanical loading, the normalized energy release rate increases when the crack is in front of a striation and decreases when the crack tip is located inside the cobalt.

Combined loading

Assuming a linear elastic behaviour of the system, the total stress intensity factor

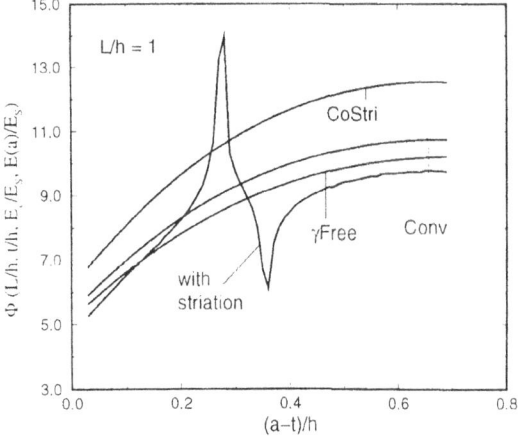

Figure 11. Thermal loading – Comparison of different grades.

K of regularly spaced cracks under combined thermomechanical loading is calculated as

$$K = K_\sigma + K_T \tag{12}$$

where K_σ is the stress intensity factor caused by applied mechanical load and K_T is that affected by the thermal load. From fracture mechanics analysis, the correlation

$$K(a) = \sqrt{\mathcal{G}\bar{E}(a)} \tag{13}$$

between the stress intensity factor $K(a)$ and the energy release rate $\mathcal{G}$ is well known. According to equation (9) and (11) the total stress intensity factor K possesses the form

$$\frac{K}{\sigma\sqrt{h}} = \left[\sqrt{\Psi} + \frac{\sigma_T}{\sigma}\sqrt{\Phi}\right]\sqrt{\frac{\bar{E}(a)}{\bar{E}_S}} \tag{14}$$

where Ψ and Φ were introduced in the previous section.

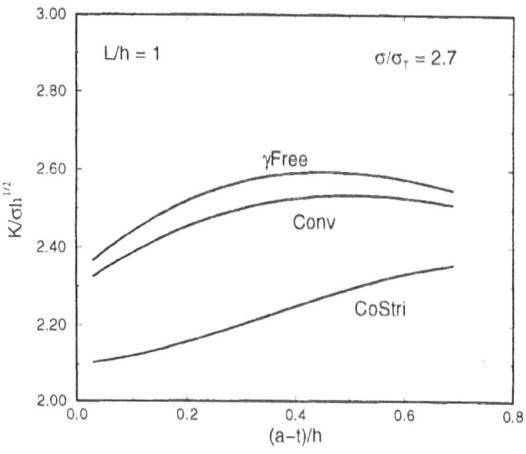

Figure 12. Combined loading – Comparison of different grades.

Results for the different grades are shown in Figure 12. The difference in the stress intensity factors is small between γFree and Conv due to the small intervall where the material data from γFree vary. CoStri distinguishes considerably from the other grades. The stress intensity factor is low, but continuously increasing. However the curves for Conv and γFree show a maximum in the stress intensity diagram.

CONCLUSION

A mesomechanical model of a coated hardmetal insert with a tough gradient surface zone was set up which considers the material properties as functions of the distance from surface. For numerical calculations with the finite element method functional gradient elements have been developed to adequately describe the continuously varying material properties in the gradient zone.

Residual stresses are generated in the insert during cooling after the coating deposition. These stresses can be estimated by a simple formula resulting from a one-dimensional model which can be solved analytically. The comparison of this analytical model and numerical calculations shows good agreement.

The thermal stresses in the coating layer are high and tensile in nature causing initial cracks perpendicular to the surface of the material. The crack/gradient interaction has been investigated for different types of gradient zones and different crack spacings in the case of thermal, mechanical and combined loadings. Crack spacing has an important influence on the crack driving force for dense crack pattern. For widely seperated initial cracks the energy release rate reaches a threshold at $L/h \geq 10$. The influence of additional Co striations in the gradient zone on the thermo-mechanical behaviour has been found to be beneficial in our initial elastic calculations.

ACKNOWLEDGMENTS

The work discussed in this paper is part of a collaborative study funded by the European Commission under Brite-Euram framework (contract No. BRE2-CT94-0620)

REFERENCES

1. H. Kolaska and K. Dreyer, Hartmetalle und ihr Einsatzfeld, DGM Fortbildungsseminar, Hannover (1992).

2. W. Schedler, "Hartmetall für den Praktiker", VDI Verlag, Düsseldorf (1988).

3. W. Koenig, K. Gerschwiler, R. Fritsch, Leistung und Verschleiss neuerer beschichteter Hartmetalle, in: "Beschichten und Verbinden in Pulvermetallurgie und Keramik," H. Kolaska, ed., VDI Verlag, Düsseldorf (1992).

4. A. Nordgren and A. Thuvander, Comb cracking of TiN, TiC and Al_2O_3-coated cemented carbide during milling of steel, Swedish Institute for Metals Research, Report no IM-2901, Stockholm (1992).

5. A. Nordgren, Influence of coating thickness in CVD-TiN coated cemented carbide upon comb cracking during milling of steel, Swedish Institute for Metals Research, Report no IM-3034, Stockholm (1993).

6. A. Nordgren and A. Thuvander, Residual stresses in CVD coated cemented carbide with structural gradient, Swedish Institute for Metals Research, Report no IM-2682, Stockholm (1990).

7. A. Nordgren and S. Jonsson, Residual stress in CVD TiN coatings on unworn and worn cemented carbide and the influence upon crack formation, Swedish Institute for Metals Research, Report no IM-3180, Stockholm (1994).

8. A. Nordgren, Influence of tailored microstructural variations in cemented carbide tools upon the propagation of cracks from surface coatings during machining of steel, Swedish Institute for Metals Research, Report no IM-2812, Stockholm (1991).

HIGH-TEMPERATURE FAILURE OF
VITREOUS-BONDED ALUMINA

C. Wolf[1] and H. Hübner

Arbeitsbereich Werkstoffphysik
Technische Universität Hamburg–Harburg
D-21071 Hamburg, Germany
[1]now at: Optische Werke G. Rodenstock
D-94209 Regen, Germany

ABSTRACT

Creep behaviour and failure characteristics of vitreous–bonded alumina as a representative material of glass–phase containing structural ceramics were studied for a wide range of grain sizes ($1 < d < 16\mu m$) and glassy phase contents ($1 < f_V < 20\%$). The time–to–failure, t_f, and the failure strain, ε_f, were found to strongly depend on the experimental variables (stress σ, temperature T) and on microstructural parameters (d, f_V). The increase in σ, f_V and d resulted in a decrease of both t_f and ε_f and caused a distinct embrittlement of the ceramics. The time–to–failure of all materials could be described by the Monkman–Grant relationship, $t_f \sim \dot{\varepsilon}_s^{-m}$, with the stationary creep rate $\dot{\varepsilon}_s$ and the Monkman–Grant exponent m. A straight line was found for each individual grain size which was independent of f_V. By including the failure strain, i.e. $t_f \cdot \dot{\varepsilon}_s^m \sim \varepsilon_f$, the data points fell on one single curve for all experimental conditions and microstructural parameters. It was concluded, therefore, that failure of all alumina ceramics investigated was controlled by the same mechanism, i.e. creep damage rather than subcritical crack growth. Grain coarsening not only lead to an increase in creep resistance but also to a reduction of the time–to–failure. Thus, both a low glassy–phase content and an intermediate grain size are recommended to obtain materials with optimum time–to–failure at elevated temperatures.

INTRODUCTION

Creep Damage

The failure of ceramic materials at high temperatures is caused by a combination of fracture mechanics and creep processes typically time and deformation-dependent in nature. The potential of a material for high-temperature applications as structural components mainly depends on its time–to–failure which typically should exceed 10^4 hours[1, 2]. Therefore, the basic damage mechanisms must be known and a precise lifetime prediction must be available if the benefits of ceramic materials are to be used in applications such as engines and gas turbines[3].

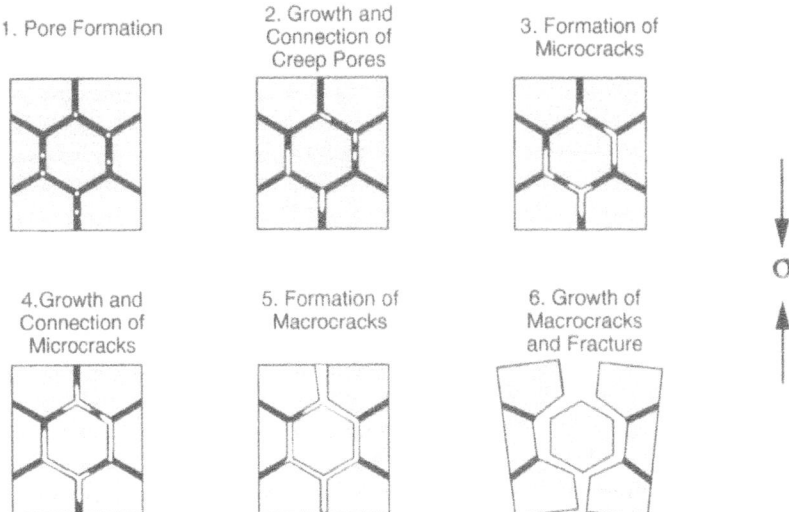

Figure 1 . Different stages of creep damage depicted schematically for a vitreous–bonded ceramic.

The development of creep damage comprises various partial steps which depend on each other. Fig. 1 schematically shows different stages of creep damage leading to final fracture. It is obvious that the different partial steps cannot be clearly separated, especially because they partially operate simultaneously. Which one of these processes will control the damage behaviour strongly depends on the applied stress and temperature as well as on the microstructure. The various partial mechanisms of Fig. 1 will not be dicussed in detail here. Rather, the reader is referred to a recent review of Chan and Page[3] where the current research on creep damage in structural ceramics is summarized.

Lifetime and Ductility

The application of ceramics as structural materials at high temperatures needs a reliable prediction of the time–to–failure, t_f. The stress dependence of t_f is usually described by a potential law,

$$t_f = B\sigma^{-a}, \tag{1}$$

where a is the stress exponent of the failure time and B is a material constant. Eq. 1 can be used in different stress regimes depending on the dominating damage mechanism[4, 5, 6]. At high stresses the time–to–failure is controlled by subcritical crack growth with $a = \nu$, ν being the stress exponent of subcritical crack growth. It typically exhibits large values ($\nu > 20$). Contrary, at low stresses the time–to–failure is determined by cavitational creep damage accumulated during deformation. Here the stress exponent a generally is smaller and comparable to the stress exponent n of the creep rate.

A common method to describe the time–to–failure controlled by creep damage is a potential law according to Monkman and Grant[7], relating t_f to the stationary or the minimum creep rate $\dot{\varepsilon}_s$:

$$t_f = C\dot{\varepsilon}_s^{-m}, \tag{2}$$

where C is a constant and m is the Monkman–Grant exponent. Sometimes a linear relationship between $\dot{\varepsilon}_s$ and $1/t_f$ is found, i.e. $m = 1$, indicating that the failure strain ε_f is a constant.

In many cases, the incorporation of ε_f in Eq. 2,

$$\frac{t_f \dot{\varepsilon}_s^m}{\varepsilon_f} = const \tag{3}$$

gives a better description of the experimental data[8]. If Eq. 3 is valid a lifetime prediction is only possible if both $\dot{\varepsilon}_s$ and ε_f are known[9].

The creep rate $\dot{\varepsilon}_s$ in Eq. 2 can be replaced by Norton's creep equation

$$\dot{\varepsilon}_s = A \frac{\sigma^n}{d^p} \exp\left(-\frac{Q_c}{RT}\right) \tag{4}$$

where A is a constant, d the grain size, and Q_c the activation energy of creep. Inserting Eq. 4 into Eq. 3 gives

$$t_f = C A^{-m} \frac{\sigma^{-mn}}{d^{-pm}} \exp\left(\frac{mQ_c}{RT}\right) \tag{5}$$

which now describes creep fracture controlled by the growth of creep pores and creep cracks[10]. Comparison of Eqs. 1 and 5 shows that the stress exponent of the failure time, a, is the product of the creep stress exponent, n, and the exponent of the Monkman–Grant relation (Eq. 2), m:

$$a = mn. \tag{6}$$

Creep Damage of Vitreous-Bonded Alumina

Comprehensive studies on creep damage of vitreous-bonded alumina were presented by Wiederhorn et al.[11, 12] and, more recently, by Fett et al.[13]. Both groups measured the time–to–failure under 4–point bending and showed that their results could be well described by Eq. 2, the exponent m having values near unity. Furthermore, Wiederhorn et al.[11, 12] found that the product $t_f \cdot \dot{\varepsilon}_s^m$ was both stress and temperature–dependent. Fett et al.[13] also investigated the stress dependence of t_f according to Eq. 1. At high stresses, the exponent a had large values between 20 and 60 which was attributed to subcritical crack growth, whereas at low stresses, a was of about 4, a figure typical of creep–induced damage[13]. Time–to–failure data of vitreous–bonded or pure alumina under compression are not known

EXPERIMENTAL DETAILS

Materials and Sample Preparation

The materials tested in this work were fabricated by sintering isostatically pressed mixtures of alumina powder* with amounts of 1, 5, 10 and 20 vol-% of a powdered silicate glass†. The glass was prepared from commercially available glass tubes by successive crushing, milling in a ball mill, sieving ($< 5\mu$m) and a final sedimentation to a particle size of $< 1\mu$m. The samples were sintered in air at temperatures between 1450 and 1650 °C up to 16 h to achieve different grain sizes.

---

*Ceralox—HPA—0.5, Condea Chemie, Germany
†Supremax, Schott Glaswerke, Germany

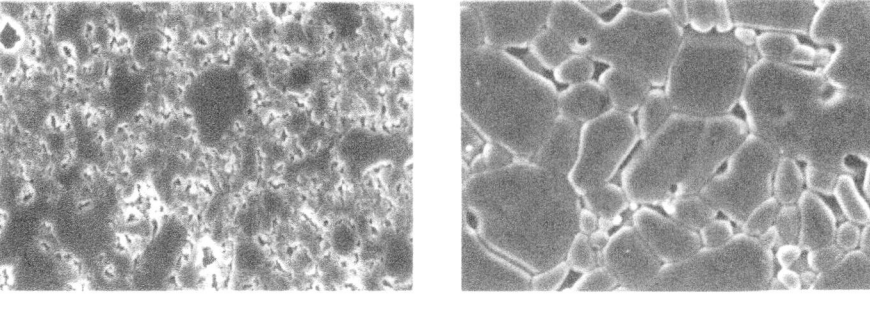

$T = 1450\ ^\circ\text{C},\ t = 0\ \text{h},\ f_V = 5\ \%;\quad$ 2μm $\quad T = 1500\ ^\circ\text{C},\ t = 16\ \text{h},\ f_V = 5\ \%;\quad$ 2μm

Figure 2 . Microstrucure of vitreous-bonded alumina.

The density of all materials was measured using the Archimedes method and was found to lie between 93 and 95 % of the theoretical. Typical microstructures of the ceramics prepared for the creep tests are shown in Fig. 2. The micrographs of Fig. 2a and 2b show microstructures obtained by different sintering schedules used to adjust different grain sizes. The glassy phase is continuously distributed around the grains. Since the glassy phase was removed by the etching procedure, the occurrence of glass pockets at triple points is obvious.

Mechanical Testing

Compressive creep tests were performed on samples with the dimension of $3.2 \times 3.2 \times 9\ \text{mm}^3$ at stresses ranging from 15 to 200 MPa and temperatures between 1100 to 1250 °C . From these tests, the stationary creep rate, $\dot{\varepsilon}_s$, the time–to failure, t_f, and the failure strain, ε_f, were determined.

RESULTS

Creep Behaviour

The creep behaviour under compression was characterized by the occurrence of the three typical creep regimes: (I) primary creep with a decreasing creep rate as a function of time or strain, respectively; (II) stationary creep with a constant creep rate, and finally (III) a pronounced tertiary creep regime with increasing creep rate leading to final rupture. Fig. 3 shows typical creep curves presented as creep rate vs. strain. As the stress is increased, the steady–state regime is decreased and tertiary creep increasingly gains importance, leading to a minimum creep rate only and to premature failure. Similar shapes of creep curves were also reported for other vitreous–bonded alumina under compression[14].

Creep curves obtained at different stresses become very similar to each other when plotted vs. a relative strain, i.e., the strain normalized to the failure strain, ε_f, as shown in the right part of Fig. 3. When expressed in units of ε_f the fractions of primary, stationary, and tertiary creep approximately have the same extensions for all stresses of the stress range investigated.

The stress dependence of the steady–state or minimum creep rate of aluminas of different grain size is shown in Fig. 4. Increasing grain size obviously leads to decreasing

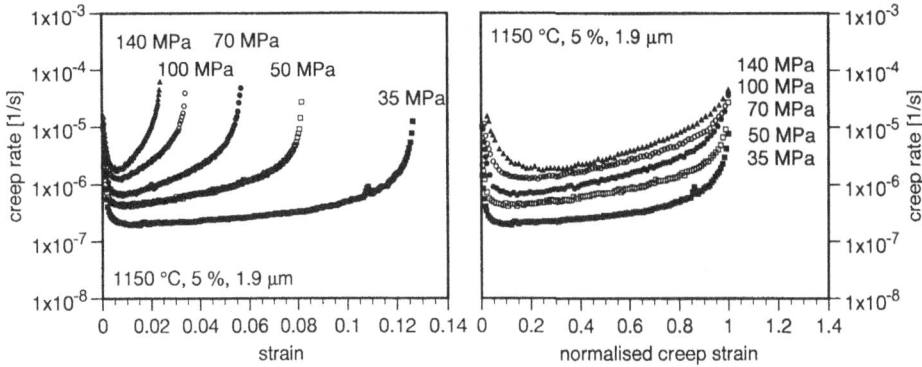

Figure 3 . Creep curves as creep rate, $\dot{\varepsilon}$ *vs.* strain, ε (left) and as creep rate, $\dot{\varepsilon}$ *vs.* normalised strain, $\varepsilon/\varepsilon_f$ (right).

creep rates and increasing stress exponents. For a detailed description of the effect of the microstructural parameters (d, f_V) and the experimental variables (σ, T), the reader is referred to Refs. (15–17).

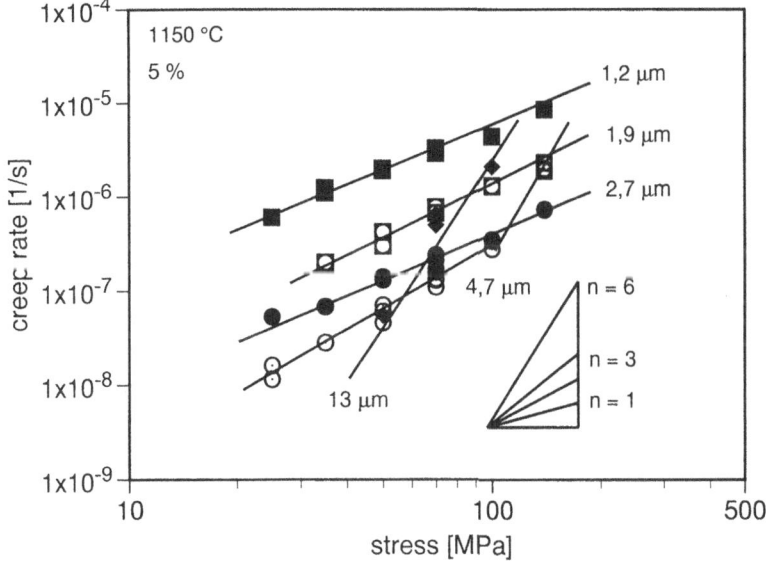

Figure 4 . Creep rate, $\dot{\varepsilon}_s$ *vs.* stress, σ for different grain sizes at a volume content of glassy phase of 5 %.

According to these references, the creep behaviour of the vitreous–bonded alumina ceramics could be characterized as follows[15, 16]: at low stresses, fine grain size and small volume content of glassy phase the deformation was viscous, the stress exponent being close to one (see also Fig. 4). The rate–controlling deformation mechanism was found to be solution–precipitation creep. With increasing stress, grain size, and volume content

503

Table 1 . Activation energy Q_c and stress exponent n for different microstructures of vitreous-bonded alumina

grain size	fine (1-3 μm)		coarse ($> 10\ \mu$m)	
$f_V =$	1 and 5 %	10 and 20 %	1 and 5 %	10 and 20 %
n (low stress)	1 — 1.5	1 — 1.5	1 — 2	1 — 3
n (high stress)	1 — 1.5	1 — 3	2 — 6	4 — 6
Q [kJ/mole]	660 — 690	570 — 600	640 — 660	580 — 660

of glassy phase the stress exponent increased to values up to 6, and the ductility was largely reduced, an effect typical of the increasing dominance of damage mechanisms.

The stress exponent n and the activation energy of creep Q_c of Eq. 4 of materials having different microstructures are summarized in Table 1. The Table also demonstrates that the stress exponent increased from about one to values up to six as the stress, grain size and volume content of glassy phase were increased. The activation energies in Table 1 were generally measured at low stresses and were shown to be characteristic for solution–precipitation creep[15].

Development of Creep Damage

The formation and growth of creep cavitaties is an important step in high-temperature failure of technical ceramics. Fig. 5 and 6 show four typical stages of the development of creep damage occurring during deformation.

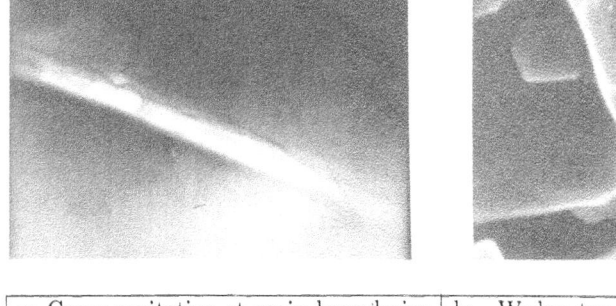

a: Creep cavitation at grain boundaries (TEM); 100nm	b: Wedge–type cavity between grain boundaries (SEM); 1µm

Figure 5 . Typical regimes of cavitation occurring during a creep test.

Fig. 5a shows cavities in a glass–filled grain boundary which are seperated by thin glass bridges. In Fig. 5b a wedge–type cavity between grain boundaries can be seen that is typical for the beginning of microcrack formation and growth. If the sample is strained further, microcracks coalesce to macrocracks, as shown in Fig. 6a for the sample surface. Finally, it was observed that most of the compressive creep samples failed by shearing as shown in Fig. 6b at a moment shortly prior to rupture.

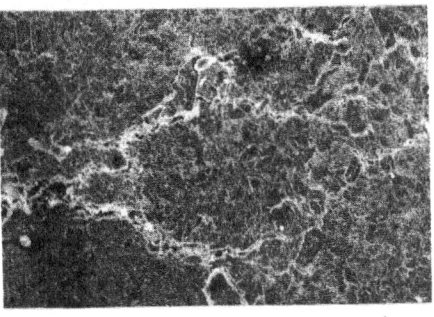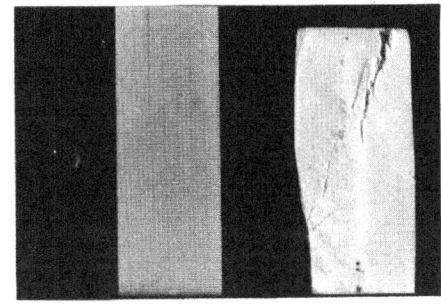

| a: Cracks at the sample surface (SEM), $[\rightarrow \sigma \leftarrow]$;
 50μm | db: Creep sample before (left) and after a creep test (right);
 2mm |

Figure 6 . Typical regimes of macroscopic damage formed during a creep test.

Time–to–Failure and Ductility

The strong effect of stress, grain size and glassy phase content on the shape of the creep curves and the stress exponent shown above also imply a strong dependence of the time–to–failure and the ductility on the experimental and microstructural parameters.

In Fig. 7 the stress dependence of the time-to-failure t_f is plotted for different grain sizes (above) and different volume fractions of the glassy phase (below). The time–to–failure obviously decreases with increasing stress following the power law described in Eq. 1, with a stress exponent a between 2.8 and about 6. The grain–size dependence of t_f can be described as follows: the time–to–failure is largest at an intermediate grain size of 2.7 μm, whereas both a decrease of grain size to 1.9 μm and an increase to 4.7 μm show only a minor effect on t_f. Materials having both finest (1.2 μm) and coarsest grain size (13 μm) exhibit the lowest level of t_f in Fig. 7 (upper part). At fine and intermediate grain sizes the stress exponent a has a value of about 3 and increases to about 5 for the coarse–grained ceramics. The effect of the volume content of glassy phase is more definite since t_f unequivocally decreases with increasing f_V.

The failure strain ε_f also was found to strongly depend on stress and microstructure as shown in Fig. 8. Generally, ε_f decreases with increasing stress, increasing grain size (Fig. 8, above), and increasing volume content of the glassy phase (Fig. 8, below). These results are in accordance with theoretical considerations of Evans et al.[18, 19], since both coarse grains and the existence of glass–filled regions lead to microstructural inhomogeneities which typically are origins of cavitation and creep cracks and therefore strongly reduce the ductility of ceramic materials.

Monkman–Grant Behaviour

The concept of characterizing the time-to–failure as a stress–dependent quantity was not fully satisfactory, since the curves shown in Fig. 7 showed a large spread for different grain sizes and volume contents of the glassy phase. A different concept of analysing failure times is that of Monkman and Grant[7], Eq. 2. To check its validity for the materials studied here, time–to–failure was plotted as a function of the stationary or the minimum creep rate $\dot\varepsilon_s$. Fig. 9 shows the data obtained over the whole range of parameters (e.g. stress, temperature, grain size, and volume content of glassy phase) investigated in the present work.

505

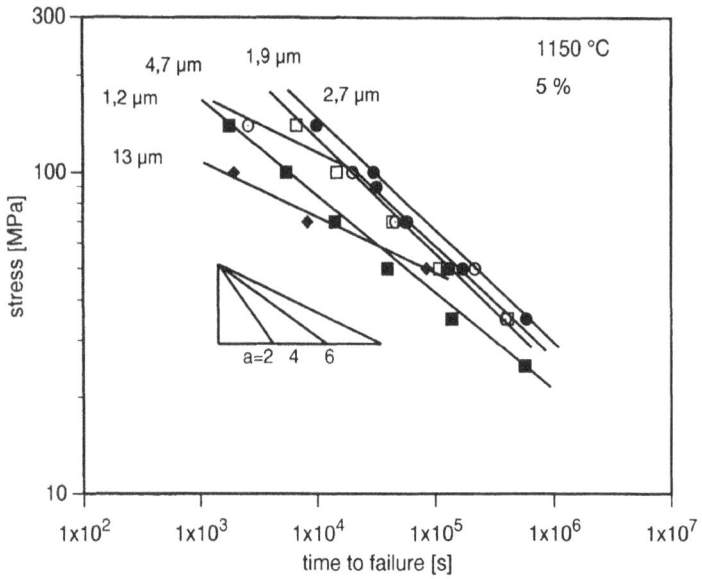

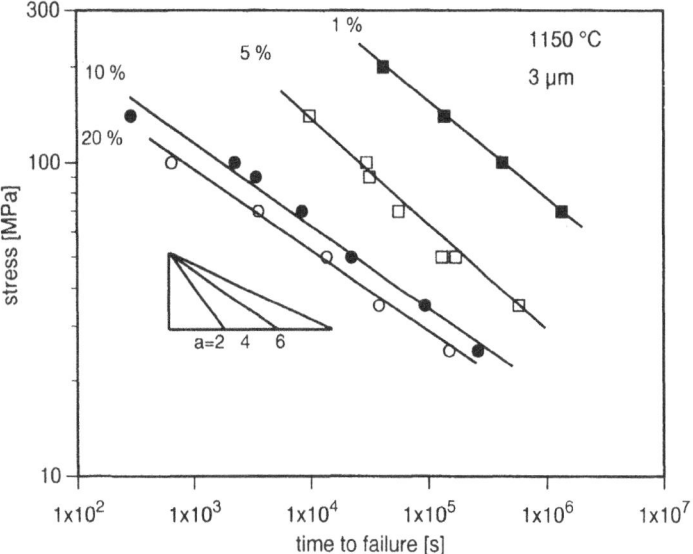

Figure 7 . Stress σ vs. time-to-failure t_f for different grain sizes at a volume content f_V of 5 % (above) and for different volume contents at a grain size of approx. 3 μm (below).

The data plotted in Fig. 9 allow a grouping of the results along straight lines for each individual grain size, but independently of stress and temperature. In comparison to Fig. 7, the data arrangement in the Monkman–Grant plot is much more homogeneous. The finding that the straight lines of Fig. 9 are independent of stress and temperature can be taken as evidence that the failure mechanism of these ceramics is mainly determined by the creep process[10, 20].

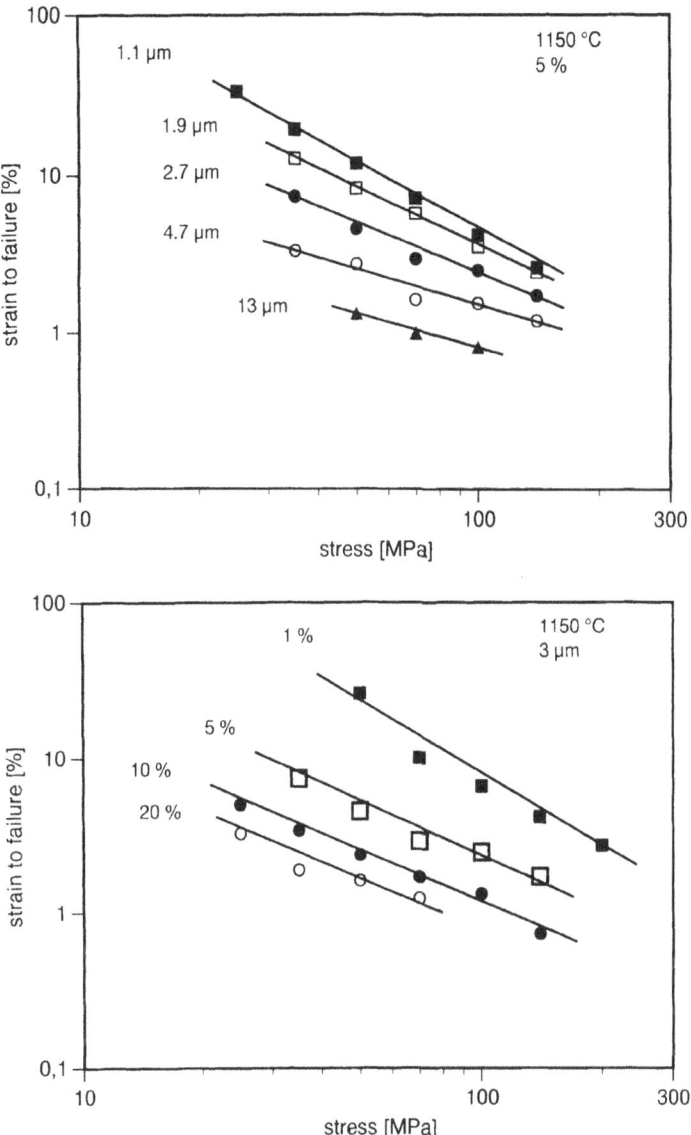

Figure 8 . Strain-to-failure ε_f *vs.* stress σ for different grain sizes at a volume content f_V of 5 % (above) and for different volume contents at a grain size of approx. 3 μm (below).

To check Eq. 6 values of the stress exponent of the creep rate, n, determined directly from creep results were compared with those calculated from the results of creep failure presented in Fig. 7 and Fig. 9, a and m, i.e. from the relationship $n = a/m$. Table 2 contains these exponents and demonstrates that the n values obtained by the different methods are nearly identical. This observtion further substantiates the interdependence between the creep and the failure processes of the alumina ceramics tested.

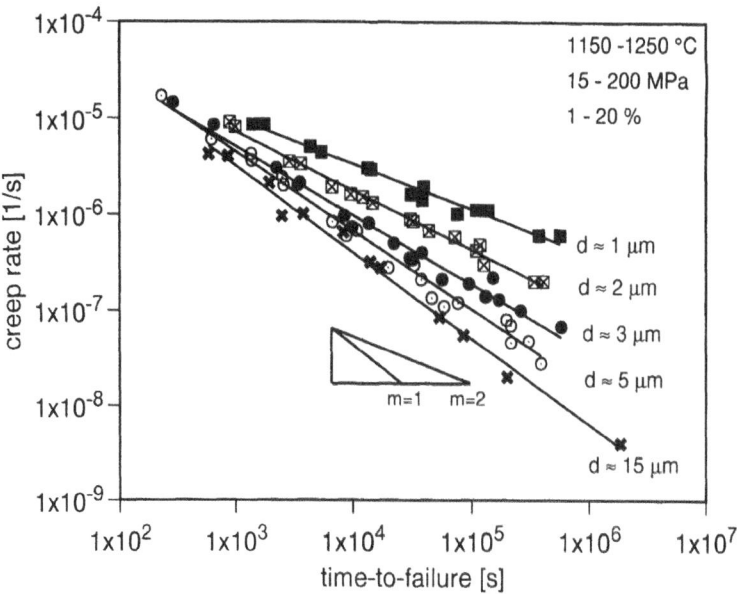

Figure 9 . Monkman-Grant plot (creep rate $\dot{\varepsilon}_s$ vs. time-to-failure t_f) for different grain sizes and glassy phase contents over the whole stress and temperature range studied in this work.

The deviation of the exponent m from unity in Fig. 9 is due to the fact that the failure strain themself depends on stress and temperature. Thus, the so-called Monkman–Grant product, e.g. $\dot{\varepsilon}_s t_f$, is not a constant but becomes dependent on the failure strain.

Table 2 . Data of the stress exponent n at a glassy phase content of 5 %, determined indirectly from the stress exponent of the time–to–failure law, a, and the exponent of the Monkman–Grant relationship, m, as well as directly from creep tests[15, 16].

grain size [μm]	1.2	1.9	2.7	4.7	13
a	3.3	3.0	2.8	3.6	5.4
m	2.2	1.8	1.7	1.2	1.0
a/m	1.5	1.7	1.6	3.0	5.4
n	1.4	1.7	1.5	2.6	5.2

The incorporation of the failure strain into the Monkman Grant relationship according to Eq. 3 is shown in Fig. 10. The figure clearly demonstrates that now all experimental data fall on a single straight line, independently of the microstructural features of the materials tested (d and f_V) and of the experimental variables employed (σ and T). The slope of this straight line is nearly unity. Allthough the log–log scale of Fig. 10 helps in smoothing the data the compact representation of the data points is surprising, especially when the great variability of the shape of the creep curves and of the time–to–failure and failure strain data is kept in mind.

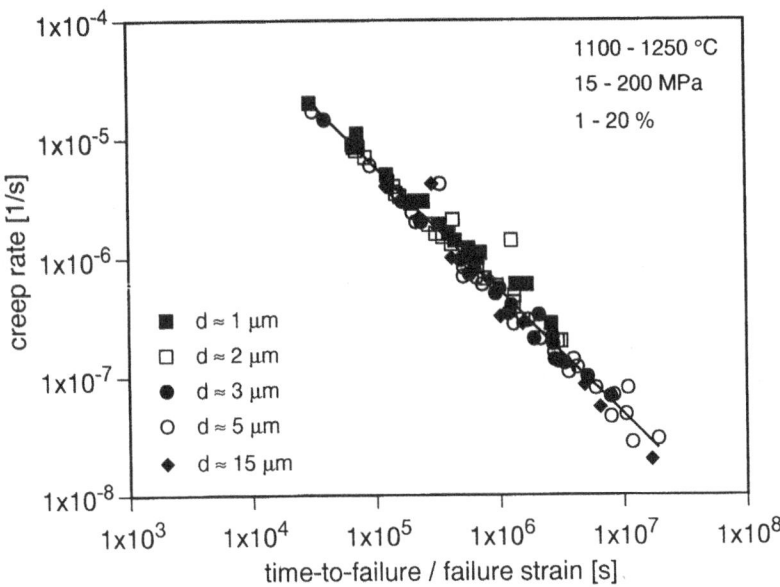

Figure 10 . Creep rate $\dot{\varepsilon}_s$ *vs.* time-to-failure divided by the failure strain t_f/ε_f for the whole range of experimental and microstructural parameters of this work.

DISCUSSION

The possibility of characterizing the failure behaviour of vitreous–bonded alumina having a wide range of grain sizes and glassy phase contents by a single curve in Fig. 10 is indicative of a close relationship between the deformation processes and those of creep damage. The agreement of the data seems only to be possible if failure of all samples was controlled by the same predominant damage mechanism. A first hint to that conclusion is the similarity of the normalized creep curves in Fig. 3b. The relative portions of the three different stages of creep deformation are approximately identical. Furthermore, the comparison of the stress exponents in Table 2 shows a direct relationship between creep and failure mechanisms. The values of the stress exponent of the time–to–failure, a, are all between 2.5 and 6, which is typical of damage by creep and not by subcritical crack growth that would demand a values above 20 and more[5, 11, 13].

The exponent of approximately one in the slope of the curve in Fig. 10 implies a linear relationship between the stationary creep rate $\dot{\varepsilon}_s$ and the average creep rate $\bar{\dot{\varepsilon}} = \varepsilon_f/t_f$. Fig. 11 schematically shows the relation between $\dot{\varepsilon}_s$ and $\bar{\dot{\varepsilon}} = \varepsilon_f/t_f$.

The ratio between the failure strain and the time–to–failure defines a creep rate which obviously is a multiple of the stationary creep rate. Furthermore, the occurrence of a single curve for all experimental data defines a constant ratio between $\dot{\varepsilon}_s$ and $\bar{\dot{\varepsilon}} = \varepsilon_f/t_f$ which should not depend on stress, temperature, grain size, and glassy phase content. Fig. 12 shows the stationary creep rate *vs.* average creep rate on a linear scale. The slope of the curve is about 0.6, i.e. the stationary creep rate lies at a level of 60 % of the average creep rate.

The type of damage mechanism which controls the time–to–failure can be identified with the help of damage mechanism maps where the failure strain ε_f is plotted *vs.* the product of stationary creep rate and time–to–failure, $\dot{\varepsilon}_s t_f$[21]. By means of this diagram

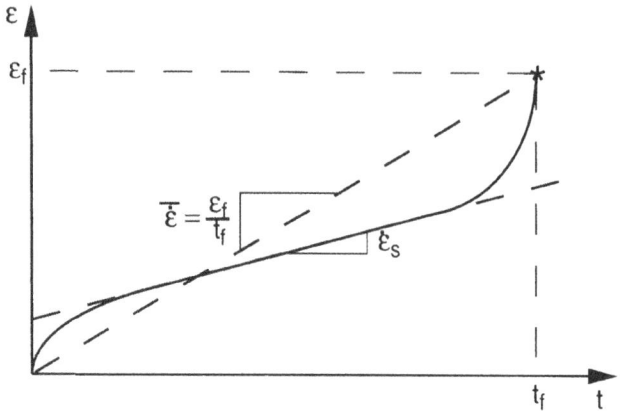

Figure 11 . Schematic of a creep curve defining the stationary creep rate $\dot{\varepsilon}_s$ and the average creep rate $\bar{\dot{\varepsilon}} = \varepsilon_f/t_f$.

a damage tolerance parameter λ was defined[21] which is given by

$$\lambda = \frac{\varepsilon_f}{\dot{\varepsilon}_s t_f}. \tag{7}$$

Eq. 7 can be interpreted as the ratio of the average creep rate to the stationary creep rate, i.e. the inverse of the slope in Fig. 12. From Fig. 12, $\lambda \approx 1.67$ is obtained which is a value typical of failure by the creep damage mechanism according to the damage mechanism map[21]. A further consequence of the constant ratio between $\dot{\varepsilon}_s$ and $\bar{\dot{\varepsilon}}$ is the self–similarity of the creep curves shown in Fig. 3b.

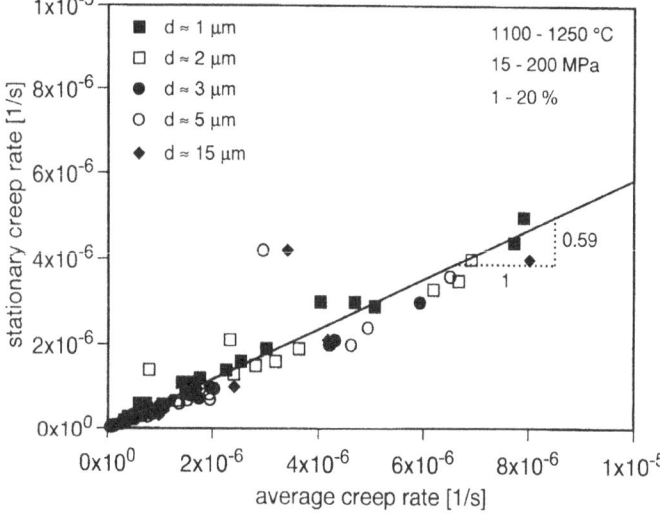

Figure 12 . Stationary creep rate $\dot{\varepsilon}_s$ vs. average creep rate $\bar{\dot{\varepsilon}} = \varepsilon_f/t_f$.

IMPLICATIONS FOR MATERIALS DESIGN

An advantage of the analysis shown in Figs. 10 and 12 respectively is the opportunity to characterize the damage behaviour of the alumina ceramics studied by only one single equation. A certain disadvantage to the prediction of the lifetime in a given application is that both the time–to–failure and the failure strain are materials parameters only available after the termination of the creep test by failure. Therefore, the use of the stress dependent time–to–failure (see Fig. 7) or the Monkman–Grant relationship (see Fig. 9) allows a less time–consuming determination of the creep lifetimes.

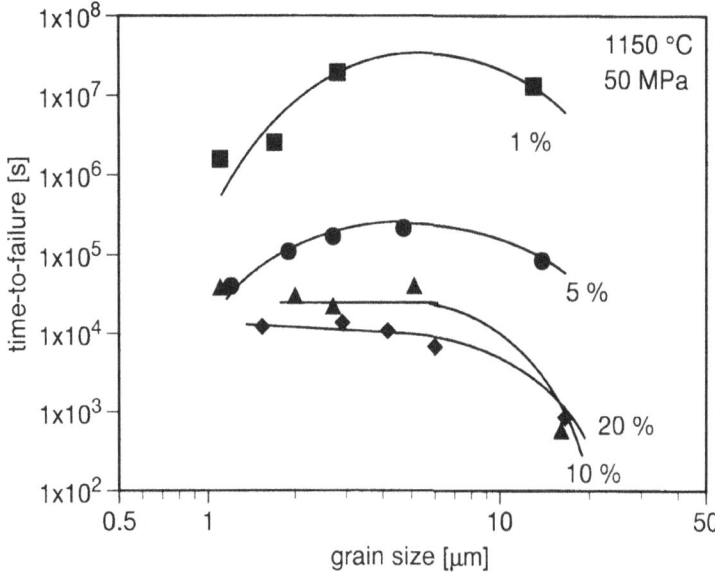

Figure 13 . Time–to–failure, t_f, vs. grain size, d, for different volume contents of glassy phase at constant stress and temperature.

From the point–of–view of optimizing the microstructure with respect to high–temperature applications, Fig. 7 has shown that an intermediate grain size and a volume content of glassy phase as small as possible will provide the longest time–to–failure. The situation of an optimum microstructure is depicted in Figs. 13 and 14 which show the time–to–failure and the time–under–creep ("straining time") conditions to reach a strain of 1 % as a function of grain size for different volume contents of glassy phase, respectively. Usually, the design of engineering structures for high–temperature applications is limited by the failure of a component or by a maximum allowable strain of a component under application. Then, the maximum service time is limited by the parameters shown in Fig. 13 and Fig. 14.

The creep rate of the vitreous–bonded alumina ceramics studied in this work has been shown to be controlled by the solution–precipitation mechanism[15, 16]. The grain–size dependence of the creep rate is characterized by $\dot{\varepsilon}_s \sim d^{-1}$ in the reaction–controlled case and by $\dot{\varepsilon}_s \sim d^{-3}$ in the diffusion–controlled case. It can be expected, therefore, that with increasing grain size, both characteristic times of Figs. 13 and 14 should also increase. Nevertheless, there exists an optimum grain size for maximum resistance to creep damage, which slightly depends on the volume content of glassy phase. At finer grain size, the creep rate is to fast for a prolonged time and at coarser grain size damage mechanisms become more predominant. On the other hand, decreasing the volume content of glassy phase generally is more beneficial to a better damage resistance since the existence of glassy phase enhances both the creep rate and the damage rate.

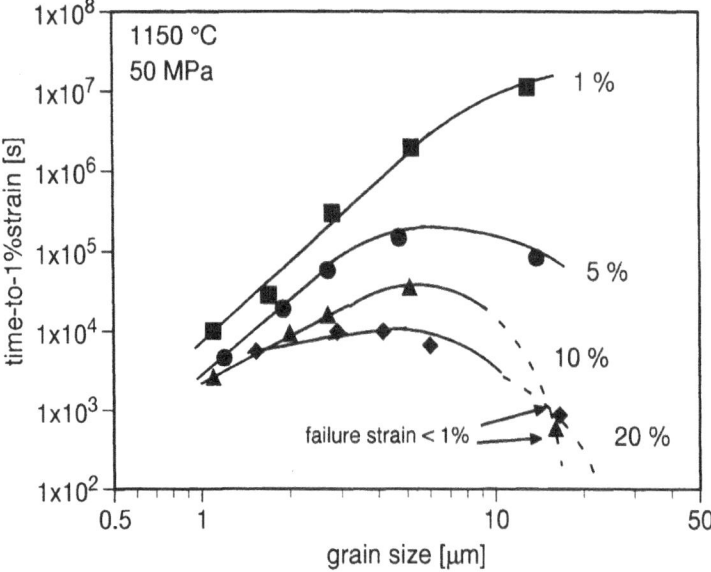

Figure 14 . Straining time to a strain of 1 %, $t_{\varepsilon=1\%}$, *vs.* grain size, d, for different volume contents of glassy phase at constant stress and temperature.

CONCLUSIONS

The creep behaviour and the failure characteristics of vitreous–bonded alumina were studied for a wide range of grain size ($1 < d < 16\mu m$) and glassy phase content ($1 < f_V < 20\%$). It was found that the time–to–failure t_f and the failure strain ε_f strongly depended on the experimental variables (stress σ, temperature T) and on microstructural parameters (d, f_V). The increase in σ, f_V and d resulted in a decrease of both t_f and ε_f and caused a distinct embrittlement of the ceramics.

The failure characteristics were expressed in terms of the stress dependence of t_f which could be described by a power–law relationship, $t_f \sim \sigma^{-a}$, where the stress exponent a had values between 2.8 and 7.1. Furthermore, the lifetime of all materials studied could be described by the Monkman–Grant relationship, $t_f \sim \dot{\varepsilon}_s^{-m}$, with the stationary creep rate $\dot{\varepsilon}_s$ and the Monkman–Grant exponent m. For each individual grain size, a single straight line was found in the Monkman–Grant plot that turned out to be essentially independent of f_V. By using the modified Monkman–Grant relationship, $t_f \dot{\varepsilon}_s^m \sim \varepsilon_f$, one single curve was obtained for all experimental conditions and microstructural parameters. From the existence of this "master curve" it was concluded that failure of all alumina ceramics investigated was controlled by the same creep damage mechanism.

It could be shown that grain coarsening not only lead to an increase in creep resistance but also to a reduction of the time–to–failure. Thus, the optimum grain size must balance these opposite influences. It was concluded, therefore, that a microstructure characterized by both a small glassy phase content and an intermediate grain size would yield a material optimized with respect to lifetime at elevated temperatures.

REFERENCES

1. A. C. F. Cocks and M. F. Ashby. Creep Fracture by Void Growth. In A. R. S. Ponter and D. R. Hayhorst, editors, *Creep in Structures (IUTAM Symp.. Leicester, UK)*. pages 368 – 386. Springer, 1980.

2. D. Carruthers and L. Lindburg. Critical Issues for Ceramics for Gas Turbines. In V. J. Tennery, editor, *Ceramic Materials Components for Engines (Proc. 3rd Int. Symp.)*, pages 1258 – 1272. Am. Ceram. Soc., 1989.

3. K. S. Chan and R. A. Page. Creep Damage Development in Structeral Ceramics. *J. Am. Ceram. Soc.*, 76(4):803 – 826, 1993.

4. R. W. Davidge. Perspectives for Engineering Ceramics in Heat Engines. *High. Temp. Technol.*, 5(1):13 – 21, 1987.

5. G. Grathwohl. Regimes of Creep and Slow Crack Growth in High-Temperature Rupture of Hot-Pressed Silicon Nitride. In R. E. Tressler and R. C. Bradt, editors, *Deformation in Ceramic Materials 2*, volume 18 of *Mater. Sci. Res.*, pages 573 – 586. Plenum, 1984.

6. B. J. Dalgleish, E. B. Slamovich, and A. G. Evans. Duality in the Creep of a Polycrystalline Alumina. *J. Am. Ceram. Soc.*, 68(11):575 – 581, 1985.

7. F. C. Monkman and N. J. Grant. An Empirical Relationship Between Rupture Life and Minimum Creep Rate in Creep Rupture Tests. *Proc. ASTM.* 56:593   605, 1956.

8. F. Dobes and K. Milicka. The Relation between Minimum Creep Rate and Time to Fractures. *Metal Sci.*, pages 382 – 384, 1976.

9. F. Povolo. Comments on the Monkman-Grant and the Modified Monkman-Grant Relationships. *J. Mater. Sci.*, 20:2005 – 2010, 1985.

10. S. M. Wiederhorn, B. J. Hockey, and T.-J. Chuang. Crepp and Creep Rupture of Structural Ceramics. In S. P. Shah, editor, *Toughening Mechanisms in Quasi-Brittle Materials*, pages 555 – 576. Kluwer Academic Publisher, 1991.

11. S. M. Wiederhorn, B. J. Hockey, R. F. Krause jr., and K. Jakus. Creep and Fracture of a Vitreous-Bonded Aluminum Oxide. *J. Mater. Sci.*, 21:810   824. 1986.

12. S. M. Wiederhorn, B. J. Hockey, and R. F. Krause jr. Influence of Microstructure on Creep Rupture. In J. A. Pask and A. G. Evans, editors, *Ceramic Microstructures +86: Role of Interfaces*, volume 21 of *Mater. Sci. Res.*, pages 795   806. Plenum Press, 1987.

13. T. Fett, M. Mißbach, and D. Munz. Failure Behavior of Al_2O_3 with Glassy Phase. *J. Europ. Ceram. Soc.*, 13:197 – 209, 1994.

14. D. R. Clarke. High-Temperatur Deformation of a Polycrystalline Alumina Containing an Intergranular Glass Phase. *J. Mater. Sci.*, 20:1321 –1332, 1985.

15. C. Wolf. *Kriechverhalten von glasphasehaltigem Aluminiumoxid*, volume 386 of *series 5*, VDI-Verlag, Düsseldorf, 1995. (PhD-Thesis, Technical University Hamburg-Harburg, 1994), in German.

16. C. Wolf and H. Hübner. Creep Mechanisms in Vitreous-Bonded Alumina. *J. Am. Ceram. Soc.*, 1995. submitted.

17. C. Wolf and H. Hübner. Creep Mechanisms in Alumina Containing a Glassy Grain Boundary Phase. In A. P. Duran and J. F. Fernandez, editors, *Engineering Ceramics*, volume 3 of *Third Euro-Ceramics*, pages 555 – 560. Faenza Editrice Iberica S. L., 1993.

18. A. G. Evans, J. R. Rice, and J. P. Hirth. Suppression of Cavity Formation in Ceramics: Prospects for Superplasticity. *J. Am. Ceram. Soc.*, 63(7-8):368 – 375, 1980.

19. A. G. Evans and W. Blumenthal. High Temperatures Failure in Ceramics. In R. C. Bradt, A. G. Evans, and D. P. H. Hasselman, editors, *Measurements, Transformations, and High- Temperature Fracture*, volume 6 of *Fractures Mechanics of Ceramics*, pages 423 – 448. Plenum Press, 1983.

20. S. M. Wiederhorn, B. J. Hockey, and D. C. Crammer. Transient Creep Behavior of Hot Isostatically Pressed Silicon Nitride. *J. Mater. Sci.*, 28:445 – 453, 1993.

21. M. F. Ashby and B. F. Dyson. Creep Damage Mechanisms and Micromechanics. In S. R. Vallori et al., editor, *Adv. Fract. Res. '84 - Proc. ICF6*, volume 1 of *3*, pages 46 – 85. Pergamon Press, 1984.

TENSILE CREEP OF SINTERED SILICON NITRIDE CERAMICS

AT ELEVATED TEMPERATURES

K. Hatanaka[1], H. Shiota[2], and K. Oshita[1]

[1]Department of Mechanical Engineering
Yamaguchi University
2557 Tokiwadai, Ube City, 755 Japan

[2]Department of Mechanical Engineering
Gifu University, 1-1 Yanagido
Gifu City, 501-1 Japan

INTRODUCTION

A good understanding of high temperature creep properties of engineering ceramics is greatly neccssitated to promote its application to high temperature-machine components. Our knowledge of high temperature creep of this material, however, is quite limited at present. One of main reasons for this is in difficulty in measuring displacement which is generated in ceramics sample loaded at elevated temperatures above 1000°C.

Most basic creep properties are obtained under tensile uniaxial loading, where precise measurement of tensile creep displacement at elevated temperatures[1~8] is required. Several displacement measurements of ceramics have been devised under tensile loading at elevated temperatures. They, however, have still some problems in both the accuracy and reliability.

The authors proposed the procedures for measuring tensile strain of silicon nitride ceramics at temperatures ranged from 1200°C to 1400°C, using the specimen with projections which work as targets and the laser-beam-type displacement measuring system in their earlier paper[9].

The tensile creep displacement was measured by using the silicon nitride ceramics specimen with projections and the laser-beam-type displacement measuring system at 1300°C in the same way as in the earlier work in the present study. Moreover, the creep deformation was numerically analyzed for the projection-accompanied specimen through finite element method. Then the relationship between stress and minimum creep strain rate was determined for silicon nitride ceramics at 1300°C.

MATERIAL AND TEST PROCEDURES

The tested material is silicon nitride ceramics which is produced by mixing Y-α-sialon and Si_3N_4 particles at ratio of 40 and 60 weight percents, respectively, and then by sintering the mixed particles in N_2-gas atmosphere at 1750℃.

Figure 1 shows the shape and dimensions of test specimen. The specimen has the four projections which work as targets for detecting the displacement by means of the laser-beam-type extensometer, as presented in the authors' earlier paper[9]. The devices for

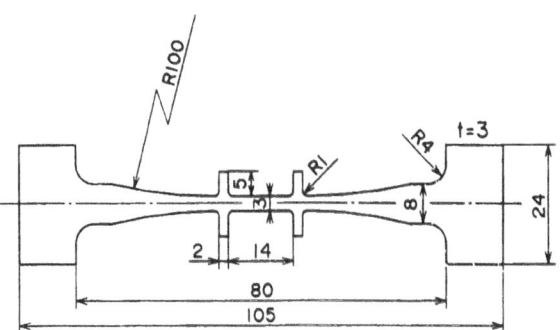

Figure 1. Shape and dimensions of test specimen.

mounting specimen were specially designed and machined from silicon carbide ceramics blocks. These were set up on the electro-hydraulic-type-test system equipped with an electric resistance type-furnace. The specimen axis was aligned for a component of bending strain to be less than 3 percent of tensile strain at room temperature, where the tensile strains detected by four gauges affixed on four side surfaces of the specimen were compared. The tensile creep test was performed at temperature, T=1300℃.

TEST RESULTS

The tensile creep displacements were measured between the two projections by means of the laser-beam-type extensometer at stresses of σ =48 and 87MPa at T=1300℃, and then they were plotted against test time t in Figs.2(a) and (b). The figures show that the quite small creep displacement of about 60 to 400 μ m can be steadily detected over period of 600 hours by the laser-beam-type displacement measuring system.

The tensile creep displacements were measured at other stresses at 1300℃ in the same way. Then the tensile creep strains ε_c were obtained from dividing them by the initial distance between the two projections. They were plotted against test time t together in Fig.3, where the final ends of the curves correspond to the failure points except the one at σ =48MPa which is interrupted at t=400 hours. According to the figure, the creep curves for quite low stresses of σ =48, 50 and 60MPa are divided into three stages of the transient, the stable and the accelerating, as in creep curves of metallic materials. By the way, the creep curves are not necessarily situated in the order of the magnitude of applied stress. This seems to be due to inequality in quality of silicon nitride ceramics, which will be stated later.

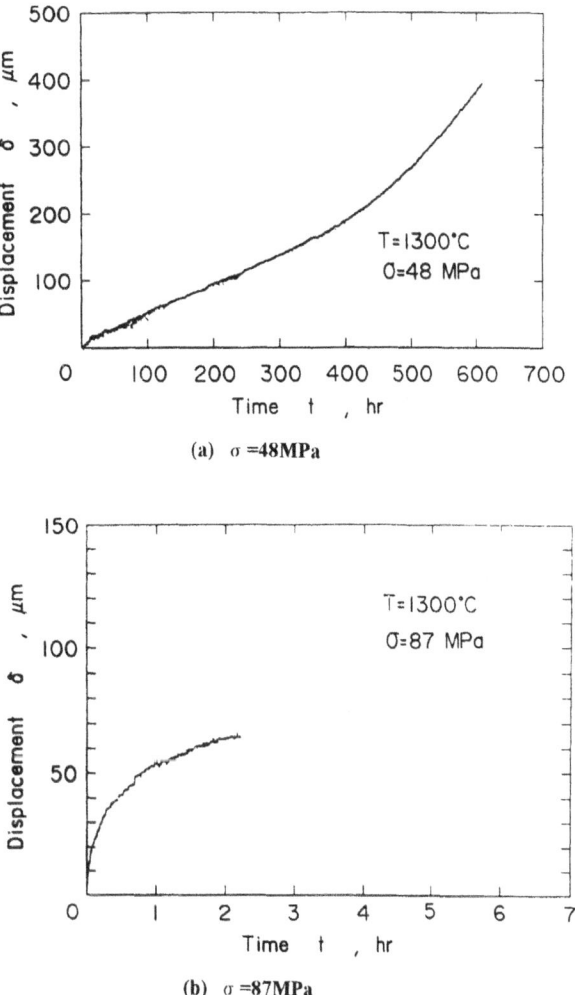

(a) σ =48MPa

(b) σ =87MPa

Figure 2. Variations of tensile creep displacement with increase in test time, which were measured at T=1300°C.

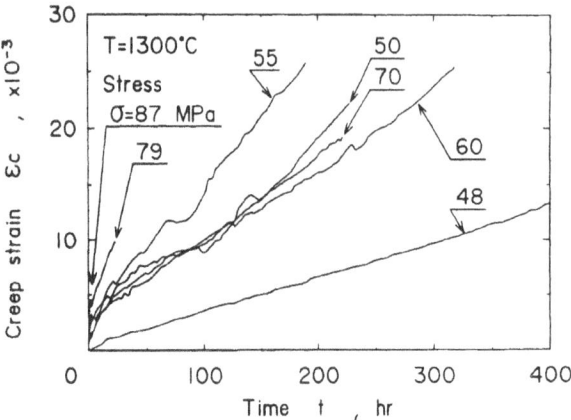

Figure 3. Tensile creep strain curves obtained from measurement by means of the laser-beam-type extensometer.

Quite large variations are observed on some creep curves presented in Fig.3. Ceramics fibers were stuffed into opening left between the loading rod and the inner circular surface of the furnace to prevent heated air from fluctuating inside the furnace. Some broken pieces of them might float in air inside the furnace and interrupt the passage of the laser beam. The transient variations arising on the creep curves seem to be concerned with such fluctuation of heated air and float of small pieces of ceramics fibers. Anyway, it should be noted that tensile creep displacement can be steadily measured over a long period at 1300°C, as shown in the creep curve for σ =48MPa, when good environmental conditions are realized.

The slopes of the creep curves were calculated at their steady stages and the minimum creep strain rates $\dot{\varepsilon}_c$ obtained at the respective stresses. It is well known that the minimum creep strain rate is related to stress by the following equation,

$$\dot{\varepsilon}_c = K\sigma^n,$$

where K is a constant and n the power exponent of stress. The minimum creep strain rates obtained from Fig.3 were plotted against stress on double logarithmic scales in Fig.4. The least square's regression analysis was made for these experimental data, and the equation,

$$\dot{\varepsilon}_c = 1.37 \times 10^{-15} \sigma^{4.09} \quad \left(sec^{-1}\right) \tag{1}$$

was determined, which is shown by the solid line in Fig.4.

The experimental data[6,8,10,11] on the relationship between $\dot{\varepsilon}_c$ and σ, which have been obtained at test temperatures around 1300°C before now are collected in Fig.5, together with the present data. According to the figure, the present $\dot{\varepsilon}_c$ versus σ plots are in the reasonable location among the other data, and the power exponent of stress of n=4.09 is also in the range from 4.0 to 6.0 reported by the other researchers.

CREEP LIFE TO FAILURE

Figure 6 shows time to failure t_f plotted against applied stress, which are scattered quite

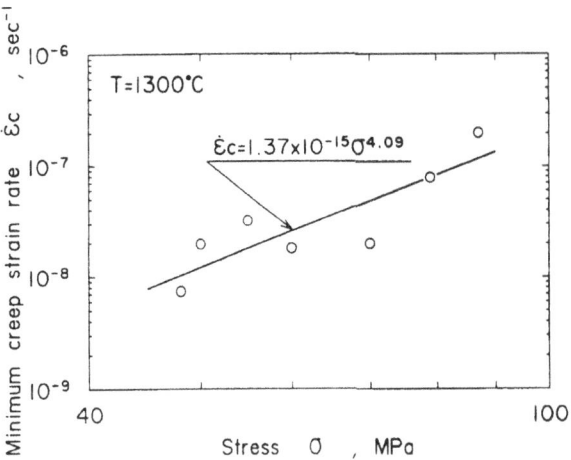

Figure 4. Minimum creep strain rate plotted against applied stress.

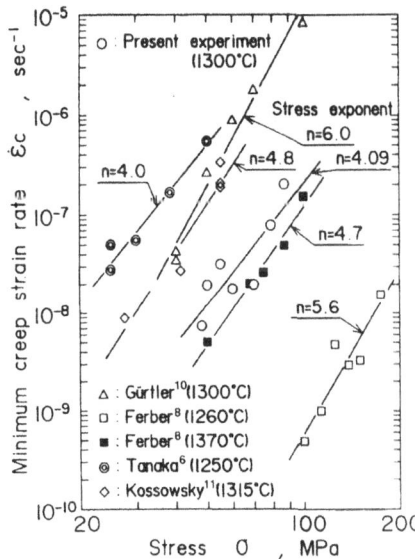

Figure 5. Experimental data on relationship between minimum creep strain rate and applied stress, which were obtained around T=1300°C before now.

widely. The solid line was determined through a least square's regression analysis for these data, and expressed by the following equation,

$$t_f = 1.74 \times 10^{15} \sigma^{-7.37} \quad (hours) \tag{2}$$

Same life data were plotted against the minimum creep strain rate in Fig.7, where the solid line represents the relationship between t_f and $\dot{\varepsilon}_c$ expressed by the equation,

$$t_f = 6.44 \times 10^{-12} \dot{\varepsilon}_c^{-1.76} \quad (hours) \tag{3}$$

The scatter in the plotted data is quite smaller in Fig.7 than in Fig.6, exhibiting that the creep

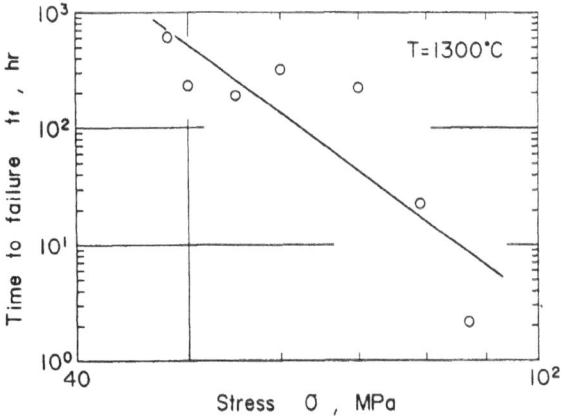

Figure 6.   Creep life time to failure plotted against applied stress.

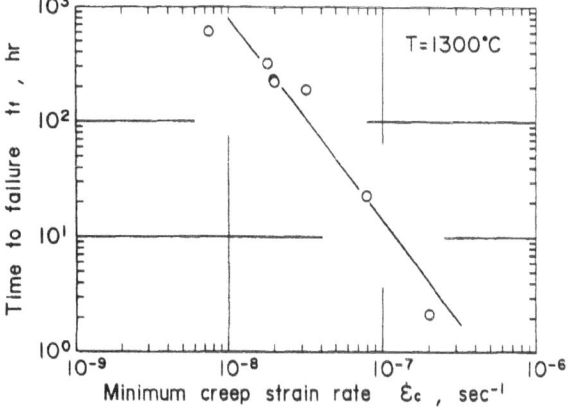

Figure 7.   Creep life time to failure plotted against minimum creep strain rate.

life to failure is more dominantly controlled by the minimum creep strain rate than the stress. It might be evidenced by such a comparison between Figs.6 and 7 that the quite large scatter observed in the t_f versus σ plots is not due to error in the measurement of the creep strain, but due to inequality in quality of the tested material, which was suggested in the previous section.

ANALYSIS

Procedures of Analysis

Nonelastic finite element analysis was made for the tensile specimen shown in Fig.1. First the stress-strain response was calculated by elastic-plastic F.E.M. analysis using the tensile stress-strain curve obtained at $1300^{\circ}C$ in our earlier paper[9]. Then this was followed by the steady creep F.E.M. analysis in which eq.(1) was employed as a constitutive equation. In the steady state creep analysis, the calculations through non-steady state creep analysis were repeated until the creep deformation reached the steady state as follows; the load increment induced by creep strain was calculated at the respective mesh nodes for every time increment predetermined, and then the equilibrium equations were constructed for these load increments at the respective nodes. The F.E.M. used in this study is the same as the one which has been employed for metallic materials, being not particular for ceramics material.

Verification of Relationship between Minimum Creep Strain Rate and Stress

The steady state creep F.E.M. analysis was performed for the specimen with four projections at applied stresses of 50, 60, 70 and 87MPa, using the constitutive equation (1). The relationships between $\dot{\varepsilon}_c$ and σ obtained from the calculation and the experimentation were compared in Fig.8. The minimum creep strain rate was calculated through the F.E.M.

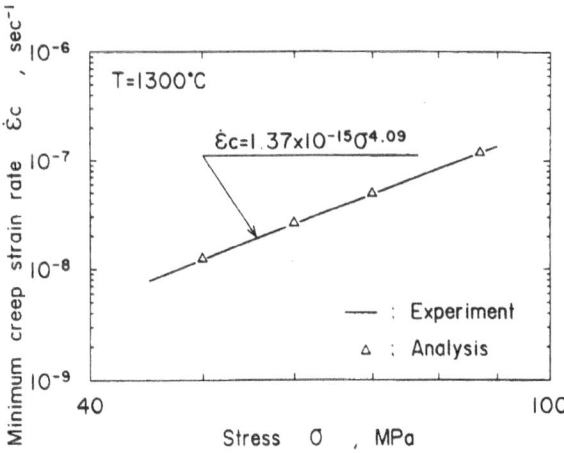

Figure 8. Comparison of the constitutive equation (1) determined from experiment and the relationship between minimum creep strain rate and stress calculated through the F.E.M. using this equation.

analysis as follows; the displacement increment between the two projections occurring for time increment $\triangle t$ at a given stress was calculated through the F.E.M., and then the average strain increment $\triangle \varepsilon_c$ was obtained from dividing this by the distance between the two projections at that time, as in estimation of the creep strain from the experimentation. The creep strain rate $\dot{\varepsilon}_c$ was determined from dividing the calculated strain increment $\triangle \varepsilon_c$ by $\triangle t$. The minimum creep strain rates calculated in this way are denoted by open triangles in Fig.8. They are in very good agreement with the solid line expressed by eq.(1). This suggests that the measured minimum creep strain rate is hardly influenced by strain concentration induced at the root of the projections. This shows that we can determine the correct relationship between $\dot{\varepsilon}_c$ and σ with practical precision without any corrections, using the tensile specimen shown in Fig.1.

Analysis of Creep Deformation of Projection-Accompanied Specimen

The progress of the creep deformation was analyzed for a quarter of the tensile specimen subjected to the stress $\sigma = 60\text{MPa}$ through the F.E.M., where eq.(1) was used as the constitutive equation. Figure 9 shows the change in distribution of the equivalent creep strain rate $\dot{\varepsilon}_c^{eq}$ with increase in time; (a) to (d) are the strain rate distributions at $t=1.0 \times 10^{-7}$, 0.65, 3.02 and 14.4 hours, respectively.

Just after the stress was applied, the high strain rate was locally developed around root of the projection, as shown in Fig.9 (a); the value of $\dot{\varepsilon}_c^{eq}$ at root of the projection is about 3.2 times as much as the one at the parallel part of the specimen. The concentration of the equivalent strain rate decreases with lapse of time due to progress of a stress relaxation occurring around root of the projection, as shown in Figs.9 (a) to (d). In addition, the domain occupied with great strain rate also diminishes as time elapses, and the magnitude of $\dot{\varepsilon}_c^{eq}$ around the root of the projection reduces from about 3.2 times to 1.6 times the one in the parallel part of the specimen as test time passes from $t=1.0 \times 10^{-7}$ to 14.4 hours, where $t=14.4$ hours are in the stage of steady state creep. The numerical calculations of the distribution in the equivalent creep strain rate at the other stresses exhibited the similar results. Thus the stress relaxation occurring in the creep process decreases the creep strain rate and constricts the size of the strain rate-concentrated region around the root of the projections.

Now, the calculated relationship between $\dot{\varepsilon}_c$ and σ is in good agreement with the one obtained from the experiment in Fig.8. This probably results from the stress relaxation effects mentioned above. The creep strain rate, however, is still slightly larger around the projection-root than in the parallel part of the specimen even at the steady creep stage. Nevertheless, the calculation coincides very well with the experimental data, as shown in Fig.8. The displacement of mesh-nodes and the strain rate at the respective node points were calculated by the F.E.M. to examine this contradiction. The calculations were made for the specimen loaded at $\sigma = 60\text{MPa}$.

Figure 10 shows the enlarged root area of the projection, where the center axis of the specimen is included. The open circles denote the locations of mesh nodes at $t=0$ hour, and they are connected with dashed line, perpendicular to the center axis of the specimen. These are referred to as the node-row hereafter. The axial displacements of the mesh-nodes at $t=14.4$ hours were calculated on a given node-row, and then the axial displacement of the mesh-node at the center axis on the same node-row was subtracted from them. The quantity obtained from such a subtraction was defined as the displacement-deviation. The displacement-deviation was plotted by the solid square marks, setting the location of the node-row at $t=0$ hour as the quantity of zero in Fig.10, where the respective solid square marks on the same row were connected with a solid line. According to Fig.10, the node-row perpendicular to the center axis of the specimen at t=0 hour bends around the root of the projection at t=14.4 hours.

Subsequently, the axial minimum creep strain rates at the mesh-nodes were calculated at σ =60MPa at t=14.4 hours. Then the minimum creep strain rate at the parallel part of the specimen was subtracted from them. This quantity was plotted by the solid square marks in Fig.11, setting the location of the initial element node denoted by a open circle mark as the basic point. According to this figure, the axial creep strain rate in the edge region around the projection-root is slightly larger than the one at the center in the parallel part of the specimen

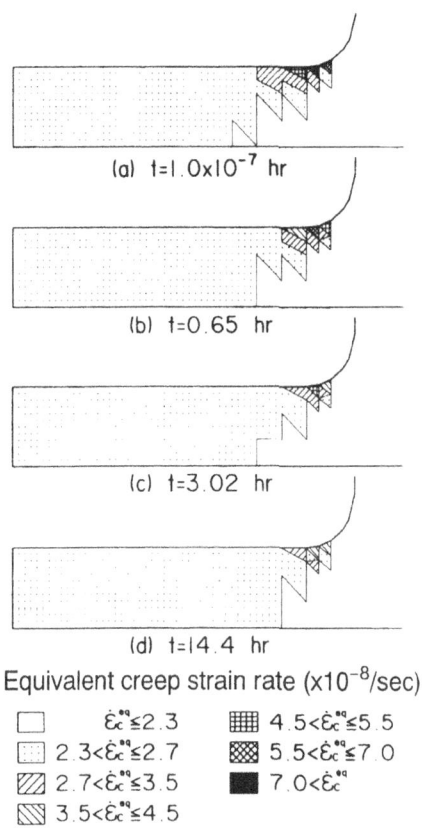

(a) t=1.0x10^{-7} hr

(b) t=0.65 hr

(c) t=3.02 hr

(d) t=14.4 hr

Equivalent creep strain rate (x10^{-8}/sec)

□	$\dot{\varepsilon}_c^{eq} \leq 2.3$	▦	$4.5 < \dot{\varepsilon}_c^{eq} \leq 5.5$
▨	$2.3 < \dot{\varepsilon}_c^{eq} \leq 2.7$	▧	$5.5 < \dot{\varepsilon}_c^{eq} \leq 7.0$
▨	$2.7 < \dot{\varepsilon}_c^{eq} \leq 3.5$	■	$7.0 < \dot{\varepsilon}_c^{eq}$
▨	$3.5 < \dot{\varepsilon}_c^{eq} \leq 4.5$		

Figure 9. Change in distribution of equivalent creep strain rate in the creep process. which was calculated through F.E.M.. A quarter of test specimen is presented.

at the steady state creep. The steady state-axial creep strain rate, however, is quite smaller in the inside region than in the edge region around the root of the projection. Meanwhile, it should be noted that the inner edge of the projection has almost the same steady creep strain rate as the one at the central part of the specimen, which is shown from the solid square marks falling nearly on the open circles at there in Fig.11.   This seems to result from that the creep strain rate is restrained from development in the edge layer by creep deformation less-developed in the inner part of the specimen around the root of the projection. The good

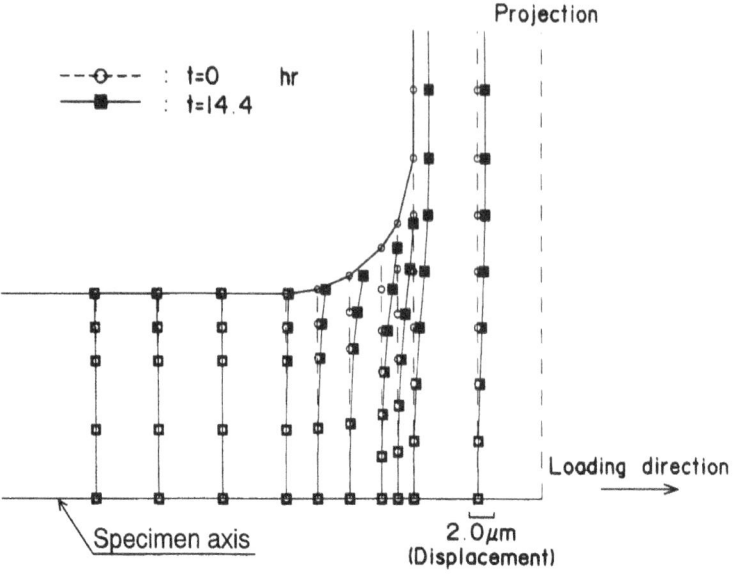

Figure 10. Distribution of axial displacement of mesh-node at steady stage of creep calculated through the F.E.M. around root of the projection. (T=1300°C, σ =60MPa)

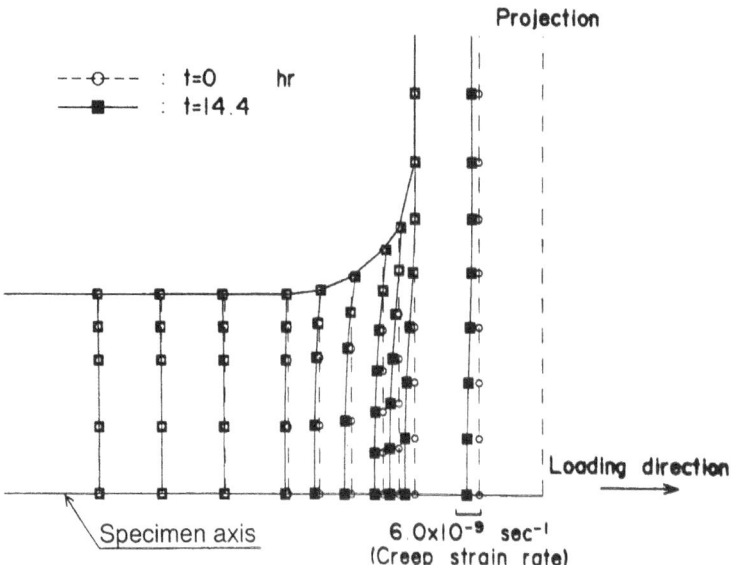

Figure 11. Distribution of steady state-axial creep strain rate at mesh-node calculated through the F.E.M.. (T=1300°C, σ =60MPa)

agreement between eq.(1) and the minimum creep strain rate versus stress curve calculated through F.E.M. using eq.(1) as the constitutive equation is understood well from such calculations. Now, we can evaluate the minimum creep strain rate of silicon nitride ceramics without any correction for the stress/strain concentrations occurring at the root of the projections, using the specimen shown in Fig.1.

The above-mentioned considerations hold for the combination of the creep constitutive equation (1) and the projection-possessing specimen shown in Fig.1. An error in the steady state-creep strain rate induced by stress/strain concentrations at the root of projection is under calculation for the different combinations of creep constitutive equation and specimen with projections.

CONCLUSIONS

Tensile creep test was performed for sintered silicon nitride ceramics at 1300°C, using the specimen with projections designed by the authors, and creep displacement was measured by means of the laser-beam-type extensometer. Then the relationship between the minimum creep strain rate and stress was determined from combining the measured creep displacement and inelastic finite element method. The main results obtained are summarized as follows:

(1) Tensile creep displacement of sintered silicon nitride ceramics was successfully measured over 600 hours at 1300°C by means of the laser-beam-type extensometer, showing the possibility of the long term-creep strain measurement of ceramics materials at test temperatures over 1000°C.

(2) The measured creep curves showed transient creep followed by steady state creep at all stress levels. Then it was found that the steady creep strain rate was expressed by the power function of the applied stress.

(3) The relationship between the steady state-creep strain rate and stress calculated through F.E.M. agreed with the power function determined between the two quantities from the experiment. It was evidenced from this that the steady state-creep strain rate could be determined from the measurement of strain between the two projections without any corrections.

REFERENCE

1. W.R. Cannon, and T.G. Langdon, Review: creep of ceramics. *J. Mater. Sci.* **18**:1(1983).
2. S.M. Wiederhorn, D.E. Roberts, and T.-J. Chuang, Damage-enhanced creep in a siliconized silicon carbide: Phenomenology, *J. Am. Ceram. Soc.* **71**:602(1988).
3. D.F. Carroll, and R.E. Tressler, Effect of creep damage on the tensile creep behavior of a siliconized silicon carbide, *J. Am. Ceram. Soc.* **72**:49(1989).
4. D.F. Carroll, S.M. Wiederhorn, and D.E. Roberts, Technique for tensile creep testing of ceramics, *J. Am. Ceram. Soc.* **72**:1610(1989).
5. M. Masuda, and M. Matsui, Fatigue in ceramics (part 4): Static fatigue behavior of sintered silicon nitride under tensile stress, *Nippon-Seramikkusu-Kyokai-Gakujutsu-Ronbunshi.* (in Japanese). **98**:83(1990).
6. T. Tanaka, N. Okabe, S. Yamamoto, H. Nakayama, A. Segawa, and T. Fujii, Strength and strain behavior in high temperature creep of fine ceramics, *J. Soc. Mat. Sci. Jpn.* (in Japanese). **39**:1692(1990).
7. T. Ohji, and Y. Yamauchi, Long-term tensile creep testing for advanced ceramics, *J. Am. Ceram. Soc.* **75**:2304(1992).
8. M.K. Ferber, and M.G. Jenkins, Evaluation of the strength and creep-fatigue behavior of hot pressed silicon nitride, *J. Am. Ceram. Soc.* **75**:2453(1992).
9. K. Hatanaka, and H. Shiota, Tensile Stress-strain response of sintered silicon nitride ceramic at elevated temperatures, *Trans. Jpn. Soc. Mech. Eng.* (in Japanese). **58**:653(1992).
10. M. Gürtler, Institut für material- und festkörperforschung , *Kernforschungszentrum Karlsruhe.* KFK-4874(1991).
11. R. Kossowsky, D. G. Miller, and E. S. Diaz, Tensile and creep strengths of hot-pressed Si₃N₄. *J. Mater. Sci.* **10**:983(1975).

DAMAGE CREEP IN SiC$_f$-MLAS COMPOSITES

Hélène Maupas, Dominique Kervadec, and Jean-Louis Chermant

Lermat, URA CNRS 1317, ISMRA
6 Bd du Maréchal Juin
14050 Caen Cedex, France

ABSTRACT

Creep in 1D and 2D SiC$_f$-MLAS composites are governed by matrix microcracking in the temperature range investigated 1273-1373K. It appears that in the case of three-point bending tests the shear stresses are significant in the case of specimens having a small gauge length/width ratio. In the stress range 40-400 MPa, the 1D fiber architecture leads to one creep mechanism which is matrix macrocracking, whereas for the 2D structure two mechanisms are needed to explain the creep in the stress range 40-250 MPa.

INTRODUCTION

It is well-known that the exceptional characteristics and behavior of ceramic matrix composites, CMCs (ceramic or glass-ceramic matrices reinforced with short or continuous ceramic fibers or whiskers), are mainly due to the role and specific properties of the fiber/matrix interfaces (or interphase(s), (see for example (1-3)). Many types of deformation and rupture processes are involved when CMCs are loaded and, or are tested in time, depending upon the relative strength, fiber/matrix cohesion, mechanical solicitation or composite architecture (1,4-8), (figure 1). Moreover for long term applications, it is necessary to have stable fibers and matrices which change as little as possible under the working conditions.

For aeronautical and space applications, many types of glass-ceramic systems have been explored for use as the matrices. They are based on tertiary or quaternary systems: LAS, MAS, CAS, ... MLAS, BMAS, YMAS, ... (9-10). Although progress is being made in improving the performance of these materials, there is at present no obvious advantage of one system over another.

The aim of this paper is to present some aspects of the creep behavior of a unidirectional (1D) and cross-ply (0-90°) (2D) MLAS glass-ceramic matrix reinforced by SiC fibers, focussing on the importance of the damage and damage micromechanisms in these composites, and on an analysis of the influence of the fiber architecture on the damage mechanisms.

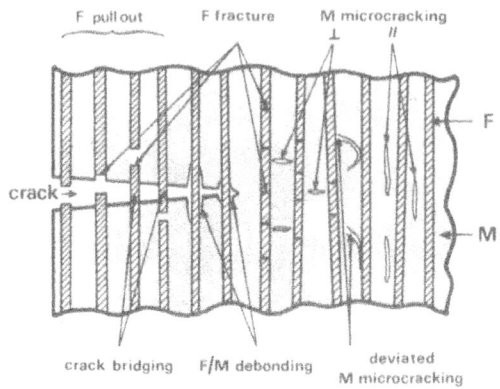

Figure 1. Different deformation and rupture micromechanisms in the case of a 1D CMC (F:fiber, M:matrix).

MATERIALS AND EXPERIMENTAL

The 1D and 2D SiC$_f$-MLAS composites investigated were produced by Aérospatiale (Etablissement de Bordeaux, France). The continuous reinforcing fibers are SiC Nicalon NLM 202 (Nippon Carbon Co, Japan). The matrix is produced by a chemical sol-gel route leading to the MLAS glass matrix of composition: $0.5MgO-0.5Li_2O-1Al_2O_3-4SiO_2$.

These composites were fabricated by winding slurry impregnated yarn onto a mandrel to form tapes. The slurry consisted of glass powder (previously mentioned), solvent and organic binder. After drying, the tapes were cut off to make plies and afterwards stacked up and densified in a hot press at elevated temperature and pressure (11). The result of this treatment is to have crystals distributed throughout the matrix which improves the conventional glass properties.

Unidirectional (1D) or cross-ply (0-90°) (2D) are obtained respectively by stacking the different plies into one or two directions. Moreover the 2D composites are symmetrical from their middle, therefore the middle 90° ply is a double layer.

If we look at the corresponding ternary diagrams MAS and LAS and pseudo-binary cordierite-spodumene diagram (12-15), several silicate phases can appear over a very narrow temperature range. Moreover many allotropic transformations can occur. It is for this reason that, in the absence of strict temperature control, different silicate phases can appear (16). X-rays, TEM and EDS analysis have shown for the 1D composites the presence of β spodumene and β cordierite, while if the matrix is fabricated at some tens of degrees higher, β eucryptite crystals and an amorphous phase are present. For 2D composites, we observe the same main crystallized phases but with a different distribution. For both types of composite, some mullite crystals are also observed (16).

Creep tests were performed using a 1380 Instron machine (Bucks, England), equipped for three-point bending, under high vacuum and for temperatures up to 2273K.

The three-point bending device was fitted with tungsten push rods and TiC knives for two possible gauge lengths of 14 or 30 mm. The temperature ranges investigated for these composites was 1073-1473K. A Sesame opening furnace (V.M.D.I., Paris, France) with tantalum heating elements was used under 10^{-5} torr.

The deformation of the specimens was followed by L.V.D.T. transducers (Penny and Giles, Christchurch, England) located outside the furnace. Therefore the machine environment must be disturbed as little as possible. Emphasis is laid on a very stable room temperature and constant cooling water flow to achieve accurate measurements (i.e. better than a micron).

Bend specimens had the following dimensions, 40 x 5 x 3 and 18 x 5 x 3 mm³, depending on the gauge length chosen. The two long edges of each samples were polished to 1μm diamond grade to facilitate observation of the damage produced in creep tests.

The present paper discusses results for 1D SiC$_f$-MLAS composites tested at temperatures between 1273-1473K and applied stresses of 50-400 MPa. The 2D SiC$_f$-MLAS results were obtained at the same temperature range, but at lower stress values, i.e. 40-260 MPa, as higher stresses led to failure upon loading.

RESULTS

Creep Curves

All SiC$_f$-MLAS creep tests were performed at constant load. The values of strain were measured using push rods. In order to compare the results for the 1D and the 2D composites, only tests with the gauge length of 14 mm will be considered. Such 1D material specimens were studied by Kervadec (17-20). The creep was measured at different stresses for each composite type. The maximum stress values possible were 250-260 and 400 MPa at 1273 and 1373K for the 2D and 1D composites; higher values led to premature failure during the loading procedure. It should be pointed out that for the same reason, at 1473K stresses are limited to 90 MPa and 250 MPa. Thus at 1473K the loadability of both composites is reduced by half compared to that at 1273K and 1373K.

For the 2D SiC$_f$-MLAS composites, the three creep curves presented in figure 2a show the variation with temperature. This paper concentrates on the results at 1373K, hence the only creep curve presented for the 1D composite is for this temperature (figure 2b). The two strain-time (ε-t) curves presented at 1373K enable a comparison of the two composites. Despite the fact that the stress values chosen are very close (250 and 300 MPa) the creep behavior has nothing in common.

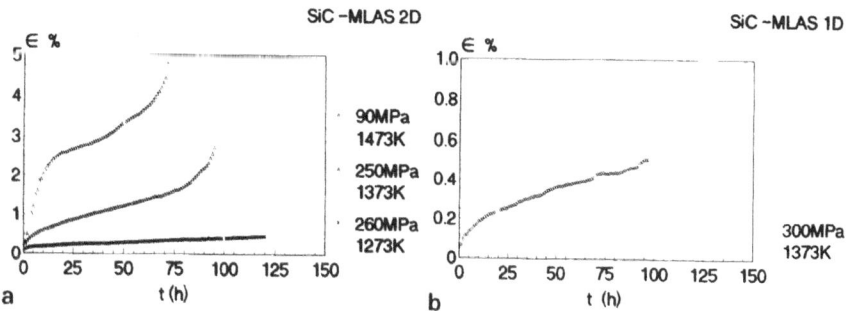

Figure 2. Strain-time curves (ε-t) for 2D SiC$_f$-MLAS (a) and 1D SiC$_f$-MLAS (b) composites.

Firstly the scale of deformation is not the same. The 2D composites deform much more: they attain a creep strain of over 1% during steady state creep, and fail at ~ 3%, whereas 1D composites only creeps 0.5% over the same period.

Secondly, the 2D creep curve shows over 100 hours the three classical creep stages. The same observation was also observed at 225 MPa. In the case of the 1D composite, no matter the applied stress from 50 to 400 MPa after 100 hours only steady state creep was observed.

Figure 2a shows the effect of temperature on the creep behavior. The loading procedure has shown that the 2D composites present the same fracture strength at 1273 and 1373K, but the creep behaviour is quite different. At 1273K the creep mechanism is perfectly stable, the strain does not exceed 0.05% in 100 hours. To illustrate this point, macrographs of two crept specimens are shown in figure 3a and 3b. The specimen tested at 1373K and 225 MPa shows a residual strain after unloading and cooling. That tested at 1273K and 260 MPa shows no residual plastic deformation.

Figure 3. Macrographs of the 2D SiC$_f$-MLAS composite tested under 1273K-260 MPa (a) and under 1373K-225 MPa (b). Both specimens have a width, W=3 mm, and were tested with the same gauge length, L=14 mm.

At 1473K the change in the 2D composite is such that the maximum sustainable stress cannot exceed 90 MPa. This macrostructural change is confirmed by microscopical investigation of creep specimens tested at 1473K. With regard to the extent of creep deformation, a comparison of the 250 MPa and the 90 MPa creep curves shows the primary stage to last longer reaching a value of ε=2.5% and the tertiary stage to end with fracture at 5%. The magnitude of these strain values indicates that at 1473K the 2D material is no longer creep resistant whatever the value of applied stress.

In order to confirm several points regarding the change of the creep behavior with temperature, the three ε-t curves (Fig. 2a) have been translated in $\dot{\varepsilon}-\varepsilon$ curves, where $\dot{\varepsilon}$ represents the instantaneous stain rate.

Strain Rate

The strain rate, $\dot{\varepsilon}$ has been calculated for each ε-t curve and plotted in figure 4a, b and c as a function of the strain. The three creep stages are clearly identified in these curves. $\dot{\varepsilon}$ first decreases, reaches a plateau value corresponding to steady state creep and then increases corresponding to tertiary creep. A change in temperature affects particularly the primary creep stage. An increase in temperature displaces the strain values to higher values, and the steady state stage starts at higher strain values, but becomes shorter. The advantage of this kind of presentations is that it enables a confirmation of earlier assumptions made interpreting ε-t curves. ε-t curves are sometimes difficult to read and the plots $\dot{\varepsilon}$ -ε enable a clear distinction between the different creep stages and the existence of a steady state stage (17, 20).

Hence after the existence of a true steady state has been confirmed, the steady-state rate, $\dot{\varepsilon}_s$, is calculated and plotted as a function of stress.

Steady-State Rate

The $\dot{\varepsilon}_s$ values, are determined from $\dot{\varepsilon}_s$-t curves. This $\dot{\varepsilon}_s$ value obtained as described above is valid if it matches the experimental results which can be expressed by the equation $\varepsilon = \varepsilon_{0s} + \dot{\varepsilon}_s t$, ε_{0s} being the primary deformation obtained if the steady state were to begin at t=0.

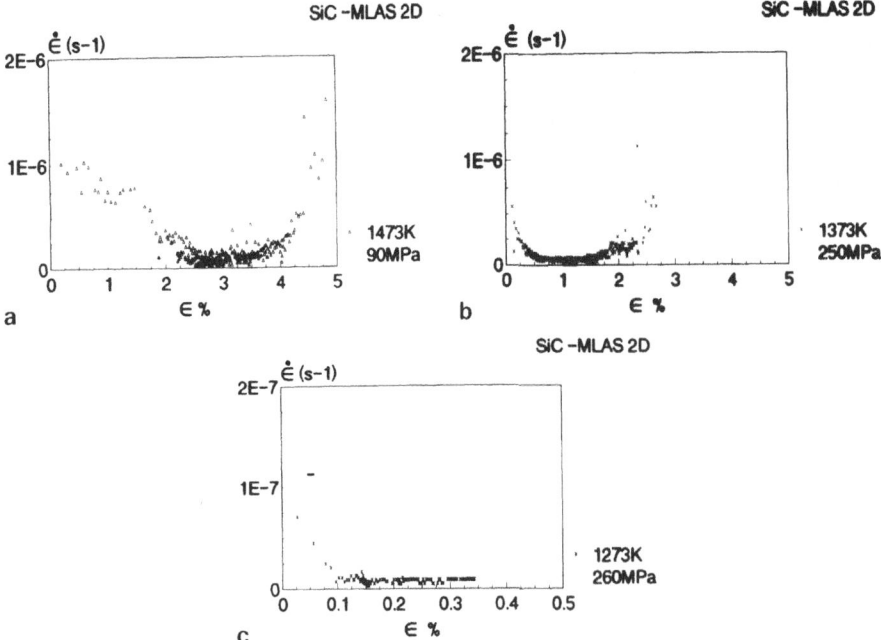

a   b   c

Figure 4. Instantaneous strain rate-strain curve ($\dot{\epsilon}$ - ϵ), corresponding to the different ϵ-t curves presented in figure 2a (the scale of the figure 4c is divided by ten compared to the others).

Traditionally $\dot{\epsilon}_s$ values obtained for one particular temperature for various applied stress values are plotted on a logarithmic scale. It is not known whether glass-ceramic composites reinforced with long fibers obey such a power law $\dot{\epsilon}_s = A\sigma^n$ (as composites are not homogeneous materials). It is just an empirical representation and does not presume any particular mechanisms. Figure 5 shows this power law plot for the 1D and 2D SiC$_f$ - MLAS at 1373K. The slopes of the various straight lines have been delineated and calculated. These values should not be compared with the n values established for the polycrystalline materials, but rather to determine the number of main mechanisms (or micromechanisms) acting within the stress range investigated. In our case both materials appear to behave differently below and above the stress value ~200 MPa. This transition is quite distinct for the 2D composites, but not for the 1D SiC$_f$-MLAS.

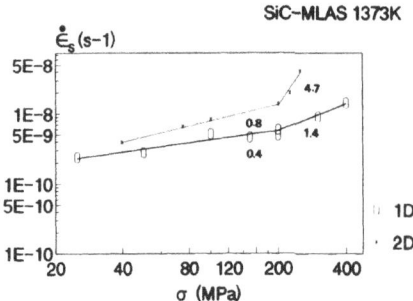

Figure 5. Steady state rate-applied stress curves ($\dot{\epsilon}_s$-σ) for both 1D and 2D SiC$_f$-MLAS composites. The numbers indicate the slopes of the sections for the double logarithmic plot.

Below, in the discussion, the reasons why the mechanisms responsible for steady state creep should or should not be separated into two regions, depending on the applied stress values will be presented (first region $\sigma \leq 200$ MPa, second region $\sigma \geq 200$ MPa) for both composites.

DISCUSSION

The arguments to be developed in the discussion will contain several points:
* the loading procedure,
* a description of the nature and propagation of the damage occurring during creep,
* the stress distribution in specimens submitted to a three-point bending test,
* the values of ε_s plotted as a function of σ.

Damage

Loading Procedure. Over the stated applied stress ranges, the two materials behave differently. Whereas for the 1D the deformation is linear elastic from 50 to 400 MPa, the 2D composites show a change in slope at about 200 MPa (figure 6).

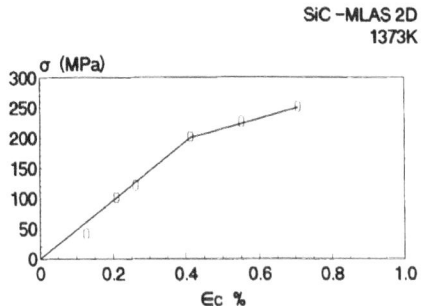

Figure 6. "Simulated" loading curve of the 2D SiC$_f$-MLAS composite obtained from the various specimens tested under stresses from 40-250 MPa. The value ε_c corresponds to the deformation at the end of the loading, and each experimental point corresponds to the chosen load for creep testing of that particular specimen.

This curve suggests that above 200 MPa, the 2D SiC$_f$-MLAS composites are damaged on loading and it is a damaged specimen which is subsequently subjected to primary creep. 1D composites are never damaged during loading and this is also the case for the 2D composites if the stress applied for the creep tests is below 200 MPa.

In order to understand which mechanisms are taking place during creep it was necessary to carry out optical and scanning microscope observations on the crept samples. For both composites a general schema of creep damage can be described as a function of the applied stress.

Creep Damage. For both composites the damage observed is essentially micro and macrocracking. Fiber/matrix decohesion and sliding are considered to act mainly in primary creep when the applied stress is redistributed on the matrix and fibers. This stress redistribution is function of the elastic and creep properties of the matrix, the fibers and the interface (21).

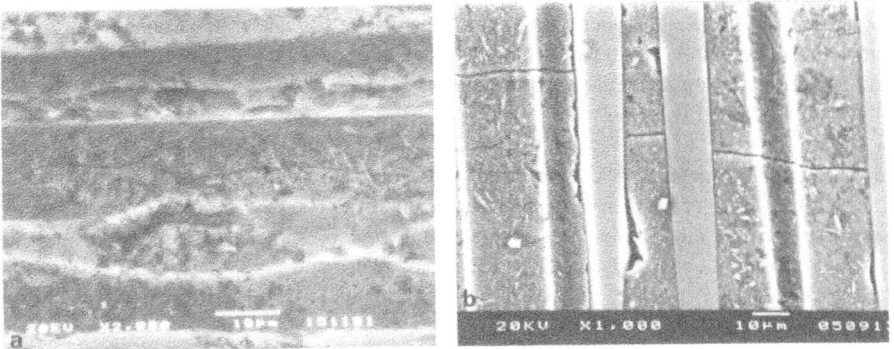

Figure 7. SEM micrographs of 1D SiC$_f$-MLAS composites creep tested at 1373K: (a) cracks parallel to the fibers; (b) perpendicular cracks located on the long edge specimen and on the side of the two inferior knives.

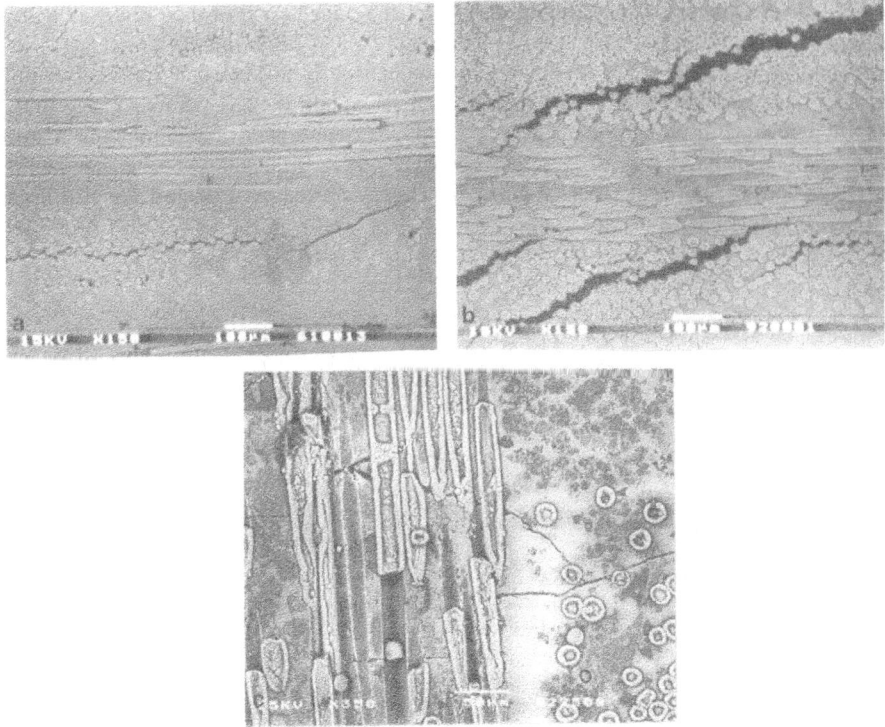

Figure 8. SEM micrographs of 2D SiC$_f$-MLAS composites creep tested at 1373K: (a) isolated cracks in the 90° plies at 100 MPa and a creep time duration test, t=190 hours; (b) at the same magnitude, damage state at 200 MPa and t=165 h; (c) at 225 MPa and t=180 h, the cracks issued from the 90° plies penetrate the 0° plies.

* *1D SiC_f-MLAS composites*

The microcracks observed in 1D SiC_f -MLAS composites were either parallel or perpendicular to the fibers. As the applied stress increases, the number of both types of microcracks increases. The microcracks perpendicular to the fibers are mainly in the region below the upper knife but on the opposite edge. Figures 7a and b give an idea of the microcrack size encountered in 1D composites at 1373K.

* *2D SiC_f-MLAS composites*

In the case of the 2D SiC_f-MLAS composites the extent of cracking is more important. All the damage is concentrated on the 90° plies and as the stress increases the cracks in the 90° plies become more and more numerous. They extend to the two nearest 0° plies and open up quite wide. The general shape of these cracks corresponds exactly to cracks propagating under a shear stress. The same kind of cracks have also been observed by Sbaizero et al. (22) in the case of four point bending tests performed on SiC_f-LAS composites at room temperature.

At low stresses the cracks are located in the double thickness 90° middle ply and in particular the main cracks are observed at the edge of the sample. Generally speaking the cracks at this stress level are quite isolated and propagate mainly in the longitudinal direction (figure 8a).

Over the period of our tests, the general crack network extension in the 90° plies is controlled by the 0° plies when the applied stress is ≤ 200 MPa. The cracks issuing from the 90° plies never cross the 0° plies (figure 8b). Over the same period of time it is only when the applied stress is greater than 200 MPa that the 0° plies begin to become damaged by cracks coming from the 90° plies (figure 8c). Simultaneously to the damage caused by the 90° plies, there is of course at almost each stress level an intrinsic mechanism damage to 0° plies (figure 9a). But compared to the size of the 90° crack plies, its contribution to the deformation is small. At high stresses, perpendicular cracks located in the same area as for the 1D composites are also visible (figure 9b). Creep induced by damage have also been observed for other CMCs like for example, SiC_f-CAS (23-26), SCS 6 SiC_f-HPSN or RBSN

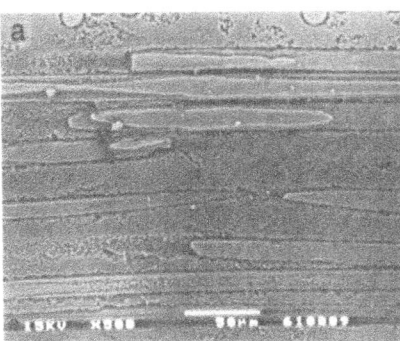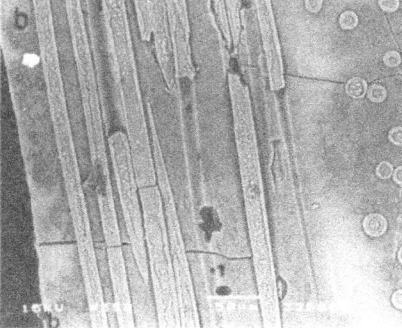

Figure 9. At 100 MPa intrinsic damage belonging to 0° plies (a); at 250 MPa cracks perpendicular to the longidutinal fibers and located in the same area as defined figure 7b are observed (b).

To understand why the main damage was a consequence of shear stresses, the stress distribution has been calculated for a sample submitted to three-point bending.

Stress Distribution: The calculations of stress distribution have been conducted using the finite element method assuming linear elasticity, and taking into account the different kinds of plies (Castem 2000 standard, LMT, ENS Cachan, France). The symmetry of the problem allows the use of only a half specimen (figure 10a).

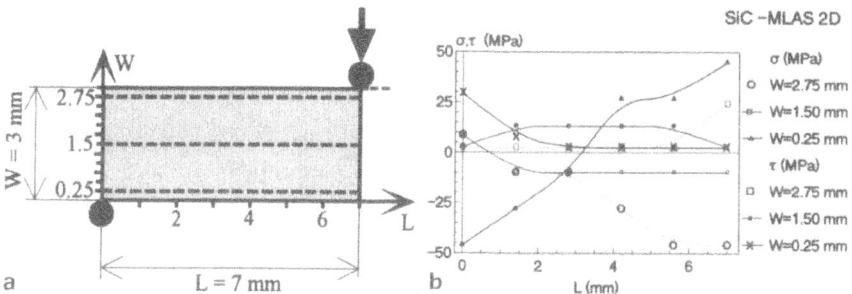

Figure 10. (a) Axes and coordinates (W,L) used to define the stress distribution occurring in one half of a three-point bending specimen; (b) σ or τ-L diagram, describing for the 2D SiC$_f$-MLAS composites the change in stresses along three lines W=0.25, 1.50 and 2.75 mm.

The change of the various normal stresses, σ (σ > 0: tensile, σ < 0: compressive), and shear stress, τ, over half a specimen is shown in figure 10b. The parameters L and W are the coordinates used to define points inside the half specimen (figure 10a).

The stress values shown values describe the change of the normal and shear stresses along three lines W=0.25 mm, W=1.5 mm (middle of the specimen) and W=2.75 mm. Additional values of W would lead to confusion. Using the same assumptions the 1D SiC$_f$-MLAS stress distribution shows the same characteristic variation. Hence the 1D SiC$_f$ -MLAS are also submitted to compressive, shear and tensile stresses. The crack network however is less extensive as for the 2D composites and quite different. The 1D parallel cracks neither open up nor propagate in the ~ 45° direction during creep. The 1D architecture is less sensitive to shear stresses.

Figure 10b shows that the shear stresses are in competition with the normal stresses. In the middle of the sample W=1.5 mm and L=2-5 mm, τ and σ have the same magnitude. Tensile stresses (σ > 0) dominate in the area defined by L=4-7 mm and W=0.25 mm. It is precisely the area in which we observed perpendicular cracks for both composites. In the middle (W=1.5 mm) and for L=0 mm tensile stresses are also greater than the shear stresses and we find here the initiation of cracks. Elsewhere there is strong competition between shear and compressive stresses. These dominate in the vicinity of the knives.

The competition between shear and compressive stresses suggest that the cracks propagate and can be nucleated under shear stresses, while both the 0° plies and the compressive stresses control the crack propagation. Thus the crack growth can be stable. These features which control the development of the damage determine the main mechanisms of creep

Steady and Tertiary Creep Stages. In the case of the 2D composites the stress range must be divided into two regions. Below 200 MPa the specimen which is subjected to creep is undamaged whereas above 200 MPa the specimen has already been damaged during loading.

A direct comparison of the 1D and 2D composites leads to the conclusion that it is not necessary to define for the 1D materials two different mechanisms responsible for the steady state at this temperature (1373K). A plot of the steady state strain, ε_s, as a function of the applied stress confirms these conclusion (figure 11). The ε_p strain is that reached at the onset of the steady state creep, whereas ε_{75} is the strain after 75 hours. In the case of both composites the deformation at 75 hours belonged to the steady state for all the stresses applied. Putting the two composites on the same scale of deformation, the 1D composites show a regular increase of ε_p and ε_{75} with stress, whereas for the 2D there is again an abrupt

535

increase about 200 MPa. The steady state corresponds to a stable crack propagation under shear and compressive stresses for both composites. When the stress level is high enough the specimen relaxes in the tensile area (W=0.25 mm, L=4-11 mm) producing perpendicular cracks.

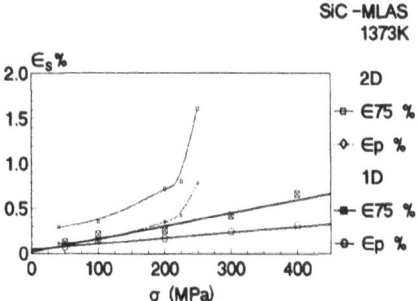

Figure 11. Deformation during the steady state as a function of the applied stress for both 1D and 2D SiC$_f$-MLAS composites.

For 2D composites creep on an already damaged sample does not modify the main mechanism proposed hitherto, but simply increases the density and the opening of the cracks over the same time period. It is for this reason that stresses above 200 MPa led to the tertiary creep within 100 hours. Thus it appears justified to define a second creep mechanism for stresses above 200 MPa, as the 2D composite submitted to creep has not the same mechanical characteristics: one could also add the contribution of the creep of the longitudinal silicon carbide fibers because for stresses above 200 MPa the 90° plies are so damaged that they leave the 0° plies bearing alone all the load. This evidence taken together led us to propose only one main mechanism for the 1D and two for the 2D (figure 12).

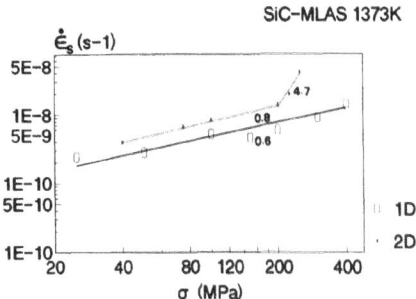

Figure 12. Steady state rate-applied stress curve ($\dot{\epsilon}_s$-σ) for both 1D and 2D SiC$_f$-MLAS composites. Only one mechanism is proposed for the 1D

The tertiary stage corresponds for the 2D materials to the damage created in the 0° plies generated by the 90° plies cracks. When the strain increases drastically (figure 2a) the frame of the composite is damaged leading towards the destruction of the structure and creep failure.

The value of the applied temperature also plays a significant role because at 1373K the microstructure and characteristics of the matrix change and thus contribute to the damage. This is the main difference between 1273 and 1373K.

536

CONCLUSION

In this paper creep results were presented on 1D and 2D SiC$_f$-MLAS composites. The main mechanism at 1273-1373K is matrix microcracking whatever the fiber architecture. For specimens tested in bending with a small gauge length/width ratio, the macrocracks propagate under shear stress and their propagation is controlled by the compressive stresses or the longitudinal fibers whatever the fiber architecture. From an accurate analysis of these creep results, it has been possible to define only one mechanism for the steady state from the ε_s - σ curves (figure 12) for the 1D composites, while for the 2D composites two mechanisms are acting depending on the stress level.

ACKNOWLEDGMENTS

Some of this work was carried out in the Groupement Scientifique GS4C "Comportements thermomécaniques des composites céramique-céramique à fibres", and supported by the CNES, the CNRS, the DRET, the MRE and the Aérospatiale, SEP and SNECMA Companies. Thanks are particularly due to Dr G. Larnac from Aérospatiale Company for fruitful discussions. We want also to cordially thank our colleague Mrs C. Rospars for the stress calculations undertaken at LMT, ENS Cachan, France.

REFERENCES

1.   K.K. Chawla, "Ceramic Matrix Composites", Chapman and Hall, 1993.
2.   D.B. Marshall, Interfaces in ceramic fiber composites, in "Ceramic Microstructures'86: Role of Interfaces", edited by J.A. Pask, A.G. Evans, Mat. Sci. Res. **21**, pp 859-868 (1987).
3.   R.J. Kerans, R.S. Hay, N.J. Pagano, T.A. Parthasarathy, The role of the fiber-matrix interface in ceramic composites, Amer. Ceram. Bull. **68** [2], 429-442 (1989).
4.   R. Warren, "Ceramic Matrix Composites", Blackie, 1992.
5.   G. Fantozzi, G. Orange, D. Rouby, Mechanical behaviour of ceramic matrix composites, Phase Transitions, **13**, 165-198 (1988).
6.   M.D. Thouless, O. Sbaizero, L.S. Sigl, A.G. Evans, Effect of interface mechanical properties on pull-out in a SiC-fiber-reinforced lithium aluminum silicate glass-ceramic, J. Amer. Ceram. Soc. **72** [4], 525-532 (1989).
7.   A.G. Evans, D.B. Marshall, The mechanical behavior of ceramic matrix composites, Acta Met. 37 [10], 2567-2583 (1989).
8.   F.E. Heredia, S.M. Spearing, A.G. Evans, Mechanical properties of continuous-fiber-reinforced carbon matrix composites and relationships to constituent properties, J. Amer. Ceram. Soc. **75** [11], 3017-3025 (1992).
9.   K.M. Prewo, J.J. Brennan, G.K. Layden, Fiber reinforced glasses and glass-ceramics for high performance applications, Amer. Ceram. Bull. **65** [2], 305-322 (1986).
10. D. Kervadec, J.L. Chermant, Les composites à matrice vitrocéramique, Rev. Comp. Nouv. Mat. **1**, 9-49 (1991).
11. G. Larnac, P. Lespade, P. Pérès, J.M. Donzac, Fiber reinforced composites. A new class of glass-ceramic materials for thermomechanical applications, in "Proc. 5th European Conference on Composite Materials" edited by A.R. Bunsell, J.F. Jamet, A. Mamiah, pp 703-708 (1992).
12. Z. Strnad, Glass-ceramic materials, liquid phase separation, nucleation and crystallization, in "Glass Science and Technology, **8**", Elsevier, pp 76-89 (1986).

13. M.D. Karkhanavala, F.A. Hummel, Reactions in the system Li_2O-MgO-Al_2O_3-SiO_2: I the cordierite-spodumene join, J. Amer. Ceram. Soc., 36 [12], 393-397 (1953).

14. E.F. Osborn, A. Muan (1961), cited by C.A. Jouenne in "Traité de Céramiques et Matériaux Minéraux", Edited by Septima, pp 185-187 (1984).

15. G.H. Beall, B.R. Karstetter, H.L. Rittler, Crystallization and chemical strengthening of stuffed β-quartz glass ceramics, J. Amer. Ceram. Soc., 50 [4], 181-190, (1967).

16. F. Doreau, H. Maupas, D. Kervadec, P. Ruterana, J. Vicens, J.L. Chermant, The complexity of the matrix microstructure in SiC-fibers reinforced glass-ceramic composites, J. Eur. Ceram. Soc. 1995, in press.

17. D. Kervadec, Comportement en fluage sous flexion et microstructure d'un SiC-MLAS 1D, Thèse de Doctorat de l'Université de Caen, 1992.

18. D. Kervadec, J.L. Chermant, Fluage et microstructure du SiCf-MLAS 1D, Rev. Comp. Mat. Avancés, 3, 173-189 (1993).

19. D. Kervadec, J.L. Chermant, Visco-elastic deformation during creep of SiCf-MLAS composite, in 6th European Conference on "Composite Materials - High Temperature Ceramic Matrix Composites", ECCM6-HTCMC, Bordeaux, France, Sept. 20-24, 1993, edited by A.R. Bunsell, A. Kelly, A. Massiah, Woodhead Pub. Ltd, pp 649-657 (1993).

20. J.L. Chermant, Creep behavior of ceramic matrix composites, Sil. Ind., (1995), in press.

21. J. Holmes, X. Wu, Elevated temperature creep behavior of continuous fiber-reinforced ceramics, in "Elevated Temperature Mechanical Behavior of Ceramic Matrix Composites" edited by S.V. Nair, K. Jakus, Butterworth - Heinneman, (1995) in press.

22. O. Sbaizero, A. Evans, Tensile and shear properties of laminated ceramic matrix composites, J. Amer. Ceram. Soc., 69 [6], 481-486 (1986).

23 C.H. Weber, J.P.A. Lofvander, A.G. Evans, Creep anisotropy of a continuous - fiber - reinforced silicon carbide / calcium aluminosilicate composite, J. Amer. Ceram. Soc., 77 [7], 1745-1752 (1994).

24. X. Wu, J.W. Holmes, Tensile creep and creep-strain recovery behavior of silicon carbide fiber/calcium aluminosilicate matrix ceramic composites, J. Amer. Ceram. Soc. 76 [10], 2695-2700 (1993).

25. J.W. Holmes, J.L. Chermant, Creep behavior of fiber - reinforced ceramic matrix composites, in 6th European Conference on "Composite Materials - High Temperature Ceramic Matrix Composites", ECCM6-HTCMC, Bordeaux, France, Sept. 20-24, 1993, edited by A.R. Bunsell, A. Kelly, A. Massiah, Woodhead Pub. Ltd, pp 633-647 (1993).

26. X. Wu, J.W. Holmes, Static and cyclic creep behavior of a SiC-fiber glass ceramic matrix composite, to appear in J. Amer. Ceram. Soc. (1995).

27. J.W. Holmes, Y. Park, J.W. Jones, Tensile creep and creep recovery behavior of a SiC-fiber Si_3N_4-matrix composite, J. Amer. Ceram. Soc. 76 [5], 1281-1293 (1993).

28. G.E. Hilmas, J.W. Holmes, R.R.T. Bhatt, J.D. Di Carlo, Tensile creep behavior and damage accumulation in a SiC-fiber/RBSN-matrix composite, in Advances in Ceramic Matrix Composites, edited by N. Bansal, Ceram. Trans. Vol. 38, pp 291-304, (1993)

29. J.N. Adami, Comportement en fluage uniaxial sous vide d'un composite à matrice céramique bidirectionel Al_2O_3-SiC, Thèse de Docteur ès Sciences Techniques, Ecole Polytechnique Fédérale de Zürich, Switzerland, (1992). This work was performed at the Institute for Advanced Materials, Joint Research Center, Petten, The Netherlands.

THE VARIATION OF INDENTATION FRACTURE
WITH TEMPERATURE IN SILICON CARBIDE,
SILICON NITRIDE AND ZIRCONIA

Marie-Odile Guillou, J. Leslie Henshall, and Robert M. Hooper

School of Engineering
University of Exeter
Exeter, Devon, EX4 4QF, UK

ABSTRACT

This paper reports the variation with temperature between 22°C and 1250°C of the indentation hardness and fracture of some advanced ceramics in polycrystalline and single crystal form. The materials tested were hot-pressed SiC, Refel SiC, single crystal SiC, hot-pressed Si_3N_4, ceria stabilized polycrystalline tetragonal ZrO_2 and calcia and yttria stabilized single crystal cubic ZrO_2. The room temperature hardnesses ranged from 8.3 GPa for the polycrystalline zirconia to 34.9 GPa for single crystal silicon carbide. In general, the relative rate of decrease in hardness with increasing test temperature was greater for the zirconias than SiC or Si_3N_4. The indentation critical stress intensity factors at room temperature varied between 0.8 $MPa.m^{0.5}$ for single crystal zirconia to 20.3 $MPa.m^{0.5}$ for the toughened zirconia. In general there was a gradual decrease in critical stress intensity with increasing test temperature. This was contrasted with the conventional K_{Ic} values which were basically temperature independent, or even increased with increasing test temperature. The ceria stabilised tetragonal zirconia exhibited a very marked decrease in toughness with increasing test temperature and Refel SiC showed a significant increase when deformation of the silicon binder phase became more prevalent above ~ 800°C.

INTRODUCTION

It is probable that the greatest usage of advanced ceramics will be in applications which take advantage of their excellent wear resistance, particularly at high temperatures, *e.g.* in grinding wheels, cutting tools, bearings, or nozzles. There are many standard and non-standard types of wear test, which are either application specific or do not necessarily relate to any potential application. It is apparent that the abrasive/erosive wear resistance of ceramics is related to their hardness and toughness, and thermal properties in some cases, although the precise form of the relationship is not well determined, *e.g.* Evans and Lawn (1975), Evans and Marshall (1981), Ajayi and Ludema (1988). In these analyses it is generally assumed that the relevant toughness parameter is the critical stress intensity factor, K_{Ic} as measured using large notched specimens, *e.g.* single edge or chevron notched beams or double torsion.

By analogy with the arguments presented that the relevant toughness associated with tensile fracture from small inclusions is not the macroscopic K_{Ic}, (Davidge and Evans, 1970; Evans, 1984), it is much more probable that in general the surface/near surface fracture processes in fully dense advanced ceramics will not be from large pre-existing voids or defects. Consequently, it is proposed that the most appropriate measure of toughness to use is that derived from indentation fracture tests, in which the loading and sampled volume are more closely similar to those encountered in applications such as those referred to above.

The purpose of this paper is to present the results of indentation hardness and fracture measurements in hot-pressed, Refel and single crystal silicon carbide, hot-pressed silicon nitride and ceria stabilized polycrystalline tetragonal zirconia and calcia stabilized single crystal cubic zirconia at temperatures up to 1250°C. The deformation and fracture mechanisms are discussed and the results compared with macroscopically notched K_{Ic} measurements.

EXPERIMENTAL PROCEDURES

Full details of the materials, specimen preparation procedures and test methods can be found in Guillou (1992). A brief outline of the salient aspects is presented herein. The ceramic test specimens were taken from the bulk of the as-supplied materials by diamond saw sectioning. The purpose of this was to avoid any near surface effects, *e.g.* residual stresses, introduced during manufacture. The specimens were then ground and polished using conventional metallographic procedures, with a final 1/4 μm diamond polish. The orientations of the single crystal specimens were determined using Laue back-reflection X-ray diffraction.

Vickers diamond pyramid indentations were used for the polycrystalline materials and (001) calcia stabilized single crystal zirconia, whereas a Berkovich diamond pyramid indentor was used for the other single crystal surfaces, since it matched the symmetry of the tested planes [(0001) for SiC and (111) for ZrO_2]. Tests up to 800°C were performed in air with an applied load of between 2.15 N and 19.8 N, and temperatures measured on the specimen surface using a type K thermocouple. The indenter was placed in contact with the surface of the ceramic test piece for approximately 20 minutes prior to performing the measured indentations to ensure that the temperature remained uniform. Above ~ 800°C diamond graphitises/oxidises in air, and hence the higher temperature tests were performed in a vacuum of better than 10^{-4} millibar, using RF heating of a graphite susceptor support. The resultant loads were in the range 99 - 132 N and the specimen surface temperatures were measured using a disappearing filament optical pyrometer. In all cases the nominal dwell time was 18 seconds, and in any particular sequence of tests, indentations were performed at the highest test temperature first, and then monotonically with decreasing test temperature.

RESULTS & OBSERVATIONS

Indentation Hardness

Polycrystalline Ceramics. The variations of the indentation hardnesses of the polycrystalline materials with temperature are shown in Figure 1. For the relatively high loads

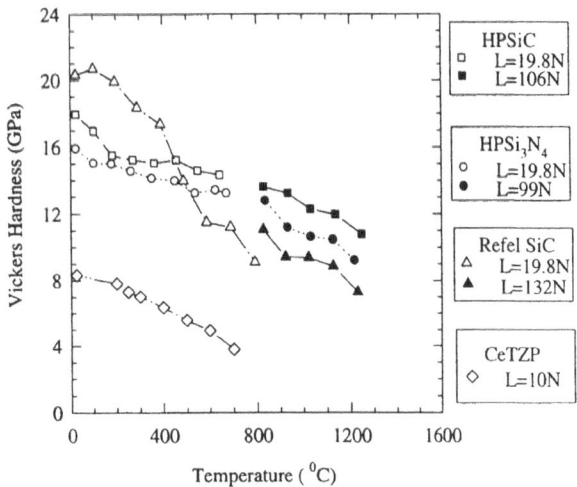

Figure 1. Variation of Vickers hardness with temperature for hot-pressed silicon carbide, HPSiC, hot-pressed silicon nitride, HPSi$_3$N$_4$, Refel SiC and ceria stabilized polycrystalline tetragonal zirconia, CeTZP. The open symbols represent tests in air and the closed symbols in vacuum.

used in this study, $i.e. > 19$ N, it would be expected that the hardness values would be independent of the applied load. It can be seen in Figure 1 that although the applied load is increased by between five and seven times, and the environment changed from air to vacuum, that there is a relatively smooth and continuous change in hardness, within experimental scatter ($<= 3\%$), between the different test methods. Refel silicon carbide is the hardest material at low temperatures ($H_V \sim 20.5$ GPa at 22°C), but with a relatively rapid decrease between 300 and 500°C as a result of softening of the silicon filler/binder phase. The hot-pressed materials have similar temperature variations, $i.e.$ decreasing in hardness by about 1/3rd between room temperature and 1250°C, with SiC harder than Si$_3$N$_4$ by approximately 2 GPa at all temperatures. The ceria stabilized tetragonal zirconia, CeTZP, is considerably softer; the hardness decreasing from 8.3 GPa at 22°C to 3.8 GPa at 700°C.

Single Crystals. The comparable temperature dependencies of the single crystal Berkovich hardnesses are presented in Figure 2. The indentations were performed on (0001) in SiC with the indenter facets parallel to $<11\overline{2}0>$ and on (001), with the facets parallel to $<100>$, and (111), with facets parallel to either $<1\overline{1}0>$ or $<11\overline{2}>$, in calcia stabilized ZrO$_2$, CaCSZ, and (111), with facets parallel to $<11\overline{2}>$, in yttria stabilized ZrO$_2$, YCSZ.

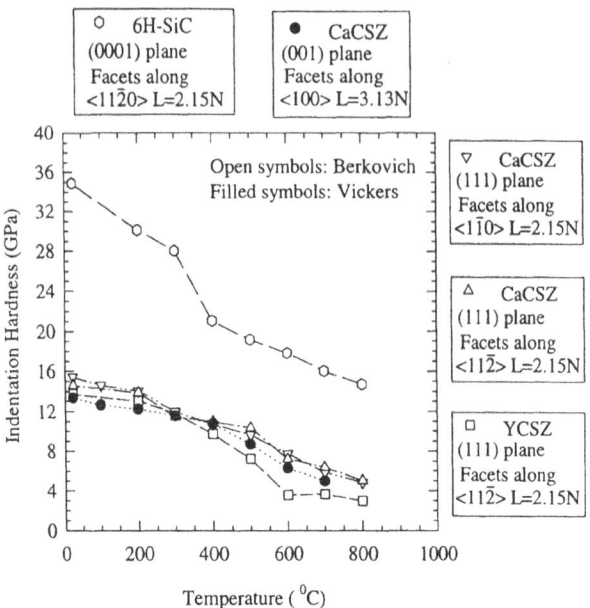

Figure 2. Variation of indentation hardness with temperature for single crystals, using either a Vickers indentor, closed symbols, for (001) plane calcia stabilized cubic zirconia, CaCSZ, or Berkovich indentor, open symbols, for (0001) plane 6H polytype silicon carbide, 6H-SiC, (111) plane CaCSZ and yttria stabilized cubic zirconia, YCSZ.

The hardness of the single crystal SiC is considerably higher at room temperature than the hot-pressed polycrystalline SiC, *i.e.* 34.9 GPa vs 17.9 GPa, but there is virtually no difference at 800°C.

The room temperature hardnesses of the single crystal zirconias lie in the range 13.4 to 15.4 GPa, which are significantly higher than the CeTZP (8.3 GPa). There is little hardness anisotropy in CaCSZ on (111) (Guillou et al., 1990). The (001) plane Vickers hardness values are slightly lower than the (111) Berkovich values, but this slight difference between the two indenter types is commonly observed. The decrease in hardness with increasing temperature is similar for all the orientations, with the hardness at 800°C being approximately 1/3rd of the room temperature value. The YCSZ has a slightly lower hardness than CaCSZ` between room temperature and 400°C, but above this temperature the decrease in hardness with increasing temperature is significantly greater. Above 600°C, the hardness becomes approximately temperature independent.

Indentation Fracture

Polycrystalline Ceramics. Even though indentation fracture testing has been analysed for many years, *e.g.* Palmqvist (1957) and Evans and Charles (1976), and is currently commonly used to determine a critical stress intensity factor using Vickers pyramidal indentations, the method of analysis of the data is not standardised. Previous work has shown (Guillou et al., 1992) that there are three main analysis equations in the literature, *i.e.* Evans (1979), Anstis et al. (1981) and Liang et al. (1990) respectively:

$$K_{Ia} = 0.631 \ L^{0.6} \ E^{0.4} \ a_v^{-0.7} \ 10^F \tag{1}$$

where $F = 1.59 - 0.34\log(c_v/a_v) - 2.02[\log(c_v/a_v)]^2 + 11.23[\log(c_v/a_v)]^3 - 24.97[\log(c_v/a_v)]^4 -16.32 \ [\log (c_v/a_v)]^5$

$$K_{Ia} = 0.0226 \ (\ E \ L \ )^{1/2} \ (\ a_v/c_v^{3/2} \ ) \tag{2}$$

$$K_{Ia} = [\ 0.3474/f(\upsilon) \ ] \ E^{0.4} \ L^{0.6} \ a_v^{-0.7} \ (\ c_v/a_v \ )^Y \tag{3}$$

where $f(\upsilon) = 14 \ [\ 1 - 8\{(4\upsilon - 0.5)/(1 + \upsilon)\}^4 \ ]$ and $Y = (\ c_v/18a_v) - 1.51$

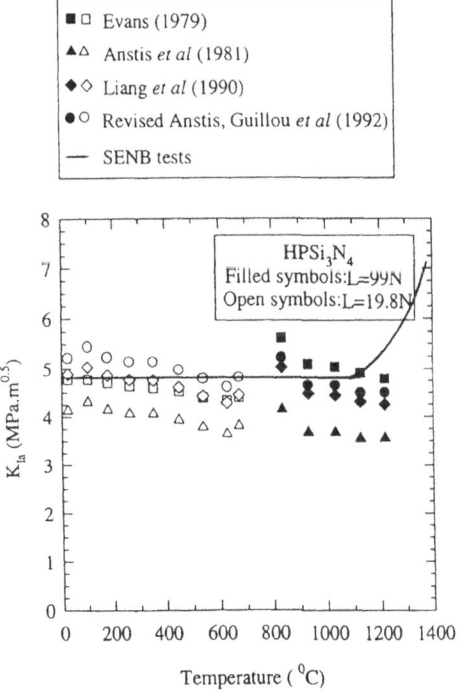

Figure 3. Comparison of the variation with temperature of the indentation critical stress intensity factor, K_{Ia}, calculated from the indentation measurements using equations (1) - (4) and single edge notched beam K_{Ic} values. The open symbols represent tests in air and the closed symbols in vacuum.

and a minor revision to the normalising constant in equation (2) has been proposed (Guillou et al., 1992) to give:

$$K_{Ia} = 0.0286 \ (\ E \ L\ )^{1/2} \ (\ a_v/c_v^{3/2}\ ) \tag{4}$$

where L is the applied load, E the Young's modulus, c_v half the tip-to-tip crack length, a_v half the indentation diagonal length and K_{Ia} is used to denote the derived critical stress intensity for crack arrest.

Figure 3 is a comparison of the K_{Ia} values calculated from the measured parameters for hot-pressed silicon nitride, $HPSi_3N_4$, between room temperature and 1250°C. The data for the temperature dependence of E were taken from Edington et al. (1975), with its value only changing slightly from 313 GPa at 22°C to 306 GPa at 672°C, and then decreasing more rapidly to be 285 GPa at 1217°C. The relevant features of the data in Figure 3 are:

(a)   there is a maximum difference of approximately 25% between the K_{Ia} values at a given temperature,

(b)   for a given set of test conditions there is a gradual decrease in K_{Ia} with increasing test temperature,

(c)   there is a distinct discontinuity in the data between the low and high load data, corresponding to test temperatures below and above 800°C respectively: this discontinuity is greatest for equation (1),

(d)   the numerical values are similar to the single edge notched beam K_{Ic} values for the same material below 1100°C (Henshall et al., 1975), but the differences are that the K_{Ic} values are essentially constant below this temperature, but then increase markedly, whereas there is a continuous slight decrease in K_{Ia} with increasing test temperature,

(e)   repeat tests were performed at 22°C after testing at the higher temperatures, which gave the same results as obtained initially.

Figure 4(a) shows a Vickers indentation and associated cracking in $HPSi_3N_4$ made at 1216°C. The cracks are well formed and relatively narrow. At higher magnifications, Figure 4(b), it can be observed that the cracks have propagated intergranularly, but with no signs of secondary cracking or localised surface deformation that would be expected to be noticeable if substantial grain boundary sliding had accompanied the indentation deformation and fracture.

The results of the indentation fracture analyses using equation (4) for HPSiC, Refel SiC and CeTZP, as well as the comparative data for $HPSi_3N_4$, are shown in Figure 5. The temperature dependent Young's moduli values for HPSiC and Refel SiC are taken from Edington et al. (1975). The value of E for CeTZP was assumed constant at 210 GPa between 22 and 700°C. The K_{Ia} results for HPSiC are similar to $HPSi_3N_4$, i.e. there is a continuous gradual decrease with increasing temperature. This is again in contrast to the K_{Ic} behaviour, which is essentially constant with increasing temperature (Henshall et al., 1979).

In contrast, the Refel SiC K_{Ia} values increase gradually with increasing temperature up to ~ 800°C, and then increase markedly, followed by a slight decrease as the test temperature is raised further. Figure 6 shows that the reason for this increase in K_{Ia} is related to the deformation and secondary cracking which occurs around the indentations at high temperatures. From Figure 7 it can be seen that the cracks primarily propagate, and eventually arrest, in the silicon binder in this material.

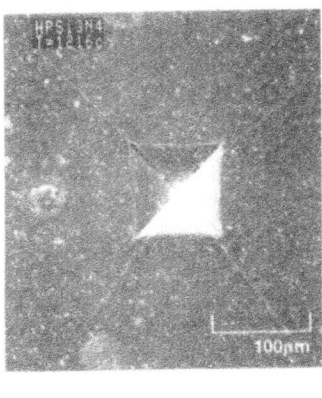

(a)

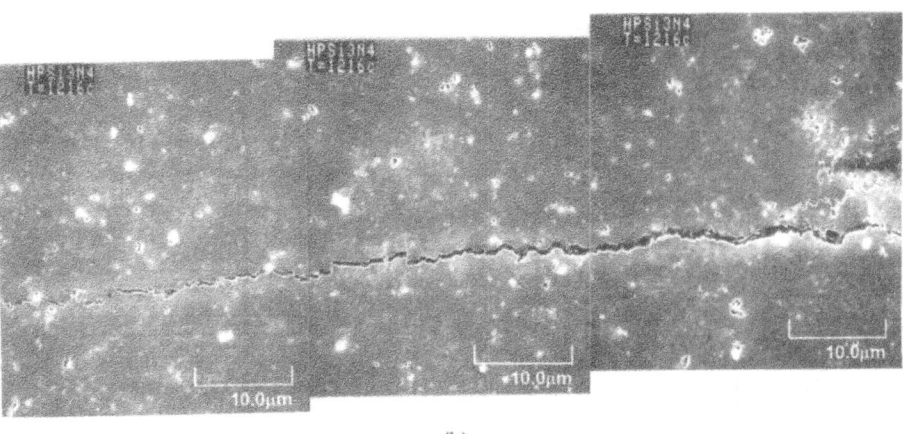

(b)

Figure 4. (a) SEM micrograph of Vickers indentation in hot-pressed silicon nitride tested at 1216°C in vacuum demonstrating general geometry and crack configurations, and (b) higher magnification SEM micrographs of one of the cracks in (a) showing that the cracks are narrow and intergranular with negligible secondary cracking or general grain boundary sliding.

The CeTZP K_{Ia} results exhibit a marked decrease between 300°C and 700°C from 7 MPa.m$^{0.5}$ to 1 MPa.m$^{0.5}$. At this relatively low load no cracking was obtained at 22°C, [although higher load tests (490 N) gave K_{Ia} = 20.3 MPa.m$^{0.5}$]. As noted by Lange (1982) and Tikare and Heuer (1991), this decrease in K_{Ia} with increasing temperature occurs as a result of the decrease in free energy between the metastable tetragonal and stable monoclinic phases. However these results would indicate that there is a significant contribution from transformation toughening at 600°C, whereas the phase diagram of Gupta and Andersson (1984) for CeTZP would suggest that the tetragonal phase is stable above ~ 500°C, and thus should not contribute to the toughness above this temperature.

By comparison with the results from HPSiC and HPSi$_3$N$_4$ it would be expected that there would probably be a slight, <10%, decrease in K_{Ia} between room temperature and 700°C in CeTZP, irrespective of any transformation effects. However this is much less than the observed *circa* twenty fold decrease.

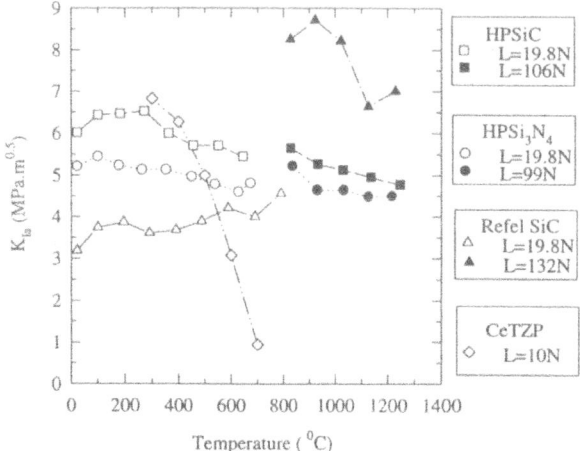

Figure 5. Variation of Indentation critical stress intensity factor, K_{Ia}, with temperature for hot-pressed silicon carbide, HPSiC, hot-pressed silicon nitride, HPSi$_3$N$_4$, Refel SiC and ceria stabilized tetragonal zirconia, CeTZP. The open symbols represent tests in air and the closed symbols in vacuum.

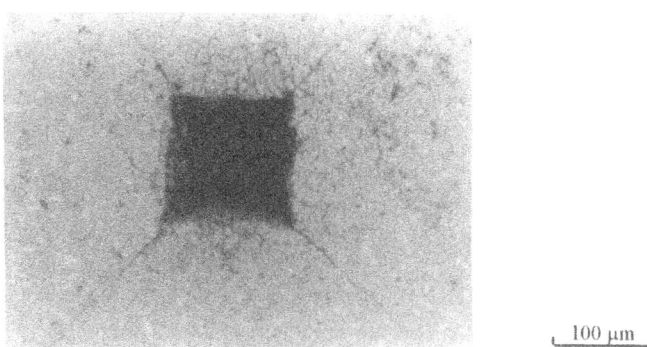

Figure 6. Optical micrograph of Vickers indentation in Refel SiC performed at 1230°C in vacuum, showing extensive deformation in the silicon phase.

546

Single Crystals. In this case the Vickers indentation data were analysed using equation (4) above, and its equivalent form (Maerky et al., 1995), was used for the Berkovich indentation data, *i.e.*

$$K_{Ia} = 0.0237 \, (E\,L)^{1/2} \, (a_b / c_b^{3/2}) \qquad (5)$$

where a_b is the facet to apex perpendicular length and c_b is the centre to tip crack length.

Figure 7. SEM micrographs of Vickers indentation in Refel SiC tested at 800°C in air. (b) is a higher magnification view of the boxed region in (a) showing that the crack arrests within the fine SiC + Si binder phase, A.

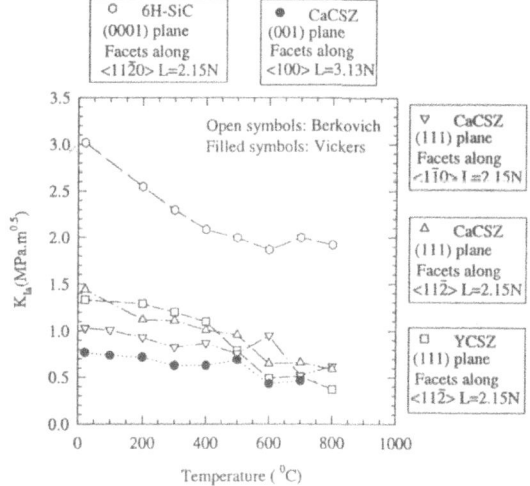

Figure 8. Variation of indentation critical stress intensity factor, K_{Ia}, with temperature for single crystals of 6H polytype silicon carbide, 6H-SiC, (001) and (111) plane calcia stabilised cubic zirconia, CaCSZ, and (111) plane yttria stabilized cubic zirconia, YCSZ. A Vickers indentor was used for the (001) plane, closed symbols, and Berkovich for (0001) and (111) planes, open symbols.

The variation of K_{Ia} with test temperature for the single crystals is shown in Figure 8. There is a marked decrease in K_{Ia} for single crystal SiC from 3.0 MPa.m$^{0.5}$ to 1.9 MPa.m$^{0.5}$ between 22°C and 800°C. This is markedly different from the results of single edge notched beam K_{Ic} values, from the same crystal, which are constant up to *ca* 600°C (Henshall et al.,

1977). It is also noteworthy that the relative decrease in K_{Ia} for the single crystal SiC is significantly greater than for hot-pressed SiC.

As can also be seen from Figure 8, there is a considerable anisotropy in K_{Ia} at room temperature in CaCSZ, which has been described and discussed previously (Guillou et al., 1990). Figure 9 shows optical micrographs of typical Berkovich indentations in CaCSZ at 300°C, 600°C and 700°C. The cracks at these temperatures propagated generally along the heights of the indentations. In Figure 9 the indenter facets are parallel to <110>, hence the traces of the cracks on (111) are <112>, and, as can be seen, the cracks lie approximately perpendicular to the indented surface. Hence these crack planes are approximately {110}. At 800°C the cracking was less regular, which would suggest that crack initiation was becoming more difficult. The anisotropy in the CaCSZ K_{Ia} values decreases with increasing test temperature. There is little difference between the YCSZ and corresponding CaCSZ K_{Ia} values below 600°C, but at the higher temperatures there is a noticeable difference.

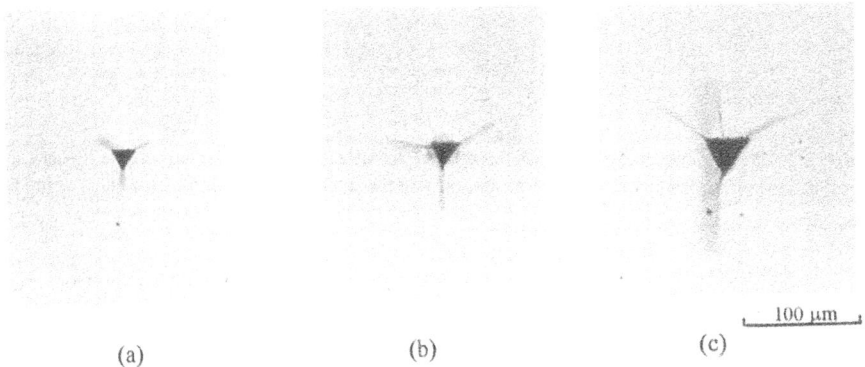

(a)   (b)   (c)

Figure 9. Berkovich indentations on (111) plane of single crystal calcia stabilized cubic zirconia at (a) 300°C, (b) 600°C and (c) 700°C. The indentor facets are parallel to <1$\bar{1}$0>.

DISCUSSION

Indentation Hardness

The room temperature indentation hardness values are generally in the order and of the magnitudes that would be expected. The only slightly surprising result is for hot-pressed SiC, which has a hardness much less than single crystal SiC. Some difference in hardness would be expected since the indentors are of differing geometries, the loads are not the same, and effects of the anisotropy in hardness will exist for the polycrystalline material. These effects however would not seem to be sufficient to explain the fact that the HPSiC hardness is only about half the single crystal value. It is also of interest to note that the Refel SiC, which contains ~ 10% free silicon, is also harder than HPSiC, whereas by comparison with cobalt-bonded carbide materials it would be expected that the hardness would be significantly lower.

The decrease in hardness with increasing test temperature again follows the trends that would be expected, with the rate of decrease being proportionally greater for the more ionically bonded zirconias than for the primarily covalently bonded silicon carbides and nitride. It is of interest to note that the rate of decrease for the polycrystalline materials is lower than for the single crystals. This may be of considerable relevance in such applications as grinding wheels, where there are high local temperatures, where it may prove to be beneficial to consider the use of polycrystalline grit rather than single crystal.

Indentation Fracture

The indentation fracture technique is commonly used at room temperature to determine critical stress intensity factors, but has been used far less frequently at elevated temperatures. From the practical viewpoint there are two main advantages of indentation fracture testing vis-à-vis conventional macroscopically notched K_{Ic} measurements. Firstly, the type of loading, and hence induced stress state, combined with the small stressed volume, is more relevant to most applications of ceramics. Secondly, the specimens are relatively straightforward to prepare and one sample can be used to obtain many results, and it is consequently possible to perform more tests at more closely spaced temperature intervals.

The results described above clearly indicate that the indentation critical stress intensity factors generally decrease with increasing test temperature. This occurs even when the conventional K_{Ic} values are either temperature independent, or increase with increasing test temperature, cf hot-pressed Si_3N_4 above 1100°C. This reinforces the fact that the use of K_{Ic} to rank materials in applications where fracture occurs under compressive loading and for small stressed volumes may be misleading.

Also, it is of relevance to note that it is sometimes assumed that an increase in the ease of plastic deformation in a material, i.e. reduced hardness, will axiomatically lead to an increase in fracture toughness. The present results clearly demonstrate that an increase in plastic deformation, i.e. decrease in hardness, may well be accompanied by a decrease in toughness.

CONCLUSIONS

This paper has reported results of the indentation hardnesses and critical stress intensity factors between room temperature and 800°C in air for hot-pressed silicon carbide and nitride, Refel silicon carbide, ceria stabilized polycrystalline tetragonal zirconia, and single crystals of silicon carbide and calcia and yttria stabilized cubic zirconias, and between 800°C and 1250°C in vacuo for the first three materials.

Hot-pressed silicon carbide and nitride show relatively little decrease of hardness with increasing temperature, i.e. ~ 30% between room temperature and 1200°C. Refel SiC, single crystal SiC and ceria stabilized TZP decrease in hardness by ~ 50% between room temperature and 800°C. The single crystal cubic zirconias show the greatest relative decrease in hardness, i.e. >65% between room temperature and 800°C. The effect of orientation on the hardness of calcia stabilized zirconia is relatively limited. The yttria stabilized zirconia decreases in hardness above 400°C to a significantly greater extent than the calcia stabilized.

The indentation critical stress intensity factors, K_{Ia}, for hot-pressed silicon carbide and nitride decrease by approximately 10 - 20% between room temperature and 1250°C. K_{Ia} for Refel silicon carbide increases gradually from 3.2 MPa.m$^{0.5}$ at room temperature to 4.6 MPa.m$^{0.5}$ at 800°C, followed by a sharp increase to > 8 MPa.m$^{0.5}$ between 800°C and 1000°C, and then a gradual decrease. The single crystal zirconia indentation critical stress intensities are markedly anisotropic at 22°C, varying between 0.8 and 1.4 MPa.m$^{0.5}$, but this anisotropy decreases with increasing test temperature such that at 800°C $K_{Ia} \sim 0.5$ MPa.m$^{0.5}$. There is a very sharp decrease in K_{Ia} for ceria stabilized TZP from ~ 20 MPa.m$^{0.5}$ at room temperature to 0.94 MPa.m$^{0.5}$ at 700°C as a result of the loss of the transformation toughening.

REFERENCES

Ajayi O.O., and Ludema, K.C., 1988, Surface damage of structural ceramics: implications for wear modelling, *Wear*, 124:237.

Anstis, G.R., Chantikul, P., Lawn B.R., and Marshall, D.B., 1981, A critical evaluation of indentation techniques for measuring fracture toughness: I, direct crack measurements, *J. Amer. Ceram. Soc.*, 64:533.

Davidge, R.W., and Evans, A.G., 1970, The strength of ceramics, *Mat. Sci. Eng.*, 6:280.

Edington, J.W., Rowcliffe, D.J., and Henshall, J.L., 1975, The mechanical properties of silicon nitride and silicon carbide, part I: materials and strength, *Powder Metallurgical Review 8*, 7:82.

Evans, A.G., 1979, Fracture Toughness: The Role of Indentation Techniques, *in:* ASTM Special Technical Publication No. 678, S.W. Freiman, ed., American Society for Testing and Materials, Philadelphia, PA.

Evans, A.G., 1984, Micromechanics of failure in brittle solids, *in:* "Microstructure and Properties of Ceramic Materials", T.S. Yen and J.A. Pask, eds., Gordon and Breach, New York.

Evans, A.G., and Charles, E.A., 1976, Fracture Toughness Determinations by Indentation, *J. Amer. Ceram. Soc.*, 59:371.

Evans, A.G., and Marshall, D.B., 1981, Wear mechanisms in ceramics, *in:* "Fundamentals of Friction and Wear of Materials", D.A. Rigney, ed., American Society for Metals, Metals Park, OH.

Evans, A.G., and Wilshaw, T.R., 1976, Quasi-static solid particle damage in brittle solids - I. observations, analysis and implications, *Acta Metall.*, 24:939.

Guillou, M.-O., Carter, G.M., Hooper, R.M, and Henshall, J.L., 1990, Hardness and Fracture Anisotropy in Single Crystal Zirconia, *J. Hard Mater.*, 1:65.

Guillou, M.-O., 1992, "Indentation Deformation and Fracture of Hard Ceramic Materials", Ph.D. thesis, University of Exeter, U.K.

Guillou, M.-O., Henshall, J.L., Hooper, R.M., and Carter, G.M., 1992, Indentation fracture testing and analysis and its application to zirconia, silicon carbide and silicon nitride ceramics, *J. Hard Materials*, 3:421.

Henshall, J.L., Rowcliffe, D.J., and Edington, J.W., 1975, The fracture toughness and delayed fracture of hot-pressed silicon nitride, *in:* "Special Ceramics 6", P. Popper, ed., British Ceramic Research Association, Stoke-on-Trent.

Henshall, J.L., Rowcliffe, D.J., and Edington, J.W., 1977, Fracture toughness in single crystal silicon carbide, *J. Amer. Ceram. Soc.*, 60:373.

Henshall, J.L., Rowcliffe, D.J., and Edington J.W., 1979, K_{Ic} and delayed fracture measurements on hot-pressed SiC, *J. Amer. Ceram. Soc.*, 62:36.

Lange, F.F., Transformation toughening, Part 5: Effect of temperature and alloy on fracture toughness, *J. Mater. Sci.*, 17:255.

Liang, K.M., Orange, G., and Fantozzi, G., Evaluation by indentation of fracture toughness of ceramic materials, *J. Mater. Sci.*, 25:207.

Maerky, C., Guillou, M.-O., Henshall, J.L. and Hooper, R.M., 1995, Indentation hardness and fracture toughness in single crystal $TiC_{0.96}$, presented at the Fifth International Conference on the Science of Hard Materials, Maui, Feb. 95, to be published.

Tikare, V., and Heuer, A.H., 1991, Temperature-dependent indentation behavior of transformation-toughened zirconia-based ceramics, *J. Amer. Ceram. Soc.*, 74:593.

Palmqvist, S., 1957, A method to determine the toughness of brittle materials, especially hard metals, *Jernkontorets Ann.*, 141:303.

ON HIGH TEMPERATURE CREEP AND DAMAGE
OF POLYCRYSTALLINE CERAMICS

H. Balke[1], W. Pompe[2], and H. Weber[2]

[1]Dresden University of Technology
 Institute of Solid Mechanics, D-01062 Dresden
[2]Max-Planck-Gesellschaft, Research Group on Mechanics
 of Heterogeneous Solids, D-01069 Dresden, Hallwachsstr. 3

INTRODUCTION

The creep life time of ceramic materials has usually a large scatter. Reasons for that are fluctuations of microstructural parameters like for instance the grain boundary energy, the diffusion constant and the geometry of the grains. All these factors influence the cavity nucleation on grain boundaries, which is the main cause of creep fracture at high temperatures. It is well known that critical stresses for cavity nucleation calculated by the theory of steady-state nucleation are unrealistically high (Evans et al., 1980). For that reason we looked for mechanisms which give rise to high stress concentrations. Sudden grain boundary sliding is often regarded as a possible mechanism for this. However there is no sufficient experimental evidence for sudden grain boundary sliding up to now. Therefore other mechanisms should be investigated to explain cavity nucleation. In the following sections we want to consider several grain boundary structures in order to investigate the influence of the grain geometry on stress amplification which may cause cavity nucleation.

POLYCRYSTALLINE STRUCTURES MODELLED
BY TWO DIMENSIONAL TESSELLATIONS

Calculations based on three–dimensional regular grain structures, done for instance by Anderson and Rice (1985), use the so-called tetrakaidecahedron (Fig. 1). But the application of these regular grain bodies implies anisotropy of the modelled material. Furthermore the handling of the three–dimensional structure is not easy.

In two dimensions (2d) the grain structure is often described by a hexagonal tessellation (Fig. 2). It has the advantage that for equal grain boundary tensions the triple junctions are in mechanical equilibrium. But these patterns do not include the randomness of the geometric parameters, and like networks consisting of tetrakaide-cahedra they contain artifical orientations. Because of their high similarity to real cross-sections two dimensional Dirichlet or Voronoi tessellations (Fig. 3) are often recommended (Riesch-Oppermann et al., 1993). They are produced in the following way (Mecke et al., 1990): First, one distributes a number of germ points over a window on a plane. Each germ point represents a future grain. The grain boundaries are drawn so that each point of a grain has a shorter distance to its germ point than to another germ point. The tessellation can be understood as a model of growing grains whose growth starts at the same time, runs off with the same velocity and comes to an end if two grains meet each other. For equal grain boundary tension the triple junctions are not in mechanical equilibrium generally. We emphasize that such a tessellation consists of straight lines connected at triple junctions.

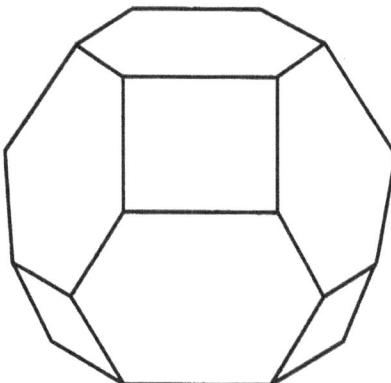

Figure 1. A three dimensional grain modelled by a tetrakaidecahedron.

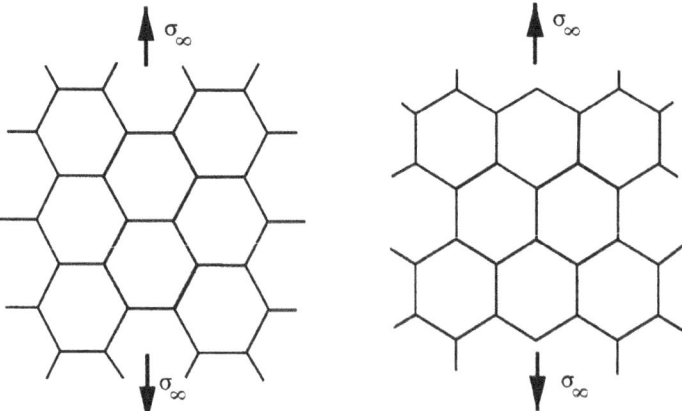

Figure 2. Hexagonal grain boundary network with different relative orientations of the external load. For the same grain boundary tension the triple junctions are in mechanical equilibrium.

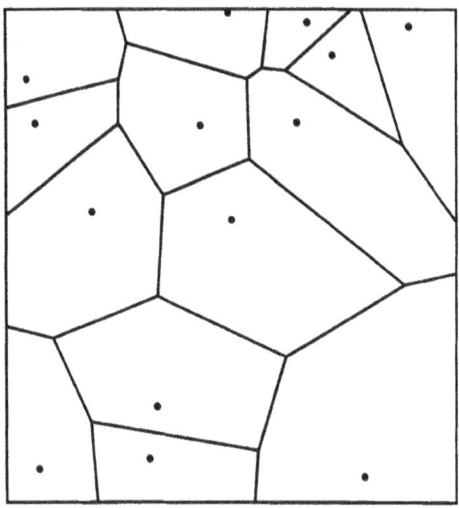

Figure 3. Dirichlet or Voronoi tessellation with 14 germ points.

DIFFUSIONAL CREEP PROCESS

At high temperatures, weak external load and small grain size, the grain boundary diffusion, the so-called Coble creep process (Coble, 1963), is the main mechanism responsible for creep and damage of one-phase ceramic materials, like pure Al_2O_3 (Gandhi and Ashby, 1979). In the following sections we want to consider the modelling of the creep process for different plane grain boundary networks. On one grain boundary of a 2d tessellation this diffusion process is described by first Fick's law

$$j(x) = -\frac{\delta_{GB} D_{GB}(T)}{\Omega k T} \nabla \mu(x),$$ (1)

where $j(x)$ and $\mu(x)$ are diffusion flux density and chemical potential at the grain boundary coordinate x, and k, T, δ_{GB} and $D_{GB}(T)$ are Boltzmann's constant, absolute temperature, grain boundary thickness and grain boundary diffusion constant. The gradient of the chemical potential (Riedel, 1987)

$$\nabla \mu(x) = -\nabla \sigma_N(x)$$ (2)

depends on the local normal stress $\sigma_N(x)$. The conservation of mass provides the continuity equation

$$\Omega \nabla j(x) + \dot{u}_N(x) = 0.$$ (3)

Combining the equations (1), (2) and (3), we get

$$\frac{\partial^2}{\partial x^2} \sigma_N(x) = -\frac{k T}{\Omega \delta_{GB} D_{GB}(T)} \dot{u}_N(x).$$ (4)

As the grains are described as rigid bodies, the relative normal displacement velocity

$\dot{u}_N(x)$ between two neighbouring grains is a linear function of the coordinate x

$$\dot{u}_N(x) = Ax + B. \qquad (5)$$

This means we consider only a steady-state process without any stress relaxation influenced by the elasticity of the material. Then the normal stress distribution $\sigma_N(x)$ has the form

$$\sigma_N(x) = -\frac{kT}{\Omega\,\delta_{GB}\,D_{GB}(T)}\left(\frac{A}{6}x^3 + \frac{B}{2}x^2 + Cx + D\right) \qquad (6)$$

with the unknown coefficients A, B, C and D. Shear stresses on the grain boundaries are not included in these considerations.

To solve the creep problem it is necessary to determine the four coefficients for each grain boundary of a tessellation. Thus we need as many conditions as unknown coefficients. These conditions are provided by the continuity of the chemical potential and the diffusion flux density at the triple junctions, the compatibility of relative grain rotation velocities and relative grain displacement velocities (Hazzledine and Schneibel, 1993) and the conditions of the static equilibrium.

Using all these relations one gets a system of equations

$$\mathbf{Q} \bullet \mathbf{V} = \mathbf{R} \qquad (7)$$

for the determination of the coefficients $\mathbf{V} = (A_1, B_1, ..., C_{n_{GB}}, D_{n_{GB}})$, where n_{GB} is the number of grain boundaries of the tessellation. The matrix $\mathbf{Q}$ contains the tessellation dependent coefficients and the vector $\mathbf{R}$ the given values of load or diffusion flux density at the external boundary of the tessellation. A dimensionless form of (7), independent of temperature and material parameter, can be used for numerical calculations (Fig. 7, Fig. 8).

The solution of the system of equations describes the creep situation at a certain time. It is not clear in which way the time development of the system goes on, because the new position of the grain boundaries after a time step and therefore also of the triple junctions is not defined by the model. If one assumes the new grain boundaries after a time step on the bisectors of the angles between neighbouring grains, there will be triangles and no triple junctions in the region where three grain boundaries meet.

The model presented in this section can be applied to different grain-boundary structures. Riedel (1987) calculated the normal stresses in a grain-boundary network of hexagonal grains for loading in direction of one symmetry axes of the grains. In his calculation stress distributions results with a maximum two times higher than the external load. Hazzledine and Schneibel (1993) applicated the creep equations to disturbed hexagonal networks. They got stress amplifications by a factor of about four related to the external load.

APPLICATION TO DIRICHLET TESSELLATIONS

For Dirichlet tessellations we get the following result: It is possible to deform these structures only by tangential displacements (Fig. 4). Reason of this specific geometric property is the existence of the dual pattern of the Dirichlet tessellation. It was shown (Balke et al., 1994, 1995) that, caused by this fact, in the case of zero normal displacement velocities on all grain boundaries there is a degree of freedom concerning the relative tangential displacement velocities. Furthermore the number of linear independent equations is $n - 1$ for n conditions of force equilibrium. This means the arrangement of grains of a Dirichlet tessellation can be deformed without external load, and by application of an external load in general direction the mechanical equilibrium demands infinitely high normal stresses on the grain boundaries.

For physical reasons –if one assumes that a real polycrystal behaves like a Dirichlet tessellation– there should be processes at the triple junction which lower these high stresses along the grain boundaries. These processes could cause higher stresses at the triple junctions.

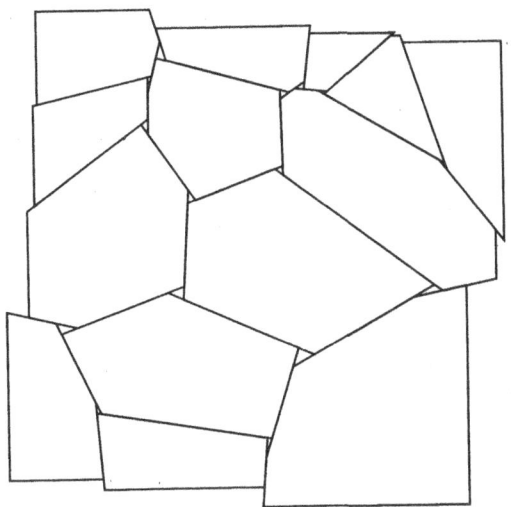

Figure 4. Dirichlet or Voronoi tessellation of Fig. 3, deformed only by relative tangential displacements. On the grain boundaries no covering occurs.

APPLICATION TO DISTURBED DIRICHLET TESSELLATIONS

If we disturb a general Dirirchlet tessellation by arbitrary displacements of the triple junctions, the degree of freedom is blocked. Application of load causes relative normal displacement velocities between neighbouring grains and normal stresses on the grain boundaries. For deviations from the Dirichlet structure (Fig. 5) shown in Fig. 6, we get results for the normal stresses which are comparable to the applied external stress (Fig. 7, 8). The magnitude of these results is of the same order as the normal stresses calculated for a hexagonal tessellation (Riedel, 1987) and for a disturbed hexagonal tessellation (Hazzledine and Schneibel, 1993). The stress amplification is two and about four in these cases, related to the external load. If a disturbed Dirichlet tessellation is modified such that it approaches a Dirichlet tessellation, normal stresses and normal displacement velocities increase (for general load direction). During this process the part of the matrix $\mathbf{Q}$ concerning the mechanical equilibrium of the grains becomes numerically singular.

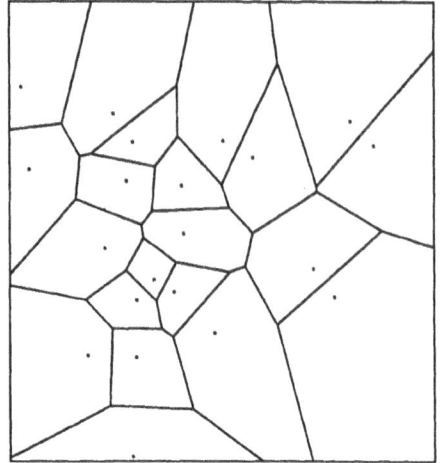

Figure 5. Dirichlet tessellation with germ points.

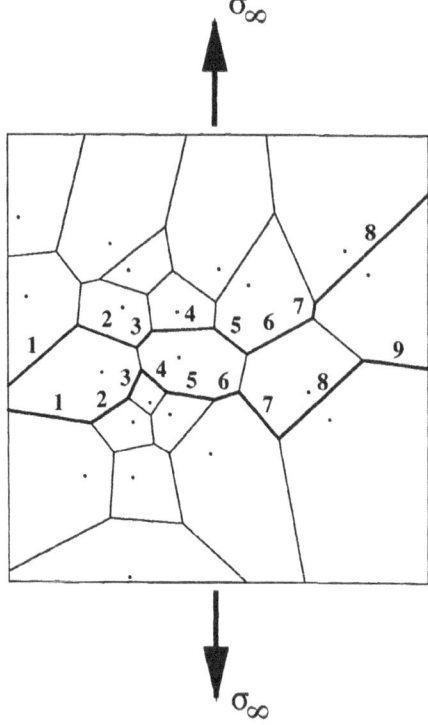

Figure 6. Disturbed Dirichlet tessellation based on Fig. 5. The germ points of the Dirichlet tessellation are marked.

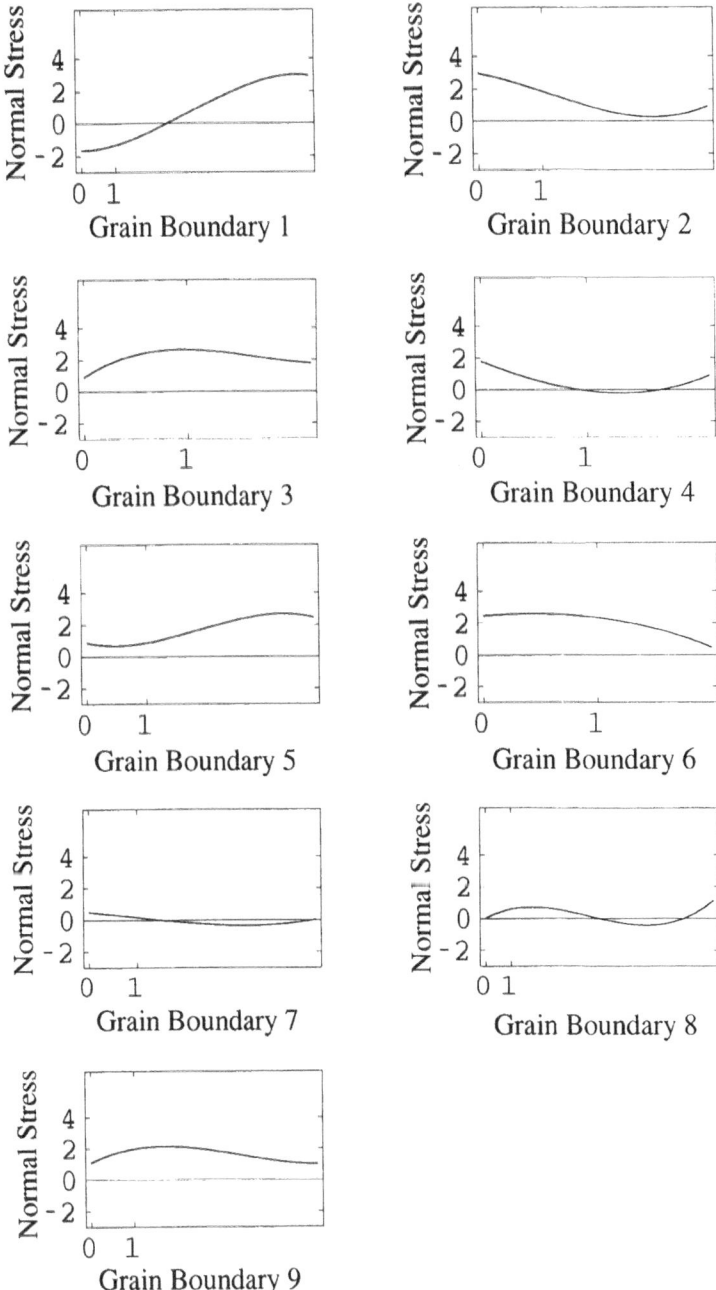

Figure 7. Normal stress distribution on the grain boundaries of the lower marked path from the left to the right. The normalisation length is the same for all grain boundaries. The direction of the external load $\sigma_\infty = 1$ is the vertical axes (Fig. 6). At the boundary the diffusion flux density is zero.

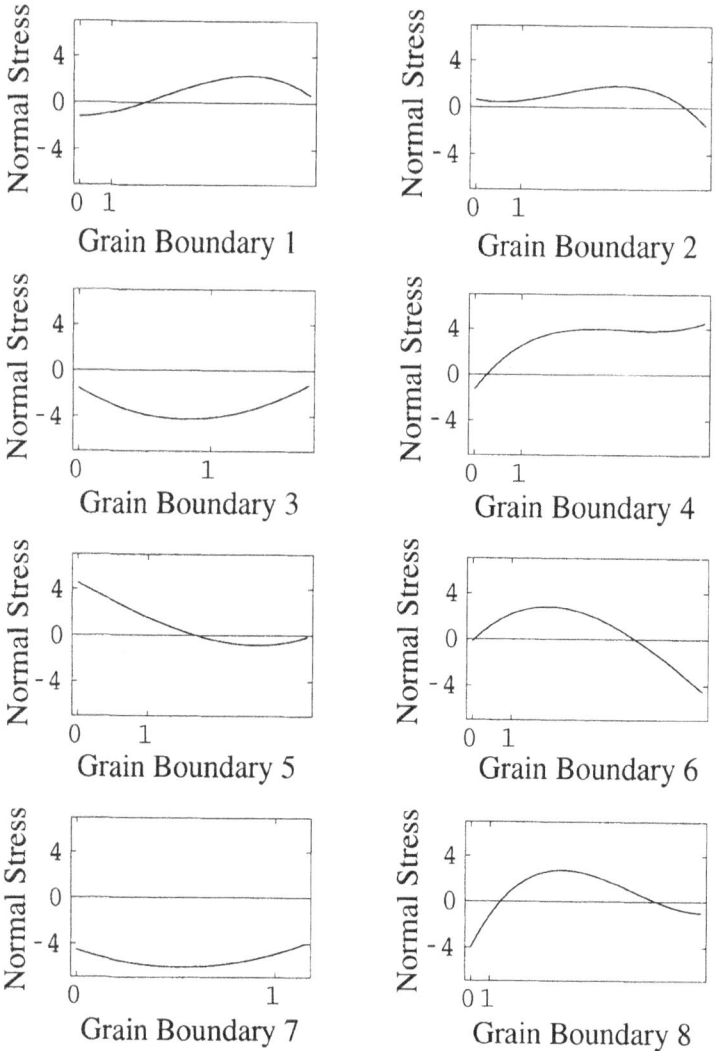

Figure 8. Normal stress distribution on the grain boundaries of the upper marked path from the left to the right. The normalisation length is the same for all grain boundaries. The direction of the external load $\sigma_\infty = 1$ is the vertical axes (Fig. 6). At the boundary the diffusion flux density is zero.

SHEAR STRESSES ON THE GRAIN BOUNDARIES

It is obviously clear that shear stresses on the grain boundaries have the tendency to suppress the sliding process on the grain boundaries caused by the degree of freedom of a Dirichlet tessellation. The question is for which viscosity one gets shear stresses so that the normal stresses on the grain boundaries are of the order of the external stress. Ashby (1972) calculated a viscosity

$$\eta = \frac{kT}{8b\delta_{GB}D_{GB}} \qquad (8)$$

resulted from the step–like character of the grain boundary, where b is the atomic size. But this viscosity (8) does not provide sufficient large shear stresses. Only for viscosities much higher than the given above we get these results. Such high viscosities can be produced only by unrealistically high roughness of the grain boundary.

RESULTS AND DISCUSSION

In this paper some models for grain structures are presented. The Coble creep mechanism was applied to the 2d structures. For common tessellations one gets result for normal stresses which are in the region of applied external stress. The Dirichlet tessellation has the property that it can be deformed without any normal displacements between neighbouring grains. For general load direction the mechanical equilibrium requires infinitely high normal stresses on the grain boundaries of these patterns. Processes at the triple junctions, which are not taken into account up to now, should lower these high stresses and cause stress amplifications at the triple junctions. The stress concentrations –on the grain boundary and at the triple junctions– could be a reason of cavity nucleation.

Shear stresses on the grain boundaries suppresses the sliding caused by degree of freedom, but a stabilisation of the Dirichlet-like structure needs an unrealistically high roughness of the grain boundary.

ACKNOWLEDGMENT

The authors would like to thank Dr. A. Brückner-Foit of the University of Karlsruhe for stimulating discussions on this topic. Furthermore we thank Dr. Riesch-Oppermann for the opportunity to use his program for the generation of Dirichlet tesselations. The work was supported by the Deutsche Forschungsgemeinschaft.

REFERENCES

Anderson, P. M., and Rice, J. R., 1985, Constrained creep cavitation of grain boundary facets, *Acta Metall.* 33:409.

Ashby, M. F., 1972, Boundary defects, and atomistic aspects of boundary sliding and diffusional creep, *Surface Science* 31:498.

Balke, H., Pompe, W., Schneider, D., and Weber, H., 1994, Bericht zum DFG-Vorhaben Po 392/1-1.

Balke, H., Pompe, W., and Weber, H., 1995, Zum Diffusionskriechen von Polykristallen mit stochastisch angeordneten Korngrenzen, *Zeitschrift Angew. Math. Mech.* 75:S219.

Coble, R. L., 1963, A model for boundary diffusion controlled creep in polycrystalline materials, *Journal Appl. Phys.* 34:1679.

Evans, A. G., Rice, J. R., and Hirth, J. P., 1980, Suppression of cavity formation in ceramics: prospects for superplasticity, *Journal Am. Ceram. Soc.* 63:368

Gandhi, C., and Ashby, M. F., 1979, Fracture-mechanism maps for materials which cleave: F.C.C., B.C.C. and H.C.P. metals and ceramics, *Acta Metall.* 27:1565.

Hazzledine, P. M., and Schneibel, J. H., 1993, Theory of Coble creep for irregular grain structures, *Acta Metall. Mater.* 41:1253.

Mecke, J., Schneider, R. G., Stoyan, D., and Weil, W. R. R., 1990, "Stochastische Geometrie", Birkhäuser Verlag, Basel.

Riedel, H., 1987, "Fracture at High Temperatures", Springer, Berlin.

Riesch-Oppermann, H., Brückner-Foit, A., Munz, D., and Winkler, T., 1993, Crack initiation, growth, and interaction in thermal cyclic loading - a statistical approach. *in* : "Behaviour of Defects at High Temperatures", ESIS 15, Ainsworth, R. A. and Skelton, R. P., Mechanical Engineering Publications, London.

PSEUDOPLASTIC DEFORMATION AND FAILURE OF Y-TZP-AL$_2$O$_3$ CERAMICS AT HIGH TEMPERATURE

V.N. Antsiferov[1], A.A. Tashkinov[2], V.E. Wildemann[2], and I.G. Sevastianova[1,2]

[1]Powder Engineering Centre with Research Institute of Problems of Powder Technology

[2]State Polytechnical University
Perm, Russia

INTRODUCTION

The study of ceramics and ceramic-base composite deformation and the development of models simulating the conditions of great deformation is an actual and important problem. An important part of it is the pseudoplastic behaviour of zirconia ceramics belonging to a complex polymorphism and structural inhomogeneity which, in combination with conditions which are favourable for superplasticity under thermal and mechanical loads, cause a substanical increase in deformation up to failure, accompanied by a phenomena of disperse failure, rotational reconstruction of the microstructure, and stable microcrack growth[1].

One of the possible ways to increase the plasticity of ceramic materials is to create conditions allowing to achieve postcritical states. In such cases there is no failure at maximum allowable stresses but, the subcritical deformation associated with the stable accumulation of microdamage causes a significant increase in critical deformations[2,3].

In the process of constructing an equilibrium diagram the object of study is a weakened zone of a test specimen. The parts of the specimen which are outside this zone· make up, together with the parts of the loading facility devices, a loading system. The results of any experiment depend on the rigidity of such a system. If the rigidity is sufficient, the equilibrium state of a material, after its passage through the maximum of the stress-strain diagram, is stable. Otherwise, the process of microdamage accumulation is unstable and the specimen fails dynamically.

As it is known, the rigidity of a body or its reverse quality, the flexibility, are the properties defining the changes of load during the movement associated with deformation. In the case of a critically soft loading, when forces are applied to a body which do not depend on its resistance, the failure corresponds to maximum possible stresses. If the loading is critically hard, equilibrium damage accumulation is possible which is reflected in the deformation diagram as its descending branch. If the flexibility of the system composed of a specimen and a test facility is not taken into account, this results in values of mechanical properties, in particular, crack resistance, which are not linked to a particular test facility and a particular specimen configuration.

The novelty of the approach in our paper is associated with the increase in the period of superplastic deformation of zirconia ceramics. Research has been performed to study and create loading conditions under which postcritical material deformation is stable until the moment when maximum failure deformation is reached.

THEORETICAL BACKGROUND

The main difficulty in the experiments is to provide sufficient rigidity of the system used to apply load to a material. For this purpose, devices were developed which increase the rigidity of standard facilities[3,4] as well as special test specimens[5,6] and special test facilities provided with fast feedback[6]. Since the success of experiments depends on the relationship between the rigidity of the loading system and that of the weakened zone, the conditions of equilibrium accumulation of damage are to be provided on the basis of special calculations of specimen dimensions[7].

The loss of stability at the postcritical stage means that the formation and the growth of cracks are becoming avalanche-like. In this case macrofailures of specimens are observed. This happens when the energy applied by the loading device and that released under the conditions of local off-loading of specimen portions become higher than is required for deformation, damage accumulation and crack formation. Any point on the descending branch of a diagram may correspond to the moment of stability loss depending on the relationship between the volume of the weakened zone and that of the specimen portion under elastic deformation and also on the rigidity of the test facility.

To evaluate the stability of zirconia ceramics postcritical deformation stability at high temperatures which is accompanied by the transformational plasticity, rotational restructuring and equilibrium growth of defects, it is necessary to analyse the balance between the energy consumption and the applied energy under the conditions of virtual increase in postcritical deformation[7]. The energy consumed consists of increments of elastic deformation energy, ΔW, and destruction work, ΔA. The energy is applied due to the work performed by external forces, ΔE. For a volume element of a material the destruction work and the increment of potential unit deformation work together make up a unit deformation work which, at any interval of deformation, may be determined as the area under the equilibrium diagram curve. At the initial stage of elastic deformation the deformation work is equal to the increment of elastic energy, i.e. the destruction work is equal to zero. For the horizontal part of the curve corresponding to the yield in the diagram practically no increment of the elastic energy is observed and the deformation work is equal to the destruction work. In this case destruction work means the magnitude of dissipation energy under the conditions of pseudoplastic deformation. At this postcritical stage the destruction work exceeds the deformation work. The steeper the plot is at the final stage of deformation, the greater is the difference. The process of destruction in the weakened zone is additionally accelerated by the liberated potential elastic deformation energy received from the other parts of a specimen.

The inequality:

$$\Delta W + \Delta A > \Delta E \qquad (1)$$

is the condition for the stability of postcritical deformation. It shows that spontaneous continuation of destruction is impossible because the applied and released energies together are insufficient to cause destruction.

Nonfulfillment of the inequality (1) corresponds to the loss of deformation stability and the dynamic destruction of the sample. From the point of view of creating the conditions under which the design rigidity of a sample is significantly lower than the rigidity of the test facility it is preferable to use bending tests. But the peculiarity of such tests is that the distribution of stresses within a specimen is rather non-uniform and it makes it difficult to interpret the experimental data by construction of stress-strain diagrams for the materials under testing. The non-linear character of the load-load point movement diagrams for three-point bending indicates that in the most loaded central zone of a specimen the accumulation of microdamage is irreversible. The greatest deviation from the proportional load-movement relationship under the conditions of elastic plastic bending is observed if the material of a sample is not plastically strengthened. As our studies show, in this case the destructive load must be by 50 % greater than the load causing plastic deformation in the samples, and non-linearity is observed on the stress-strain diagram. The lower degree of load increase and the horizontal or the descending branch of a plot testify that there is post-critical deformation in the weakened zone of a sample.

EXPERIMENTAL DETAILS

For our studies Y-TZP-Al$_2$O$_3$ ceramics were selected which had been produced from ultradispersed powder subjected to laser treatment[8]. The materials were produced by the method of cold moulding followed by heat treatment at 1800 °C. The test samples having prism shape, 3.5 x 5 x 40 mm in size, were ground and their sharp edges were chamfered. The samples were tested at 1600 °C by the three-point bending method with the distance between the supports being 25 mm. The speed of the crosspiece movement was 0.16 mm/min.

RESULTS AND DISCUSSION

The studies allowed to determine the conditions of equilibrium post-critical deformation in the local zone of a sample:

$$2Ah_b{}^3 + C < \theta\varepsilon_b,$$
$$A = b\varepsilon_t (E - E_1) [1 - \varepsilon_t{}^2/(3\varepsilon_b{}^2)] + 2/3 \cdot b\varepsilon_b (E_1 + D) - bD\varepsilon_p \quad (2)$$
$$c = 2/3 \cdot bD\varepsilon_b h^3$$

where θ characterises the rigidity of a loading system, ε_t and ε_b are deformations corresponding to the yield point and the ultimate strength of a material, E is the modulus of elasticity, E_1 is the modulus of strengthening, and D is the modulus of strength loss. The latter two characteristics determine the slope of stress-strain diagram branches corresponding to plastic and post-critical deformations, b is the width of the beam rectangular cross section, h is half of the cross section height, ε_p is the maximum obtainable deformation for the given material corresponding to the decrease of resistance to zero. The distance h_b from the beam central axis to the border of the post-critical deformation zone can be determined from the bending moment equation:

$$M = Ah_b{}^2 + bD\varepsilon_p h^2 - c/h_b \quad (3)$$

An analysis of the stability conditions shows that if the rigidity of a test facility is sufficient to start the process of post-critical deformation, there will be no loss stability if loading is continued.

High-temperature tests involving samples of Y-TZP-Al$_2$O$_3$ ceramics allowed to construct load-deflection diagrams with clearly defined pseudo-plastic deformation branches. A typical deformation diagram with a descending branch is shown in Fig. 1. It indicates that at the post-critical stage of deformation the dissipation processes are in equilibrium. The gentle slope of the descending branch testifies that TZP-Al$_2$O$_3$ ceramic material is highly tough and that its tendency to dynamic failure is low. The bending strength at 1600 °C was 170 to 175 MPa. This indicates that the mechanical properties of the ceramics are high and their superplastic deformation margins are wide. The experimental sample before and after testing is shown in Fig. 2. The level of superplastic deformations attained was 30 %.

Studies were also conducted to determine the effect of surface machining on the properties of samples. The tests gave a complete load-deflection diagram with a descending branch, as shown in Fig. 3. To determine the stability of post-critical deformation, off-loading was performed at a speed of test facility crosspiece movement of 0.34 mm/min.

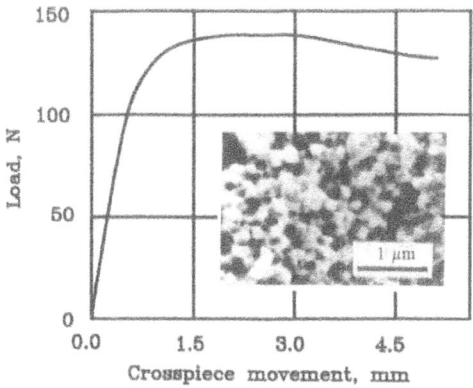

Figure 1. Diagram of superplastic deformation of a ceramic sample at 1600 °C.

Additional surface machining of samples caused the development of surface flaws and damage to the samples which was manifest in a steeper slope on the diagram. But in this case the conditions for the stability of post-critical deformation also applied. The equilibrium character of dissipative processes in the structure of surface-machined samples is confirmed by their mechanical behaviour under the conditions of off-loading and second loading. The experimental samples preserved their integrity when the test facility was stopped at moments corresponding to any point on the descending branch of the stress-strain diagram and after subsequent off-loading and removal from the facility. The results of the study show that the post-critical deformation of samples with surface flaws causes a two-fold increase in the superplastic deformation margin.

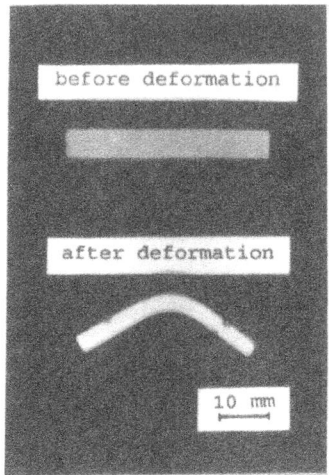

Figure 2. Superplasticity of Y-TZP-Al$_2$O$_3$ ceramics at 1600 °C.

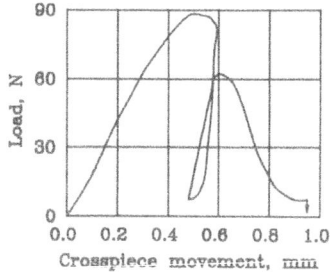

Figure 3. Complete diagram of ceramic sample deformation with off-loading at post-critical stage at 1600 °C.

SUMMARY

The theoretical studies conducted have allowed to determine the mechanism of post-critical superplastic deformation of zirconia ceramics. Conditions have been determined under which post-critical deformation could be achieved during bending tests. The results of high-temperature tests of Y-TZP-Al$_2$O$_3$ ceramics show that the superplastic behaviour of materials is possible and that great deformation of ceramic materials can be achieved if the conditions for stable dissipation processes in the structure of material are provided. The experiments show that achievement of the post-critical deformation stage increases more than twice the fracture strain of superplastic ceramic materials.

REFERENCES

[1] F. Wakai, Superplasticity of Ceramics, Ceramics International, 17:153 (1991).

[2] S.D. Volkov, G.I. Dubrovina, and Yu.P. Sokovnin, On the Theory of Failure Stability of Technical Materials, Strength of Materials, 2:3 (1978).

[3] A.A. Lebedev and I.G. Chausov, Material Testing Facility Constructing Complete Equilibrium Deformation Diagrams, Strength of Materials, 12:104 (1981).

[4] Yu.V. Sokolkin and V.E. Wildemann, Postcritical Deformation and Failure of Composite Materials, Mechanics of composite materials, 29:163 (1993).

[5] A.A. Lebedev, I.G. Chausov, and Yu.L. Evetsky, Methods for the Construction of Complete Sheet Material Deformation Diagrams, Strength of Materials, 9:29 (1986).

[6] S.D. Volkov, Yu.P. Guskov, and V.I. Krivospitskaya et al., Experimental Functions of Alloy Steel Resistance under Tension and Torsion, Strength of Materials, 1:3 (1979).

[7] V.E. Wildemann, Yu.V. Sokolkin, and A.A. Tashkinov, Boundary Problems of Continual Failure Mechanics, Perm (1992).

[8] V.N. Antsiferov, I.G. Sevastianova, and V.I. Ovchinnikova et al., The Peculiarities of Ultradispersed State Formation in Ceramic Powders of ZrO$_2$-Y$_2$O$_3$-Al$_2$O$_3$ System, Refractories, 11:12 (1994).

EFFECT OF REMANENT POLARIZATION LEVEL ON MACROCRACK

EXTENSION IN LEAD ZIRCONATE TITANATE (PZT)

Michael D. Hill and Grady S. White

Ceramics Division, NIST
Gaithersburg, MD 20899

Isabel K. Lloyd

Engineering Materials Department
University of Maryland, College Park
College Park, MD

INTRODUCTION

Many applications of piezoelectric actuator materials inherently incorporate large numbers (10^7 to 10^{10}) of loading cycles at appreciable strains. Therefore, reliability of these materials under cyclic loading conditions is of paramount concern. However, the most commonly used piezoelectric actuator, PZT, has been observed to fail during cyclic loading at stress levels below the failure stress measured by monotonic loading both at room temperature[1] and at an elevated temperature[2] of 180 °C. This suggests the existence of cyclic failure mechanisms in addition to those known to occur during monotonic loading. We examined the high power, resonator material, PZT-8, to investigate damage processes during cyclic loading. PZT-8 was designed for use in high frequency (10 kHz to 100 kHz), high-power applications such as resonators and ultrasonic motors. A consequence of resonant operation is component heating (up to 200 °C)[2] due to electrical and mechanical loss. Partial depolarization has been observed as a result of resonance stress cycling in the temperature range 150 °C $\leq$ T $\leq$ 200 °C.[3] It is the purpose of this research to examine the effect of partial depolarization on macrocrack extension in poled polycrystalline PZT-8.

In previous work on fully poled ferroelectric materials, including PZT, Vickers indentations were used to provide stable, characterizable macrocracks.[4] It was observed that for Vickers indentation with a diagonal placed along the symmetry axis, crack lengths parallel to the poling direction were shorter than those perpendicular to the poling direction.[4,5,6] Although indentation cracks in fully poled material have been examined, there has never been a systematic study of the effect of thoroughness of poling on the indentation crack lengths in any ferroelectric material. Since partial depolarization has been shown to occur (high temperature/high stress) in PZT-8 with or without indentations[3], a knowledge of the relation between polarization level and macrocrack extension is necessary before the reliability of PZT-8 in service may be estimated.

PROCEDURE

The material used in this work was PZT-8, a polycrystalline ceramic with tetragonal symmetry, 1μm to 2 μm equiaxed grains, and a Curie temperature, T_c, of 300 °C. The material was received in 6.5 mm x 12.4 mm x 70.0 mm billets, poled perpendicular to the longitudinal axis and with electrodes applied. Vickers indentations were placed on a polished PZT surface using a load of 0.5Kg, and crack lengths were determined by inspection in an optical microscope. Eight indentations were measured to obtain crack length values.

ELECTRICAL RESONANT LOADING

In addition to mechanical loading, electrical loading was also investigated. As has been described previously[2,3,7], electrical cycling was performed by application of 50 V parallel to the poling direction at the specimen longitudinal resonance frequency (24 kHz). During loading, the temperature of the sample increased to a steady state value between 170 °C and 190 °C. A schematic of the electrical loading appears on Figure 1.

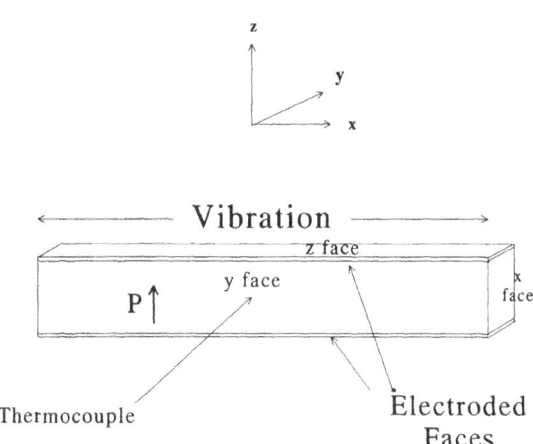

Figure 1. Schematic of electrical resonant cycling procedure showing the three oriented planes (x, y and z).

Electrical Poling

Since as received samples had been poled by the supplier, to investigate the effect of thoroughness of poling on mechanical properties, it was first necessary to depole the material. This was accomplished by short circuiting the billet and heating it to 400 °C for

12 h. Samples were re-poled by immersion in oil and simultaneous application of electric fields at 22 ° C or 115 °C for 10 min. Samples with a range of different remanent polarization levels were obtained.

Measurement of Material Coefficients

The compliance term at constant electric field perpendicular to the poling direction s_{11}^E, was measured by determination of the fundamental resonance frequencies (f_r) for the transverse vibration mode and application of the relation[8]:

$$s_{11}^E = (4\rho l^2 f_r^2)^{-1}$$

Here, ρ is the sample density (7.6 ± 0.1 g/ml by the Archimedes method) and l is the sample length. The resonance frequency was located with an impedance analyzer. Since the material was poled perpendicular to the longest dimension (Figure 1), it was possible to obtain only the s_{11}^E using the impedance analyzer.

The transverse piezoelectric strain coefficient, d_{31}, was calculated[8] from the fundamental transverse resonance and anti-resonance frequencies and the sample capacitance, as measured by an impedance analyzer. Due to the geometry of the samples, the d_{31}, rather than the d_{33}, was measured.

The dielectric constant was determined from the capacitance measured at 1kHz using an impedance analyzer. Unclamped relative dielectric constants were obtained from the capacitance by the expression:

$$K_{ij}^\sigma = \epsilon_{ij}^\sigma/\epsilon_o = (Ct/\epsilon_o A)$$

where C is the capacitance in picofarads, ϵ_{ij} is the permittivity, ϵ_o is the dielectric constant in a vacuum (8.85 x 10^{-12} F/m), t is the sample thickness and A is the electrode area. K_{33}^σ and K_{11}^σ were the unclamped relative dielectric constants parallel and perpendicular to the poling direction respectively.

Strain measurements were performed by application of an ac field (3 MV/cm) across the sample and measurement of the resultant displacement with an Linear Variable Displacement Transducer (LVDT) connected to a lock-in amplifier. A resolution of 3 nm was achievable with this system.

The displacement (strain) perpendicular to the poling direction was measured on heating to 450 °C with a commercial dilatometer[b]. Samples were heated at a rate of 10 °C/min., and data were collected in intervals of 5 °C.

RESULTS AND DISCUSSION

Effect of High Temperature Cyclic Loading on Crack Extension

Three cubes, 4.0 mm x 4.0 mm x 4.0 mm, were cut from: i.) the high stress/temperature region of an electrically cycled specimen (1 h at 180 °C) ii.) a poled virgin sample and iii.) a thermally depoled sample. Indentations were placed on each of the three orthogonal faces of the cubes (shown in Figure 1). Table 1 shows that the crack lengths were anisotropic (shorter parallel to poling direction) in the virgin samples and isotropic in electrically cycled or depoled samples. Electrical cycling changed the crack length symmetry from the transverse isotropic (∞m) of poled polycrystalline ferroelectrics to the isotropic symmetry of depoled material. Crack lengths for cycled material were similar to those of depoled material, with lengths close to the average of those in the perpendicular and parallel directions in the poled material.

Table 1. Indentation Crack Lengths for X, Y and Z faces (all lengths in μm)

		Poled (Virgin)	180 °C Cycled	Depoled
X face	//P	88.8 ± 10.0	106.5 ± 5.7	98.1 ± 7.8
	⊥P	135.1 ± 6.1	105.1 ± 8.7	102.8 ± 9.0
Y face	//P	81.3 ± 11.5	106.3 ± 9.2	101.2 ± 6.8
	⊥P	123.8 ± 8.9	101.5 ± 6.8	107.7 ± 11.7
Z face	//y dir.	100.4 ± 6.7	105.0 ± 3.2	104.5 ± 10.2
	⊥y dir.	101.1 ± 7.8	94.4 ± 6.7	102.9 ± 7.3

There is considerable disagreement regarding the physical mechanism causing the indentation crack length anisotropy. Previous investigators suggested that it was associated with macroscopic poling residual stress fields interacting with the cracks from the Vickers indentation[4], while others claim it was related to toughening by 90° domain wall motion (ferroelastic toughening)[5,6]. In the first case, the short crack length parallel to the poling direction would be caused by a macroscopic compressive residual poling stress acting on cracks parallel to the poling direction; similarly, the long crack length perpendicular to the poling direction would be caused by a macroscopic tensile poling-induced stress acting perpendicular to the poling direction. In the second case, ferroelastic domain wall motion in front of an advancing crack parallel to the poling direction would dissipate mechanical energy, and shield the crack tip from the applied load. This shielding would increase the fracture toughness parallel to the poling direction, resulting in shorter indentation crack lengths, while simultaneously decreasing the fracture toughness perpendicular to the poling direction, resulting in longer indentation crack lengths.

Effect of Remanent Polarization on Indentation Crack Lengths

To investigate the interplay of polarization level and indentation crack length anisotropy, depoled samples were then subsequently poled at various electric fields to span a range of remanent polarizations, aged at 22 ° C for 7 days, and indented (on y face in Figure 1). Figure 2 shows the indentation crack length plotted vs. the piezoelectric d_{31} coefficient, an estimate of the remanent polarization. Weakly poled (low d_{31}) material showed no detectable indentation crack length anisotropy with crack lengths similar to depoled material. For well poled (high d_{31}) material, crack length anisotropy was observed, with the cracks parallel to the poling direction shorter and cracks perpendicular to the poling direction longer than the depoled cracks. Samples experiencing 180 °C electrical cycling show final d_{31} levels close to the d_{31} value (Figure 2) joining the two regions.

To explain Figure 2, it is instructive to examine the electrical poling process. Electrical poling occurs when unit cell dipole axes of a polycrystalline material are aligned by the application of large electric fields (2 MV/m to 5 MV/m). For a randomly oriented tetragonal material, electrical poling occurs in two stages; 180° domain wall motion and 90° domain wall motion. An 180° domain wall is a boundary between volumes of oppositely poled unit cells. During poling, 180° domain walls move, orienting individual unit cell dipoles with an applied field. Complete 180° dipole alignment results in the annihilation of all 180° domain walls, with no resultant change in unit cell orientation and no residual stress[9]. During *poling*, 180° domain wall motion is nearly complete before 90° domain wall motion begins[10].

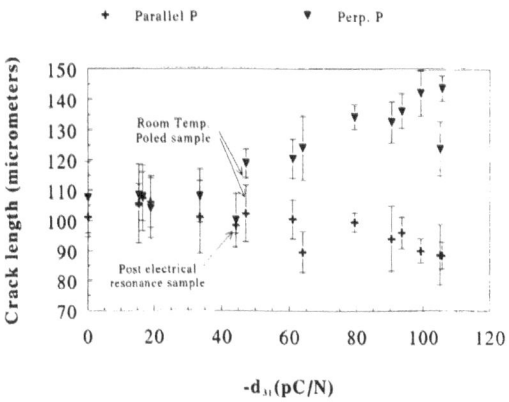

Figure 2. Effect of thoroughness of poling (as measured by d_{31}) on the 0.5kg indentation crack length anisotrophy in PZT-8. Electrically cycled specimen and the room temperature poled specimen are specifically marked.

A 90° domain wall is a boundary between volumes of material whose tetragonal c axes are oriented 90° apart (ferroelastic twin). Movement of a 90° domain wall changes the orientation of the tetragonal c axis by 90°, resulting in a macroscopic change in sample dimensions after the field is removed (remanent strain). After poling from an unpoled state, 90° domain wall motion causes expansion in the poling direction and contraction perpendicular to the poling direction. However, much of the strain which would result from 90° domain wall motion is clamped by the mechanical constraint of neighboring grains, resulting in a residual stress at grain boundaries. This residual stress would be anisotropic due to the preferred c-axis orientation caused by the poling field. During *depoling*, 90° domain wall motion would occur more readily than 180° domain wall motion since the former is aided by residual grain boundary stresses, which force 90° domain walls to reposition themselves to minimize this stress.

It is our contention is that the threshold d_{31} observed in Figure 2 above was the polarization level where there is complete 180° dipole alignment (180° domain walls have been annihilated) but where the 90° domain walls are positioned so as not to contribute to the total polarization. The following evidence supports this assertion:

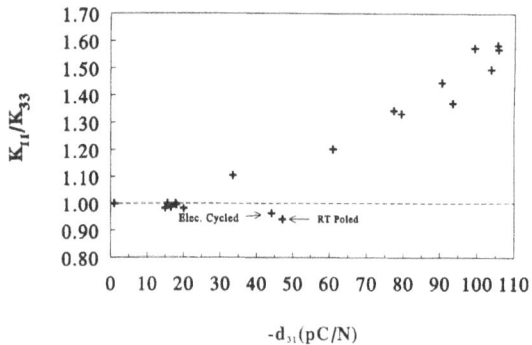

$-d_{31} (pC/N)$

Figure 3. Effect of thoroughness of poling (d_{31}) on the unclamped dielectric anisotropy ratio ($\epsilon_{11}{}^\circ/\epsilon_{33}{}^\circ$) in PZT-8. Electrically cycled specimen and the room temperature poled specimen are specifically marked.

It has been reported[11] that the unclamped permittivity ratio ($\epsilon_{11}{}^\circ/\epsilon_{33}{}^\circ$) is a measure of the 90° domain wall motion contribution to the polarization. Figure 3 shows the unclamped permittivity ratio ($\epsilon_{11}{}^\circ/\epsilon_{33}{}^\circ$) vs. d_{31} for the partially poled samples. As in Figure 2, two regions are shown, connected at a $d_{31} \approx -44$ pC/N. For low values of d_{31}, where the permittivity ratio ($\epsilon_{11}{}^\circ/\epsilon_{33}{}^\circ$) is close to 1.0, only 180° domain walls have moved in response to the poling field. At larger d_{31} levels, where 90° domain wall motion contributes to the polarization, the ($\epsilon_{11}{}^\circ/\epsilon_{33}{}^\circ$) increases with d_{31}. Therefore, Figure 3 shows results consistent with the existence of a transition from polarization by 180° domain wall motion to polarization by 90° domain wall motion at a $d_{31} \approx -44$ pC/N.

Figure 3 also shows that the $\epsilon_{11}{}^\circ/\epsilon_{33}{}^\circ$ of the high temperature resonance cycled sample was very close to 1.0, suggesting that an effect of the 180° resonance cycling is partial depolarization by removing the 90° domain wall contribution to the remanent polarization.

For poling PZT-8, temperatures around 100 °C are required to move 90° domain walls at any applied field[12]. Room temperature poling will therefore pole the material only by 180° domain wall motion. Partially poled material, in which *only* 180° domain wall motion occurred was obtained by application of a very large electric field (5 MV/m) at 22 °C. It was intended that this sample have complete 180° domain orientation with a minimal 90° contribution to the remanent polarization. Figure 4 shows a strain vs. field curve for this sample and no remanent strain (and no remanent 90° domain wall motion) with no applied field was observed. This sample had a $d_{31} = -47.2 \pm 0.7$ pC/N and $\epsilon_{11}{}^\circ/\epsilon_{33}{}^\circ$ =0.94 $\pm$ 0.05[1] similar to the high temperature cycled sample (Figures 2 and 3). The similarity in the d_{31} and $\epsilon_{11}{}^\circ/\epsilon_{33}{}^\circ$ values for the two samples supports the assertion that in the post-resonated sample (T $\approx$ 180 °C), the contribution to the final d_{31} was exclusively due to the 180° dipole alignment.

---

[1]All ± numbers represent standard uncertainty.

Figure 5 shows transverse strain (perpendicular to the poling direction) vs. temperature curves for a thermally depoled, a fully poled and a 180 °C resonated sample. The displacement was measured *in situ* as each sample was heated at 10 °C/min. to a depoled state above T_c (≈300 °C) at 450 °C. The fully poled material (Figure 5) showed a larger

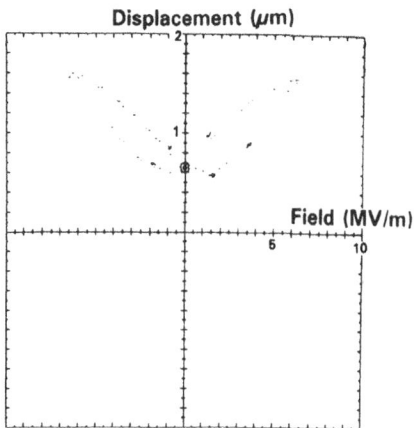

Figure 4. Room Temperature (22 °C) Mechanical strain vs. Electric Field loop for thermally depoled PZT-8. No remanent strain (strain at 0 field) was observed even at large electric fields.

lateral strain than either the depoled sample or the high temperature resonated sample. After cooling to room temperature, we observed a final strain of 1.4×10^{-3} in the fully poled material, while the other samples showed no change in the dimensions. The results are consistent with the expected depoling behavior in the two specimens. While the isotropic depoled sample was expected to show only normal thermal expansion, the fully poled sample was expected to exhibit an additional transverse expansion, consistent with the relaxation of the 90° domain wall motion contribution to the poled state (reversal of the poling strain). The transverse expansion curve of the electrically resonated sample coincides with that of the depoled material, indicating that on heating to a depoled condition, the 90° domain wall relaxation was negligible; all of the 90° domain wall contribution to polarization had relaxed previously due to the resonant cycling. This is further evidence that high temperature cyclic loading relaxed the 90° domain walls and that piezoelectric response observed after high temperature cyclic loading was solely due to the 180° domain alignment.

All of this evidence strongly suggests that high temperature electrical cycling causes the 90° domain walls, displaced during poling, to relax to their original pre-poled positions. The evidence also suggests that the anisotropy of the indentation crack lengths was an effect of 90° domain wall displacement during poling.

CONCLUSIONS

Room temperature indentations on material previously cycled at resonance (T≈180 °C) did not show the crack length anisotropy observed in fully poled material and the crack lengths in both directions were close to those of depoled specimens. Relative to fully poled material, high temperature electrical cycling caused an increase in the crack lengths parallel to the poling direction and a decrease in the crack lengths perpendicular to the poling direction. A study of crack length anisotropy for a range of poled specimens revealed two regions of behavior: 1.) Weakly poled material showing no crack length anisotropy with crack lengths similar to depoled material. 2.) Well poled material showing

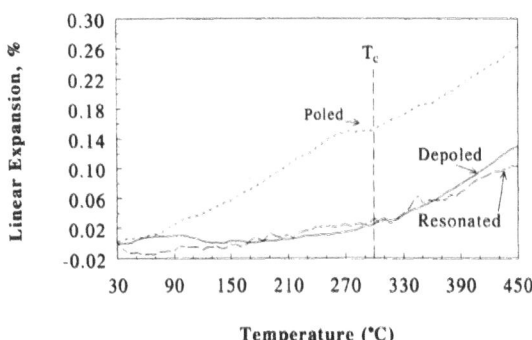

Figure 5. Dilatometry measurement of strain (as % linear expansion) perpendicular to the poling direction as a function of temperature for fully poled, depoled, and high temperature resonated PZT-8 specimens.

crack length anisotropy, with the cracks parallel to the poling direction shorter and perpendicular cracks longer than cracks in depoled material. The "threshold" (d_{31}) connecting these two regions was close to the d_{31} of electrically cycled samples (-44.2 pC/N). Dielectric anisotropy and remanent strain measurements suggest that the "threshold" d_{31} was that poling level remaining when all poling residual stresses due to 90° domain orientation were allowed to relax. The resultant residual stress relief was responsible for the change in indentation crack behavior after electrical cycling. Furthermore, the crack length anisotropy in poled ferroelectrics results from the 90° domain wall contribution to the remanent polarization.

ACKNOWLEDGMENTS

The authors would like to acknowledge Ed Fuller at NIST, Craig Near of Morgan Matroc and Eric Cross, Qiming Zhang and Paul Moses of the Pennsylvania State University for the helpful discussions. Partial support from the Office of Naval Research is gratefully appreciated.

REFERENCES

1 R.C. Pohanka, P.L. Smith and J. Pasternak, "The Static and Dynamic Strength of Piezoelectric Materials," Ferroelect. **50** pp 286-91 (1983)

2 G.S. White, A.S. Raynes, S.W. Freiman and Mark Vaudin; "Fracture Behavior of Cyclically Loaded PZT," J. Amer. Ceram. Soc. **77** [10] pp 2603-8 (1994)

3 M.D. Hill, G.S. White, C.S. Hwang and I.K. Lloyd, "Cyclic Damage in PZT," accepted for publication by the J. Amer. Ceram. Soc.

4 T. Yamamoto, H. Igarashi and K. Okazaki, "Dielectric, Electromechancal, Optical and Mechanical Properties of Lanthanum-Modified Lead Titanate Ceramics," J. AMer. Ceram. Soc. **66** [5] pp 363-6 (1983)

5 G.G. Pisarenko, V.M. Chushko, and S.P. Kovalev; "Anisotropy of Fracture Toughness of Piezoelectric Ceramics," J. Am. Ceram. Soc., **68** [5] pp 259-65 (1985)

6 K. Mehta and A. V. Virkar; "Fracture Mechanisms in Ferroelectric-Ferroelastic Lead Zirconate Titanate [Zr:Ti = 0.54:0.46] Ceramics," J. Am. Cer. Soc., **73** [3] pp 567-74 (1990)

7 R. Gerson, S.R. Burlage and D. Berlincourt, "Dynamic Tensile Strength of a Ferroelectric Ceramic," J. Acoust. Soc. Am. **33** [11]
1483-5 (1961)

8 IEEE Standard on Piezoelectricity ANSI/IEEE Std. 176-1987 1988

9 D. Berlincourt and H.A. Krueger, "Domain Processes in Lead Zirconate Titanate and Barium Titanate Ceramics," J. Appl. Phys. **30** [11] pp 1804-10 1959

10 L.E. Cross, "Ferroelectric Ceramics: Tailoring Properties for Specific Applications," in Ferroelectric Ceramics, Birkhauser-Verlag, Basel pp 1-85 (1994)

11 D. Berlincourt, "Variation of Electroelastic Constants of Polycrystalline Lead Zirconate Titanate with Thoroughness of Poling," J. Acoust. Soc. Amer. **36** [3] pp 515-20 (1964)

12 Morgan Matroc Inc. - Specifications Booklet

13 M.D. Hill - Unpublished Results

AUTHORS

T. Akatsu
Tokyo Institute of Technology, Yokohama, JAPAN

R.A. Andrievski
Institute for New Chemical Problems, Russian Academy of Sciences, Chernogolovka, Moscow Region, RUSSIA

M. Anglada
Universidad Politécnica de Cataluna, Barcelona, SPAIN

V.N. Antsiferov
Research Institute of Problems of Powder Technology, Perm, RUSSIA

S. Arakawa
Nagaoka University of Technology, Nagaoka, JAPAN

H.-A. Bahr
Max-Planck-Society, Research Group on Mechanics of Heterogeneous Solids, Dresden, GERMANY

U. Bahr
Technical University Dresden, Dresden, GERMANY

H. Balke
Technical University Dresden, Dresden, GERMANY

S.M. Barinov
High Tech Ceramics Research Centre, Russian Academy of Sciences, Moscow, RUSSIA

H. Basoalto
University of London, London, UNITED KINGDOM

F. Billi
University of Rome, Rome, ITALY

R. Biswas
Exeter University, Exeter, UNITED KINGDOM

R.C. Bradt
University of Alabama, Tuscaloosa, AL, USA

W.A.M. Brekelmans
Eindhoven University of Technology, Eindhoven, THE NETHERLANDS

R. Brook
University of Oxford, Oxford, UNITED KINGDOM

A. Brückner-Foit
Universität Karlsruhe, Karlsruhe, GERMANY

J.W. Cao
Toyohashi University of Technology, Toyohashi, JAPAN

J.A. Celemin
Polytechnic University of Madrid, Madrid, SPAIN

I.-W. Chen
University of Michigan, Ann Arbor, MI, USA

J.-L. Chermant
LERMAT, ISMRA, Caen, FRANCE

J. Chevalier
INSA-GEMPPM, Villeurbanne, FRANCE

T.-J. Chuang
National Institute of Standards and Technology, Gaithersburg, MD, USA

N. Claussen
Technische Universität Hamburg-Harburg, Hamburg, GERMANY

R. Dal Maschio
Università di Trento, Trento, ITALY

R. Danzer
Montanuniversität Leoben, Leoben, AUSTRIA

J.H.P. de Vree
Eindhoven University of Technology, Eindhoven, THE NETHERLANDS

G. de With
Philips Research Laboratories, Eindhoven, THE NETHERLANDS

K.Y. Donaldson
Virginia Polytechnic Institute and State University, Blacksburg, VA, USA

L.J.M.G. Dortmans
Centre for Technical Ceramics, Eindhoven, THE NETHERLANDS

M. Drissi-Habti
LERMAT, ISMRA, Caen, FRANCE

J. Dusza
Institute of Materials Research, Slovak Academy of Sciences, Kosice, SLOWAKIA

M. Elices
Polytechnic University of Madrid, Madrid, SPAIN

580

M. Engineer
University of Michigan, Ann Arbor, MI, USA

M. Enoki
The University of Tokyo, Tokyo, JAPAN

G. Fantozzi
INSA-GEMPPM, Villeurbanne, FRANCE

R. Fernández
Universidad Politécnica de Cataluna, Barcelona, SPAIN

T. Fett
Forschungszentrum Karlsruhe, Karlsruhe, GERMANY

B. Fiedler
Technische Universität Hamburg-Harburg, Hamburg, GERMANY

M. Gallmann
Fraunhofer-Institut für Werkstoffmechanik, Freiburg, GERMANY

A. Gerbatsch
Max-Planck-Society, Research Group on Mechanics of Heterogeneous Solids, Dresden,
GERMANY

R.J. Gettings
National Institute of Standards and Technology, Gaithersburg, MD, USA

A. Ghosh
Philips, Ann Arbor, MI, USA

C.B. Gilpin
California State University, Long Beach, CA, USA

V.T. Golovchan
Institute for Superhard Materials, National Academy of Sciences, Kiev, UKRAINE

G. Grathwohl
University of Bremen, Bremen, GERMANY

D.J. Green
The Pennsylvania State University, University Park, PA, USA

O. Grigor'ev
Institute for Problems of Materials Science, National Academy of Sciences, Kiev, UKRAINE

M.-O. Guillou
University of Exeter, Exeter, UNITED KINGDOM

F. Guiu
University of London, London, UNITED KINGDOM

T. Hansson
Nagaoka University of Technology, Nagaoka, JAPAN

D.P.H. Hasselman
Virginia Polytechnic Institute and State University, Blacksburg, VA, USA

K. Hatanaka
Yamaguchi University, Tokiwadai, Ube City, JAPAN

J.C. Hay
University of Houston, Houston, TX, USA

J.R. Hellmann
The Pennsylvania State University, University Park, PA, USA

C.H. Henager
Battelle Pacific Northwest Laboratories, Richland, WA, USA

J.L. Henshall
Exeter University, Exeter, UNITED KINGDOM

D. Hertel
Universität Karlsruhe, Karlsruhe, GERMANY

M.D. Hill
National Institute of Standards and Technology, Gaithersburg, MD, USA

M. Hoffman
Technische Hochschule Darmstadt, Darmstadt, GERMANY

T. Hollstein
Fraunhofer-Institut für Werkstoffmechanik, Freiburg, GERMANY

M. Holzherr
Fraunhofer-Institute for Materials Physics and Surface Engineering, Dresden, GERMANY

R.M. Hooper
University of Exeter, Exeter, UNITED KINGDOM

H. Hübner
Technische Universität Hamburg-Harburg, Hamburg, GERMANY

P. Hvizdos
Institute of Materials Research, Slovak Academy of Sciences, Kosice, SLOWAKIA

T. Inoue
Osaka National Research Institute, Osaka, JAPAN

S. Ivanov
Institute for Problems of Materials Science, National Academy of Sciences, Kiev, UKRAINE

D.S. Jacobs
University of Michigan, Ann Arbor, MI, USA

B.-K. Jang
The University of Tokyo, Tokyo, JAPAN

M.G. Jenkins
University of Washington, Seattle, WA, USA

Q. Jiang
University of Nebraska-Lincoln, Lincoln, NE, USA

W. Kanematsu
National Industrial Research Institute of Nagoya, Nagoya, JAPAN

V. Kartuzov
Institute for Problems of Materials Science, National Academy of Sciences, Kiev, UKRAINE

M. Kawahara
Tokyo Metropolitan University, Tokyo, JAPAN

D. Kervadec
LERMAT, ISMRA, Caen, FRANCE

B.-N. Kim
Tokyo Metropolitan University, Tokyo, JAPAN

S. Kimura
Yamanashi University, Yamanashi, JAPAN

G. Kirchhoff
Fraunhofer-Institute for Materials Physics and Surface Engineering, Dresden, GERMANY

T. Kishi
The University of Tokyo, Tokyo, JAPAN

H. Kishimoto
Toyota Technological Institute, Nagoya, JAPAN

H.-J. Kleebe
University of Bayreuth, Bayreuth, GERMANY

H.N. Ko
Nakanihon Automotive College, Gifu, JAPAN

A.S. Kobayashi
University of Washington, Seattle, WA, USA

D. Koch
University of Bremen, Bremen, GERMANY

Y. Kodama
National Industrial Research Institute of Nagoya, Nagoya, JAPAN

U. Köpke
Technische Universität Hamburg-Harburg, Hamburg, GERMANY

Y. Koshka
Institute for Problems of Materials Science, National Academy of Sciences, Kiev, UKRAINE

V. Kostopoulos
Commission of the European Communities, Joint Research Centre, Petten,
THE NETHERLANDS

S.P. Kovalev
Institute for Problems of Strength, National Academy of Sciences, Kiev, UKRAINE

D. Kovar
Carnegie Mellon University, Pittsburgh, PA, USA

J.J. Kübler
Swiss Federal Laboratories for Materials Testing and Research, Dübendorf, SWITZERLAND

K. Kubo
Setsunan University, Neyagawa, JAPAN

M. Kuntz
University of Bremen, Bremen, GERMANY

R. Kurth
Forschungszentrum Jülich, Jülich, GERMANY

V.I. Kushch
Institute for Superhard Materials, National Academy of Sciences, Kiev, UKRAINE

A.G. Lanin
The Institute of the Scientific Industrial Association "Lutch", Podolsk, Moscow Region,
RUSSIA

F.F. Lange
University of California, Santa Barbara, CA, USA

E. Lara-Curzio
Oak Ridge National Laboratory, Oak Ridge, TE, USA

B.F. Lawlor
Forschungszentrum Jülich, Jülich, GERMANY

S.-H. Lee,
Ssangyong Cement Industrial Co., Ltd., DaeJeon, KOREA

Z. Lencès
Institute of Inorganic Chemistry, Slovak Academy of Sciences, Bratislava, SLOWAKIA

M. Li
University of London, London, UNITED KINGDOM

Z. Li
Watkins-Johnson, Scotts Valley, CA, USA

G. Liu
University of Illinois, Urbana, IL, USA

S.-Y. Liu,
University of Michigan, Ann Arbor, MI, USA

L. Llanes
Universidad Politécnica de Cataluna, Barcelona, SPAIN

J. LLorca
Polytechnic University of Madrid, Madrid, SPAIN

I.K. Lloyd,
University of Maryland, College Park, MD, USA

Y. Lu
Virginia Polytechnic Institute and State University, Blacksburg, VA, USA

T. Lube
Montanuniversität Leoben, Leoben, AUSTRIA

E.H. Lutz
LWK-Plasmakeramik GmbH, Gummersbach, GERMANY

L. Ma
University of California, Los Angeles, CA, USA

T. Mangialardi
Rome University, Rome, ITALY

A. Martin
Polytechnic University of Madrid, Madrid, SPAIN

S.W. Martz
The Pennsylvania State University, University Park, PA, USA

M. Matsui
NGK Insulators Ltd., Mizuho, Nagoya, JAPAN

Y. Matsuo
Tokyo Institute of Technology, Tokyo, JAPAN

R. Matt
University of Karlsruhe, Karlsruhe, GERMANY

H. Maupas
LERMAT, ISMRA, Caen, FRANCE

F. Mignard
INSA-GEMPPM, Villeurbanne, FRANCE

T. Miyajima
National Industrial Research Institute of Nagoya, Nagoya, JAPAN

Y. Miyashita
Nagaoka University of Technology, Nagaoka, JAPAN

D. Munz
Universität Karlsruhe, Karlsruhe, GERMANY

N. Murayama
National Industrial Research Institute of Nagoya, Nagoya, JAPAN

Y. Mutoh
Nagaoka University of Technology, Nagaoka, JAPAN

M.M. Nagl
Universidad Politécnica de Cataluna, Barcelona, SPAIN

H. Naito
Tokyo Metropolitan University, Tokyo, JAPAN

T. Nishida
Kyoto Institute of Technology, Kyoto, JAPAN

K. Ogawa
Kyoto University, Kyoto, JAPAN

N. Ogawa
Nihon University, Koriyama, JAPAN

H.-K. Oh
Ssangyong Cement Industrial Co., Ltd. DaeJeon, KOREA

C. Olagnon
INSA-GEMPPM, Villeurbanne, FRANCE

K. Ono
Nihon University, Koriyama, JAPAN

K. Oshita
Yamaguchi University, Tokiwadai, Ube City, JAPAN

M.-J. Pan
The Pennsylvania State University, University Park, PA, USA

A.E. Paolini
Rome University, Rome, ITALY

J.Y. Pastor
Polytechnic University of Madrid, Madrid, SPAIN

S.D. Peteves
Commission of the European Communities, Joint Research Centre, Petten,
THE NETHERLANDS

G. Pezzotti
Toyohashi University of Technology, Toyohashi, JAPAN

W. Pfeiffer
Fraunhofer-Institut für Werkstoffmechanik, Freiburg, GERMANY

I. Pflugbeil
Max-Planck-Society, Research Group on Mechanics of Heterogeneous Solids, Dresden,
GERMANY

J.P. Piccola
Boeing Commercial Airplane Group, Seattle, WA, USA

G.G. Pisarenko
Institute for Problems of Strength, National Academy of Sciences, Kiev, UKRAINE

W. Pompe
Max-Planck-Society, Research Group on Mechanics of Heterogeneous Solids, Dresden,
GERMANY

H. Prielipp
Technische Universität Hamburg-Harburg, Hamburg, GERMANY

G.D. Quinn
National Institute of Standards and Technology, Gaithersburg, MD, USA

M.J. Readey
Sandia National Laboratories, Albuquerque, NM, USA

A. Reichl
Innomess GmbH, Marl, GERMANY

H. Riedel
Fraunhofer-Institut für Werkstoffmechanik, Freiburg, GERMANY

J. Rödel
Technische Hochschule Darmstadt, Darmstadt, GERMANY

J. Rohde
Max-Planck-Institut für Metallforschung, Stuttgart, GERMANY

M. Rombach
Fraunhofer-Institut für Werkstoffmechanik, Freiburg, GERMANY

D. Rouby
INSA-GEMPPM, Villeurbanne, FRANCE

E. Rudnayová
Institute of Materials Research, Slovak Academy of Sciences, Kosice, SLOWAKIA

P. Sajgalik
Institute of Inorganic Chemistry, Slovak Academy of Sciences, Bratislava, SLOWAKIA

S. Sakaguchi
National Industrial Research Institute of Nagoya, Nagoya, JAPAN

M. Sakai
Toyohashi University of Technology, Toyohashi, JAPAN

C. Santulli
Commission of the European Communities, Joint Research Centre, Ispra, ITALY

T.A. Sarkisyan
Mendeleev University of Chemical Technology of Russia, Moscow, RUSSIA

S. Schmauder
University of Stuttgart, Stuttgart, GERMANY

G.A. Schneider
Technische Universität Hamburg-Harburg, Hamburg, GERMANY

I.B. Sevostianov
Max-Planck-Society, Research Group on Mechanics of Heterogeneous Solids, Dresden,
GERMANY

I.G. Sevastianova
State Polytechnical University, Perm, RUSSIA

V.M. Sglavo
Università di Trento, Trento, ITALY

J.K. Shang
University of Illinois, Urbana, IL, USA

V.Y. Shevchenko
High Tech Ceramics Research Centre, Russian Academy of Sciences, Moscow, RUSSIA

H. Shiota
Gifu University, Gifu City, JAPAN

G. Sines
University of California, Los Angeles, CA, USA

S. Sodeoka
Osaka National Research Institute, Osaka, JAPAN

E. Sommer
Fraunhofer-Institut für Werkstoffmechanik, Freiburg, GERMANY

M. Steen
Commission of the European Communities, Joint Research Centre, Petten,
THE NETHERLANDS

R.W. Steinbrech
Forschungszentrum Jülich, Jülich, GERMANY

M. Sternitzke
University of Oxford, Oxford, UNITED KINGDOM

R.N. Stevens
University of London, London, UNITED KINGDOM

Y.J. Stockmann
Forschungszentrum Jülich, Jülich, GERMANY

M. V. Swain
University of Sydney, Sydney, AUSTRALIA

A.A. Tashkinov
State Polytechnical University, Perm, RUSSIA

J. Tatami
Tokyo Institute of Technology, Tokyo, JAPAN

M. Tilscher
Forschungszentrum Karlsruhe, Karlsruhe, GERMANY

D.K. Tran
University of Washington, Seattle, WA, USA

R.E. Tressler
The Pennsylvania State University, University Park, PA, USA

A. Ueno,
Toyota Technological Institute, Nagoya, JAPAN

K. Ueno,
Osaka National Research Institute, Osaka, JAPAN

M.A.J. van Gils
Centre for Technical Ceramics, Eindhoven, THE NETHERLANDS

A. Vojta
Max-Planck-Society, Research Group on Mechanics of Heterogeneous Solids, Dresden,
GERMANY

S. Wakayama
Tokyo Metropolitan University, Tokyo, JAPAN

R.J. Wakeman
Exeter University, Exeter, UNITED KINGDOM

Y. Wang
University of Nebraska-Lincoln, Lincoln, NE, USA

Z. Wang
National Institute of Standards and Technology, Gaithersburg, MD, USA

P.D. Warren
Oxford University, Oxford, UNITED KINGDOM

H. Weber
Max-Planck-Society, Research Group on Mechanics of Heterogeneous Solids, Dresden,
GERMANY

H.-J. Weiss
Fraunhofer Institute for Materials Physics and Surface Technology, Dresden, GERMANY

H. Weitzing
Technische Universität Hamburg-Harburg, Hamburg, GERMANY

E. Weppelmann
Fraunhofer-Institut für Werkstoffmechanik, Freiburg, GERMANY

G.S. White
National Institute of Standards and Technology, Gaithersburg, MD, USA

K.W. White
University of Houston, Houston, TX, USA

V.E. Wildemann
State Polytechnical University, Perm, RUSSIA

M. Wittling
CSIRO Division of Applied Physics, Lindfield, AUSTRALIA

C. Wolf
Technische Universität Hamburg-Harburg, Hamburg, GERMANY

W. Wong-Ng
National Institute of Standards and Technology, Gaithersburg, MD, USA

Y.Y. Yang,
Universität Karlsruhe, Karlsruhe, GERMANY

E. Yasuda
Tokyo Institute of Technology, Yokohama, JAPAN

K. Yasuda
Tokyo Institute of Technology, Meguro-ku, Tokyo, Japan

M. Yoda
Nihon University, Koriyama, JAPAN

C.-T. Yu
University of Washington, Seattle, WA, USA

A.L. Yurkov
Mendeleev University of Chemical Technology of Russia, Moscow, RUSSIA

B. Zickgraf
Max-Planck-Institut für Metallforschung, Stuttgart, GERMANY

A. Zolochevsky
Kharkov Technical University, Kharkov, UKRAINE

INDEX

The manufacturer's authorised representative in the EU is Springer
Nature Customer Service Centre GmbH, Europaplatz 3, 69115 Heidelberg,
Germany. If you have any concerns regarding our products, please
contact ProductSafety@springernature.com

Printed and bound by CPI Group (UK) Ltd, Croydon, CR0 4YY
23/04/2026
02095607-0011